THE TSUNAMI THREAT - RESEARCH AND TECHNOLOGY

Edited by **Nils-Axel Mörner**

The Tsunami Threat - Research and Technology
Edited by Nils-Axel Mörner

Published by InTech
Janeza Trdine 9, 51000 Rijeka, Croatia

Publishing Process Manager Iva Lipovic

Technical Editor Teodora Smiljanic

Cover Designer InTech Design Team

Additional hard copies can be obtained from orders@intechopen.com

The Tsunami Threat - Research and Technology, Edited by Nils-Axel Mörner
p. cm.
ISBN 978-953-307-552-5

For EU product safety concerns:
IN TECH d.o.o., Prolaz Marije Krucifikse Kozulić 3, 51000 Rijeka, Croatia,
info@intechopen.com or visit our website at intechopen.com.

Meet the editor

Nils-Axel Mörner took his PhD in Quaternary Geology at Stockholm University in 1969. He was head of a personal institute at Stockholm University and the Swedish National Council on Paleogeophysics & Geodynamics (P&G) from 1991 up to his retirement in 2005. He has written hundreds of research papers and a number of books. He is a global traveller and has undertaken field studies in all continents. Several students have taken their doctoral degree at the P&G institute, which became an international centre for paleoseismics, neotectonics, paleomagnetism and global sea level changes. He was president of the INQUA Neotectonics Commission (1981-1989) and president of the INQUA Commission on Sea Level Changes and Coastal Dynamics (1999-2003). From 1996 he has been deeply engaged in paleotsunami research in Sweden and the Maldives. In 2008 he was awarded the Golden Condrite of Merit (from Algarve University) "for his irreverence and contribution to our understanding of sea level change".

Contents

Preface

Just after 9.30 in the morning of Sunday, November 1, 1755 (All Saints Day), Lisbon was struck by a very large earthquake. Buildings collapsed, roaring sounds filled the air, particles were thrown up in the air obscuring the Sun, fire started and an immense tsunami wave broke in over land. A terrible disaster had occurred, often known as *"the Disaster at Lisbon"*. Large parts of the city were in ruins and possibly up to 100,000 persons were killed. *"Take care of the living and bury the dead"* was the laconic advice to the dumbfounded King José I from his minister Sebastião José Carvalho e Melo, later to become known as Marquis de Pombal, one of the true heroes of the tragic event. The prime area of disaster was the coastal sections of Portugal, Spain and Morocco. The tsunami wave was also recorded in Brazil and the West Indices. The disaster had wide intellectual impacts, too. The philosopher Leibniz had claimed (1714) that *"this is the best of worlds"*. In Candide (1759), Voltaire used the "Destruction of Lisbon" as the final evidence against Leibniz's idea, and let the world become far from ideal, rather close to hell. Some saw the disaster as an act of God and claimed that Lisbon therefore was not to be rebuilt, an idea strongly contested by Voltaire. Lisbon was, of course, rebuilt. The Lisbon 1755 event was probably the first well-recorded tsunami in the history of tsunami research. Over 250 years have passed since the event. Still, it seams, we have learned very little. A new event may well reoccur "at any time" in the same area. Today we are very much more vulnerable for extended disastrous effects, however, the number of people has multiplied several times, buildings have crept very close to the sea, and the beaches are crowded with people. An effective warning system should have been installed, and a disaster plan should have been constructed. It should have been a central issue for the sake of precaution. Still, this is not yet the case.

On December 26, 2004 (the Boxing Day), the Indian Ocean was stuck by a disastrous tsunami event that killed, at least 227,898 people. It was initiated by the Sumatra-Andeman Island Earthquake of a magnitude of M 9.3 on the Richter scale. This earthquake set up a series of tsunami waves that spread away from the epicentre at a rate of about 700 km/hour. When the waves reached coastal areas they started to break in over the coasts. The first sign of the approaching tsunami was an initial withdrawal of water. This phase lasted for some minutes, then came the disastrous breaking wave followed by its back-swash phase. The world was shocked and amazed. This time we seem to have learned a lesson (just read this book). Today there is an Indian Ocean tsunami warning system and agencies are established to guide in case of a new event and to support in the field.

These two examples of exceptionally large tsunami events both refer to submarine earthquakes. Volcanism, submarine slides and impacts may also generate tsunamis.

The Krakatoa multiple volcanic eruptions occurred on August 26–27, 1883. The 4th eruption at 10.02 on August 27 set up a very intensive and destructive tsunami with a death toll in the order of 40,000 persons. The maximum wave height was 35 m and it was still 2.4 m high when it reached Jakarta after 2.5 hours.

The Storegga submarine slide at about 7000 BP set up a tsunami, which is well recorded along the Norwegian coasts (e.g. Bondevik et al., 1997) and in Scotland (e.g. Dawson et al., 1988). Even terrestrial slides that go into a lake or the sea set up local tsunamis.

The huge impact that is held to have hit the Earth at the Cretaceous/Tertiary (C/T) boundary is likely to have set up a very large tsunami that had disastrous effects on the fauna and flora at the end of the Cretaceous (e.g. Bourgeous et al., 1988). Recently, tsunamis have also been claimed to occur on Mars as a function of impacts (Mahaney et al., 2010).

Even boulders that fall from a vertical cliff into the sea (or a lake) may set up a minor local tsunami wave. When Charles Darwin and his group on January 20, 1832 were watching the glaciers on the opposite side of Beagle Channel, a big iceberg fell into the sea. It set up a large and rapidly moving wave, which threw their boat up on land and nearly killed them (Darwin, 1839). Today we know that this was a local tsunami.

This book, we may say, is another sign of the new awareness of the tsunami phenomenon, which is a natural process in Earth's dynamics (Mörner, 2010). Tsunamis have always been a part of Earth's dynamics and will always continue to be. What we as humans can do is to undertake precautional efforts like installation of warning systems and establishment of effective disaster planning. All this calls for a basic knowledge about the phenomenon as such, and this is the reason why this book is published today.

The title of the book was chosen to provide a direct message that we are dealing with a *"threat"* to humanity in the form of a phenomenon called tsunami, which in the past caused a lot of damage and suffering. The Indian Ocean tsunami in 2004 with a death toll of 229,866 persons was number 13 or 12 on the terrible list of death toll per natural disaster event (Mörner, 2010, Table 2).

This book includes 33 separate papers, which have been arranged in three main sections; (1) hazard and vulnerability, (2) observational records and (3) technology. All these papers present a well of information, partly novel, which will hopefully act as useful, not to say necessary, base-document both for the understanding of the problem and the threat, and for the argumentation for wise decisions in order to stay well prepared at future events to come.

Stockholm, December 2010

Nils-Axel Mörner
Paleogeophysics & Geodynamics
Sweden

Bondevik, S., Svensen, J.I. & Mangerud, J., 1997. Tsunami sedimentary facies deposited by the Storegga tsunami in shallow marine basins and coastal lakes, western Norway. *Sedimentology,* 44: 1115-1131.
Bourgeois, J., Hansen, T.A., Wiberg, P.L. & Kauffman, E.G., 1988. A tsunami deposit at the Cretaceous–Tertiary boundary in Texas. *Science,* 241: 567-570.
Darwin, C., 1839. *Voyage with H.M.S. Beagle* (with later editions, 1845 and 1860).
Dawson, A.G., Long, D. & Smith, D.E. 1988. The Storegga Slide: Evidence from eastern Scotland for a possible tsunami. Marine Geology, 82: 271-276.
Mahaney,W.C., Dohm, J.M., Costa, P. & Krinsley, D.H., 2010. Tsunamis on Mars: Earth analogues of projected Martian sediment. *Planetary and Space Science,* 58: 1823-1831.
Mörner, N.-A., 2010. Natural, man-made and imagined disaster. From the Editor's desk, *Disaster Advances,* 3(1): 3-5.

Part 1

Hazard and Vulnerability

1

The Asian Tsunami's Havoc and Death Toll: Nature's Wrath or Human Shortsightedness?

Zvy Dubinsky[1], Osnat Chomsky[2] and Steve Brenner[3]
[1]*The Mina & Everard Goodman Faculty of Life Sciences*
Bar-Ilan University, Ramat-Gan 52900
[2]*National Institute of Oceanography, Israel Oceanographic and Limnological Research, Tel Shikmona, P.O. Box 8030, Haifa 31080*
[3]*Department of Geography and Environment, Bar-Ilan University*
Ramat-Gan 52900
Israel

1. Introduction

The unprecedented devastation caused by the December 2004 tsunami (Reuters, 2005) was followed by environmentalists' claims that the human toll and overall impact were aggravated by disregard for conservation of coral reefs and mangrove forests (Channel News Asia, 2005; Cripps, 2005; Mangrove Action Project, 2005). Corals surround tropical shorelines (Spalding et al., 2001) and beyond them, on land, grow, in most cases, dense belts of mangroves (Chapman, 1977). These twin barriers are thought to protect shorelines from waves and may be capable of mitigating tsunami impact (Fig. 1). According to reports in the mass media, reefs and mangroves sheltered people from the tsunami's impact, whereas in the hardest hit areas in terms of human casualties, coral reefs and mangroves had been removed in order to make room for shrimp and fish mariculture (Kaban, 2005). Here we present an attempt to evaluate the potential of the combination of coral reefs and mangroves to attenuate tsunamis, and validate the claims regarding the protective function of reefs and mangroves. Special attention will be given to the extent of the recent destruction of reef and coral ecosystems in the region afflicted by the tsunami, and the importance of their conservation and of remediation and rehabilitation efforts (Kaban, 2005).
Tsunamis have ravaged coastal communities since the earliest extant myths and written records (Table 1). The most prominent among these was the one generated by the volcanic eruption of Thera (Santorini), thought to have wiped out the Minoan civilization (Galanopoulos, 1960; Pararas-Carayannis, 1973), and sending survivors as far east as Canaan, where they settled as the biblical "Sea people" or Philistines, who competed with the Hebrews for what was to become the "Holy Land". There are historical records of tsunamis with death tolls ranging from no victims to estimates of over 300,000 in the last one (Table 1).
The Sumatra tsunami of December 2004 was the first to be witnessed via television in virtually every home, bringing the destructive potential of oceanic waves to unprecedented public awareness. In the wake of this tsunami, there have been several reports in the mass media regarding the reduction in the human death toll and overall destruction in areas that

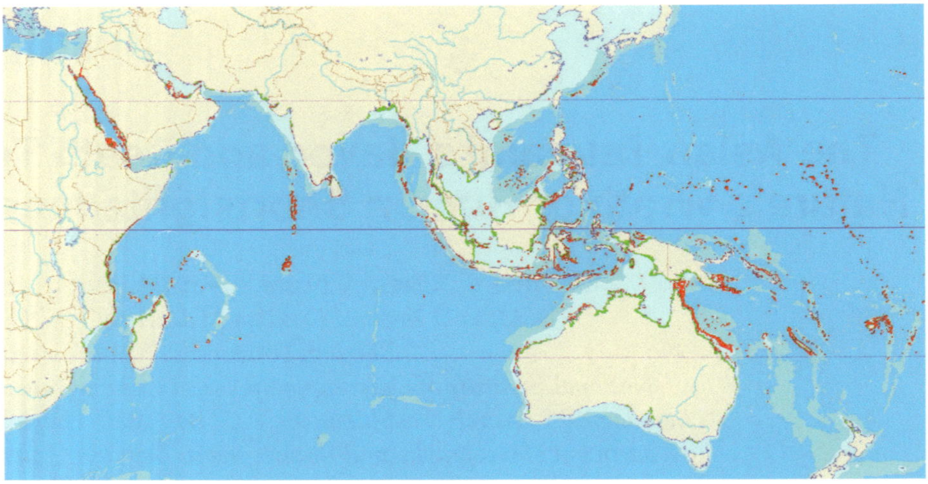

Fig. 1. World distribution of coral reefs and mangroves. Key: Coral = red, Mangrove = green
After www.unep-wcmc.org/marine/data/coral_mangrove/marine_maps_main.html

Principal areas	**Magnitude**	**Year**	**Fatalities**
Crete-Santorini, Ancient Greece	?	1600 ± 50 BC	~100,000
Tokaido-Nankaido, Japan	8.4	1707	30,000
Portugal, Morocco	8.5	1755	60,000
Ryukyu Trench, Japan	7.4	1771	13,486
South China Sea	7.0	1782	40,000
Kyushu Island, Japan	6.4	1792	15,030
Sumatra region	8.7	1833	36,000
Sumatra region	8.5	1861	Thousands
Northern Chile	8.5	1868	25,674
Krakatau, Indonesia	?	1883	36,500
Sanriku, Japan	7.6	1896	26,360
Tokyo	8.3	1923	140,000
Northeastern coast, Japan	8.9	1933	3,064
Japan	?	1944	1,251
Shikoku, Japan	8.0	1946	1,330
Fukui, western Japan	7.1	1948	3,769
Hokkaido	8.2	1952	8,233
Kamchatka Peninsula, Russia	8.2	1952	Considerable loss of life
Chile, Hawaii, Japan, and the Philippines	9.5	1960	500-2,300
Philippines		1976	5,000
Kobe, Western Japan	7.2	1995	1,800
Papua New Guinea	7.1	1998	3,000
Indian Ocean	9.0	December 26, 2004	305,276

Table 1. Historical records of major tsunamis

were sheltered from the wave by intact reefs and mangrove belts (Chapman, 1977; Spalding et al., 2001; Channel News Asia, 2005; Cripps, 2005; Mangrove Action Project, 2005). These reports were eventually substantiated by a field survey of Sri Lankan mangrove sites, whose destruction increased tsunami damage (Dahdouh-Guebas et al., 2005).

The global distributions of coral reefs and mangroves are almost identical: they skirt shorelines between 30 degrees north and south of the equator. These ecosystems are linked by complementary environmental requirements, and also by interactions among physical, chemical, and biological processes. Furthermore, the destruction of one endangers and eventually destroys the other (Kuhlman, 1998).

Coral reefs are biogenic, calcium carbonate, marine structures, based on the symbiosis between corals and unicellular microalgae, the zooxanthellae (Brandt, 1883). The algae contribute high-energy photosynthate whereas the host animal, an avid predator on zooplankton, provides the symbionts with metabolic wastes rich in nitrogen and phosphorus. It is this association that allows corals to outcompete seaweeds in the warm, transparent, nutrient-poor "blue deserts" of the tropics. Coral reefs extend over thousands of kilometers from the surface down to 100 m, usually decreasing in vigor below 30 m. This depth distribution is dictated by the exponential decrease in underwater light, upon which the algae depend (Achituv & Dubinsky, 1990). In recent decades, rising ocean temperatures have been implicated in bleaching events in which corals lose their symbionts and die (Hoegh-Guldberg, 2004). In addition, terrigenous nutrients and sediments disrupt the host–symbiont association, leading to the replacement of corals by other benthic communities. Over 70% of the world's reefs have been destroyed or are in great danger (Wilkinson, 2002), (Fig. 2). In the areas worst hit by the recent tsunami, the overriding cause of reef destruction (Fig. 3) was the spread of shrimp ponds and other mariculture enterprises, as well as tourist recreational facilities.

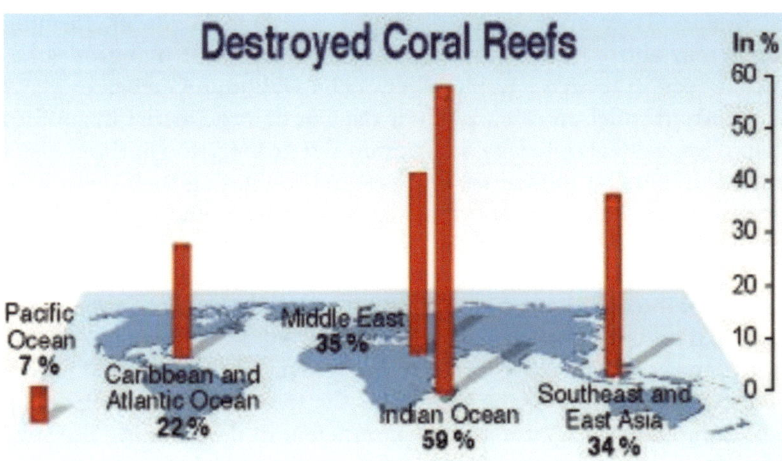

Fig. 2. Global distribution of coral reef destruction
www.unep.org/ vitalwater/marine.htm

Mangroves dominate 75% of tropical coastlines between 25° N and 25° S (Duke et al., 2002). These coastal forests are made up of some forty species of trees and shrubs adapted to life in high-salinity waters, where they are rooted in anaerobic mud. They depend on aerial roots, or pneumatophores, supplying their roots with oxygen (Hanagata et al., 1999). The

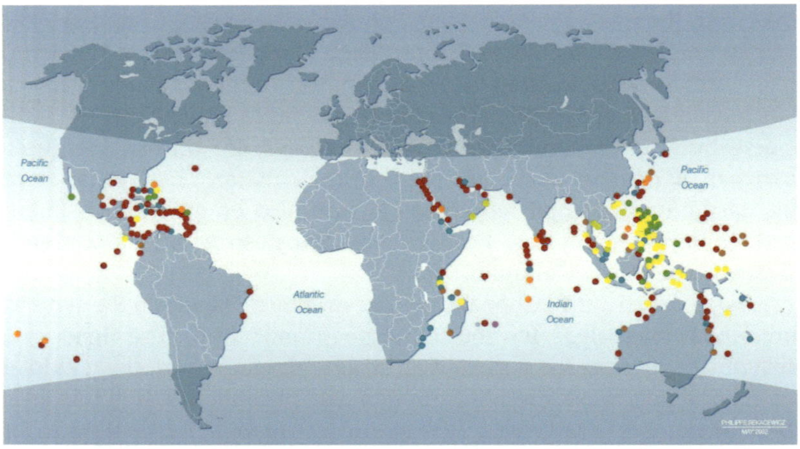

Fig. 3. Main causes for reef destruction: tourism; poison fishing; overexploitation; sedimentation; coral harvesting; dynamite fishing; pollution
After Bryant et al., Reefs at Risk; a Map-Based Indicator tom the World's Coral Reefs, World Resources Institute (WRI), Washington DC, 1998

mechanisms allowing them to retain water against the osmotic gradient posed by seawater still elude full understanding (Zimmermann et al., 2002). To increase chances of survival, in many mangrove species, fruits germinate while still connected to the mother plant, and when they fall off, they become planted in the soft soil (Hogarth, 1999). Mangrove forests may reach a width of several hundred meters and a height of up to 40 m.

Mangroves act as a filter, preventing sediment from reaching the reef and abrading the delicate coral tissues. They also absorb dissolved nutrients in runoff, forming a biofilter protecting reefs from eutrophication. Indeed, the destruction of mangroves has led to the death of adjacent reefs (Dubinsky & Stambler, 1996; Dahdouh-Guebas et al., 2005). In yet another recent study (Danielsen *et al.*, 2005) of the Cuddalore District in southeastern India based on satellite images validated by surveys on the ground, the authors also confirm the role of mangroves in attenuating tsunami waves and mitigating their destructive potential. Their data are in agreement with experiments, which have shown that 30 trees per 100 square meters suffice to dissipate most of a tsunami's energy (Hiraishi & Harada, 2003). More recently, Yanagisawa et al. (2009) used satellite images and field measurements to assess the damage of the mangrove forests caused by the 2004 Asian tsunami along a stretch of the coast of Thailand. They also used a numerical model to simulate the effects of the tsunami on the mangroves and the role of the mangroves in reducing the inundation depth and concluded that in addition to the density of the tree coverage, the tree stem diameter and the initial inundation depth are crucial parameters in determining the survival rate of the trees as well as the mangroves ability to reduce the inundation depth. The mangrove roots serve as the reef's nurseries, where juveniles of myriad reef creatures find shelter from predators. Coral reefs weaken surf, reducing the removal of soil and the uprooting of mangroves and their seedlings.

The human need for land, firewood, and material for construction has acerbated the global climate change-driven decline of corals (Buddemeier et al., 2004) and mangroves (Table 2). However, our burgeoning appetite for seafood, and human greed have probably been the driving force in recent decades behind the clearcutting of mangroves and their replacement, as well as that of reefs, by various mariculture enterprises and recreational villages. In Southeast Asia, this destruction has led to losses of 70-90% of the reefs and mangroves (Wilkinson, 2002).

It is of great interest to explore the potential of the twin barriers, mangrove forests and coral reefs, to attenuate destructive waves and prevent disaster. We present here an initial attempt of physically modeling and evaluating the attenuation of tsunami waves by reefs and their further weakening by mangrove belts.

While it is clear that both the coral reefs and the mangroves play a role in reducing the devastating effects of tsunami run-up and inundation of the coastal area, most of the evidence has been anecdotal and taken from the mass media or from eyewitness reports. Furthermore, the primary measure of tsunami damage is the number of casualties, which may be more a function of land use and demographics than the force of the tsunami itself. Therefore, in order to quantitatively assess the potential mitigating effects of coral reefs and mangroves, we have conducted a series of simulations with an established wave model (Chen et al., 2000; Kennedy et al., 2000), which has been tested for a wide variety of surface wave types including tsunamis. The goal of our experiments is to demonstrate the role of the bulk properties of the combined coral reef - mangrove system as a barrier that can mitigate the potential damage of the tsunami by reducing the inundation depth. We do not consider or focus on the small scale details of the structure of the reef or the mangrove forest as was done in other studies (e.g., Yanagisawa et al., 2009).

2. Model description and setup

The model we have chosen is the fully nonlinear, Boussinesq surface wave model FUNWAVE (Chen et al., 2000; Kennedy et al., 2000) based on the work of Wei et al. (1995). FUNWAVE is capable of simulating waves in the open sea as well as in the nearshore zone, and has been tested extensively for a wide range of surface wave types, including tsunamis (Chen et al., 2000; Watts et al., 2003). The model is based on the shallow water equations which consist of equations for the conservation of mass (i.e., the free surface height) and the horizontal momentum with parameterizations added to account for frictional dissipation and wave breaking. Further details can be found in the references mentioned above. For this study we use the two dimensional version of the model.

Since our goal is to assess the potential role of coral reefs and mangroves in reducing the impact of tsunamis in general, rather than trying to simulate the specific details of the 2004 Asian tsunami, we focus on simulations of an idealized case, with bathymetry typical of an ocean basin, and initial conditions typical of a tsunami that might be induced by an earthquake of a magnitude 8-9 or more. The model domain consists of a 400-km-long by 200-km-wide channel. The bathymetry includes a 300-km-long abyssal plain with a constant depth of 4000 m, a 40-km-wide continental slope over which the seafloor rises to 200 m, followed by a continental shelf also 40 km wide. The beach above the mean water level continues with the same slope as the shelf until reaching an elevation of 25 m, remaining flat until the end of the domain. The bathymetry is uniform in the cross-channel direction, and a sponge zone is imposed along all four boundaries. The grid spacing is 250 m.

For all experiments, the initial free surface is level and at rest, except in the area of the earthquake. Here we assume that displacement takes the form of a 100-km-wide delta function in which half of the zone is lifted and half is depressed. The magnitude is similar to that estimated for the recent Sumatra tsunami, with instantaneous, initial upward and downward displacements of +7.2 m and –3.5 m, respectively (Yalciner et al., 2004).

Period covered	Loss, %	Region and country
	Africa	
To 1980s[a]	50	Angola
To 1980s	60	Ivory Coast
To 1980s	50	Gabon
To 1980s	70	Guinea-Bissau
1971-88	4	Kenya
To 1980s	60	Tanzania
	Latin America	
1983-90	-6 (gain)	Costa Rica
1983-90	8	El Salvador
1983-90	72	Guatemala
1970s to 1992	65	Mexico
1983-90	67	Panama
1982-92	25	Peru
	Asia	
To 1986	20	Brunei
To 1980s	55	Indonesia
To 1992-93	74	Malaysia
To 1992-93	75	Myanmar
To 1980s	78	Pakistan
1918 to 87-88	67	Philippines
To 1993	84	Thailand
To 1993	37	Vietnam
	Oceania	
To 1992-93	8	Papua New Guinea

[a] Unless otherwise stated, the decline is relative to the earliest pre-destruction records

Table 2. Mangrove loss in selected countries with available data from the World Resources Institute 1991, 2001 (Adeel & Pomeroy, 2002)

A coral reef can affect wave propagation and development by presenting an abrupt change in bathymetry and through increased bottom roughness. To model the former effect, at any grid point where a reef is present, the bottom depth is replaced by a step that rises to 1 m below the surface. This was done for the three grid lines between the 11 and 14 m isobaths. In FUNWAVE, wave energy dissipation is simulated by friction with the sea floor which is represented by a quadratic bottom drag formulation (i.e., a drag coefficient multiplied by the velocity magnitude and the relevant velocity component). The default value of the dimensionless, bottom drag coefficient is 0.001. To model the roughness effect, we arbitrarily assumed that the bottom drag coefficient at the reef grid points is the default value multiplied by a constant factor. A range of multiplicative factors between two and twenty was tested. The changes in the wave height over and beyond the reef were not overly sensitive to the values tested and, thus, we chose a factor of ten as representative.

Mangroves present a permeable obstacle through which the wave can continue to propagate, but the densely-packed roots and trees will lead to more rapid wave-energy dissipation. Here, too, we model this effect through an increase of the drag coefficient. Field studies of the impact of mangroves on short-period (wind) waves (Mazda et al., 1997) as well as wave tank models of tsunamis (Hiraishi & Harada, 2003) show that the increased rate of energy dissipation across mangrove or other tropical tree forests can be expressed as an equivalent drag coefficient that can be several orders of magnitude larger than usual. Other investigators studied the dependence of the drag coefficient as a function of the tree density for longer period waves. Thus for tidal scale flow Mazda et al. (2005) estimated the drag coefficient to be in the range of O(0.1) - O(10) while for a tsunami Teh et al. (2009) estimated values of O(1). Thus, in our simulations at grid points where mangroves are assumed to be present, we multiplied the default drag coefficient by a factor of 200, thus giving a value of 0.2 which is at the lower end of the values estimated by Mazda et al. (2005) and somewhat smaller than the values of Teh et al. (2009). The mangroves extend from a water depth of 2.5 m to a land elevation of 3.75 m, thereby covering a 6-grid-point, 1.5-km-wide coastal strip.

3. Model results

Here we present results from four simulations for our typical rectangular ocean basin with various combinations of fringing reefs and mangroves and an initial surface displacement representative of the recent Sumatra earthquake. The simulations are designated as: 1) NONE - a smooth, sandy bottom only; 2) REEF - with a 750 m wide reef in the form of a bathymetric step rising abruptly from 12 m to 1 m below the surface; 3) MANGROVE - with a 1.5 km wide mangrove forest centered on the still water coastline in which the bottom drag is significantly increased; and 4) BOTH - with a reef and a mangrove forest. The results are summarized in Figure 4, where we show the time records of virtual wave gauges located at four points along the centerline of our channel. The points are: (A) immediately seaward of the coral reef; (B) immediately behind the reef; (C) at the mean water level coastline, which is roughly midway through the mangrove; and (D) immediately landward of the mangrove.

At point A, there was a clear separation between the simulations with and without the reef as the primary wave approached and passed the station. Until minute 42, the NONE and MANGROVE curves were coincident as were the REEF and BOTH curves. In the two simulations with the reef, the shoaling effect of the reef led to a maximum wave height on the seaward side that was more than 40% larger than in the cases without the reef, i.e., 11.7 m and 8.3 m, respectively. As the secondary wave approached, the curves from the four simulations began to separate and fall into two new groups - with and without mangroves. The mangroves apparently reflected a certain amount of the wave energy back to the sea, thereby producing a higher secondary wave than in the no-mangrove cases. Behind the reef, at point B, the dissipating effect of the reef was evident. For the two simulations with the reef, the maximum primary wave height was reduced by 25%, reaching only 7.1 m as compared to 9.4 m for the two cases without the reef. Also, as at point A, the influence of wave energy reflection by the mangroves can be seen with the appearance of a significant secondary wave only in the two simulations that included the mangroves.

Point C is the mean still-water coastline. From here we began to see noticeable differences between the results of the simulations. Not surprisingly, the highest wave occurred in the NONE simulation. Based on additional simulations run without the reef and mangrove, we found that an important parameter that controls the extent of run up and inundation is the slope of the beach. However, since our interest here is to examine the role of the reef and the mangrove, we only present the results with the same bottom slope in all experiments. The order of the maximum wave heights above-ground for the four simulations is 11.8 m, 9.8 m, 8.5 m, and 7.3 m for NONE, REEF, MANGROVE, and BOTH, respectively.

Finally, point D is located four grid points landward of the coast (one point past the mangrove). As at point C, the NONE simulation produced the highest wave, with a height of 8.4 m above-ground. The additional energy dissipation due to the reef reduced the wave height by 26%. In contrast, the mangrove forest was very effective in blocking most of the wave energy, leading to a 93% reduction in wave height. When both the reef and mangroves were included, the tsunami was completely attenuated and did not emerge from the forest. From our results, it is clear that while both coral reefs and mangroves contribute to mitigating the destruction caused by tsunamis, it is the latter that provide the most effective protection.

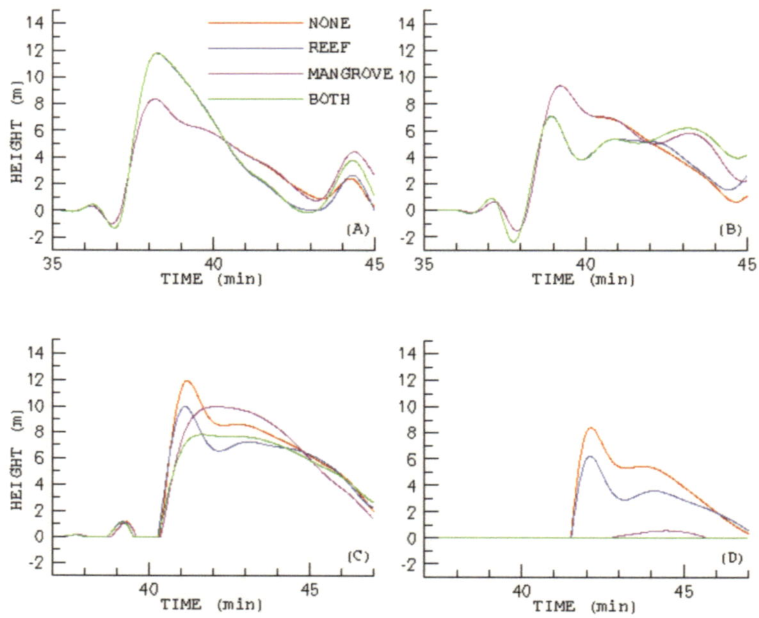

Fig. 4. Time series from wave gauges located at points (A) seaward of the reef, (B) immediately behind the reef, (C) still-water coastline, which is mid-way through the mangrove, and (D) immediately behind the mangrove. For the sea points (A) and (B), the values shown are the wave heights relative to the mean water level while for the land points (C) and (D), the values shown are the water levels relative to the land elevation. At point A, the curves for the primary wave for the NONE and MANGROVE cases are identical as are the curves for the REEF and BOTH cases. For the BOTH case the tsunami never reaches point D, beyond the mangrove, as indicated by the horizontal line

4. Conclusion

Based on our results and the results of other field and modeling studies, we can conclude with reasonable confidence that the numerous anecdotal media reports describing the protection afforded by intact coral reefs and mangrove belts from the 2004 Asian tsunami are most likely true. Indeed, the paired reef-mangrove belt can significantly reduce and, in some cases, totally dissipate tsunami-like wave energy, thereby reducing the destruction of property, and most importantly, greatly reducing the human death toll. In summary, the conservation of coral reefs and adjacent mangrove forests, the remediation of damaged ones, and the massive planting of new mangroves, will most likely contribute significantly to the reduction and, perhaps, prevention of destruction on the scale of the Sumatra tsunami, in addition to protecting shorelines and contributing to the preservation of our planet's vanishing biodiversity. The importance of preserving and planting mangrove forests is summarized in the post-tsunami document "The Tsunami and Coastal Wetlands - Recommendations for Action" (2005).

5. Acknowledgements

We would like to thank Prof. James T. Kirby from the Center for Applied Coastal Research, University of Delaware, for providing the FUNWAVE code and user's manual.

6. References

Achituv, Y. & Dubinsky, Z. (1990). Evolution and zoogeography of coral reefs, In: *Coral Reefs: Ecosystems of the World*, Dubinsky, Z. (Ed.), pp. 1-10, Elsevier, Amsterdam

Adeel, Z. & Pomeroy, R. (2002). Assessment and management of mangrove ecosystems in developing countries. *Trees*, 16, 235-238

Brandt, J. (1883). Über die morphologische und physiologishe Bedeutung des Chlorophylls bei Tieren. *Zool. Stn. Neapol.*, 4, 191-302

Bryant, D.; Burke, L.; McManus, J. & Spalding, M. (1998). *Reefs at Risk; a Map-based Indicator of Threats to the World's Coral Reefs*, World Resources Institute (WRI), Washington DC

Buddemeier RW; Kansas Geological Survery; Kleypas, J.A.; National Center for Atmospheric Research & Aronson, R.B. Dauphin Island Sea Lab. (February 2004). *Coral Reefs & Global Climate Change: Potential Contributions of Climate Change to Stresses on Coral Reef Ecosystems Prepared for the Pew Center on Global Climate Change.*

Channel News Asia (2005). Tsunami calamity highlights key protective role of coral, mangroves. Posted on January 6, 2005 07:24 PM. Filed under: Environment

Chapman, V.J. (ed.) (1977). *Wet Coastal Ecosystems, Ecosystems of the World*, Vol. 1. Elsevier, Amsterdam

Chen, Q.; Kirby, J.T.; Dalrymple, R.A.; Kennedy, A.B. & Chawla, A. (2000). Boussinesq modeling of wave tranformation, breaking, and runup, II: 2D. *Journal of Waterway, Port, Coastal and Ocean Engineering*, 126, 48-56

Cripps, S., Director of the Global Marine Programme at the environment group WWF International (2005). Terradaily Paris (AFP) Jan 06, 2005, Terra Wire

Dahdouh-Guebas, F.; Jayatissa, L.P.; Di Nitto, D.; Bosire, J.O.; Lo Seen, D. & Koedam, N. (2005). How effective were mangroves as a defence against the recent tsunami? *Current Biology*, 15, R443-R447

Danielsen, F., Sorensen, M.K., Olwig, M.F. *et al.* (2005). The Asian tsunami: a protective role for coastal vegetation. *Science*, 310, 643

Dubinsky, Z. & Stambler, N. (1996) Eutrophication, marine pollution and coral reefs. *Global Change Biology*, 2, 511-526

Duke, N.C.; Yuk King Lo, E. & Sun, M. (2002). Global distribution and genetic discontinuities of mangroves - emerging patterns in the evolution of *Rhizophora. Trees*, 16, 65-79

Galanopoulos, A. G. (1960). Tsunamis observed on the coasts of Greece from antiquity to present time. *Annali de Geofisica*, X111, 371-386

Hanagata, N.; Takemura, T.; Karube, I. & Dubinsky, Z. (1999). Salt/water relationships in mangroves. *Israel Journal of Plant Sciences*, 47, 63-76

Hiraishi, T. & Harada, K. (2003). Greenbelt Tsunami Prevention in South-Pacific Region, Report of the Port and Airport Research Institute 42, 1-23. Available at http://eqtap.edm.bosai.go.jp/useful_outputs/report/hiraishi/data/papers/greenbelt.pdf

Hoegh-Guldberg, O. (2004). Coral reefs in a century of rapid environmental change. *Symbiosis*, 37, 1-31

Hogarth, P.J. (1999). *The Biology of Mangroves*, Oxford University Press, Oxford

Kaban, M.S., Indonesia's Forestry Minister (2005). Reuters, Jakarta, Indonesia. Updated: 8:40 a.m. ET Jan. 25, 2005

Kennedy, A.B.; Chen, Q.; Kirby, J.T. & Dalrymple, R.A. (2000). Boussinesq modeling of wave tranformation, breaking, and runup, I: 1D. *Journal of Waterway, Port, Coastal and Ocean Engineering*, 126, 39-47

Kuhlman, D.H.H. (1998). The sensitivity of coral reefs to pollution. *Ambio*, 17, 13-21

Mangrove Action Project, Tsunami Action Alert (January 4, 2005). http://www.earthisland.org/map/tsunami.htm

Mazda, Y.; Magi, M.; Kogo, M. & Hong, P.N. (1997). Mangroves as a coastal protection from waves in the Tong King delta, Vietnam. *Mangroves and Salt Marshes*, 1, 127-135

Mazda, Y.; Kobashi, D. & Okada, S. (2005). Tidal-scale hydrodynamics within mangrove swamps. *Wetlands Ecology and Management*, 13, 647-655

Pararas-Carayannis, G. (1973). The waves that destroyed the Minoan empire. *Sea Frontiers*, 19, 94-106

Reuters (21 Feb 2005). 05:55:00 GMT

Spalding, M.D.; Ravilious, C. & Green, E.P. (2001). *UNEP-WCMC World Atlas of Coral Reefs*, University of California Press, California

Teh, S.Y.; Koh, L.H.; Liu, P.L.-F.; Ismail, A.I.M. & Lee, H.L. (2009). Analytical andnumerical simulation of tsunami mitigation by mangroves in Penang, Malaysia. *J. Asian Earth Sci.*, 36, 38-46

The Tsunami and Coastal Wetlands - Recommendations for Action Special Session on Tsunami and Coastal Wetlands, Asian Wetland Symposium (9 February 2005). Global Environment Centre, 2nd Floor, Wisma Hing, No. 78, SS2/72, 47300 Petaling Jaya, Selangor DE, Malaysia

Watts, P.; Grilli, S.T.; Kirby, J.T.; Fryer, G.J. & Tappin, D.R. (2003). Landslide tsunami case studies using a Boussinesq model and a fully nonlinear tsunami generation model. *Natural Hazards and Earth System Sciences*, 3, 391-402

Wei, G.; Kirby, J.T.; Grilli, S.T. & Subramanya, R. (1995). A fully nonlinear Boussinesq model for surface waves, Part I: Highly nonlinear unsteady waves. *Journal of Fluid Mechanics*, 294, 71-92

Wilkinson, C. (Ed.) (2002). *Status of Coral Reefs of the World*, Australian Institute of Marine Sciences

Yalciner, A.C.; Taymaz, T.; Kuran, U.; Pelinovsky, E. & Zaitsev, A. (2004). Model studies on December 26, 2004 Indian Ocean tsunami. http://yalciner.ce.metu.edu.tr/sumatra/

Yanagisawa, H.; Ksohimura, S.; Goto, K.; Miyagi, T.; Imamura, F.; Ruangrassamee, A. & Tanavud, C. (2009). The reduction effects of mangrove forest on a tsunami based on field surveys at Pakarang Cape, Thailand and numerical analysis. *Estuar. Coast. Shelf Sci.*, 81, 27-37

Zimmermann, U.; Wagner, H.-J.; Heidecker, M.; Mimietz, S.; Schneider, H.; Szimtenings, M.; Haase, A.; Mitlohner, R.; Kruck, W. & Hoffmann, R. (2002). Implications of mucilage on pressure bomb measurements and water lifting in trees rooting in high-salinity water. *Trees - Structure and Function*, 16, 100-111

2

On the Need for a Tsunami Warning System in the North East Atlantic Area (Gulf of Cadiz)

Baptista M. A.[1,3], Omira R.[1,2], Matias L.[1,4], Miranda J. M.[1], Annunziato A.[5], Carrilho F.[4] and Kaabouben F.[2]
[1]*Instituto Dom Luiz University of Lisbon, CGUL, IDL, Lisbon,*
[2]*University Ibn Tofail, Faculty of Sciences, Kénitra,*
[3]*Instituto Superior de Engenharia de Lisboa,*
[4]*Instituto de Meteorologia, Lisbon,*
[5]*Joint Research Center, European Commission,*
[1,3,4]*Portugal*
[2]*Morocco*
[5]*Italy*

1. Introduction

Tsunami events are described in the historical documents of coastal countries in the North East Atlantic area but the need for a Tsunami Early Warning System was recognized only after the Indian Ocean tsunami on December 2004. The huge devastation and loss of lives caused by this unique event point out the attention of worldwide authorities to the need to implement a global tsunami warning system and the need for long term preparation of populations in danger to tsunamis was also recognized.

The key elements of a Tsunami Warning System (TWS) include: tsunami hazard and risk assessment, seismic monitoring and earthquake detection, sea level monitoring and tsunami detection, dissemination of warnings and mitigation programs and public awareness. Tsunami hazard assessment is a key element of all TWS. This assessment requires knowledge of: (i) past tsunami occurrences such as historical and paleo-tsunamis and (ii) possible earthquake-tsunami sources, their likelihood of occurrence and their effects along the threatened coasts. The compilation of tsunami catalogs and inundation mapping constitute the main components of tsunami hazard assessment. In order to complete tsunami hazard assessment it is desirable to compute the probability that a tsunami of a particular size will strike a particular section of coast (IOC, 2008).

The operational components of a TWS are the seismic and the tsunami detection networks. The existence and operation of a good seismic network is essential to the performance of a TWS due to the fact that seismic waves are recorded almost immediately after the onset of an earthquake as they travel through the earth much faster than tsunami waves propagate through the ocean. Once the earthquake is identified, the warning centres use sea level data to confirm that a tsunami was generated or, if there are no changes in sea level, to cancel the alert messages. In those regions where ocean bottom sensors (tsunameters) are in place the generation of a tsunami can be confirmed well before its arrival to the coast.

The impact of tsunamis may be considerably reduced if timely warnings are issued to the endangered population and if these populations are prepared well in advance to react properly to a tsunami warning, especially in areas where tsunami travel times to shore are rather short. In the Chile 2010 case several saved their lives thanks to the historical memory of the previous event of 1960 and escaped to high levels as they felt the earthquake and/or saw the sea level retracting.

In this chapter we present the efforts to assess the tsunami hazard towards the construction of an effective tsunami warning system in the Gulf of Cadiz, which could work as the regional tsunami watch centre for the entire North East Atlantic region. These efforts include the compilation of the tsunami catalogs, identification/evaluation of tsunami sources, site-specific inundation mapping for some of the most sensitive areas and the design for a reliable tsunami detection network based upon the existent coastal tide gage network in Southwest Iberia and Northwest Morocco and the installation of a minimum number of tsunameters that enssure an acceptable protection for the endangered populations. Finally we present the modus operandi of the Portuguese Tsunami warning system.

2. Tsunami events in the North East Atlantic

Tsunami catalogs constitute the primary tool for understanding tsunami hazard specific area and must be able to enlighten the cause and impact of each event along the study areas. The building of a catalog is mainly based upon the interpretation of the historical accounts that can provide information on intensity and frequency of occurrence in a specific region for a specific time period. Due to the fact that written reports just cover a short period of time the extension of the time span of the catalog is only possible through investigation on paleo-tsunami deposits.

Until last century tsunami descriptions were found in historical reports and in earthquake catalogs as a secondary effect of earthquakes and/or volcanic activity.

The earliest tsunami catalog, in Europe, appeared in 1962 (Ambraseys, 1962). The first attempt to compile an European catalogue including both the Atlantic and Mediterranean areas was concluded in mid to late nineties in the framework of GITEC (Genesis and Impact of Tsunamis on the European Coasts), and GITECTWO (Genesis and Impact of Tsunamis on the European Coasts - Tsunami warning and observation) projects, this database was updated within TRANSFER (Tsunami Risk ANd Strategies For the European Region) project.

In the aftermath of the Indian ocean tsunami, with the implementation of a global tsunami warning system and the division of the ocean in four regions: Pacific, Indian, Caribean and North East Atlantic & Mediterranean, new databases for each regions are now available at the IOC-UNESCO tsunami website. The database for the North Atlantic and Mediterranean is designated as NEAMTWS database (http://www.ioc-tsunami.org/images/stories/File/neamtws.zip).

According to this database we can distinguish three main sub-regions in the North East Atlantic: the Atlantic (AT) area extending from the north coast of Africa towards France, the North Sea (NS) between the United Kingdom and the Scandinavian Peninsula and the Norwegian Sea (NWS).

There are three main sources/causes of tsunamis: submarine earthquakes, landslides (aerial or submarine) and volcanic eruptions. Tsunami events recorded or identified in the North East Atlantic region are mainly due to the first and second mechanisms.

Northern Europe events, in the North sea and Norwegian sea areas, are rather infrequent and mainly local events. The main source of tsunamis are those generated in the fjords by gravitative landslides generated in the fjords (Tinti, 1993) such as the April, 7th, 1934 that caused gigantic waves in the Tafjord (NEAMTWS database).

The tsunamis generated in the area designated above as AT are mainly caused by submarine earthquakes generated in the complex plate boundary domain extending from the Azores Islands towards the western Mediterranean. The average convergence velocity between the Nubia and Eurasia plate boundary is circa 5 mm/year close to Iberia. This value is compatible with large return periods for very strong tsunamigenic earthquakes.

The largest landslide generated tsunami in this region was caused by the Storegga submarine landslide on the continental slope of Norway about 8000 BP (before present). According to (Bondevik et al., 2003) a slide of circa 3500 km^2 slide into the ocean and caused a tsunami that generated onshore deposits in Norway, Faroe Islands and Scotland. The study of tsunami deposits in western Norway suggest a 10-12m run up.

In the historical period the largest tsunami was the November, 1st, 1755, generated by a strong magnitude earthquake offshore Iberia. It devastated the Iberian Peninsula and north Morocco Atlantic coasts, causing great damage and casualties. It was observed all over the North Atlantic in central and south America. The magnitude of the earthquake was recently re-evaluated from the macro-seismic field as M_w=8.5±0.3 (Solares & Arroyo, 2004).

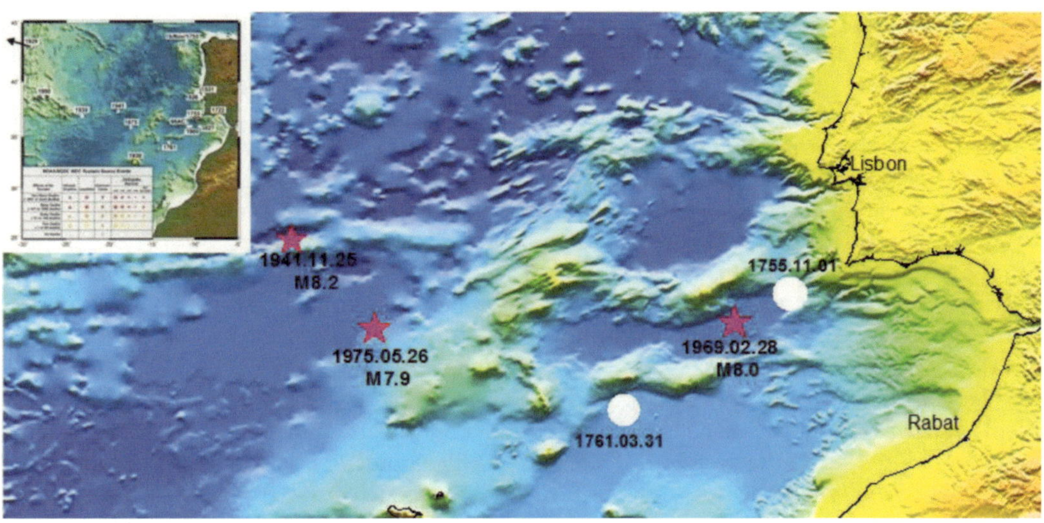

Fig. 1. Great tsunamigenic earthquakes in the North East Atlantic

Reports describe 10-15m waves at Cape St. Vincent and along the Gulf of Cadiz. In Lisbon the number of casualties, due exclusively to the tsunami, is estimated as close to 900 and the inundation distance, in Lisbon downtown, is estimated as 250m (Baptista, 1998). The effects of the tsunami are described in Madeira, Cadiz and Cornwall, and the agitation of closed waters was observed in faraway places like Scotland (see Chambers, 1757). The tsunami crossed the whole Atlantic Ocean being observed in some of the Caribbean Islands, approximately 8 hours after the main shock (Urban, 1755).

Since the installation of the tide-gage networks in the area some tsunamis, generated by submarine earthquakes, were recorded: the 1939 local event in the Azores, the

November,25th, 1941, the February, 28th, 1969 and the May,26th, 1975 events. Two exceptions should mentioned: the 1929 tsunami generated by Grand Banks submarine landslide and the 1930 local tsunami that occurred in Madeira Island caused by an aerial landslide (Baptista & Miranda, 2009).

In figure 1 we present the location of known tsunami events in the North East Atlantic area.

3. Earthquake sources and tsunami scenarios

The western segment of the Eurasia-Nubia plate boundary extends between the Azores toward the western Mediterranean. In the Azores the interplate domain is rather complex as consequence of the small spreading velocity, though it generates spreading along the Terceira and Pico-Faial axes (Miranda et al., 1998; Fernandes et al., 2007). In the area between 24°W and 19°W it is supposed to follow a prominent morphological feature, the Gloria Fault in an almost pure transcorrent way. To the east of 19°W the interplate domain is morphologically complex, characterized by a series of huge ridges and seamounts, as the Gorringe Bank, the Coral Patch and Ampère seamounts. These features delimitate morphologically depressions such as the Horseshoe and Tagus abyssal plains.

The Gorringe Bank is a large uplifted block of oceanic lithosphere approximately 180 km long and 60-70 km wide, trending N55E. The Bank itself is a huge morphological high that reaches 25 m below sea level at Gettysburg and 60 m close to Ormonde seamounts, respectively. This lithospheric block is usually interpreted as an almost continuous section of oceanic crust and upper mantle (Bergeron & Bonnin, 1991) and is associated with a large isostatic anomaly (ca. 300 mGal), which indicates the presence of a thick, high-density body close to the surface (Bergeron & Bonnin 1991; Sartori et al., 1994). For a long time this anomaly has been interpreted as a sign of the lack of isostatic equilibrium (Bergeron & Bonnin, 1991) and, thus, a potential seismogenic feature.

The Horseshoe Abyssal Plain is an elongated feature bounded by the Ampère and Coral Patch seamounts to the south, the Gorringe Bank and Tagus abyssal plain to the north, the Madeira Tore rise to the west and the Iberian continental margin to the east. In its deeper part the crust is around 15 km thick. The two major faults identified in this zone are the Horseshoe Fault (Zitellini et al., 2004) and the Marques de Pombal Fault (Zitellini et al., 2001).

The Horseshoe Fault is a reverse fault oriented perpendicular to the present day kinematic displacement of Nubia with respect to Iberia. It is one of the most important tectonic accidents identified in the area but, while close to the source area of the 28th February 1969 earthquake, the geometry of the fault does not match the structure as mapped by seismostratigraphy.

The Marques de Pombal Fault (Zitellini et al., 2001) is a large active compressive tectonic structure located 100 km offshore SW Cape S. Vicente. It displays a pronounced drag fold on the fault hanging-wall and the height of the escarpment is taller in the north where it reaches 1.2 km.

The Gulf of Cádiz developed as a result of interaction between the southern end of Iberia paleo-margin, the westward displacement of the Gibraltar arc, and the convergence between the Africa and Eurasia plates. The tectonic mechanisms responsible for its modern features imply two different structural behaviors in the Gulf of Cadiz: in the eastern part, from the Horseshoe Abyssal Plain towards the Gibraltar Arc, a mixture of thin skinned and thick skinned tectonics comprising westward thrusting of the upper sedimentary units and the

whole crust respectively; and in the western part, towards the Gorringe Bank and Coral Patch, a thick skinned tectonics comprising the whole crust, and possibly the upper mantle too, by means of high angle reverse faults (Medialdea et al., 2004).
East of Gibraltar Straight lays the Alboran Sea, between the Betic Cordillera in southern Iberia and the Rif-Tell fold and thrust belt in northern Morocco. During the early and middle Miocene the Alboran domain underwent extension and thinning of a previously thickened crust. This geodynamic environment led to the formation of an eastward-thickned wedge of deformed sediments west of Gibraltar (Torelli et al., 1997, Gutscher et al., 2002).
In order to design tsunami scenarios, all the above mentioned earthquake sources were geometrically simplified into Typical faults" (Lorito et al., 2008) (see figure 2 a-b). These typical faults represent the maximum credible earthquake (MCE) scenarios that could generate large tsunamis in the Gulf of Cadiz region.

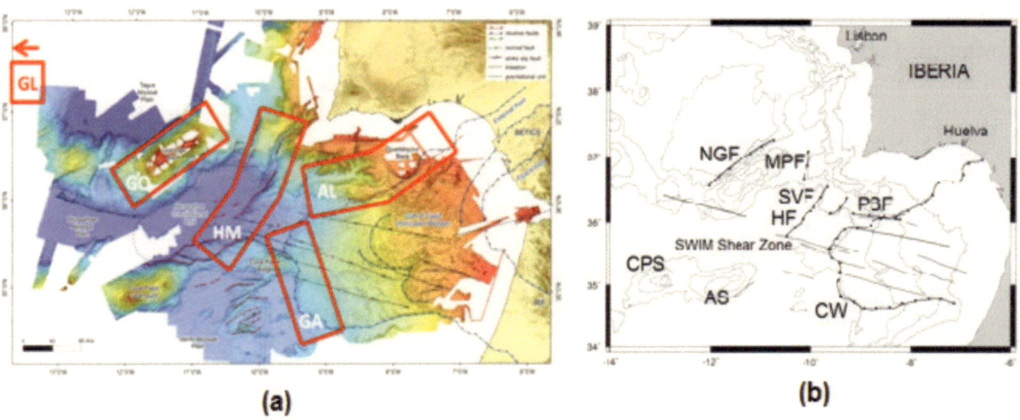

Fig. 2. (a) Earthquake/tsunami source zones (b) Typical Faults

4. Sites-specific inundation mapping

The most obvious hazard from a tsunami is the flooding or the inundation. Damages on coastal areas due to the impact of tsunami waves are the direct result of three factors: erosion, inundation and impact on structures. Tsunami inundation is measured as the maximum horizontal distance reached inland by the water and run-up as the greatest vertical height reached by the waves. Thus, the impact of a tsunami depends strongly on the morphology of the target coasts. A tsunami impacting coastal cliffs can exhibit high run-up with little inundation, whereas a tsunami impacting a flat beach can produce great inundation distance and small run-up.
Site-specific inundation mapping is recognized as an important tool in tsunami hazard assessment. Detailed deterministic modeling based on most credible scenario best serves the purpose of coastal engineering to develop effective measures to protect coastal regions against tsunami threat (Tinti & Armigliato, 2003). Mapping inundation due to the most credible earthquake/tsunami scenarios provide an estimation of the areas that could be flooded if a similar event occurs. The information contained in these maps can be used by emergency planners and civil authorities to draw evacuation maps and to delineate evacuation strategies.

This section is addressed to predict tsunami inundation in various specific sites in the countries surrounding the Gulf of Cadiz. Deterministic tsunami hazard approach based upon particular source scenarios and tsunami hydrodynamic modelling has been employed. The initial condition used in the propagation code is the sea surface elevation due to the vertical displacement of the ocean bottom. Assuming the incompressibility of the water, the initial sea surface deformation is instantaneous and equals the ocean bottom deformation. The computation of the sea bed deformation is based on the elastic dislocation theory computed through Okada's formula (1985) and a discretized dip-directed slip distribution along the fault plane is adopted, following the smooth closure condition that is described in Geist & Dmowska (1999). The initial deformation of the ocean surface is used as initial condition of the tsunami propagation model.

The tsunami propagation model, used in this study, is based on the shallow water wave theory. The numerical code is based upon COMCOT code (Liu et al., 1998), which uses an explicit leap frog finite difference scheme to solve the equations of continuity and momentum on a dynamically coupled system of nested grids. Due to the size of the physical domain of interest in this area the Coriollis's effect can be neglected and it is not considered here. The non linear shallow water equations in Cartesian coordinates are expressed through equations 1, 2 and 3.

$$\frac{\partial \eta}{\partial t}+\frac{\partial P}{\partial x}+\frac{\partial Q}{\partial y}=0 \tag{1}$$

$$\frac{\partial P}{\partial t}+\frac{\partial}{\partial x}\left(\frac{P^2}{H}\right)+\frac{\partial}{\partial y}\left(\frac{PQ}{H}\right)+gH\frac{\partial \eta}{\partial x}+\tau_x=0 \tag{2}$$

$$\frac{\partial Q}{\partial t}+\frac{\partial}{\partial x}\left(\frac{PQ}{H}\right)+\frac{\partial}{\partial y}\left(\frac{Q^2}{H}\right)+gH\frac{\partial \eta}{\partial y}+\tau_y=0 \tag{3}$$

Where η represents the free surface elevation, h is the still water depth, H is the total water depth, $u=(u_x,u_y)$ is the velocity vector, P and Q are the volume flux components in x and y directions respectively; τ_x and τ_y are the bottom frictional terms. H=h+ η is the total water depth.

A system of nested grids is used with increasing resolutions towards the coast in order to ensure a good description of bathymetric and topographic features close to coastal areas. In this grid system the inner (higher resolution) grid has a grid cell size and time step smaller than the outer/adjacent grids. These variations in time step and grid size are used to ensure the stability of the computations. In our case grids of 800m x 800m and 50m x50m are used in the source region (open ocean) and coastal areas respectively. An intermediate grid of 200m x 200m is used to ensure computation stability in the transition between open ocean and coastal areas grids. In order to compute the wave's propagation over dry land (inundation) the code uses the "moving boundary condition" (Wang and Liu, 2006).

In figure 3a–e, we present the initial sea surface displacements computed for each considered submarine earthquake scenario in the Gulf of Cadiz region. In spite of the scenarios' magnitude, which is relatively smaller (8.2) than the 1755 earthquake, the the initial sea surface displacement can be as high as 7m for the average slip values considered (further details in Omira et al., 2009).

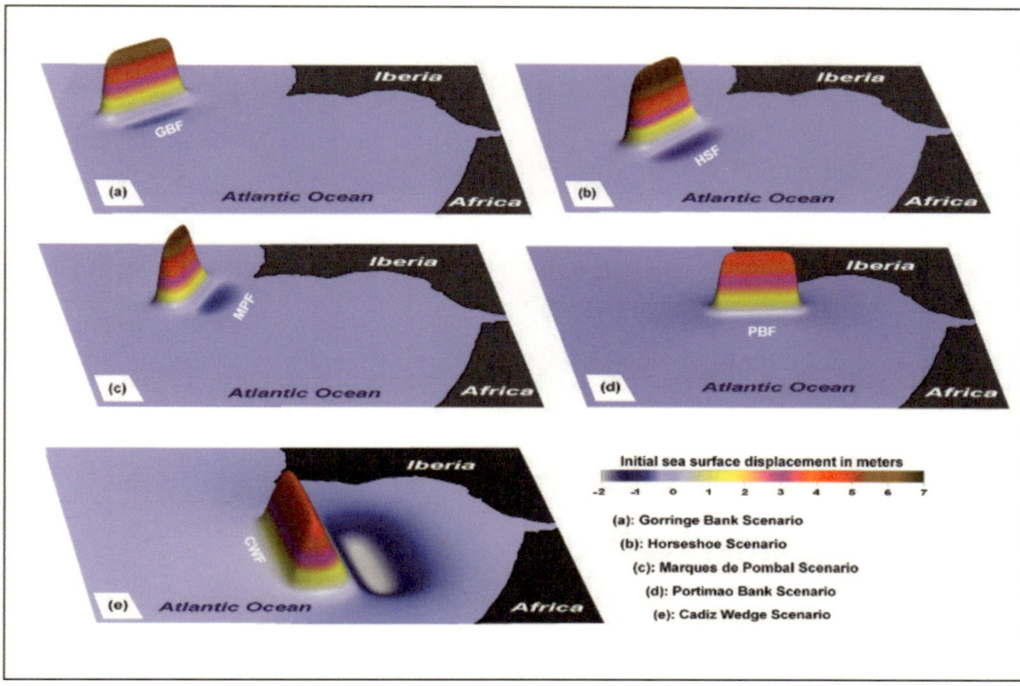

Fig. 3. Initial sea surface perturbation for each tsunami scenario in the Gulf of Cadiz; (a) Gorring Bank scenario; (b) Marques de Pombal scenario; (c) Horseshoe scenario; (d) Portimao Bank scenario; (e) Cadiz Wedge scenario.

Results of tsunami simulations are presented in figure 4. We compute a global MWH (maximum wave height) map in which maximum values of wave height are extracted along the coastline, taking in consideration all the 5 established earthquake scenarios (Fig.4-a). The spatial distribution of the MWH along coasts of Portugal, Spain and Morocco shows values varying from 1m up to 6m. Examination of these results indicates a concentration of tsunami energy in the South of Portugal and in the Northwest of Morocco. This is due to two main reasons: (i) the fault orientation and (ii) the bathymetry. At first order, the tsunami amplitude is maximal in the direction perpendicular to the fault strike, but when the wave reaches the near-shore the bathymetry affects the wave-form and the shallow water amplifies the wave amplitude.

Four selected sites are considered to predict the tsunami inundation: Sagres and Lagos-Portimao in the south coast of Portugal, Huelva located in the SW of Spain and Casablanca in the NW of Morocco (Fig. 4).

For each test areas we compute the maximum flow depth distribution due to the 5 MCE scenarios. Here we present only the worst inundation scenario at each site (Fig 4b-e). For both Sagres and Lagos-Portimao sites the worst tsunami impact is obtained for the Marques de Pombal fault, while for Huelva and Casablanca sites the worst flooding cases correspond to the of Cadiz Wedge and Horseshoe scenarios, respectively. These results are in agreement with the recent studies focused in the tsunami impact in the Gulf of Cadiz region (Omira et al., 2009; Omira et al., 2010b (in press); Lima et al., 2010).

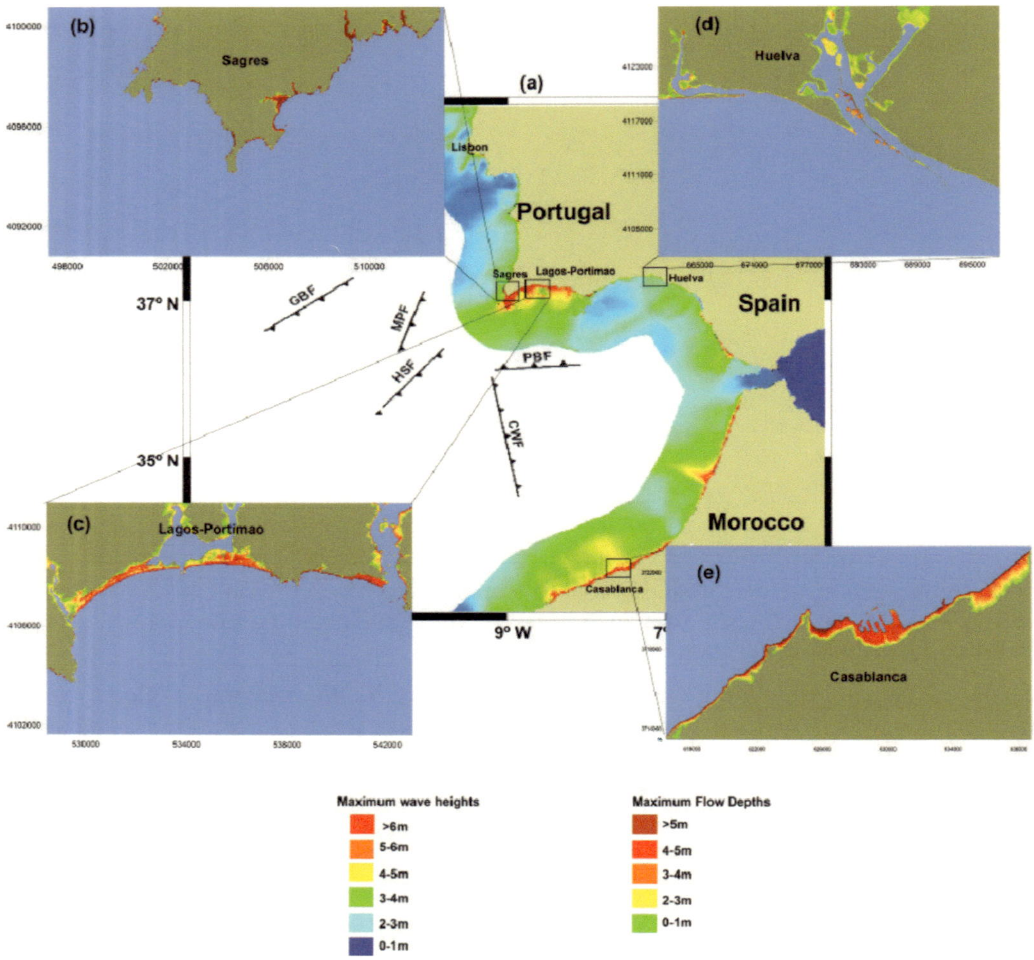

Fig. 4. (a) MWH along the Gulf of Cadiz coasts considering the 5 tsunamigenic scenario (adapted from, Omira et al. 2009).Worst case inundation maps for sites-specific: (b) Sagres site; (c)Lagos-Portimao site(adapted from Omira et al., 2010b(in press)); (d)Huelva site (adapted from Lima et al., 2010) and (e)Casablanca Site.

Predicted tsunami inundation in different sites-specific (Fig.4b-e) shows the high level of vulnerability of coastal areas of Portugal, Spain and Morocco. This vulnerability increases with the high density of occupation and the presence of coastal structures that are not configured and maintained in ways that effectively reduce the risk of exposure to the threat of tsunami.

5. The existing tsunami detection network and its potential improvements

The present sea-level detection network includes solely coastal tide stations from Portugal mainland; Azores and Madeira (cf. Fig.5). No ocean bottom sensors are in place and these are essential for tsunami detection before they hit the coast. The tsunami travel times to coastal points in the Gulf of Cadiz are rather short, 20-30 minutes, and this shows the utmost importance of deploying tsunami ocean bottom sensors.

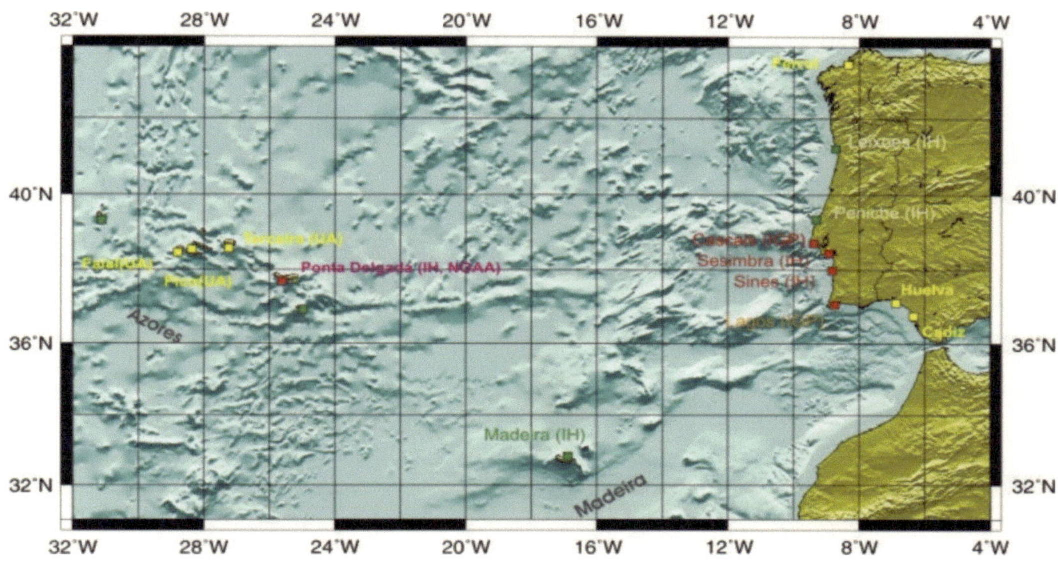

Fig. 5. Coastal tide gage stations connected to Instituto de Meteorologia (in Lisbon).

Recently Omira et al. (2009) published a study on the design of the "optimum" offshore sea-level network for the area. The study takes into account the tsunami radiation pattern and the tsunami travel time to the coast, based upon worst credible earthquake scenario in the area, and focus on maximizing warning times and minimizing the number of tsunameters to be deployed. The main conclusion of this study points to the installation of three DART-like sensors with a spacing of 110 km that are required to constitute the deep ocean tsunami monitoring network (Fig. 6).

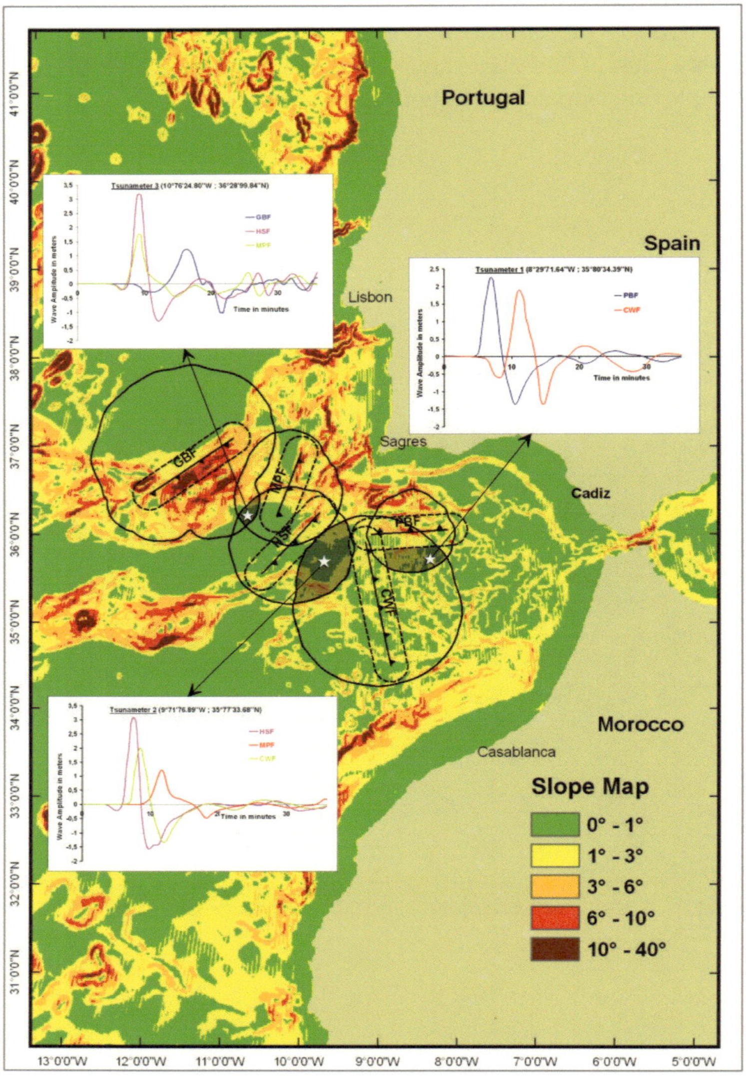

Fig. 6. Offshore tsunameter network for the Gulf of Cadiz. White stars are the recommended position of the ocean bottom sensors. Graphs illustrate waveforms recorded by the proposed stations; only waves detected in the time interval (0, 10)minutes are considered for tsunami detection in order to maximize the warning time (from Omira et al. 2009)

Figure 7 displays the warning time along the coasts of the Gulf of Cadiz area considering all five credible tsunami scenarios and taking into account the established tsunameters network. The proposed design provides a minimum advance warning time of 7.0 to 15.4 min for the first threatened coastline.

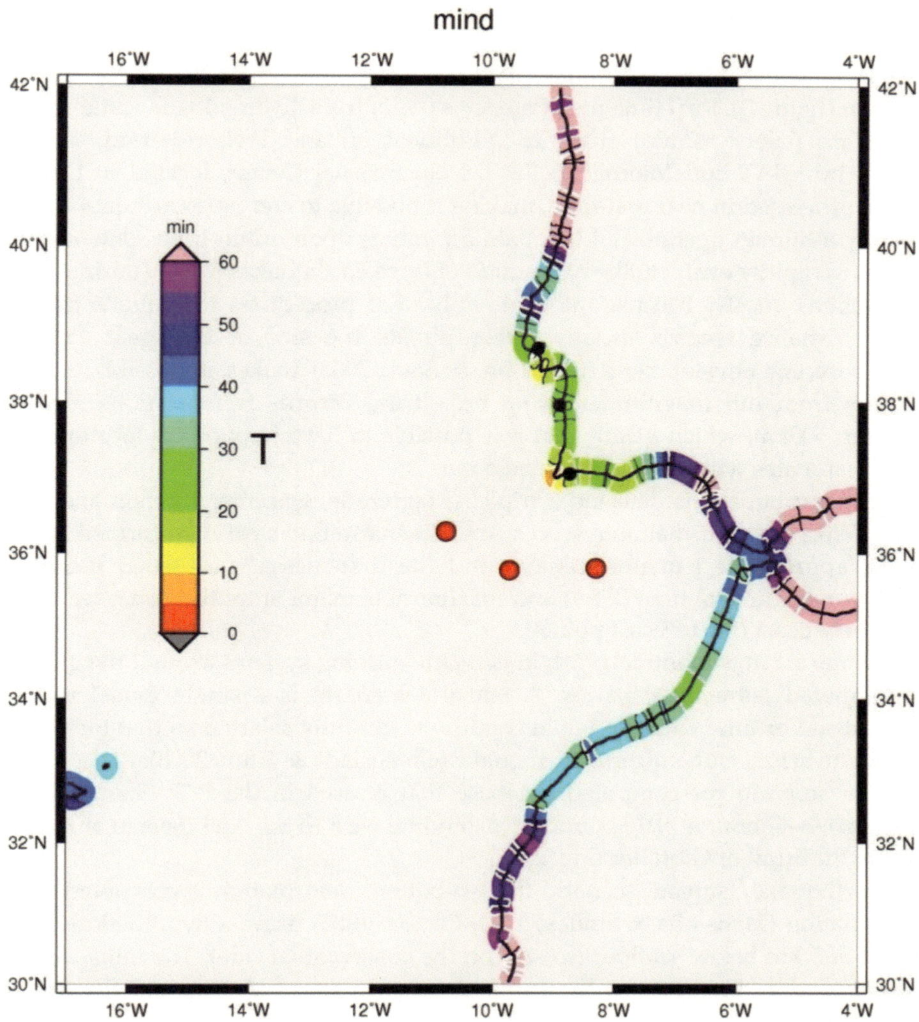

Fig. 7. Red dots correspond to tsunameters network; color scale represents the warning time

6. Modus Operandi of the PtTWS

The development of the Portuguese Tsunami Warning System follows a sequential data collection and analysis, from the origin of the tsunamigenic earthquake to the issuing of messages to the Portuguese Civil Protection authorities, as indicated schematically in figure 8. The PtTWS includes three main components: the seismic detection, the tsunami detection/analysis and the issue of warnings/alerts. In Portugal, the Instituto de Meteorologia (IM) is the only national institution operating on a 24x7 basis that is also responsible for the Portuguese seismic network, which makes IM the natural candidate to host the Portuguese system. IM is the Portuguese National Tsunami Focal Point as regards the NEAMTWS.

The seismic technology is the first one to be used in the detection of a possible tsunamigenic earthquake. In recent times there was a significant enhancement in the seismic network coverage around the Azores-Gibraltar plate boundary, and in particular for the Gulf of Cadiz region (figure 3). Real time data (latency <10sec) from 45 broadband stations and near real-time data (latency~2min) from 28 additional enhanced short-period stations are transmitted by VSAT and Internet to the IM Operational Center, located in Lisbon. The records are processed in near real-time, making it possible to compute earthquake locations, validated by a human operator, in less than 5 minutes upon origin time. One of the major problems is to rapidly evaluate the magnitude of large earthquakes (Mw>7) using data from regional stations, mostly because the commonly used procedures to evaluate magnitudes from short distance records usually underestimate the size of the great earthquakes. However, a recent development (Bormann & Saul, 2008) makes it possible to compute reliable Mw from mB magnitudes using broadband records from stations at distances starting from 500km, which means that it is possible to have hypocentre location and Mw magnitude estimates within the mentioned 5 min.

Each time an earthquake is detected a triplet (Magnitude, epicenter location and depth) is computed. With this information a search in the scenario database is performed in order to choose the appropriate tsunami scenario and the first message is issued including the estimated tsunami arrival time (ETA) and maximum tsunami amplitude on selected sites at the Portuguese coast (the forecast points).

The state of the art in tsunami forecast, in tsunami warning systems around the globe, relies on pre-computed tsunami scenarios. A tsunami scenario is a single model run that is calculated ahead of time with the initial conditions carefully selected so that they are likely to represent an actual tsunamigenic earthquake (Greenslade & Titov, 2008).

The tsunami scenario pre-computed database that is used in the PtTWS covers the area along the Azores-Gibraltar plate boundary extending west to the Archipelago of Azores and eastward to the Strait of Gibraltar (Fig. 2).

For each earthquake/tsunami scenario the sea-bottom deformation is computed using the Okada's equation (Mansinha & Smiley, 1971; Okada, 1985) using a hypothetical top of the fault depth of 5 km below seafloor (to stay on the conservative side). The initial sea-surface elevation is used as the input to the tsunami propagation model; should the earthquake depth be greater, correction factors are applied to take that into account. The propagation model uses Mader's SWAN non linear shallow water equation model; mass and momentum conservation equations in two dimensions are solved over the calculation space. The initialization of the calculation domain is performed taking into account the size of the initial disturbance (this means that for smaller earthquakes the calculation domain is smaller and the cell size smaller).

The bathymetric grid used in tsunami propagation was generated from a compilation of multisource depth data from multibeam surveys, in the area of interest, digitalized bathymetric charts of different scales. The different data were merged on a unique database and all data transformed to WGS84/UTM coordinates (fuse 29). The final scenario database contains scenarios calculated one point every half a degree, all magnitudes from 6.5 to 8.75 (0.25 interval) in a total of over 6000 scenarios.

The next level of decision in the PtTWS is taken with the help of the Tsunami Analysis Tool (TAT). TAT is being developed at the Joint Research Centre of the European Commission

(JRC) to assist the tsunami warning centre (TWC) operator in deciding if a tsunami has been generated or not, in case of a large enough seismic event.

It is well known that only a few of the large earthquakes do generate tsunamis. It is then essential to confirm (or to exclude) the generation of a tsunami using the sea level observations. Through TAT it is possible to compare in real-time the sea level observations and the tsunami waveforms for the selected scenario, allowing for a fast evaluation of the generation of the tsunami (Fig. 8). This process allows the updating of messages to the Civil Defense and emergency authorities (multiple languages messages may be generated).

It is important to note that the operational procedures planned for the PtTws based on tectonically credible scenarios (worst case) extend the basic requirements defined for the NEAM Regional Tsunami Warning Centres (RTWCs) where a simple distance and magnitude criteria is proposed. However, the recommend message thresholds are to be adopted. The current alerting procedure defined for NEAMTWS is to send an advisory message if the maximum tsunami amplitude exceeds 0.2 m and to send a watch message if that amplitude is above 0.5 m. Below the 0.2 m threshold Information messages will be delivered to the Portuguese Civil Defense in case of a significant earthquake or in case any size earthquake is felt close to the coast. In addition, dedicated tsunami messages are planned to be broadcast to the responsible of coastal sensible infrastructures like harbors.

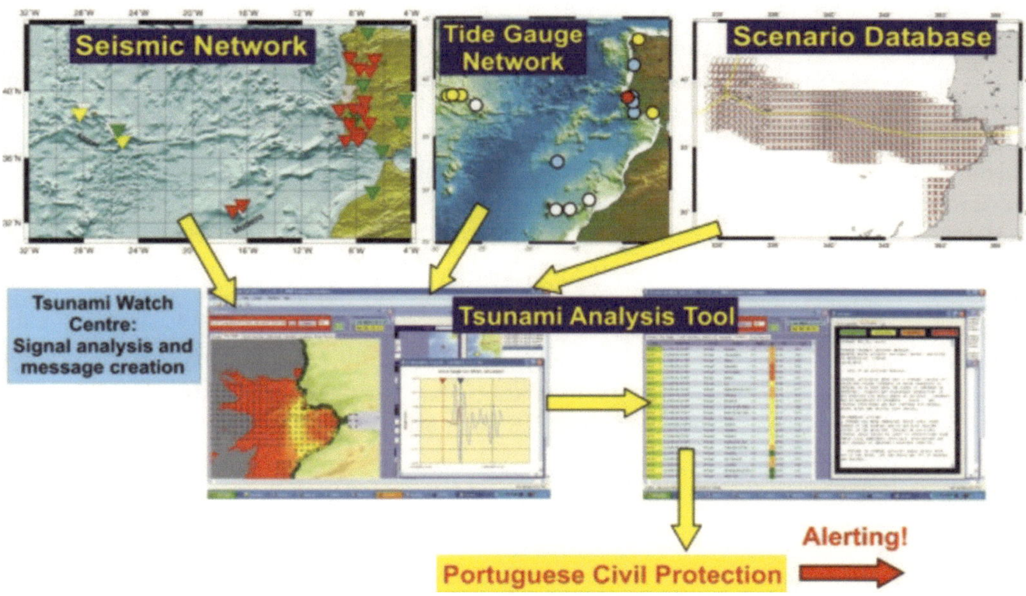

Fig. 8. Main operational components of the PtTWS (from Annunziato et al., 2009)

One of the most important activities of the PtTWS operators is the detection of the tsunami event. This is achieved through the analysis of the sea level measurements - bottom pressure sensors, if available and tidal sensors - to check whether they deviate from the usual tidal trend (this tidal filtering is performed automatically by the TAT using a configurable moving average). By comparing the real signal with the expected calculated value for that location it is possible to anticipate the alert level and be able to say what would occur in other locations (tsunami forecast). In case the expected calculated trend does not reflect the

real trend, it is possible either to adjust the calculation with a numerical factor or use a different scenario that better represents the current data. This may happen because the initial estimates of the epicenter location, fault depth and magnitude are affected by a large uncertainty and thus the assumed scenario could not reflect the real tsunamigenic source.
TAT contains also dedicated sections for running training simulations which may be very useful for preparing the operators that will be sitting 24/7 at the control panel and need to take appropriate decisions. The simulation mode allows using historical measured events (in Atlantic or other world areas) which are injected at the appropriate time to be compared with simulations from the scenario database.

7. Conclusion

The first three sections of this chapter have been addressed to tsunami hazard evaluation through historical events compilation, tsunamigenic sources determination and site-specific inundation mapping. Outcomes of this studies showed clearly that this is a sensitive area to tsunami occurrence and also that the tsunami phenomenon presents a real menace for the surrounding countries: Morocco, Portugal and Spain. Only the implementation of an end-to-end tsunami early warning system for North East Atlantic region reassures the coastal communities and contributes to their safety in case of tsunami occurrence.
Since there is no offshore tsunami sea-level detection network implemented in the region, we presented, a proposal of tsunameters network for the area. This strategy recommended that 3 DART-like stations should be deployed. Their installation locations have been determined so as to ensure a maximum WT, as well as maximum coverage of tsunami potential hazard areas. The approach provides 7.0 to 15.4 min as a minimum advance warning time for the first threatened coastline. This time may not be sufficient for a global evacuation procedure. This is the reason why awareness campaigns and exercises should be implemented in such region, explaining in particular to the inhabitants not to wait official tsunami warning in case of large felt earthquakes, but to immediately move to higher ground or vertically evacuate in a concrete building, as demonstrated by recent Chile tsunami event of 2010.
Due to the fact that tsunamis are considered infrequent in this region of the globe, although their impact can be disastrous, the implementation of an early warning system in such areas, where the last catastrophic event happened 250 years ago turns out to be a difficult task.

8. References

Ambraseys, N.N. (1962). Data for the investigation of the seismic sea waves in the eastern Mediterranean. *Bull. Seism. Soc. Am.* 52, 4, 895–913.

Annunziato, A.; Carrilho, F.; Matias, L.M.; Baptista, M.A. & Omira, R. (2009). Progress in the establishment of the Portuguese Tsunami Warning System. *EMSC (European Mediterranean Seismological Centre) Newsletter*. April 2009. ISSN 1607-1980

Baptista, M. A.; Heitor, S.; Miranda, J. M.; Miranda, P. & Mendes Victor, L. (1998). The 1755 Lisbon tsunami; evaluation of the tsunami parameters. *J. Geodynamics,* 25, 143-157.

Baptista, M.A. & Miranda, J.M. (2009). Revision of the Portuguese catalog of tsunamis. *Nat. Hazards Earth Syst. Sci.*, 9, 25–42.

Bergeron, A. & Bonnin, J. (1991). The deep structure of Gorringe Bank (NE Atlantic) and its surrounding area. *Geophys. J. Int.*, 105, 491-502.

Bondevik , S.; Mangerud, J.; Dawson, S.; Dawson, A. G. & Lohne, O. (2003) Record-breaking height for 8000 years-old tsunami in the North Atlantic. *EOS*, 84, 289–300.

Bermann, P. & Saul, J. (2008). The new IASPEL standard broadband magnitude mB. *Seism. Res. Lett.*, 79,5, 698–706.

Chambers, C. (1757). Another Account of the Same Earthquake at Madeira in a Letter from Charles Chambers to his Father, in London, dated at Madeira, Novem. I, 1755, Communicated by Lewis Crusius, D. D. F. R. S., Lett. XVIII, for the Year 1756, Philosophical Transactions, Vol. XLIX, Part II, London.

Crawford, D.A. & Mader, C.L. (1998). Modelling asteroid impact and tsunami. *Science of Tsunami Hazards*, 16, 1, 21 -30.

Dawson, A.G., Lockett, P. & Shi S. (2004). Tsunami hazards in Europe. *Environment International*, 30, 577- 585.

Fernandes, R.M.S., Miranda, J.M.; Meijninger, B.M.L.; Bos, M.S.; Noomen, R.; Bastos, L.; Ambrosius, B.A.C. & Riva, R.E.M. (2007). Surface Velocity Field of the Ibero-Maghrebian Segment of the Eurasia-Nubia Plate Boundary. *Geophys. J. Int.*, 169, 1, 315-324.

Geist, E. L. & Dmowska, R. (1999). Local tsunamis and distributed slip at the source. *Pure Appl. Geophys.*, 154, 485-512.

Greenslade, D. & Titov, V. (2008). A Comparison Study of Two Numerical Tsunami Forecasting Systems. *Pure appl. geophys.* 165, 1991–2001

Gutscher, M-A.; Malod, J.; Rehault, J-P.; Contruci, I.; Klingelhoefer, F.; Mendes-Victor, L. & Spakman, W. (2002). Evidence for active subduction beneath Gibraltar. *Geology*, 30, 12, 1071–1074.

IOC (2008). Tsunami glossary. http://ioc3.unesco.org/itic/files/tsunami_glossary_sm-all.pdf

Lima, V.V.; Miranda, J.M.; Baptista, M.A.; Catalão, J.C.; Gonzalez, M.; Otero, L.; Olabrrieta, M.; Álvarez-Gómez, J.A. & Carreño, E. (2010). Impact of a 1755 like tsunami in Huelva Spain. *Nat. Hazards Earth Syst. Sci.*, 10, 139-148.

Liu, PL-F.; Woo, S-B.; Cho, Y-S. (1998). Computer programs for tsunami propagation and inundation. Cornell University, New York.

Mansinha. L; Smylie, DE. (1971). The displacement field of inclined faults. *Bull. Seism. Soc. Am.*, 61,1433–1440

Medialdea, T.; Vegas, R.; Somoza, L.; Vázquez, J.T.; Maldonado, A.; Díaz-Del-Río, V.; Maestro, A.; Córdoba, D. & Fernández-Puga, M.C. (2004). Structure And Evolution Of The "Olistostrome" Complex Of Thegibraltar Arc In The Gulf Of Cádiz (Eastern Central Atlantic): Evidence From Two Long Seismic Cross-Sections. *Marine Geology*, 209, 173–198

Miranda, J. M., Baptista, M.A., Terrinha, P. & Matias L. (2008). Tsunamigenic source areas for Portugal mainland, Iberia, Oral Communication, session Tsunami Early Warning Systems and Tsunami Risk Mitigation in the European-Mediterranean Region. *31 General Assembly of the European Seismological Commission*. September 2008, Crete, Greece.

Okada, Y. (1985). Surface deformation due to shear and tensile faults in a half-space. *Bull. Seism. Soc. Am.*, 75:1135–1154

Omira, R.; Baptista, M. A.; Matias, L.; Miranda, J. M.; Catita, C.; Carrilho, F. & Toto, E. (2009). Design of a Sea-level Tsunami Detection Network for the Gulf of Cadiz. *Nat. Hazards Earth Syst. Sci.*, 9, 1327-1338..

Omira, R.; Baptista, M. A.; Miranda, J. M.; Toto, E.; Catita, C.& Catalao, J. (2010a). Tsunami vulnerability assessment of Casablanca-Morocco using numerical modelling and GIS tools. *Natural Hazards*, 54,75-95.

Omira, R.; Baptista, M. A. & Miranda, J. M. (2010b, in press). Evaluating tsunami impact on the Gulf of Cdaiz coast (Northeast Atlantic). *Pure and Applied Geophysics*.

Sartori, R.; Torelli, L.; Zitellini, N.; Peis, D. & Lodolo, E. (1994). Eastern segment of the Azores-Gibraltar line (central-eastern Atlantic): An oceanic plate boundary with diffuse compressional deformation. *Geology*, 22, 555-558.

Solares, J. M. & Lopez-Arroyo, A. (2004). The great historical 1755 earthquake: Effects and damage in Spain. *J. Seismol..*, 8, 275- 294.

Terrinha, P.; Matias, L.; Vicente, J.; Duarte, J.; Luís, J.; Pinheiro, L.; Lourenço, N.; Diez, S.; Rosas, F.; Magalhães, V.; Valadares, V.; Zitellini, N.; Mendes-Víctor, L. & MATESPRO Team (2009) - Morphotectonics and Strain Partitioning at the Iberia-Africa plate boundary from multibeam and seismic reflection data. *Marine Geology* , 267, 3-4, 156-174.

Tinti, S. (1993). The project GITEC: a European effort to address tsunami risk, in P. A. Merriman and C. W. A. Browitt (eds.), Natural Disaster, Protecting Vulnerable Communities. *Thomas Telford, London, pp. 147–155.*

Tinti, S. & Armigliato, A. (2003). The use of scenarios to evaluate the tsunami impact in southern Italy. *Marine Geology*, 199, 221-243.

Torelli, L.; Sartori, R. & Zetillini, N. (1997). The giant chaotic body in the Atlantic Ocean off Gibraltar: new results from a deep seismic reflection survey. *Mar. Petrol. Geol.* 14, 125-138.

Urban, S. (1755). The Gentleman's Magazine, printed by: D. Henry, D. and Cave, R., St. John' Gate, London, 554–564.

Wang, X. and Liu, P. L.-F.: COMCOT User Manual - Version 1.6., School of Civil and Environmental Engineering, Cornell University, Ithaca, NY 14853, USA, available at: http://ceeserver.cee. cornell.edu/pll-group/doc/comcot user manual v1, .pdf, 2006a.

Zitellini, N.; Mendes, L.; Cordoba, D.; Danobeitia, J. J.; Nicolich, R.; Pellis, G.; Ribeiro, A.; Sartori, R. & Torelli, L. (2001), Source of the 1755 Lisbon earthquake and tsunami investigated, *Eos Trans. AGU*, 82,26, 285- 291.

Zitellini, N.; Rovere, M.; Terrinha, P.; Chierici, F.; Matias, L. & BIGSETS team (2004). Neogene through Quaternary tectonic reactivation of SW Iberian passive margin. *Pure Appl. Geophys.*, 161, 565- 587.

Zitellini, N.; Gràcia, E.; Matias, L.; Terrinha, P.; Abreu, M. A.; DeAlteriis, G.; Henriet, J. P.; Dañobeitia, J. J.; Masson, D. G.; Mulder, T.; Ramella, R.; Somoza, L. & Diez, S. (2009). The quest for the Africa-Eurasia plate boundary west of the Strait of Gibraltar., Earth and Planetary Science Letters, 280, 1-4, 15, 13-50.

3

Detection of Tsunamis from Changes in Ocean Surface Roughness

Benjamin D. Hamlington[1], Oleg A. Godin[2,3]
Vladimir G. Irisov[2,4] and Robert R. Leben[1]
[1]Colorado Center for Astrodynamics Research, University of Colorado at Boulder
[2]Cooperative Institute for Research in Environmental Sciences
University of Colorado at Boulder
[3]NOAA Earth System Research Laboratory, Physical Sciences Division, Boulder
[4]ZelTechnology LLC, Boulder
USA

1. Introduction

The need for a reliable system for early tsunami detection and warning was made painfully clear by the over two hundred thousand lives lost to the tsunami generated by the 9.3 magnitude Sumatra-Andaman earthquake that swept across the Indian Ocean on 26 December 2004 (Stein & Okal, 2005; Lay et al., 2005; Titov et al., 2005). The tsunami claimed the lives of over 220,000 people and despite a lag of up to several hours between the earthquake and arrival of the tsunami in some locations, the majority of victims were given little or no warning of the impending threat. While tsunamis occur much more frequently in the Pacific Ocean and a tsunami warning system has been in place in the region for many years, no such system was in place in the Indian Ocean and the communications infrastructure was not adequate for issuing widespread warnings at the time of the Sumatra-Andaman tsunami.

An early and dependable assessment of a tsunami threat requires detection of the tsunami wave in the open ocean away from the shore (Lautenbacher, 2005; Levin & Nosov, 2005; Bernard et al., 2006; Schindele et al., 2008). In the open ocean, however, the wave amplitude of the tsunami is small (generally less than one meter) and it is only as it approaches the shore that the tsunami rapidly grows in amplitude. Given the expansiveness of the ocean, sensors capable of detecting the tsunami must have very broad coverage. In addition to detecting the tsunami early enough to provide adequate warning, the method of detection must be reliable with few false warnings. If coastal populations go to great lengths to move to safe areas only to find out later such an evacuation was unnecessary, they may be less likely to heed warnings in the future.

By complementing traditional seismic data and point measurements as provided by the Deep-Ocean Assessment and Reporting of Tsunamis (DART) buoys network (Gonzalez et al., 2005; Bernard et al., 2006), satellite observations of tsunami manifestations can potentially improve the accuracy and timeliness of tsunami forecasts (Levin & Nosov, 2005; Synolakis & Bernard, 2006; Geist et al., 2007; Wei et al., 2008; Behrens et al., 2008), increase the lead time of tsunami warnings, decrease the probability of false alarms (Walker, 1996; Dudley and Lee, 1998; Godin et al., 2004; Nagai et al. 2007), and help avoid unnecessary evacuations (Dudley and Lee, 1998; Bernard et al., 2006).

Satellites have detected gravity waves induced by tsunamis in the ionosphere and the potential to use global positioning system satellites (GPS) for early tsunami detection is being explored (Artru et al., 2005; Occhipinti et al., 2006). Satellite altimeters have also sampled several tsunamis over the past two decades. The satellite altimetry sea surface height (SSH) measurements of the Sumatra-Andaman tsunami were used by a number of authors to study the properties of the tsunami, its propagation and scattering from the coastline as well as to improve characterization of the seismic source of the tsunami, and to verify numerical models (Fine et al., 2005; Kulikov et al., 2005; Smith et al., 2005; Song et al., 2005; Titov et al., 2005; Ablain et al., 2006; Hirata et al., 2006; Kumar et al., 2006; Fujii & Satake, 2007; Gower, 2007; Hayashi, 2008; Hoechner et al., 2008; Sladen & Hebert, 2008). Detection of earlier weaker tsunamis in less extensive satellite altimetry SSH records is discussed by Okal et al. (1999) and Zaichenko et al. (2005).

Although measurements of SSH can provide definitive detection of sufficiently large tsunamis, the spatial coverage and temporal resolution of satellite altimeters are not suitable for forming the basis of a system for the early detection of tsunamis. Satellite altimeters provide measurements only along their ground tracks and generally require ten days to obtain near global coverage. The chances are remote of actually observing a tsunami early enough to warn coastal inhabitants. Of the tsunami manifestations in the deep ocean, variations in ocean surface roughness are the most relevant and promising to detect tsunamis from space provided that these factors can be revealed by orbiting active (scatterometers) and passive (radiometers) scanning and microwave sensors, which have broad coverage of hundreds of kilometers across the satellite ground track.

Tsunami-induced variations in surface roughness away from the shore were first observed in visible light originating from a tsunami approaching Oahu in 1994 (Walker, 1996; Dudley and Lee, 1998). These variations were given the name "tsunami shadows" and appear as extended darker strips on the ocean surface along a tsunami front. Formation of the tsunami shadows as areas with a different root mean square (RMS) surface slope has been explained theoretically as a result of air-sea interaction; specifically tsunami-induced perturbations in the wind velocity close to the ocean surface that are predicted to be much larger than currents in the tsunami wave (Godin, 2003, 2004; Rowan, 2004). Later theoretical studies (Godin, 2005; Troitskaya & Ermakov, 2008) corroborated these conclusions.

The first definitive measurements of the tsunami effect on sea surface height and radar backscattering strength (a measure of ocean surface roughness) in the open ocean were obtained from satellite altimeters during passage of the 2004 Sumatra-Andaman tsunami. In this study we concurrently employ radar backscattering strength and SSH data obtained by satellite altimeters for the Sumatra-Andaman event as well as for three other tsunamis. Through statistical analyses of multiple years of satellite altimeter observations, we definitively demonstrate that the Sumatra-Andaman tsunami induced distinctive variations in ocean surface roughness and tentatively demonstrate that three other tsunami-induced variations in ocean surface roughness.

The sections below are organized as follows: In section 2, we outline a theory explaining the magnitude and spatial structure of tsunami-induced variations in ocean surface roughness. In section 3, we discuss our analysis of ocean surface roughness variations by introducing the data used in this study and explaining the statistical randomization tests. Section 4 provides results of the randomization tests for the 2004 Sumatra-Andaman tsunami as well as three other weaker tsunamis and demonstrates that the tsunami causes distinct, detectable changes in ocean surface roughness. In section 5, we discuss the feasibility of using variations in ocean surface roughness for the early detection of tsunami waves. Section 6 provides a summary and discussion of further research on this topic.

2. Theory of tsunami-induced wind velocity perturbations

Long surface gravity waves in the ocean modulate short gravity and gravity-capillary waves, and change ocean surface roughness through the interaction of short waves with near-surface currents and variations of near-surface wind-induced by long waves (Hara & Plant, 1994; Troitskaya, 1994; Kudryatsev et al., 1997; Cohen & Belcher, 1999; Godin & Irisov, 2003). Modulation due to currents is negligible for a tsunami in the deep ocean (Godin, 2003, 2004). Tsunami-induced variations in ocean surface roughness away from the shore result from variations in wind velocity that accompany tsunami waves, and have been predicted to be much larger than currents in the tsunami wave (Godin, 2003, 2004, 2005). Godin (2003, 2004, 2005) found that significant variations in the mean wind velocity arise from the generation of viscous waves in the atmosphere by coherent large-scale motion of the ocean surface in a tsunami wave. However, the magnitude of the surface roughness modulations and the position of areas with increased and decreased roughness are sensitive to the choice of a closure hypothesis for turbulence in the atmospheric boundary layer. Although any theoretical explanation of tsunami-induced surface roughness variations is still tentative, measuring these variations in the open ocean can provide insight into the physics of the interaction of fast surface waves with turbulent wind.

As outlined in Godin et al. (2009), there is no universally accepted model of airflow over fast sea waves. Using assumptions made in (Godin, 2005), in the presence of a monochromatic tsunami wave, the wind speed relative to the ocean surface retains a logarithmic profile up to a few tens of meters above the surface. The effective wind speed depends on characteristics of the tsunami and differs from the background wind speed by the factor:

$$M = 1 - \frac{\kappa a c}{H u_* \ln \beta} \tag{1}$$

where κ is the von Karman constant, u_* is the friction velocity, H is the height of the background logarithmic boundary layer, a is the SSH change due to the tsunami, c is the tsunami phase speed,

$$\beta = \frac{\kappa u_* T}{2\pi z_0} \tag{2}$$

z_0 is the roughness length, and T is the tsunami period.

For a monochromatic tsunami wave, the effective wind speed varies periodically in time with the SSH change. As long as the relaxation time of wind waves is much smaller than the tsunami period, the time dependence of the effective wind speed can be disregarded when determining characteristics of the ocean surface roughness that correspond to a given instantaneous value of the wind modulation M. Variations in the radar backscattering strength at nadir, σ_0, resulting from the tsunami-induced wind variations can be found in the Modified Chelton-Wentz algorithm (Witter & Chelton, 1991) or the Freilich-Challenor algorithm (Freilich & Challenor, 1994) by comparing the σ_0 values that correspond to the background and effective wind speeds. In section 5, computed estimates of the maximum and minimum radar backscattering strength variations will be compared to measurements taken by the Jason-1 satellite altimeter during the 2004 Sumatra-Andaman tsunami.

3. Statistical analysis of ocean surface roughness variations

3.1 Data

Satellite altimetry provides concurrent measurements of SSH and σ_0. While not the focus of our study, SSH measurements allow us to identify the location of the leading edge of the tsunami. Detection of tsunamis in SSH measurements has been demonstrated in several previous studies (Okal et al., 1999; Ablain et al., 2006), although weak tsunamis generally remain obscured by background ocean variability. By utilizing satellite altimetry data, we are able to identify the location of the tsunami's leading edge in SSH measurements and test the concurrently measured σ_0 values for tsunami-induced changes in ocean surface roughness.

While initial studies have focused on the 2004 Sumatra-Andaman tsunami, the statistical analysis described below can be applied to any tsunami event occurring within the modern altimetry era. The Radar Altimeter Database System (RADS) was used to search and collect historical altimeter records. RADS allow the user to quickly search and obtain records from several satellite altimeters (Naeije et al., 2000). With the poor spatial coverage and temporal resolution provided by satellite altimeters, the chances of an altimeter sampling a tsunami are remote. Through an extensive search of the past 17 years of satellite altimetry data, we have identified four tsunami events on which we will focus our attention: 1992 Nicaraguan tsunami, 1995 Chile tsunami, 2004 Sumatra-Andaman tsunami, and 2010 Chile tsunami.

3.2 Statistical randomization tests

Ocean surface roughness is influenced by diverse phenomena in the ocean and atmosphere, including wind gusts, currents, internal gravity waves, and oceanographic fronts. The resulting σ_0 variability can far exceed the expected tsunami-induced variations. To determine whether the σ_0 variations observed were indeed caused by a tsunami and whether tsunami signals can be reliably extracted from σ_0 data, data with and without the tsunami present must be compared.

Tsunamis have several distinctive spatio-temporal characteristics that aid in the retrieval of the tsunami signal from the "noise" arising from other geophysical processes. Perhaps the most distinctive attribute of tsunami-induced roughness variations is their propagation speed relative to the ocean bottom. However, satellite altimeters only provide a "snapshot" of the ocean surface, and thus the propagation speed of the tsunami cannot be used to identify tsunami-induced features in satellite altimeter data. Instead, we systematically utilize spatial filtering (Powell & Leben, 2004) to suppress σ_0 variations that are unrelated to tsunamis.

To determine if σ_0 variations were induced by the passage of a tsunami, statistical randomization tests (Edgington, 1995) were performed to compare data with and without the tsunami present. One thousand 3.2°-windows (each containing 64 points of data) were randomly selected from the area of the ocean through which the tsunami passed. Mean σ_0 values were subtracted in each window to calculate the σ_0 anomaly. The RMS values and the number of zero crossings were calculated for the σ_0 anomaly in each window and compared to the respective values in the 3.2°-window containing the leading edge of the tsunami. The RMS σ_0 anomaly characterizes the strength of the surface roughness variations, while the number of zero crossings serves as a measure of the spatial scale of the variations. If the tsunami-induced variations were distinctive and unique, we would expect the window containing the leading edge of the tsunami to have both a higher RMS and a greater number of zero crossings than found in the 1,000 randomly selected windows.

In addition to the randomization tests using the RMS σ_0 anomaly and zero crossings, a spectral approach is implemented. In each randomly selected 3.2°-window, the σ_0 and SSH data were detrended by removing the linear fit computed using least-squares. Windows with unphysically large SSH variations of more than 150 cm from the median were excluded if randomly selected. The Fourier spectrum of the detrended data was normalized to account for the high variability of the high-frequency part of the spectrum. The Fourier components with spatial scales from 90 km to 300 km, scales representative of the tsunami, were summed. To quantify the correlation between the σ_0 and SSH anomalies, the Fourier spectrum of their cross-correlation function was calculated in the range from 90 km to 300 km. The spectral measure of the σ_0 anomaly correlation with the SSH anomaly was computed for each of the 1,000 randomly selected windows and compared to the spectral measure computed for the window containing the leading edge of the tsunami. If the spectral measure of the window containing the tsunami signal is larger than the spectral measure computed for the randomly selected windows, we can determine that the σ_0 variations during the tsunami were unique and attributable to the passage of the tsunami.

4. Results

4.1 2004 Sumatra-Andaman tsunami

Four satellite altimeters overflew the 2004 Sumatra-Andaman tsunami during its propagation across the Indian Ocean. Envisat, Geosat Follow-On (GFO), Jason-1, and TOPEX/Poseidon measured the tsunami at times ranging from two hours to seven hours after the tsunamigenic earthquake occurred. Of these four satellite altimeters, Jason-1 provides the earliest observations of the Sumatra-Andaman tsunami and has the most extensive records. Jason-1 encountered the leading edge of the tsunami 1 h 53 minutes after the earthquake (Ablain et al., 2006; Gower, 2007) at about 5°S in the Indian Ocean heading northeast on ascending pass 129 of cycle 109 (Fig. 1). Since the satellite launch in 2001, Jason-1 has collected an extensive set of high-quality SSH and σ_0 data, allowing us to characterize the variability of the radar backscattering strength under various atmospheric conditions without the presence of the tsunami.

The leading front of the tsunami is contained in the window between 6°S and 2°S, with the spatial extent of the segment on the order of the tsunami wavelength. The tsunami signal is clearly present in this window and is well above the noise level in SSH records (Fig. 2A). Data quality in this window is high and there are few data points excluded by quality controls. Radar backscattering strengths measured in both the Ku- and C- microwave frequency bands in the vicinity of the leading front of the tsunami show up to 1 dB variations, which are not present in measurements along the same pass of the cycles before and after the tsunami (Fig. 2B, C). Using equations 1 and 2, and assuming the height of the background logarithmic boundary layer to be H = 50-70 m (Garratt, 1994), we calculate the

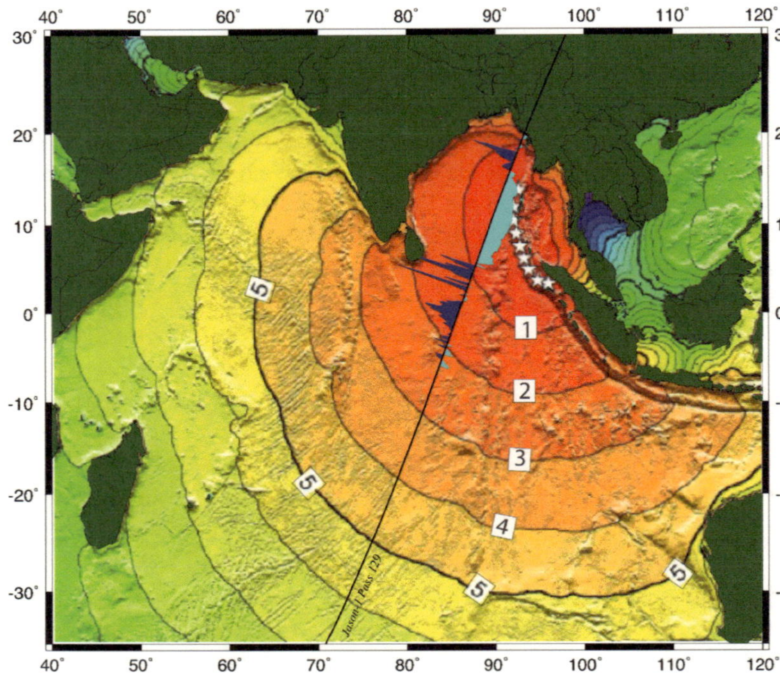

Fig. 1. Jason-1 satellite altimeter overflight of the Sumatra-Andaman tsunami. The Jason-1 ground track and C-band σ_0 data for pass 129 of cycle 109 is superimposed on contours of the tsunami leading wave front at hourly intervals after the earthquake. White stars show the location of the tsunami wave sources. (Tsunami wave front graphic is provided courtesy of the National Geophysical Data Center/NOAA)

tsunami period to be T = 35-45 min (Gower, 2007), and a maximum and minimum SSH anomaly of 0.6-0.7m and -(0.4-0.5m). We obtain values of about 1 dB for the maximum variation of the radar backscattering strength that are consistent with the values in Figs. 2B and C. A more detailed comparison of the observed to predicted σ_0 variations, however, is not possible because of uncertainty in knowledge of the local meteorological parameters and the tsunami spectrum, and the high sensitivity of the tsunami-induced changes in the surface roughness to poorly known environmental parameters, such as the background wind speed. Despite this difficulty, the σ_0 variations can be attributed to the tsunami by performing statistical randomization tests using historical Jason-1 σ_0 data and comparing to the σ_0 data collected during the passage of the tsunami. The Jason-1 σ_0 data obtained during the Sumatra-Andaman tsunami passage were discussed by Troitskaya & Ermakov (2005; 2008), but they did not compare the data to analogous data in the absence of the tsunami.

To perform the randomization tests, one thousand 3.2°-windows centered between 20°S and 10°N were randomly selected in the tropical Indian Ocean from cycles 1-174 of Jason-1. The data were processed as outlined in section 3.2, and the RMS values and the number of zero crossings were calculated for the σ_0 data in each window and compared to the respective values in the window covering 5.6°S to 2.4°S along pass 129 of cycle 109. Only a few percent of the randomly selected windows simultaneously have equal or larger values of both the RMS σ_0 anomaly and the number of zero crossings. The statistical significance of the hypothesis that surface roughness variations with and without the tsunami are not substantially different is 3.15% and 0.93% when estimated using the Ku- and C-band σ_0 data, respectively (Fig. 3).

As described in section 3.2, we also performed randomization tests implementing a spectral approach. Each 3.2°-window was processed and filtered as outlined in section 3.2, and the Fourier spectrums were computed for the σ_0 anomalies and cross-correlation function of the SSH and σ_0 anomalies. The spectral measures of these anomalies were compared to those computed for the window covering 5.6°S to 2.4°S along pass 129 of cycle 109. The probability distribution densities of the spatially filtered σ_0 anomaly, as measured in the C-band (Fig. 4a) and the σ_0 - SSH correlation (Fig. 4b), show that the σ_0 variability equalled or exceeded its level in the tsunami event in 1.7% of the cases, while the correlation level in the tsunami event was uniquely large. For the σ_0 data obtained in the Ku-band, the corresponding significance level is 6.8% for the σ_0 variability (Fig. 4c); the σ_0 - SSH correlation remains uniquely large in the tsunami event (Fig. 4d). Thus, the randomization tests provide strong support for the hypothesis that the σ_0 variations and hence the underlying surface roughness variations observed by Jason-1 during the passage of the Sumatra-Andaman tsunami passage, were caused by the tsunami.

In addition to detecting the leading wave front of the tsunami where the amplitude of the wave is often the largest, it is important to be able to detect the tsunami away from

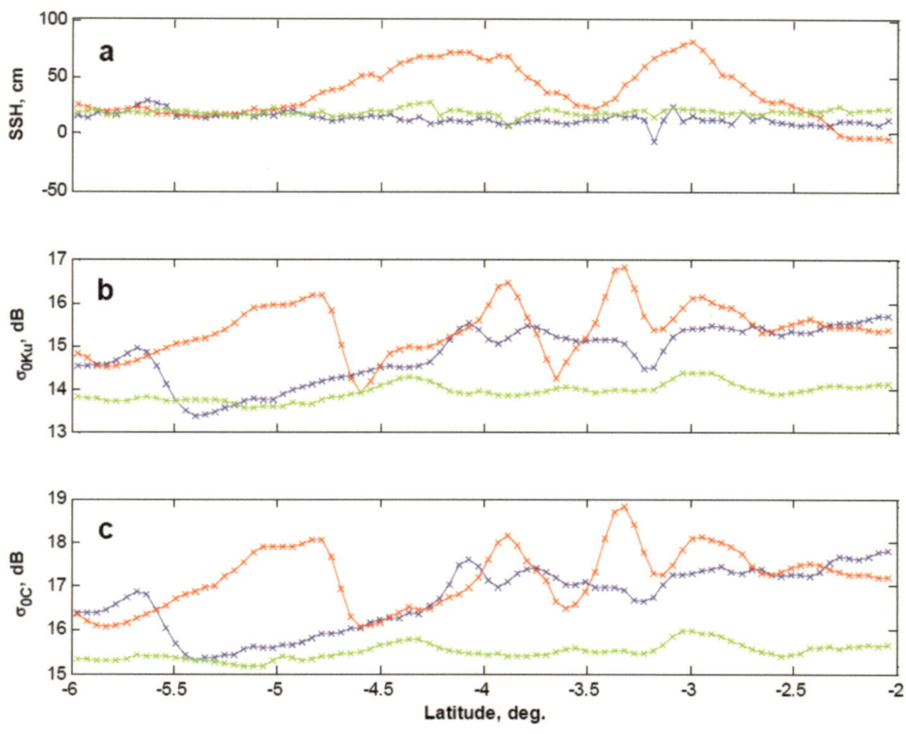

Fig. 2. Jason-1 data for pass 129 from 6°S to 2°S shown for the cycles before the tsunami (blue), coincident with the tsunami (red) and after the tsunami (green). (a) Sea surface height. (b) Ku-band radar backscattering strength. (c) C-band radar backscattering strength

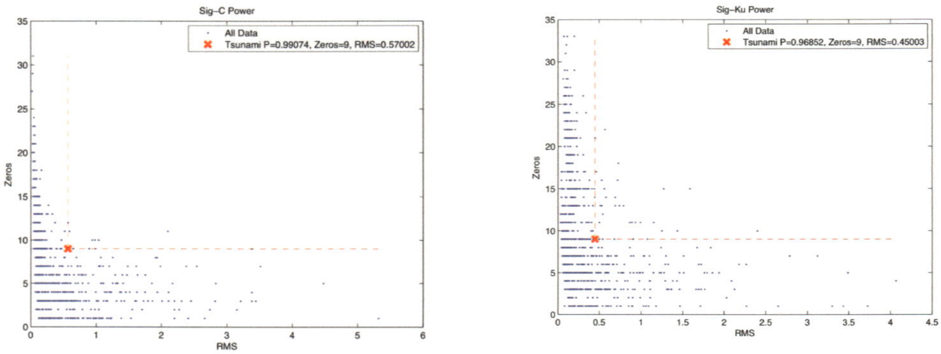

Fig. 3. Randomization tests of the radar backscattering strength data. (a) Ku-band σ_0 data. (b) C-band σ_0 data. Statistical significance of the hypothesis that surface roughness variations with and without the tsunami are not substantially different is found to be 3.15% and 0.93% for the Ku-band and C-band measurements respectively

the leading edge where the magnitude of the tsunami has been diminished. We applied the same data processing techniques to 3.2° non-overlapping windows, which extend from 2.4°S (where the window test above ends) to the north along the Jason-1 pass 129 of cycle 109. Windows that contained data gaps too extensive to apply the processing and filtering techniques were excluded from the subsequent randomization tests. The other two windows that were used for testing ranged from 2.5°S to 0.7°N and from 0.8°N to 4.0°N. For the window beginning at 2.5°S, only 9.1% and 9.7% of the randomly selected windows simultaneously have equal or larger numbers of both RMS σ_0 anomalies and the number of zero crossings in the Ku- and C-bands, respectively. For the window beginning at 4.0°N, only 6.4% and 2.2% of the randomly selected windows had greater RMS σ_0 anomaly and more zero crossings for the Ku- and C-bands, respectively. Similarly, for the randomization test using the spectral approach, the window beginning at 2.5°S yielded a spectral measure for the σ_0 - SSH correlation that was uniquely large. The same analysis could not be applied to the window beginning at 0.8°N due to data gaps.
These additional randomization tests away from the leading wave front of the tsunami reinforce the evidence of exceptional features in radar backscattering strength in the presence of a tsunami. Furthermore, these results demonstrate that our data processing algorithms allow detection of the tsunami manifestations in the radar backscattering strength for various tsunami waveforms and can successfully discriminate between regions where the tsunami is and is not present.

4.2 2010 Chile tsunami

Detection of tsunami manifestations in the Jason-1 σ_0 records was simplified by the extraordinary strength of the Sumatra-Andaman tsunami. Other weaker tsunamis provide a much more rigorous test of our detection algorithm. As a result of excellent satellite altimeter coverage, a good candidate for analysis is the 2010 Chile tsunami that was generated by an Mw 8.8 earthquake on February 27th, 2010. Tsunami waves hit coastal towns in Chile with substantial wave heights. Although a warning was generated for the entire Pacific region, the tsunami did not significantly affect other areas such as Hawaii, New Zealand, Australia, or Japan. There were multiple satellite altimeters (Jason-1, Jason-2

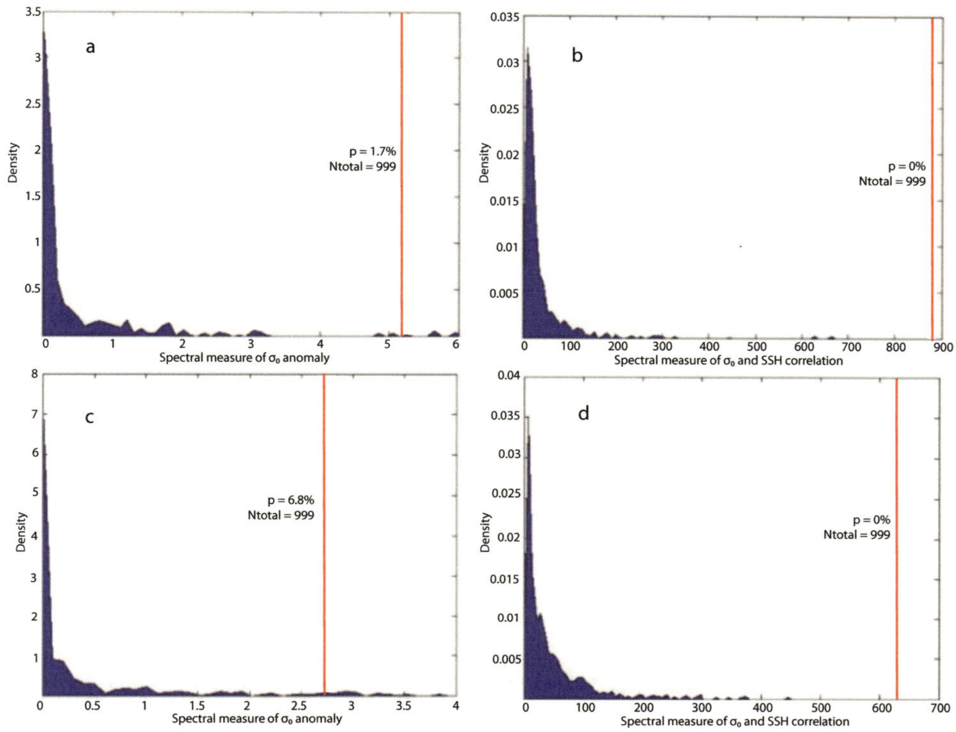

Fig. 4. Randomization tests of the spectral measures of the radar backscattering strength anomaly and its correlation with sea surface height anomaly as observed by Jason-1 in the vicinity of the tsunami leading front. (a) Randomization tests of the spatially filtered C-band σ_0 anomaly. (b) Randomization tests of the co-spectrum of the spatially filtered C-band σ_0 and sea surface height anomalies. (c) Randomization tests of the spatially filtered Ku-band σ_0 anomaly. (d) Randomization tests of the co-spectrum of the spatially filtered Ku-band σ_0 and sea surface height anomalies

and Envisat) that overflew the tsunami wave field shortly after the earthquake occurred. Resulting largely from the weak tsunami and small wave amplitude in the open ocean, the tsunami signal was only definitively identified in one pass from these three altimeters despite the excellent altimeter coverage. Using a simple filtering technique to remove variations in SSH not resulting from the tsunami, the leading wave front of the tsunami is positively identified about 7.5 hours after the generation of the tsunami in pass 143 of Jason-1 cycle 300 around 15°S. The passes from the cycles 299 and 301 (before and after the one coincident with the tsunami) were averaged together, smoothed and then subtracted from pass 143 of Jason-1 cycle 300. The filtered signal agrees reasonably well with the Method of Splitting Tsunami (MOST) model results obtained from NOAA/PMEL/Center for Tsunami Research (Synolakis et al., 2008) for the Chile tsunami, and confirms the location of the leading edge near 15°S (Fig. 5). The spatial extent of the leading edge, however, differs between the filtered Jason-1 signal and the results from the MOST model, likely resulting from the oblique sampling of the tsunami by Jason-1. In other words, the Jason-1 ground track of pass 143 did not enter very far into the tsunami wave field and stayed close to the leading front of the tsunami as the wave propagated across the Pacific Ocean. Fig. 6A shows

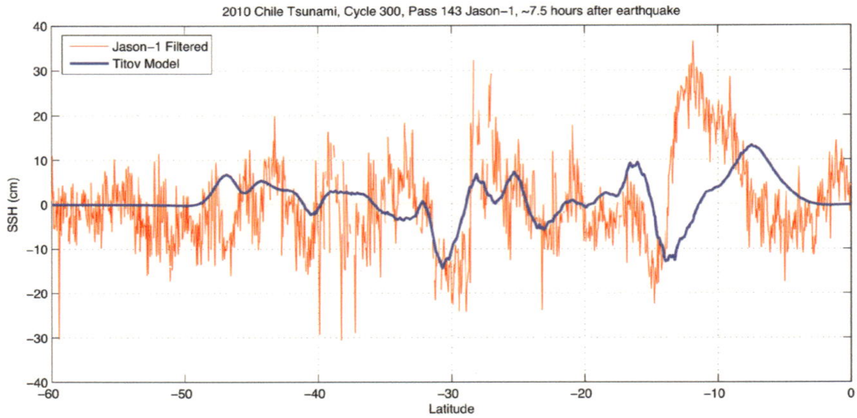

Fig. 5. Filtered SSH data from pass 143 of cycle 300 of Jason-1, occurring about 7.5 hours after the earthquake. The filtered signal is in agreement with the results from the MOST model and appears to confirm the location of the leading wave front of the tsunami around 15°S

the MOST model results at the time Jason-1 entered the tsunami wave field, while Fig. 6B shows the model results at the time the satellite altimeter departed the tsunami wave field. As a result of the oblique sampling of Jason-1, a small time difference between reality and model could result in a considerable change in the location and spatial extent of the leading edge of the tsunami.

The leading front of the tsunami is contained in the window between 16°S and 12°S, with the spatial extent of the segment on the order of the tsunami wavelength. The tsunami signal is clearly present in this window with amplitude of greater than 20 cm (Fig. 5 & 7A). Data quality in this window is high and there are few data points excluded by quality controls. Radar backscattering strengths measured in both the Ku and C microwave frequency bands in the vicinity of the leading front of the tsunami do not show variations as strong as those observed for the 2004 Sumatra-Andaman tsunami, particularly in the C-band. This is not surprising given the relative strengths of the two tsunamis.

To perform the randomization tests, one thousand 3.2° windows centered between 50°S and 10°N were randomly selected in the Pacific Ocean from cycles 1-305 of Jason-1. The data was processed as outlined in section 3.2, and the RMS values and the number of zero crossings were calculated for the σ_0 data in each window and compared to the respective values in the window covering 15.5°S to 12.3°S along pass 143 of cycle 300. The statistical significance of the hypothesis that the surface roughness variations with and without the tsunami are not substantially different is 0.4% and 20.8% when estimated using the Ku- and C-band σ_0 data, respectively. While the randomization test on the Ku-band radar backscattering strength yielded positive identification at the 1% significance level, 30 zero crossings and an RMS σ_0 anomaly of only 0.20 were found for the window containing the leading front of the tsunami. The high significance level obtained from the randomization test is primarily a result of the large number of zero crossings. Given the physical characteristics of the tsunami, we would expect fewer zero crossings. Without knowledge of the wind speed at the time of the tsunami, however, the expected RMS σ_0 anomaly is unknown. The C-band radar backscattering strength for the same window contained 10 zero

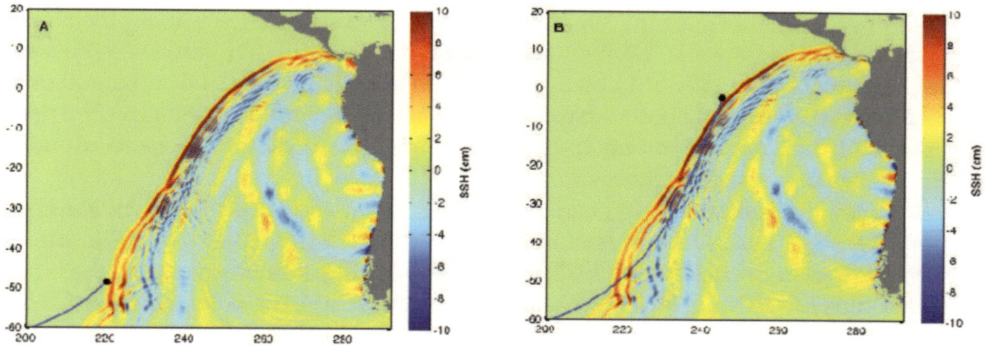

Fig. 6. MOST model results at the time Jason-1 entered the tsunami wave field (according to the model) (A) and at the time Jason-1 departed the tsunami wave field (B). The oblique sampling of the leading front of the tsunami by Jason-1 provides a possible explanation for the discrepancy between model results and Jason-1 data as seen in Fig. 5

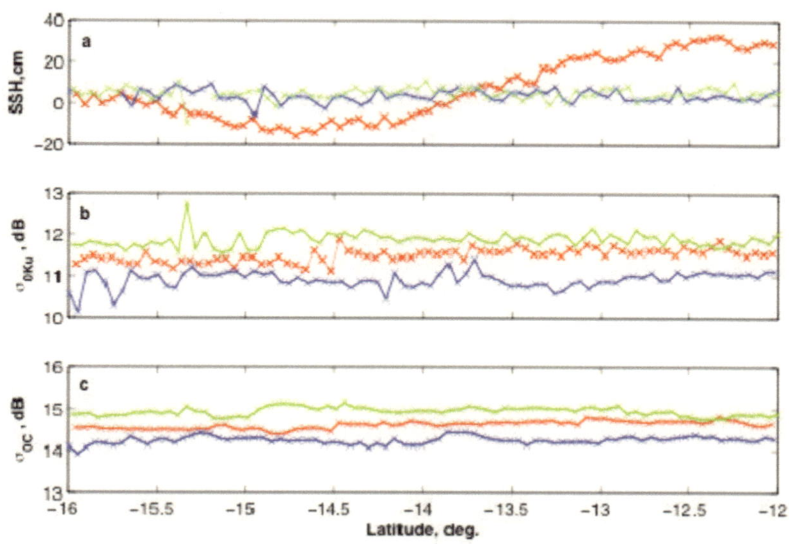

Fig. 7. Jason-1 data for pass 143 from 16°S to 12°S shown for the cycles before the 2010 Chile tsunami (blue), coincident with the tsunami (red) and after the tsunami (green). (a) Sea surface height. (b) Ku-band radar backscattering strength. (c) C-band radar backscattering strength

crossings and RMS σ_0 anomaly of 0.12, producing poor results for the randomization tests. The randomization tests utilizing the spectral approach were not conducted because the lack of noticeable variations in the σ_0 anomaly and weaker SSH anomaly signal are unlikely to produce positive results.

4.3 1992 Nicaragua tsunami and 1995 Chile tsunami

While we have positively detected the tsunami-induced variations in ocean surface roughness for the 2004 Sumatra-Andaman tsunami, we have only tentatively identified the tsunami-induced variations for the 2010 Chile tsunami, raising questions about the ability to detect weaker tsunami signals. Okal et al. (1999) studied seven other tsunami events using satellite altimetry. Using only SSH measurements, one tsunami (1992 Nicaragua Tsunami) was positively detected and another tsunami (1995 Chile tsunami) was tentatively detected. Using the statistical randomization tests, we can attempt to detect the tsunami-induced variations in ocean surface roughness.

The 1992 Nicaragua tsunami was generated by a "tsunami earthquake," i.e., an earthquake that produces an unusually large tsunami relative to the earthquake magnitude due to a slow rupture (Kanamori & Kikuchi, 1972). The ERS-1 satellite altimeter sampled the tsunami wave field on passes 523 and 525 of cycle 87. Both tracks sampled the tsunami about 3.5 and 5.5 hours after the tsunamigenic earthquake, respectively. Okal et al. (1999) determined that the tsunami signal could not be detected in pass 523 and focused their attention on 525. ERS-1 entered the tsunami wave field around 17°S, with wave amplitudes seen in the SSH of less than 10 cm (Fig. 8). Variations in the radar backscattering strength were around 0.5 dB.

To perform the randomization tests, one thousand 3.2° windows between 40°S and 20°N were randomly selected in the Pacific Ocean from cycles 83-101 of ERS-1. The data was processed as outlined in section 3.2, and the RMS values and the number of zero crossings were calculated for the σ_0 data within each window and compared to the respective values in the window covering 17.0°S to 13.8°S along pass 525 of cycle 87. The statistical significance of the hypothesis that the surface roughness variations with and without the tsunami are not substantially different is 2.6% when estimated using the Ku-band σ_0 data. This suggests a positive identification of the tsunami signal from variations in ocean surface roughness.

As mentioned above, Okal et al. (1999) tentatively identified the 1995 Chile tsunami from SSH measurements. A relatively large earthquake generated a tsunami with run-ups of 2 m in the Marquesas Islands. The TOPEX/Poseidon satellite altimeter sampled the tsunami wave field along two passes (230 and 232) of cycle 105, 3 and 5 hours after the earthquake occurred. The tsunami signal was not detected in pass 230, but there was tentative identification of the tsunami in pass 232. Okal et al. (1999) estimate that TOPEX/Poseidon entered the tsunami wave field around 36°S and exited the wave field around 25°S. SSH wave amplitudes in this region were found to be only 10 cm with maximum variations in the radar backscattering strength of approximately 0.4 dB (Fig. 9). Additionally, comparing the SSH data from the cycles before and after the one coincident with the tsunami show very little variation in SSH from cycle to cycle. When using these cycles to filter the target cycle, only a small signal with amplitude of much less than 10 cm remains near 27°S (Fig. 10).

We performed the randomization tests by randomly selecting one thousand 3.2°-windows centered between 50°S and 10°N in the Pacific Ocean from cycles 50-200 TOPEX/Posedion. The data was processed as outlined in section 3.2, and the RMS values and the number of zero crossings were calculated for the σ_0 data in each window and compared to the respective values in the window covering 29.0°S to 25.8°S along pass 232 of cycle 105. The statistical significance of the hypothesis that the surface roughness variations with and without the tsunami are not substantially different is 2.5% when estimated using the Ku-band σ_0 data. This suggests a positive identification of the tsunami signal from the variations in ocean surface roughness. Given the relatively small signal found in the SSH data, however, this identification remains tentative.

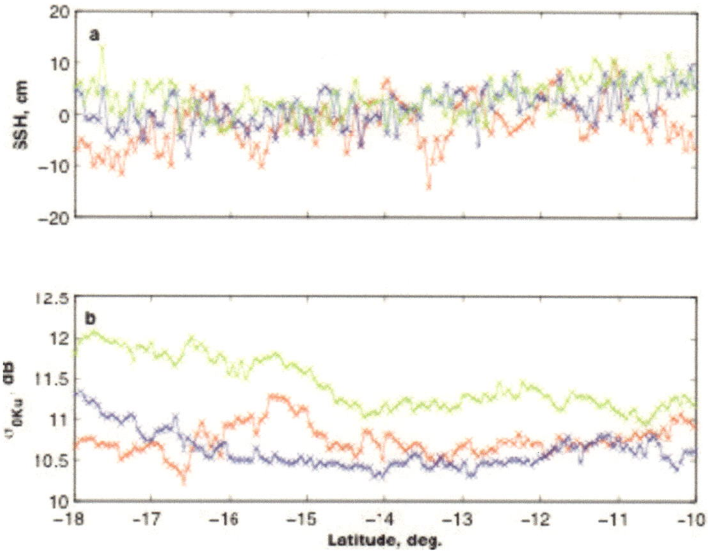

Fig. 8. ERS-1 data for pass 525 from 18°S to 10°S shown for the cycles before the 1992 Nicaragua tsunami (blue), coincident with the tsunami (red), and after the tsunami (green). (a) Sea surface height. (b) Ku-band radar backscattering strength

5. Feasibility of tsunami early detection and warning

The use of radar backscattering strength measurements for the detection of a tsunami in the open ocean has been positively demonstrated for the 2004 Sumatra-Andaman tsunami and tentatively demonstrated for the 1992 Nicaragua tsunami, 1995 Chile tsunami, and 2010 Chile tsunami. Satellite altimeters provide concurrent measurements of SSH and σ_0 allowing for the identification of the leading edge in SSH and subsequent testing of this leading edge using randomization tests on σ_0 measurements. While the four tsunamis mentioned above were well sampled by satellite altimeters, historically satellite altimeter measurements of tsunamis are uncommon. Even with several satellites on orbit, nadir-pointing satellite altimeters do not provide the ground track coverage necessary to ensure sampling of a tsunami wave field. Furthermore, as the results above show, even if timely sampling of the tsunami was available in real time, separating the tsunami signal from the background ocean variability is difficult. The use of satellite altimeters as the foundation for an early detection system is not plausible.

While radar backscattering measurements from satellite altimeters are only available along one-dimensional lines traced by the nadir ground track, two-dimensional images of tsunami-induced changes in ocean surface roughness could be obtained by using microwave radiometers and radars already on orbit. As outlined in section 2, Godin et al. (2009) present a model for calculating the tsunami-induced changes in ocean surface roughness, providing a factor directly related to SSH that corresponds to the modulation of background wind speed resulting from the passage of a tsunami. The ability to detect the tsunami from such a two-dimensional image is largely dependent on the variability and strength of the background wind field at the time of the tsunami passage. With a constant background wind of physically realistic magnitude, the tsunami-induced changes in ocean

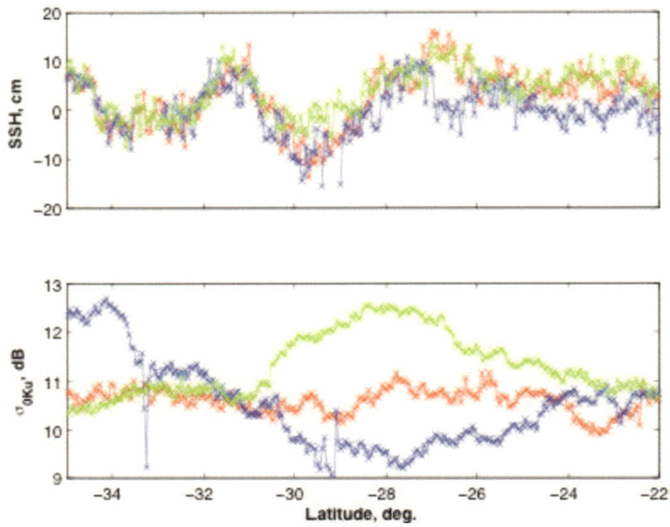

Fig. 9. TOPEX/Posedion data for pass 232 from 35°S to 22°S shown for the cycles before the 1995 Chile tsunami (blue), coincident with the tsunami (red), and after the tsunami (green). (a) Sea surface height. (b) Ku-band radar backscattering strength

surface roughness are apparent (Fig. 11A). Values shown are computed from equations 1 and 2, assuming a constant background wind of 3 m/s and a SSH profile derived from the Jason-1 measurements for the 2004 Sumatra-Andaman tsunami (the two-dimensional SSH values were not measured directly by Jason-1, but are representative of the magnitudes during the Sumatra-Andaman tsunami). When using the QuikSCAT wind speeds from December 26, 2004 (the day of the Sumatra-Andaman tsunami), however, the tsunami-induced σ_0 changes are obscured by the variability of the background wind field (Fig. 11B). As in the one-dimensional case, it will likely be necessary to find an appropriate method for

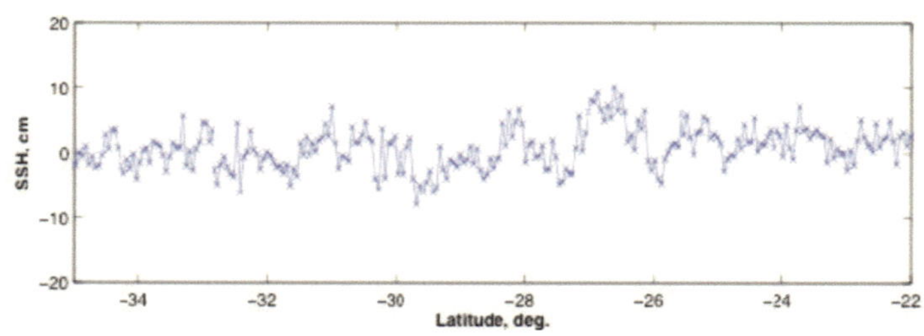

Fig. 10. TOPEX/Posedion data for pass 232 from 35°S to 22°S filtered using cycles before and after the cycle containing the tsunami. The amplitude of the signal is greatly reduced as a result of the similarity between the three signals, as seen in Fig. 9A

filtering the data to allow separation of the tsunami-induced signal from the background wind variability and measurement noise.

To date, searches of two-dimensional images of ocean surface roughness measurements from instruments already on orbit have not yielded any positive identification of a tsunami-induced signal, partially due to poor sampling coincident with the events. Nevertheless, it is possible that sufficient satellite sampling of the tsunami wavefront and other physical characteristics of the tsunami could be utilized in the detection of a tsunami using changes in ocean surface roughness. Multiple two-dimensional images of the same region obtained with relatively short time separation could allow for the use of the rapid propagation speed and large spatial extent of the tsunami in the open ocean to aid in the early detection of the tsunami signal. Lognonn

6. Summary and conclusions

Satellite altimeters provide the chance to study the effects of a tsunami wave in the open ocean through concurrent measurements of the sea surface height and the radar backscattering strength. Availability of the SSH data allows one to compare statistical properties of the radar backscattering strength when there is and there is not a tsunami wave present, without having to use tsunami source and propagation models. Using satellite altimeter observations, we have demonstrated that tsunamis in the open ocean cause distinct, measurable changes in ocean surface roughness. We have shown this to be true definitively for the 2004 Sumatra-Andaman tsunami and tentatively true for the 2010 Chile tsunami, 1992 Nicaragua tsunami, and 1995 Chile tsunami.

Although the feasibility of tsunami detection from changes in ocean surface roughness has been demonstrated using measurements from satellite altimeters, the practical issue of optimal retrieval of a tsunami signal from other sources of ocean roughness measurements remains an open question. Use of radar backscattering measurements from satellite altimeters would be impractical for tsunami detection and early warning purposes because of the limited number of operational satellite altimeters. Even if the data could be processed quickly enough to be useful, the temporal resolution and spatial coverage of nadir pointing altimeter measurements is not adequate for tsunami detection and warning. The tsunami-

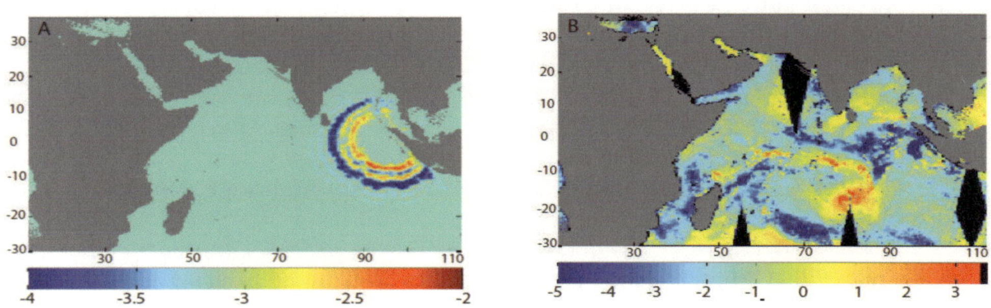

Fig. 11. Using the model presented in Godin et al. (2009) and equations 1 & 2, the two-dimensional field of radar backscattering strength (dB) is computed for A) a constant background wind of 3 m/s, and B) a background wind field obtained from QUIKSCAT on December 26, 2004. SSH data used for computing the values has been derived from the Jason-1 measurements of the 2004 Sumatra-Andaman tsunami

induced surface roughness variations, however, are likely to be observable with other types of space and airborne sensors. Unlike the sea surface height, which is measured at nadir points along the satellite ground track, variations in ocean surface roughness can potentially be measured over wider swaths with side-looking radars and scanning microwave radiometers. The broader surface coverage of these sensors suggests that they are more promising for early tsunami detection and may be an important component in a future global system for tsunami detection and warning. Further research is required to demonstrate tsunami detection with such instruments and, as seen in Fig. 9, appropriate analysis techniques will have to be developed to extract the tsunami signal from such data. We anticipate that our study will stimulate the development of data processing algorithms and microwave sensors for the identification of tsunami-induced ocean surface roughness changes to complement or enhance existing regional tsunami detection and early warning systems and contribute to a future global system.

7. References

Ablain, M., Dorandeu, J., Le Traon, P.-Y., and Sladen A., (2006). High resolution altimetry reveals new characteristics of the December 2004 Indian Ocean tsunami, *Geophys. Res. Lett.*, 33, L21602, doi:10.1029/2006GL027533.

Artru, J., Lognonné, P., Occhipinti, G., Crespon, F., Garcia, R., Jeansou, E., Murakami, M. (2005). Tsunami detection in the ionosphere, *Space Res. Today*, 163, 23-27.

Behrens, J., Androsov, A., Harig, S., Klaschka, F., Mentrup, L., Pranowo, W. S., Cui, H. Y., Schroter, J., and Hiller, W. (2008). Design and performance testing of a multi-sensor quick assimilation technique for tsunami early warning in the GITEWS simulation system, *Geophys. Res. Abstr.*, 10, EGU2008-A-01878.

Bernard, E. N., Mofjeld, H. O., Titov, V., Synolakis, C. E., and Gonzalez, F. I., (2006). Tsunami: scientific frontiers, mitigation, forecasting and policy implications, *Phil. T. R. Soc. A*, 364, 1989–2007, doi:10.1098/rsta.2006.1809.

Dudley, W. C. and Lee, M. (1998). Tsunami!, University of Hawaii Press, Honolulu, 5 pp., 302–303, 321–322.

Edgington, E. S. (1995). Randomization Tests, Marcel Dekker, New York.

Fine, I. V., Rabinovich, A. B., and Thomson, R. E. (2005). The dual source region for the 2004 Sumatra tsunami, *Geophys. Res. Lett.*, 32, L16602, doi:10.1029/2005GL023521.

Freilich, M. H. and Challenor, P. G. (1995). A new approach for determining fully empirical altimeter wind speed model functions, *J. Geophys. Res.*, 99, 25051–25062.

Fujii, Y. and Satake, K. (2007). Tsunami source of the 2004 Sumatra-Andaman earthquake inferred from tide gauge and satellite data, *B. Seismol. Soc. Am.*, 97, 192–207, doi:10.1785/0120050613.

Garratt, J. R. (1994). The Atmospheric Boundary Layer, Cambridge University Press, Cambridge.

Geist, E. L., Titov, V. V., Arcas, D., Pollitz, F. F., and Bilek, S. L. (2007). Implications of the 26 December 2004 Sumatra-Andaman earthquake on tsunami forecast and assessment models for great subduction-zone earthquakes, *B. Seismol. Soc. Am.*, 97, 249–270, doi:10.1785/0120050619.

Godin, O. A. (2003). Influence of long gravity waves on wind velocity in the near-water layer and feasibility of early tsunami detection, *Dokl. Earth Sci.*, 391, 841–844.

Godin, O. A. (2004). Air-sea interaction and feasibility of tsunami detection in the open ocean, *J. Geophys. Res.*, 109, C05002, doi:10.1029/2003JC002030.

Godin, O. A. (2005). Wind over fast waves and feasibility of early tsunamidetection from space, in: Frontiers of Nonlinear Physics, edited by: Litvak, A., Inst. Appl. Phys., Nizhny Novgorod, 210–215.

Godin, O. A. and Irisov, V. G. (2003). A perturbation model of radiometric manifestations of oceanic currents, *Radio Sci.*, 38, 8070, doi:10.1029/2002RS002642.

Godin, O.A, Irisov, V.G., Leben, R.R., Hamlington, B.D., Wick, G.A. (2009). Variations in sea surface roughness induced by the 2004 Sumatra-Andaman Tsunami, *Nat. Hazards Earth Syst. Sci.*, 9, 1135-1147.

Gonzalez, F. I., Bernard, E. N., Meinig, C., Eble, M. C., Mofjeld, H. O., and Stalin, S. (2005). The NTHMP tsunameter network, *Nat. Hazards*, 35, 25–39, doi: 10.1007/s11069-004-2402-4.

Gower, J. (2007). The 26 December 2004 tsunami measured by satellite altimetry, *Int. J. Remote Sens.*, 28, 2897–2913, doi:10.1080/01431160601094484.

Hara, T. and Plant, W. J. (1994). Hydrodynamic modulation of short wind-wave spectra by long waves and its measurement using microwave backscatter, *J. Geophys. Res.*, 99, 9767–9784.

Hayashi, Y. (2008). Extracting the 2004 Indian Ocean tsunami signals from sea surface height data observed by satellite altimetry, *J. Geophys. Res.*, 113, C01001, doi:10.1029/2007JC004177.

Hirata, K., Satake, K., Tanioka, Y., Kuragano, T., Hasegawa, Y., Hayashi, Y., and Hamada, N. (2006). The 2004 Indian Ocean tsunami: Tsunami source model from satellite altimetry, *Earth Planets Space*, 58, 195–201.

Hoechner, A., Babeyko, A. Y., and Sobolev, S. V. (2008). Enhanced GPS inversion technique applied to the 2004 Sumatra earthquake and tsunami, *Geophys. Res. Lett.*, 35, L08310, doi:10.1029/2007GL033133.

Kanamori, H. and Kikuchi, M. (1993). The 1992 Nicaragua Earthquake: a slow tsunami earthquake associated with subducted sediments, *Nature*, 361, 714-716.

Kudryavtsev, V. N., Mastenbroek, C., and Makin, V. K. (1997). Modulation of wind ripples by long surface waves via the air flow: a feedback mechanism, *Bound.-Lay. Meteorol.*, 83, 99–116.

Kulikov, E. A., Medvedev, P. P., and Lappo, S. S. (2005). Satellite recording of the Indian Ocean tsunami on December 26, 2004, *Dokl. Earth Sci.*, 401A, 444–448.

Kumar, B. P., Kumar, R. R., Dube, S. K., Murty, T., Gangopadhyay, A., Chaudhuri, A., and Rao A. D. (2006) Tsunami travel time computation and skill assessment for the 26 December 2004 event in the Indian Ocean, *Coast. Eng. J.*, 48, 147–166.

Lay, T., Kanamori, H., Ammon, C. J., Nettles, M., Ward, S. N., Aster, R. C., Beck, S. L., Bilek, S. L., Brudzinski, M. R., Butler, R., DeShon, H. R., Ekstrom, G., Satake, K., and Sipkin, S. (2005). The Great Sumatra-Andaman Earthquake of 26 December 2004, Science, 308, 1127–1133, doi:10.1126/science.1112250.

Lautenbacher, C. C. (2005). Tsunami warning systems, *The Bridge*, 35, 21–25.

Levin, B. W. and Nosov, M. A. (2005). Physics of Tsunamis and Kindred Phenomena, Janus-K, Moscow.

Naeije, M. , Schrama, E., and Scharroo, R. (2000). The Radar Altimeter Database System project RADS. *Proc. of the IEEE 2000 Int. Geoscience and Remote Sensing Symp. (IGARSS 2000)*, Honolulu, HI, IEEE, 487-490.

Nagai, T., Kato, T., Moritani, N., Izumi, H., Terada, Y., and Mitsui, M. (2007). Proposal of hybrid tsunami monitoring network system consisted of offshore, coastal and on-site wave sensors, *Coast. Eng. J.*, 49, 63–76.

Occhipinti, G., Lognonné, P., Kherani, A., Hebert, H. (2006). 3D Waveform modeling of ionospheric signature induced by the 2004 Sumatra tsunami, *Geophys. Res. Lett.*, doi:10.1029/2006GL026865.

Okal, E. A., Piatanesi, A., and Heinrich, P. (1999) Tsunami detection by satellite altimetry, *J. Geophys. Res.*, 104, 599–615.

Powell, B. S. and Leben, R. R. (2004) An optimal filter for geostrophic mesoscale currents from along-track satellite altimetry, *J. Atmos. Ocean. Tech.*, 21, 1633–1642.

Rowan, L. (2004). Tsunami and its shadow, *Science*, 304, p.1569.

Schindele, F., Loevenbruck, A., and Hebert, H. (2008). Strategy to design the sea-level monitoring networks for small tsunamigenic oceanic basins: the Western Mediterranean case, *Nat. Hazards Earth Syst. Sci.*, 8, 1019–1027, http://www.nat-hazards-earth-syst-sci.net/8/1019/2008/.

Sladen, A. and Hebert, H. (2008) On the use of satellite altimetry to infer the earthquake rupture characteristics: application to the 2004 Sumatra event, *Geophys. J. Int.*, 172, 707–714, doi:10.1111/j.1365-246X.2007.03669.x.

Smith, W. H. F., Scharroo, R., Titov, V. V., Arcas, D., and Arbic, B. K. (2005) Satellite altimeters measure tsunami, *Oceanography*, 18, 11–13.

Song, T. Y., Ji, C., Fu, L.-L., Zlotnicki, V., Shum, C. K., Yi, Y., and Hjorleifsdottir, V. (2005) The 26 December 2004 tsunami source estimated from satellite radar altimetry and seismic waves, *Geophys. Res. Lett.*, 32, L20601, doi:10.1029/2005GL023683.

Stein, S. and Okal, E. A. (2005) Speed and size of the Sumatra earthquake, *Nature*, 434, 581–582, doi:10.1038/434581a.

Synolakis, C. E. and Bernard, E. N. (2006). Tsunami science before and beyond Boxing Day 2004, *Philos. T. Roy. Soc. A.*, 364, 2231- 2265, doi:10.1098/rsta.2006.1824.

Synolakis, C. E., Bernard, E. N., Titov, V. V., Konoglu, U., Gonzalez, F. I. (2008). Validation and verification of tsunami numerical models, *Pure Appl. Geophys.*, 165(11-12), 2197-2228.

Titov, V., Rabinovich, A. B., Mofjeld, H. O., Thomson, R. E., and Gonzalez, F. I. (2005). The global reach of the 26 December 2004 Sumatra tsunami, *Science*, 309, 2045–2048, doi:10.1126/science/1114576.

Troitskaya, Y. I. (1994). Modulation of the growth rate of short surface capillary-gravity wind waves by a long wave, *J. Fluid Mech.*, 273, 169–187.

Troitskaya, Y. I. and Ermakov, S. A. (2005). Recording of the December 26, 2004 tsunami in the open ocean based on variations in radar scattering section, *Dokl. Earth Sci.*, 405A, 1384–1387.

Troitskaya, Y. I. and Ermakov, S. A. (2008) Manifestations of the Indian Ocean tsunami of 2004 in satellite nadir-viewing radar backscattering variations, *Int. J. Remote Sens.*, 29, 6361–6371, doi:10.1080/01431160802175348.

Walker, D. A. (1996). Observations of tsunami "shadows": A new technique for assessing tsunami wave heights?, *Science of Tsunami Hazards*, 14, 3–11.

Wei, Y., Bernard, E. N., Tang, L., Weiss, R., Titov, V. V., Moore, C., Spillane, M., Hopkins, M., and Kanoglu, U. (2008). Realtime experimental forecast of the Peruvian tsunami of August 2007 for US coastlines, *Geophys. Res. Lett.*, 35, L04609, doi:10.1029/2007GL032250.

Witter, D. L. and Chelton, D. B. (1991). A geosat altimeter wind speed algorithm and a method for altimeter wind speed algorithm development, *J. Geophys. Res.*, 96, 8853–8860.

Zaichenko, M. Y., Kulikov, E. A., Levin, B. V., and Medvedev, P. P. (2005). On the possibility of registration of tsunami waves in the open ocean with the use of a satellite altimeter, *Oceanology*, 45, 194–201.

4

Overtopping and Run-Up Hazards Induced by Solitary Waves and Bores

Tom. E. Baldock and Damitha Peiris
School of Civil Engineering, University of Queensland
Australia

1. Introduction

Solitary waves are wave forms that consist of a single wave, rather than waves that form part of a series of regular waves or random waves, the latter being typical of ocean wind and swell waves. A solitary wave that propagates with constant form has a unique shape for a given water depth and height, but in general, single waves of arbitrary shape are referred to as solitary waves. Single waves that have broken can be regarded as solitary bores. While the exact solitary wave shape has long been used to represent tsunami, Madsen et al. (2008) show that this is not likely to be usually the case and more complex wave shapes result. However, tsunami waves often closely resemble a solitary wave type and such waves still form the basis of most tsunami modelling. Tsunami are generated by impulsive geophysical events on ocean bed, and take the form of long waves with small steepness in the open ocean (Synolakis and Bernard, 2006). As witnessed in the 2004 Indian Ocean Tsunami, at the shore, the leading waves of a tsunami may steepen sufficiently to break and form very long surf bores.

The impact of tsunami bores may be very different to that from non-breaking or standing solitary waves, and this may require different disaster management strategies and evacuation plans, particularly when considering the initial impact and first few minutes after the wave arrival. Hall and Watts (1953) provided the first study of such waves with later laboratory studies of bore run-up providing further details on the speed of the run-up in comparison to theory (Yeh, 1991). In addition to tsunami, inundation of coastal zones by overwash is a major flooding hazard in many regions. In this instance, the overwash may be a result of run-up and overtopping on the sub-aerial beach, or as a result of overwash from waves in shallow water as the tide level inundates a barrier. In these cases, the overwash may be driven by surf zone bores or long wave surges that arise from wave groups in the shallow water or surf zone. The leading part of the long wave surge typically exhibits a solitary wave shape, as shown by Baldock (2006) and Nielsen and Baldock (2010). This overland flow in the run-up zone on beaches is frequently termed swash, and comprises an uprush (landward) and backwash (seaward) phase.

This work considers these issues, and presents the results of recent laboratory experiments measuring the overtopping flows from solitary waves and long bores. The results are discussed in the context of recent run-up theory, and used to further verify that theory for application to long bores, solitary waves and potentially initial tsunami impact. In addition, using the recent solutions, some estimates of the fluid forces, maximum inundation depths and the potential human safety hazard arising during the run-up are presented. This chapter is organised as follows: section 2 provides an overview of previous work on tsunami, overwash, and overtopping of wave-run-up, together with an outline of the model for bore run-up (Shen and Meyer, 1963; Peregrine and Williams, 2001; Guard and Baldock, 2007). Section 3 describes the experimental setup, data collection and wave conditions in the experiments. Results and comparisons with the Guard and Baldock (2007) model are presented in section 4. Application of the model to derive flow forces and a hazard assessment in the run-up zone are discussed in section 5, with final conclusions in section 6.

2. Background

2.1 Previous studies

An extensive summary of the literature on tsunami propagation and impact is given by Synolakis and Bernard (2006). Here, the focus is on the run-up and overtopping flow, which has not been as extensively studied. Traditional tsunami models assume non-breaking waves impacting at the shoreline. In addition to the single positive "hump" of a classical solitary wave, typical tsunami also exhibit N-shaped wave forms, with either a leading depression (LDN wave) or a leading elevation (LEN wave) and the latter are primarily considered here. Additionally, tsunami waves, or the leading positive waves in a tsunami wave train, may make landfall in the form of broken waves or bores, which impact coastal defenses, beaches and lead to the initial overwash or overtopping of coastal dunes and seawalls. Eventually, the large mass of water in the main following tsunami wave overtakes the initial bore-driven run-up, generally leading to further inundation. However, initially, for the first few minutes, the impact of the tsunami may be dominated by the run-up from broken waves or bores. This initial period is important in the context of human safety on the immediate foreshore and in terms of warning systems and evacuation strategies. It is also relevant to the potential impact forces on structures, particularly if the run-up picks up debris along the coastline (Yeh, 2006). Furthermore, in the event that warnings are non-existent or too late for inland evacuation, the initial impact and human hazard are important in designing vertical evacuation strategies.

Early tsunami observations were frequently interpreted as turbulent bores (Synolakis and Bernard, 2006). Peregrine (1966; 1967) formulated the depth-averaged or non-linear shallow water (NLSW) equations that describe both the propagation and run-up of such bores. Further work by Hibberd and Peregrine (1979) provided numerical solutions for the overland or swash flows for long bores. This complemented the earlier theoretical work of Shen and Meyer (1963), later adapted by Peregrine and Williams (2001) as an analytical solution for the hydrodynamics in the swash zone. Synolakis (1987) provided solutions for the run-up of non-breaking solitary waves. However, as noted by Synolakis and Bernard (2006), run-up is still neglected in many tsunami models, which can lead to estimates of run-up that differ by factors of 2. In addition, if the run-up is not properly included in the modeling, accurate estimates of flow depths and flow velocities cannot be produced. The local bathymetry and most landward beach slope experienced by the wave is a controlling factor in the run-up and inundation (Synolakis and Bernard, 2006) and hence the wave amplitude at the original shoreline (dry beach) is important in determining the final run-up behaviour. However, most historical observations are of the maximum run-up or maximum landward penetration of tsunami waves, although post-tsunami survey data from more recent events does include inundation depths.

The solutions of the NLSW equations are applicable to describe the run-up of long bores, from which overtopping flow volumes can be determined, together with the hydrodynamics in the inundation zone. For non-breaking waves, the analytical solitary wave solutions are relevant (e.g. Carrier et al., 2003), but to the authors' knowledge have not been applied to describe overwash or overtopping volumes. Previous work has considered the classical analytical solution of Shen and Meyer (1963) to describe bore run-up and overtopping due to run-up, or swash overtopping (Peregrine and Williams, 2001). However, recent new solutions to the NLSW (Guard and Baldock, 2007) show that the Shen and Meyer solution is not conservative for engineering design, and that it significantly underestimates flow depths and overtopping flow volumes. Given that the usual criteria for human safety during flood events is based on a product of water depth and flow velocity, and that the forces on structures are proportional to the momentum flux, a product of the velocity squared and depth, the underestimation of flow depths by the traditional model can lead to significant underestimation of potential hazards.

Numerical modelling of wave overwash over steep coastal structures has used a wide range of techniques, from non-linear shallow water wave models (e.g. Kobayashi and Wurjanto, 1989; Dodd, 1998) to Boussinesq models (e.g. Stansby, 2003) to Navier-Stokes solvers (e.g. Ingram et al., 2009). A number of empirical formula are also available (Goda, 2009), although these are more applicable for sequences of periodic or random waves. However, these lose the transparency of the swash models that are controlled by the nearshore boundary conditions. On natural beaches, sand dunes and beach berms provide the first line of coastal defence, and overtopping leads to flooding of the backshore as well as the transport and deposition of marine sand and saline water. For extreme conditions, when storm surge elevations exceed the berm crest, wave overtopping may combine with a steady flood flow (Hughes and Nadal, 2009), and berm rollover and breaching of the barrier may occur. Overviews of these processes are given by Kraus et al. (2002) and Donnelly et al. (2006). Similar processes may occur due to tsunami overtopping and drawdown.

Guard and Baldock (2007) and Guard et al. (2005) proposed a new model for the run-up of long bores and leading tsunami bores. The models are written in non-dimensional form, which enables them to be applied easily to different beach slopes and run-up conditions. This model provides different solutions for different boundary conditions at the original still water line, with Shen and Meyer (1963) as a special case. The model boundary condition represents different fluxes of mass and momentum into the run-up zone, which in turn control the flow depths and magnitude of the overtopping flows if the run-up exceeds the beach length. Similar controls on the mass and momentum flux into the run-up zone could be expected for solitary-type non-breaking waves.

In natural conditions, the run-up experiences a truncated beach if the run-up exceeds the beach crest, dune crest or structure crest, and then inundation by overtopping occurs. While some previous laboratory experiments have investigated the influence of the boundary conditions on the flow depth, no previous data on overtopping of either non-breaking solitary waves or breaking solitary bores is available. Further, the theory of Guard and Baldock (2007) has not been applied to overtopping conditions, nor to estimate impact forces and hazards. This work presents some recent experiments on this issue and compares the observations with the model results. Subsequently, the model is used to illustrate how the varying boundary conditions will influence the hazards in the run-up zone in terms of flow depths and flow forces. The data and model are presented in non-dimensional form to make the results applicable over a wide range of scales.

2.2 Guard and Baldock (2007) swash model

The Shen and Meyer (1963) model is one of a number of possible solution to the nonlinear shallow-water equations (Peregrine and Williams, 2001; Pritchard and Hogg, 2005), and represents an asymptotic description of the flow close to the wave tip for a single and specific boundary condition. The solution corresponds to conditions for a dam-break wave up a slope, where the water behind the dam is initially stationary. If applied to the swash zone, this solution neglects the mass and momentum of the flow behind the incident bore front, which in many cases is very significant. Recognising this, Guard and Baldock (2007) developed new numerical solutions to the non-linear shallow-water equations that include the initial mass and momentum behind the incident bore front. An analytical form of these solutions was later developed by Pritchard *et al.* (2008), and which verified the previous numerical scheme. A brief description of the model follows, focusing on the influence of the model boundary condition on the flow depth in the run-up zone and the duration of the inflow across the original shoreline.

The model solves a non-dimensional form of NLSWE written in characteristic form as (Peregrine and Williams, 2001):

$$\begin{aligned} &\frac{\partial h}{\partial t}+\frac{\partial (uh)}{\partial x}=0 \\ &\frac{\partial u}{\partial t}+u\frac{\partial u}{\partial x}+\frac{\partial h}{\partial x}+1=0 \end{aligned} \tag{1}$$

where u and h are flow velocity and depth, respectively. The dimensionless variables are scaled as follows:

$$x=\frac{x^*\sin\gamma}{A} \quad t=t^*\sin\gamma\sqrt{\frac{g}{A}} \quad h=\frac{h^*\cos\gamma}{A} \quad u=\frac{u^*}{\sqrt{gA}} \tag{2}$$

where dimensional variables are starred and A=R/2, where R is the maximum run-up elevation above the initial elevation. For most natural beaches, cosγ can be taken equal to one, and sinγ ≈γ. Substituting for the non-dimensional shallow water wave speed,

$$c=\sqrt{h} \tag{3}$$

these equations may be rearranged to obtain:

$$\begin{aligned} &\left(\frac{\partial}{\partial t}+(u+c)\frac{\partial}{\partial x}\right)(u+2c+t)=0 \\ &\left(\frac{\partial}{\partial t}+(u-c)\frac{\partial}{\partial x}\right)(u-2c+t)=0 \end{aligned} \tag{4}$$

Using the characteristic variables $\alpha(x,t)$ and $\beta(x,t)$ gives:

$$\frac{d}{dt}\alpha=0 \quad \text{on} \quad \frac{dx}{dt}=u+c \tag{5}$$

$$\frac{d}{dt}\beta = 0 \quad \text{on} \quad \frac{dx}{dt} = u - c \tag{6}$$

where

$$\begin{aligned} \alpha(x,t) &= u + 2c + t \\ \beta(x,t) &= u - 2c + t \end{aligned} \tag{7}$$

The solutions and model results are controlled by the value of $\alpha(t)$ on the seaward swash boundary and the Shen-Meyer solution corresponds to a constant value of $\alpha=2$, which results in very shallow inundation depths. The key change introduced by Guard and Baldock (2007) allowed the value of the characteristic variable on the seaward boundary, α, to vary over time, such that:

$$\alpha(x_b, t_b) = 2 + kt_b \tag{8}$$

where $k=0$ corresponds to the Shen-Meyer solution and k is expected to lie in the range $0<k<1.5$, with $k=1$ corresponding to a uniform incident bore. The solutions in non-dimensional form are independent of beach slope, γ, and run-up amplitude (2A). Consequently, the Shen and Meyer solution is identical for all run-up events (there is only a single solution for $k=0$), but the Guard-Baldock model allows different boundary conditions for different incident wave types or conditions, by specifying different values for k.
Varying the model conditions on the seaward boundary, by varying $\alpha(t)$ on the incoming characteristics, results in deeper and less asymmetric swash flows than predicted by the Shen-Meyer solution. This physical significance of the characteristic values α and β is shown by the non-dimensional volume flux, q, and momentum flux, M, per unit length of wave crest which are:

$$\begin{aligned} q = uh &= \frac{1}{32}(\alpha + \beta - 2t)(\alpha - \beta)^2 = q^* \frac{\cos\gamma}{\sqrt{gA^3}} \\ M = u^2 h &= \frac{1}{64}(\alpha + \beta - 2t)^2 (\alpha - \beta)^2 = M^* \frac{\cos\gamma}{\rho g A^2} \end{aligned} \tag{9}$$

Here, the starred variables are again dimensional, and the scaling follows that of Peregrine and Williams (2001), where A is half of the vertical run-up elevation above the still water shoreline, i.e. R=2A.
The interpretation in terms of the overall fluid physics is that solutions for different $\alpha(t)$ correspond to different mass and momentum flux at the seaward boundary. By varying k, and consequently $\alpha(t)$, the incoming mass and momentum fluxes at the seaward boundary can differ for each swash event, which results in different flows within the swash zone. Increases in k represent more sustained inflow conditions and lead to solutions that have a longer duration of inflow across the original shoreline, later times of flow reversal, increased depths in the swash, and a more symmetric velocity field between uprush and backwash. Figures 1 and 2 illustrate how the flow depths, the flow volume and the time of flow reversal are influenced by the choice of k. The time of flow reversal at the seaward boundary of the swash varies significantly for solutions between $k=0$ and $k=1$; at the seaward swash boundary flow reversal occurs after one-quarter of the whole swash cycle for $k=0$, whereas

for k=1 flow reversal occurs approximately half-way through the swash cycle (Guard and Baldock, 2007). This leads to much greater volumes of water entering the swash zone than predicted by the Shen-Meyer solution, which was originally observed on the basis of overtopping experiments using monochromatic waves by Baldock et al. (2005). Thus, the specified boundary condition, as defined by k, controls the overall asymmetry of the flow field, the flow depth (figure 2) and the quantity of fluid that can be expected to overtop a truncated beach.

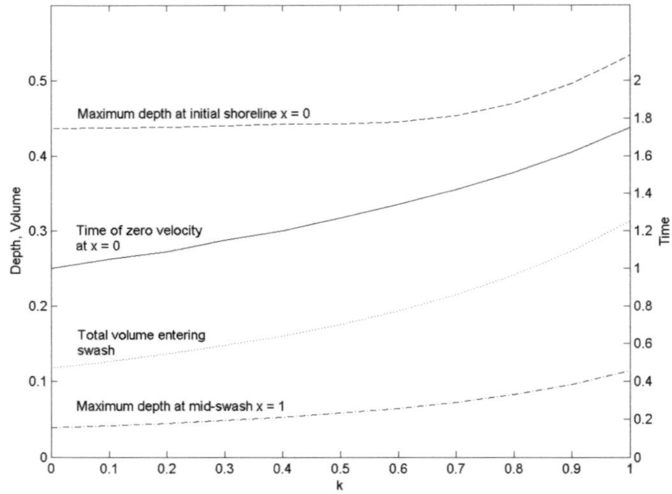

Fig. 1. The dependence of key swash hydrodynamic parameters on the swash boundary condition, parameterised through k. After Guard and Baldock (2007).

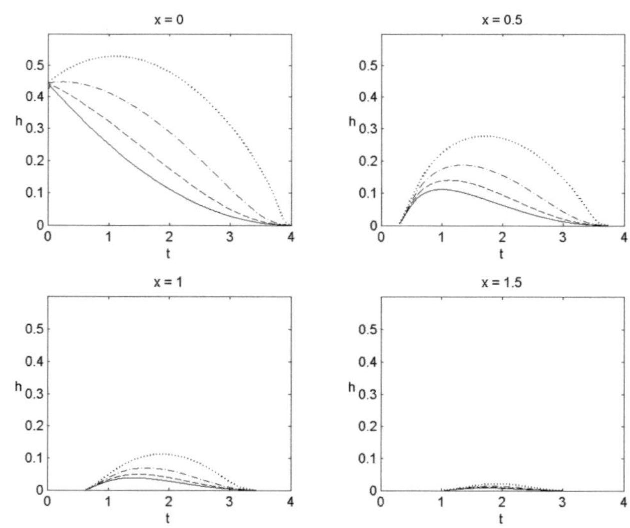

Fig. 2. Time-history of flow depth at different cross-shore locations indicated. Solid line, k=0 (SM63); Dashed-dotted line, k=1/3; Dashed line, k=2/3; dotted line, k=1. x varies from 0 at the SWL to 2 at the run-up limit. After Guard and Baldock (2007).

3. Experimental setup

3.1 Wave flume and instrumentation

The overtopping experiments were conducted in a 0.85 m wide, 0.75 m deep and 28 m long wave flume in the Gordon Mckay Hydraulics Laboratory at the University of Queensland. The bathymetry comprises of a 10.5m long horizontal section from the wavemaker to the toe of a uniform long sloping beach of gradient 0.11. The sloping beach was constructed in two parts; a fixed lower section below the water line and an adjustable beach with removable panels above the still water line. Figure 3 and figure 4 illustrates the general arrangement, where the origin of the horizontal coordinate is at the still water line (SWL) and positive onshore. Two configurations of the bed were used, firstly a smooth painted marine plywood bed, and secondly, with the bed covered by sand paper (0.2 mm grain size). The removable panels on the upper beach can be replaced by an overtopping tank sunk within the bed at

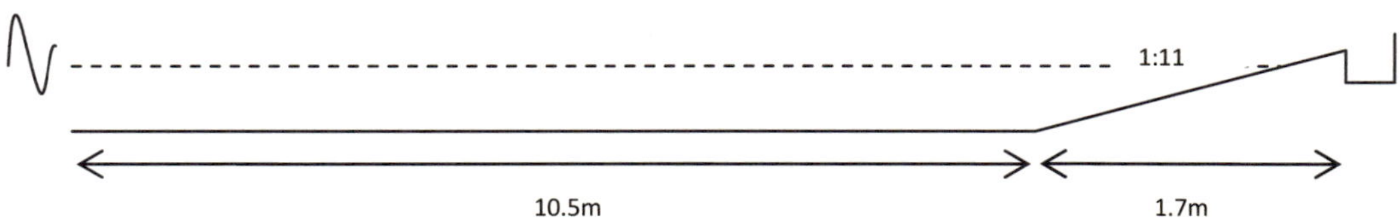

Fig. 3. Schematic of wave flume and overtopping tank for one tank position.

Fig. 4. Photo of upper beach panels, overtopping tank and instrumentation.

different elevations relative to the still water line and the maximum run-up. A special tank was designed, which encompassed the middle two-thirds of the flume to avoid wall effects, and which also included thin "wing" walls to prevent flow from entering from the sides (figure 4). The tank is 0.55m in length, 0.45m in width, and 0.195m deep, and results in a truncated beach at the seaward or offshore edge of the tank. The tank edge has very little effect on the uprush flow since the flow at the edge is super-critical for the majority of the uprush (Peregrine and Williams, 2001) and as shown later.

The experiments were performed with the overtopping tank located at six locations along the beach, ranging from 0.05-0.26m above the still water line. The water depth over the horizontal section of the flume was also varied, which additionally changes the tank positions relative to the SWL.

The tank position is written in non-dimensional form, such that the non-dimensional position, or truncation point for the beach, is given by $E=2z/R$, where R is the maximum vertical run-up excursion for each individual wave condition, and z is the elevation of the tank edge relative to the still water line (SWL). Hence, E ranges from zero, at the SWL, to two, at the run-up limit, for each particular test. This follows the Peregrine and Williams (2001) scaling. R was determined for each wave condition by running the identical experiment with a non-truncated beach without the tank, again for both the smooth and sandpaper bed surfaces. Overtopping volumes were measured in the tank using an ultrasonic distance sensor to measure the surface elevation change between the start and end of the test, with a calibrated conversion function to account for small differences in tank area with water surface elevation. The hydrodynamics were obtained from ultrasonic displacement sensors and an Acoustic Doppler Velocimeter (ADV), mounted at different locations along the flume. However, for the present work, only the ultrasonic displacement sensor data are used; the ADV and one ultrasonic sensor provide hydrodynamic boundary condition data for future numerical modelling purposes.

3.2 Wave conditions

Solitary type waves and solitary broken waves (bores) were generated using a computer-controlled hydraulic piston wave generator which has stroke lengths of up to 1.2m to generate long bores. The waves were generated by applying a known voltage signal to the wave paddle, which were generated using either error functions or using the method developed to generate solitary waves as per Goring (1979) and Baldock et al. (2009). The wave height and wavelength is controlled by varying the stroke (distance moved) and speed of the wavemaker. This has the effect of altering the hydrodynamics of each wave case, particularly the duration of the inflow into the swash zone, and therefore corresponds to waves which generate different swash boundary conditions or waves that are described by different values of k. The wavemaker motion and resulting wave forms are very repeatable, enabling the use of multiple tank positions and repeat runs of the same wave form. Two wave types were used in this study; waves that broke at the shore (shore breaks), and breaking waves that broke prior to reaching the SWL forming well-developed bores. This distinction was made through visual observation for each case. Altogether, 19 different wave conditions were used, repeated for four different water depths and six different (in absolute elevation) tank positions. For each wave condition, this provides measurements of the overtopping at 15-20 different truncation points between the SWL and maximum run-up.

4. Experimental results

4.1 Run-up

This section presents measurements of the run-up for the different wave conditions and surface roughness. The data are non-dimensionalised using the offshore water depth, following Hall and Watts (1953) and later researchers. Figure 5 illustrates the variation of the maximum run-up elevation, R, over the smooth bed with the maximum inundation depth or bore height measured at the SWL, H_b. Non-dimensionalising by the offshore water depth does not fully remove the depth-dependency from the data, showing that the maximum run-up is not a linear function of offshore water depths. However, the differences for water depths of 15-26cm are relatively small, and it is only for the smallest water depth that a significant difference occurs. R/d is well correlated with H_b/d, and the coefficient of proportionality ranges from 3-4, i.e. $R=CH_b$, where $3<C<4$ for these data. If the wave height at the toe of the beach is chosen for the correlation instead of H_b, the data for R/d tend to collapse more onto a single line (i.e. a linear depth dependency), but the scatter in the data is significantly greater. The relationship between run-up and offshore or nearshore wave height is therefore similar too, but not exactly the same as the theoretical and observed behaviour of non-breaking solitary waves, which show R/d is proportional to $(H/d)^{5/4}$ (Synolakis, 1987).

The behaviour is very similar on the rough slope (figure 6), and the coefficients of proportionality do not change significantly, ranging from 3-3.7. This is consistent with previous run-up measurements and suggests that the frictional effect at the wave tip is quite small. Even though the relative roughness has changed by an order of magnitude, the increased roughness reduces the run-up by only 5-10%. The non-dimensional run-up coefficients are however different from that observed by Baldock et al. (2009), which was

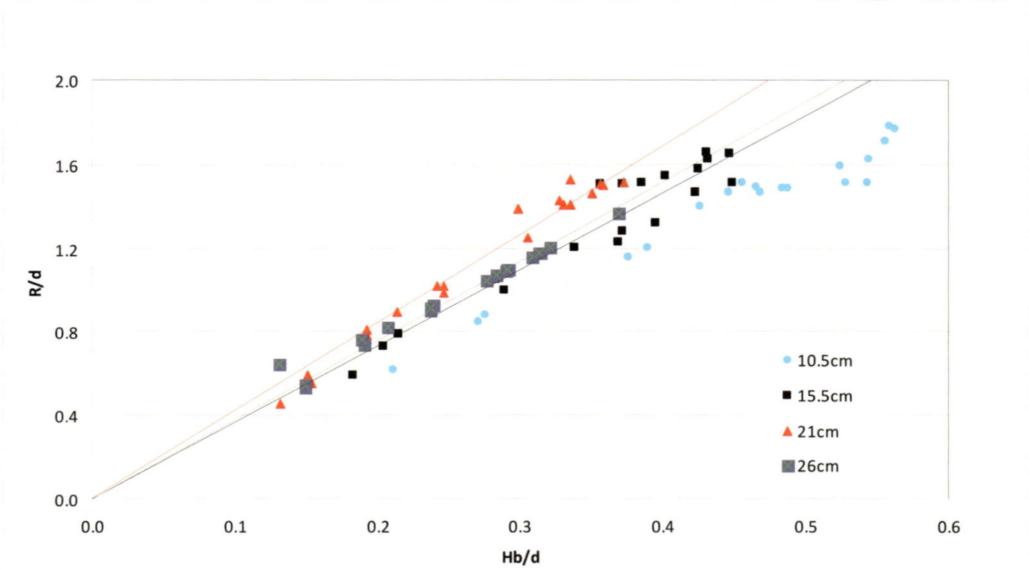

Fig. 5. Non-dimensional run-up versus maximum non-dimensional wave elevation at the still water shoreline. Smooth beach. Legend indicates water depths for each data series.

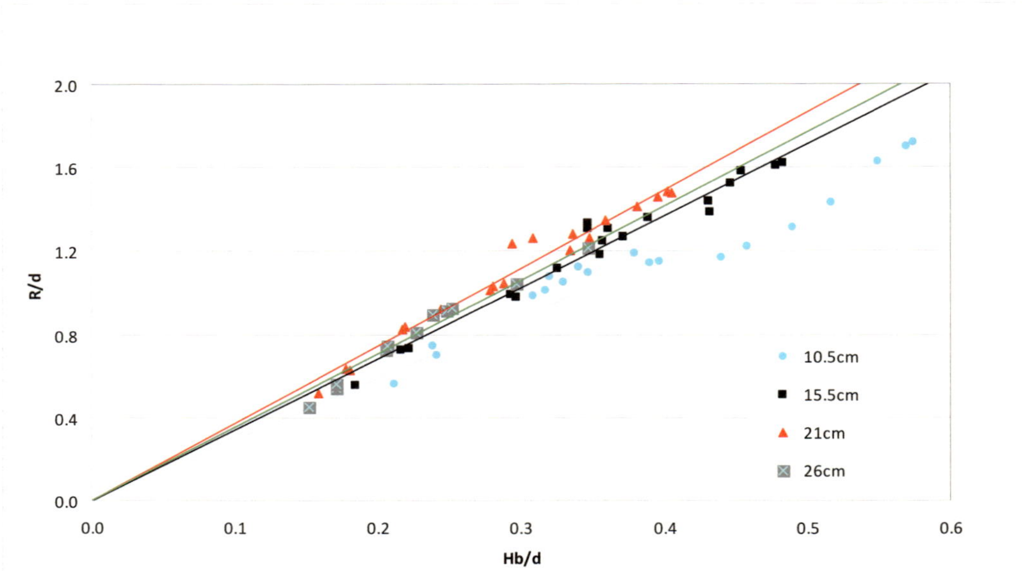

Fig. 6. Non-dimensional run-up versus maximum non-dimensional wave elevation at the still water shoreline. Rough beach. Legend indicates water depths for each data series.

C=2.6, where upper beach was a milder gradient than the offshore beach gradient. The lack of sensitivity of the run-up to the change in roughness is for two reasons; firstly, at laboratory scale Reynolds numbers, the friction factors are not expected to vary significantly with changes in relative roughness, as shown on classical diagrams, e.g. the Moody diagram; and secondly, the run-up tip does not travel as a solid body, but is constantly overrun by the flow behind (Barnes and Baldock, 2010). The latter effect further reduces the influence of friction on the thin run-up tip. Based on the Guard and Baldock (2007) model, the maximum depth at the SWL is about one-quarter of the run-up elevation (see figure 1), or alternatively, $R \approx 4H_b$ for $k \approx O(1)$, so the data is consistent with the theory, allowing for frictional effects. No obvious difference in run-up trend was observed between shore breaks and developed bores.

4.2 Overtopping

The type of incident wave is illustrated in figure 7, which shows the time-history of the water surface elevation at the SWL-beach intersection (original shoreline) for 3 different incident waves, W1-W3. The wavemaker stroke length and speed varies for each case, leading to a weak shore break (W1), a strong shore break (W2) and a fully developed bore (W3), respectively. The product of velocity and flow depth during the inflow to the swash zone varies for each case, which controls the overtopping rate higher on the beach face at the truncation point. This variation cannot be predicted by the Shen and Meyer (1963) solution, since all incident waves have the same boundary condition.

The measured overtopping volume per unit width of beach, V*(E), is non-dimensionalised following Peregrine and Williams (2001) as follows:

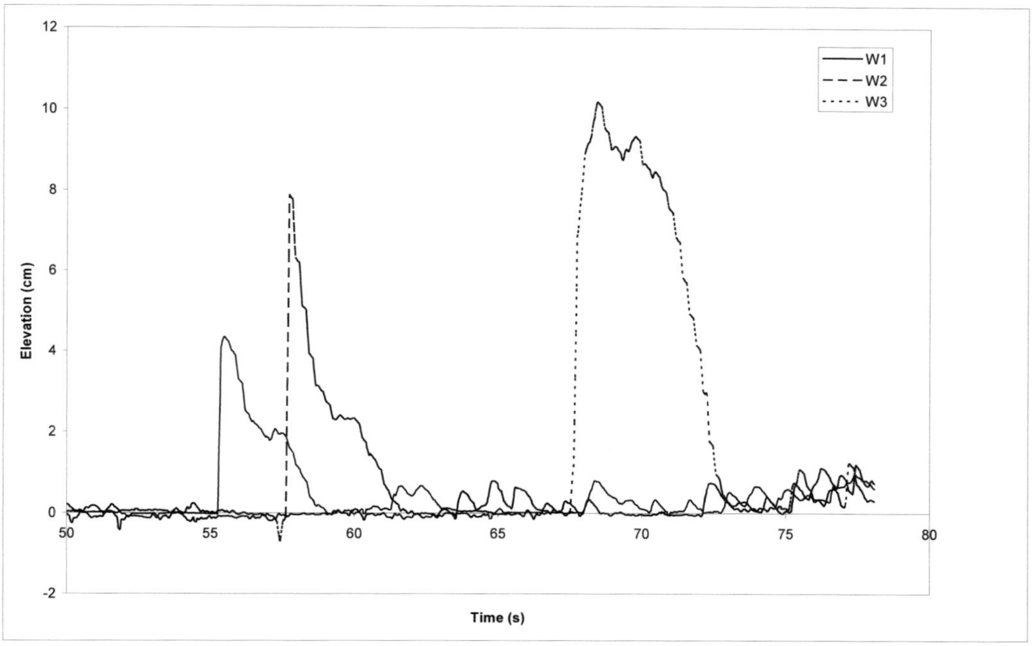

Fig. 7. Water surface elevation at the original still water shoreline for different wave types, corresponding to a weak shore break (W1), a strong shore break (W2) and a fully developed bore (W3).

$$V(E) = V^*(E)\sin(2\gamma) / 2A^2 \tag{10}$$

The non-dimensional overtopping volumes are plotted versus the relative truncation elevation , E, of the beach, where E=2z/R such that E=0 at the SWL and E=2 at the run-up limit for each wave condition. Figures 8 and 9 illustrate the data for shore break waves and fully developed bores, respectively. The theoretical overtopping derived by Peregrine and Williams (2001), itself based on the swash theory of Shen and Meyer (1963), and the empirical overtopping curve of Baldock et al. (2005) are also shown.

A distinct difference is observed between the two wave types. Waves that break at the shore generally result in smaller overtopping dimensionless volumes, and the relationship between V(E) and truncation point is quite different from that for bores. For bores, the overtopping rates increase rapidly as E decreases, whereas for the waves breaking at the shore the rate of increase is less sensitive to E. Further, the overtopping rate for the developed bores is multi-valued for the same value of E, in contrast to the shore break conditions which shows a relatively consistent trend with E, and in contrast to the Peregrine and Williams solution. This is not scatter in the data, as shown below, and is consistent with the theory of Guard and Baldock (2007). If data for a single constant offshore water depth is considered in isolation, these differences between the shore breaks and developed bores are even more marked (figure 10). The non-dimensional overtopping rate for shore breaks becomes insensitive to E, whereas for the developed bores the overtopping is not solely determined by E, and that it is again multi-valued for constant E.

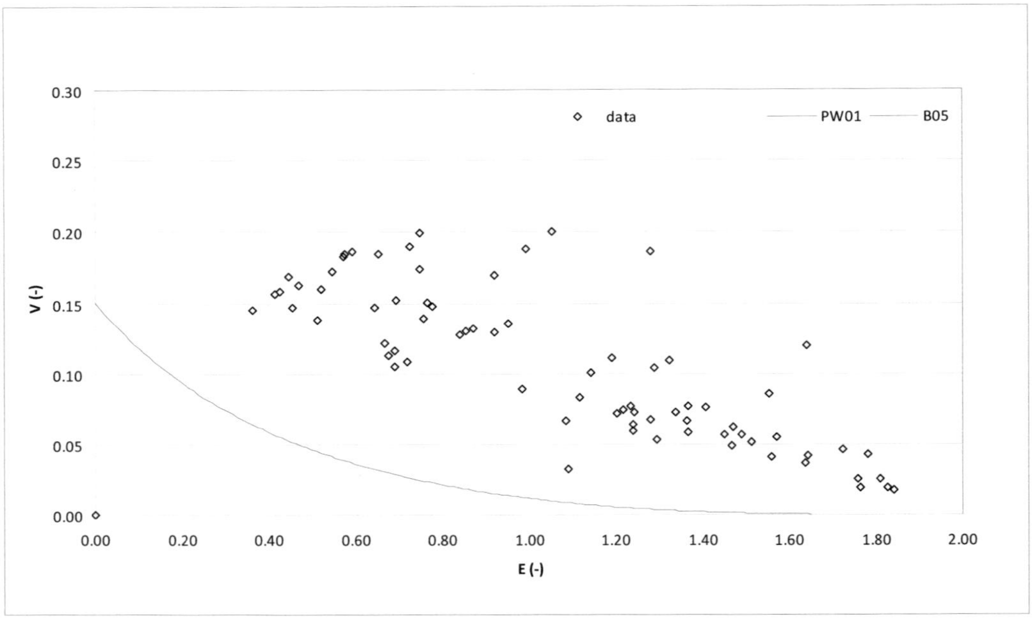

Fig. 8. Non-dimensional overtopping volume versus non-dimensional truncation point. Shore break waves and smooth beach. Legend indicates theoretical curve (solid line) of Peregrine and Williams (2001) and empirical curve (dashed line) of Baldock et al. (2005)

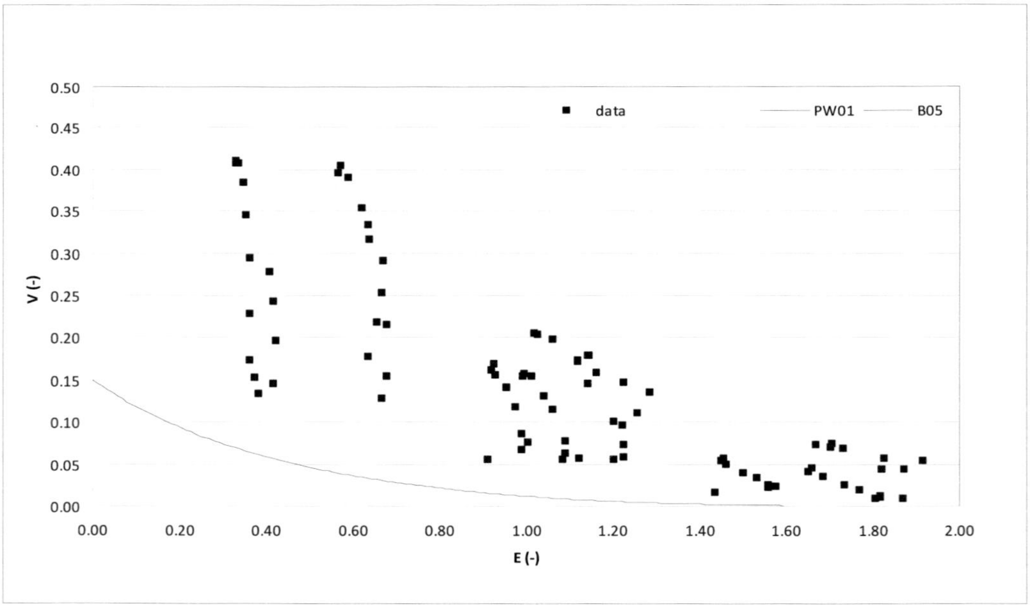

Fig. 9. Non-dimensional overtopping volume versus non-dimensional truncation point. Developed bores and smooth beach. Legend indicates theoretical curve (solid line) of Peregrine and Williams (2001) and empirical curve (dashed line) of Baldock et al. (2005)

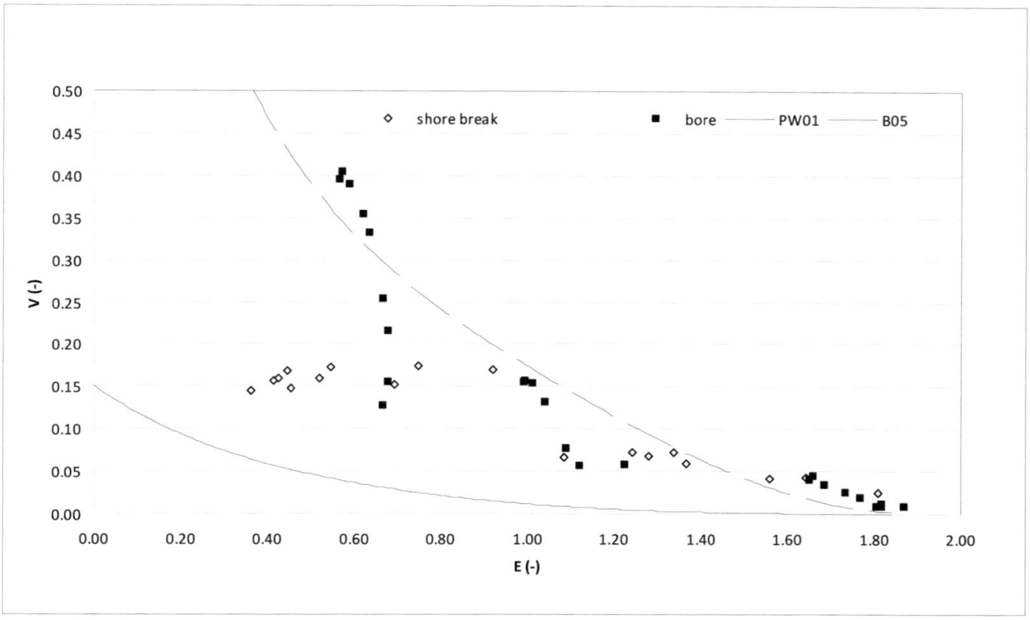

Fig. 10. Non-dimensional overtopping volume versus non-dimensional truncation point. Shore breaks and developed bores, smooth beach, depth=15.5cm. Legend indicates theoretical curve (solid line) of Peregrine and Williams (2001) and empirical curve (dashed line) of Baldock et al. (2005).

For the developed bores, V(E) is a monotonic function of E for large E, consistent with the asymptotic solutions of the NLSWE close to the run-up limit. However, for smaller values of E, both the dimensional and the non-dimensional overtopping rates are very dependent on the incident bore conditions. Hogg et al. (2010) have further developed semi-analytical theoretical solutions to account for the variation of the overtopping rates as k is varied in the Guard and Baldock (2007) swash model. It is of interest that for E>1, the overall trend of V(E) is similar for the two different wave types. For the rough beach the trends are very similar, and as a function of E, the non-dimensional overtopping volumes do not decrease significantly (not shown).

The non-dimensional overtopping volumes obtained from the Guard and Baldock (2007) solution, and ignoring the influence of the edge on the flow, are shown in figure 11. These calculations determine the overtopping volume from the product $V(E)=uh$ at the truncation point, but do not account for the influence of the truncation point on the flow further offshore. However, the edge does not influence the flow until the flow drops to critical conditions at the edge and Peregrine and Williams (2001) show that the supercritical overtopping volume and the total overtopping volume are similar for E>0.4 and very similar for E>1. Hence, some differences are expected between the full solution and the approximate calculations for small E. The full solution for k=0 and the approximate solution are in very good agreement for E>0.4, which shows the influence of the edge is indeed minor for most practical applications. Note that E<0.4 corresponds to locations in the lower (most seaward) fifth of the swash zone, i.e. z/R<0.2.

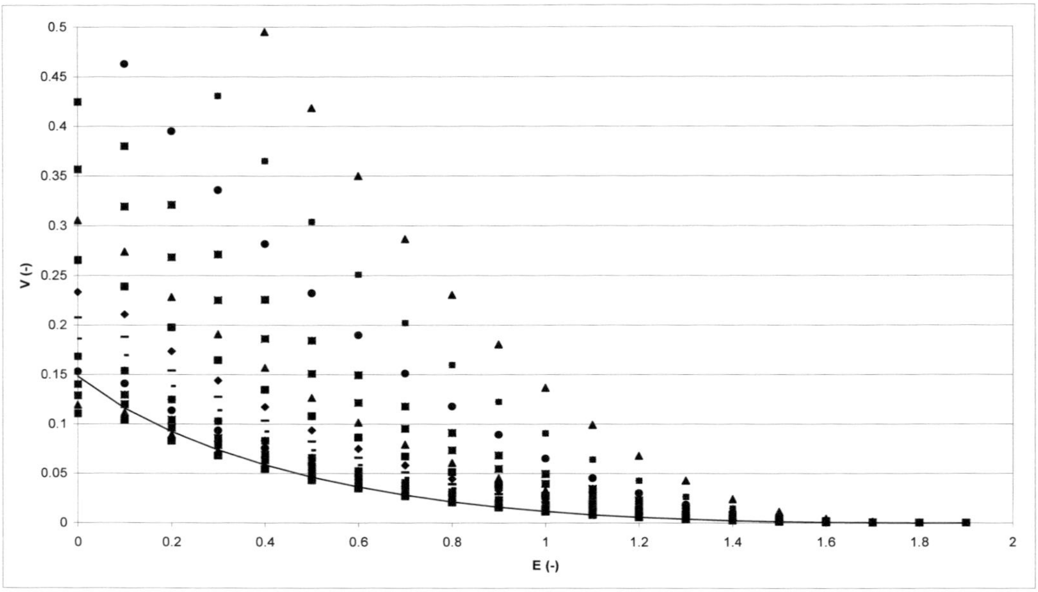

Fig. 11. Non-dimensional overtopping volume versus non-dimensional truncation point based on Guard and Baldock (2007) solution and ignoring the influence of the edge on the flow. Symbols show influence of varying k, from k=0 to k=1.5 at intervals of k=0.1. Solid line is theoretical curve of Peregrine and Williams (2001).

For k=0, it appears that the presence of the edge increases the volume flux past a given elevation in comparison to the solution for a non-truncated beach. This might be expected on the basis of a reduction in pressure at the free overfall and the formation of a classical M2 flow profile seaward of the edge, which reduces the gradient of the adverse (offshore dipping) water surface slope further offshore and hence reduces the influence of gravity in slowing the uprush flow.

The different symbols in figure 11 at constant values of E show the influence of varying k, or varying the incident mass and momentum flux in the incoming bore. The predicted overtopping volume increase by a relatively constant factor of about 2.6 as k increases from k=0 to k=1, and increase further to O(10) for k=1.5. More importantly, the predicted multiple values of V(E) at constant E are entirely consistent with the observations and in complete contrast to the monotonic solution predicted by Peregrine and Williams (2001). Comparing figure 11 with figure 9 shows that the model underestimates the measured non-dimensional overtopping volumes to some extent for E>1. However, the range of variation in the predicted and measured overtopping at a constant value of E is very similar, as illustrated in figure 12. The predicted overtopping volumes for k=0.9 and k=1.4 encompass the range of data in these experiments. This is consistent with the previous estimates of k for developed bores (Guard and Baldock, 2007) and from field observations of surf zone bores and swash run-up (Power et al., 2009).

It should be noted that the model is based on inviscid flow conditions. Therefore, the absolute values of the measured non-dimensional overtopping rates are sensitive to the influence of friction, which will have the greatest effect on the thin run-up tip and hence the value of R. For a real swash, the maximum run-up is expected to be reduced by a greater

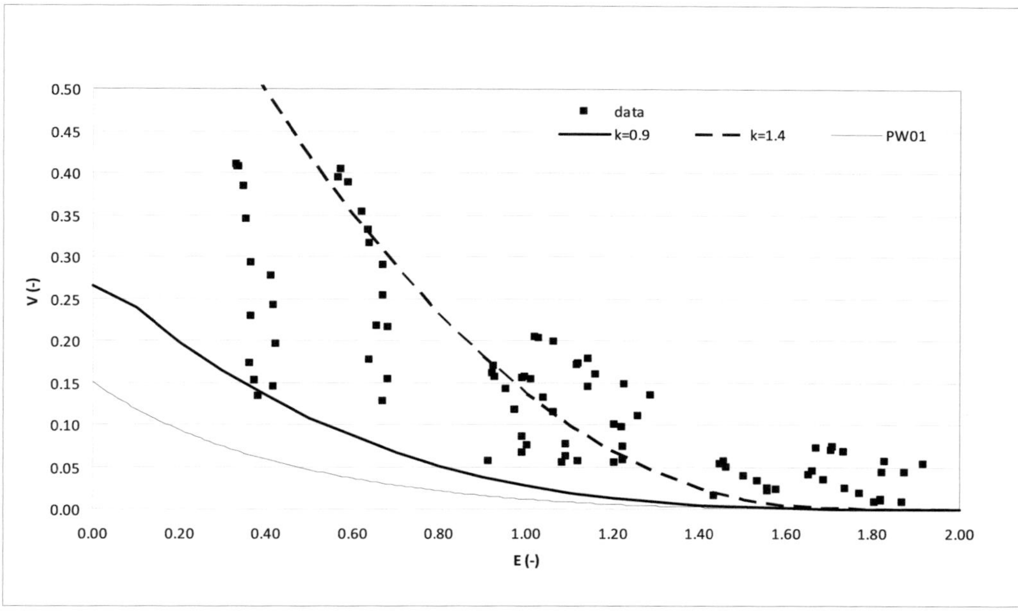

Fig. 12. Non-dimensional overtopping volume versus non-dimensional truncation point compared to range expected from Guard and Baldock (2007) solution. Developed bores and smooth beach. Thin solid line is theoretical curve of Peregrine and Williams (2001), thick solid line and thick dashed line correspond to solutions for k=0.9 and k=1.5, respectively.

extent than the depth and flow velocity at the edge, since frictional effects are greatest in the shallow flow depths at the wave tip. Thus, R is reduced in the experiments in comparison to the inviscid solution, whereas the dimensional overtopping volume is probably not reduced to such an extent. The non-dimensionalisation therefore increases the measured V(E) compared to the inviscid solution, as shown by comparing figures 9 and 11. Alternatively, this is equivalent to the value of E for the measured data being slightly too large, since they are based on the measured run-up, not the idealised inviscid run-up, which will be larger.

5. Run-up hazards

5.1 Personal safety

The ability of people to maintain a secure footing in flood flow is determined by both the flow depth and the flow velocity. For characterising flood risks to personal safety, a simple and widely used parameter is based on the product of water depth and flow velocity exceeding a critical value, typically, u^*h^*>0.5-1 (Abt et al., 1989; Ramsbottom et al., 2003). The dimensionless value of this parameter is

$$uh = \frac{u^* h^* \cos\gamma}{A\sqrt{gA}} \tag{11}$$

This is plotted in figures 13a&b, for uprush and backwash, respectively, and again for a range of k values, from which values of u^*h^* can be determined for any location in the run-up zone for any particular bore run-up amplitude.

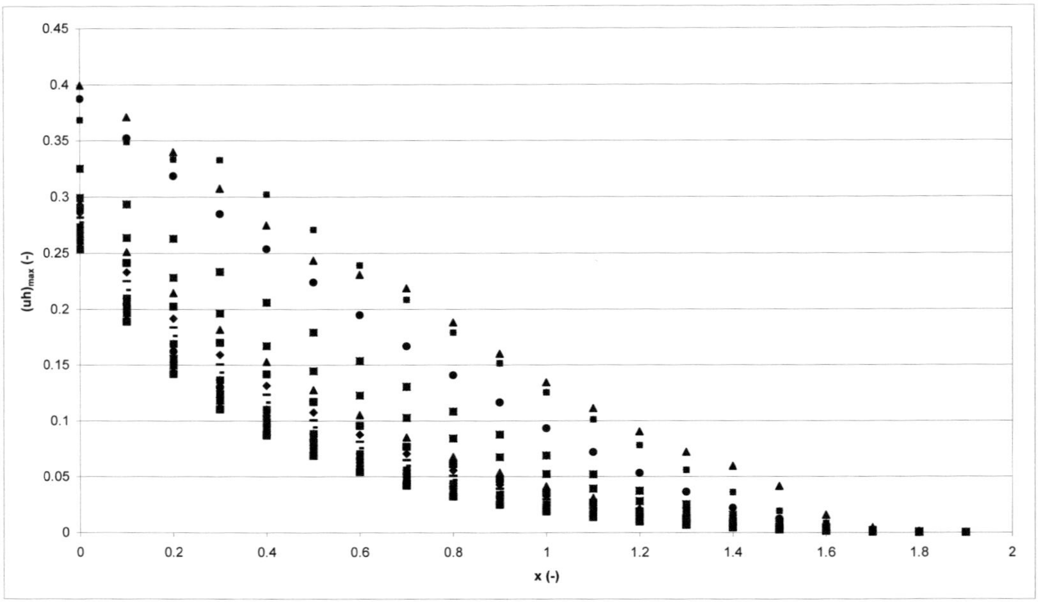

Fig. 13a. Non-dimensional hazard values during uprush versus non-dimensional distance based on Guard and Baldock (2007) solution. Symbols show influence of varying k, from k=0 to k=1.5 at intervals of k=0.1.

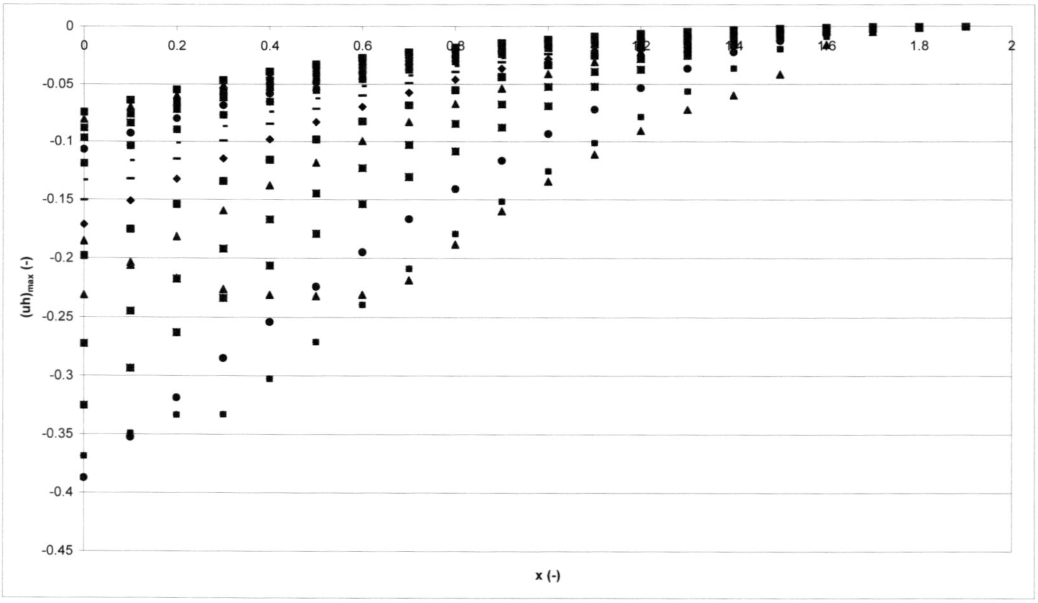

Fig. 13b. Non-dimensional hazard values during backwash versus non-dimensional distance based on Guard and Baldock (2007) solution. Symbols show influence of varying k, from k=0 to k=1.5 at intervals of k=0.1.

The hazard during uprush and backwash is approximately equal, although the resulting danger from being carried seaward is probably greater than that of being carried landward. The influence of the boundary conditions and inflow is again very important, and the Shen and Meyer solution is clearly not conservative. Variations in k lead to changes in the values of uh and u^*h^* of a factor 2-6, with greatest variation in the mid-part of the run-up zone.

5.2 Fluid forces

A similar analysis can be performed for the fluid forces, which for surface piercing structures are proportional to the momentum flux, u^2h, as given in (9), and assuming a constant drag coefficient. The maximum flow velocity within the swash zone does not vary as much as the depth varies as the parameter k is changed for different inflow conditions. This is because the maximum velocities are governed by the speed of the advancing and receding shoreline, and the Guard and Baldock solutions converge asymptotically to the Shen and Meyer solution close to the shoreline. Hence, the variation in the forces during uprush and backwash are not as great as those for the water depth (figures 14a&b). Nevertheless, variations in k still lead to variations in the likely fluid forces of a factor 2 in the uprush and a factor 4 in the backwash. Forces in the backwash are a little higher than in the uprush. The key point, however, is that the Shen and Meyer solution is not conservative and predicts the minimum hazard that can be expected during bore run-up, as opposed to the maximum hazard, which is likely to be much better described by solutions with $1<k<1.5$. Yeh (2006) provided a similar analysis for the Shen and Meyer (1963) swash solution (i.e. equivalent to k=0). Here, figures 14a&b, in conjunction with (9), enable the values of $u^{2*}h^*$ to be determined for any location in the run-up zone for any particular bore run-up amplitude and different values for k.

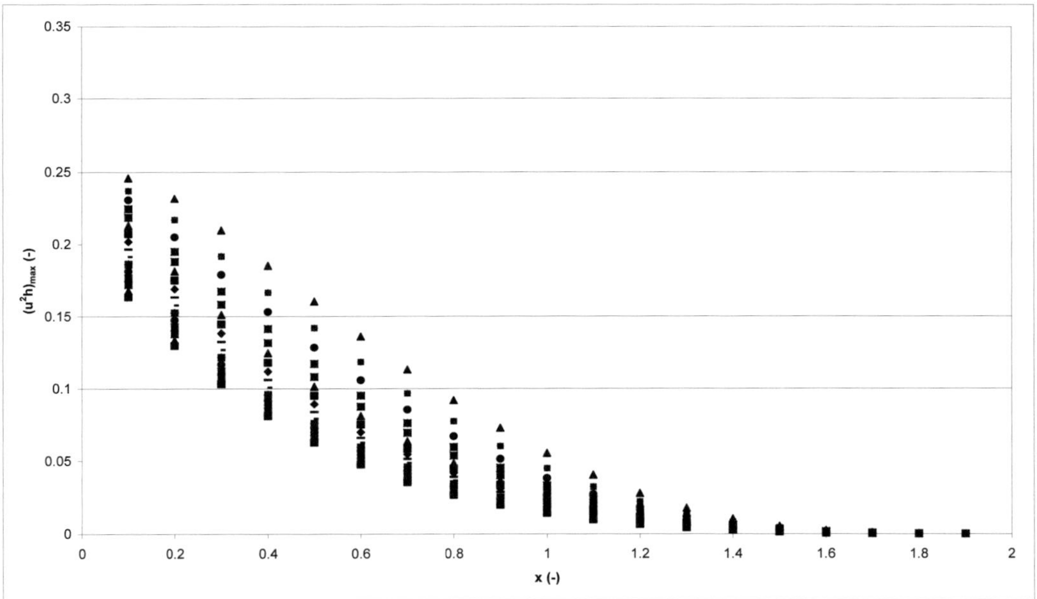

Fig. 14a. Non-dimensional force during uprush versus non-dimensional distance based on Guard and Baldock (2007) solution. Symbols show influence of varying k, from k=0 to k=1.5 at intervals of k=0.1.

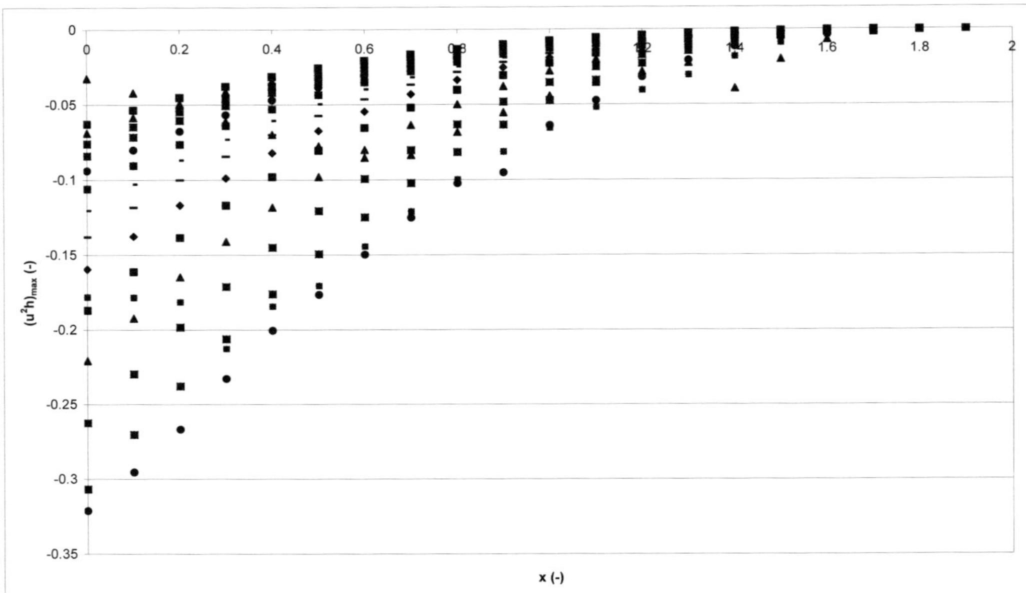

Fig. 14b. Non-dimensional force during backwash versus non-dimensional distance based on Guard and Baldock (2007) solution. Symbols show influence of varying *k*, from *k*=0 to *k*=1.5 at intervals of *k*=0.1.

6. Conclusion

New experimental data on the run-up and overtopping of solitary waves and solitary bores has been presented. The data are discussed with regard to the classical solutions to the non-linear shallow water wave equations and more recent theory. The results show that the Shen and Meyer (1963) swash solution is not conservative in terms of predicting overtopping or overwash flow volumes on beaches and mildly sloping structures. The actual overtopping volumes may be an order of magnitude greater. Further, that solution underestimates hazards in the run-up zone. More recent solutions (Guard and Baldock, 2007) show good agreement with the overtopping data, and demonstrate the difference in flow conditions that arises from different conditions at the seaward swash boundary. This may explain the considerable degree in the variation of overtopping rates in other data sets, previously attributed to scatter. Hence, a range of severity of the hazard can be expected, depending on the wave form. The results are discussed in terms of their applicability to tsunami modelling and evacuation and emergency management strategies.

7. References

Abt, S.R., Wittler, R.J., Taylor, A. and Love, D.J., 1989. Human stability in a high flood zone. *Water Resources Bulletin*, 25(4): 881-890.

Baldock, T.E., 2006. Long wave generation by the shoaling and breaking of transient wave groups on a beach. *Proceedings of the Royal Society A-Mathematical Physical and Engineering Sciences*, 462(2070): 1853-1876.

Baldock, T.E., Cox, D., Maddux, T., Killian, J. and Fayler, L., 2009. Kinematics of breaking tsunami wavefronts: A data set from large scale laboratory experiments. *Coastal Engineering*, 56(5-6): 506-516.

Baldock, T.E., Hughes, M.G., Day, K. and Louys, J., 2005. Swash overtopping and sediment overwash on a truncated beach. *Coastal Engineering*, 52(7): 633-645.

Barnes, M.P. and Baldock, T.E., 2010. A lagrangian model for boundary layer growth and bed shear stress in the swash zone. *Coastal Engineering*, 57(4): 385-396.

Carrier, G.F., Wu, T.T. and Yeh, H., 2003. Tsunami run-up and draw-down on a plane beach. *J. Fluid Mech.*, 475: 79-99.

Dodd, N., 1998. Numerical model of wave run-up, overtopping, and regeneration. *Journal of Waterway Port Coastal and Ocean Engineering-Asce*, 124(2): 73-81.

Donnelly, C., Kraus, N. and Larson, M., 2006. State of knowledge on measurement and modeling of coastal overwash. *Journal of Coastal Research*, 22(4): 965-991.

Goda, Y., 2009. Derivation of unified wave overtopping formulas for seawalls with smooth, impermeable surfaces based on selected clash datasets. *Coastal Engineering*, 56(4): 385-399.

Goring, D.G., 1979. Tsunamis - the propagation of long waves on to a shelf, California Institute of Technology, Pasadena, 337 pp.

Guard, P.A. and Baldock, T.E., 2007. The influence of seaward boundary conditions on swash zone hydrodynamics. *Coastal Engineering*, 54(4): 321-331.

Guard, P.A., Baldock, T.E. and Nielsen, P., 2005. General solutions for the initial run-up of a breaking tsunami front, International Symposium on Disaster Reduction on Coasts, Melbourne.

Hall, J.V. and Watts, J.W., 1953. Laboratory investigation of the vertical rise of solitary waves on impermeable slopes. , Tech. Memo. 33, Beach Erosion Board, USACE.

Hibberd, S. and Peregrine, D.H., 1979. Surf and run-up on a beach: A uniform bore. *J. Fluid Mech.*, 95(2): 323-345.

Hogg, A.J., Baldock., T.E. and Pritchard, D., 2010. Overtopping a truncated planar beach. Journal of Fluid Mechanics, in press.

Hughes, S.A. and Nadal, N.C., 2009. Laboratory study of combined wave overtopping and storm surge overflow of a levee. *Coastal Engineering*, 56: 244-259.

Ingram, D.M., Gao, F., Causon, D.M., Mingham, C.G. and Troch, P., 2009. Numerical investigations of wave overtopping at coastal structures. *Coastal Engineering*, 56(2): 190-202.

Kobayashi, N. and Wurjanto, A., 1989. Wave overtopping on coastal structures. *Journal of Waterway Port Coastal and Ocean Engineering-Asce*, 115(2): 235-251.

Kraus, N.C., Militello, A. and Todoroff, G., 2002. Barrier breaching processes and barrier spit breach, stone lagoon, california. *Shore and Beach*, 70: 21-28.

Madsen, P.A., Fuhrman, D.R. and Schaeffer, H.A., 2008. On the solitary wave paradigm for tsunamis. *Journal of Geophysical Research-Oceans*, 113(C12).

Nielsen, P. and Baldock, T.E., 2010. I-shaped surf beat understood in terms of transient forced long waves. *Coastal Engineering*, 57(1): 71-73.

Peregrine D.H., 1967. Long waves on a beach. *Journal of Fluid Mechanics*, 27: 815-830.

Peregrine and H., D., 1966. Calculations of development of an undular bore. *Journal of Fluid Mechanics*, 25: 321-330.

Peregrine, D.H. and Williams, S.M., 2001. Swash overtopping a truncated plane beach. *J. Fluid Mech.*, 440: 391-399.

Power, H.E., Palmsten, M., Holman, R.A. and Baldock, T.E., 2009. Remote sensing of swash zone boundary conditions using video and argus, Coastal Dynamics 2009. ASCE, Tokyo.

Pritchard, D., Guard, P.A. and Baldock, T.E., 2008. An analytical model for bore-driven run-up. *Journal of Fluid Mechanics*, 610(1): 193.

Pritchard, D. and Hogg, A.J., 2005. On the transport of suspended sediment by a swash event on a plane beach. *Coastal Engineering*, 52(1): 1-23.

Ramsbottom, D., P. Floyd and Penning-Rowsell., E., 2003. Flood risks to people phase 1, r&d technical report, fd2317defra DEFRA / Environment Agency, Flood Management Division, London, UK, London.

Shen, M.C. and Meyer, R.E., 1963. Climb of a bore on a beach. Part 3. Runup. *J. Fluid Mech.*, 16: 113-125.

Stansby, P.K., 2003. Solitary wave run up and overtopping by a semi-implicit finite-volume shallow-water boussinesq model. *Journal of Hydraulic Research*, 41(6): 639-647.

Synolakis, C.E., 1987. The runup of solitary waves. *J. Fluid Mech.*, 185: 523-545.

Synolakis, C.E. and Bernard, E.N., 2006. Tsunami science before and beyond boxing day 2004. *Philosophical Transactions of the Royal Society A-Mathematical Physical and Engineering Sciences*, 364(1845): 2231-2265.

Yeh, H., 2006. Maximum fluid forces in the tsunami runup zone. *Journal of Waterway Port Coastal and Ocean Engineering-Asce*, 132(6): 496-500.

Yeh, H.H., 1991. Tsunami bore runup. *Natural Hazards*, 4: 209–220.

5

Major Tsunami Risks from Splay Faulting

Mohammad Heidarzadeh
Assistant professor, Department of Civil Engineering, Tarbiat Modares University
Tehran,
Iran

1. Introduction

One of the main lessons learned from the great 2004 Sumatra-Andaman earthquake was the fact that tsunami generation process due to large subduction earthquakes is rather complicated. Hence, modelling tsunamis by assuming a simple rupture on a megathrust may not account for actual variations of observed tsunami runups in the near field. From this viewpoint, the 2004 Indian Ocean tsunami was a milestone in tsunami research in that it clearly showed the effect of secondary tsunami sources on intensifying the near-field tsunami heights. The phenomena that are triggered by the main subduction earthquakes and locally contribute to tsunami in addition to the main slip on the subduction zone are known as secondary tsunami sources. The most important secondary sources are submarine landslides, whose effect was mainly evidenced during the 1992 Flores Island tsunami (Synolakis and Okal, 2005; Hidayat et al., 1995), and splay fault branching which was observed during some large subduction earthquakes such as the 1946 Nankai tsunami (Cummins and Kaneda, 2000), 1960 Chilean and 1964 Alaskan tsunamis (Plafker, 1972), and most recently during the 2004 Sumatra-Andaman earthquake and tsunami (Sibuet et al., 2007).

Here, in this chapter we focus on splay faults which are known as one of the important secondary tsunami sources and were responsible for a large part of tsunami deaths during past tsunamis. Splay faults, sometimes known as imbricate faults, are steeply-dipping thrust faults which branch upward from the subduction zone to the seafloor. As splay faults often have steep dip angles, they are capable of producing large seafloor deformation which can significantly increase tsunami runup heights in the near-field. Figure 1 schematically shows a splay fault which branches from the plate boundary. As shown, an abrupt increase in seafloor uplift happens at the location of the splay fault. It is clear that this enhanced seafloor uplift will cause larger tsunami wave heights in the near field.

With this background, it is clear that tsunami hazard assessment without taking into account the effect of possible splay fault branching due to large subduction zone earthquakes may result in underestimating the actual existing tsunami hazards. Hence, in this chapter we study the consequences of splay faulting on tsunami waves in the near-field. First, we make a review of some actual splay faulting cases. Then, the characteristics of splay fault branching from the main plate boundary will be discussed. In the next section, numerical modeling of tsunami will be performed to investigate the effect of splay faulting on tsunami wave heights in the near-field. Finally, we make some practical recommendations on how to take into account the effect of splay faulting for tsunami hazard assessment.

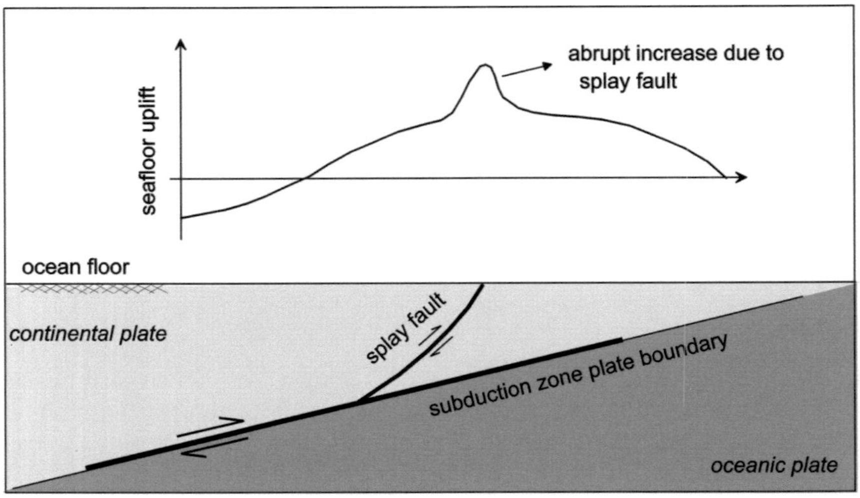

Fig. 1. Sketch showing a splay fault branching upward from a subduction zone plate boundary (bottom) along with the variation of seafloor uplift due to the earthquake (top).

2. Characteristics of splay fault branching

Although we do not intend in this chapter to study the dynamics of splay fault branching from geological and seismological point of views, but basic information on splay faulting can be useful in view of tsunami hazard assessment.

Many studies show that splay faults exist in most subduction zones in the world (Sykes and Menke, 2006; Ryan and Scholl, 1989). According to Sykes and Menke (2006), splay faults are common in most modern accretionary prisms, which grow as sediments are added from the upper plate. These thrust faults in accretionary prisms may rupture due to great subduction zone earthquakes. Fukao (1979) suggested that rupture propagation onto splay faults within accretionary wedges is one of the main mechanisms for generating large tsunamis. According to Park et al. (2002), as splay faults are relatively weak zones within accretionary prisms, it is likely that these weak zone to be repeatedly selected for rupture propagation of subduction earthquakes.

In fact, different studies showed that the total slip during a large megathrust earthquake can be partitioned between the subduction-zone plate boundary and splay faults within the accretionary wedge (e.g., Baba et al., 2006). The amount of slip that transfers from the plate boundary onto splay faults during large subduction earthquakes and the pattern of slip partitioning between them can be an important issue in view of tsunami hazard assessment because the vertical seafloor uplift due to splay faults is relatively larger resulting in large tsunamis. However, the pattern of slip partitioning between subduction zone and splay faults seems casual (Park et al., 2002; Baba et al., 2006). In addition, it is not known if splay faults can rupture occasionally by themselves or not (Sykes and Menke, 2006).

Despite this, some authors made efforts to study the complex phenomenon of splay fault branching during large subduction earthquakes. Cooke (1997) studied the effect of frictional strength variations near fault tips on the pattern of splay fault branching. The result of his work is shown in Fig. 2 indicating that gradual changes in frictional strength may produce broad zones of splay fractures, whereas abrupt changes in frictional strength produce single splay fractures. Kame et al. (2003) studied the effects of pre-stress state, rupture velocity, and branch angle on dynamic fault branching. Their study showed that the pre-stress state has a significant effect on the most favored direction for dynamic branching. Kame et al. (2003) showed that the enhanced dynamic stressing of a rapidly propagating rupture could nucleate failure on a fault which would not necessarily be judged the most favorably oriented one based on the pre-stress state. In summary, Kame et al. 's (2003) study revealed that the tendency of a fault to branch depends on the orientation of the local pre-stress field relative to that of the main fault, the rupture velocity, and the angle between the main and the branching faults.

Fliss et al. (2005) studied the relationship between fault branching and rupture directivity. Wang and Hu (2006) presented a new theory to explain accretionary prisms in subduction earthquake cycles. This theory provides a conceptual framework for investigating the evolution history of accretionary prisms and forearc basins and related phenomena like splay-faulting during great earthquakes and its role in accommodating deformation and generating tsunami.

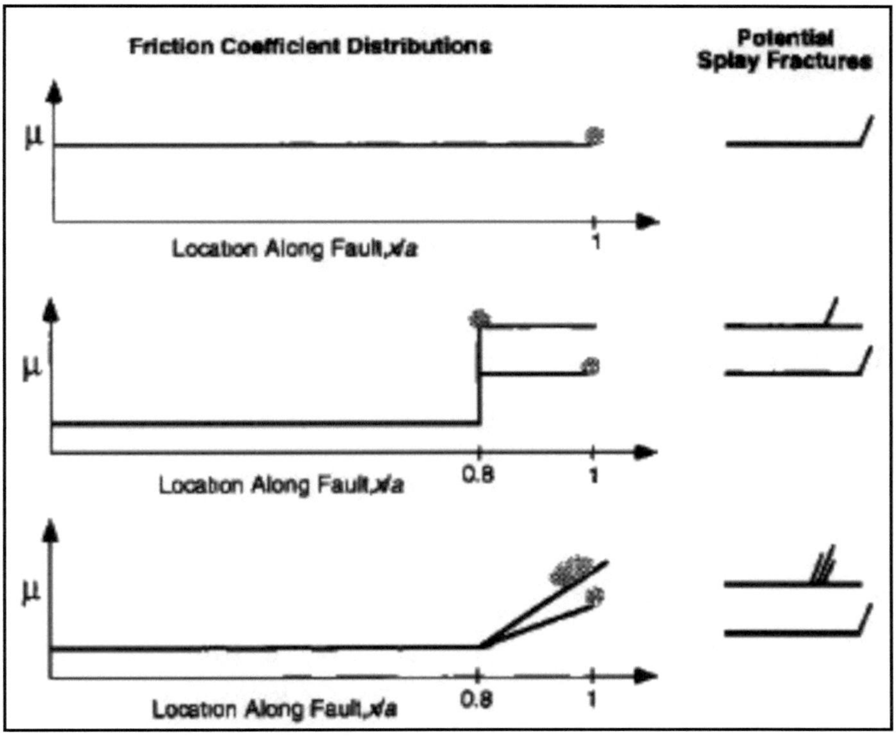

Fig. 2. The effect of frictional strength variations near fault tips on the pattern of splay fault branching. Uniform friction coefficient produces single fractures (top and middle graphs). Linear increases in friction coefficient can produce broad zones of splay fractures near the fault tip (bottom graph). Region of potential splay cracks is shaded (After Cooke, 1997).

Others who studied the mechanism of splay fault branching are: Bhat et al. (2007), Schulson et al. (1999), and Paliakov et al. (2002).

3. Review of some splay faulting cases

The pronounced effect of splay faulting on tsunami wave heights in the near-field was evidenced during some large tsunamis like 1946 Nankai earthquake and tsunami, 1960 Chilean and 1964 Alaskan earthquakes and tsunamis, and most recently during the 2004 great Sumatra-Andaman earthquake and tsunami.

The seafloor deformation caused by the 1964 Alaskan earthquake along with the tectonic setting of the Aleutian subduction zone is shown in Fig. 3. As can be seen, the 1964 Alaskan earthquake with a moment magnitude of *Mw* 9.2 (Kanamori, 1977), was associated with a major splay faulting at the location of the Fatton Bay Fault which locally increased the seafloor uplift by a factor of 2. The seafloor uplift was around 12 m at the location of the splay fault branching (Fig. 3-top) whereas its maximum value was around 6 m at the adjacent regions. It is clear that such a large seafloor deformation can cause a catastrophic tsunami because the maximum near-filed tsunami runup is a direct function of the seafloor deformation at the earthquake source (Synolakis, 2003). In other words, the larger the vertical seafloor deformation in the earthquake source, the larger the tsunami runup heights that is produced.

The other well-known case of splay fault branching during a large subduction earthquake is the 1960 Chilean earthquake and tsunami which is the largest ever recorded earthquake with a moment magnitude of *Mw* 9.5 (Kanamori, 1977). The rupture mechanism of this earthquake was thoroughly studied by Plafker (1972). A summary of the seafloor vertical uplift caused by the earthquake and the tectonic setting of the Peru-Chile subduction zone is shown in Fig. 4. It is clear in Fig. 4 that the seafloor uplift is significantly larger in the vicinity of the splay fault compared to that of the adjacent regions.

Other cases of splay fault branching are the 1946 Nankai (*Mw* 8.3) and 1944 Tonankai (*Mw* 8.1) earthquakes both originated at the Nankai subduction zone, southwest Japan. Studies show that perhaps the Nankai subduction zone is susceptible to splay fault branching and many splay faults exist in this subduction zone (Cummins and Kaneda, 2000 ; Sunagawa and Hayashi, 2007; Cummins et al., 2001). The presence of splay faults at the location of Nankai subduction zone was studied by Park et al. (2002) using seismic reflection data (Fig. 5). The location of splay faults are shown in Fig. 5 by green arrows.

We may point out the great 2004 Sumatra-Andaman earthquake and tsunami as another example. Plafker et al. (2007) and Sibuet et al., (2007) presented evidence that the 2004 Indian Ocean tsunami was associated with a splay fault originating at the interplate fault plane which increased the tsunamigenic effects.

4. Modelling the effect of splay fault branching on tsunami waves

In this section, the effect of splay faulting in a subduction zone on tsunami waves will be studied using numerical modelling of tsunami waves. As a case study, the Makran subduction zone (MSZ) at the north-western Indian Ocean will be considered for our modelling efforts.

As shown in Fig. 6, the MSZ is formed by the northward subduction of the Arabian plate beneath the Eurasian one. This zone extends east from the Strait of Hormoz in Iran to near

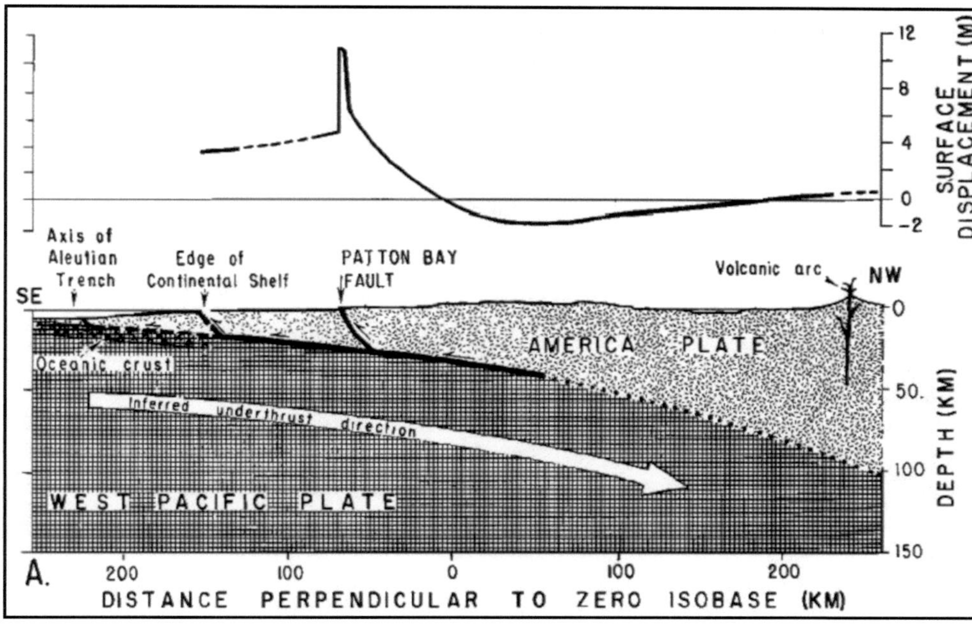

Fig. 3. Seafloor deformation caused by the 1964 Alaskan earthquake (top), and branching of a splay fault at the Patton Bay Fault location (bottom). (After Plafker, 1972).

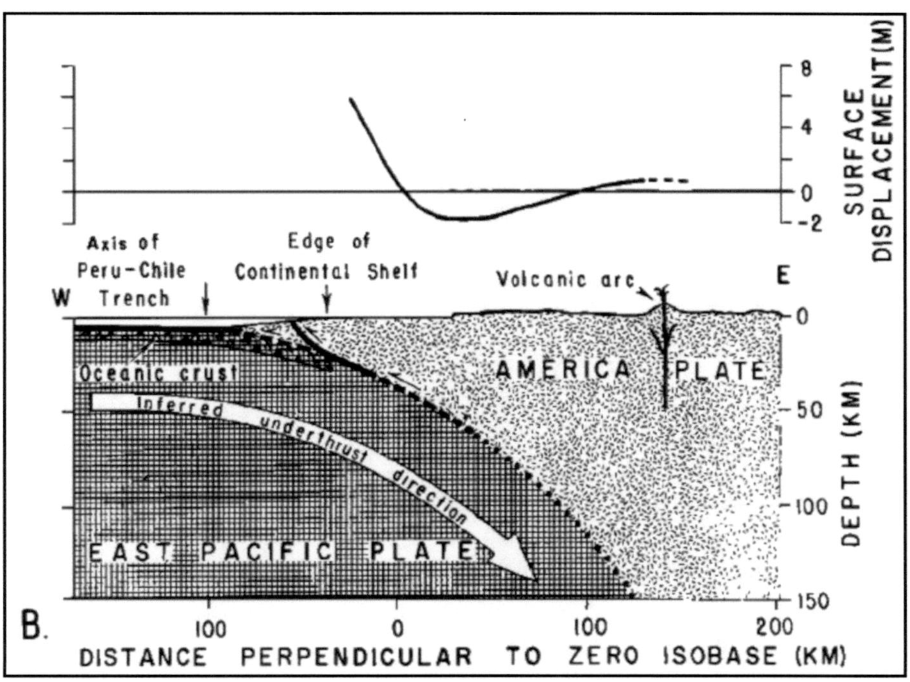

Fig. 4. Seafloor deformation caused by the 1960 Chilean earthquake (top), and branching of a splay fault (bottom). (After Plafker, 1972).

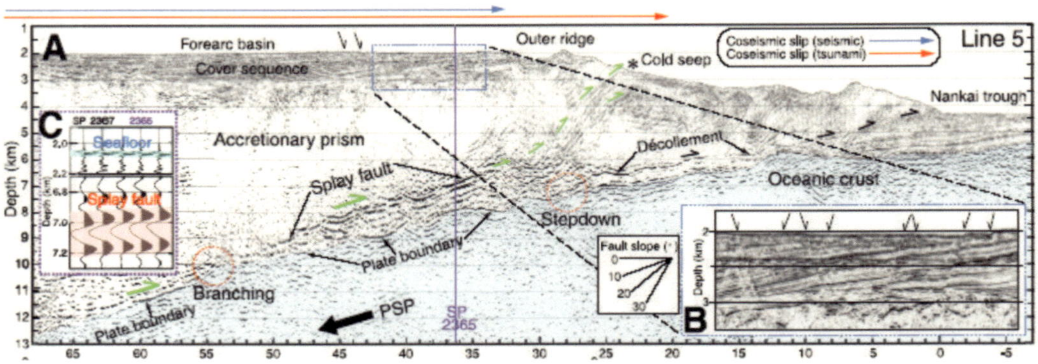

Fig. 5. Multichannel seismic profiles obtained at the location of Nankai subduction zone showing the splay faults (A). Green and black arrows show motions of the splay fault slip and the décollement or normal fault, respectively. Locations of both the splay fault's initial branching and the décollement stepdown to the top of the oceanic basement are marked in red dotted circles. Note active normal faults (inset B) cutting the well-stratified, landward tilting cover sequence and reverse polarity reflection (inset C) of the splay fault at ~7km depth around shot point (SP) 2365. (After Park et al., 2002).

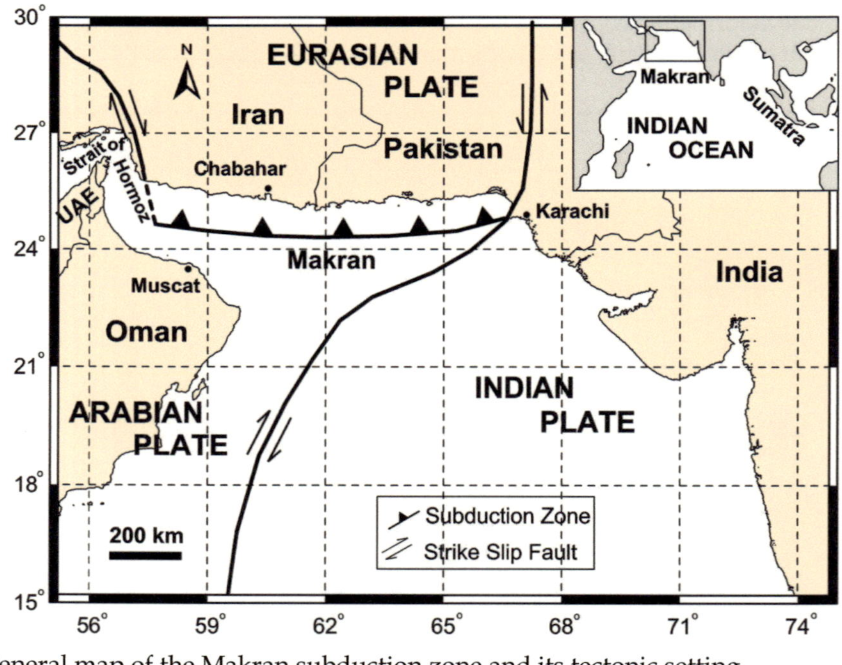

Fig. 6. General map of the Makran subduction zone and its tectonic setting.

Karachi in Pakistan with a length of about 900 km. The reasons that we have chosen the MSZ for our case study are twofold: firstly, Heidarzadeh et al. (2008) presented evidence that a splay fault may have been responsible for the huge runup heights observed in the near field during the Makran earthquake and tsunami of November 27, 1945, and secondly, Mokhtari et al. (2008) showed that many splay faults are present in the MSZ using 2D seismic reflection profiles. Figure 7 is an example of the 2D seismic reflection profiles presented by Mokhtari et al. (2008). In addition, the Makran region has one of the largest accretionary wedges on the earth with an extreme sediment thickness of about 7 km indicating a high probability for splay fault branching due to large subduction earthquakes. Therefore, it can be seen that splay fault branching is a matter of concern in the MSZ and it is likely that this phenomenon occur during large subduction earthquakes in this region. As a result, the possible effect of splay fault branching on tsunami waves should be taken into account for any tsunami hazard assessment in this region.

Here, to quantitatively evaluate the effect of splay faulting on tsunami waves, we numerically model a large earthquake and the consequent tsunami in the MSZ twice: with and without considering a splay fault branching. In the following subsections, the details of our modellings will be presented.

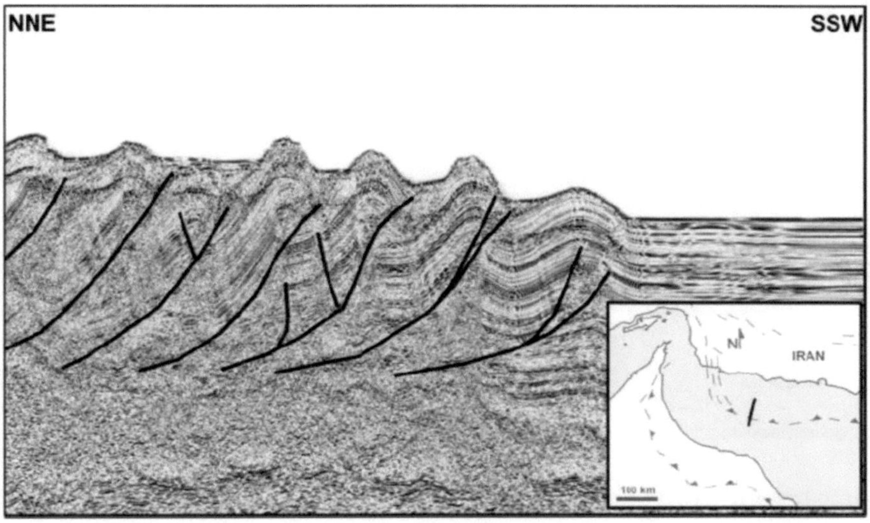

Fig. 7. An example of 2D seismic reflection profiles obtained at the Makran subduction zone (MSZ). The thick lines show the locations of splay faults. The inset at the bottom-right shows the location of the profile in the MSZ (After Mokhtari et al., 2008).

4.1 The earthquake scenario and tsunami source modelling

To realistically choose the scenario earthquake for our tsunami simulations, we first review the history of large earthquake occurrence in the MSZ. This will help understand which parts of the MSZ are susceptible to fail into large-size earthquakes.

Historical earthquakes in the MSZ were studied by some authors including Quittmeyer and Jacob (1979), Page et al. (1979), Ambraseys and Melville (1982), and Byrne et al. (1992). A review of these studies indicates that the Makran region experienced at least seven large earthquakes (M>7) in the past 500 years rupturing the plate boundary in four different segments as shown in Fig. 8. Also, the information of these large earthquakes is summarized in Table 1. As indicated in Fig. 8 by a question mark, the event of 1483 in the western Makran is uncertain, and some authors believe that the western Makran is entirely aseismic.

As the earthquake scenario in this study, we suppose that the blocks A, B, and C shown in Fig. 8 are ruptured simultaneously. This scenario was previously used by Okal and

Synolakis (2008). The reason behind this is the study by Stein and Okal (2007) who proposed that the maximum earthquake size expected from a subduction zone depends on the length of a continuous fault system along a convergent plate boundary. This continuous segment is about 500 km (blocks A, B, and C in Fig. 8) for the MSZ as the segmentation of this subduction zone was confirmed by Byrne et al. (1992).

The tsunami modeling process can be divided into three parts: generation, propagation, and runup (Synolakis, 2003). Generation modeling forms the first stage in the modeling of tsunami, and includes the calculation of the initial disturbance of ocean surface due to the earthquake-triggered deformation of the seafloor using seismic parameters.

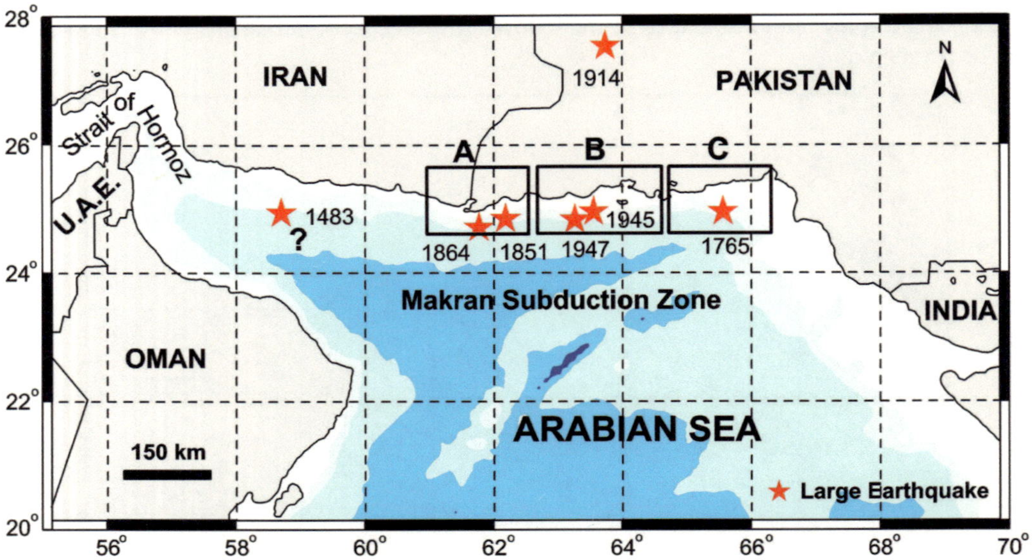

Fig. 8. Large historical earthquakes in the Makran subduction zone (stars) showing the different segments of the plate boundary which ruptured due to large earthquakes (modified from Byrne et al., 1992 and Okal and Synolakis, 2008).

No.	Date (yyyy-mm-dd)	Latitude (°N)	Longitude (°E)	Ms	Mw	Intensity (MM)	Focal depth (km)
1	1483-??-??	24.90	57.90			10	
2	1765-??-??	25.40	65.80			8-9	
3	1851-04-19	25.10	62.30			8-9	
4	1864-??-??	25.12	62.33			6-8	
5	1914-??-??	29.70	63.80	7.0			
6	1945-11-27	24.50	63.00		8.1		25.0
7	1947-08-05	25.10	63.40		7.6		35.0

Table 1. Summary of the information of the past large earthquakes in the MSZ. Abbreviations are: MM, Modified Mercalli; Ms, Surface wave magnitude; Mw, earthquake moment magnitude; °N, degree North; and °E, degree East.

The algorithm of Mansinha and Smylie (1971) was used to calculate the seafloor deformation. This algorithm is based on seismic parameters that include the strike, dip, and slip angles, the amount of slip, the dimensions of the ruptured area (length and width), and the earthquake depth. Empirical equations of Wells and Coppersmith (1994) were used to estimate the corresponding earthquake magnitude of the fault rupture followed by calculation of the rupture width and surface displacement. We validated the predictions of the empirical equations using the available seismic parameters of the 1945 event. The only instrumentally recorded large earthquake at Makran is the event of 1945 (Mw 8.1) rupturing approximately one-fifth of the plate boundary and causing about 7 m of slip (Byrne et al., 1992). Table 2 (first row) presents the source parameters for the scenario earthquake that we evaluate in this study. As shown in Table 2, the scenario earthquake features a seismic moment of about 1.95×10^{22} N m (1.95×10^{29} dyne × cm).

For the case that the earthquake is associated with splay fault branching, a hypothetical splay fault inspired by the 1946 Nankai and 1960 Chilean earthquakes was assumed to branch during the scenario earthquake. Table 2 (rows 2 and 3) presents the seismic parameters of the splay fault and the main plate boundary slips for this new earthquake. We note that the seismic moment of this earthquake was kept unchanged and was the same as that of the first scenario (Table 2- last column). Therefore, due to the slip on the splay fault, the maximum slip on the main plate boundary was reduced.

Using the seismic parameters presented in Table 2 for the two cases of with and without splay fault branching, tsunami source modelling is performed whose results are shown in Fig. 9. The maximum seafloor uplift due to the scenario earthquake was about 4.5 m without splay fault branching (solid line- Fig. 9-top) which was raised to 7.2 m due to the presence of the splay fault. However, it should be noted that Fig. 9 shows that the increase in the amount of the seafloor uplift is limited to the areas close to the splay fault.

Case	Mw	Slip type	L	W	D	H	δ	λ	φ	Earthquake moment (N.m) +
Without SF	8.6	PB slip	500	100	13	25	7	90	265	1.95×10^{22}
With SF	8.6	PB slip	500	100	12	25	7	90	265	1.95×10^{22}
		SF slip	100	50	10	15	30	90	265	

Table 2. Source parameters of the scenario earthquake with and without splay faulting in the MSZ. Abbreviations are: SF, Splay Fault; Mw, earthquake moment magnitude; PB, Plate Boundary; L, rupture Length in km; W, rupture Width in km; D, fault Displacement in m; H, earthquake depth in km; δ, dip angle in degrees; λ, slip angle in degrees; and φ, strike angle in degrees. The rigidity of the earth is considered to be 3×10^{10} N m^{-2}.

4.2 Tsunami modelling

The numerical model TUNAMI-N2 was used for modelling tsunami propagation and coastal amplifications. The model was developed by Nobuo Shuto and Fumihiko Imamura of the Disaster Control Research Center in Tohoku University (Japan) through the Tsunami Inundation Modeling Exchange (TIME) program (Goto et al., 1997). TUNAMI-N2 is one of the key tools for numerical modeling of wave propagation and coastal amplification of

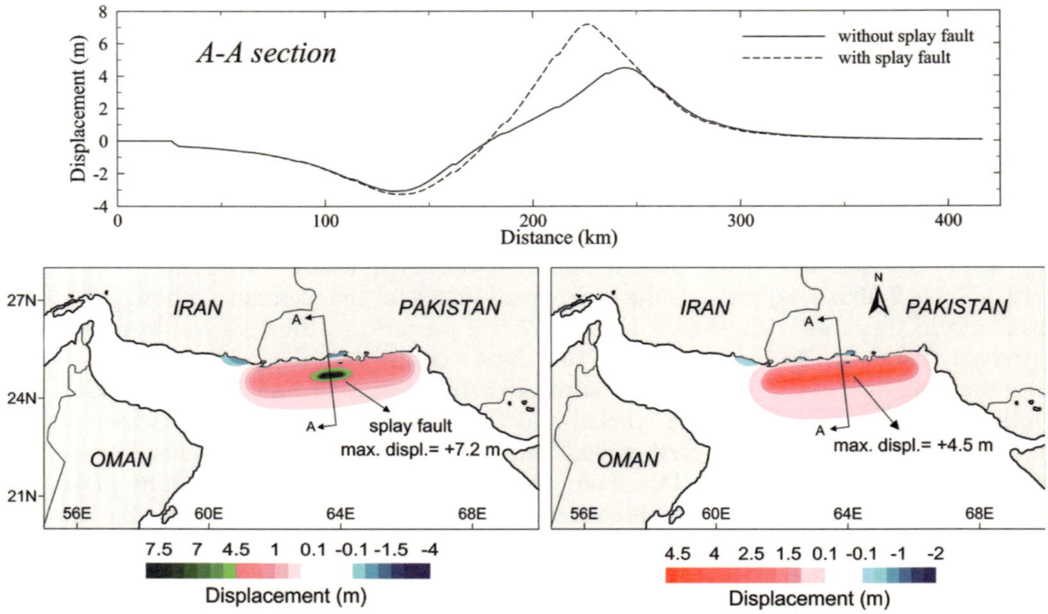

Fig. 9. Results of tsunami source modelling for the cases of without (right panel) and with (left panel) splay fault branching.

tsunami in relation to different initial conditions (Yalciner et al., 2004). A similar methodology is used in the numerical model MOST (Method Of Splitting Tsunami) developed by Titov and Synolakis (1997). TUNAMI-N2 and MOST are nonlinear shallow water codes that have yielded satisfactory agreement with laboratory and field data (Yeh et al., 1996).

We used a 833 × 444 grid and 369852 grid points in our computational domain. The time step was 1.0 s to satisfy the stability condition. The duration of tsunami propagation was 3 h. Bathymetry data provided through the GEBCO digital atlas (General Bathymetric Chart of the Oceans) was applied in this study (IOC et al., 2003). As runup modeling is not applicable to large computational domains, we calculated the maximum positive tsunami heights along the coast which give a reasonable approximation of the runup heights (Tinti et al., 2006).

Snapshots of tsunami propagation for both cases are shown in Fig. 10 at different times of 10, 30, and 60 minutes after the earthquake. Also, to evaluate the effect of splay faulting on tsunami wave heights, the distribution of tsunami wave heights along the north coast of the MSZ (i.e., the southern coast of Iran and Pakistan) are shown in Fig. 11 for the both cases. Results of tsunami modeling (Fig. 11) reveal that possible splay faulting during large mega-thrust earthquakes in the MSZ can locally increase the maximum tsunami wave height by nearly a factor of 2. Based on Fig. 11, the maximum simulated wave height is about 12 m in the vicinity of the splay fault while it was about 6 m at the same location in the previous simulation. Similar to the seafloor uplift pattern, the effect of splay fault branching on the distribution of tsunami wave height is localized to the vicinity of the splay fault. In other words, the effect of possible splay fault branching on tsunami is limited to the near field. As shown in Fig. 10, no significant difference can be observed in tsunami snapshots for the two cases of with and without splay faulting in the far field.

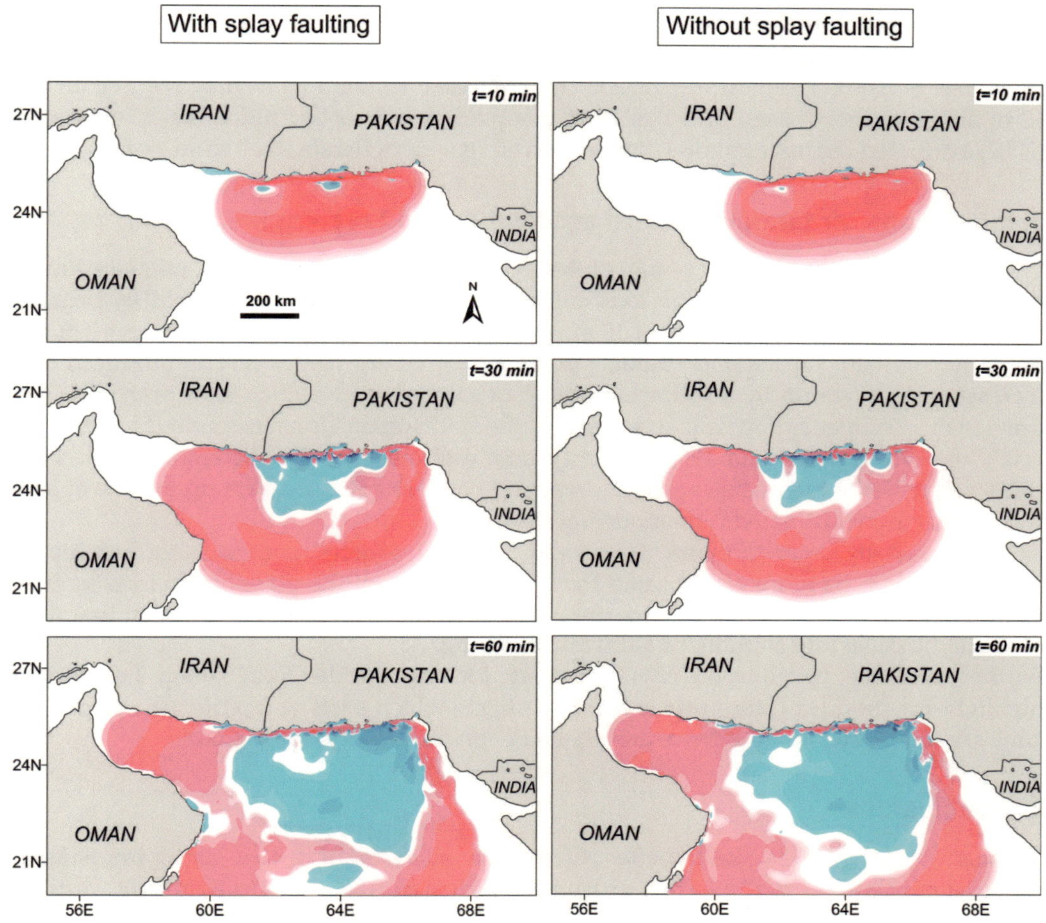

Fig. 10. Snapshots of tsunami simulations due to the two cases of without (right columns) and with (left columns) splay fault branching at different times.

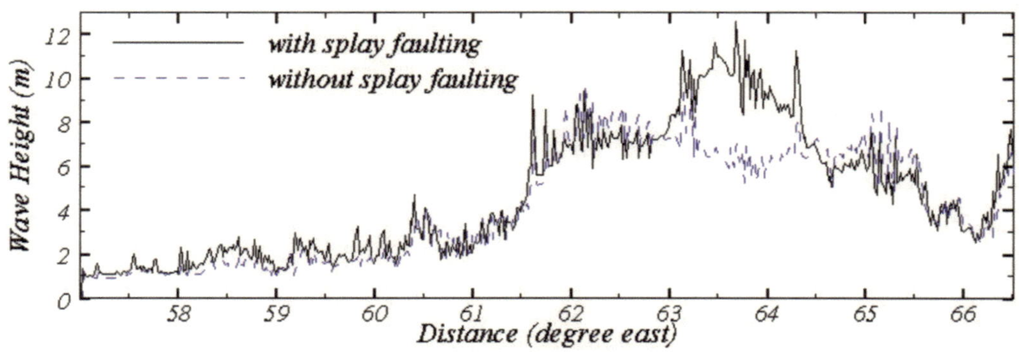

Fig. 11. Distribution of tsunami wave heights along the north coast of the MSZ (i.e., the southern coasts of Iran and Pakistan) due to the two cases of with and without tsunami.

The above results are consistent with the observations of tsunami runup during some actual large tsunami. For example, Plafker et al. (2007) reported that the peak runup was as high as 39 m west of Banda Aceh during the 2004 Indian Ocean tsunami while they were about 5 to 12 m along the north coast and 7 to 20 m along the west coast of Indonesia. Plafker et al. (2007) attributed the huge runup height observed in west of Banda Aceh to splay faulting.

5. Indications for tsunami hazard assessment and recommendations

We showed that the branching of a hypothetical splay fault from the plate boundary during large subduction earthquakes can locally increase the maximum wave height by nearly a factor of 2. In our modeling, a slip of 10 m was supposed on the splay fault. It is evident that the larger the slip on the splay fault, the larger the runup height that is produced. The increase of local runup by a factor of about 2 or more due to a splay fault branching was previously documented by field surveys, e.g., the 2004 Indian Ocean tsunami (Plafker et al., 2007), the 1960 Chilean and 1964 Alaskan tsunami (Plafker, 1972). Therefore, for planning purposes, it seems reasonable and conservative to consider that local runup may be at least twice the amount estimated from modeling studies.

As discussed by some authors previously, tsunami sources due to large subduction earthquakes can be more complicated than a simple slip on the plate boundaries as other phenomena like splay fault branching are likely to occur. Therefore, it is essential that these effects to be taken into account for tsunami modelling.

Furthermore, the pronounced effect of splay faulting on the local runup height may highlight the need for future research to investigate which parts of a particular subduction zone are more likely to experience this phenomenon during large earthquakes.

6. Conclusions

To quantitatively investigate the effect of possible splay faulting on intensifying tsunami wave heights, a large earthquake and tsunami (Mw 8.6) was numerically modelled in the Makran subduction zone for two cases of with and without splay fault branching. In this study, a hypothetical splay fault was used whose seismic parameters were inspired by the 1946 Nankai and 1960 Chilean earthquakes. Modelling of a hypothetical splay fault showed that it can locally increase the maximum wave height by nearly a factor of 2. Hence, for planning purposes, we propose a safety factor of 2 for the tsunami wave heights obtained from regular tsunami simulations.

7. References

Ambraseys, N.N., & Melville, C.P. (1982). A history of Persian earthquakes. Cambridge University Press, Cambridge, Britain, 218 pp.

Baba, T., Cummins, Ph.R., Hori, T., & Kaneda, Y. (2006). High precision slip distribution of the 1944 Tonankai earthquake inferred from tsunami waveforms: Possible slip on a splay fault, *Tectonophysics,* 426, 119–134, doi:10.1016/j.tecto.2006.02.015.

Bhat, H.S., Olives, M., Dmowska, R., & Rice, J.R. (2007). Role of fault branches in earthquake rupture dynamics, *Journal of Geophysical Research,* 112, B11309, pp. 1-16.

Byrne, D.E., Sykes, L.R., & Davis, D.M., (1992). Great thrust earthquakes and aseismic slip along the plate boundary of the Makran subduction zone. *Journal of Geophysical Research* , 97, B1, pp. 449–478.

Cooke, M.L. (1997). Fracture localization along faults with spatially varying friction. *Journal of Geophysical Research,* 102, B10, pp.22425-22434.

Cummins, Ph.R. ; & Kaneda, Y. (2000). Possible splay fault slip during the 1946 Nankai earthquake. *Geophysical Research Letters,* 27, 17, pp. 2725–2728.

Cummins, Ph.R., Hori, T., & Kaneda, Y. (2001). Splay fault and megathrust earthquake slip in the Nankai Trough, *Earth Planets Space,* 53, 243–248.

Fliss, S., Bhat, H.S., Dmowska, R., & Rice, J.R. (2005). Fault branching and rupture directivity, Journal of Geophisical Research, 110, B06312, pp. 1-22.

Fukao, Y. (1979). Tsunami earthquakes and subduction processes near deep-sea trenches. *Journal of Geophysical Research,* 84, pp. 2303–2314.

Goto, C., Ogawa, Y., Shuto, N., & Imamura, F. (1997). Numerical method of tsunami simulation with the leap-frog scheme (IUGG/IOC Time Project), IOC Manual, UNESCO, No. 35.

Kame, N., Rice, J.R., & Dmowska, R. (2003). Effects of prestress state and rupture velocity on dynamic fault branching, Journal of Geophysical Research, 108, B5, pp. 1-21.

Kanamori, H. (1977). The energy release in great earthquakes. *Journal of Geophysical Research,* 82, 20, pp. 2981-2987.

Mansinha, L., & Smylie, D.E., (1971). The displacement field of inclined faults. *Bulletin of the Seismological Society of America,* 61, 5, pp. 1433–1440.

Mokhtari, M., Abdollahie, I., & Khaled Hessami, H. (2008). Structural elements of the Makran region, Oman sea and their potential relevance to tsunamigenisis. *Natural Hazards,* 47, pp. 185–199.

Heidarzadeh, M., Pirooz, M.D., Zaker, N.H., Yalciner, A.C., Mokhtari, M., & Esmaeily, A. (2008). Historical tsunami in the Makran subduction zone off the southern coasts of Iran and Pakistan and results of numerical modeling. *Ocean Engineering* 35, 8 & 9, pp. 774–786.

Hidayat, D. ; Barker, J.S., & Satake, K. (1995). Modeling the seismic source and tsunami generation of the December 12, 1992 Flores Island, Indonesia, earthquake. *Pure and Applied Geophysics,* 144, 3 &4, pp. 537-554.

IOC, IHO, & BODC (2003), Centenary edition of the GEBCO digital atlas, published on CD-ROM on behalf of the Intergovernmental Oceanographic Commission and the International Hydrographic Organization as part of the general bathymetric chart of the oceans, British oceanographic data centre, Liverpool.

Okal, E.A., & Synolakis, C.E. (2008). Far-field tsunami hazard from mega-thrust earthquakes in the Indian Ocean. *Geophysical Journal International* , 172, 3, pp. 995–1015.

Page, W.D., Alt, J.N., Cluff, L.S., Plafker, G. (1979). Evidence for the recurrence of large-magnitude earthquakes along the Makran coast of Iran and Pakistan. *Tectonophysics,* 52, pp. 533–547.

Park, J.-O., Tsuru, T., Kodaira, S., Cummins, Ph.R., & Kaneda,Y. (2002). Splay fault branching along the Nankai subduction zone, Science, 297, pp. 1157-1160.

Poliakov, A.N.B., Dmowska, R., & Rice, J.R. (2002). Dynamic shear rupture interactions with fault bends and off-axis secondary faulting, *Journal of Geophysical Research,* 107, B11, pp. 1-18.

Plafker, G. (1972). Alaskan earthquake of 1964 and Chilean earthquake of 1960: implications for arc tectonics. *Journal of Geophysical Research*, 77, pp. 901–923.

Plafker, G., Ward, S.N., Nishenko, S.P., Cluff, L.S., Coonrad, J., & Syahrial, D. (2007). Evidence for a secondary tectonic source for the cataclysmic tsunami of 12/26/2004 on NW Sumatra. *Seismological Society of America*, Annual meeting, Kona, Hawaii, April 11–13. Abstract.

Quittmeyer, R.C., & Jacob, K.H. (1979). Historical and modem seismicity of Pakistan, Afghanistan, northwestern India, and southeastern Iran. *Bulletin of the Seismological Society of America*, 69, 3, pp. 773–823.

Ryan, H.F., & Scholl, D.W. (1989). The evolution of forearc structures along an oblique convergent margin, central Aleutian Arc. *Tectonics*, 8, pp. 497–516.

Schulson, E.M., Iliescu, D., & Renshaw, C.E. (1999). On the initiation of shear faults during brittle compressive Failure: A new mechanism, *Journal of Geophysical Research*, 104, B1, pp. 695-705.

Sibuet, J.C., et al. (2007). 26th December 2004 great Sumatra–Andaman earthquake: Co-seismic and post-seismic motions in northern Sumatra. *Earth and Planetary Science Letters*, 263, pp. 88–103.

Stein, S., Okal, E.A. (2007). Ultralong period seismic study of the December 2004 Indian Ocean earthquake and implications for regional tectonics and the subduction process. *Bulletin of the Seismological Society of America*, 97, 1A, pp. S279–S295.

Sykes, L.R., & Menke, W. (2006). Repeat times of large earthquakes: implications for earthquake mechanics, *Bulletin of the Seismological Society of America*, 96, 5, pp. 1569–1596.

Sunagawa, Y., & Hayashi, D. (2007). Development of splay faults in the Nankai Accretionary prism, *Bull. Fac. Sci.,Univ. Ryukyus*, 84 , pp. 15 -31.

Synolakis, C.E. (2003). Tsunami and seiche. In: *Earthquake Engineering Handbook*, CRC Press, W. F Chen., & C. Scawthorn (Eds.), Chapter 9, pp. 1–90.

Synolakis, C.E. ; & Okal, E.A. (2005). 1992-2002: Perspective on a decade of post-tsunami surveys, in: Tsunami ed. by K. Satake, *Advances in Natural and Technological Hazards*, 23, pp. 1–30.

Tinti, S., Armigliato, A., Manucci, A., Pagnoni, G., Zaniboni, F., Yalciner, A.C., and Altinok, Y. (2006). The generating mechanisms of the August 17, 1999 Izmit Bay (Turkey) tsunami: Regional (tectonic) and local (mass instabilities) causes. *Marine Geology*, 225, pp. 311–330.

Titov, V.V., and Synolakis, C.E. (1997). Extreme inundation flow during the Hokkaido-Nansei-Oki tsunami. *Geophysical Research Letters*, 24, pp. 1315–1318.

Wang, K., & Hu., Y. (2006). Accretionary prisms in subduction earthquake cycles: The theory of dynamic Coulomb wedge, *Journal of Geophysical Research*, 111, B06410, pp. 1-16.

Wells, D.L., & Coppersmith, K.J. (1994). New empirical relationships among magnitude, rupture length, rupture width, rupture area, and surface displacement. *Bulletin of the Seismological Society of America*, 84, 4, pp. 974–1002.

Yalciner, A.C., Pelinovsky, E., Talipova, T., Kurkin, A., Kozelkov, A., and Zaitsev, A. (2004). Tsunamis in the Black Sea: comparison of the historical, instrumental, and numerical data. *Journal of Geophysical Research*, 109, 12, pp. 2003–2113.

Yeh, H., Liu, P., & Synolakis, C.E. (1996). Long wave runup models. World Scientific Publication Company, London, 403 pp.

6

Forecasting Landslide Generated Tsunamis: a Review

Marcello Di Risio, Paolo De Girolamo and Gian Mario Beltrami
University of L'Aquila (DISAT-LIAM)
Italy

1. Introduction

When earthquakes occur, suddenly bottom deformations will induce water response that is almost instantaneous. Then, bottom displacements affect wide areas and generate transient waves (tsunamis) usually characterized by a relatively small height with respect to their length, i.e. by a small steepness. Indeed, just after the earthquake, the water free surface mimics the final bottom deformations, that can be of the order of some meters, and small amplitude long waves start to propagate resulting in a potential transoceanic devastation (Synolakis et al., 2002).

When either subaerial or submerged landslides occur, the displacements will take place on both larger temporal-, and smaller spatial-scale. The deformations are of the order of hundreds of meters and the generated waves are quite different from those induced by submarine earthquakes. Hence, landslide tsunamis tend to be a local phenomenon, although extreme. Since two centuries ago, submarine failures were related to "sea waves without earthquake" when great waves were observed without any account of earthquakes (e.g. Mallet & Mallet, 1858; Milne, 1898; Montessus de Ballore, 1907). In an illuminating reasoning Mallet & Mallet (1858) argued that "great underwater slippage takes place [...] the effect upon the surface of the sea is at the same moment to originate a positive and a negative wave. [...] The magnitude of the wave raised is dependent upon that of the mass of solid material that has suddenly changed its place, upon the depth of water in which the slippage has occurred, upon the rapidity of the transposition, and in minor degree upon the form and material of the portion of the bank that has slipped."

One of the more meaningful event occurred at Lituya Bay (Alaska, July 9, 1958), when a subaerial landslide triggered by an earthquake entered the water at the Gilbert Inlet and generated an impulse wave (Miller, 1960). The induced run-up was the highest known: the water rose up to 524 m at the slope in front of the landslide, then the waves propagated into the Bay and eventually radiated offshore through the Bay inlet with minimum effects outside the Bay itself. Another example, related to submarine failures, is the well documented tsunami generated by an underwater slump triggered by a magnitude 7 earthquake on July 17, 1998 close to the Sissano Lagoon, Sandaun Province, Papua New Guinea (see Synolakis et al., 2002). The tsunami struck about 30 km wide area, i.e. rather small, by inducing a maximum run-up of about 15 m and killing over 2100 people.

Landslide generated impulse waves can occur not only at the sea boundaries, but also in enclosed basins (i.e. reservoirs and lakes). The event that took place at the Vajont Valley (Italy) on October 9, 1963 is perhaps the more sorrowful one. A subaerial landslide of about

270 millions of cubic meters fell into the Vajont reservoir and generated impulse waves that ran up to about 235 m. The waves overtopped the concrete arc dam, flooding the village of Longarone, and causing 1901 casualties (e.g. Panizzo et al., 2005b).
When geological evidence can predict the probable occurrence of landslides, engineering tools are needed in order to forecast the features of the induced free surface perturbation.
At this end, the phenomenon of a landslide tsunami can be qualitatively separated in four different phases, each characterized by its own physical features:

(i) a solid or granular mass starts to move from one of the boundary of a water body;

(ii) an impulsive perturbation is generated as a consequence of the interaction with the water surface;

(iii) the generated perturbation propagates into the water body;

(iv) the propagated perturbation interacts with water body boundaries (i.e. coastlines).

Then, modeling of landslide hazards requires information and data from seismology, geology, geotechnical engineering (i-ii) and hydrodynamics (ii-iii-iv), resulting in a challenging task (Synolakis & Bernard, 2006).
Figure 1 depicts, although qualitatively, the phenomena. While the first phase (i) is intimately related to geological events, the other phases (ii-iii-iv) encompass challenging hydraulic topics such as the water entry problem, the propagation of the generated water-surface perturbation, as well as the run-up and overtopping at the boundaries.
More in details, when the landslide reaches the free surface (Figure 1, generation area, also referred to as the splash zone in the case of subaerial landslide, Walder et al., 2003), a perturbation is generated and energy exchange between landslide and water takes place. The induced waves quickly leave the generation area and propagate in the near field. The near field features of such a perturbation depend upon the characteristics of the landslide (volume, velocity, underwater travel time, density, porosity, shape of the front, slope angle, etc...). Then the waves propagate in the far field where frequency and directional energy dispersion, refraction and diffraction occur, depending upon water body configuration and waves features. Finally, the perturbation reaches the water body boundaries and it runs up and floods coastal areas, often resulting in great damage and danger for human activities. In the case of reservoirs normal modes (seiches) can be excited and the impulsive perturbation impinges the dam and, for extreme events, overtops or destroys it by flooding downstream.
The last two phases (iii-iv) tend to coexist, influencing each other. When the landslide occurs directly at the coast, impulse waves both radiate seaward and propagate alongshore. The complex interaction that exists between the generation and the propagation mechanisms has therefore to be taken into account. In such a case trapped waves can be triggered by the source of the tsunami and propagate along the coast by inducing large wave run-up observed in some real cases (Ursell, 1952; Liu & Yeh, 1996; Liu et al., 1998; Johnson, 2007).
The present chapter aims at reviewing the available engineering tools that can be used to estimate the features of generated waves and their effects on boundaries, i.e. wave run-up and, in the case of reservoirs, dam overtopping.
The topic is intimately related to experimental investigations carried out in the past aimed at defining empirical formulations. Indeed, when forecasting of impulse waves features and effects is needed in a particular real case, it is possible to (i) build the specific hydraulic model, in which the topography of the area of the probable landslide and of the water body are reproduced (e.g. Davidson & McCartney, 1975; Chaudhry et al., 1983; Fritz et al., 2001) or (ii) refer to the results of well known basic research experiments performed under idealized

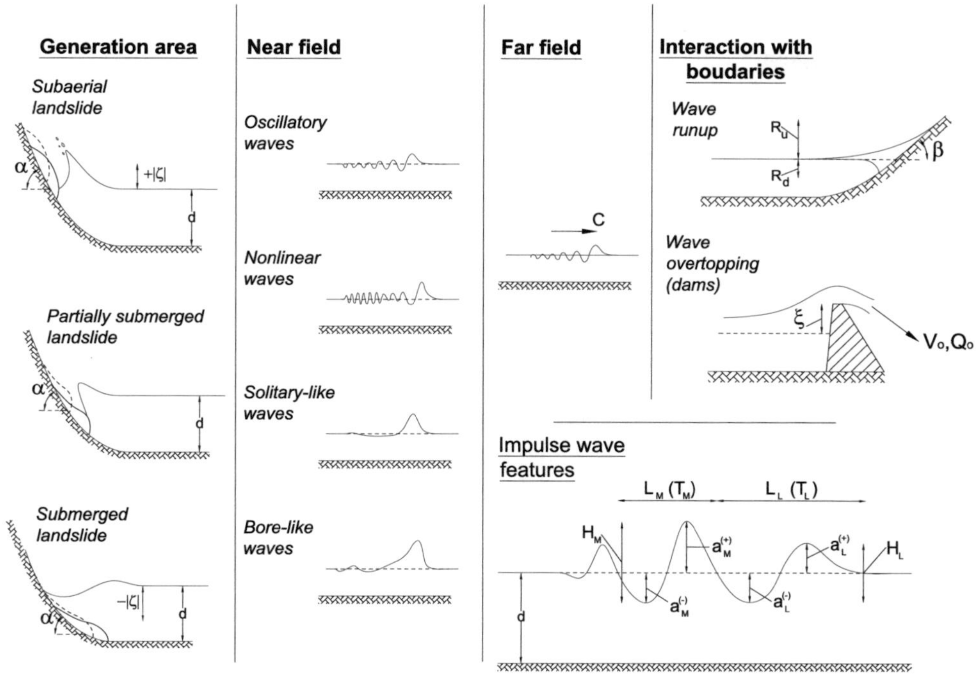

Fig. 1. Qualitative sketch of landslide induced impulse waves.

assumptions, both two-dimensional and three-dimensional, intended to be representative of real cases (e.g. Huber, 1982; Panizzo et al., 2005a; Di Risio et al., 2009a;b).

The chapter is structured as follows. The next section deals with the generation and propagation of impulse waves induced by both underwater and subaerial landslide. The phenomena are qualitatively described and empirical formulations suitable for wave features forecasting are reported on the basis of past researches. The succeeding sections briefly describe the interaction of impulse waves with water body boundaries, i.e. wave run-up and overtopping. Concluding remarks close the chapter. Due to the limited number of pages, only generation and propagation of impulse waves are described in depth.

2. Impulse waves generation and propagation

This section deals with experimental investigations aimed at analyzing physical phenomena occurring in the generation area and in the near- and far-field. In the former, landslide energy is transferred to water, whereas in the near- and far-field the waves start to propagate and radiate in the water body.

The first experimental task that has to be addressed is impulse waves reproduction in physical models. The first studies were carried out by Russell (1838; 1845). This autor used a vertical falling box in order to generate free surface transient perturbations similar to the "large solitary elevation" (christened "wave of translation") induced by a sudden stop of a boat as observed by Sir Scott Russell along a channel.The box height was higher than the water depth and this impulse waves generation method was used by many authors (e.g. Wiegel et al., 1970; Noda, 1970; Monaghan & Kos, 2000) and is often referred to as "Scott Russell's wave generator".

Some authors (e.g. Panizzo et al., 2002; Di Risio, 2005; Di Risio & Sammarco, 2008; Yim et al., 2008) used a falling box characterized by an height lower than the water depth. In this way, clearly, a return flow occurs over the top of the box and the generator may be addressed as "pseudo-Scott Russells wave generator". Another option to simulate the impulse wave generation by a landslide was to use either a vertical (e.g. Noda, 1970) or an inclined (e.g.Miller, 1970; Iwasaki, 1983; Cea et al., 2004) moving paddle. Several studies have been aimed at providing analytical methods able to define paddle displacements for the generation of cnoidal and solitary waves, similar to impulse waves (e.g. Goring, 1978; Synolakis, 1990; Kattel & Eric, 2002; Malek-Mohammadi & Testik, 2010).

However, the preceding methods allow to study the impulse wave propagation, but lack in studying the landslide-water energy exchange. Then, a series of experimental investigations involved the use of solid boxes sliding down inclines (e.g. Wiegel et al., 1970; Heinrich, 1992; Watts, 1998; 2000; Watts et al., 2000; Walder et al., 2003; Panizzo et al., 2005a; Sue et al., 2006; Sue, 2007; Enet & Grilli, 2007; Ataie-Ashtiani & Nik-Khah, 2008a;b; Sæ levik et al., 2009) or small calcareous stones (de Carvalho & do Carmo, 2007). Nevertheless, the shape of the boxes has to be defined preliminarily, and the deformations and porosity of real landslides cannot be reproduced in the model.

Watts et al. (2005) used a validated 2D fully non-linear potential-flow numerical model in order to evaluate the influence of the shape of an underwater solid landslide, and therefore to address the concern about the shape of the solid boxes. These authors found that the higher the spreading of Gaussian shape, the lower the amplitude of generated waves. Then, Watts et al. (2005) indicate that semi-elliptical shaped rigid body represents the worst case scenarios, at least for underwater landslides (Watts, 2000; Enet & Grilli, 2007; Di Risio et al., 2009a;b). For the case of subaerial landslides, Ataie-Ashtiani & Nik-Khah (2008a) suggested that the slide shape does not affect substantially the generated waves. However, comprehensive and parametric experimental investigations aimed at evaluating the influence upon the generated waves of the slope of the landslide front (γ) have not been carried out so far (at least to our knowledge). The topic was partially tackled by Kamphuis & Bowering (1970).

In order to reproduce the deformations of the landslide at model scale, some earlier investigations employed sand bags (e.g. Davidson & McCartney, 1975). Huber (1980; 1982) succesively used granular material with the aim of fully reproducing both deformations and porosity of landslides. More recently, a series of two-dimensional experiments were performed at the Laboratory of Hydraulics, Hydrology and Glaciology of Zurich (VAW-ETH) by using granular landslides with varying geometry and energy (e.g. Fritz et al., 2003a;b; 2004; Zweifel et al., 2006; Heller & Hager, 2010b). The only three-dimensional experimental investigation with granular landslide is related to an attempt to reproduce the Lituya Bay event (Fritz et al., 2009). However, when granular landslides are used, scale effects will be encountered, mainly related to surface tension and fluid viscosity. This was in particular observed by Heller et al. (2008) when the water depth is lower than 0.20 m. As a general rule, scale effects tend to reduce the generated wave amplitudes, whilst their influence on wave celerity is found relatively small.

For a seek of synthesis, experimental investigations were carried out with solid boxes falling vertically (Scott Russell's wave generator) or sliding along inclines with different shapes (semi-elliptical, triangular, parallelepiped), with deformable sand bags and with granular materials. When comparing solid and granular materials induced waves, for subaerial landslide it seems that there is agreement that solid blocks induce higher waves. In particular, Zweifel (2004) showed that, at least for two-dimensional water bodies, granular landslides induce lower waves with respect to that induced by solid blocks. The differences in terms of induced wave amplitude increase as the landslide Froude number (F) increases ($= v_s/\sqrt{gd}$, being v_s the landslide impact velocity and $\sqrt{gd}$ the shallow water wave celerity, d the water depth and g the gravitational acceleration):

$$\frac{a_{Mb}/d - a_M/d}{a_{Mb}/d} = 1 - 0.26\text{F} \tag{1}$$

where a_M is the maximum wave amplitude for granular landslide and a_{MB} for solid blocks (for wave parameters symbols see Figure 1). As discussed by Heller & Kinnear (2010a), the differences arise from the influence of porosity (being zero for solid blocks), slide front angle (being fixed for solid blocks and time varying for granular landslides), transition between slope toe and channel bottom. Furthermore the blockage ratio ($= w/w_f$) can influence induced waves. Indeed, for two-dimensional experiments with granular material, the landslide model fill the whole channel width (w_f), whereas solid blocks width (w) has to be lower than channel one in order to allow it to slide the incline.

On the other hand, in the case of underwater landslides, the differences between granular landslides and solid blocks are controversial. Some numerical simulations suggested that intense slide deformation in shallow water, even if crudely represented, induce higher waves (Grilli & Watts, 2005). Some experiments were conducted with deformable underwater landslide (Fleming et al., 2005; Ataie-Ashtiani & Nik-Khah, 2008b). It was observed that both sand bags and granular underwater landslide induce lower waves if compared with solid blocks ones (Ataie-Ashtiani & Nik-Khah, 2008b). In the authors' opinion this aspect should be investigated. Furthermore, differences can arise from the transition between slope toe and channel bottom. It was observed that a sudden stop of solid underwater blocks induce higher waves if compared with a smooth transition allowing the landslide model to decelerate naturally at the horizontal bottom (Sue et al., 2006; Koo & Kim, 2008).

In the following subsections, forecasting formulae for subaerial and underwater landslide induced waves are reviewed.

2.1 Subaerial landslides generated waves

A qualitative description of the phenomena can be found in Liu et al. (2005) and Di Risio et al. (2009a). When landslide enters the water body, it pushes ahead the fluid and a leading positive seaward radiating wave is generated. Once the landslide becomes totally submerged, the water is initially depressed by generating a trailing wave through. Strong alongshore free surface gradients occur in the generation area resulting in converging flows that collide and rebound along the centreline of the landslide. The rebound is the responsible of a large positive wave radiating offshore. This section deals with the problem of forecasting the features of the generated and offshore-radiated waves.

Earlier experiments were conducted in two-dimensional wave flumes facing the problem of impulse wave propagation arising from water surface disturbances ((Prins, 1958), i.e. related to the classical problem of Cauchy-Poisson, Lamb, 1932, †238 - 240). In that study the leading wave types were inferred from experimental observations, i.e. oscillatory linear waves, solitary-like and cnoidal waves as well as bore-like waves (see Figure 1), depending on the geometrical features of wave disturbance, i.e. its height and length. However, the surface disturbances were static, then Miller (1970) extended the work by Prins (1958) by using a moving vertical paddle with the aim of gaining insight about the relationship between leading wave type and paddle Froude number. Then Wiegel et al. (1970) used the

Scott Russell's wave generator and Noda (1970) related the wave type to landslide Froude number, i.e. the ratio between the falling box velocity (v_s) and shallow water waves celerity ($\sqrt{gd}$). Furthermore, for the case of horizontal moving paddle, he provided an empirical formulation for maximum wave crest amplitude based on his linear analytical modeling (Table 1, abbreviation NODA1970). In the meantime, Kamphuis & Bowering (1970) provided the first empirical formulations based on their own experiments (Bowering, 1970). The experiments consisted in a tray sliding down an incline. They proposed a dimensional analysis, by defining the relationships f between any dependent quantity (in dimensionless form, Π):

$$\Pi = f\left(\mathrm{F},\mathrm{M},\mathrm{G},\mathrm{X},\mathrm{S},t\sqrt{\frac{g}{d}},\gamma,\alpha,p\right) \tag{2}$$

where F is the already defined landslide Froude number ($= v_s/\sqrt{gd}$), M is the dimensionless two-dimensional landslide volume ($= V_s/wd^2$, V_s the landslide volume, w landslide width, d water depth), G the specific gravity ($= \rho_s/\rho$, ρ_s the landslide bulk density, ρ water density), X the dimensionless propagation distance ($= x/d$, x the distance from impact point), t the time, g the gravitational acceleration, γ the landslide front slope, α the slope angle of the incline to horizontal, p the landslide porosity. Experimental investigation showed that the generated maximum wave height H_M is strongly affected by the dimensionless volume (M) and landslide Froude number (F). The leading wave period T_l is affected only by the dimensionless distance. Empirical formulations are reported in Tables 1 and 2 (abbreviation KB1970).

The first three-dimensional experimental investigation was carried out by Huber (1980; 1982). He performed several two-dimensional experiments by employing a granular landslide model. Then the experimental results were compared with a limited number of three-dimensional experimental findings in order to estimate a reduction parameter capable of taking into account the three-dimensional effects due to lateral dispersion. Empirical correlations were provided only on a tabular form, but Huber & Hager (1997) re-analyzed the same experimental data and a series of empirical formulation were defined, for both two- and three-dimensional water bodies (HH1997 in Table 1). They found that, for two-dimensional water body, the main effect on wave height is related to the slope angle (α): the lower the slope angle, the lower the wave height. Furthermore, increasing dimensionless landslide volume (M) and specific gravity (G) and decreasing of dimensionless distance (X) result in increasing wave height. Nevertheless, they simulated fast landslides, then the influence of landslide Froude number was not addressed. Three-dimensional water bodies lateral dispersion are modeled by a squared cosine function involving the wave propagation direction θ (measured with respect to the direction of landslide motion). After few years, granular material was used to simulate landslides within a long series of two-dimensional experiments at the wave flume of VAW-ETH (Zurich). The experimental observations were used to characterize the water flow in the near- and far-field (Fritz et al., 2003a), as well as the impact crater features (Fritz et al., 2003b), scale effects influence on experimental results (Heller et al., 2008), wave types (Fritz et al., 2004). Also, a series of empirical formulations were defined (Fritz et al., 2004; Zweifel et al., 2006; Heller & Kinnear, 2010a). By means of large scale PIV measurements they found that fluid velocities exist always after the landslide impact, then when dealing with numerical simulations, free surface does not suffice and also velocity fields have to be considered. They observed that the flow is substantially irrotational except close to the interface between granular material and water and in the rear of the landslide where turbulent flow occurs. These observations confirm the assumptions made in using of mathematical method, either numerical or analytical, based on Laplace equation. Of particular interest for our purposes is the empirical formulations proposed by this research group. Indeed, different empirical formulations were provided with different experimental ranges and by considering several dimensionless groups (ETH2004, ETH2006 and ETH2010 in Tables 1 and 2). The influences of (i) landslide Froude number (F), (ii) dimensionless landslide thickness (S $= s/w$, being the landslide thickness s measured perpendicular to the incline), (iii) specific gravity (G) and (iv) incline slope (α) are addressed: the higher the values of M, S, G, the higher the induced wave height and period; the lower the value of α (i.e. less steep inclines), the higher the induced wave height and period. It has to be stressed that the influence of the incline slope seems to be in contrast with the three-dimensional results showed by Huber & Hager (1997) that observed increasing wave height for increasing α.

Other experimental investigations were carried out by using solid landslides. Walder et al. (2003), on the basis of an interesting scaling analysis of the Euler equations, demonstrated the importance of the dimensionless underwater travel time τ ($= t_s / \sqrt{gd}$, being t_s the underwater travel time), as confirmed analytically by Di Risio & Sammarco (2008), the dimensionless landslide volume M, the landslide Froude number F and of the incline slope α. They proposed empirical formulations for maximum wave height in the near field involving the dimensionless underwater travel time and landslide volume (W2003 in Table 1). It was shown that such a relationships can model also the experimental data collected during past researches (Bowering, 1970; Huber, 1980). In this way, knowledge of underwater travel time is needed, and they related it to dimensionless landslide length $\tilde{L}$ ($= L/d$, being L the landslide length measured along the incline). However the authors report in the paper the values of empirical coefficients based on their own data only. Furthermore they related the near field wavelength to underwater landslide, needed to estimate the extension of the near field, i.e. about three times the wavelength. It has to be stressed that the estimation of underwater travel time, and in turn wavelength and near field extension, did not involve the slope angle α. Synthetically, the experimental work of Walder et al. (2003) showed that wave height increases for decreasing underwater travel time (then decreasing dimensionless landslide length) and increasing dimensionless landslide volume. It has to be stressed that the blockage ratio for those experiments was quite less than unity (i.e. equal to 0.53), therefore experimental data could be somewhat affected by some reduction (Heller & Kinnear, 2010a).

Further two-dimensional tests aimed at providing empirical formulations were performed by Di Risio (2005). He performed a series of experimental tests in the Environmental and Maritime Hydraulics Laboratory (LIAM) of the University of L'Aquila, by using a pseudo-Scott Russell's wave generator (i.e. being the height of the box lower than water depth) in a two-dimensional wave flume. He provided empirical formulations (LIAM2005-2D in Table 1), based on the dimensional analysis by Kamphuis & Bowering (1970), equation (2), by keeping constant the porosity ($p = 0$), the slope angle ($\alpha = 90°$), the landslide front slope ($\gamma = 90°$) and the specific gravity ($G = 1.36$). This study represent an extension of the work by Kamphuis & Bowering (1970), even if only for vertical landslides. Moreover he performed a series of axial symmetric experiments by using a vertical falling solid body (a kind of cylindrical Scott Russell's wave generator) and by providing further empirical formulations (LIAM2005-AX in Tables 1 and 2). Similar behavior of wave features were found for both two-dimensional and axial symmetric water bodies: the higher the landslide Froude number (F) and the dimensionless landslide thickness (S) and the shorter the dimensionless distance (X or R $= r/d$, being r the distance from impact point in three-dimensional water body), the higher the maximum and leading wave height and crest amplitude. For the wave period it was found that it is slightly affected by the landslide Froude number.

Ref. **(Abbr.)**	$\mathbf{H_M/d}$	$\mathbf{a_M/d}$	$\mathbf{T_M\sqrt{g/d}}$
Noda (1970) (NODA1970)	=	$1.32F$	=
Kamphuis & Bowering (1970) (KB1970)	$F^{0.7}(0.31+0.20\log M)+$ $+0.35\exp(-0.08X)$	=	=
Huber & Hager (1997) (HH1997-2D)	$0.88G^{0.25}M^{0.5}X^{-0.25}\sin\alpha$	=	=
Huber & Hager (1997) (HH1997-3D)	$1.76G^{0.25}M^{0.5}R^{-0.67}\sin\alpha\cos^2(0.67\theta)$	=	=
Walder et al. (2003) (W2003)	=	$1.32(\tau/M)^{-0.68}$ with $\tau=4.5\tilde{L}^{0.5}$	=
Fritz et al. (2004) (ETH2004)	=	$0.25F^{1.4}S^{0.8}$	=
Panizzo et al. (2005) (LIAM2005-3D)	$0.102\Sigma^{0.572}F^{0.297}R^{-0.440}(\sin\alpha)^{-0.286}\cdot$ $\cdot\exp(0.6\cos\theta)$	=	$3.010\Sigma^{0.279}F^{0.145}R^{0.17}(\sin\alpha)^{-0.04}$
Di Risio (2005) (LIAM2005-2D)	$0.897S^{0.642}F^{0.531}X^{-0.273}$	$0.572S^{0.832}F^{0.398}X^{-0.215}$	$5.638S^{0.078}F^{0.053}X^{0.374}$
Di Risio (2005) (LIAM2005-AX)	$0.264S^{1.535}F^{0.527}R^{-0.828}$	=	$11.2S^{0.847}R^{0.110}$
Zweifel et al. (2006) (ETH2006)	=	Near field: $0.33FS^{0.5}M^{0.25}G^{0.25}$ Far field: $2S^{0.5}\tanh\left(0.25F^{1.5}M^{0.5}G^{0.5}X^{-0.5}\right)$	=
Ataie-Ashtiani & Nik-Khah (2008) (AANK2008)	=	$\left(0.398+0.076M^{1.27}F^{2.54}\right)\tau^{-0.26}\cdot$ $\cdot M^{0.26}\tilde{L}^{-0.125}S^{0.125}X^{-0.48}$	$\left(4.14+3.88M^2F^4\right)\tau^{-0.114}\cdot$ $\cdot M^{0.114}\tilde{L}^{0.1}S^{-0.1}X^{0.16}$
Heller & Hager (2010) (ETH2010)	Near field: $0.56F^{0.8}M^{0.2}G^{0.2}S^{0.4}(\cos 0.9\alpha)^{0.4}$ Far field: $0.75F^{0.8}M^{0.2}G^{0.2}S^{0.4}(\cos 0.9\alpha)^{0.4}X^{-0.27}$	Near field: $0.44F^{0.8}M^{0.2}G^{0.2}S^{0.4}(\cos 0.9\alpha)^{0.4}$ Far field: $0.60F^{0.8}M^{0.2}G^{0.2}S^{0.4}(\cos 0.9\alpha)^{0.4}X^{-0.27}$	Near field: $9F^{0.5}M^{0.13}G^{0.13}S^{0.25}(\cos 0.9\alpha)^{0.25}$ Far field: $9F^{0.25}M^{0.06}G^{0.06}S^{0.13}(\cos 0.9\alpha)^{0.13}X^{0.31}$

Table 1. Empirical formulations for subaerial landslide generated waves. Highest wave.

Ref. **(Abbr.)**	$\mathbf{H_l/d}$	$\mathbf{a_l/d}$	$\mathbf{T_l\sqrt{g/d}}$
Kamphuis & Bowering (1970) (KB1970)	=	=	$11+0.225X$
Fritz et al. (2004) (ETH2004)	=	=	$4.92F^{0.5}M^{0.2}X^{0.33}(c/\sqrt{gd})^{-1}$
Panizzo et al. (2005) (LIAM2005-3D)	$0.090\Sigma^{0.381}F^{0.198}R^{-0.810}(\sin\alpha)^{-0.404}\cdot$ $\cdot\exp(1.37\cos\theta)$	$0.025\Sigma^{0.343}F^{0.178}R^{-0.800}(\sin\alpha)^{-0.544}\cdot$ $\cdot\exp(1.60\cos\theta)$	$6.960R^{0.18}\exp(0.23\cos\theta)$
Di Risio (2005) (LIAM2005-AX)	$0.268S^{1.711}F^{0.553}R^{-0.934}$	=	$9.43S^{0.508}F^{0.112}R^{0.177}$

Table 2. Empirical formulations for subaerial landslide generated waves. Leading wave.

Ref. (Abbr.)	**Landslide model**	F	M	G	Σ	$X, (R)$	S	α (°)	θ (°)	γ (°)
Kamphuis & Bowering (1970) (KB1970)	Solid	0.9–3.1	0.04-0.79	2.7	=	7.2-74.3	=	20-90	=	75.5-101.5
Huber & Hager (1997) (HH1997-2D)	Granular	0.53-3.69	0.03-2.60	1.33	=	5-100	=	28-50	=	=
Huber & Hager (1997) (HH1997-3D)	Granular	0.53-3.69	0.03-2.60	1.33	=	5-30	=	28-50	0-90	=
Walder et al. (2003) (W2003)	Solid	1.0-4.1	0.14-3.24	2.9	0.28-2.94	8-20	0.24-1	11.2-19.5	=	32
Fritz et al. (2004) (ETH2004)	Granular	1.08-4.66	0.07-6.19	1.62	=	0-25	0.07-0.60	45	=	=
Panizzo et al. (2005) (LIAM2005-3D)	Solid	1.0-2.2	=	2.2	0.04-0.68	1.3-15.1	0.11-0.45	16-36	0-90	90
Di Risio (2005) (LIAM2005-2D)	Solid	0.29-2.58	0.09-4.17	1.36	0.28-12.50	1.7-89.2	0.22-2.50	90	=	90
Di Risio (2005) (LIAM2005-AX)	Solid	0.29-2.64	0.03-0.90	1.36	0.30-8.96	1.7-52.6	0.87-4.78	90	=	90
Zweifel et al. (2006) (ETH2006)	Granular	1.08-4.89	0.04-1.36	0.95-2.64	=	0-49	0.08-1.13	45	=	=
Ataie-Ashtiani & Nik-Khah (2008) (AANK2008)	Solid	Not Declared	9-240	1.9	0.02-0.18	0-7.2	0.1-0.4	15-60	=	60-90
Heller & Hager (2010) (ETH2010)	Granular	0.86-6.83	0.05-2.98	0.59-1.72	=	0-59	0.09-1.64	30-90	=	=

Table 3. Experimental ranges for empirical formulations of Tables 1-2.

Similar qualitative behavior was observed by Panizzo et al. (2005a) in the case of three-dimensional water bodies (LIAM2005-3D in Tables 1 and 2). They used a tray sliding along an incline that generated impulse waves in a wave tank. The slope of the incline was adjustable, so they investigated also the influence of α on generated waves. Actually, they followed the scaling analysis proposed by Walder et al. (2003), but their results showed that the underwater travel time depends on the same dimensionless groups proposed by Kamphuis & Bowering (1970). They found that the generated wave height increases as the landslide Froude number (F) and the dimensionless landslide front area ($\Sigma = ws/d^2$) increase. Furthermore they observed that the wave height grows as the incline slope (α) becomes smaller, therefore in contrast with the formulation by Huber & Hager (1997).
Ataie-Ashtiani & Nik-Khah (2008a) showed that the landslide shape does not strongly affect the wave height. They proposed their own empirical formulations (AANK2008 in Table 1) that work well also for experimental data by Bowering (1970), Huber & Hager (1997) and Walder et al. (2003). It has to be noted that the data collected by Huber & Hager (1997) is related to granular landslide models that should result in lower wave height if compared with experiments performed with solid blocks (Zweifel, 2004; Heller & Kinnear, 2010a). The doubt can be partially solved by observing that the experiments performed by Walder et al. (2003) and Ataie-Ashtiani & Nik-Khah (2008a) are characterized by a blockage ratio ($= w/w_f$, being w_f the wave flume width) equal to 0.526 and about 0.12 respectively. Indeed, as indicated by Heller & Kinnear (2010a), decreasing blockage ratio tends to reduce generated wave height.

Inspection of empirical formulations (see Tables 1 and 2) reveals that wave height (or wave crest amplitude) increases as dimensionless Froude number (F), dimensionless landslide volume (M) or dimensionless landslide front area (Σ), dimensionless landslide thickness (S) and specific gravity (G) increase. As far as the slope of the incline (α) is concerned, all the formulations but the one by Huber & Hager (1997) indicated that the lower the slope, the higher the wave height. Wave height attenuation is larger in three-dimensional and axial symmetric water bodies with respect to two-dimensional geometries. The landslide front slope angle (γ) influence was investigated only by Kamphuis & Bowering (1970) and Heller & Hager (2010b). When the angle that the front of landslides forms with water ($= \pi - \alpha - \gamma$) decreases, wave height will increases too. In general, granular landslides seems to generate lower waves as indicated by Zweifel (2004) and landslide shape does not affect significantly the induced wave height (Ataie-Ashtiani & Nik-Khah, 2008a).
The induced wave period is strongly related only to dimensionless distance from generation area (X or R): the higher the distance, the higher the wave period, i.e. wave dispersion occurs. The influences of landslide Froude number (F), the slope of the incline (α), the specific gravity (G) and the dimensionless landslide volume (M) seems to be slight.
It has to be noticed that for the use of some empirical formulations (i.e. W2003 and AAKK2008) the dimensionless underwater travel time value (τ) is needed. The only available formulations are those proposed by Walder et al. (2003) and Panizzo et al. (2005a):

$$\tau = 4.5\sqrt{\tilde{L}} \quad ; \quad \tau = 0.43\Sigma^{-0.27}F^{-0.66}(\sin\alpha)^{-1.32} \tag{3}$$

As observed by Walder et al. (2003), the first of equations (3), should be used when landslide motion is manly affected by frictional forces (i.e. Coulumbic friction) and hydrodynamic drag is less important (Savage & Hutter, 1989).

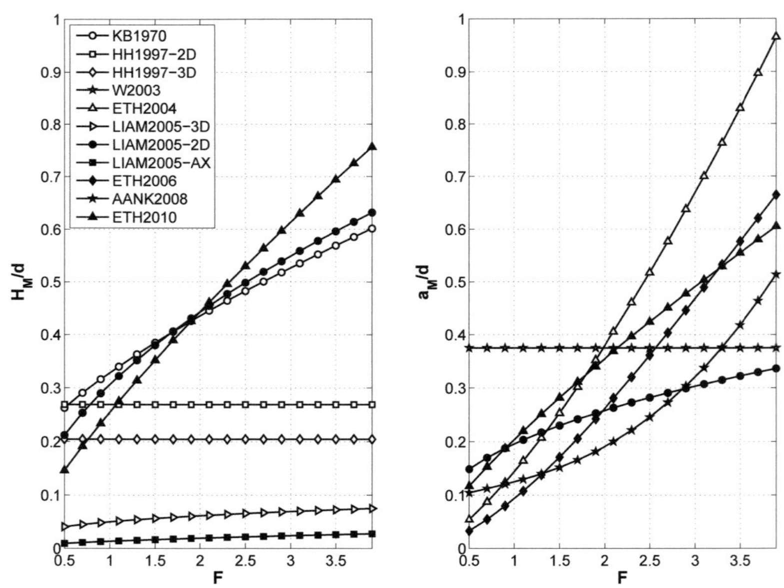

Fig. 2. Computed values of relative maximum wave height (H_M/d) and maximum wave crest amplitude (a_M/d) for varying landslide Froude number F (M = 0.5, G = 1.4, Σ = 0.5, X = R = 5, S = 0.5, $\alpha = 45°$, $\theta = 0°$)

Figure 2 compares the computed maximum wave height (left panel) and maximum wave crest amplitude (right panel) with the empirical formulations described above (Tables 1). It has to be noted that HH1997-3D and LIAM2005-3D are based on three-dimensional experiments, LIAM2005-AX on axial symmetric water bodies, whereas the others refer to two-dimensional investigations. In order to compare the formulations, the curves are plotted together by keeping fixed some parameters ($M = 0.5, G = 1.4, \Sigma = 0.5, X = R = 5, S = 0.5, \alpha = 45°, \theta = 0°$). It has to be stressed that some formulations are applied outside their own experimental ranges (see Table 3). Some considerations arise from figure inspection. Maximum wave height is considered first. As expected, the lowest wave height is estimated by LIAM2005-AX and LIAM2005-3D. The three-dimensional results are higher than the axial symmetric ones, due to (i) the wave directivity that is present for the experiments of Panizzo et al. (2005a) and totally absent in axial symmetric geometries (Di Risio, 2005) and (ii) as axial symmetric experiments were carried out with vertical falling solid boxes ($\alpha = 90°$). For landslide Froude number lower than about 2.0, empirical formulation ETH2010, based on granular material experiments, gives lower maximum wave height than LIAM2005-2D and KB1970, based on solid blocks experiments. However the differences decrease with increasing landslide Froude number, somewhat in contrast with Zweifel (2004). Furthermore, for landslide Froude number higher than about 2.0, granular material experiments give higher maximum wave height than solid blocks ones. The higher the value of dimensionless landslide thickness (S), the higher the landslide Froude number from which granular landslide models give higher wave height. However it has to be noted that landslide Froude number experimental ranges of LIAM2005-2D and KB1970 are 2.6 and 3.1 respectively.
As far as the maximum wave crest amplitude is concerned (Figure 2, right panel), it can be noted that the ETH's empirical formulations give the higher values for high landslide Froude numbers, i.e. granular material seems to induce higher crest elevation. For landslide Froude number lower than about 2.0, the empirical formulation proposed by Walder et al. (2003) on the basis of solid blocks experiments gives the highest value. For higher values of F, ETH2004-2006-2010, based on granular landslide experiments, give higher values than the solid boxes. It could be argued that the deformability of granular slide plays a role in generating waves for high landslide Froude number due to increasing of landslide front area that, as indicated by Panizzo et al. (2005a), induces an increasing of wave height. This aspect should be addressed in order to clarify the applicability of empirical formulations.

2.2 Underwater landslides generated waves

Physical features of underwater landslide generated waves are significantly different from that induced by subaerial landslides described in the previous section. As observed experimentally by Enet & Grilli (2007) in a three-dimensional wave tank, when underwater landslides occur, a depression of free surface takes place above the initial location of the slide. Then it rebounds to generate elevation perturbation radiating offshore as a leading elevation N-wave (Tadepalli & Synolakis, 1994) followed by trailing smaller waves. The rebound wave also propagates shoreward and is reflected by the shore, inducing wave run-up.
One of the first experimental works was carried out by Wiegel (1955) who used solid boxes, either falling vertically or sliding down an incline, in order to generate impulse waves in a two-dimensional wave flume. It was observed that the wave amplitude increases as the submerged weight increases and the initial submergence of the slide ($|\zeta|$) decreases. The wave period is related to the length of the slide and on the slope of the incline: the flatter the incline (the lower the slope α), the greater the wave period.

As the landslide starts to move when it is totally submerged, the features of generated waves are likely to be strongly related to the landslide motion. This aspect was addressed by the key work of Watts (1998). His experimental work was devoted at providing a method aimed at defining empirical formulation giving characteristic wave amplitude at a particular distance away from initial location of a submerged landslide. Based on scaling analysis of Euler equations, similar to Walder et al. (2003), he provided the following dimensional relationship:

$$\frac{\eta}{x_0} = f\left(\frac{b}{|\zeta|}, \alpha, \mathrm{Z}, \mathrm{HA}_0\right) \tag{4}$$

where η is the characteristic wave amplitude, b is the horizontal dimensions of the landslide, $|\zeta|$ the initial submergence, α the slope of the incline, already used for subaerial landslide, Z the submergence number ($= x_0 \sin\alpha/|\zeta|$), related to the characteristic length of landslide motion (x_0), and HA_0, the Hammack number ($= t_0 \sqrt{gd}/b$), is related to the characteristic time of the landslide motion (t_0). It has to be noted that, compared to the subaerial landslide generated waves problem, the landslide Froude number does not suffice to describe the phenomena. In particular, some measurements of the characteristic of the landslide motion are needed (i.e. x_0 and t_0). Based on theoretical analysis of solid block motion proposed by Pelinovsky & Poplavsky (1996), it was showed that characteristic length x_0 and time t_0 of landslide motion can be related to the initial acceleration a_0 and terminal velocity u_t:

$$x_0 = u_t^2/a_0 \qquad ; \qquad t_0 = u_t/a_0 \tag{5}$$

that, in turn, are related to the dynamical forces acting on the landslide during its motion (i.e. Coulumbic friction force, added mass force and hydrodynamic drag). Then the influence of initial acceleration and terminal velocity upon the induced wave is emphasized. Furthermore, the Hammack number represents a dimensionless near field wavelength (Watts, 2000). Indeed, characteristic wavelength λ_0 can be expressed in terms of the characteristic time of landslide motion ($= t_0 \sqrt{gd}$), and the Hammack number is the ratio between λ_0 and the horizontal landslide length b. This result is similar to the analytical one by Tuck & Hwang (1972) who showed that waves dispersion can be neglected (i.e. the near field) up to a distance of $b/(\tan\alpha)^2$ when an underwater disturbance of scale length b occurring on a sloping beach that forms an angle α to the horizontal generates an impulse waves train. Hence, the key result of the work by Watts (1998) is the formal coupling of the landslide motion to the generated waves (non-dimensional wave curve). Based on this formalism, Grilli & Watts (2005) and Watts et al. (2005) performed a series of two-dimensional numerical simulations aimed at providing empirical formulations giving the wave amplitude in the near field. The shape of the underwater landslide was selected on the basis of some preliminary simulations (Grilli & Watts, 2005) that indicated the semi-elliptical landslides as the worst case scenarios. The empirical formulation reads as follows (hereinafter referred to as W2005-2D, Watts et al., 2005):

$$\eta_{2D} = \frac{u_t^2}{a_0}\,(0.0574 - 0.0431 \sin\alpha)\,\mathrm{S}_b \left(\frac{L \sin\alpha}{|\zeta|}\right)^{1.25} \{1 - \exp\left[-2.2\,(\mathrm{D}-1)\right]\} \tag{6}$$

where S_b is a different definition of dimensionless landslide thickness ($= s/b$) if compared to that used for subaerial landslide. Formulation (6) is valid within the following ranges: $5° < \alpha < 30°$, $0.06 < b/d < 1.5$, $0.008 < \mathrm{S}_b < 0.2$, $1.46 < \mathrm{D} < 2.93$. In order to provide also three-dimensional wave amplitude, the following relationship was proposed by the same authors (hereinafter referred to as W2005-3D):

$$\eta_{3D} = \eta_{2D}\left(\frac{w}{w+\lambda_0}\right) \tag{7}$$

where w is the landslide width and λ_0 is the characteristic wavelength($= t_0\sqrt{gd}$).
The three-dimensional experimental work by Enet & Grilli (2007) confirmed the importance of initial acceleration and terminal velocity and successfully validated the empirical formulation provided by Watts et al. (2005).
Based on three-dimensional experiments, Ataie-Ashtiani & Nik-Khah (2008b) and Najfi-Jilani & Ataie-Ashtiani (2008) proposed further empirical formulations aimed at forecasting wave features in both the near field and the far field. The width of the wave tank was about 12 times the landslide width, and the waves were free to propagate also in the rear of the slope that did not span on the whole cross section of the tank. Here we report only the formulation for near field wave amplitude. Ataie-Ashtiani & Nik-Khah (2008b) and Najfi-Jilani & Ataie-Ashtiani (2008) used a series of dimensionless groups slightly different from that proposed by Watts et al. (2005) and expressed the near field wave amplitude as follows (hereinafter referred to as NJAA2008):

$$\eta = \frac{u_t^2}{a_0}\left(a_1 S_b^2 + a_2 S_b + a_3\right)\left(\frac{|\zeta|}{L}\right)^{b_1 \ln(S_b)+b_2} \tag{8}$$

whose parameters a_i and b_i are reported in Table 4. Formulation (8) is valid within the following ranges: $15° < \alpha < 60°$, $0.008 < S_b < 0.8$, $0.26 < |\zeta|/L < 0.86$, $D = 1.9$.
Figure 3 shows the computed values against the varying relative submergence ($b/|\zeta|$) and slope incline (α). It has to be stressed that for underwater landslide the higher the slope (the steeper the incline), the higher the wave amplitude, in contrast with subaerial generated waves.
It has to be noted that empirical formulations for underwater landslide generated waves were developed for rigid landslide and the effect of deformability and porosity is controversial. Watts & Grilli (2003) suggested that landslide deformation, at least at early stage, does not affect its motion, but preliminary numerical simulations of Grilli & Watts (2005) showed that landslide deformability is able to increase amplitude and affect waveform of generated waves. On the contrary, experimental investigations of Ataie-Ashtiani & Nik-Khah (2008b) showed that the slide deformations cause an average reduction of about 20-30% in energy conversion if compared with rigid slide. The problem should be addressed more in details.

2.3 Impulse waves propagation alongshore

The first attempt to analyze wave propagation alongshore was performed in the case of an ideal circular island (Yeh et al., 1994; Briggs et al., 1995a; Liu et al., 1995; Cho & Liu, 1999; Cho et al., 2004). Motivated by the extreme inundation observed in the rear side of Babi Island (Indonesia) during an earthquake induced tsunami attack (December 12, 1992), the experiments consisted in the study of inundation map around a circular island undertaken

a_1	$1.44976(\sin\alpha)^3 - 2.02732(\sin\alpha)^2 + 0.71456(\sin\alpha) - 0.04967$
a_2	$-0.00692(\sin\alpha)^3 + 0.14119(\sin\alpha)^2 - 0.04103(\sin\alpha) + 0.00308$
a_3	$0.00580(\sin\alpha)^3 - 0.00913(\sin\alpha)^2 + 0.00309(\sin\alpha) + 0.00019$
b_1	$9.68278(\sin\alpha)^3 - 12.68411(\sin\alpha)^2 + 4.64830(\sin\alpha) - 0.65307$
b_2	$19.23503(\sin\alpha)^3 - 27.66358(\sin\alpha)^2 + 12.49724(\sin\alpha) - 3.72730$

Table 4. Parameters of equation (8)

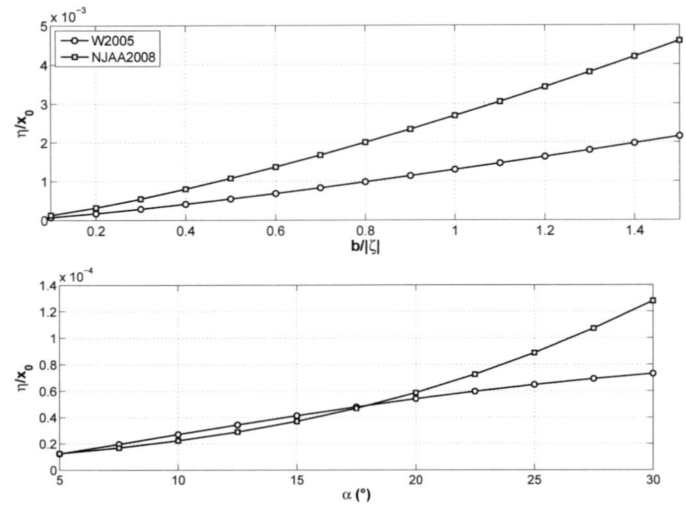

Fig. 3. Computed values of near field wave amplitude (η) normalised with characteristic length of landslide motion (x_0). Upper panel: $\alpha = 30°$, $S_b = 0.1$, $D = 1.9$. Lowe panel: $S_b = 0.1$, $b/|\zeta| = 0.1$, $D = 1.9$.

to the attack of a solitary wave propagating from offshore. Here we want to pay attention to impulse wave propagation when impulse waves are generated directly on the shore. Indeed, landslide generated waves can remain trapped close to the coast due to refraction phenomena and edge waves occur (Ursell, 1952; Lynett & Liu, 2005; Johnson, 2007). Ursell (1952) showed that waves on straight coast can remain trapped close to the shoreline depending on the beach slope (edge wave modes). In particular, he showed that the number of possible edge waves modes n_M has to satisfy the following relationship

$$n_M \leq \frac{1}{2}\left(\frac{\pi}{2\alpha} - 1\right). \tag{9}$$

Then, the higher the beach slope, the lower the energy trapped close to the coast. It has to be noted that perfect trapping was demonstrated not possible in a polar-symmetric topography, i.e. for ideal conical island (Renzi & Sammarco, 2010).

Liu et al. (1998) generated wave trains with a hinged wave-maker moving at the undisturbed shoreline of a straight coast (Liu & Yeh, 1996, $\alpha = 15°$, $\cot\alpha = 3.7$). Those experiments showed that perturbations occurring at the coast generate dispersive waves, even if in very shallow water. On the basis of standard spectral analysis they argued that near the shoreline the Stokes edge waves (zero-th edge wave mode) dominates and wave energy trapping occur for a long distance. Then, Liu et al. (2005) experimentally reproduced landslides sliding down an incline ($\alpha = 26.6°$, $\cot\alpha = 2$) in a wave tank. The tank was about 4 times the landslide width (triangular section body). They used experimental data to validate a RANS-VOF numerical model and they observed that the sidewalls affected the waves propagating alongshore (i.e. induced wave run-up) just after about 1.5 seconds. After Lynett & Liu (2005) performed a series of numerical simulations in the case of elliptical solid landslides sliding down a sloping beach, hence characterized by a straight coast. By using of landslide motion law proposed by Watts (1998), they used the numerical results to provide empirical formulations suitable to forecast run-up along the coast for both subaerial and submerged landslides. More in particular, they investigated the role of a series of dimensionless parameters: (i) dimensionless landslide thickness ($S_\zeta = -\tilde{s}/\zeta$, being the origin of ζ axis located at the undisturbed shoreline, positive upward and $\tilde{s}$ the maximum landslide thickness measured along the vertical direction); (ii) landslide steepness ($\mu = 2\pi\tilde{s}/b = k_s\tilde{s}$); (iii) symmetry of the landslide ($\kappa = b_f/b_b$, being b_f and b_b the horizontal distances from the location of maximum landslide thickness forward and backward respectively, $b = b_f + b_b$); (iv) aspect ratio ($A = b/w$); (v) specific gravity (G) and (vi) beach slope (α). They defined a "near-shore" far field, where "edge waves become important", whereas the near field wave pattern is strongly related to the source of impulse waves. In particular they showed that the near field extends up to about 5 times the landslide width for the subaerial case, whereas in the case of underwater landslide, also the aspect ratio of the body plays a key role in the extension of the near field:

$$\left(\frac{y_{nf}}{w}\right)_S \simeq 5 \quad ; \quad \left(\frac{y_{nf}}{w}\right)_U \simeq 5\left(\frac{L}{W}\right) \tag{10}$$

For the underwater case, Lynett & Liu (2005) provided empirical formulations for the maximum run-up (R_u) and run-down (R_d) immediately landward the landslide centreline:

$$\frac{R_u}{\tilde{s}} \simeq 0.30 S_\zeta^{1.16} \mu^{-1.40} A^{0.83} \kappa^{-0.015} G^{0.80} (\tan\alpha)^{1.71} \tag{11}$$

$$\frac{R_d}{\tilde{s}} \simeq -0.23 S_\zeta \mu^{-1.19} A^{0.87} \kappa^{-0.230} G^{0.62} (\tan\alpha)^{1.14} \tag{12}$$

At the near-far field division, the run-up reads as follows:

$$\frac{R_u}{\tilde{s}} \simeq 0.07 S_\zeta \mu^{-6/5} A^{5/6} \kappa^{-1/4} G^{4/3} (\tan\alpha)^{4/3} \tag{13}$$

As the edge waves propagate alongshore, dispersion effects can induce a secondary run-up peak. Numerical simulations of Lynett & Liu (2005) showed that in some cases this secondary peak can exceed the run-up observed just landward of the landslide centreline. This secondary peak was observed to be located at:

$$\frac{y_{sp}}{w} \simeq 0.6 S_\zeta^{-1} \mu^{6/5} A^{-1} \kappa^{1/3} G^{-1/5} (\tan\alpha)^{-6/5} \tag{14}$$

where the run-up can be estimated by the following relationship:

$$\frac{R_u}{\tilde{s}} \simeq 0.17 S_\zeta^{4/3} \mu^{-8/5} A^{5/6} \kappa^{-1/3} G^{3/4} (\tan\alpha)^{5/3} \tag{15}$$

In the case of subaerial landslide, only the run-up at the near-far field division was provided:

$$\frac{R_u}{\tilde{s}} \simeq 0.1 S_\zeta^{-1/4} G \tag{16}$$

Validity ranges of the dimensionless relationships proposed by Lynett & Liu (2005) follow: $0.125 < |S_\zeta| < 1.0$, $0.01 < \mu < 0.25$, $0.7 < \kappa < 1.7$, $1 < A < 10$, $1.8 < G < 3.2$, $1.9° < \alpha < 5.7°$.
It can be observed that the simulated beach slope ranged from about 2° ($\cot\alpha = 30$) up to about 6° ($\cot\alpha = 10$). These values represent typical values of coastal areas, but some steeper slopes can exist where landslides are likely to occur. Such a case occurred at the volcanic island of Stromboli on December 30, 2002 when two landslides generate impulse waves that propagate along the coast of the island itself (Tinti et al., 2005a;b). The slope of the island flank

is about 18° ($\cot\alpha = 3$), then quite larger than the slopes simulated by Lynett & Liu (2005). This event motivated a series of three-dimensional experimental works aimed at reproducing the wave generation by semi-elliptical landslides sliding down plane beaches with straight coasts (Di Risio et al., 2009a) and flanks of conical islands (Di Risio et al., 2008; 2009b; Molfetta et al., 2010).

Di Risio et al. (2009a) carried out a series of experiments in a three-dimensional wave tank where a plane beach was placed. Then the experimental layout was similar to the cases simulated by Lynett & Liu (2005), but with a steeper beach. Along the beach a semi-elliptical landslide model (G = 1.83) were left to slide along the incline. Both partially submerged and subaerial landslide were reproduced (i.e. $0.28 < S_\zeta < \infty$, being $S_\zeta = \infty$ when landslide center mass is located at the undisturbed shoreline elevation). The tank width was about 27 times landslide width, quite larger than previous experimental investigations (Liu et al., 2005; Enet et al., 2003; Enet & Grilli, 2007). Nevertheless the tank was short in the offshore direction and front sidewalls also reflected generated waves. The experimental layout allowed to observe the near field wave pattern, just landward to the landslide, and the propagation alongshore of the leading wave before the sidewalls reflection contaminated the induced waves. During the experiments the secondary run-up peak was located and the maximum run-up was located at about two times the landslide width away the centreline of the landslide rather than directly landward the landslide. At the near-far field division defined by Lynett & Liu (2005) they observed that the dimensionless run-up induced by the leading wave was equal to about $R_u/s \simeq 0.18$, whilst the maximum run-up was $R_u/s \simeq 0.28$, by keeping in mind that the latter could be affected by sidewalls reflection. Although outside the validity range, the formulation (16) gives a value equal to 0.25 ($S_b = 0.28$), quite similar to the observed value. The work by Di Risio et al. (2009b) was aimed at reproducing impulse waves propagation around a conical island, similar to that of Briggs et al. (1995a), placed at the centre of a large wave tank (50 m long, 30 m wide, 3 m high at the Research and Experimentation Laboratory for Coastal Defence - LIC - of the Technical University of Bari, Italy). The landslide model was exactly the same used by Di Risio et al. (2009a). During the experiments, the induced run-up along the coast was measured by means of special gauges directly embedded into the island flanks. Only subaerial landslides were reproduced and the role of undisturbed shoreline radius was investigated: the higher the undisturbed shoreline radius, the higher the induced run-up. The large dimensions of the wave tank allowed to observe the propagation of waves along the whole circular coastline and dispersive features of wave packets were observed, being the maximum run-up induced by the first wave near the generation area, by the second wave up to a curvilinear distance alongshore equal to about 8 times the landslide width, then the third wave induces the maximum run-up and so on. Although based on relatively few data, Di Risio et al. (2008) provided a series of empirical formulations giving the run-up induced by the impulse waves around the island by the first four waves of the generated train:

$$\frac{R_u^{(1)}}{s} = 1.049\left(\frac{w}{r_0}\right)^{-0.470}\left(\frac{\tilde{y}}{w}\right)^{-2.021} \tag{17}$$

$$\frac{R_u^{(2)}}{s} = 0.386\left(\frac{w}{r_0}\right)^{-0.066}\exp\left[\left(-0.131\frac{\tilde{y}}{w}\right)\right] \tag{18}$$

$$\frac{R_u^{(3)}}{s} = 0.011\left(\frac{w}{r_0}\right)^{-0.945}\exp\left[\left(0.007\frac{\tilde{y}}{w}\right)^2\right] \tag{19}$$

$$\frac{R_u^{(3)}}{s} = 0.011\left(\frac{w}{r_0}\right)^{-0.341} \exp\left[\left(0.001\frac{\tilde{y}}{w}\right)^2\right] \tag{20}$$

where $\tilde{y}$ is measured along the undisturbed shoreline (i.e. it is curvilinear coordinate) with radius equal to r_0. Validity ranges of the above relationships follow: $0.15 < w/r_0 < 0.19$, $1 < \tilde{y}/w < 20$. To date, these are the unique empirical formulations based on experimental data aimed at forecasting wave run-up on island coastlines when a landslide occurs directly on the flank of the island itself.

3. Impulse wave run-up

Once impulse waves are generated by landslides, they propagate and interact with water body boundaries. The waves run up the coastal boundaries and induce even large inundations.
Since the earlier work by Hall & Watts (1953) and Camfield & Street (1970), a long series of experiments have been carried out in order to forecast the run-up induced by a single impulse wave. Almost all the experiments dealt with solitary waves (breaking or non-breaking). This is due to the fact that solitary wave has been recognized so far as the proper model to describe tsunami approaching shoreline. Actually, Madsen & Schaffer (2010) has recently shown that this is not realistic for a geophysical tsunami.
For non-breaking solitary waves, Synolakis (1987) found that non-linear and linear shallow water equations for a sloping beach (angle to the horizontal β) connected to an horizontal bottom have the same solution given by:

$$\frac{R_u}{d} = 2.831\sqrt{\cot\beta}\left(\frac{H}{d}\right)^{1.25} \tag{21}$$

where H is the wave height. The run-up law given by equation (21) is consistent with the Green law (Lamb, 1932, §185), valid for the propagation of waves in a canal of gradually varying rectangular section. This law states the free surface elevation (η) is inversely proportional to mean water depth, i. e.:

$$\eta \propto d^{-1/4} \tag{22}$$

The relationship (21) is valid for non-breaking solitary waves, then a kind of breaking criterion is needed in order to check its validity:

$$\frac{H}{d} > 0.8183\,(\cot\beta)^{-10/9}. \tag{23}$$

However, this is not a proper breaking criterion as based on the limit at which the Jacobian of hodograph transformation used by Synolakis (1987) becomes singular. It was noted that the assumptions inherent in the shallow-water wave theory are violated before the Jacobian becomes singular, then it cannot be viewed as the actually point where the waves break but a validity limit for analytical results (Meyer, 1986a;b). Indeed, numerical simulations of Grilli et al. (1997) showed that solitary waves break during climbing on sloping beach if

$$\frac{H}{d} > 16.9\,(\tan\beta)^2. \tag{24}$$

The analytical method of Synolakis (1987) has been used by Li & Raichlen (2001) that retained higher order terms in the initial condition. Then a corrected run-up law for non-breaking solitary waves has been defined:

$$\frac{R_u}{d} = 2.831\sqrt{\cot\beta}\left(\frac{H}{d}\right)^{5/4} + 0.293\,(\cot\beta)^{3/2}\left(\frac{H}{d}\right)^{9/4} \tag{25}$$

A first attempt to provide a reliable formulation for run-up forecasting induced by other waves than solitary one was made by Tadepalli & Synolakis (1994). They provided analytical relationships giving induced run-up in the case of leading-elevation and leading-depression generalized, isoscele and double N-waves. The maximum run-up was showed to be induced by leading-depression isosceles and double N-waves given by:

$$\frac{R_u}{d} = 3.86\sqrt{\cot\beta}\left(\frac{H}{d}\right)^{1.25} \quad ; \quad \frac{R_u}{d} = 4.55\sqrt{\cot\beta}\left(\frac{H}{d}\right)^{1.25} \tag{26}$$

respectively. It has to be noted that the resulting run-up is higher than the one induced by the solitary wave.

Nevertheless many other researches still preferred to focus on the run-up induced by solitary waves by making use of different approaches. Li & Raichlen (2003) proposed a simple energy model that gives a reliable estimation of the run-up resulting from breaking solitary waves by taking into account also the shape of the run-up tongue:

$$\frac{R_u}{d} = \frac{1 - E_b/E_i}{1.5k_t}\left(\frac{H}{d}\right) \tag{27}$$

where E_b and E_i are the dissipated energy by breaking processes and incident solitary wave energy respectively and k_t is the shape factor of the run-up tongue ($k_t = 0.33$ for a triangular run-up tongue, $k_t = 0.12$ based on numerical simulation by Li, 2000, $k_t = 0.18$ based on experimental data). For the ratio E_b/E_i numerical simulations by Li (2000) have been used by Li & Raichlen (2003):

$$\frac{E_b}{E_i} = C\left[A\ln\left(\cot\beta\right) + B\right] \tag{28}$$

with

$$A = -0.470\left(\frac{H}{d}\right) + 0.534 \quad ; \quad B = 2.165\left(\frac{H}{d}\right) - 1.154 \quad ; \quad C = 0.190\left[\ln\left(\frac{H}{d}\right)\right] + 0.969. \tag{29}$$

Gedik et al. (2005) performed an experimental investigation aimed at including the influences of sand grain diameter (d_s) for sloping sandy beach, also in the case of armored beaches (armor units diameter d_{n50}):

$$\frac{R_u}{d} = 4\cdot10^{-4}\left(\frac{H}{d_s}D\cot\beta\right)^{0.921} \quad ; \quad \frac{R_u}{d} = 5\cdot10^{-3}\left(\frac{H}{d_{n50}}D\cot\beta\right)^{0.954} \tag{30}$$

Hughes (2004a) used experimental and numerical data of past studies (Hall & Watts, 1953; Li, 2000; Briggs et al., 1995b; Synolakis, 1986; 1987) in order to define empirical formulation for breaking and non-breaking solitary wave run-up based on the wave momentum flux parameter proposed by Hughes (2004b).

It has to be noted that the formulations cited above are almost all related to solitary waves. However, when impulse waves are generated by landslides, especially in enclosed basins (i.e. reservoirs), the waveform is likely to be related to other wave types (e.g. Panizzo et al., 2005a). Recently, on the basis of the definition of representative solitary wave period (Kobayashi & Karjadi, 1994), it has been demonstrated that the role of the relative wave height (H/d) for solitary wave run-up (relative to wave height R_u/H) is replaced by the dimensionless wave period ($T\sqrt{g/d}$) for oscillatory waves, whilst the role of beach slope remains unchanged

(Fuhrman & Madsen, 2008; Madsen & Fuhrman, 2008). Then, a general relationship that gives impulse wave run-up can be formally expressed as follows:

$$\frac{R_u}{H} = a_1 \left(\frac{H}{d}\right)^{a_2} \left(T\sqrt{\frac{g}{d}}\right)^{a_3} (\tan\beta)^{a_4}. \tag{31}$$

For solitary wave run-up $a_3 = 0$, whereas for sinusoidal (linear) waves $a_2 = 0$. Run-up law proposed by Synolakis (1987) reads $a_1 = 2.831, a_2 = +1/4, a_3 = 0$ and $a_4 = -1/2$. Consistently, Green Law can be recasted by keeping $a_1 \neq 0, a_2 = +1/4, a_3 = 0, a_4 \neq 0$.
Muller (1995) performed a series of experimental tests providing the following empirical formulation:

$$\frac{R_u}{H} = 1.25 \left(\frac{H}{d}\right)^{0.25} \left(\frac{H}{\lambda}\right)^{-0.15} \left(\frac{\pi}{2\beta}\right)^{0.20} \tag{32}$$

The experimental ranges of the above relationship follow: $0.011 < H/d < 0.521$, $0.001 < H/\lambda < 0.013$, $18.4° < \beta < 90°$. It has to be stressed that the wavelength λ estimation is needed in order to use the formulations (32). Hence, it can be used only if also wave celerity estimate is available, for instance by using solitary wave celerity (Boussinesq, 1872).
In order to overcome the shortcomings of relationship (32), Di Risio (2005) performed a series of two dimensional tests by providing the following formulations for wave run-up:

$$\frac{R_u}{H} = 1.37 \left(\frac{H}{d}\right)^{0.51} \left(T\sqrt{\frac{g}{d}}\right)^{0.47} (\sin\beta)^{0.26} \tag{33}$$

This formulation is consistent with theoretical reasoning of Fuhrman & Madsen (2008), equation (31), and it is the first empirical formulation involving the wave period. Experimental ranges of relationship (33) follow: $0.18 < H/d < 0.70$, $7.48 < T\sqrt{g/d} < 15.60$, $22° < \beta < 84°$. Equation (33) can be used also for solitary waves by defining the representative wave period T_S as the time interval during which the normalized solitary wave surface elevation (η/H) exceeds some small threshold level (Kobayashi & Karjadi, 1994; Fuhrman & Madsen, 2008):

$$T_S = 1.03\sqrt{\frac{16d^2}{3Hg(1 + H/d)}}. \tag{34}$$

4. Impulse wave overtopping

If the induced run-up exceeds the dam free-board (ξ) the water will overtop the structure, and floods the downstream areas. In this case, overtopping volume and overtopping flow forecasting can be precious in modeling downstream flooding (e.g. Risley et al., 2006).
To date, some analytical solutions exist for bore waves (e.g. Peregrine & Williams, 2001; Baldock et al., 2005). The only researches dealing with dam overtopping due to impulse waves are those of Muller (1995) and Di Risio (2005). The former performed a series of two-dimensional experiments by reproducing the effect of both the dam slope and crest width. The following empirical formulation was given for the overtopping volume per unit width:

$$\frac{V}{d^2} = 1.45\tilde{\kappa} \left(\frac{H}{d}\right)^{1.33} \left(T\sqrt{\frac{g}{d}}\right)^{0.44} \left(1 - \frac{\xi}{R_u}\right)^{2.2} \tag{35}$$

For small dam crest width (relative to impulse wave height) the coefficient $\tilde{\kappa}$ is equal to 0.53, 0.61 and 0.66 for β equal to 90°, 45° and 18.4° respectively. It has to be noted that equation (35) can be used once one has estimated the wave run-up. Muller (1995) provided also the time duration $\tilde{t}_0$ of the overtopping event, even if only in the case of $\xi = 0$, then giving an estimate of maximum overtopping discharge per unit width $q_0 = V_0/\tilde{t}_0$:

$$\tilde{t}_0 = 4\left(T\sqrt{\frac{g}{d}}\right)^{0.44} \tag{36}$$

It can be observed that the overtopping discharge ($q_0 = V_0/t_0$) does not depend upon wave period. Experimental ranges of equation (35) follow: $0.019 < H/d < 0.488, 9.0 < T\sqrt{g/d} < 21.0$. In a similar way, Di Risio (2005) performed two-dimensional experiments giving the following empirical formulation:

$$\frac{V}{d^2} = 0.0084\left(\frac{H}{d}\right)^{3.18}\left(T\sqrt{\frac{g}{d}}\right)^{1.58}\left(\frac{\xi}{d}\right)^{-1.62}(\sin\beta)^{-0.61} \tag{37}$$

with the experimental ranges $0.18 < H/d < 0.70$, $7.48 < T\sqrt{g/d} < 15.60$, $22° < \beta < 84°$, $0.09 < \xi/d < 1.42$.

5. Concluding remarks

A comprehensive review of landslide generated wave forecasting has been presented. The review covers the overall phenomenon, from wave generation up to wave propagation both alongshore and offshore and impulse wave interaction with coasts and dam (i.e. run-up and overtopping).
The main aim of the work is to review empirical formulations useful to practically forecast generated wave features (i.e. maximum wave height, crest amplitude and period; leading wave height, crest amplitude and period) both in the near and far field. Furthermore attention has been paid to the propagation alongshore when landslides generate waves directly at the coast and edge waves modes can be triggered. The interaction of impulse waves with water body boundaries has been described in terms of wave run-up and dam overtopping.
Some aspects related to generation phenomena have to be addressed. In particular, the applicability of empirical formulations based on solid landslide experiments should be investigated with respect to the role of deformations and porosity on wave generation, especially in the case of underwater landslide that seems to be controversial.

6. Acknowledgments

The physical modeling performed by the authors and described in the present chapter were funded by the RID (National Dam Office of the Italian Government), by the MIUR (Italian Minister of Formation, University and Research) and by Dipartimento Nazionale della Protezione Civile (Italian Department for Civil Protection). The authors wish to thanks all the people who participate to the experimental investigations. In particular Mr. M. Nardi and L. Matergia, the LIAM technicians, are acknowledged. Furthermore Eng. M. G. Molfetta, F. Aristodemo, A. Romano, G. Bellotti and A. Panizzo are acknowledged for their cooperation during experimental investigations and data analysis on conical island and planar beach.

7. References

Ataie-Ashtiani, B. & Nik-Khah, A. (2008a). Impulse waves caused by subaerial landslides, *Environ. Fluid Meach.* 8(3): 263–280, doi:10.1007/s10652–008–9074–7.

Ataie-Ashtiani, B. & Nik-Khah, A. (2008b). Laboratory investigations on impulsive waves caused by underwater landslide, *Coast. Eng.* 55(12): 989–1004, doi: 10.1016/j.coastaleng.2008.03.003.

Baldock, T., Hughes, M., Daya, K. & Louysa, J. (2005). Swash overtopping and sediment overwash on a truncated beach, *Coast. Eng.* 52(7): 633–645, doi: 10.1016/j.coastaleng.2005.04.002.

Boussinesq, J. (1872). Theorie des ondes et des remous que se propagent le long d'un canal rectangulaire horizontal, en communiquant au liquide contenu dans ce canal des vitesses sensiblement pareilles de la surface au fond, *J. Math. Pures Appl.* 17: 55–108.

Bowering, R. J. (1970). *Landslide generated waves: a laboratory study*, Master's thesis, Queen's University, Kingston, Ontario, Canada.

Briggs, M. J., Synolakis, C. E., Harkins, G. S. & Green, D. R. (1995a). Laboratory experiments of tsunami runup on a circular island, *Pure Appl. Geophys.* 144(3/4): 569–593, doi: 10.1007/BF00874384.

Briggs, M. J., Synolakis, C. E., Harkins, G. S. & Hughes, S. A. (1995b). Large-scale, three-dimensional laboratory measurements of tsunami inundation, *Tsunamis: Progress in Prediction, Disaster Prevention and warning.*, Kluver Academic Publishers, pp. 129–149, ISBN: 0-7923-3483-3.

Camfield, F. E. & Street, R. L. (1970). Shoaling of solitary waves on small slopes, *J. of the Waterways and Harb. Div., Proc. of the Am. Soc. of Civil Eng.* 95(WW1): 1–22.

Cea, L., Ferreiro, A., Vazquez-Cendon, M. E. & Puertas, J. (2004). Experimental and numerical analysis of solitary waves generated by bed and boundary movements, *Int. J. Numer. Meth. Fl.* 46(8): 793–813, doi: 10.1002/fld.774.

Chaudhry, M. H., Mercer, A. G. & Cass, D. (1983). Modeling of slide-generated waves in a reservoir, *J. Hydraul. Eng.-ASCE* 109(11): 1505–1520, doi: 10.1061/(ASCE)0733-9429(1983)109:11(1505).

Cho, Y. S. & Liu, P. L.-F. (1999). Crest-length effects in nearshore tsunami run-up around island, *J. Geophys. Res.* 104(C4): 7907–7913, doi: 10.1029/1999JC900012.

Cho, Y.-S., Park, K.-Y. & Lin, T.-H. (2004). Run-up heights of nearshore tsunamis based on quadtree grid system, *Ocean Eng.* 31(8–9): 1093–1109, doi: 10.1016/j.oceaneng.2003.10.011.

Davidson, D. D. & McCartney, B. L. (1975). Water waves generated by landslides in reservoirs, *J. of the Hydraul. Div., Proc. of the Am. Soc. of Civil Eng.* 101(HY12): 1489–1501.

de Carvalho, R. F. & do Carmo, A. J. S. (2007). Landslides into reservoirs and their impacts on banks, *Environ. Fluid Mech.* 7(6): 481–493, doi: 10.1007/s10652-007-9039-2.

Di Risio, M. (2005). *Landslide generated impulsive waves: generation, propagation and interaction with plane slopes. An experimental and analytical study*, PhD thesis, University of Roma Tre, Rome.

Di Risio, M. & Sammarco, P. (2008). Analytical modeling of landslide-generated waves, *J. Waterway, Port, Coastal, Ocean Eng.* 134(1): 53–60, doi: 10.1061/(ASCE)0733-950X(2008)134:1(53).

Di Risio, M., Bellotti, G., Panizzo, A., Molfetta, M. G., Aristodemo, F., De Girolamo, P., Pratola, L. & Petrillo, A. F. (2008). Tsunamis generated by landslide along the coast of a conical island: new three-dimensional experiments, *Proc. of 31st International Conference of*

Coastal Engineering, Vol. 2, pp. 1396–1408, doi: 10.1142/9789814277426_0116.

Di Risio, M., Bellotti, G., Panizzo, A. & De Girolamo, P. (2009a). Three-dimensional experiments on landslide generated waves at a sloping coast, *Coast. Eng.* 56(5–6): 659–671, doi: 10.1016/j.coastaleng.2009.01.009.

Di Risio, M., De Girolamo, P., Bellotti, G., Panizzo, A., Aristodemo, F., Molfetta, M. G. & Petrillo, A. F. (2009b). Landslide-generated tsunamis runup at the coast of a conical island: new physical model experiments, *J. Geophys. Res.* 114, C01009, 16 pp, doi: 10.1029/2008JC004858.

Enet, F., Grilli, S. T. & Watts, P. (2003). Laboratory experiments for tsunami generated by underwater landslides: comparison with numerical modeling, *Proc. of the Thirteenth Int. Offshore and Polar Eng. Conf., Honolulu, Hawaii, USA, May 25–30*, pp. 372–379.

Enet, F. & Grilli, S. T. (2007). Experimental study of tsunami generation by three-dimensional rigid underwater landslides, *J. Waterway, Port, Coastal, Ocean Eng.* 133(6): 442–454, doi: 10.1061/(ASCE)0733-950X(2007)133:6(442).

Fleming, J. G., Walters, R. A., Sue, L. P. & Nokes, R. I. (2005). Experimental design for solid block and granular submarine landslides: a unified approach, *Tsunamis: case studies and Recent Developments*, Springer, K. Satake (ed.), pp. 259–277, doi: 10.1007/1-4020-3331-1_16.

Fritz, H. M., Hager, W. H. & Minor, H. E. (2001). Lituya bay case: rockslide impact and wave run-up, *Science of Tsunami Hazards* 19(1): 3–22.

Fritz, H. M., Hager, W. H. & Minor, H. E. (2003a). Landslide generated impulse waves. 1. Instantaneous flow fields, *Exp. Fluids* 35(6): 505–519, doi: 10.1007/s00348-003-0659-0.

Fritz, H. M., Hager, W. H. & Minor, H. E. (2003b). Landslide generated impulse waves. 2. Hydrodynamic impact craters, *Exp. Fluids* 35(6): 505–519, doi: 10.1007/s00348-003-0660-7.

Fritz, H. M., Hager, W. H. & Minor, H. E. (2004). Near field characteristics of landslide generated impulse waves, *J. Waterway, Port, Coastal, Ocean Eng.* 130(6): 287–302, doi: 10.1061/(ASCE)0733-950X(2004)130:6(287).

Fritz, H. M., Mohammed, F. & Yoo, J. (2009). Lituya bay landslide impact generated mega-tsunami 50th anniversary, *Pure Appl. Geophys.* 166(1–2): 153175, doi: 10.1007/s00024-008-0435-4.

Fuhrman, D. R. & Madsen, P. A. (2008). Surf similarity and solitary wave runup, *J. Waterway, Port, Coastal, Ocean Eng.* 134(3): 195–198, doi: 10.1061/(ASCE)0733-950X(2008)134:3(195).

Gedik, N., Irtem, E. & Kabdasli, S. (2005). Laboratory investigation on tsunami run-up, *Ocean Eng.* 32(5–6): 513–528, doi: 10.1016/j.oceaneng.2004.10.013.

Goring, D. G. (1978). *Tsunamis. The propagation of long waves onto a shelf.*, PhD thesis, California Institute of Technology, Pasadena, California.

Grilli, S., Svendsen, I. & Subramanya, R. (1997). Breaking criterion and characteristics for solitary waves on slope, *J. Waterway, Port, Coastal, Ocean Eng.* 123(3): 102–112, doi: 10.1061/(ASCE)0733-950X(1997)123:3(102).

Grilli, S. T. & Watts, P. (2005). Tsunami generation by submarine mass failure. I: modeling, experimental validation, and sensitivity analyses, *J. Waterway, Port, Coastal, Ocean Eng.* 131(6): 283–297, doi: 10.1061/(ASCE)0733-950X(2005)131:6(283).

Hall, J. V. & Watts, J. W. (1953). Laboratory investigation of the vertical rise of solitary waves on impermeable slopes, *Technical Memorandum 33*, Beach Erosion Board, US Army Corps of Engineers.

Heinrich, P. (1992). Nonlinear water waves generated by submarine and aerial landslides, *J. Waterway, Port, Coastal, Ocean Eng.* 118(3): 249–266, doi: 10.1061/(ASCE)0733-950X(1992)118:3(249).

Heller, V., Hager, W. H. & Minor, H.-E. (2008). Scale effects in subaerial landslide generated impulse waves, *Exp. Fluids* 44(5): 691–703, doi: 10.1007/s00348-007-0427-7.

Heller, D. & Kinnear, R. D. (2010a). Discussion of "Experimental investigation of impact generated tsunami; related to a potential rock slide, western norway" by G. Sælevik, A. Jensen, G. Pedersen [Coastal Eng. 56 (2009) 897–906], *Coast. Eng.* 57(8): 773–777, doi: 10.1016/j.coastaleng.2010.02.008.

Heller, D. & Hager, W. H. (2010b). Impulse product parameter in landslide generated impulse waves, *J. Waterway, Port, Coastal, Ocean Eng.* 136(3): 145–155, doi: 10.1061/(ASCE)WW.1943-5460.0000037.

Huber, A. (1980). Schwallwellen in seen als floge von felssturzen (reservoir impulse waves caused by rockfall), *Technical Report Mitteilung 47*, Laboratory of Hydraulics, Hydrology and Glaciology, Swiss Federal Institute of Technology.

Huber, A. (1982). Impulse waves in swiss lakes as a results of rock avalanches and bank slides. experimental results for the prediction of the characteristic numbers of these waves., *Comminsion Internationale des Grands Barrages, 14 Congres des Grand Barrages, Rio de Janeiro, 1982*, pp. 311–390.

Huber, A. & Hager, W. H. (1997). Forecasting impulse waves in reservoirs, *Comminsion Internationale des Grands Barrages, 19 Congres des Grand Barrages, Florence, 1997*, pp. 993–1005.

Hughes, S. A. (2004a). Estimation of wave run-up on smooth, impremeable slopes using the wave momentum flux parameter, *Coast. Eng.* 51(11-12): 1085–1104, doi: 10.1016/j.coastaleng.2004.07.026.

Hughes, S. A. (2004b). Wave momentum flux parameter: a descriptor for nearshore waves, *Coast. Eng.* 51(11-12): 1067–1084, doi: 10.1016/j.coastaleng.2004.07.025.

Iwasaki, S. (1983). Experimental study of a tsunami generated by a horizontal motion of a sloping bottom, *Bull. of the Earthq. Res. Inst.* 57: 239–262.

Johnson, R. S. (2007). Edge waves: theories past and present, *Phil. Trans. R. Soc. A* 365(1858): 2359–2376, doi: 10.1098/rsta.2007.2013.

Kamphuis, J. W. & Bowering, R. J. (1970). Impulse waves generated by landslide, *Proc. of 12th Coastal Engineering Conference*, pp. 575–588.

Kattel, G. & Eric, B. (2002). Accuracy of solitary wave generation by a piston wave maker, *J. of Hydraul. Res.* 40(3): 321–331, doi: 10.1080/00221680209499946.

Kobayashi, N. & Karjadi, E. A. (1994). Surf-similarity parameters for breaking solitary-wave runup, *J. Waterway, Port, Coastal, Ocean Eng.* 120(6): 645–650, doi: 10.1061/(ASCE)0733-950X(1994)120:6(645).

Koo, W. & Kim, M.-H. (2008). Numerical modeling and analysis of waves induced by submerged and aerial/sub-aerial landslides, *KSCE J. Civ. Eng.* 12(2): 77–83, doi: 10.1007/s12205-008-0077-1.

Lamb, H. (1932). *Hydrodynamics*, 6th edn, Dover Publications, Inc., ISBN: 0-486-60256-7.

Li, Y. (2000). Tsunamis: non-breaking and breaking solitary wave runup, *Technical Report KH-R-60*, W.M. Keck Laboratory of Hydraulics and Water Resource, California Institute of Technology, Pasadena, California.

Li, Y. & Raichlen, F. (2001). Solitary wave runup on plane slope, *J. Waterway, Port, Coastal, Ocean Eng.* 127(1): 33–44, doi: 10.1061/(ASCE)0733-950X(2001)127:1(33).

Li, Y. & Raichlen, F. (2003). Energy balance model for breaking solitary wave runup, *J. Waterway, Port, Coastal, Ocean Eng.* 129(2): 47–59, doi: 10.1061/(ASCE)0733-950X(2003)129:2(47).

Liu, P. L.-F., Cho, Y. S., Briggs, M. J., Kanoglu, U. & Synolakis, C. E. (1995). Runup of solitary waves on a circular island, *J. Fluid Mech.* 302: 259–285, doi: 10.1017/S0022112095004095.

Liu, P. L.-F. & Yeh, H. H. (1996). The generation of edge waves by a wave-maker, *Physics of Fluids* 8(8): 2060–2065, doi: 10.1063/1.869008.

Liu, P. L.-F., Yeh, H., Lin, P., Chang, K. T. & Cho, Y. S. (1998). Generation and evolution of edge-wave packets, *Physics of Fluids* 10(7): 1635–1657, doi: 10.1063/1.869682.

Liu, P. L.-F., Wu, T.-R., Raichlen, F., Synolakis, C. E. & Borrero, J. C. (2005). Runup and rundown generated by three-dimenional sliding masses, *J. Fluid Mech.* 536: 107–144, doi: 10.1017/S0022112005004799.

Lynett, P. & Liu, P. L.-F. (2005). A numerical study of the run-up generated by three-dimensional landslides, *J. Geophys. Res.* 110, C03006, 16 pp., doi: 10.1029/2004JC002443.

Madsen, P. A. & Fuhrman, D. R. (2008). Run-up of tsunamis and long waves in terms of surf-similarity, *Coast. Eng.* 55(3): 209–223, doi: 10.1016/j.coastaleng.2007.09.007.

Madsen, P. A. & Schaffer, H. A. (2010). Analytical solutions for tsunami runup on a plane beach: single waves, N-waves and transient waves, *J. Fluid Mech.* 645: 27–57, doi: 10.1017/S0022112009992485.

Malek-Mohammadi, S. & Testik, F. Y. (2010). New methodology for laboratory generation of solitary waves, *J. Waterway, Port, Coastal, Ocean Eng.* 136(5): 286–294, doi: 10.1061/(ASCE)WW.1943-5460.0000046.

Mallet, R. & Mallet, J. W. (1858). *Fourth report upon the facts and theory of earthquake phenomena*, Transaction of the British Association for the Advancement of Science, 1852 to 1858, Taylor and Francis, Red Lion Court, Fleet Street.

Meyer, R. E. (1986a). On the shore singularity of water-waves. I. The local model, *Physics of Fluids* 29(10): 3152–3163, doi: 10.1063/1.865967.

Meyer, R. E. (1986b). On the shore singularity of water-waves. II. Small waves do not break on gentle beaches, *Physics of Fluids* 29(10): 3164–3173, doi: 10.1063/1.865968.

Miller, D. J. (1960). Giant waves in lituya bay, alaska, Geological Survey Professional Paper 354-C. US Government Printing Office, Washington DC.

Miller, R. L. (1970). Prediction curves for waves near the source of an impulse, *Proc. of 12th Coastal Engineering Conference*, pp. 609–624.

Milne, J. (1898). *Earthquakes and other Earth movements*, Paul, Trench, Trubner & Co., London.

Molfetta, M. G., Di Risio, M., Bellotti, G., Romano, A., Pratola, L., De Girolamo, P. & Damiani, L. (2010). Tsunamis generated by landslides along the coast of a conical island: a new set of three-dimensional experiments, *Coastlab2010*, Accepted Abstract.

Monaghan, J. J. & Kos, A. (2000). Scott Russell's wave generator, *Phys. Fluids* 12(3): 622–630, doi: 10.1063/1.870269.

Montessus de Ballore, F. (1907). *La science seismologique*, A. Colin, Paris.

Muller, D. (1995). Auflaufen und uberschwappen von impulswellen an talsperren (run-up and overtopping of impulse waves at dams), *Technical Report Mitteilung 137*, Laboratory of Hydraulics, Hydrology and Glaciology, Swiss Federal Institute of Technology.

Najfi-Jilani, A. & Ataie-Ashtiani, B. (2008). Estimation of near-field characteristics of tsunami generation by submarine landslide, *Ocean Eng.* 35(5–6): 545–557, doi: 10.1016/j.oceaneng.2007.11.006.

Noda, E. K. (1970). Water waves generated by landslides, *J. of the Waterways and Harb. Div.,*

Proc. of the Am. Soc. of Civ. Eng. 96(WW4): 835–855.

Panizzo, A., Bellotti, G. & De Girolamo, P. (2002). Application of wavelet transform analysis to landslide generated waves, *Coast. Eng.* 44(4): 321–338, doi: 10.1016/S0378-3839(01)00040-0.

Panizzo, A., De Girolamo, P. & Petaccia, A. (2005a). Forecasting impulse waves generated by subaerial landslide, *J. Geophys. Res.* 110, C12025, 23 pp., doi: 10.1029/2004JC002778.

Panizzo, A., De Girolamo, P., Di Risio, M., Maistri, A. & Petaccia, A. (2005b). Great landslide events in italian artificial reservoir, *Natural Hazards and Earth System Science* 5: 733–740, SRef–ID: 1684–9981/nhess/2005–5–733.

Pelinovsky, E. & Poplavsky, A. (1996). Simplified model of tsunami generation by submarine lanslides, *Phys. Chem. Earth* 21(1–2): 13–17, doi: 10.1016/S0079-1946(97)00003-7.

Peregrine, D. H. & Williams, S. M. (2001). Swash overtopping a truncated plane beach, *J. Fluid Mech.* 440: 391–399, doi: 10.1017/S002211200100492X.

Prins, J. E. (1958). Characteristics of waves generated by a local disturbance, *Trans. Am. Geoph. Un.* 39(5): 865–874.

Renzi, E. & Sammarco, P. (2010). Landslide tsunamis propagating around a conical island, *J. Fluid Mech.* 650: 251–285, doi: 10.1017/S0022112009993582.

Risley, J. C., Walder, J. S. & Denlinger, R. P. (2006). Usoi dam wave overtopping and flood routing in the Bartang and Panj Rivers, Tajikistan, *Natural Hazards* 38(3): 375–390, doi: 10.1007/s11069-005-1923-9.

Russell, J. S. (1838). Report of the committee on waves, *Rep. 7th Meeting British Association Advanced Science, Liverpool, 1837*, pp. 417–496.

Russell, J. S. (1845). Report on waves, *Rep. 14th Meeting British Association Advanced Science, York, 1844*, pp. 311–390.

Sæ levik, G., Jensen, A. & Pedersen, G. (2009). Experimental investigation of impact generated tsunami; related to a potential rock slide, western norway, *Coast. Eng.* 56(9): 897–906, doi: 10.1016/j.coastaleng.2009.04.007.

Savage, S. B. & Hutter, K. (1989). The motion of a finite mass of granular material down a rough incline, *J. Fluid Mech.* 199: 177–215, doi: 10.1017/S0022112089000340.

Sue, L. P., Nokes, R. I. & Walters, R. A. (2006). Experimental modeling of tsunami generated by underwater landslides, *Science of Tsunami Hazards* 24(4): 267–287.

Sue, L. P. (2007). *Modelling of tsunami generated by submarine landslides*, PhD Thesis, University of Canterbury, Christchurch, New Zealand.

Synolakis, C. E. (1986). *The runup of long waves*, PhD thesis, California Institute of Technology, Pasadena, California.

Synolakis, C. E. (1987). The runup of solitary waves, *J. Fluid Mech.* 185: 523–545, doi: 10.1017/S002211208700329X.

Synolakis, C. E. (1990). Generation of long waves in the laboratory, *J. Waterway, Port, Coastal, Ocean Eng.* 116(2): 252–266, doi: 10.1061/(ASCE)0733-950X(1990)116:2(252).

Synolakis, C. E., Bardett, J. P., Borrero, J. C., Davies, H. L., Okal, E. A., Silver, E. A. & Sweet, S. Tappin, D. R. (2002). The slump origin of the 1998 papua new guinea tsunami, *Proc. R. Soc. Lond. A* 458(2020): 763–789, doi: 10.1098/rspa.2001.0915.

Synolakis, C. E. & Bernard, E. N. (2006). Tsunamis science before and beyond boxing day 2004, *Phil. Trans. R. Soc. A* 364(1845): 2231–2265, doi: 10.1098/rsta.2006.1824.

Tadepalli, S. & Synolakis, C. E. (1994). The run-up of n-waves on sloping beach, *Proc. R. Soc. Lond. A* 445: 99–112.

Tinti, S., Manucci, A., Pagnoni, G., Armigliato, A. & Zaniboni, F. (2005a). The 30 december

2002 landslide-induced tsunamis in stromboli: Sequence of the events reconstructed from the eyewitness accounts, *Nat. Hazards Earth Syst. Sci.* 5: 763–775, SRef–ID: 1684–9981/nhess/2005–5–763.

Tinti, S., Maramai, A., Armigliato, A., Graziani, L., Manucci, A., Pagnoni, G. & Zaniboni, F. (2005b). Observations of physical effects from tsunamis of december 30, 2002 at stromboli volcano, southern italy, *Bull. Volcanol.* 68(5): 450–461, doi: 10.1007/s00445-005-0021-x.

Tuck, E. O. & Hwang, L. S. (1972). Long wave generation on a sloping beach, *J. Fluid Mech.* 51(3): 449–461, doi: 10.1017/S0022112072002289.

Ursell, F. (1952). Edge waves on a sloping beach, *Proc. R. Soc. Lond. A* 214(1116): 569–593, doi: 10.1098/rspa.1952.0152.

Walder, J. S., Watts, P., Sorensen, O. E. & Janssen, K. (2003). Tsunamis generated by subaerial mass flows, *J. Geophys. Res.* 108(B5), 2236, 19 pp., doi: 10.1029/2001JB000707.

Watts, P. (1998). Wavemaker curves for tsunamis generated by underwater landslides, *J. Waterway, Port, Coastal, Ocean Eng.* 124(3): 127–137, doi: 10.1061/(ASCE)0733-950X(1998)124:3(127).

Watts, P. (2000). Tsunami features of solid block underwater landslides, *J. Waterway, Port, Coastal, Ocean Eng.* 126(3): 144–152, doi: 10.1061/(ASCE)0733-950X(2000)126:3(144).

Watts, P., Imamura, F. & Grilli, S. (2000). Comparing model simulations of three benchmark tsunami generation cases, *Science of Tsunami Hazards* 18(2): 107–123.

Watts, P. & Grilli, S. T. (2003). Underwater landslide shape, motion, deformation, and tsunami generation, *Proc. of the Thirteenth International Offshore and Polar Engineering Conference, Honolulu, Hawaii, USA, May 25–30*, pp. 364–371.

Watts, P., Grilli, S. T., Tappin, D. R. & Fryer, G. J. (2005). Tsunami generation by submarine mass failure. I: predictive equations and case studies, *J. Waterway, Port, Coastal, Ocean Eng.* 131(6): 298–310, doi: 10.1061/(ASCE)0733-950X(2005)131:6(298).

Wiegel, R. L. (1955). Laboratory studies of gravity waves generated by the movement of a submerged body, *Trans. Am. Geoph. Un.* 36(5): 759–774.

Wiegel, R. L., Noda, E. K., Kuba, E. M., Gee, D. M. & Tornberg, G. F. (1970). Water waves generated by landslides in reservoir, *J. of the Waterways and Harb. Div., Proc. of the Am. Soc. of Civ. Eng.* 96(WW2): 307–333.

Yeh, H., Liu, P., Briggs, M. & Synolakis, C. (1994). Propagation and amplification of tsunamis at coastal boundaries, *Nature* 372: 353–355, doi: 10.1038/372353a0.

Yim, S. C., Yuk, D., Panizzo, A., Di Risio, M. & Liu, P.-F. (2008). Numerical simulation of wave generation by a vertical plunger using rans and sph models, *J. Waterway, Port, Coastal, Ocean Eng.* 134(3): 143–159, doi: 10.1061/(ASCE)0733-950X(2008)134:3(143).

Zweifel, A. (2004). *Impulswellen: Effekte der Rutschdichte und der Wassertiefe*, PhD thesis, ETH Zurich, Zurich (in German).

Zweifel, A., Hager, W. H. & Minor, H.-E. (2006). Plane impulse waves in reservoirs, *J. Waterway, Port, Coastal, Ocean Eng.* 132(5): 358–368, doi: 10.1061/(ASCE)0733-950X(2006)132:5(358).

7

The Vulnerability of a City – Diagnosis from a Bird's Eye View

Hannes Taubenböck
German Aerospace Center (DLR),
German Remote Sensing Data Center (DFD)
Germany

1. Introduction

When the tsunami in the Indian Ocean on 26 December 2004 hit the city of Banda Aceh on the island of Sumatra, Indonesia, neither the city administration nor its inhabitants, nor national or international organisations were prepared. Approximately 60.000 of the 260.000 inhabitants died, leaving other 30.000 homeless and causing an enormous impact on the local economy.

In the aftermath of this event tsunami early warning system were developed and are operated today (e. g. the German Indonesian Tsunami Early Warning System - GITEWS (Lauterjung, 2005)). However, the problem of earthquake or tsunami prediction in a deterministic sense has not been solved yet (Zschau et al, 2002). Thus, an end-to-end tsunami early warning system includes not only the tsunami warning, but also the assessment of vulnerability, perception studies, evacuation modeling, eventually leading to technical requirements for monitoring stations and recommendations for adaptation and mitigation strategies (Taubenböck et al., 2009a).

In this study we address several specific questions on the capabilities of one discipline - remote sensing - for diagnosing the multi-faceted and complex vulnerability of a city:

- Which remotely sensed data sets are appropriate analyzing vulnerability in highly complex urban landscapes?
- What capabilities and limitations does urban remote sensing have regarding mapping, analysis and assessment of risks and vulnerability?
- How can interdisciplinary approaches extend the applicability of earth observation?

In general, earth observation serves as an independent and relatively cost-effective data source. Its distant view enables clarity - a principle especially true for remote sensing (Dech & Messner, 2005). The bird's eye view provides an up-to-date and areas-wide spatial overview and at the same time capturing the small-scale heterogeneity of urban landscapes. Especially in explosively sprawling and changing urban areas of developing countries remote sensing often serves as the only reliable data source.

In recent years, airborne and spaceborne remote sensing systems and image analysis techniques have developed to an extent where civil and commercial earth observation (EO) instruments can contribute significantly in supporting the management of major technical and natural disasters as well as humanitarian crisis situations (Voigt et al., 2007).

The use of various satellite data sets and remote sensing methods for post catastrophe damage detection was and still is topic of several studies (Eguchi, et al., 2000; Huyck et al., 2003, Saito & Spence, 2003; Pesaresi, Gerhardinger & Haag, 2007). As example Gusella, Huyck & Adams (2008) used an object-oriented approach to count the number of collapsed buildings after the earthquake in Bam, Iran.
An operational service at DLR's German Remote Sensing Data Center (DFD), the "Center for Satellite Based Crisis Information" (ZKI) supports decision-makers and relief organisations. Its function is the rapid acquisition, processing and analysis of satellite data and the provision of satellite-based information products during and after natural and environmental disasters, for humanitarian relief activities, as well as in the context of civil security (www.zki.dlr.de).
In recent years, the research also focused on assessing vulnerability and risk of systems before an expected event aiming at the development of strategies of preparedness (Schneiderbauer, 2007). Earth observation data are widely used for mapping or up-dating building inventory data (Miura & Midorikawa, 2006; Sarabandi et al, 2008) as input for multi-criteria GIS-based vulnerability studies (Rashed & Weeks, 2003).
The work presented here is embedded in the Numerical Last Mile Tsunami Early Warning and Evacuation Information System (Last-Mile) project (Birkmann et al., 2008a; www.last-mile-evacuation.de). The city of Padang, Indonesia, has been indicated as one of the most plausible localities for a tsunami of disastrous proportions in the near future (Borrero et al., 2006; McCloskey et al., 2010). Off the coast of Padang, the Sunda Arc marks an active convergent plate boundary. Thus, the city is located in a zone of extreme high probability to severe earthquakes and potential triggered tsunamis. According to the catalogue prepared for tsunamis in the Indian Ocean, which includes about ninety tsunamis, 80% of the tsunamis are from Sunda arc region, where on an average, tsunamis are generated once in three years (Rastogi and Jaiswal, 2006). Padang is the capital city of the Sumatera Barat province in West Sumatra, and with almost one million inhabitants the third largest city on the island of Sumatra, Indonesia.
The complex and dynamic urban system of Padang features a high concentration of population, infrastructure and economic values. Padang features supra-regional relevance with an international airport, a port as well as binding to the rail network. Thus, the city possesses a central economical role for the coastal region and the mountainous back-country. It is located directly on the coast and partially sited beneath the mean sea level.

2. Vulnerability – a conceptual framework

Vulnerability is a term of such broad use as to be almost useless for careful description at the present, except as a rhetorical indicator of areas of greatest concern (Timmermann, 1981). A review of the literature indicates a number of contrasting definitions of what vulnerability means, as well as numerous conflicting perspectives on what should or should not be included within the broad assessment of vulnerability in cities. Definitions and concepts on 'vulnerability' evolve from different schools and disciplines - e. g. social scienes (Bohle, 2001; Cannon et al., 2003), business sciences (Briguglio, 2003), natural sciences (Correll et al., 2001), climatology (IPCC, 2001), engineering (Correira et al, 1987) or disaster management (Blaikie et al, 1994). The various disciplines and perspectives on vulnerability show the multidimensionality of the concept. Holistic concepts try to incorporate its multidimensionality (Bogardi & Birkmann 2004; Füssel, 2005; Taubenböck et al, 2008). For this paper, we adapt the definition of the UN: Subject to the considered period of time, the geographic location (system) and a scale-dependent reference system, vulnerability is the condition determined by physical, social, economic and environmental factors or processes, which increase the susceptibility of a community to the impact of hazards (UN/ISDR, 2004). Vulnerability is also part within the higher-ranking risk concept. The UN (2004) determined that the risk to a particular system has two factors. One factor is the "hazard" itself, which is a potentially damaging physical event, phenomenon, or human activity that is characterized by its location, intensity, frequency, and probability. The second factor is "vulnerability," which denotes the relationship between the severity of hazard and the degree of damage caused. Thus, risk results from a future interplay of a hazard and the various components defining vulnerability. The effect of a natural hazard on the objects or people of a particular area exemplify the complex interrelationships and emerging domino effects.

However, we are still dealing with a paradox: we aim to measure vulnerability, yet we cannot define it precisely (Birkmann, 2006). The various approaches, concepts and definitions are on an abstract level and objectively quantification or measurement thus is a complex task. Indicator systems are usually used to resolve the abstract term 'vulnerability' into measurable parameter (Bollin & Hidajat, 2006).

Components specifying vulnerability include the physical, demographic, social, economic, ecological and political aspects contributing and adding up to the holistic conceptual idea (Table 1). This conceptual idea adapted for this study uses a meta-framework applicable to various systems (e.g., urban areas), various elements or attributes within a system (buildings, people, environmental services, etc.), various scales (local, regional, national, global), and various hazards (earthquake, tsunami, landslides, climate change, etc.). In general the various components described above suggest the complexity of dynamic interactions and interdependencies evolving in a system in case of a disastrous event. This highlights the need for multidisciplinary scientific approaches to converge on the problem of holistic identification of vulnerability (Taubenböck et al., 2008).

To converge on the problem of assessing the complexity of risk and vulnerability, the rather abstract components of this meta-framework need further partitioning. Specifications are dependent on the considered system and the type of hazard. In our case, indicators are developed for the example of an "urban system prone to a tsunami hazard". A consistent systematization itemizes the components into a set of measurable indicators contributing to all the various stages of the risk management cycle. As example, higher-ranking physical vulnerability is hard to determine, but one of many indicators such as number of affected buildings, enables quantification of the situation. An indicator is defined as a variable, which is a representation of an attribute, such as quality and/or characteristics of a system (Gallopin, 1997). The quality of the indicator is determined by its ability to indicate the characteristics of a system which are relevant to the underlying interest determined by the goal or guiding vision (Birkmann, 2006b). The result is an arbitrary number of measurable indicators contributing aspects to the associated component and to the ranking in the holistic conceptual framework (Table 1).

The overall conceptual framework serves as the basic outline to identify capabilities and limitations of remote sensing data and methods to contribute to the holistic concept.

Meta – Framework		Itemization based on the considered hazard, system, time and scale		
Conceptual Framework	**Components**	**Causes**	**Indicators / Variables**	**Index**
Risk – HAZARD	Natural hazard, human threat, phenomenon	Earthquakes, volcanic eruptions, floods, droughts, landslides, cyclones, tropical storms, terrorist attacks, etc.	Magnitude, intensity, spatial exposure, probability of occurrence, duration, time	Indexing of indicators
Risk – HAZARD	Secondary threats, aftereffects	Landslides, tsunamis, fires, etc.	Height, slope, orientation, soil type, spatial exposure, etc.	
Risk – Vulnerability (Exposure x Susceptibility / Coping Capacity)	Physical Vulnerability	Location	Accessibility, distances, etc.	
		Structural exposure	Number of structures, built-up density, building height, building material and construction type, roof type, building age, urbanization rate, sealed areas, open spaces, etc.	
		Critical infrastructure	Street- and infrastructure network, public transport, communication lines, pipelines, supply, lifelines, etc.	
	Demographic Vulnerability	Population structure	Total population, population density distribution, day- and night-time distribution, age pattern, etc.	
		Population development	Population growth rates, migration rates, etc.	
	Social Vulnerability	Social status	Education, public awareness, health, social network, gender, etc.	
		Accessibility to and supply of local facilities	Hospitals, schools, fire brigades, shelters, etc.	
	Economic Vulnerability	Individual financial potential	Per-capita income, insurance, property, unemployment rate, etc.	
		Governmental potential	Local relief budget, gross national product, help programmes and organisations, inflation, Human Poverty Index (HPI), etc.	
	Political Vulnerability	Decision structure	Political system, willingness, early warning systems, crisis and information management, etc.	
	Ecological Vulnerability	Natural resources	Water supply and balance, agriculture, forests, etc.	

Table 1. A hierarchical holistic framework conceptualizing hazards, vulnerability and risk to derive a selection of measurable indicators for a specified system (adapted from Taubenböck et al., 2008)

3. Diagnosis of a city's vulnerability to tsunami hazards using earth observation

3.1 Remotely sensed data sets

With respect to the holistic idea of vulnerability and risk concepts the question arises why remotely sensed data sets should be used addressing this topic? In general, because remote sensing provides spatially consistent data sets that cover large areas with both high spatial resolution and high temporal frequency. In particular in developing countries where geo-information is rarely available earth observation is often the only up-to-date, area-wide and independent data source and thus indispensable.

However, the capabilities of various sensors reach far beyond the obvious benefits available at popular platforms like Google Earth or Bing. They provide reflective responses all along the electromagnetic spectrum which enables detection of objects or patterns of the earth's surface and their condition. Remote sensing platforms and sensors cover many spatio-temporal dimensions, with a flexible repetition rate and in various scales ranging from spatially detailed analysis on single buildings or building block level to global studies on continental scales for various urban applications (Taubenböck & Dech, 2010):

Very high resolution multispectral optical satellite data from i. e. GeoEye I, WorldView I & II, Ikonos (cp. Fig. 3), Quickbird, CARTOSAT, SPOT or RapidEye feature a geometric resolution ranging from 41 cm to 5 m, which is feasible for the small-scale objects typical in urban environments. A highly detailed spectral coverage of the electromagnetic spectrum by hyperspectral sensors like the airborne sensors HyMap, AVIRIS (or in the near future the German satellite EnMAP) enable derivation of surface materials or land surface temperature in high geometric resolution from 2m - 30m. Laser Altimeter (LIDAR) is also an optical remote sensing technology for highly detailed profiles of 3D elevations of the earth's surface; stereo images can also be used for that purpose. In terms of temporal analysis, optical sensors such as Landsat (since 1972), SPOT (1986), or IRS (1988) enable monitoring and detection of changes with reduced spatial resolution.

In addition to optical systems, SAR antennas operate almost independently of meteorological conditions and solar illumination. Thus, they are especially in often cloudy areas of high importance. There are, at present several SAR sensors in space as well as data from past missions offering a broad and global observation of the planet (e.g., ERS-2, RadarSat, Envisat, TerraSAR-X, and the space shuttle) in different frequencies, polarizations, and geometric resolutions. In addition, interferometric digital elevation model (DEM) can be derived, such as the X- and C-band data from the Shuttle Radar Topography Mission (SRTM) are used to analyze the terrain situation. It is a surface model with a pixel-spacing of 25 meters, and thus enables a rather coarse overview. Future missions, like TanDEM-X will provide a higher geometric detail of up to 12 meters. Even aerial acquisitions are possible due to the full-time imaging potential of radar. Furthermore, new radar satellites such as TerraSAR-X, CosmoSkyMed, and Advanced Land Observing Satellite (ALOS) enable the extraction and analysis of urban structures based on geometric resolutions up to 1 m (Roth et al, 2005).

Satellites from EUMETSAT enable to monitor current weather situations to include predictions of heavy rainfall or cyclones, etc.; however, this field of application is not part of our study.

3.2 Mapping

The data and images presented and described above are *data* not *information*. The strength of remote sensing with its synoptic overview allows independent, fast, up-to-date, area-wide and relatively cost-effective transformation of data (or images) into information (Taubenböck & Dech, 2010). Making use of a vast amount of methodologies - e.g. statistical-, neural-, fuzzy classifiers - for automatic information extraction for particular data sets (Mather, 2004) this transformation leads to application-driven products. As example, we used an object-oriented, multi-level, hierarchical classification approach (Taubenböck et al, 2010) to derive the land-cover of Padang from Landsat as well as Ikonos data in combination with a digital surface model (DSM). Beyond the spectral information within the remotely sensed data, the fuzzy logic-based object-oriented approach allows to use shape, neighbourhood, hierarchical and context information (cp. Fig. 1).

A pixel-based classification approach extracting urban and non-urban structures from a single TerraSAR–X image has also been applied (Esch et al., 2010). The urban footprint is extracted by analysing the speckle characteristics of a TerraSAR-X scene based on an estimation of the local coefficient of variation and the fading texture of the whole scene. The basic concept is to extract reliable urban features as seed points, which are represented by bright point scatterers (corner reflectors) for the urban footprint classification. The 'urban seeds' depict especially the reflection information of vertical structures. Urban seeds are extracted from intensity and texture, derived from a two-tiered speckle divergence calculation with a 9x9 and a 35x35 window. To derive the urban footprint, analysis techniques are used to densify and generalize the detected urban seeds (Esch et al., 2010).

For multi-temporal urban growth analysis the urban footprint classifications from Landsat of the years 1989 and 2000 and from TerraSAR-X of 2010 were combined. Post-classification comparison was found to be the most accurate procedure and had the advantage of indicating the nature of the changes (Mas, 1999).

Fig. 1 shows various results: Multi-temporal urban change detection of a multi-sensoral approach using Landsat and TerraSAR-X data of the coastal urban area of Padang (Fig 1a). The result allows detecting, quantifying and localizing spatial urbanization over time on urban footprint level. The result displays extensive urban sprawl as well as re-densification in the urban center in Padang over the last two decades. The geo-spatial product allows a first assessment of areas of high vulnerability due to natural hazards.

For mapping the complexity of the urban environment, multi-sensoral data processing of high resolution optical satellite data in combination with a high resolution airborne digital surface model allows the derivation of a 3D city model (Fig. 1d). The results are eight classes mapping the urban morphology—houses, streets, sealed areas, grassland, trees, wetland, bare soil, and water. This is e.g. relevant for identification and quantification of the building stock, built-up density, floor-space index, building heights, vegetation fraction, infrastructure, or undeveloped areas (Fig 1b & c). In terms of vulnerability an inventory of the urban morphology is an essential information basis for urban planning and risk reduction strategies.

3.3 Analysis

Beyond the capabilities of mapping the city remotely sensed data sets and results can be used for analyzing the urban structure as well as the natural conditions in and around the urban landscape in the context of hazards and vulnerability.

Regarding hazards, e. g. the inclination of the slope is computed using a digital elevation model (SRTM) with a tangent equation taking neighbouring pixel height values into account. The result projects the spatial distribution of the steepness of slopes, which serve as

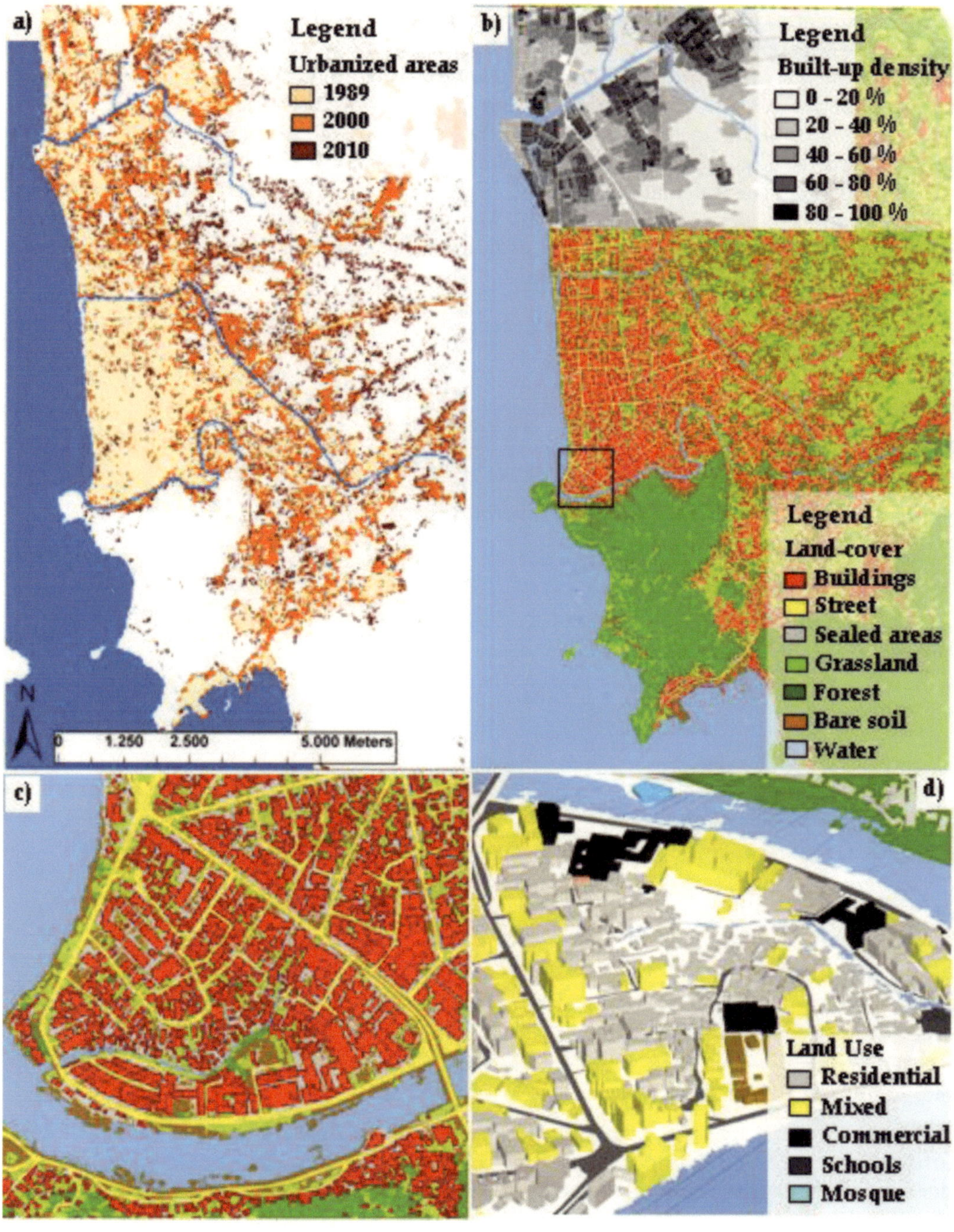

Fig. 1. Classification of multi-sensoral remotely sensed data a) Change detection from 1989 until 2010 on urban footprint level b) Land cover classification on high geometric resolution and built-up density derivation on block level c) Detail of the land-cover classification d) 3-D city model and integration of land use

one component to identify areas at categorical risk for landslides as secondary threat of an earthquake. Of course, steepness is not solely decisive for the occurrence of landslides, but it does provide a spatial containment to possible areas. Figure 2a displays critical slopes with an inclination of 10 degrease and higher along the coastal area of Padang.
The height information in combination with the orientation of the slopes calculated from the DEM is also used to approximately assess affected areas in case a tsunami wave hits the urban coastline. The calculation is based on the coastline extracted from a land cover classification from Ikonos data (cp. Fig 1b). The DEM allows a coarse identification of areas prone to a potential tsunami wave. Figure 2a indicates areas lower than 4 meter above sea level and thus of highest spatial exposure.

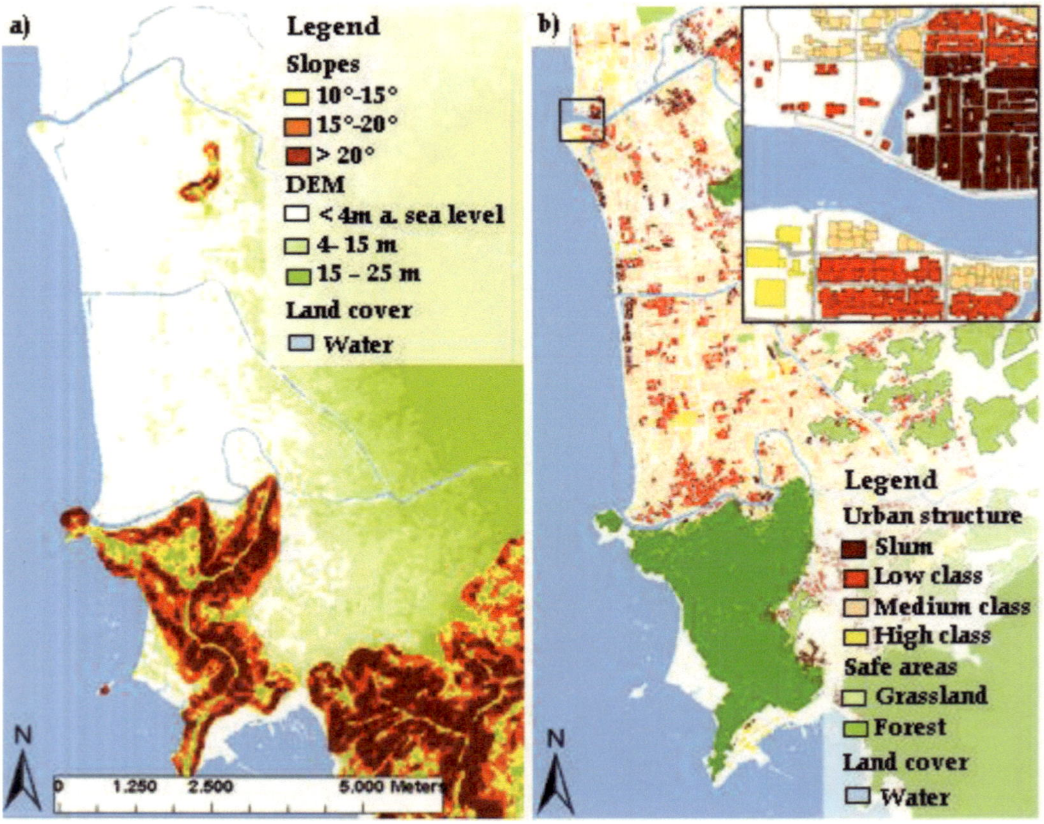

Fig. 2. Physical parameters - Structural types

Risk and vulnerability is an inherently spatial problem. Cities contain physically different urban structures. Subdividing the complex urban landscape into homogeneous units aims at deriving meaningful entities for vulnerability assessment. Therefore the land-cover classification (cp. Fig. 1b) and the physical parameters, such as built-up density, location, alignment or heights, are used for a semantic classification of urban structure types. As example, the building density is displayed in figure 1b for a part of the city of Padang. The idea of semantic classification aims at a first assumed evaluation of vulnerability between physically homogeneous sectors within the city's landscape.

The approach for semantic classification is generic, aiming at transferability on any urban area throughout the world with similar physical parameters available. Therefore, we used descriptive statistic values as Quartile (Q1, Q3), Median (Med) or Mean (M) to subdivide the different classes. The six resulting semantic classes are defined by the following statistical borders. An example, we classify 'informal settlements' using built-up density values higher than the third quartile (Q3) of the complete spectrum of the built-up density values classified in Padang. Analogous, the buildings of informal settlements are assumed to be smaller than Q3 and lower than Q1 of the particular spectrum of values classified for the city of Padang. Utilizing these statistical parameters the classification is not affected by cultural or regional characteristics of urban morphology occurring worldwide. With respect to this methodology, only the semantic nomenclature has to be adjusted on the particular structures and locations.

Figure 2b displays six different urban structure types- 'informal settlements (slums)', 'suburbs', 'low class areas' (LC), 'middle class areas' (MC) and 'high class areas' (HC). As an example, slum areas are defined by the highest built-up density measured within the urban environment with mostly one storey buildings, with the smallest buildings sizes. With the nomenclature 'slum' the classification by solely physical parameters sets a first hint at building quality and thus, stability and also in the direction of assumed socioeconomic relevance.

3.4 Assessment

The city, seen as a human product, is the physical and architectonic reflection of the society that created it (Gonzalez & Medina, 2004). Thus, the data and products indirectly contain additional information. The products presented above refer to land cover, which is defined as the physical material on the surface, while land use refers to the human activity that takes place on, or makes use of that land. The fundamental problem for remote sensing is that while there is often a relatively simple and direct relationship between land cover type and detected spectral reflectance, the same is seldom true for land use. Nevertheless, physical parameters like building sizes and heights, roof types, etc. often indirectly correlate to the usage of the buildings.

As figure 3 visualizes the buildings classified as industry or university clearly differ in alignment, size and shape from buildings classified as residential areas. Thus, the physical parameters of the urban morphology can be used for systematic detection of probable non-residential areas. The type of usage, like in this case industry or university needs to be integrated from external knowledge, like e. g. field surveys or data bases from the city.

Furthermore it is obvious, that the knowledge on this physical and semantic information on urban morphology (building size, height, occupancy, location, semantic class) allows indirect conclusions about the population distribution, even on its spatiotemporal shift within the course of a day.

On the basis of the punctual information of population data (1000 households) collected through fieldwork (UNU-EHS, 2008) the mathematical concept is based on a bottom-up extrapolation. For the different semantic classes the particular punctual information is averaged to get a characteristic number of inhabitants per m^2 per semantic class. The results for the residential areas are listed in table 2. In addition, the survey included other occupancies, like industry, offices or markets. These results are displayed in table 3. With these external data sets the extrapolation is calculated with respect to semantic classes and the knowledge of living spaces per individual house in the particular areas.

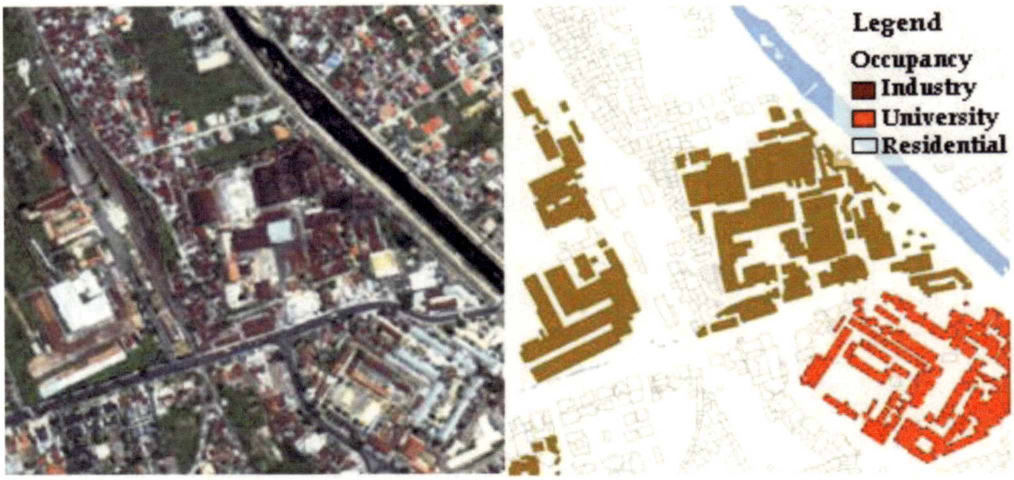

Fig. 3. Structural differences between various occupancies of buildings visualized in Ikonos imagery and in the classification result

Semantic Classes	Living space per inhabitant (Li)	
	day	night
class 1: "SLUM"	49.2 m^2/p	26.1 m^2/p
class 2: "LC"	93.3 m^2/p	47.5 m^2/p
class 3: "MC"	133 m^2/p	54.5 m^2/p
class 4: "SUBURB"	48.2 m^2/p	31.5 m^2/p
class 5: "HC"	-	-
class 6: "residuals"	58 m^2/p	41 m^2/p
Without semantic classes	93.3 m^2/p	47.0 m^2/p

Table 2. Mean living space for various semantic classes of residential areas

Figure 4 shows a bottom-up distribution of punctual population data for the entire city of Padang. Figure 4a displays population at the morning per building – e. g. localizing high population density in market areas, schools, etc.. Figure 4b shows the population flux between day- and night-time and reveals a spatial shift of vulnerability, e. g. the highly prone market areas at day-time are almost empty and thus less vulnerable at night-time. Beyond the indirect assessment of time-dependent population assessment the semantic classification already hints at socio-economic parameters of the people residing in these areas. As an example, densely populated areas generally correlate to high poverty and low quality housing, economic and social components of vulnerability.

A study proved that also a correlation between physical urban morphology and socioeconomic parameters of the people is basically given (Taubenböck et al., 2009b). The sample parameters "income per month" and "value of property" were related to the various semantic classes. The results confirm the hypothesis that the physically classified "informal settlements" and "suburbs" represent lowest income and property values. With slight variations – a linear correlation can not be detected in such a complex environment – a consistent rising of income levels is revealed to "low, middle and high class" areas. Thus, remote sensing may provide a cost-effective method to reduce, but not replace extensive ground data collection and provide spatial value-adding via interdisciplinary cooperation.

Semantic Classes	**Working spaces per inhabitant (Li)**	
	day	**night**
office, bank, and other commercial usage	20 m^2/p	1000 m^2/p
shopping mall, market	10 m^2/p	100 m^2/p
industrial facilities	10 m^2/p	30 m^2/p
harbour, railway station, sports facility	20 m^2/p	100 m^2/p
mosque	-	1000 m^2/p
school, university	10 m^2/p	1000 m^2/p

Table 3. Mean space for various semantic classes of commercially used areas

Besides interdisciplinary contributing to the demographic, social or economic sphere of the vulnerability framework, physical stability analysis of the structures is an essential step to identify physical vulnerability. In fast changing, complex and small-scale urban environments the identification and localization of vulnerable as well as stable structures is therefore the basic element to implement mitigation strategies and to reduce vulnerability. With almost 100.000 buildings within the study area (cp. Figure 1 & 2) surveying of the complete building stock is out of reach for time and cost reasons.

An interdisciplinary analysis of civil engineering and remote sensing aims at overcoming this problem: On the one hand civil engineering allows a highly detailed stability analysis of individual buildings by an extensive house by house inspection. For Padang, 500 buildings have been surveyed gathering data on their physical characteristics like e. g. height, material, existence of columns, foundation of main structures, existence of stirrup or/and main beam or damage due to previous earthquakes. In combination with physical tests, like the hammer test on supporting pillars, the stability of the structures was assessed individually. An additive approach was applied for the various physical parameters and a normalized vulnerability (stability) index calculated (100 = stable structure; 0 instable structure). The sample selection of the surveyed buildings aimed at a complete coverage of differing housing types -from small, low height shacks to large, high-rise building types, equally distributed within the entire city area of Padang. 500 buildings out of almost 100.000 are comparably few. The challenge consists of a value-adding combination of civil engineering and remote sensing.

The capabilities of remote sensing are area-wide, up-to-date, and quantitative derivation of physical characteristics of the complete building stock. On the other hand the capabilities of remote sensing allow area-wide, up-to-date, and quantitative derivation of physical characteristics of the complete building stock, with comparatively to in-situ surveys limited knowledge on structural parameters. Structural parameters from remote sensing are limited to roof colour and type, height and size. The high variability of roof colours does not enable any correlation to the stability of the structures. Thus, we plotted the stability of the surveyed buildings to their size combined with height (volume of the structure) against each other (Fig. 5).

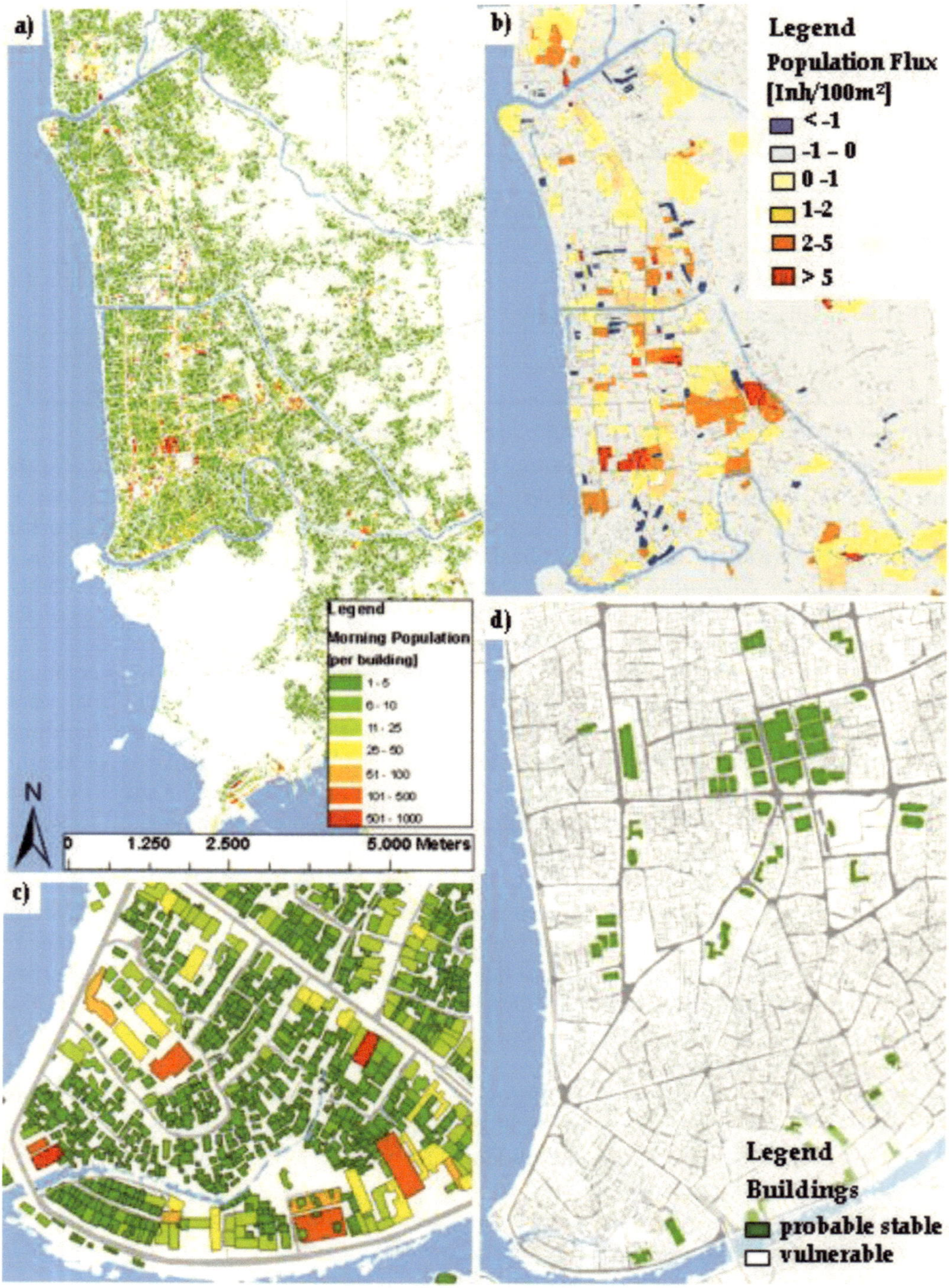

Fig. 4. Population distribution in dependence of daytime, population flux and assessment of buildings suitable for vertical evacuation

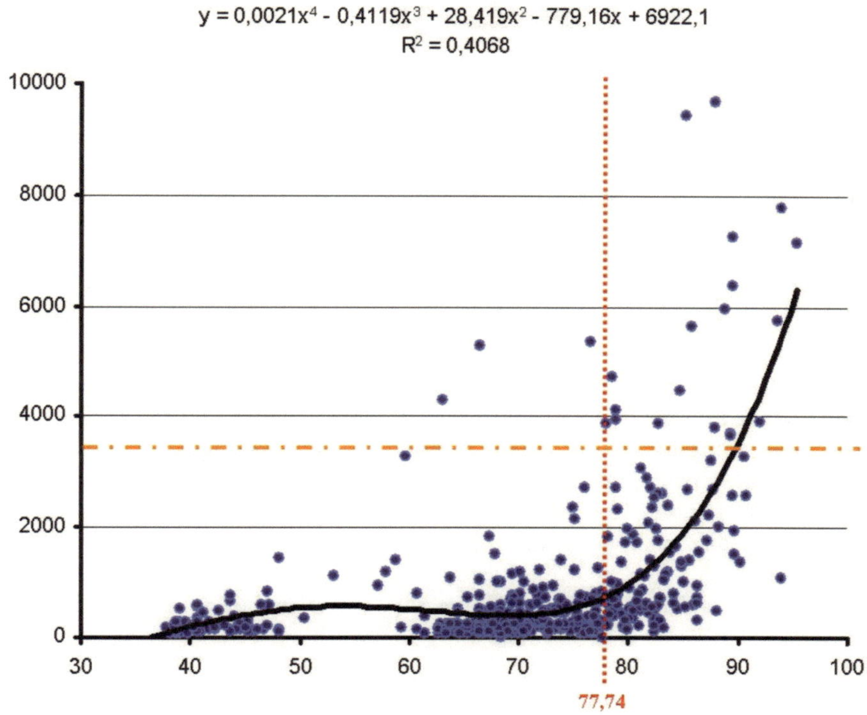

Fig. 5. Correlation of building volumes derived from remote sensing with the vulnerability analysis of the particular buildings from a civil engineering survey

The result does not show an obvious correlation, but reveals that buildings with a higher stability (vulnerability value) predominantly show a higher building volume. In our case we identified an 85.71 % probability for a stable structure for buildings with a volume higher than 3500 m^3. The extrapolation on the complete building stock using the 3-D city model identified 183 buildings with a high probability to be appropriate for vertical evacuation. The interdisciplinary analysis thus allows reducing the amount of buildings to be surveyed for a final official designation as vertical evacuation shelter from 100.000 to 183, which is feasible for surveys.

3.5 Accuracy assessment

The synoptic overview of remote sensing products in the previous chapters show area-wide and spatially highly detailed information extraction, but the accuracy of cadastral data sets is not achieved.

On urban footprint level, the large amount of mixed spectral information in such a coarse ground resolution - caused by multiple land covers within the lowest granularity of one pixel - the accuracy is limited. But for the requirement of mapping the large city footprint, its spatial dimension and the spatial developments over the years, the Landsat and TerrSAR-X images provide enough information for an assessment of urban change (cp. Fig 1). The accuracy of every classification result was assessed with 250 randomly distributed pixels. Due to missing ground truth data at historic times, the accuracy was assessed visually by comparing classification results to the Landsat data. Thus, this assessment of accuracies already includes uncertainties. Even so, the high overall accuracies range from 86% to 93% correctly classified pixels. For TerraSAR-X data a comparison with the high resolution settlement mask revealed an accuracy of the urban footprint classification in 2010 of 79.2 % (Taubenböck et al, 2011).

On highest resolution the implemented automatic object-oriented, hierarchical classification methodology showed an accuracy of 84.2% correctly detected houses. Manual revision enabled to raise the accuracy to 97%. The remaining land cover classes of bare soil, grassland, forest, sealed areas, streets, water range between 80% and 95% accuracy. Consequently, the inferred products of the land cover classification for analysis of the urban landscape or the surface (e. g. building density, vegetation fraction, etc.) inherit the errors from the classification results. The accuracy of the land use classification has been tested with field observations. As the algorithm applies residential usage by default and changes the occupancy only if the described structural components apply, accuracy assessment has been carried out for non-residential buildings. The result reveals that the indirect derivation of non-residential occupancies is with 92.4 % correctly identified structured highly accurate. But, remote sensing shows clearly limitations regarding the amount of not detected non-residential areas, due to e. g. business, office or working activities in buildings structurally similar to residential neighbourhoods. Thus, the producer's accuracy is with 47.4 % comparatively low.

The accuracy for the indirect assessment of population distribution, the population flux or the building stability from remotely sensed products has been assessed as well. For the population distribution, accuracy assessment was performed by means of 500 remaining buildings from a field survey not used for the calibration of the extrapolation algorithm. The assessment shows the compliance of field work results from the questionnaire with the calculated population per house based on the extrapolation algorithm. Table 5 presents a comparison between extrapolation results: In one section is the accuracy achieved by extrapolation using spatial specification by semantic classes as listed. This shows the accuracy of the remaining field survey data for a specific semantic class with the extrapolated data. In the other section, the extrapolation based on the mean value of living space per person for all surveyed residentially used buildings is presented as comparison.

The extrapolation using spatial structuring of the complex urban environment results in significant higher accuracies than averaging the entire city.

Classes	Semantic classes	STD	Mean value	STD
class 1 - SLUM	94.21 %	1.91	63.08 %	2.43
class 2 - LMC	89.48 %	3.32	73.11 %	3.95
class 3 - MC	83.82 %	6.79	80.03 %	6.13
class 4 - SUBURB	87.56 %	7.82	79.56 %	9.12
class 5 - HMC	-	-	-	-
class 6 - residuals	74.27 %	9.41	68.09 %	10.24
Complete (AOI)	86.73 %	5.71	-	-
Without spatial differentiation (AOI)	-	-	72.12 %	9.34

Table 5. Accuracy assessment of the extrapolation algorithm and comparison of the approach utilizing semantic classes and the approach utilizing a mean value

For the building stability assessment a probability of 85.71% has been calculated that the extrapolation applies to stable shelters for vertical evacuation. However, this accuracy assessment should not lead to a definite designation of the identified potential buildings as vertical shelters, but as a containment of possibilities.

4. The earthquake in Padang on the 30th September 2009

On September 30, 2009, a severe earthquake took place in the Indian Ocean with a magnitude of 7.6 and several aftershocks. The epicentre was registered about 50 km north-eastern of Padang in a depth of 85 km. Heavy shocks caused the collapse of many buildings and bridges, fires broke out and major parts of the technical infrastructure failed. More than 770 people died, more than 2100 are injured (information of October 2, 2009). A second earthquake with a magnitude of 6.6 occurred hours later on October 1 near the town of Sungai Penuh. In the capital city of Padang, 383 people were killed, public services such as electricity and water were disrupted, and major markets and transport routes were destroyed or badly damaged (Mercy Corps, 2010).

Here as well, hundreds of houses were damaged, two people died and a large number became injured or homeless. Strong rain make rescue work difficult and caused landslides in some areas of the affected region. The International Charter on Space and Major Disasters was triggered to provide post-disaster satellite imagery for damage mapping and to support the aid response (www.zki.dlr.de).

The rapid mapping activities of ZKI of the DLR were substantially supported with the manifold available data sets from the pre-disaster phase. While usually damages are mapped using post-disaster data on a raster basis of 250 meters, the preparation within the Last-Mile consortium enabled to now deliver fast and reliable information on individual building level. The combination of the 3-D city model, the population distribution, the street network, the knowledge on safe areas, etc. with post-disaster remote sensing data, allowed within a few hours and days to provide a fast and more detailed spatial and quantitative information for rescue teams than ever before in the history of remote sensing. Figure 6 shows an overview on partly or completely damaged structures after the earthquake on 30th of September 2009 using post-disaster satellite data. Additionally the location of critical infrastructures such as hospitals or schools is displayed.

5. Interdisciplinary outreach within the conceptual framework

The holistic idea of the risk and vulnerability framework (cp. Table 1) reveals that earth observation data and methods are especially appropriate analyzing spatial exposure due to natural hazards as well as physical vulnerability indicators. In the assessment part indirect correlations with external data sets from social sciences or civil engineering revealed the capability of interdisciplinary approaches to overcome the limitations of one single discipline.

The combination of geo-spatial information products from remote sensing in combination with data, methods and results of inundation modelling allows high precision risk identification, localization and evaluation (Goseberg & Schlurmann, 2009). The combination of the physical urban morphology with socio-economic parameters allows statements on socio-economic patterns, and thus susceptible areas (Setiadi et al., 2010). The time-dependent knowledge on population distribution is a vulnerability indicator itself, but can additionally serve in combination with the knowledge on the street network and the width of streets, etc. to

Fig. 6. Rapid mapping post-disaster analysis using pre- and post disaster satellite data and geo-spatial products (www.zki.dlr.de)

model evacuation scenarios (Borrero et al., 2006; Lämmel, Grether & Nagel, 2010). The combination of area-wide and up-to-date available physical parameters of the building stock derived from remote sensing in combination with civil engineering (Leone et al. 2007; Münich et al., 2006) can also support the latter approach with the actual distribution of vertical shelters.

The multidimensional information on physical, demographic, social or economic parameters in combination with land surface information like elevation, slopes, location, etc. can support urban growth modelling substantially (Cheng & Masser, 2003; Hill & Lindner, 2010). The anticipation of urban dynamics serves as a valuable planning tool for sustainable development, due to its capability to base informed decisions on various scenarios.
For sustainable urban planning the variety of geo-spatial results need a standardized format to be integrated in an easy to handle geo-information-system (Wahl, Schnabel & Klein, 2008). Thus, the results can be visualized and used for quantitative analysis for specific problems. The higher-ranking goal is supporting the process of urban planning with the provision of objective, quantitative and area-wide data for more informed decisions.

6. Conclusion and outlook

The study has demonstrated that earth observation data, methods and products can significantly support identification, localization and evaluation of vulnerability and risks in cities. The main findings and results address the three questions we defined earlier in the introduction.
In general, remote sensing enables an independent, up-to-date, geometrically and thematically highly resolved analysis of the physical urban environment as well as the assessment of indirect parameters like e.g. the population distribution. High resolution sensors are especially valuable due to the small-scale heterogeneous structure typical for urban environments. Furthermore, interdisciplinary analysis allows value-adding such as the assessment of building stability with methods of civil engineering.
The conceptual framework shows the multi-disciplinarity and multi-dimensionality of vulnerability and risk. Remote sensing can only contribute bits and pieces to the holistic concept being one essential scientific discipline especially strong in the derivation of physical vulnerability indicators. Indirect correlation methods in interdisciplinary approaches allow contributions to demographic, social or economic vulnerability parameters. Thus, we need to overcome the problem of marginal or isolated questions failing to integrate them on higher-ranking goals without collaborating with other research disciplines.
Furthermore the spatial exposure on various hazards, in our case, the assessment of potential inundation areas of a tsunami or landslide prone areas can be assessed using earth observation data. .
These vulnerability studies and the geo-spatial products are of essential importance in an actual disaster case. The rapid mapping products were delivered within hours and days after the earthquake in Padang in 2009 with a never before seen thematic and geometrical resolution, due to their availability beforehand. Thus, rapid crisis intervention and knowledge generation needs this preparedness for the highly detailed statements.
Earth observation data and methods deliver objective and quantitative geo-spatial results or products with accuracies mostly above 80 % and sometimes even higher. Thus, remote sensing has the capability to provide an objective basis for informed decisions needed for sustainable urban planning or the development of risk mitigation strategies. But these earth observation products are not yet established in the decision-making process, acceptance needs to be found by users, politicians and a legal foundation needs to be found for juristic acceptance.

Within a different study of Mery Corps (2010) after the earthquake in Padang in 2009, perhaps the most important finding of the survey is that six months after the earthquake, many people continue living in unsafe housing. Rebuilding rates are extremely low compared to the number of damaged houses. This underlines, that transforming knowledge into action is a major challenge for success to rebuild resilient city. Thus, future work needs to include the last step of bringing information gained from science into action. The scientific results are valueless if they do not transform into practical value.

7. References

Birkmann, J. (2006): *Measuring vulnerability to promote disaster-resilient societies: Conceptual frameworks and definitions, in: Measuring Vulnerability to Natural hazards – Towards Disaster Resilient Societies,* edited by: Birkmann, J., New York, United Nations University, 9–54, 2006.

Birkmann, J., Dech, S., Goseberg, N., Lämmel, G., Moder, F., Nagel, K., Oczipka, M., Schlurmann, T., Setiadi, N., Siegert, F., Strunz, G. and Taubenböck, H.: *Numerical Last-Mile Tsunami Early Warning and Evacuation Information System,* in: International Conference on TsunamiWarning (ICTW), Bali, Indonesia, 12 pp., 12–14 November 2008a.

Blaikie, P., Cannon, T., Davis, I. & Wisner, B. (1994): *At Risk: Natural Hazards, People's Vulnerability and Disasters.* London: Routeledge. S. 320.

Bogardi, J. and Birkmann, J.: *Vulnerability Assessment: The First Step Towards Sustainable Risk Reduction,* In: *Disaster and Society – From Hazard Assessment to Risk Reduction,* edited by: Malzahn, D. and Plapp, T., Logos Verlag Berlin, Berlin, 75–82, 2004.

Bohle H.-G.: *Vulnerability and Criticality: Perspectves from Social Geography,* IHDP Update 2/2001, Newsletter of the International Human Dimensions Programme on Global Environmental Change, 1–7, 2001.

Bollin, C., Hidajat, R. (2006): *Community-based risk index: Pilot implementation in Indonesia.* In: Birkmann, J. (2006): *Measuring Vulnerability to Natural hazards – Towards Disaster Resilient Societies.* New York, United Nations University. S. 524.

Borrero, J. C.; Sieh, K.; Chlieh, M. & Synolakis, C. E. (2006): *Tsunami inundation modeling for western Sumatra.* Proceedings of the National Academy of Science of the United States of America, 2006, 103, 19673–19677. PNAS online publication, available at http://www.pnas.org

Brakenridge, G. R., Anderson, E., Nghiem, S. V., Caquard, S., Shabaneh, T. B. (2003): *Flood warnings, flood disaster assessments, and flood hazard reduction: The roles of orbital remote sensing.* Proceedings of the 30th International Symposium on Remote Sensing of Environment.

Briuglio, L. (2003): *Methodological and Practical Considerations for Constructing Socio-Economic Indicators to Evaluate Disaster Risk.* Institute of Environmental Studies, University of Columbia, Manizalez, Colombia. Programme on Information and Indicators for Risk Management. IADBECLAC- IDEA.

Cannon, T., Twigg, J., and Rowell, J. (2003): *Social Vulnerability: Sustainable Livelihoods and Disasters, Report to DFID Conflict and Humanitarian Assistance Department (CHAD) and Sustainable Livelihoods Support Office,* http://www.proventionconsortium.org/themes/default/pdfs/CRA/DFIDSocialvulnerability.pdf.

Chen, P., Liew, S. C., and Kwoh, L. K. (2005). *Tsunami damage assessment using high resolution satellite imagery: a case study of Aceh, Indonesia.* Proceedings, International Geoscience and Remote Sensing Symposium (IGARSS), Seoul, South Korea.

Cheng, J. & Masser, I. (2003): Urban *growth pattern modelling: a case study of Wuhan city, PR China.* In: Landscape and Urban Planning, 62, 2003, 199 - 217.

Correira, F., Santos, M. & Rodrigues, R. (1987): *Engineering Risk in Regional Drought Studies.* In: L. Duckstein and E.J. Plate, (eds.), *Engineering Reliability and Risk in Water Resources*, Dordrecht/Boston: Martinus Nijhoff, S. 61-86.

Correll, R., Cramer, W., and Schellnhuber, H.-J.: Potsdam sustainability days, Symposium on "Methods and Models of Vulnerability Research, Analysis and Assessment", 2001.

Davidson, R. (1997): *An Urban Earthquake Disaster Risk Index,* The John A. Blume Earthquake Engineering Center, Department of Civil Engineering, Stanford University, Stanford, Report No. 121, 1997.

Dech, S. & R. Messner (2005): *Berge aus dem All.* Frederking & Thaler, München.

Eguchi, R.T., Huyck, C.K., Houshmand, B., Shinozuka, M., Yamazaki, F., Matsuoka, M. and Ulgen, S. (2000): *The Marmara Turkey Earthquake: Using Advanced Technology to Conduct Earthquake Reconnaissance.* Research Progress and Accomplishments, 1999-2000, MCEER-00-SP01, Multidisciplinary Center for Earthquake Engineering Research, University at Buffalo.

Esch. T, Thiel M., Schenk A., Roth A., Mehl H., Dech S. (2010): *Delineation of Urban Footprints from TerraSAR-X Data by Analyzing Speckle Characteristics and Intensity Information,* In: IEEE - Transactions on Geoscience and Remote Sensing. vol. 48, no. 2, pp. 905–916, Feb. 2010.

Füssel, H. M. (2005): *Vulnerability in climate change research: A comprehensive conceptual framework.* University of California, International and Area Studies - Breslauer Symposium No. 6.

Gonzalez, R. R. & Medina, J. S. (2004): *Theories, Models and Urban Realities. From New York to Kathmandu.* In: Dela 21, pp. 64-81.

Goseberg, N. & Schlurmann, T. (2009): *Enhanced hazard mapping on a medium-resolved numerical grid for the city of Padang, West Sumatra.* In: Journal of Ship Technology 5, July, Nr. 2, S. 13-2,1.

Gusella, L., Huyck, C. & Adams, B.J. (2008): *Counting the number of collapsed buildings using an object-oriented analysis: Case study of the 2003 Bam earthquake,* Remote Sensing for Resilient Multi-hazard Disaster Response, Volume II, MCEER: Buffalo.

Hill, A.; Lindner, C. (2010): *Simulating informal urban growth in Dar es Salaam, Tanzania – A CA-based land-use simulation model supporting strategic urban planning.* In: Koch, A.; Mandl, P. (Eds.): Modeling and simulating urban processes, Lit Verlag, Münster (forthcoming)

Huyck, C. H., Adams, B.J., Cho, S., Chung, H.-C., Eguchi, R. T. (2005): *Towards Rapid Citywide Damage Mapping Using Neighborhood Edge Dissimilarities in Very High-Resolution Optical Satellite Imagery – Application to the 2003 Bam, Iran.* In: Earthquake Spectra, December 2005, Vol. 21, Issue S1, pp. S255-S266

IPCC (2001): *Impacts, Adaption, and Vulnerability*, Published for the Intergovernmental Panel on Climate Change, University Press, Cambridge, ISBN 0-521-80768-9, 2001.

Lämmel, G., Grether, D. & Nagel, K. (2010): *The representation and implementation of time-dependent inundation in large-scale microscopic evacuation simulations.* In: Transportation Research Part C: Emerging Technologies. vol. 18, pp. 84-98.

Lauterjung, J. & Rudloff, A. (2005): *GITEWS – The German Indonesian Tsunami Early Warning System,* EOS T. Am. Geophys. Un., 86(52), Fall Meet. Suppl., 2005.

Leone, F., Denain, J.-C., Vinet, F., Bachri, S., (2007): *Analyse spatiale des dommages au bâti de Banda Aceh (Sumatra, Indonésie): contribution à la connaissance du phénomène et à l'élaboration de scénarios de risque tsunami.* In: Lavigne, F., Paris, R., in press. *Le tsunami du 26 décembre 2004 en Indonésie - Rapport scientifique du programme TSUNARISQUE (2005-2006).* Délégation Interministérielle pour l'aide Post-Tsunami (DIPT), Ambassade de France en Indonésie, Centre National de la Recherche Scientifique: 356p.

Mas, J.-F. (1999): *Monitoring land-cover changes: A comparison of change detection techniques.* International Journal of Remote Sensing, 20(1), 139–152.

Mather, P. M. (2004): *Computer Processing of Remotely-Sensed Images:* An Introduction. Third Edition John Wiley. p. 292.

McCloskey, J., Lange, D., Tilmann, F., Nalbant, S., Bell, A., D. Hillman Natawidjaja & Rietbrock, A. (2010): *The September 2009 Padang Earthquake.* Nature Geoscience, Correspondence.

Mercy Corps (2010): *Padang houses by house – A Padang housing and vulnerability assessment,*http://prattinfographicsworkshop.files.wordpress.com/2010/08/2010-july-29-final-english-padang-housing-and-vulnerability-assessment-hires.pdf

Miura., H. and Midorikawa, S. (2006): *Updating GIS building inventory data using high-resolution satellite images for earthquake damage assessment,* Earthquake Spectra, 22(4): 151-168.

Münich, J., Taubenböck, H., Stempniewski, L., Dech, S., Roth, A. (2006): *Remote sensing and engineering: An interdisciplinary approach to assess vulnerability in urban areas,* First European Conference on Earthquake Engineering and Seismology. Geneva, Switzerland. p. 10 on CD-ROM.

Pesaresi, M. , Gerhardinger, A. & Haag, F.(2007): *Rapid damage assessment of built-up structures using VHR satellite data in tsunami-affected areas,* In: International Journal of Remote Sensing, 28: 13, 3013 — 3036

Post, J., Wegscheider, S., Mück, M., Zosseder, K., Kiefl, R., Steinmetz, T., & Strunz, G. (2009): *Assessment of human immediate response capability related to tsunami threats in Indonesia at a sub-national scale,* Nat. Hazards Earth Syst. Sci., 9, 1075-1086, doi:10.5194/nhess-9-1075-2009, 2009.

Rashed, T. & Weeks, J. (2003): *Assessing vulnerability to earthquake hazards through spatial multicriteria analysis of urban areas,* International Journal of Geographical Information Science, 17(6), 547–576.

Rastogi, B. K. and Jaiswal, R. K. (2006): *A Catalogue of Tsunamis in the Indian Ocean,* Science of Tsunami Hazards, 25(3), 128–143, 2006.

Roemer, H.; Kaiser, G.; Sterr, H. & Ludwig, R. (2010): *Using remote sensing to assess tsunami-induced impacts on coastal forest ecosystems at the Andaman Sea coast of Thailand. Nat.* Hazards Earth Syst. Sci., vol. 10, 729-745.

Saito, K., Spence, R. & Foley, T. (2005): *Visual damage assessment using high-resolution satellite images following the 2003 Bam Iran earthquake,* Earthquake Spectra, Special Issue 1, Vol 21, 2005, pp. 309-318.

Saito, K & Spence, R. (2003): *Using high-resolution satellite images for post earthquake building damage assessment: A study following the 26.1.01 Gujurat earthquake,* Earthquake Spectra.

Sarabandi, P., Kiremidjian, A., Eguchi, R. T. and Adams, B.J., (2008) *Building inventory compilation for disaster management: Application of remote sensing and statistical modeling,* Technical Report Series MCEER-08-0025, MCEER: Buffalo.

Setiadi, N., Taubenböck, H., Raupp, S. & Birkmann, J. (2010): *Integrating socio-economic data in spatial analysis: An exposure analysis method for planning urban risk mitigation.* 15th International Conference on Urban Planning and Regional Development in the Information Society (REALCORP), Vienna, Austria. GeoMultimedia 2010. pp. 367-374.

Schneiderbauer, S. (2007): *Risk and Vulnerability to Natural Disasters – from Broad View to Focused Perspective,* PhD-Thesis, Freie Universit¨at Berlin, pp. 121, available at: http://www.diss.fu-berlin.de/2007/498/indexe.html .

Taubenböck, H., Esch, T., Felbier, A., Roth, A. & Dech, S. (2011): *Pattern-based accuracy assessment of an urban footprint classification using TerraSAR-X data.* In: IEEE Geoscience and Remote Sensing Letters. Vol. 8; No. 2; pp. 5; March 2011.

Taubenböck H. & Dech, S. (2010): *Remote Sensing and Urban Planning: A Common Future?* In: V1 Magazine - Promoting Spatial Design for a Sustainable Tomorrow. http://www.vector1media.com/articles/features/13931-remote-sensing-and-urban-planning-a-common-future-

Taubenböck, H., Esch, T., Wurm, M., Roth, A. & Dech, S. (2010): *Object-based feature extraction using high spatial resolution satellite data of urban areas.* In: Journal of Spatial Science, vol. 55, no. 1., pp. 117-133.

Taubenböck, H., Goseberg, N., Setiadi, N., Lämmel, G., Moder, F., Oczipka, M., Klüpfel, H., Wahl, R., Schlurmann, T., Strunz, G., Birkmann, J., Nagel, K., Siegert, F., Lehmann, F., Dech, S., Gress, A., Klein R. (2009a): *Last-Mile preparation for a potential disaster – Interdisciplinary approach towards tsunami early warning and an evacuation information system for the coastal city of Padang, Indonesia.* In: Nat. Hazards Earth Syst. Sci.. Vol. 9, pp. 1509-1528. http://www.nat-hazards-earth-syst-sci.net/9/1509/2009/nhess-9-1509-2009.html

Taubenböck, H., Wurm, M., N. Setiadi, N. Gebert, A. Roth, G. Strunz, J. Birkmann & S. Dech (2009b): *Integrating Remote Sensing and Social Science – The correlation of urban morphology with socioeconomic parameters.* In: Urban Remote Sensing Joint Event, Shanghai, China. pp. 7, CD-ROM.

Taubenböck, H., Post, J., Roth, A., Zosseder, K., Strunz, G. & Dech, S. (2008): *A conceptual vulnerability and risk framework as outline to identify capabilities of remote sensing.* In: Nat. Hazards Earth Syst. Sci.. Vol. 8, no. 3, pp. 409-420. http://www.nat-hazards-earth-syst-sci.net/8/409/2008/ nhess-8-409-2008.html

Timmermann, P. (1981): *Vulnerability, Resilience and the Collapse of Society,* No. 1 in Environmental Monograph, Toronto: Institute for Environmental Studies, University of Toronto.

Turner, B. L., Kasperson, R. E., Matson, P. A., McCarthy, J. J., Corell, R. W., Christensen, L., Eckley, N., Kasperson, J. X., Luers, A., Martello, M. L., Polsky, C., Pulsipher, A., and Schiller, A. (2003): *A framework for vulnerability analysis in sustainability science,* P. Natl. Acad. Sci. USA, 100(14), 8074–8079, 2003.

Turker, M. & Cetinkaya, B. (2005): *Automatic detection of earthquake-damaged buildings using DEMs created from pre- and post-earthquake stereo aerial photographs,* In: International Journal of Remote Sensing, 26, pp. 823–832.

United Nations (1991): *Mitigating Natural Disasters: Phenomena, Effects, and Options: a Manual for Policy Makers and Planners,* New York: UNDRO (United Nations Disaster Relief Organization), 1991.

United Nations Development Programme (UNDP) (2004): *Reducing Disaster Risk – A challenge for development – A Global Report,* ISBN 92-1-126160-0, New York, USA, 2004.

United Nations/ISDR (International Strategy for Disaster Reduction) (2004): *Living with Risk: A Global Review of Disaster Reduction Initiatives,* United Nations International Strategy for Disaster Reduction, UN Publications, Geneva, Switzerland, 2004.

United Nations University Institute for Environment and Human Security (UNU-EHS) (2008): *Household Survey on Socio-Economic Vulnerability to Tsunamis in Padang City.*

Voigt, S., Kemper, T., Riedlinger, T., Kiefl , R., Scholte, K., & Mehl, H. (2007): *Satellite image analysis for disaster and crisis-management support,* IEEE Transactions on Geoscience and Remote Sensing, 45(6), 1520–1528.

Wahl, R., Schnabel, R. & Klein, R. (2008): *From Detailed Digital Surface Models to City Models Using Constrained Simplification*. Photogrammetrie, Fernerkundung, Geoinformation (2008):3(207-215)

Zschau, J., Isikara, M., Ergünay, O, Yalcin, M.N., Erdik, M. (2002): *Towards an earthquake early warning system for the megacity Istanbul*. Early Warning Systems for Natural Disaster Reduction, edited by J.Zschau and A.Küppers, pp.433-440, Springer.

8

Resistance of Flexible Emergent Vegetation and Their Effects on the Forces and Runup due to Waves

Sundar Vallam, Murali Kantharaj and Noarayanan Lakshmanan
Indian Institute of Technology Madras
India

1. Introduction

Field surveys of post great Indian Ocean tsunami of 2004 revealed that coastal vegetation can greatly improve the safety of coastal infrastructure. Buildings exposed to direct attack of the giant wave suffered greater damage than those fronted by vegetation. Thus, understanding the effect of vegetation on the hydrodynamic impact of such long waves on the coastal structures has become an important issue among the coastal engineering community. In the backdrop of such requirement, a comprehensive experimental program was taken up to investigate on the dynamic pressures exerted on a wall due to cnoidal waves in the presence and absence of a patch of elastic vertical cylinders. The patch of vertical cylinders is expected to replicate typical characteristics of vegetation of emerged type.

The resistance offered by vegetation to flow is essentially of drag in nature (Fischenich, 2000 and Darby, 1999) and is generally termed as, Green Belt effect. However, the nature of drag may be largely influenced by the hydro-elastic interaction of the current with the vegetal stems (Freeman, 1997). As such, stiffness of the vegetation appears to be one of the major parameters that govern the net drag offered to the flow. The percentage of reduction in the energy due to vegetation is also governed by the average diameter of the members within the vegetation, its height, distribution, density and type (Darby, 1999; Fischenich, 2000). In the past, several researchers have studied the flow characteristics past vegetal patches in open channels in order to evaluate friction factors for quantifying the resistance. However, measurements and quantification of resistance in the practical range of flow and vegetal parameters applicable to coastal protection are rather limited due to problems relating to scaling of all the fluid-structure physics.

1.1 Previous research

Frictional losses caused by momentum transfer through vegetated zone of a channel were the focus of most of the earlier investigations. Through a series of experiments using four species of coniferous trees in water and air, Fathi-Moghadam and Kouwen (1997) presented a model for estimating the drag coefficient of vegetation. Using this model, effects of flow depth and land slope on the *Darcy friction factor (f)* was investigated by Fathi-Moghadam (2006). Numerical and experimental works have been performed by Kutija *et al.* (2003) to analyse and formulate the drag effects induced by vegetation. A vegetation index (E_{vi}) was proposed as one of the leading parameters to account for the effects of leaf density, shape and rigidity of individual trees. Meijer and Van Velzen (1999), Nepf (1999) and Dudley et al. (1998) reported the dependency of flow resistance on the density of the vegetation within the group and on the bending stiffness of the species. A formula for evaluation of vegetation resistance was proposed by Righetti and Armanini (2002) based on an analytical two layer model for sparsely distributed bushes under uniform flow conditions. The analytical model was verified with the experimental data.

The problem of vegetal resistance to flow can be categorised into submerged vegetation and emergent vegetation as claimed by (Liu et al., 2003). Since this study focuses on emergent vegetation, a concise review of relevant studies follows. Kouwen and Fathi-Moghadam (2000) and Jarvela (2004) investigated the variation in f due to vegetation in an open channel and concluded that the fundamental vegetation properties to be considered in establishing an equation for resistance are leaf density, shape, flexibility and manner of deflection of the tree species. Jarvela (2002) investigated the influence of the type, density and placement of vegetation, flow depth and velocity on friction losses. Large variations in the f, with depth of flow, velocity and vegetative density were observed. Further, the transition between submerged and partially submerged flow condition was observed to be at a depth of about 80% of the undeflected plant height.

Freeman *et al.* (2000), focussing mainly on *Manning's n,* argued that the equations, parameters, and methods developed by Ree and Crow (1977) and Kouwen and Li (1980) for a combined density and blockage of heavy ground cover and grasses do not produce satisfactory results. A methodology to determine flow resistance coefficients for submerged and partially submerged shrubs and woody vegetation was proposed. Data for developing the method were extracted from the study of Werth (1997), in which twenty natural plant species with both homogenous and mixed plant spacing were investigated with and without leaves. Unfortunately, a density measure such as leaf area was not recorded. Separate empirical regression equations were developed for the submerged and partially submerged cases revealing the modulus of elasticity to be a critical parameter, which can be estimated from the ratio of the un-deflected plant height to the stem diameter H/d, through field measurements. The approach of Freeman *et al.* (2000) does not directly take into consideration the deformation of foliage, as the stiffness modulus is a property of the stems. It appears not to be reasonable to assume that E can be related to only H/d, which implies that the differences in flexibility between species can be neglected. Later, Fathi-Moghadam (2006) proposed an empirical model to estimate the flow resistance coefficients (including the Darcy-Weisbach friction factor f and Manning's n) for flow inside emergent flexible vegetation.

The cantilever beam theory was used to compute the deflection of the flexible vegetation. In most of the earlier studies, the individual stems of the vegetation were simulated by a group of slender cylinders of constant height and diameter for the tandem configuration (Meijer and Van Velzen 1999). The results from experimental and numerical studies on the attenuation of incident wave energy by coastal forest including mangroves carried out by Hamzah *et al.* (1999) and Harada and Imamura (2006) are similar. The details of a pilot project on the integrated coastal vegetation in Sri Lanka are reported by Tanaka (2009). It is obvious that attention by researchers worldwide has been towards the understanding of the interaction of waves with vegetation.

A brief review of literature on the run-up due to long waves followed by pressures on vertical wall is discussed below. Hall and Watts (1953) proposed an empirical formula for run-up on an impermeable slope with an angle of 45^0 due to solitary waves. Carrier and Greenspan (1958) solved the shallow water equations analytically obtaining solutions for the prediction of the long wave run-up over smooth plane beaches. Synolakis (1986) investigated the run-up of non-breaking long waves theoretically by applying a power law to predict solitary wave run-up on a smooth plane beach. Other notable contributions on run-up of long waves over slopes are due to Li and Raichlen (2001), Carrier *et al.* (2003) and Hughes (2004).
An empirical formula of Goda (1974) to calculate pressures on vertical walls is widely adopted. Other salient contributions are due to Führböter (1993), and Müller and Whittaker (1993). It was generally concluded that sloping walls experience more load than vertical walls. Through experimental studies, Furukawa *et al.* (1997) demonstrated that the wave dissipation depends on forest density and the diameter of the tree trunks. The literature review thus indicates that information on wall pressures fronted by vegetation, in particular due to long waves, is scanty. This aspect in addition to the steady flow past vegetation is focussed through a well-controlled experimental program. Vegetative parameters, including the green belt width, stem diameter, spacing and their rigidity are varied resulting in a holistic view of the present wave vegetation interaction problem.
The effect of two prominent configurations of plantations, viz., tandem and staggered, in attenuating the wave induced pressures is studied. The vegetative parameters such as the width of the green belt (*BG*), diameter of the individual stems (*D*), their spacing (*SP*) were varied for an overall understanding of the steady floe or wave-vegetation interaction problem. Cnoidal waves were considered in the wave flume experiments in order to replicate the shallow water behaviour of tsunami as much as possible. The range of *Ursell parameter* considered for this purpose is 18 to 700.

2. Experimental program

2.1 Vegetation parameters

The mechanical characteristics of vegetation have a considerable effect on the resistance to flow of water. For tall vegetation, these properties include leaf density, shape and the general flexibility of tree species. To estimate the f due to flow through emerging and flexible vegetation, an index to account for the effects of the vegetation's mechanical characteristics is required. Since, the trees could be characterised similar to cantilever beams, the mechanical characteristics may be brought in through the resonance frequency of the first mode of vibration, namely the natural frequency of the vegetal stems. The natural frequency could be related to the '***Vegetation Index***' (Fathi-Moghadam 2007). The resonance frequencies are denoted as f_j (with $j = 1, 2, 3, \ldots, n$, where f_1 is the fundamental or base natural frequency and $f_2, \ldots, n$ are higher modes of natural frequencies). For a linear and homogeneous beam, f_j depends upon the beam's length l, mass per unit length m, second moment of inertia I, modulus of elasticity E, as well as a dimensionless parameter λj which in turn is a function of beam geometry and the boundary conditions under which it is tested. The mathematical relationship between the resonance frequency and the above variables may be written as (Timoshenko and Gere 1961),

$$f_j = \frac{\lambda_j^2}{2\pi}\left(\frac{EI}{ml^4}\right)^{0.5} \tag{1}$$

The above parameter characterizes the height, mass (including leaf density) and the moment of inertia of a tree. Herein, the beam length l is taken to be the height of the vegetal stem. According to Fathi-Moghadam (2007), on transferring the measurable parameters to the right side, the relationship between known and unknown group of variables involving the natural frequency of first mode, f_1, will be:

$$E_{vi} = f_1^2 \left(\frac{m_s}{l} \right) \tag{2}$$

where the total mass of the tree $m_s = ml$, and E_{vi} is called the **"vegetation index"** of a tree species. Measuring the height, mass and fundamental natural frequency of the tree, the vegetation index is estimated by the above equation.

2.2 Modelling of vegetation

Motivated by the need to study the hydro-elastic interaction of the flow with that of the vegetal stems, a suitable material for model tests had to be identified to represent coastal vegetation and mangroves in the real world. Yang and Choi (2009) indicated that the main effect of vegetal flexibility is the increase of Reynolds shear stresses and the intensity of turbulence in the vertical direction. Hence, reproducing the hydro-elastic behaviour is the key in understanding the effects of vegetation on pressures and run-up. One of the key parameter to study the hydro-elastic interaction of the flow with the vegetal stem to represent real-world coastal vegetation is Young's modulus E. Common timber has an E value in the range from 10.05 to 15GPa. Mangrove has maximum E values of 20.03GPa [Gan *et al.* (2001)].

To cover the wide range of E, a reference value of 14GPa was assumed for field condition, serving as a basis of the lab tests. In order to scale down the prototype values and identify the model material, assuming that the flow is due to storm surges and tsunamis, and considering also the experimental ranges of flow (including velocities and flow depth), a guiding scale ratio of 1:40, i.e., a Scale Factor (SF) of 40 was adopted. This would mean that the E value for the model material should be about 0.35Gpa, which is quite difficult to identify. The most practical option for achieving the above is to consider EI as a single parameter and scaling the rigidity. For the present study the material considered is Polyethylene with an E value of about 3.8 Gpa which is about ten times that of the required value. Since the rigidity is modelled herein, the above said variation could be compensated with the variation in I of the material. Having chosen the model material, the typical prototype dimensions of vegetal stems are considered in the range of 100mm - 400mm. Accordingly, using the Froude model law of scaling, the rigidity with the scale factor of $[SF]^5$ - the base diameter for model vegetation will fall in the range of 1.65 - 5.5 mm. The base diameter as arrived above only ensures that the bending action of the vegetal stems is properly scaled. Thus herein, the advantage of modelling rigidity rather than Young's modulus is clearly brought out. However, for accurate hydrodynamic interaction of the flow and stems, the hydrodynamic forces need to be correctly scaled as per Froude's Law. Considering Morison's force regime (Chakrabarti, 1983) the corresponding (to the above range) diameters for the exposed part of vegetal stems were arrived at In order to ensure that the change in diameter does not affect the bending behaviour of the model vegetation, a small clearance is provided between the base and the section, where the diameter changes from a lower to a higher value. It is to be mentioned that, the diameter at root is used only for calculation of EI of the vegetal stems.

2.3 Experimental parameters

The parameters that have been considered here include the size, elastic modulus of the vegetal stems, configuration of the vegetal patch or green belt, as well as the spacing between the individual plants within the green belt etc. Quantification of friction factor and energy loss in steady uniform flow is of primary interest and has practical relevance as far as long waves are concerned. For the measurement of pressures and forces on the structures mounted at a distance from the shoreline/reference line and run-up on mild slope of 1:30, a number of wave and beach parameters were considered in the wave flume experiments. The following variables are of interest for both currents and waves for investigating the response function such as resistance, force or run-up.

$$\text{Vegetal Response} = f\,(\rho, g, h, H, T, B_s, D, D_b, BG, f_1, l, SP, E, L, \beta, R_u, V), \tag{3}$$

where, ρ- mass density of water; g- gravitational acceleration; h- depth of water in the flume; H-wave height; T- Wave period; B_s- Width of the structure; D- diameter of the vegetation; D_b- diameter at the root of the vegetation; BG- width of green belt; f_1 –frequency of first mode of the vegetal stem; l- height of the vegetation; SP- centre to centre spacing between vegetation; E - modulus of plant stiffness; L- Wave length; β- Beach slope; R_u- Run-up; V– Flow velocity; The h refers to, h_s- depth of water at the toe of the structure; h_{avg}- average depth of water on the up-stream and down stream of flow in open channel.

The variables seen in equation (3) are grouped as per Buckingham's Pi theorem. The parameters thus obtained are tabulated in Table 1 in which, U_{max} is the velocity V and L_o is the deep water wave length. The non dimensional quantities investigated in the study are Darcy's f, and Manning's n in steady uniform flow and wave run-up R_u/H, pressures P^* and forces F^* on structures. They are designated as below.

$$f = \frac{8H_f gh}{LV^2} \tag{4}$$

$$F^* = \frac{F_{max}}{0.5 * \rho g H^2 B_s} \tag{5}$$

$$P^* = \frac{P_{max}}{H} \tag{6}$$

2.4 The test set-ups

Test facility for steady-uniform flow

The experiments were conducted in a horizontal, impermeable and rigid glass walled flume 10m long, 0.5m wide and 0.6m deep in the Hydraulics Laboratory at Indian Institute of Technology Madras, India. A head tank at one end served as a reservoir for the discharge into the flume, in which the water level could be maintained at a constant level by adjusting the inflow valve and the outflow weir. Closely spaced slender cylindrical members representing vegetation (with E=3.8E+09 N/m^2) were positioned at a distance of 6 m from one end of the flume in tandem configuration. On the down stream end, an over flow weir was used to adjust the desired flow depth. The water depth was recorded on the upstream as well as on the downstream with a measuring scale fixed along the sides of the glass flume. The velocity on the upstream and downstream of the vegetation was measured with Bikini type propeller velocity meter.

The experiments were conducted for three different depths of flow (210 mm, 180 mm and 80 mm) for the tandem configuration considered herein. For each of above depths, three different widths of green belt of 250 mm, 625 mm and 1000 mm were employed for the tests. Two different diameter of vegetation of 10 mm and 3 mm for each of the green belt as mentioned earlier were adopted. Furthermore, for each of the two diameters considered, two different spacing (37.5 mm and 75 mm) were selected. Thus, for the present study, 12 different experimental set ups were considered. The schematic representation of the vegetation and flow parameters considered is shown in Fig.1 and typical configurations of vegetation for D = 10 mm, SP = 37.5 mm and 75 mm and BG = 1.0m, 0.625m and 0.250m is shown in Fig.2. The details of the flume and the experimental set-up for the tests with steady flow past vegetation are shown in Fig.3. The range of vegetation and flow parameters covered under the present study is provided in Table.1.

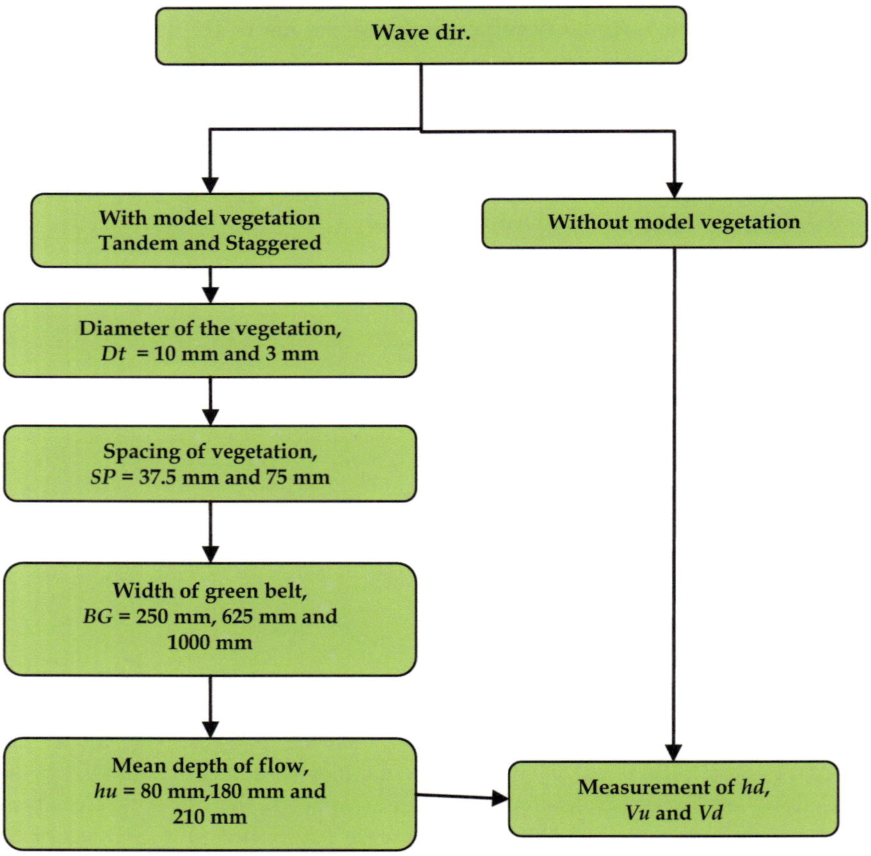

Fig. 1. Experimental Program for open channel flow experiments

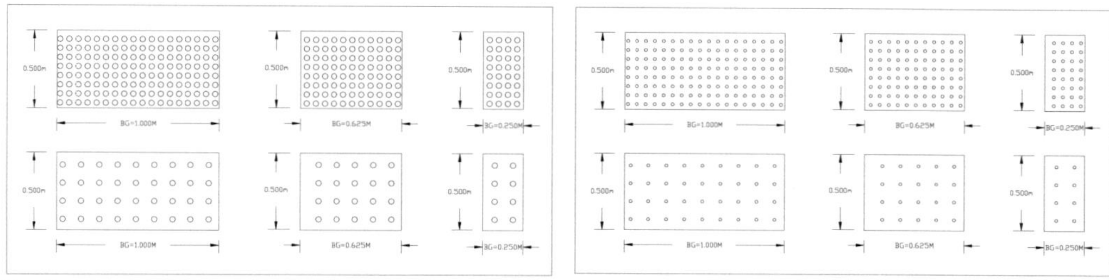

Tandem Configuration:
SP=0.0375 m and 0.075 m; D = 0.01 m.

Tandem Configuration:
SP=0.0375 m and 0.075 m; D = 0.003 m.

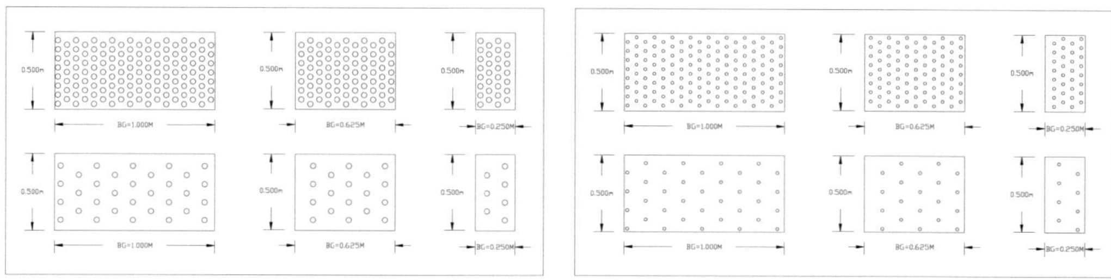

Staggered Configuration:
SP=0.0375 m and 0.075 m; D = 0.01 m.

Staggered Configuration:
SP=0.0375 m and 0.075 m; D = 0.003 m.

Fig. 2. Different configurations of model stems used in open channel and wave flume tests.

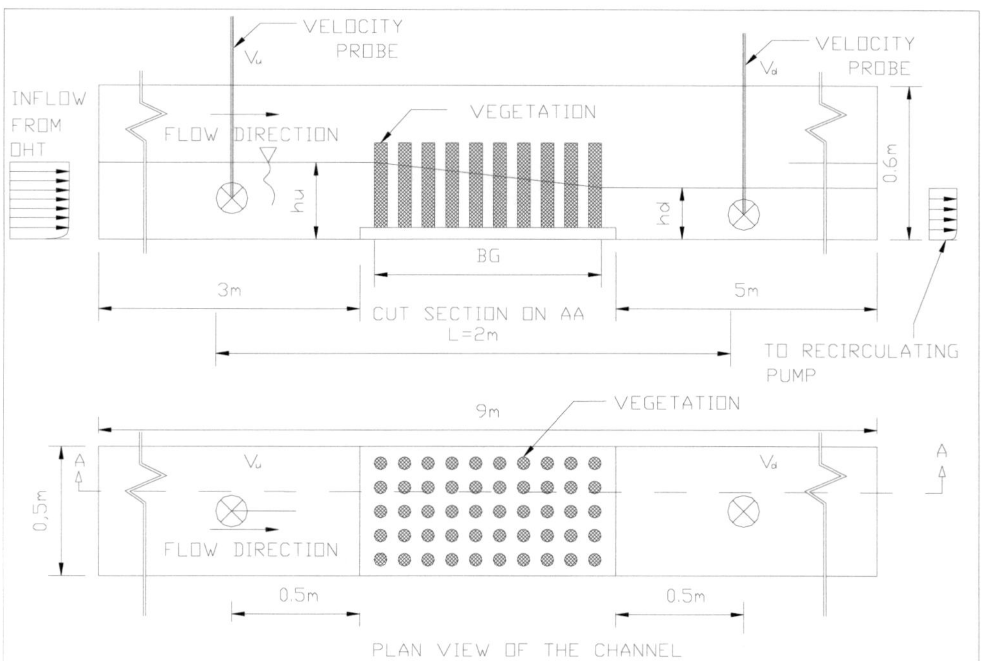

Fig. 3. Schematic diagram of test set-up in open channel for Tandem configuration.

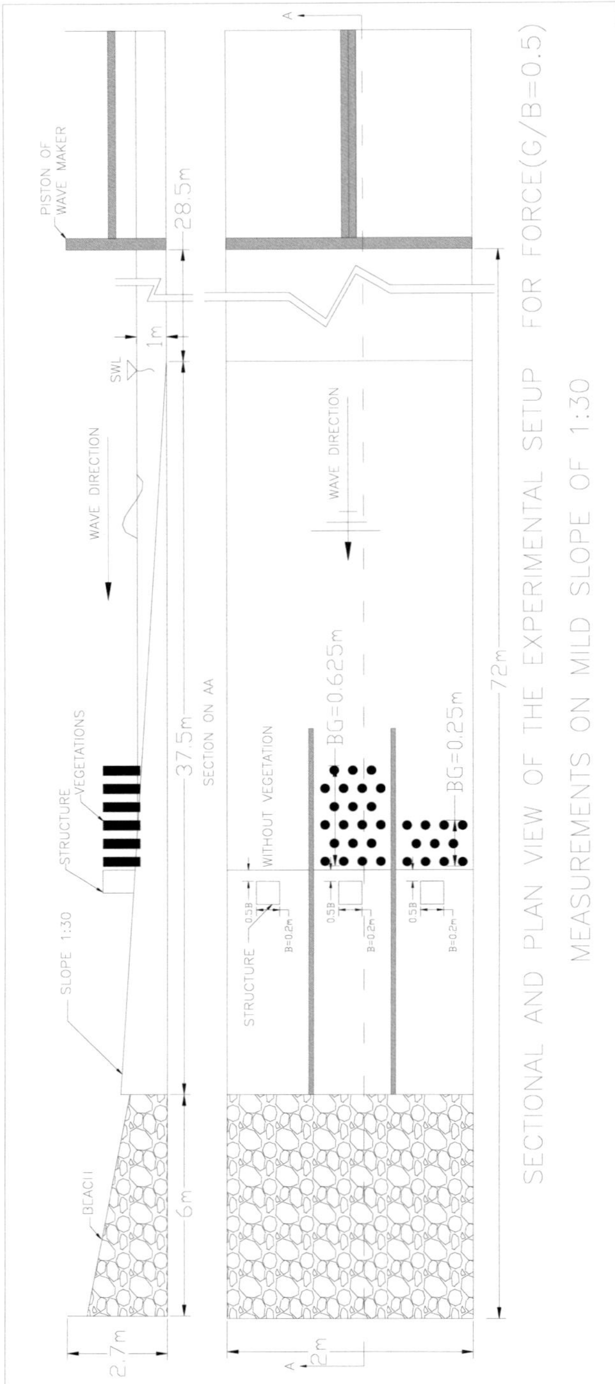

Fig. 4 (a). Experimental setup in the wave flume for measuring forces on model buildings fronted by vegetation.

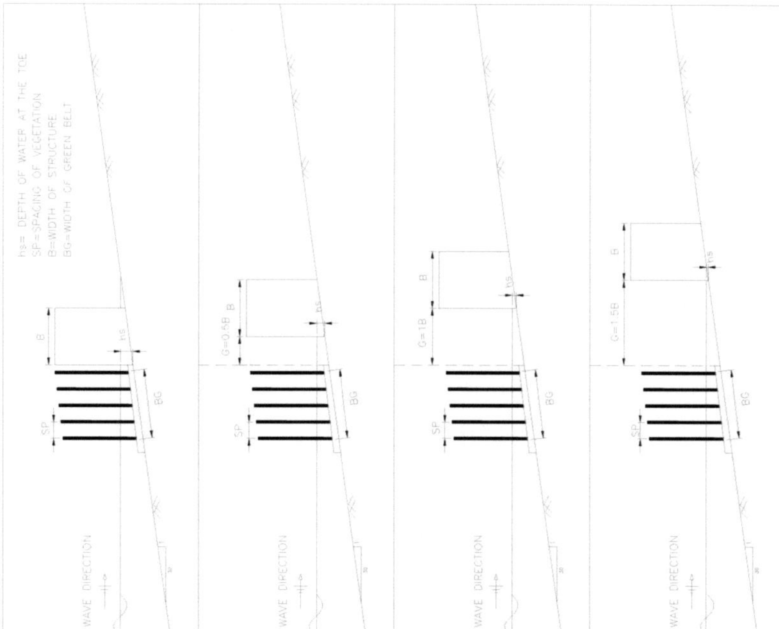

Positions of model building for tests with model vegetation

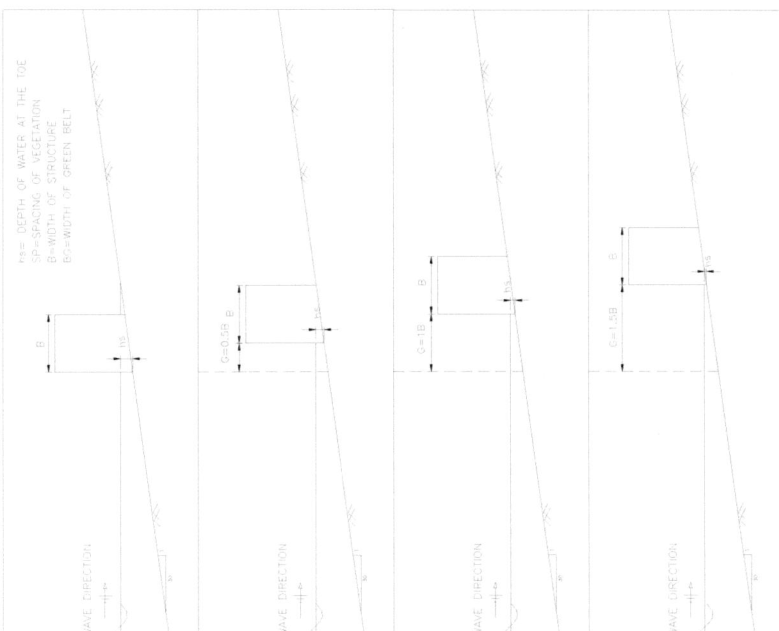

Positions of model building for tests without model vegetation

Fig. 4 (b). Conditions for which forces due to waves on a model building were measured.

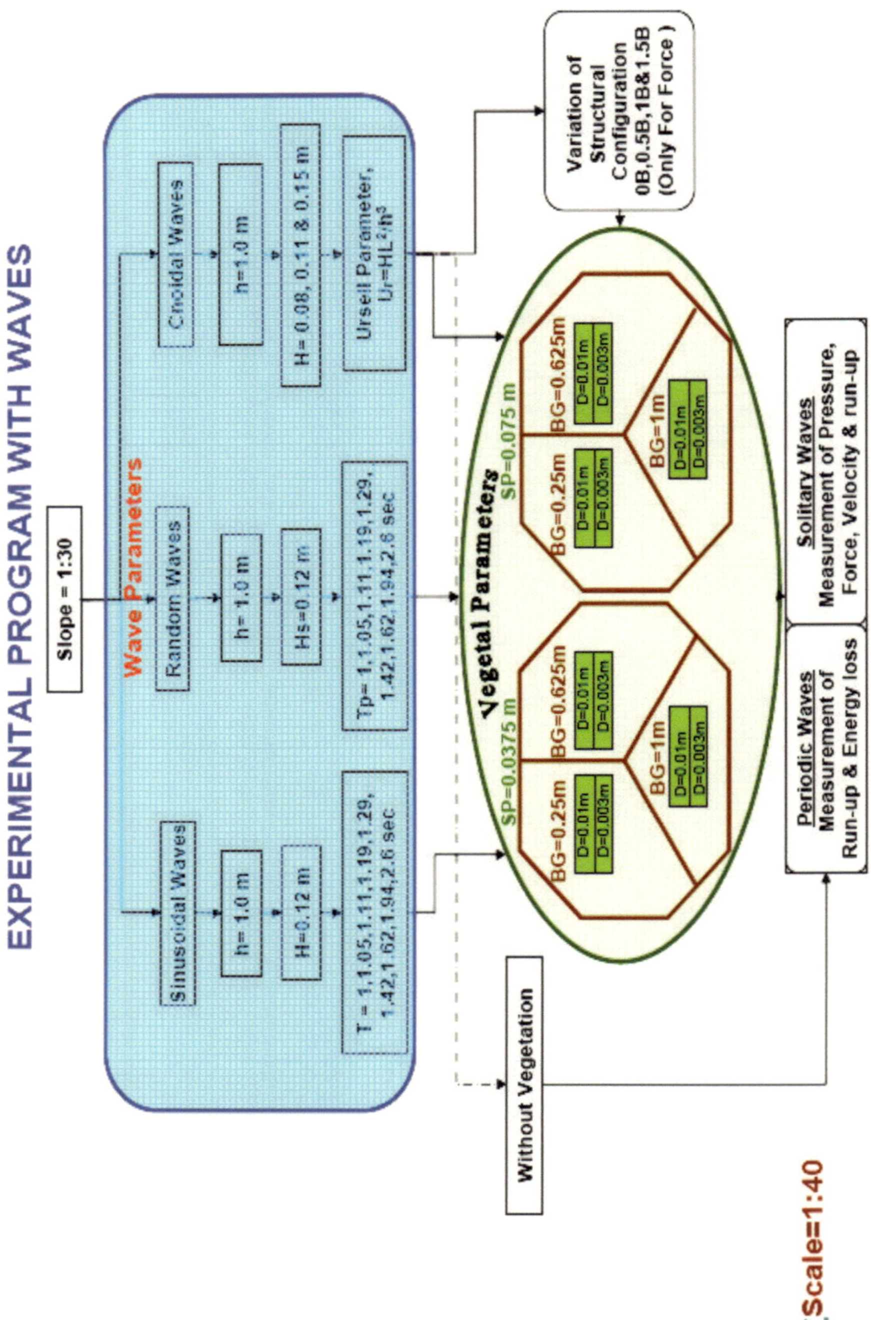

Fig. 4 (c). Experimental Programs for wave flume investigations for pressure, runup and force measurement.

BG=0.25 m and D = 0.003 m. *BG=0.25 m and D = 0.01 m.*

Fig. 5a. Typical views of flow gradient in the open channel tests (flow from left to right)

Fig. 5b. Typical views of model setup for run-up measurement (left) and force measurement (right). The model buildings could be seen with the instrumentation..

Froude Number , $F_r = \left(\dfrac{V_{avg}}{\sqrt{g * h}}\right)$	0.23 to 0.60
Relative spacing, $R_s = \left(\dfrac{SP}{D}\right)$	3.75,7.5,12.5 and 25
Velocity of ratio, $V_r = \left(\dfrac{V_{avg}}{f_1 D}\right)$	2.19 to 83
Frequency parameter, $f_z = \dfrac{1}{T}\left(\sqrt{\dfrac{h_{avg}}{g}}\right)$	0.247 to 1.42
Reynolds number, $\mathrm{Re} = \left(\dfrac{V_{avg} * h_{avg}}{\upsilon}\right)$	13693 to 146179
Vegetation-Flow Parameter, $VFP = \dfrac{EI(BG/D)}{\rho H l^3 V_{avg}^2 (SP/D)}$	0.001to 14.37
Vegetal parameter, $a = BG{\cdot}SP/D^2$	94 to 8333

Table 1. Range of dimensionless parameters covered in the present study

Among all the aforesaid parameters, the Vegetation-Flow Parameter, *VFP,* is newly introduced to obtain a unique parameter that combines the effects of structural rigidity of the individual model stem, flow and vegetation parameters. It is prudent to now recall the fact that, the earlier investigators have used the Vegetation Index as a parameter to classify vegetation (Fathi-Moghadam, 2007). However, it is to be mentioned that the Vegetation Index does not include parameters relating to rigidity *EI,* vegetation parameters *BG, SP/D* and the flow parameters. Hence, it is proposed that the present parameters derived from this work will be more relevant and comprehensive in view of the hydro-elastic processes being considered.

Flume facility for test in Waves

The present experimental study was carried out in a wave flume of 50 m long, 2.0 m wide and 2.7 m deep, in Department of Ocean Engineering, Indian Institute of Technology Madras, India. A computer controlled wave maker is installed at one end of the flume and on the other end; a rubble mound wave absorber is placed to minimize the reflection from the end of the flume. The water depth in the flume can be varied from 0.3 m to 2.0 m. For the present study, the water depth was maintained as 1.0 m and hence, the wave maker was operated in the piston mode. The wave generating system is capable of generating Sinusoidal, random waves, cnoidal and solitary waves. The wave maker can be operated either in piston mode or, in hinged mode primarily for the generation of deep-water waves and fully computer controlled using Wave Synthesizer. As the number of tests to be carried out is voluminous and in order to save time for testing, simultaneous testing of different configurations were aimed at without sacrificing the quality of testing. The flume of 2m width was partitioned longitudinally into three parts each of width of 0.66 m.

In the present experimental program, the wave elevation, wave run-up, pressures and forces on the structures as well as the velocity before and after the Green belt were measured using a set of instruments in the wave flume. The host computer that is used for controlling the wave maker is also used for acquiring the response from the different instruments used in this study. Then the calibration coefficients for each instrument have been directly input into the data collection software for directly obtaining the physical quantities. The calibration of all the instruments is done for each set of experiments to minimize the calibration errors in the experiments.

Propeller type velocity meters were used to measure the velocity on the upstream and downstream of the Green belt for the tests in the open channel. The wave probe used for the present tests comprises of two thin, parallel stainless steel electrodes. When immersed in water, the change in the conductivity of the instantaneous water column between the two electrodes is measured. This change is proportional to the variation of the water surface elevation between the electrodes. A beam type single component force transducer was used to measure the inline wave force on the model structure in the wave flume. The load cell had a maximum capacity of 200 N with an accuracy of 0.1 N and had strain gauges in half-bridge configuration. Similar to the principle of the wave gauge, the run-up probe was fabricated to measure the run-up on the outer cylinder. The run-up probe consists of two parallel stainless steel wires of 1 mm diameter spaced 10 mm apart and are fixed at specified locations on the beach slope. When immersed in water, the electrodes measure the conductivity of the instantaneous water volume between them. The conductivity change is proportional to the variation in the water surface elevation. Static calibration of run-up probe has been carried out to obtain the calibration constants.

Test Procedure

The open channel tests were manually carried out by adjusting discharges and measuring flow depths using height gauges and velocities using vane type flow meters. The flow depths h_u and h_d and mean flow velocities V_u and V_d have been measured for all the test combinations. Each of the above tests has been repeated 3 times for ensuring repeatability of measurements and only the mean values have been taken for further processing. These measurements have been used to arrive at the f for various flow as well as vegetative parameters. Snap shots recorded during the tests are shown in Fig.5a. The pictures bring out the hydraulic gradients observed during the tests. The resistance to the flow due to the presence of vegetation is dependent on the gradient between upstream and downstream flows. In general, it is observed that higher *D* and *BG* results in a higher gradient, the details of which for all the tests carried out are reported and discussed later. In the present study, the first modal frequency, f_1, was obtained by free vibration tests conducted after fixing strain gauges in a half bridge on the root of the model vegetal stem.

In the case of the tests in the wave flume, all the test signals acquired through an amplifier were filtered through a 20Hz low pass filter to remove the sampling ripples. A Personal Computer was used for the generation of waves and for the simultaneous acquisition of signals from the sensor pick-ups. The electrical signals were acquired using quartz clock controlled sampling and converted to digital form using a 12bit A/D (Analog to Digital) converter. The signals from various instruments were acquired with a sampling frequency of 40Hz. Sufficient time gap was allowed between successive runs to restore calm water condition in the wave flume.

Initially, tests without the vegetal patches were carried out in each of the flume compartments. The run–up measurements have been carried out without the wall (used for pressure measurements) and the model building (used for force measurements). The tests with Cnoidal waves were carried out in two water depths of 0.9 and 1 m, The run-up, R_u was thus measured for both the cases i.e., presence and absence of vegetation in order to quantify the role of wave height on the run-up (R_u/H). In addition, to the measurement of pressures on a wall in absence and presence of vegetation, the tests on the measurement of forces due to Cnoidal waves on a structure positioned at a distance, *G* of 0.0B, 0.5*B*, 1.0*B* and 1.5*B*, from the shoreline/reference line that was carried out under the second phase of the wave flume experiments. A model structure of dimension, 0.2 m*0.2 m and height of 0.3 m was mounted on a cantilever type load cell.

The depth of water at the toe, *h* of the beach slope was maintained at 1m. The testing parameters for both waves and the vegetation are the same as that adopted for the tests with pressures as discussed earlier. The test set-up for the measurement of forces on the structure (in absence of vegetation and with Staggered configuration of vegetation for any two *BG*) for G/B = 0, 0.5, 1.0 and 1.5. Typical experimental set-up used for the tests in wave flume for measuring the wave forces on a model building is shown in Fig.4a. The conditions for which the forces acting on a model building were measured in presence and absence of vegetation are projected in Fig. 4b. The complete details of the experimental investigations carried out are concisely summarised in Fig.4c.

3. Vegetal resistance and resulting runup, pressure and force

3.1 Resistance of flexible emergent vegetation in steady uniform flow

From the literature it is observed that a number of authors have investigated the resistance characteristics of emergent vegetation in the past under steady uniform flow while a few have also considered the flexibility of vegetation. However, a comprehensive study considering more appropriate modelling of vegetal flow interaction is found to be warranted. The present effort focuses on the effect of the flexibility of vegetation on the resistance offered to the flow. The flexibility, being more of a material property, should be directly related to the Young's Modulus *E* of the vegetal material as found from a standard bending experiment of a simply supported beam with point loads and without involving the flow parameters. Furthermore, rigidity is defined as a product of *E* and the moment of inertia *I*. By doing this, the indirect way of establishing *E* as done by some of the authors (Freeman 1997; Jarvela 2004) has been avoided. Further, the limitation of the validity of *E* or the empirical model being a function of un-deflected plant height and the depth of flow is overcome.

In the present flow vegetation interaction, the resistance to the fluid flow is considered to be characterised by a number of friction factors. The resistance coefficients that are in use are the *Manning's n, Darcy Weisbach friction factor, f* and the *Chezy's constant C*. There are also a few studies where the vegetal drag has been investigated. The Darcy *f* has been quantified in the past through measurements in open channel experiments (Unny and Kouwen, 1969; Petryk and Bosmajian, 1975; Chen, 1976 and Werth, 1997) using various vegetation models. In general, it may be noted that the interaction between vegetation and flow has not been modelled appropriately and the elastic characteristics of vegetation has not been given due consideration. However, the previous investigators reported that quantifying the resistance using *f* has several advantages such as (i) the flow being taken past enclosed / partially enclosed vegetal volumes, (ii) ease of measurements in the laboratory, (iii) possibility of quantification of wall effects in open channel, etc. As such, the present work aims to investigate the Darcy's *f and Manning's n* for various flow as well as vegetative parameters.

In the present study, the resistance is investigated with respect to Reynolds' number (*Re*), relative spacing (*SP/D*), reduced velocity (V_r) and Vegetation-Flow Parameter (VFP) of the vegetal patch. The vegetation-flow parameters considered in the present study not only accounts for the flexural rigidity of the vegetation but also for the width of the green belt *BG*, flow depth *h* and the diameter of the vegetation *D*. As experiments were carried out over a wide range of flow as well as vegetative parameters, determination of *f and n* irrespective of the depth of flow and the un-deflected plant height to stem diameter is made possible. However, herein the results on the variation of *f* alone are projected.

Determination and quantification of resistance due to Vegetation

Utilising the Bernoulli's equation, the loss of head due to friction can be determined as,

$$H_f = \frac{V_u^2 - V_d^2}{2g} + (h_u - h_d) \tag{7}$$

V, h and *g* in the above equations are defined earlier and the suffix *u* and *d* correspond to the respective variables on upstream and downstream of flow respectively. However, H_f in the above equation also includes the contribution of flume walls and the bed. The energy loss due only to vegetation $H_{f\,(veg)}$ may be calculated by deducting the energy losses due to side and bottom walls $H_{f(wall)}$, from the total energy losses $H_{f(veg+wall)}$

$$H_{f(veg)} = H_{f(veg+wall)} - H_{f(wall)} \tag{8}$$

$$H_{f(veg+wall)} = \left\{ \left(\frac{V^2_{u(veg+wall)} - V^2_{d(veg+wall)}}{2g} \right) + (h_{u(veg+wall)} - h_{d(veg+wall)}) \right\} \tag{9}$$

$$H_{f(wall)} = \left\{ \left(\frac{V^2_{u(wall)} - V^2_{d(wall)}}{2g} \right) + (h_{u(wall)} - h_{d(wall)}) \right\} \tag{10}$$

$$H_{f(veg)} = \left\{ \left(\frac{V^2_{u(veg+wall)} - V^2_{d(veg+wall)}}{2g} \right) + (h_{u(veg+wall)} - h_{d(veg+wall)}) \right\} - \left\{ \left(\frac{V^2_{u(wall)} - V^2_{d(wall)}}{2g} \right) + (h_{u(wall)} - h_{d(wall)}) \right\} \tag{11}$$

Herein, $V_{u(veg+wall)}$ is upstream velocity measured in the presence of vegetation, $V_{u(wall)}$ is upstream velocity measured in the absence of vegetation; $h_{u(veg+wall)}$ is depth of flow upstream of the vegetation; $h_{u(wall)}$ is depth of flow at the upstream location in the absence of the vegetation; $V_{d(veg+wall)}$ is velocity measured with vegetation on the downstream; $V_{d(wall)}$ is velocity measured without vegetation on the downstream; $h_{d(veg+wall)}$ is depth of flow down stream of the vegetation; $h_{d(wall)}$ is depth of flow downstream in the absence of the vegetation; $V_{d(veg+wall)}$ is velocity measured with vegetation on the downstream. It is to be noted that the above approach consistently eliminates the presence of wall effects in the measurements, irrespective of the location of upstream and downstream measurement points. A similar exercise has been carried out for the Manning's *n*, however, the details of which are not reported. Snap shots recorded during the tests are shown in Fig.5. The pictures bring out the hydraulic gradients observed during the tests.

Friction factors and their quantification

Froude model law is usually employed for studying open channel hydraulics. However, it the dependence of *f* with respect to *Re* is considered up-front for the purpose of validation of the results with that in the literature. Jarvela (2002) has measured the *f* for sedges for a wide range of *Re* in a flow depth of 0.2m, the results of which are superposed with the present results for *f*, for flow depth ranging between 0.19m and 0.2m, in Fig.6a. This comparison validates the measurements with that of Jarvela (2002) in the range of *Re* considered. The above said variation have been grouped for three different values of h_{avg}/R_h of 1.5, 1.7 and 1.9 and superposed in Fig.6b, suggesting that the *f* has a strong correlation with the non-dimensional flow. From the results it is seen that the *f* is sensitive to the variation h_{avg}/R_h for lower *Re*. The friction factor appears to increase drastically by about 50%-100% as the h_{avg}/R_h increases by about 100%.

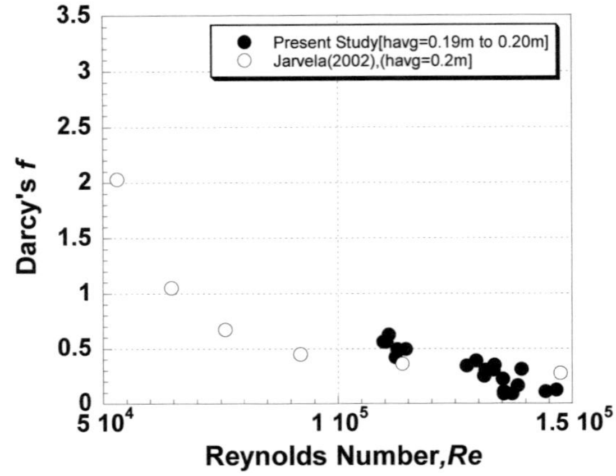

Fig. 6 a. Comparison of variation of *f* with *Re*-(present results with Jarvela(2002)

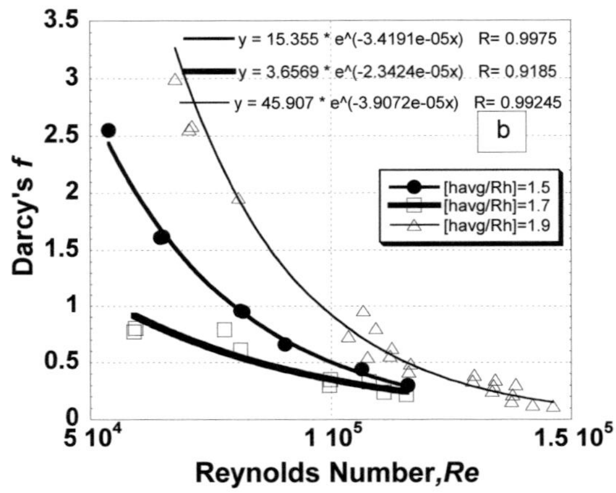

Fig. 6 b. Variation of *f* with Re as a function of relative flow depth.

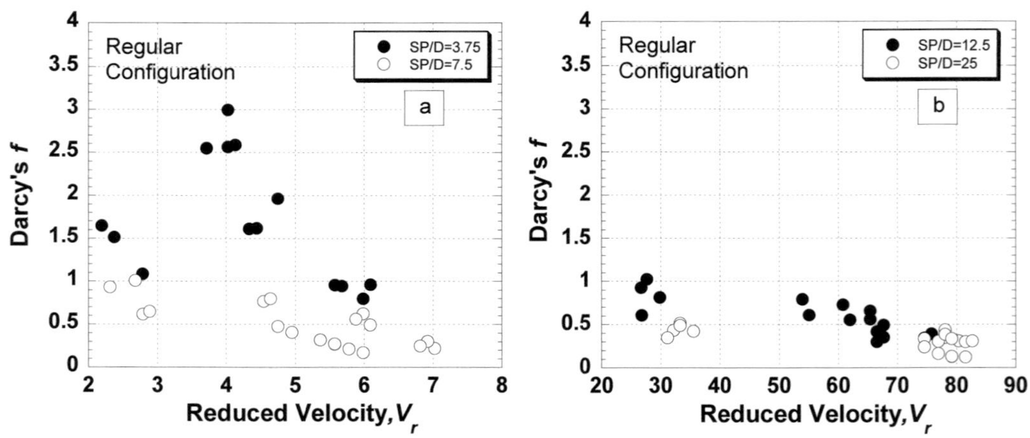

Fig. 7. Variation of Darcy's *f* with V_r for different ranges of *SP/D* in tandem configuration.

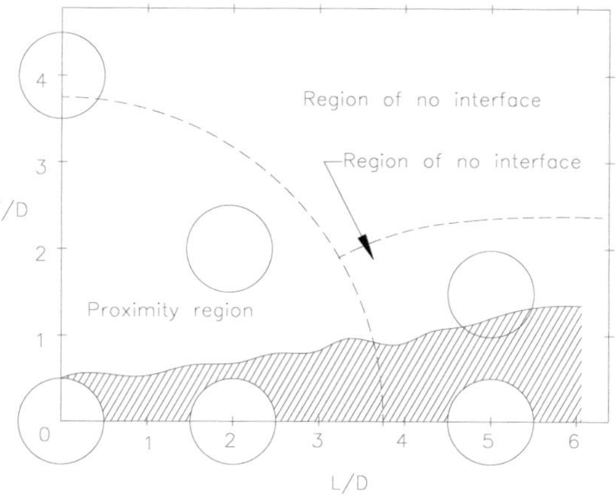

Fig. 8. Interface Region in flow around cylinders (after Zdravkovich and Medeiros, 1991).

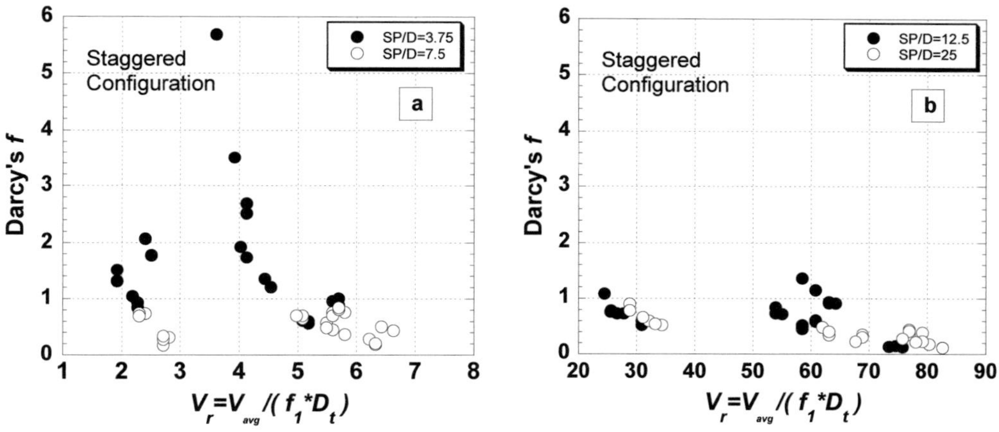

Fig. 9. Variation of Darcy's f with V_r for different ranges of SP/D in Staggered configuration.

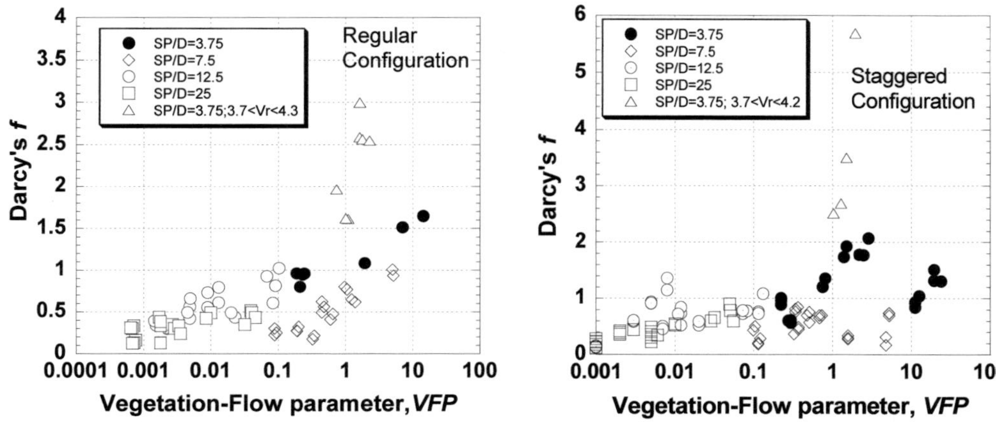

Fig. 10. Variation of Darcy friction factor f with Vegetation-Flow Parameter for different ranges of SP/D in Tandem configuration.

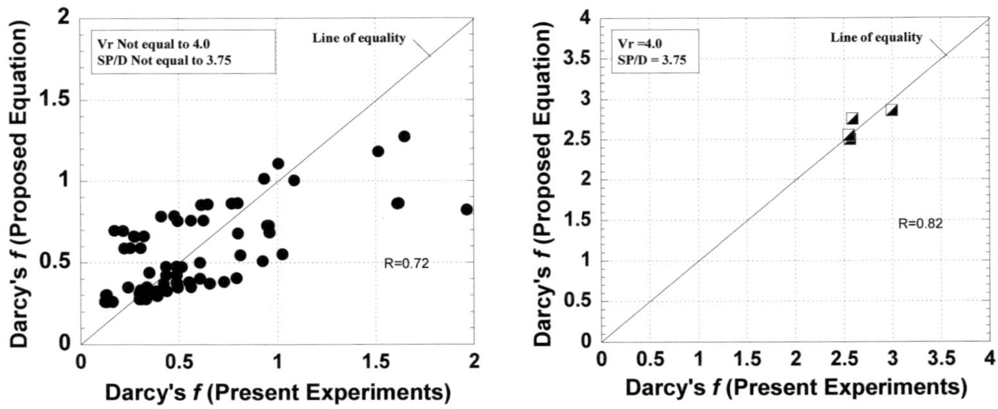

Fig. 11. Darcy's *f* - Comparison of present proposed equation with present measurements. When proximity and wake effects are presents, large friction is observed.

Proposed empirical formula (no wake and proximity effects)	Proposed empirical formula (with wake and proximity effects)
$f_{\mathrm{Re}\,g} = 1.005 * \left(\frac{EI*(BG/D)}{\rho HV^2 l^3 (SP/D)}\right)^{0.1680} \left(\frac{V_{avg}}{\sqrt{gh_{avg}}}\right)^{0.2479}$	$f_{\mathrm{Re}\,g} = 7.488 * \left(\frac{EI*(BG/D)}{\rho HV^2 l^3 (SP/D)}\right)^{-0.2654} \left(\frac{V_{avg}}{\sqrt{gh_{avg}}}\right)^{1.099}$
$f_{Stg} = 0.463 * \left(\frac{EI*(BG/D)}{\rho HV^2 l^3 (SP/D)}\right)^{0.0789} \left(\frac{V_{avg}}{\sqrt{gh_{avg}}}\right)^{-0.5211}$	$f_{Stg} = 0.196 * \left(\frac{EI*(BG/D)}{\rho HV^2 l^3 (SP/D)}\right)^{1.4136} \left(\frac{V_{avg}}{\sqrt{gh_{avg}}}\right)^{-3.098}$
$n_{\mathrm{Re}\,g} = 0.0422 * \left(\frac{EI*(BG/D)}{\rho HV^2 l^3 (SP/D)}\right)^{0.1332} \left(\frac{V_{avg}}{\sqrt{gh_{avg}}}\right)^{-0.2217}$	$n_{\mathrm{Re}\,g} = 0.229 * \left(\frac{EI*(BG/D)}{\rho HV^2 l^3 (SP/D)}\right)^{0.01812} \left(\frac{V_{avg}}{\sqrt{gh_{avg}}}\right)^{1.26}$
$n_{zz} = 0.0466 * \left(\frac{EI*(BG/D)}{\rho HV^2 l^3 (SP/D)}\right)^{0.0641} \left(\frac{V_{avg}}{\sqrt{gh_{avg}}}\right)^{-0.0865}$	$n_{zz} = 0.0279 * \left(\frac{EI*(BG/D)}{\rho HV^2 l^3 (SP/D)}\right)^{0.8016} \left(\frac{V_{avg}}{\sqrt{gh_{avg}}}\right)^{-1.554}$
$E_{L(Reg)} = 0.4369 * \left(\frac{EI*(BG/D)}{\rho h_u V_{avg}^2 l^3 (SP/D)}\right)^{0.2038} \left(\frac{V_{avg}}{\sqrt{gh_{avg}}}\right)^{0.5413}$	$E_{L(Reg)} = 4.25 * \left(\frac{EI*(BG/D)}{\rho h_u V_{avg}^2 l^3 (SP/D)}\right)^{0.2357} \left(\frac{V_{avg}}{\sqrt{gh_{avg}}}\right)^{3.0015}$
$E_{L(Stg)} = 0.171 * \left(\frac{EI*(BG/D)}{\rho h_u V_{avg}^2 l^3 (SP/D)}\right)^{0.1028} \left(\frac{V_{avg}}{\sqrt{gh_{avg}}}\right)^{-0.3949}$	$E_{L(Stg)} = 0.099 * \left(\frac{EI*(BG/D)}{\rho h_u V_{avg}^2 l^3 (SP/D)}\right)^{1.2722} \left(\frac{V_{avg}}{\sqrt{gh_{avg}}}\right)^{-1.7077}$

Table 2. Proposed formula for *f*, *n* and E_L

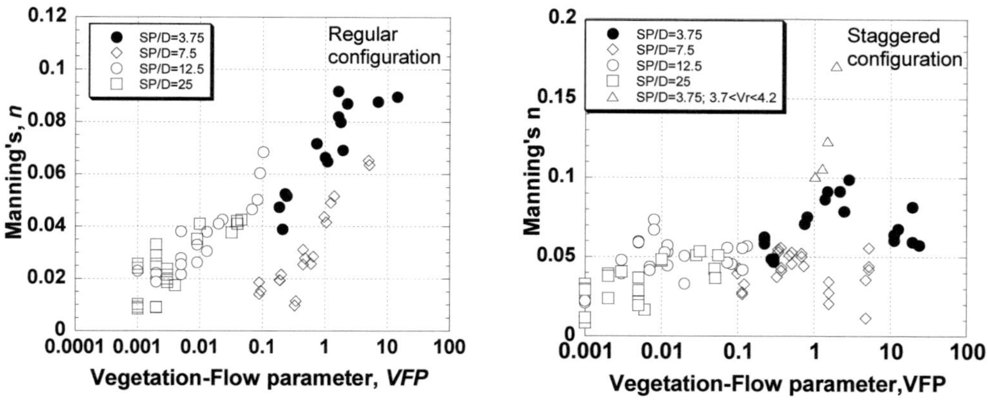

Fig. 12. Variation of Manning's n with Vegetation-Flow Parameter for different ranges of *SP/D*.

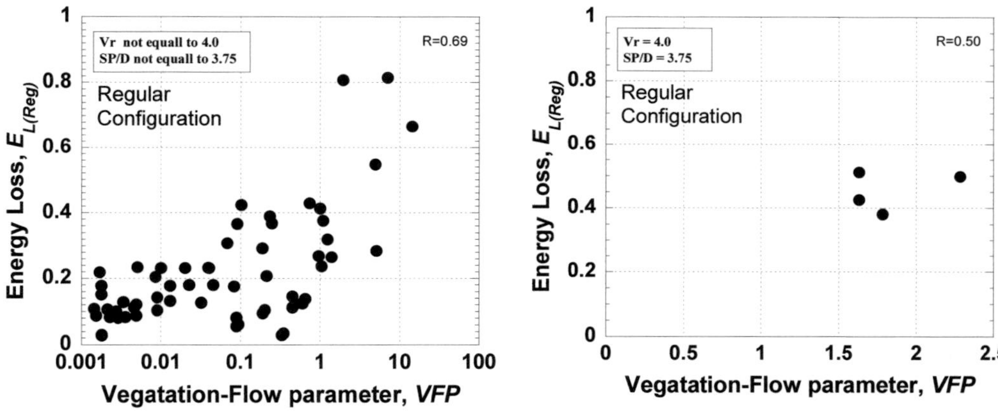

Fig. 13. Representation of Manning's n with Vegetation-Flow Parameter for different ranges of *SP/D*.

Dependence of 'f' on Reduced Velocity 'V_r' and Relative Spacing 'SP/D'

Considering the number of parameters involved in the resistance offered by vegetal stems needs an in-depth understanding of the possible non-dimensional groups that could dictate specific regions of the flow-vegetation interaction. Detailed analysis indicate that the reduced velocity (V_r) and vegetation-flow parameter to be important parameters.

The variation of f with Reduced Velocity V_r for different ranges of Relative spacing, *SP/D* is investigated in Fig.7 for tandem configuration of vegetation. The range of experimental data for V_r is $2.0< V_r <8.0$ and $20< V_r <90$. It is found that the f has a strong correlation with V_r and *SP/D*. The f is consistently close to or less than 1.0 for values of V_r other than about 4.0 and *SP/D* other than 3.75. The f reduces with an increase in the V_r for *SP/D* greater than 3.75. For *SP/D* =3.75, the f is found to increase for V_r in between 3 and 4, thereafter, the f in general is found to decrease. Two distinct processes that have been observed in the experiments are responsible for this distinct behaviour of f. These are the Wake and Proximity induced vibrations of the model stems. The Wake induced vibrations of Vortex Induced Vibrations (VIV) is expected to have a significant effect in the flow-vegetation

interaction and care has been taken to keep the elasticity and flow variables with in the range of requisite non-dimensional parameters, the most important being the Reduced Velocity, V_r. In the present experimental parameters, V_r is defined as $V_{avg}/(f_1 D_t)$ as described earlier. Since the *Strouhal Number,*($S_t = Df_s/V_{avg}$) that is about 0.25, does not appear in the definition of V_r, it is expected that *"lock-in"* (matching of Vortex shedding frequency with the natural frequency of the model vegetal stem) will take place around V_r =4.0. When the lock-in occurs, the vortex street downstream of any structural member will be oscillated with the most possible structural vibrations. It may be concluded that, higher the structural vibrations, better the energy dissipation leading to an increase in *f*. This is observed in this case to be as high as 200% more than the non lock-in regime resulting in values of *f* as high as 3.0. However, the same is not observed for other values of *SP/D* considered in the study. This can be explained with reference to the proximity effect. The proximity effect is the effect of presence of neighbouring cylinders in the flow, in the transverse direction. The combined interaction of Proximity and Wake induced oscillations has been investigated by Zdravkovich and Medeiros (1991). For benefit of the readers, a figure describing this effect is reproduced in Fig.8 based on the original publication. It is seen that, as long as the *SP/D* is about 3.8, the combined Proximity and Wake effects will be present in the structural response. Hence, we obtain the highest measured *f* for the combination of *SP/D* =3.75 and V_r ~4.0. It is to be noted that this refers to densely spaced vegetation. In the experiments, this is the lowest value of relative spacing that has been considered. Specific mention needs to be made to the fact that, closer spacing may not be possible in real situations pertaining to vegetation in flow or Greenbelt. In the case of sparsely spaced vegetation, for other *SP/D* (7.5, 12.5 & 25.0) and V_r >20, the *f* is found to be consistently around or less than 1.0, which is obvious as no lock-in or a proximity effect is expected for this combination of parameters.
It is usually expected that the staggered placement of vegetal stems would be more effective in offering resistance to flow (Fig.9). The staggered placement is seen to give about 20% improved resistance when compared to tandem configuration.

Vegetation Flow Parameter 'VFP' and its effect on resistance

Relating *EI, BG, l,* V_{avg}, ρ, h_{avg} and *SP/D*, the Vegetation-Flow Parameter is defined as,

$$\mathrm{VFP} = \left(\frac{EI * (BG / D)}{\rho h_u l^3 V_{avg}^2 (SP / D)} \right) \tag{12}$$

The plot showing variation of *f* with *VFP* is shown in Fig.10 for both tandem and staggered configuration. For lower rigidity and higher *SP/D*, the *f* is less than 1.0. A general trend of higher *f* for higher *VFP* is clearly visible, meaning that, higher Relative Rigidity will provide higher resistance. For the range of up to *VFP* = 0.8, the *f* value is always less than 1.0. The critical range of *VFP* is about 1.0. At this *VFP*, very high values of *f* (up to above 3) may be encountered due to the combined effect of Wake and Proximity induced oscillations. Further beyond *VFP* =1, an increasing trend in *f* is once again observed. For a maximum value of *VFP* ~15, the maximum *f* value of 1.5 is observed. In the critical range of *VFP* =1, a steep increase in *f* with *VFP* is observed from 1.5 to 3.0 with a maximum *f* occurring at *VFP* ~1.5. The forgoing discussion confirms that the new vegetal parameter *VFP,* is able to reasonably scale the structural rigidity with reference to the vegetation and flow parameters. Hence this shall be used as a key parameter in greenbelt design.

Empirical relationship for f

The aforementioned discussion indicates that the vegetal resistance or drag depends itself on many variables. With the results presented above as the background, a multiple regression analysis with suitable variables provides the most useful relationship which could be used is applications. For this purpose the *Fr* and *VFP* were selected. The *Fr* determines hydraulic regime of the flow while the *VFP* determines the hydro-elastic scale. Hence, the parameters shall appear in the relationship. However, the data was split into two categories. The first category pertaining to all the data except the range of *SP/D* =3.75 and V_r ~4.0. The second category pertaining to the range of *SP/D* =3.75 and V_r ~4.0. These classifications were necessary to avoid unwanted scatter in the fit and provide best possible regression coefficient. The new empirical equations thus obtained are given in eqs.24 and eq.25 for the two categories of data.
While *SP/D* ≠ ~3.75 and V_r ≠ ~4.0

$$f_{tandem} = 1.005 * \left(\frac{EI * (BG/D)}{\rho h_u V_{avg}^2 l^3 (SP/D)} \right)^{0.1680} \left(\frac{V_{avg}}{\sqrt{g h_{avg}}} \right)^{0.2479} \tag{13}$$

While *SP/D* = ~3.75 and V_r = ~4.0

$$f_{tandem} = 7.488 * \left(\frac{EI * (BG/D)}{\rho h_u V_{avg}^2 l^3 (SP/D)} \right)^{-0.2654} \left(\frac{V_{avg}}{\sqrt{g h_{avg}}} \right)^{1.099} \tag{14}$$

The above relationships represent the data within a maximum standard deviation of 10%. These relationships may be used with in the range of $0.25 \leq F_r \leq 0.60$. Further, it is observed that all the parameters are physically reasonably related to the *f*. In order to ascertain the quality of the fit presented above, the relationship between measured and predicted *f* values for a given set of experimental parameters for the above equations are presented in Fig.11. It is once again noticed that most of the data fall around the line of equality for both the data ranges. It is proposed that the above empirical models may be used in practical application of Green belt / Bio shield.

Manning's coefficient 'n' for vegetal patch.

Similar to '*f*', the Manning's n could also be studied in terms of *Re* and Reduced Velocity V_r, and similar discussion could be made. Following the procedures adopted for *f*, the dependence of Manning's *n* with the newly proposed Vegetation Flow Parameter is brought out in Fig. 12. The comparison shows an excellent correlation for any given *SP/D* (7.5-25). The Manning's *n* varies between 0.01 and 0.065. However, for the case, where proximity effect is present, the minimum Manning's *n* observed is about 0.04 which increases to about 0.09 towards the experimental regime of wake effects. The wake effects are not very much pronounced in the case of Manning's *n* in tandem configuration. It is observed that for the range of Vegetation Flow Parameter studied, the Manning's *n* has a linear relationship with the Vegetation Flow Parameter. Similar to '*f*', the newly proposed empirical relationship for Manning's *n* in terms of F_r and *VFP* are given in Table.2. These relationships represent the data within a maximum standard deviation of 12%. These relationships may be used with in the range of $0.25 \leq Fr \leq 0.60$. Further, it is observed that all the parameters are physically related to the *n*.

The Manning's n has been computed for staggered configuration from the experimental measurements using the same approach outlined earlier. The superiority the staggered configuration over the tandem configuration is brought out once again in the above figure. The increase in n when the proximity and wake effect are present is about 50-60% occurring at the value of *VFP* of about 1.0. For the other scenarios, a 20-30% increase is evident. Once again, the empirical relationships for evaluating the 'n' value for staggered configuration are given in Table.2. It is found that these relationships are better than that proposed by Freeman *et al.* (2000) as demonstrated by Noarayanan et al. (2009). It is also demonstrated that the presently proposed vegetal parameter, VFP, is superior to the earlier vegetal parameters.

3.2 Relationship of energy loss with vegetation-flow parameter

The f and n are the most important parameters for vegetal drag or head loss. However, their direct use in the design of Greenbelt is not quite straightforward. One needs to have a fair understanding of the energy loss and its direct relation to the vegetal and flow parameters in order to perform a design calculation for Green belt. To this end, the normalized Energy Loss $E_{L(Reg)}$ and $E_{L(Stg)}$, with reference to the upstream energy content is considered as against the Vegetation Flow Parameter *VFP* so that a direct relation between the Greenbelt parameters and drop in energy within the Greenbelt may be obtained. The variations of $E_{L(Reg)}$ as a function of *VFP* for tandem and staggered configurations are displayed in Figs.13a and Fig.13b respectively. The $E_{L(Tan)}$ is found to increase from about 0.1 at *VFP* ~0.001 to 0.8 at *VFP* 1.5. It is to be recalled that the maximum f and n have been observed at the same value of *VFP* as reported in earlier figures. Beyond *VFP* =1.5, the $E_{L(Tan)}$ reduces and is about 0.30 at *VFP*~15. Furthermore, the effects of wake and proximity interactions are once again brought out in the $E_{L(Tan)}$ also as the variation of $E_{L(Tan)}$ for *SP/D*=3.75 show a clearly superior performance than the other combinations. Similar trend is observed in the Energy Loss with *VFP* for the vegetations in the Staggered configuration. The energy loss is about 150 - 200% more (0.4 - 0.9) for *SP/D*=3.75 while, the range of $E_{L(Tan)}$ for higher *SP/D* is between 0.05-0.4. The mechanisms of wake and proximity effects are discussed earlier and hence are not repeated here.

An attractive proposition is to obtain a direct relationship between $E_{L(Reg)}$ and the independent parameters concerning the flow and Greenbelt. Once again, the multiple regression procedure is extended for $E_{L(Reg)}$ and other parameters. However, in this case, to keep the relationship simple only F_r and *VFP* are used in the regression procedure.

While $SP/D \neq \sim 3.75$ and $V_r \neq \sim 4.0$

$$E_{L(Reg)} = 0.4369 * \left(\frac{EI * (BG/D)}{\rho h_u V_{avg}^2 l^3 (SP/D)} \right)^{0.2038} \left(\frac{V_{avg}}{\sqrt{g h_{avg}}} \right)^{0.5413} \quad (15)$$

While $SP/D = \sim 3.75$ and $V_r = \sim 4.0$

$$E_{L(Reg)} = 4.25 * \left(\frac{EI * (BG/D)}{\rho h_u V_{avg}^2 l^3 (SP/D)} \right)^{0.2357} \left(\frac{V_{avg}}{\sqrt{g h_{avg}}} \right)^{3.0015} \quad (16)$$

Similar relationship for the Staggered configuration are presented below to obtain a direct relationship between $E_{L(Stg)}$ and the independent parameters concerning the flow and

Greenbelt. Once again, the multiple regression procedure is extended for $E_{L(Stg)}$and other parameters, yielding the following relationships.
While $SP/D \neq \sim 3.75$ and $V_r \neq \sim 4.0$

$$E_{L(Stg)} = 0.171 * \left(\frac{EI * (BG/D)}{\rho h_u V_{avg}^2 l^3 (SP/D)} \right)^{0.1028} \left(\frac{V_{avg}}{\sqrt{g h_{avg}}} \right)^{-0.3949} \tag{17}$$

While $SP/D = \sim 3.75$ and $V_r = \sim 4.0$

$$E_{L(Stg)} = 0.099 * \left(\frac{EI * (BG/D)}{\rho h_u V_{avg}^2 l^3 (SP/D)} \right)^{1.2722} \left(\frac{V_{avg}}{\sqrt{g h_{avg}}} \right)^{-1.7077} \tag{18}$$

3.3 Application of the model

A typical sample application of the present empirical mode for a real field situation is demonstrated herein based on the newly proposed empirical relationship for $E_{L(Reg)}$ eq.16. The different parameters in the field obtained based on design requirements are provided in Table.3.The procedure used for obtaining the values in the above table is highlighted below.

Design Example

Design a green belt for a storm surge of 5m, in such a way that the surge height may be reduced by half on its lee side. The available width of beach may be taken to be 150m. Assume an *Fr* of 0.4 for the flow induced by surge of height of 5m, assumed to reduce on downstream to 2.5m. The width of green belt has to be less than or equal to 150 m.

The following are the assumptions:

1. Froude number is assumed to be 0.40 (can be obtained from the ingress velocity in the field).
2. Fully grown diameter beachfront vegetation is taken to be 0.4m and (2/3) of this diameter is used for design.
3. Vegetal parameters: E = 9.0 GPa (common Oak) to 18.5 GPa (Red Mangroves)
   ρ = 1100 kg/m3
   SP/D = 6.0 (for design)
   = 4.0 (for fully grown condition)

Calculation procedure:

Using the above data and assumptions, the variables in *VFP* are derived and applied in eq.12. The design $E_{L(Reg)}$ is obtained based on given data as 0.36 with a factor of uncertainty of 1.5. Finally, the green belt parameter *BG* is obtained from eq.15 and reported in the above table.
The attractiveness of eq.15 and the usefulness of the newly proposed *VFP* are brought out by the above design example. While the said equation is simple to use, it relates the progressive loss of energy within the Greenbelt as a function of *VFP* which contains the vegetal parameters, flow parameters and greenbelt parameters. Hence, for a required E_l, one could obtain the *BG* and *SP/D*. Secondly; since $E_{L(Reg)}$ is proportional to *VFP* we get a higher *BG* for a weaker species (like Oak) and lesser *BG* for a stronger species (like Mangrove) which is physically understandable. In other words, realistic design of Greenbelt is possible with the proposed equations which include the *VFP* parameter.

Properties	Oak tree[Light]	Casurina[Medium]	Mangrove[Heavy]
E	9Gpa	14Gpa	18.5Gpa
Dt	0.267 m	0.267 m	0.267 m
λ	1	1	1
f_1	0.3037	0.37	0.43
Vr	51.9	41.6	36.2
SP/Dt	6	6	6
RR	3.87	4.02	3.98
Design BG	**120 m**	**80 m**	**60 m**

Table 3. Design example-Comparison of design for various species.

3.4 Engineering benefits of vegetation on beaches

This part of the chapter is purposed in bringing out the beneficial effects of vegetation on beaches. It is common to presume that the presence of vegetal patches on beach fronts would help protect the beach. Numerous studies have been reported in the literature, wherein, the friction factors or drag coefficient due to vegetation is reported.

Pressure reduction on a wall protected by vegetal patch

The variation of measured non-dimensional pressure, $P^* = p_{max}/H$ during the impingement of wave, with vegetal parameter (α) and relative water depth (h_s/gT^2). Herein, the vegetal parameter (α) is defined as ($BG{\cdot}SP/D^2$). Typical plots showing the variations of dimensionless pressure along the vertical wall, for $h_s/(gT^2) = 0.0004$ and various α, are shown in Fig. 14. As the wall is exposed to shallow water waves the variations in the dynamic pressures along its depth is found to be insignificant for any of the two configurations considered. Further, the results indicate, that the pressures at all the three wall ports reduce in presence of vegetation. Note that the pressure reduces from about 10% to a maximum of about 35%. Further, the pressure reduction is observed to be almost twice for staggered vegetation. The pressures measured by the pressure port near the free surface at $z/h_s = -0.05$ is lower than that of the middle pressure transducer, while that close to the bottom experiences the least pressure. However a pressure reduction at the top transducer is not expected. Note that the top pressure port is close to the free surface and above the lowest location of the wave crest, leading to a pressure relief in this region as the wave runs up the wall. Hence, the trend in pressure variation along a wall as reported. Typical variations of P^* versus α for $z/h_s = -0.45$ and two values of $h_s/(gT^2)$ shown in Fig. 15 indicate that the least pressure is experienced by wall-fronted by staggered vegetation. The variation of P^* with $h_s/(gT^2)$ for all the pressure port locations and $\alpha = 375$ in Fig. 16 shows that it decreases with an increase of $h_s/(gT^2)$ as pressure exerted by long waves is higher, which was found consistent for the other α tested.

The results on the wall pressures revealed that the staggered configuration has a clear advantage because the pressure reduction is found to be almost twice. This is due to the continuous dissipation from the sea-front to the lee-side of vegetation caused by staggered vegetation which is absent for tandem vegetation. This aspect was confirmed by Kothyari *et al.* (2009) through experiments in open channel flow. It is observed that the placement of vegetation has a major role in drag offered by the stems in the flow. As a result more energy is dissipated in case of staggered placement, leading to reduced pressures.

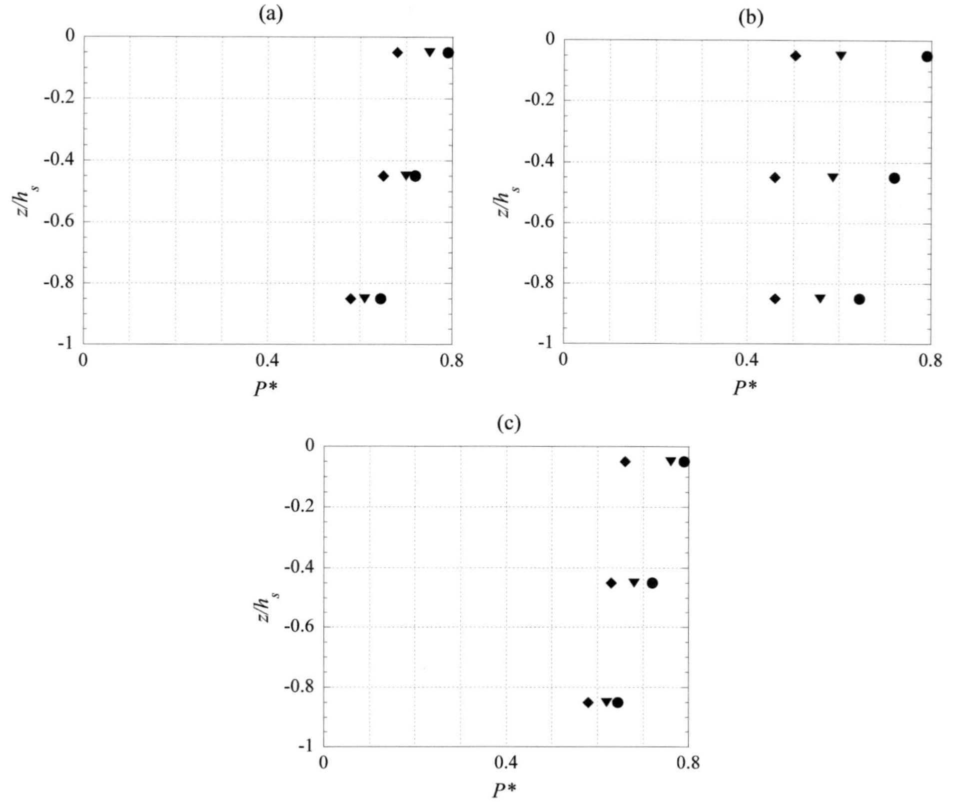

Fig. 14. Variation of $P^* = P_{max}/H$ with z/h_s for H=0.11 m and h_s/gT^2=0.0004 with (●) No, (♦) staggered and (▼) tandem vegetation for α = (a) 1041, (b) 2083, (c) 2604

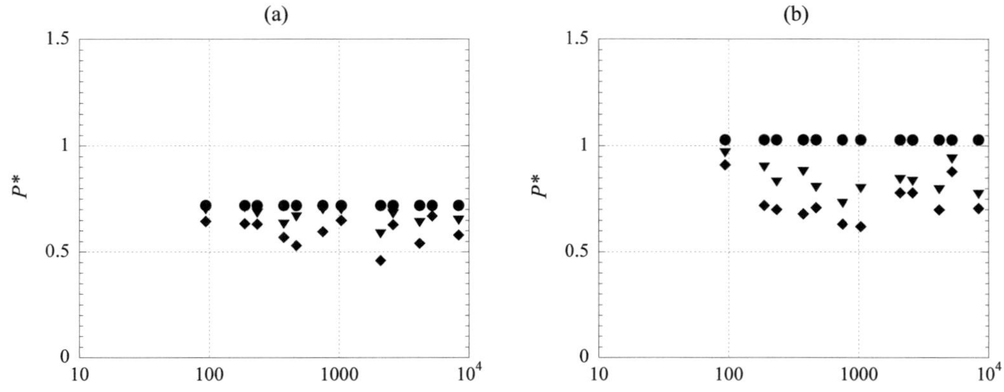

Fig. 15. Variation of $P^* = P_{max}/H$ with α at z/h_s =–0.45 for h_s/gT^2 = (a) 0.0004, (b) 0.0001 (with (●) No, (♦) staggered and (▼) tandem)

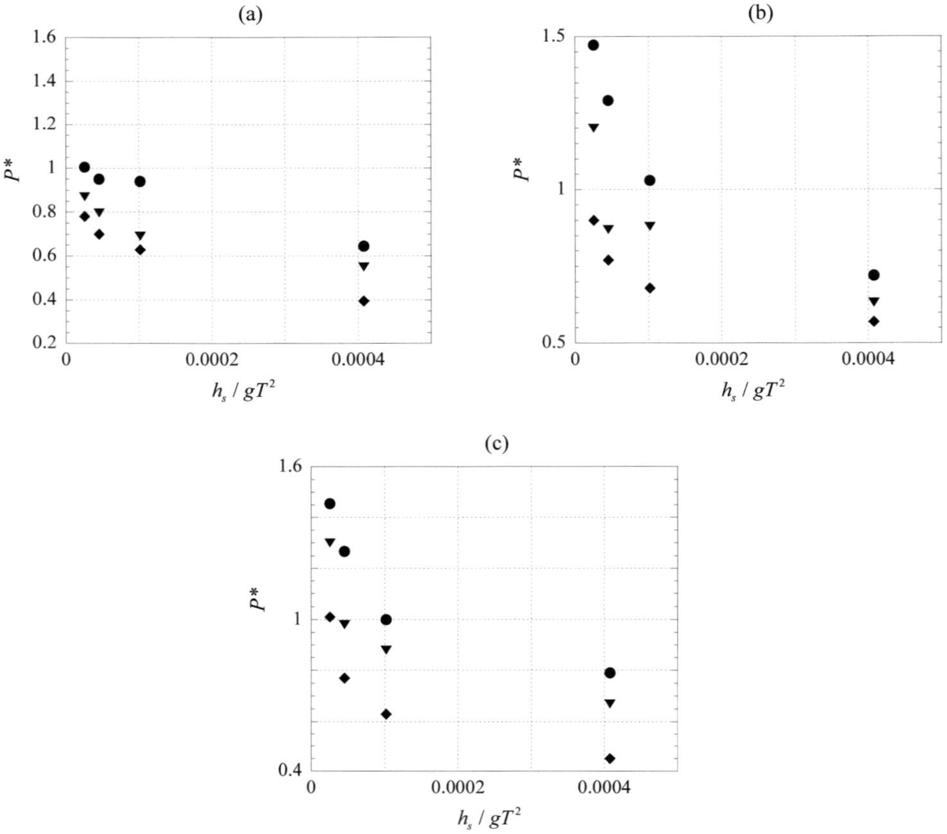

Fig. 16. Variation of $P^* = P_{max}/H$ with h_s/gT^2 for α =375 and z/h_s = (a) –0.85, (b) –0.45, (c) –0.05, (with (●) No, (♦) staggered and (▼) tandem)

The parameter α appears to be reasonable to quantify the effect of vegetal placement. Increasing SP/D corresponds to an increase in the spacing between the individual stems within the green belt, leading to an inferior green belt, for which BG is wider. It was observed that a dense vegetal placement reduces pressure as $\alpha < 750$ with the percentage reduction in pressure increase with an increase in the green belt width by about 25 to 30%. For $\alpha > 750$, i.e. a larger spacing between the stems, a marginal drop in the mean percentage of pressure reduction of about 22 to 30% was observed. From the time histories recorded by the centre pressure sensor, equations through a regression analysis for pressure reduction due to staggered and tandem vegetations were obtained as

$$P_{StaggeredVeg} = 1.244\left(\frac{h_s}{gT^2}\right)^{-0.3232} \cdot \alpha^{0.0062} \tag{19}$$

$$P_{TandemVeg} = 0.092\left(\frac{h_s}{gT^2}\right)^{-0.4635} \cdot \alpha^{0.0874} \tag{20}$$

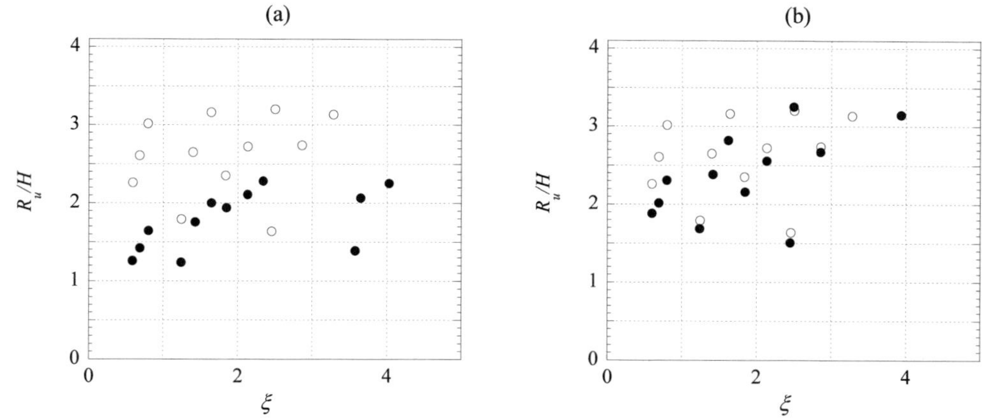

Fig. 17. Variation of R_u/H with ξ for α = (a) 375, (b) 750, Symbols (with (•) No, (♦) staggered and (▼) tandem

Run up on beaches fronted by vegetal patch

The tests on the measurement of wave run-up on the slope were conducted only for staggered vegetation for which the pressures on the lee-side wall were less than that for tandem vegetation, as reported above. Only typical results on the variations of R_u/H with ξ in presence and absence of vegetation for two values of α are shown in Fig. 17 for *BG*= 1 m. It is seen that R_u/H increases with an increase in ξ. The plots also show variations of wave run-up on a slope without vegetation. A significant reduction of run-up is observed in presence of staggered vegetation.

The non-dimensional run-up increases linearly as the surf similarity parameter increases. The percentage reduction of run-up due to waves in the presence of the vegetation is about 40% for the higher surf similarity parameter. Even if the width of vegetation with respect to the water depth is less, the attenuation is dominant due to the flow passing through a medium of lower porosity. The R_u/H on the slope decreases in the presence of green belt, the rate of reduction being significant to an extent of about 60% for Cnoidal waves for higher values of ξ, that is, for long waves propagating over a slope. The tests with long waves exhibits that high degree of reduction in the run-up can be achieved even with sparse spacing (larger porosity) provided the greenbelt is wider. For larger width of green belt, the run-up is independent of the porosity, as shown using the *SP/D* parameter, of the vegetation. The percentage reduction of run-up is more for lesser *BG* of lesser porosity. The effect of vegetation is to reduce the R_u/H_s by about 40 %. This is valid for all the Vegetal parameter investigated.

Attenuation of forces on model buildings

Only salient results on the variations of dimensionless force, F^* (= $\left[\left(F_{max}\right)/\left(0.5*\rho*g*H^2*B_s\right)\right]$) are presented herein in order to clearly bring out the prime aspects of attenuation of wave forces on typical model building fronted by vegetation. The variations of F^* with the vegetation-flow parameter *VFP* is plotted in Fig.18 for the value of α=375 and for *G/B* values of 0, 0.5, 1, and 1.5. To bring out clarity on the effect of vegetation, the results obtained for forces in its presence and absence are superposed. For the case *G/B*=0 (recall, G/B is gap

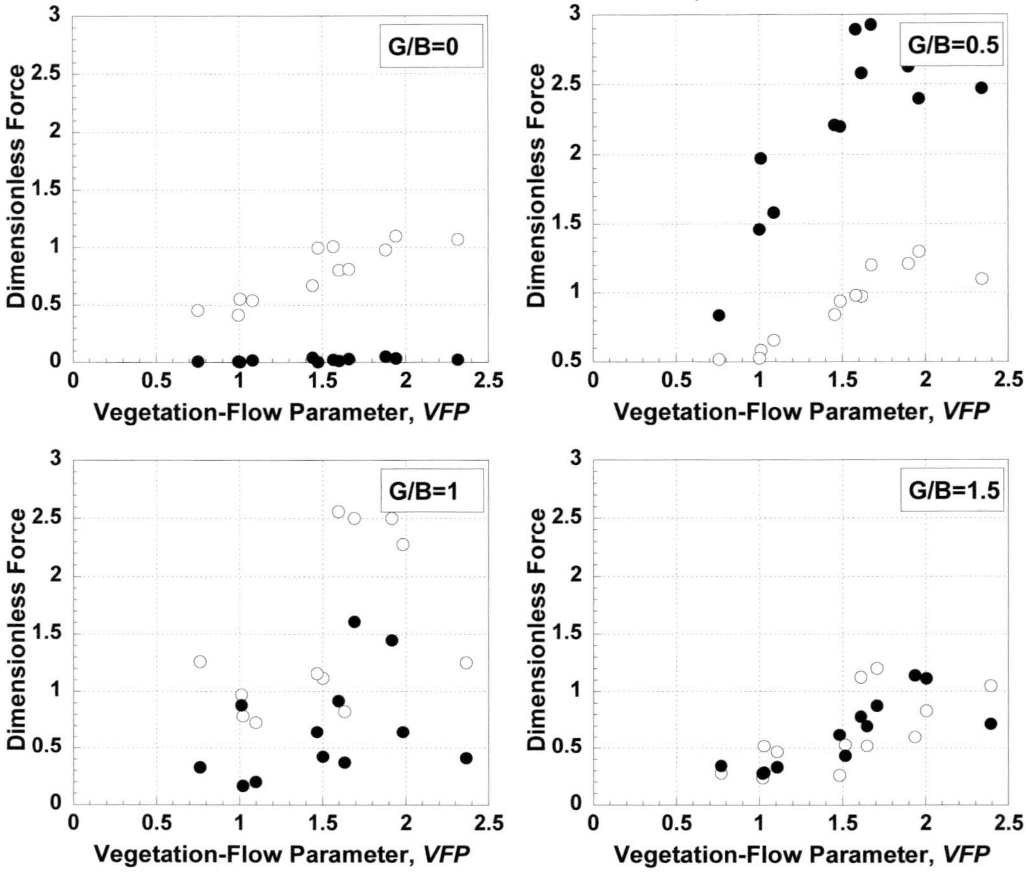

Fig.18. Variation of Dimensionless Force with Vegetation-Flow Parameter for ($BG*SP/D^2$)=375 (O-No Veg and ●-Veg)

between rear face of vegetal patch and the front face of building); the maximum dimensionless force in the presence of vegetation is found to be about 80 - 95% less than that with vegetation. As *VFP* increases the force on the structure in the absence of vegetation is found to increase, whereas, this effect is not found to be significant when vegetation is present in front of the structure. For the case of *G/B*=0.5, the forces are clearly seen to increase for the structure fronted by vegetation, the rate of increase being higher with an increase in *VFP*. For the case of *G/B* = 1, although a similar trend in its variation as observed for *G/B*=0.5 is seen, the rate of increase in the force is found to much lesser. For the last case of *G/B* = 1.5, the forces in the presence of vegetation is found to be less compared to that obtained for the earlier two cases. It is observed that for the gap ratios <1.5, there is significant standing wave formation in the space between vegetation and building leading to the increase in the force. In addition, the kinetic energy of the reformed wave on the lee side of the vegetation may be more incase of *G/B* = 0.5 while this effect reduces as *G/B* increases. Hence most favorable location for the structure is adjacent to the green belt (*G/ B* =0) or away from the green belt by more than 1.5*B*. The maximum value of the dimensionless force is 1.2, 3, 2.6 and 1.3 for *G/B*=0, 0.5, 1, 1.5 respectively.

It is also observed that the F^* has a linear relationship with *VFP* for a given *[BG*SP/D²]*. Further, most of the points pertaining to G/B = 1.5 shows that the forces are mostly lower incase of buildings fronted by vegetation that the case of no vegetation. Hence it may be concluded that the value of G/B > 1.5 will yield beneficial effects incase of bio-shield. However the force reduction incase of G/B>1.5 could not be quantified as such a scenario was not included in the present experimental program. Considering the fact that all experiments are conducted for high Ursells parameter ($U_r \geq 180$), the above discussion indicates that G/B=0 and G/B>1-5 will be certainly beneficial against disastrous waves.

4. Summary and conclusions

A comprehensive experimental program was carried out for investigation of the interaction of emergent vegetation of different characteristics with uniform steady open channel flow as well as with waves of high Ursell's parameter. The study involved measurement of energy loss in the experiments in open channel. In addition, measurement of pressures on a vertical wall and forces on a model building, both placed at the shoreline and wave run up on the beach with a slope of 1:30, have been carried out in a wave flume.

An essential aspect of the present study is that the hydro elastic regime of the present problem, pertaining to plantations along open coasts, has been modelled using Froude law. The effects of canopy and leaf area, wind, surface roughness of the vegetal stems etc are not considered.

The resistance due to vegetation in a steady-uniform flow have been presented and discussed in terms of Darcy's-Weisbach friction factor f and Manning's friction co-efficient n in both Regular and Staggered configuration of elastic cylindrical stems emerging the water surface. The friction factors have been investigated in terms of *Re*, *Fr*, Reduced Velocity (V_r) and vegetation-flow parameter (*VFP*). The Vegetation-Flow Parameter is a newly proposed non-dimensional parameter to represent the relative elasticity of vegetal stems with respect to flow characteristics. Based on the above studies, new semi empirical relationships are proposed for f, n and E_L (energy loss) for all the flow regimes tested in terms of *Fr* and *VFP*.

It is concluded from the wave flume experiments that the Staggered configuration of the vegetation can serve as a better attenuator of waves, thus leading to lesser pressures on a vertical wall on its leeside for all the flow and vegetal parameters considered in the study. The pressures reduce on the wall by about 15-30% for regular vegetal configuration and by about 30-60% for Staggered vegetal configurations.

The non-dimensional run-up increases linearly as the surf similarity parameter increases. The percentage reduction of run-up due to waves in the presence of the vegetation is about 40% for the higher surf similarity parameter. R_u/H on the slope decreases in presence of Green Belt, the rate of reduction being significant to an extent of about 60% for Cnoidal waves for higher values of ξ, that is, long waves propagating over a slope. For larger width of green belt, the run-up is independent of the porosity, as shown using the *SP/D* parameter, of the vegetation. The percentage reduction of run-up is more for lesser *BG* of lesser porosity. The effect of vegetation is to reduce the R_u/H_s by about 40 %. This is valid for all the Vegetal parameter investigated.

The results on the variation of forces on a structure positioned at different distances from the green belt and subjected to the action of Cnoidal waves are presented and discussed in detail. The parameter, *[BG*SP/D²]* is found to be quite useful in understanding the effects of placement of vegetation in combination with the width of green belt on the forces, F^*. On the other hand *G/B*=0 or *G/B*>1.5 has been found to be suitable for the green belt to be effective in reducing theforces on the structure on its lee side, while *G/B*=0.5 has been found to cause adverse effect.

5. References

Carrier, G.F. and Greenspan, H.P. (1958). Water waves of finite amplitude on a sloping beach. *Journal of Fluid Mechanics*, 4, 97-109.

Carrier, G.F., Wu, T.T. and Yeh, H. (2003). Tsunami run-up and draw-down on a plane beach. *Journal of Fluid Mechanics*, 475 (1), 79-99.

Chen,C.I. (1976). Flow resistance in board shallow grassed channels. *Journal of the Hydraulics Division*, ASCE 102 (3), pp 307-322.

Chakrabarti, S.K. (1983). Hydrodynamics of offshore structures. Computational Mechanics Publications, Southampton Boston.

Darby, S. E. (1999). Modeling effect of riparian vegetation on flow resistance and flood potential. *J. Hydraul. Eng.*, 125. pp 443–454.

Dudley, F.S.J., Abt, S.R., Bonham, C.D., Watson, C.C., and Fischenich, J.C. (1998). Evaluation of flow-resistance equations for vegetated channels and floodplains. *Technical Report EL-98-2. U.S. Army Engineer Research and Development Center, Vicksburg, MS.*

Fathi-Moghadam, M. and Kouwen, N., (1997). Nonrigid, nonsubmerged, vegetative roughness on floodplains. *Journal of HydraulicEngineering* 123 (1), 51–57.

Fathi-Moghadam, M. (2006). Effects of land slope and flow depth on retarding flow in non-submerge vegetated lands. *J. Agron.ANSI.*5 (3), pp 536-540.

Fathi-Moghadam, M. (2007). Characteristics and mechanics of tall vegetation for resistance to flow, *African jr of biotechnology* Vol.6 (4), pp 475-480.

Fischenich, C. (2000). Resistance due to vegetation, *EMRRP Technical Notes, ERDC TN-EMRRP-SR-07, US Army Engineer Research and Development Center, Vicksburg, MS.*

Freeman, G. E. (1997). Analysis and prediction of plant stiffness modulus for selected plants. *Federal Contract No. DACW-39-97-M-1413, U.S. Waterway Experiment Station, U.S.Army Corps of Engineers*, Vicksburg, Miss.

Freeman, G.E., Rahmeyer, W. H. and Copeland, R. R. (2000). Determination of resistance due to shorts and woody vegetation. *Rep.ERDC/CHL TR-00-25, Engineer Research and Development Center,U.S. Army Corps of Engineers.*

Fuehrboeter, A. (1993). Wave loads on seadikes and seawalls. In: Abbott, M.B., Price, W.A. (Eds.), Coastal,Estuarial and Harbour Engineer's Reference Book. Chapman & Hall, London, pp. 351–367.

Furukawa, K., Wolanski, E., Mueller, H. (1997). Currents and sediment transport in mangrove forests. *Estuarine and Coastal Shelf Science* 44(3), 301-310.

Gan, K.S., Lim, S.C., Choo, K.T., Jantan, M.D. (2001). Timber notes-medium hardwoods 4. *Timber Technology Bulletin* 21, 1-7.

Goda, Y. (1974). New wave pressure formula for composite breakwater, *Proceedings of the 14th International Conference on Coastal Engineering, Copenhagen, Denmark* (1974) pp. 1702–1720.

Hall Jr, J.V. and Watts, G.M. (1953). Laboratory Investigation of the vertical rise of solitary waves on impermeable slopes Techincal Memorandum No.33, Beach Erosion Board, U.S.Army corps of Engineers, Washington, DC.

Hamzah, L., Harada, K., Imamura, F. (1999). Experimental and numerical study on the effect of mangrove to reduce tsunami. *Tohoku J. Nat. Disaster Sci.* 35(2), 127-132.

Harada, K. and Imamura, F. (2005). Effects of coastal forest on tsunami hazard mitigation - A preliminary investigation. Tsunamis, Case Studies and Recent Developments, Springer, Netherlands, pp. 279-292.

Hughes, S.A. (2004). Estimation of wave run-up on smooth, impermeable slopes using the wave momentum flux parameter, *Coastal Engineering*, Vol 51, pp- 1085-1104.

Jarvela,J. (2002). Determination of flow resistance of vegetated channel banks and flood plains. In River flow 2002,(Eds.Bousmar,D.and Zach,Y.) Lisse,Swets Zeitlinger, pp 311-318.

Järvelä, J. (2004). Determination of flow resistance caused by non-submerged woody vegetation, *Int. J. River Basin Manage.* 2 (2004) (1), pp. 61–70.

Kothyari, U.C, Hayashi, K., Hashimoto, H. (2009). Drag coefficient of unsubmerged rigid vegetation stems in open channel flows. *J. Hydraulic Res.* 47(6), 691-699.

Kouwen, N. and Li, R.M. (1980). Biomechanics of vegetative channellinings. *Journal of the Hydraulics Division*, ASCE 106 (6), 1085–1103.

Kouwen, N. and Fathi-Moghadam, M. (2000). Friction factors for coniferous trees along rivers. *Journal of Hydraulic Engineering*, ASCE, 126(10), pp 732-740.

Kutija, V., Hong, H.T.M., (1996). A numerical model for assessing the additional resistance to flow introduced by flexible vegetation. *J. Hydraulic Res.* 34(1), 99-114.

Li, Y. and Raichlen, F. (2001). Solitary wave run-up on plane slopes., *Journal of Waterway Port, Coastal and Ocean Engineering*, Vol. 127, pp. 33–44.

Liu, P., Lynett. and Wu, T. (2002). Modeling wave run-up with depth integrated equations, *Journal of Waterways, Port, Coastal and Ocean Engineering, ASCE* 46, pp. 89–107.

Meijer, D.G. and Van Velzen, E.H. (1999). Prototype-scale flume experiments on hydraulic roughness of submerged vegetation, *Proceedings of XXVIII AIHR Conference, Graz (A), Sept. 1999.*

Mueller and Whittaker, (1993). G.U. Muller and T.J.T. Whittaker, An investigation of breaking wave pressures on inclined walls, *Ocean Engineering* 20 (4) (1993), pp. 349–358.

Nepf, H.M. (1999). Drag, turbulence and diffusion in flow through emergent vegetation. *Water Resources. Res,*. 35 (2), pp 479–489.

Noarayanan, L., Murali, K. and Sundar, V. "Wave attenuation by flexible vegetation on a mild slope"Proc. International Conference in Ocean Engineering-2-5 Feb 2009 IIT Madras, INDIA, pp 1157-1169

Petry.K.S. and Bosmajian,G.B. (1975). Analysis of flow through vegetation. *Journal of hydraulic division*.ASCE 101(7), pp 871-884.

Ree, W.O. and Crow, F.R. (1977). Friction factors for vegetated waterways of small slope. *Agricultural Research Service* ARS-S-151. U.S. *Department of Agriculture, Washington, D.C.*

Righetti, M., and Armanini, A. (2002) Flow resistance in open channel flows with sparsely distributed bushes." *J. Hydrol.*, 269_1-2_, 55-64.

Sarpkaya, T. and Isaacson, M. (1979) Mechanics of wave forces on offshore structures, Van Nostrand reinhold company , New york, pp-420(to be removed)

Synolakis, C.E. (1986). The run-up of solitary waves. Journal of Fluid Mechanics vol. 185. Cambridge University Press, United Kingdom, pp. 523-545.

Tanaka, N. (2009). Vegetation bio-shields for tsunami mitigation: Review of effectiveness, limitations, construction, and sustainable management. *J. Landscape and Ecological Eng.* 5(1), 71-79.

Timoshenko, S.P. and Gere, J.M. (1961). *Theory of elastic stability, McGraw Hill, New York.*

Werth, D. (1997). Predicting flow resistance due to vegetation in flood-plains, *PhD thesis, Utah State University, Logan, Utah.*

Yang, W. and Choi,S.U (2009). Impact of stem flexibility on mean flow and turbulence structure in open-channel flows with submerged vegetation. Jl. Hydraulic Res. 47 (4), 445-454.

Zdravkovich, M.M. and Medeiros, E.B. (1991). Journal of Wind Engineering and Industrial Aerodynamics. *Elsevier Science Publishers* B.V.,Amsterdam - Printed in The Netherlands, 38 pp 197-211.

9

Effectiveness and Limitations of Vegetation Bioshield in Coast for Tsunami Disaster Mitigation

Norio Tanaka
Graduate School of Science and Engineering, Saitama University
Japan

1. Introduction

Tsunamis are one of the most awful natural disasters that can cause catastrophic damage to both human life and socioeconomic property. When a tsunami reaches a coast, it may appear as a rapidly falling or rising tide or a series of breaking waves. When a tsunami reaches the shore, part of the tsunami wave is reflected offshore, but most of it slows down, increases in height, runs up towards the land with huge energy, and causes massive destruction on the coast and in the hinterlands.

Mitigation, preparedness, and escape are representative countermeasures to protect human lives and infrastructure facilities from natural hazards. Mitigation techniques are broadly categorized in two ways. These are artificial methods (hard solutions) and natural methods (soft solutions utilizing a natural buffer zone of coastal vegetation, sand dunes, lagoons, or coral reefs). Artificial methods are mainly kinds of sea walls (huge embankments, tsunami gates) that can be constructed on the coastal area for any predicted tsunami height. However, the construction costs of the artificial methods can be very high, which restricts development in many cases. Some developed countries like Japan, which has frequent tsunamis or storm surge threats, have employed such techniques. On the other hand, the use of coastal vegetation as a natural method for disaster mitigation was also discussed in Japan more than 100 years ago (Honda, 1898).

Recently, natural methods have been widely understood to reduce the tsunami energy remarkably, although they cannot completely stop the tsunami itself, and their effectiveness depends on the magnitude of the tsunami and the type of vegetation structure (Tanaka et al., 2007). Shuto (1987) analyzed the effects and limitations of coastal vegetation from historical records of tsunamis that occurred in Japan. Especially, since the 1998 Papua New Guinea tsunami (Dengler & Preuss, 2003), many researchers have begun investigating the effects of coastal vegetation on tsunami mitigation using water flume experiments (Harada & Imamura, 2000), field investigations and numerical simulations (Hiraishi & Harada, 2003). After the Indian Ocean tsunami on 26 December 2004, further research was conducted to elucidate the effectiveness of coastal vegetation (Danielsen et al., 2005; Kathiresan & Rajendran, 2005, 2006; Dahdouh-Guebas et al., 2005; Harada & Imamura, 2006; Tanaka & Sasaki, 2007; Tanaka et al., 2006a, 2006b, 2007; Sasaki et al., 2007; Nandasena et al., 2007) in order to minimize infrastructure damage and protect human lives. Danielsen et al. (2005) pointed out that the deterioration and clearing of mangroves and other types of coastal vegetation along many coastlines has increased their vulnerability to storm and tsunami damage. Many field investigations after the huge tsunami suggest that establishing or strengthening greenbelts of mangroves and other coastal forests may play a key role in reducing the effect of future extreme events. Therefore, natural methods that employ coastal vegetation together with other natural features such as sand dunes or lagoons have been studied because they require relatively little capital investment compared to artificial measures, provide human-friendly beach fronts, and enhance inter-relationships with other ecological systems.

For optimal planning of the natural systems and their maintenance, the effects of vegetation on the reduction of tsunami energy are to be quantitatively elucidated. This requires knowledge of the limitations of coastal forests in tsunami mitigation and consideration of how to maintain the forest as a natural disaster buffer zone. The objectives of this article are to review the important points about the 1) effectiveness and vulnerability of coastal forests against tsunami revealed by field investigations, 2) implications of results obtained from numerical simulations, 3) breaking conditions of coastal tree species and forests, and 4) future establishment and management of coastal vegetation.

2. Effectiveness and vulnerability of coastal forests against tsunami revealed by field investigations

2.1 Effectiveness of coastal forests in tsunami mitigation

Dengler & Preuss (2003) investigated the disaster caused by the 1998 Papua New Guinea tsunami and found that *Casuarina* trees presented relatively greater resistance than palm trees. However, they pointed out the need to conduct further research in understanding the interaction of trees, roots, and high-flow water regimes in order to utilize vegetation as a tsunami abatement measure because significant scouring occurred in the areas where the root systems were undermined, and in some cases, trees caused additional damage to their surroundings in the fast-moving water. This demerit was also pointed out by Shuto (1987).

Kathiresan & Rajendran (2005) concluded that the presence of mangroves reduced the human death toll along the Tamil Nadu coast of southeast India, although Kerr et al. (2006) argued that all mechanistically important factors should be addressed in considering the issue. However, the buffering behavior of mangroves has been examined in many other post-tsunami investigations. Dahdouh-Guebas et al. (2005) showed by cluster analysis that the man-made structures located directly behind the most extensive mangroves were less damaged, while Danielsen et al. (2005) reported that a dense mangrove of *Rhizophora* spp. and *Avicennia* spp. (density = 14-26 tree trunks per 100 m^2) contributed to decrease damage in 96% of surveyed cases in India. Exposed villages suffered the highest levels of damage, and those behind mangroves experienced intermediate levels of damage.

Tanaka et al.(2007) showed some examples that coastal vegetation could reduce the tsunami velocity, and water depth by comparing damage situations with/without vegetation in Sri Lanka and Thailand. Fig.1 (a) shows a schematic of the tsunami damage at Kalutara, Sri Lanka. The *Cocos nucifera* trees growing along the coast (Line A) were 3-7 m in height and 3 m apart, but some of them were damaged, and houses behind the trees were damaged up to

100 m from the coast. *Pandanus odoratissimus* and *Casuarina equisetifolia* were found as coastal vegetation at Line B, 500 m south of Line A and near the mouth of the Kalu Ganga (river). The vegetation width in the tsunami direction was 20 m and 40 m for *P. odoratissimus* and *C. equisetifolia*, respectively. At the interface, *P. odoratissimus* and *C. equisetifolia* were mixed for a distance of about 20 m. The tsunami height 60 m inland from the coast was 0.6 m, and the houses located within this area were not as damaged because the tsunami height was low compared to its height at Line A. It is supposed that the tsunami current was repelled by the vegetation because they played a role in blocking the current.

Fig.1(b) shows an example how the growth stage affects tsunami inundation. At Line A, the tsunami, about 3 m high at the coast, swept about 1.8 km inland. It passed through 200 m of *C. equisetifolia* vegetation. The forest does not contribute the tsunami mitigation. The trunk diameter of the trees was about 0.5-1 m, but the forest density is very low. At Line B, *C. equisetifolia* is younger and the density is larger in comparison with Line A. The tsunami broke young *C. equisetifolia*, the trunks of which were about 0.07 m in diameter. However, the broken trees and branches were accumulated in front of the *C. equisetifolia* of which trunk diameter is larger than 0.1m. It had a role to repell the current.

2.2 Effects of forest density, tree-trunk diameter, and stand structure of trees

Field surveys in Sri Lanka and Thailand after the Indian Ocean tsunami showed that older (larger diameter) *Casuarina equisetifolia* belts on the coast withstood the tsunami but failed to provide good protection (Tanaka et al., 2006a, 2006b, 2007). Tanaka et al. (2007) showed that tree growth and forest density can have a significant effect on tsunami mitigation because trees with larger trunk diameters require more space (lower tree density) between them to grow. It suggests that trunk diameter and density effect, important parameters for estimating vegetation drag, cannot be discussed independently. Tanaka et al. (2007) reported that the relationship between the crown height and tsunami height is also important in terms of the drag characteristics of broad-leaved trees because they have branches with large diameters. The United Nations Food and Agriculture Organization (FAO) (2007) pointed out that active forest management is required to produce variously-aged stands of trees with a range of sizes and with branches at all levels to enhance the potential for mitigation, particularly in smaller tsunamis, because the mitigation potential of *Pinus* spp. and *Casuarina* spp. declines with age due to self-thinning (density is decreased) and because tree crown height exceeds the tsunami water depth, etc.

A combination of different tree species is recommended in a buffer forest (Tanaka et al., 2007, 2008b, 2009). *Pandanus odoratissimus* (Fig.2a) was observed to grow under the shade of the taller *Cocos nucifera* and *Casuarina equisetifolia* (Fig.2b). In field observations, two layers of vegetation, *P. odoratissimus* and *C. equisetifolia*, in the vertical direction exhibited a strong potential to decrease the damage behind the vegetation cover (Tanaka et al., 2007), but a combination of *P. odoratissimus* and *C. nucifera* had little effect because they leave wide gaps vertically. *C. equisetifolia* is recommended as an appropriate tree species for a mixed forest. If it does not break, it traps debris. Considering the other roles, i.e. trapping debris and providing something to climb to escape or a soft-landing place for people, that coastal vegetation plays (Shuto, 1987; Tanaka et al. 2006b, 2007), a two-layer forest of *P. odoratissimus* and dense *C. equisetifolia* should be planted and preserved near the coast, and other broad-leaved trees should be grown behind this buffer forest (towards inland), as shown in Fig. 3 (Tanaka et al., 2007).

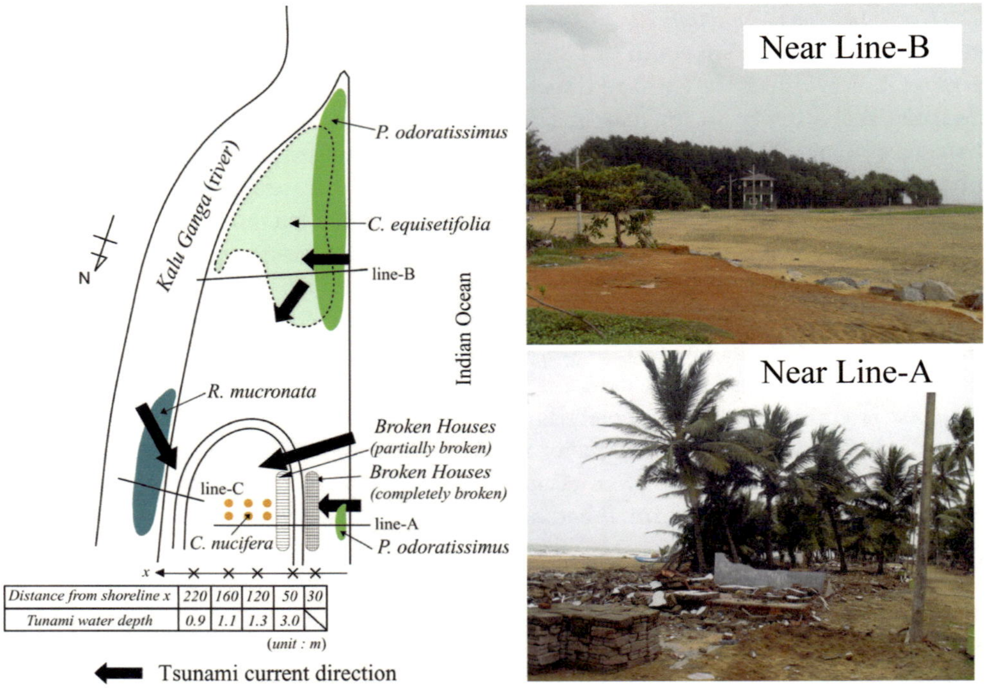

(a)

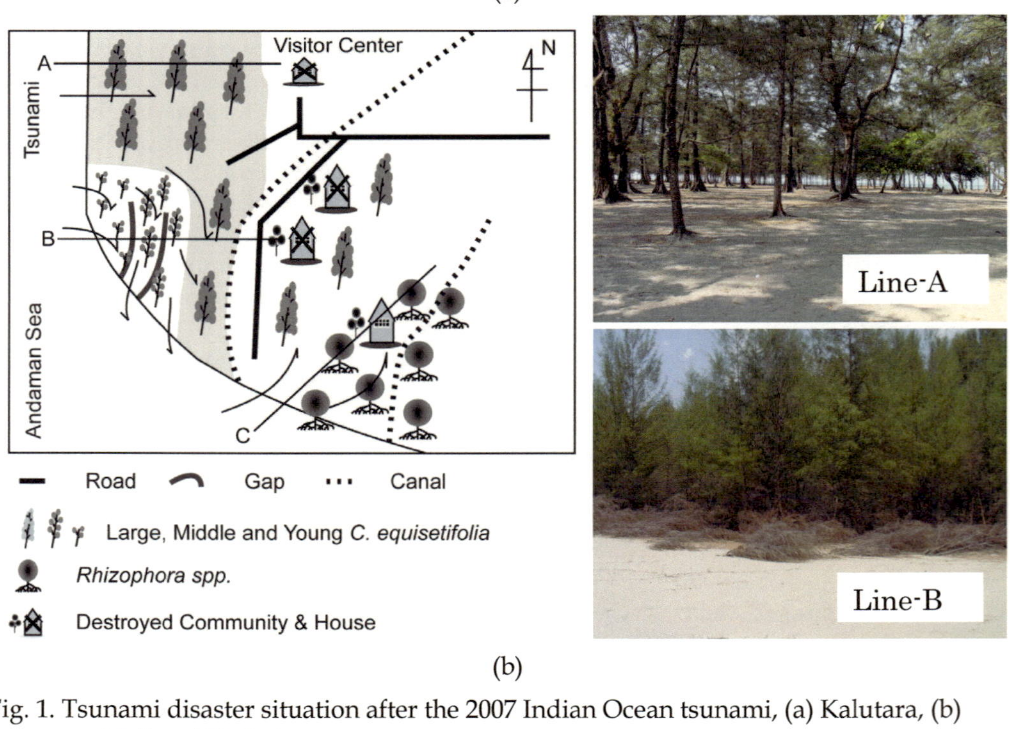

(b)

Fig. 1. Tsunami disaster situation after the 2007 Indian Ocean tsunami, (a) Kalutara, (b) Laem son National Park

(a)

(b)

Fig. 2. Effective coastal vegetation, (a) *Pandanus odoratissimus*, and (b) *Casuarina equisetifolia* in tropical countries (undergrowth bushes is young *P. odoratissimus*)

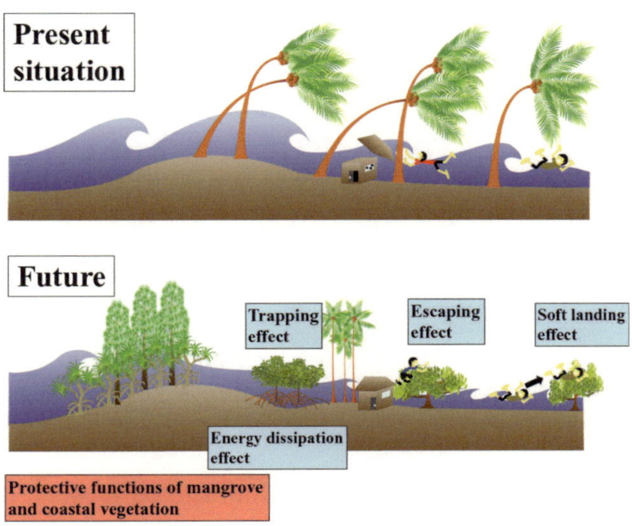

Fig. 3. Functions of coastal vegetation for tsunami disaster mitigation (modified from Tanaka et al.(2007))

3. Implications from numerical simulations

3.1 Effectiveness of coastal vegetation for reducing velocity and runup height of tsunamis

A densely grown coastal vegetation belt of *P. odoratissimus* for reducing the tsunami energy was quantitatively analyzed by an enhanced one-dimensional numerical model that included variations of topography and tsunami characteristics (Nandasena et al., 2008b). The drag and inertia forces were assumed as the total resistance generated by the vegetation. Fig.4 shows an example of their simulation. It can be observed that the wave reflection was dominant at the vegetation front. Vegetation also reduces the velocity, water depth inside forest, and tsunami front arrival. Expecting this kind of vegetation effect, many researches conducted numerical simulations. The model output by Hiraishi & Harada (2003) suggested a more than 90% reduction in maximum tsunami flow pressure by a 100 m-wide forest belt when the tree density is very high (30 tree trunks per 100 m^2). Model results obtained by Harada & Imamura (2006), and Tanaka et al. (2006a, 2008b) for various types of coastal vegetation, including mangroves, were very similar. Tanaka et al. (2006a) modelled the relationship of species-specific differences in drag coefficient to tsunami height. The species differences in mitigating tsunami water depth, velocity, and tsunami arrival time behind a forest in relation to tsunami height were found to be quite large, and *P. odoratissimus* was considered more effective in reducing the velocity and water depth of the current through the forest than other common trees, including the coconut tree (*C. nucifera*) and a kind of mangrove (*Avicennia alba*) using a numerical simulation (Tanaka et al., 2006a). These results point out the importance of preserving or selecting appropriate species to act as tsunami barriers and offer sufficient shoreline protection. For *P. odoratissimus*, extensive numerical simulations were conducted and effects of the vegetation has been quantitatively evaluated with many different tsunami conditions by Thuy et al.(2010b)

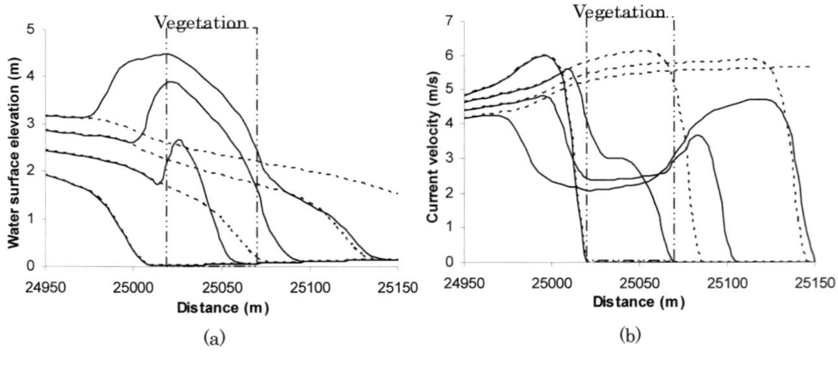

Fig. 4. Spatial variation of (a) water depth, and (b) current velocity for tsunami periods of 10 min (modified from Nandasena et al.(2008b)). Line: with vegetation, dashed-line: without vegetation. Flow is from left to right

3.2 Vulnerability of sand dunes, coastal vegetation, and coral reefs: Open gap problem

A gap in the coastal zone is reported to increase risks and potential damage behind the forest. Gaps in sand dune vegetation are due to natural or artificial causes, such as construction of access roads to a beach or for sand mining, the mouths of rivers, mangrove channels opening onto the sea (FAO, 2007), and coral reefs for constructing headlands or embayments (Fernando et al., 2005). The water flow through the gaps is accelerated as it moves into the constriction (FAO, 2007; Fernando et al., 2005, 2008; Mascarenhas & Jayakumar, 2008; Thuy et al., 2009). When the gap is narrow, it increases the velocity immediately behind the gap, although the water depth will actually decrease in most cases (Nandasena et al., 2008a). The areas behind a coastal forest can still be protected from the tsunami (FAO, 2007; Thuy et al., 2008), but gaps increase the hazards in the gap line (Fernando et al., 2005, 2008; Cochard et al., 2008). As the gap width increases, the flow velocity at the end of the open gap first increases, reaches a maximum, and then decreases (Fig.5), while the run-up height increases monotonously. The maximum velocity in the condition by Thuy et al.(2009) is 1.7 times the maximum velocity without a coastal forest (Fig.6).

Although the danger at the gap has been pointed out, large number of coastal vegetation belt were still constructed with the gap after the Indian Ocean tsunami. Fig.7 shows the planting situation in eastern coast in Sri Lanka. A lot of *Casuarina* forests are planted there but some forests has supposed to make dangerous zone behind or around the forest although the tsunami damage in most of the area are supposed to be mitigated. As it is not realistic to consider a coastal forest without any gap in the barrier, careful planning is required in the design of an actual coastal forest to incline the gap direction away from the tsunami current direction or to stagger it to reduce the water velocity through the gap (Thuy et al., 2009).

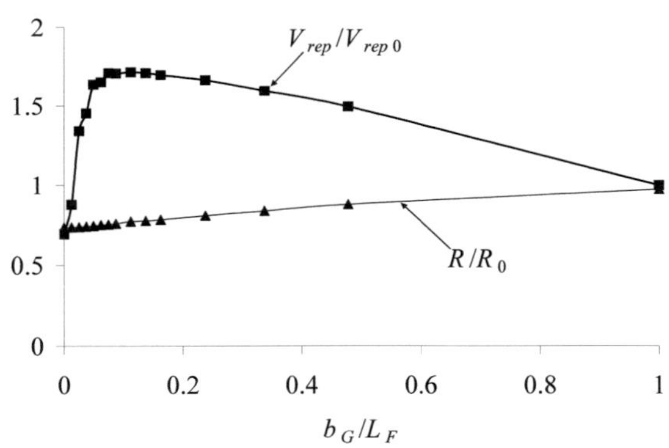

Fig. 5. Variation of representative velocity behind a gap (V_{rep}/V_{rep0}) and run-up height (R/R_0) in relation to the ratio of a gap width(b_G) and channel width(L_F). Subscript 0 indicates the case of no vegetation. (modified from Thuy et al.(2009))

4. Breaking conditions of coastal tree species and forest due to tsunami force

4.1 Damage to single tree

Tanaka et al.(2007) reported the damage situation of tress in detail (Fig.8) Main trunks larger than 0.1 m in diameter were seldom broken by 5-10 m tsunami, except for *Rhizophora apiculata* (Kang Island in Thailand), which was broken by 8 m tsunami, *Rhizophora mucronata* (Medilla in Sri Lanka), by 6 m tsunami, *C. equisetifolia* (Ban Thale Nok in Thailand) by 10 m tsunami, and *P. odoratissimus* (Hikkaduwa, Medilla in Sri Lanka) by 5-6 m tsunami. The

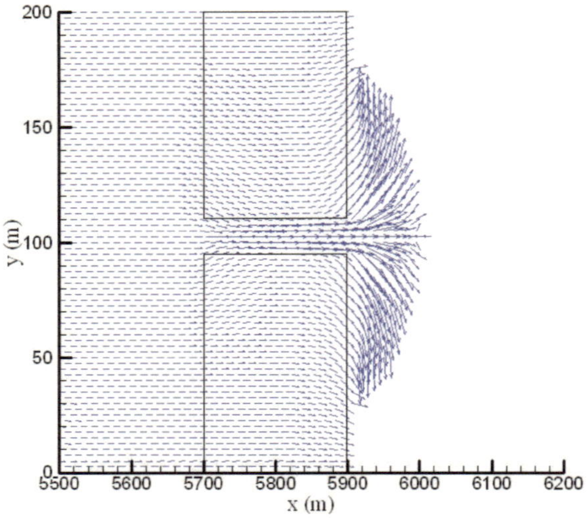

Fig. 6. Flow pattern around a gap (modified from Thuy et al.(2009))

Fig. 7. Planting situation in eastern coast in Sri Lanka, (a) aerial photo (from Google Earth), (b) straight gap inside forest (photo by Methsiri Samarakoon)

pattern of uprooting was related to the strength of the substrate and was usually observed at the front line of the vegetation (Fig. 8(a), (c)). *P. odoratissimus* and *R. apiculata*-type trees have many aerial roots, and the moment by drag force can be shared by the aerial roots. Thus, they were able to withstand a less than 5 m tsunami, even with debris attached to the aerial roots and additional force applied. However, if the drag moment exceeded the threshold for the breaking moment when the tsunami water was high, the trunks were broken just above the aerial roots (Fig. 8(b),(d)).

Tanaka et al. (2006b, 2008b) conducted tree breaking tests in situ using small diameter trunks or branches and showed that the breaking moment curve of representative tree species (Fig.9) can be classified into three parts by elasticity and habitat as

1. the curve for coastal vegetation (ex. *P. odoratissimus*),
2. the curve for elastic trees
3. the curve for trees with hard trunks (ex. *C. equisetifolia*).

Fig. 8. Damaged tree patterns, (a) erosion and overturning of *Casuarina equisetifolia*, (b) tree trunk breakage of *Pandanus odoratissimus*, (c) uprooted *Cocos nucifera*, (d) tree trunk breakage of *C. equisetifolia* and *Rhizophora* spp.

Although the breaking test was conducted in Sri Lanka before the 2006 Java tsunami (Tanaka & Sasaki, 2007), the breaking condition coincided well with the actual damage observed at the sites of the Java tsunami, except for *C. equisetifolia* (Tanaka et al., 2009). *C. equisetifolia* does not have aerial roots, but its diameter becomes greater at ground level than at breast height as the tree grows. Thus, its trunk was rarely broken. Most of the damage to *C. equisetifolia* was due to uprooting. This indicates that the breaking condition is not well expressed by the breaking test using *C. equisetifolia* with small trunk diameters. Further study of the breaking or uprooting conditions for *C. equisetifolia* is recommended with real-scale breaking or uprooting tests like those of Peltola et al. (2000) because *C. equisetifolia* is expected to be an effective species for dissipating tsunami energy in Sri Lanka, Thailand (Tanaka et al., 2006a, 2006b, 2007; FAO, 2007), and India (Mascarenhas & Jayakumar, 2008).

Tanaka & Sasaki (2007) examined the height at which *P. odoratissimus* and *C. nucifera* trees were broken in two tsunamis (the 2004 Indian Ocean and 2006 Java tsunamis) and found that the threshold of tsunami water depth at which a tree trunk broke was about 80% of the tree height for *P. odoratissimus* (Fig.10). *C. nucifera* was seldom damaged except for small trees, but when it was toppled, the critical tsunami height was about 80%, the same as *P. odoratissimus*. Because the breaking point of *P. odoratissimus* is 1-2 m high, the moment by drag at that point is smaller than that of a tree without aerial roots. Although the threshold tsunami water depth for breaking *P. odoratissimus* is similar to that for *C. nucifera*, the trunk

strength of *P. odoratissimus* is not as great and it receives a large drag force from the tsunami current when the tsunami water depth is greater than the height of its aerial roots.
The studies by Tanaka & Sasaki (2007) and Tanaka et al. (2008b) elucidated the breaking condition for a single tree when the scouring around the tree is not severe. However, the breaking condition is not applicable for critical tree damage in the fringe area of a forest, especially the frontal vegetation on sand dunes where the substrate is not hard and severe scouring occurs.

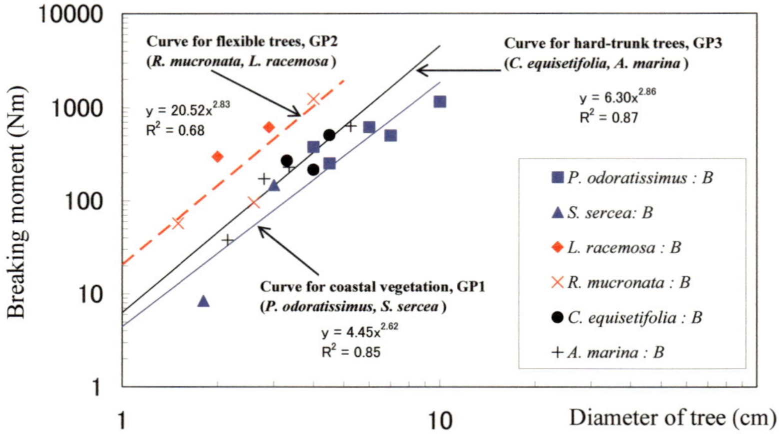

Fig. 9. Breaking moment curve of representative coastal tree species in tropics (modified from Tanaka et al.(2009))

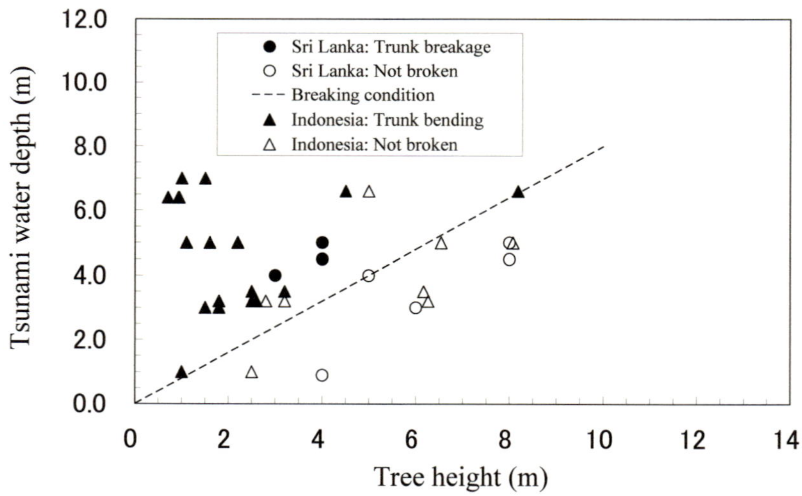

Fig. 10. Tree breaking situation in relation to tsunami water depth and tree height (modified from Tanaka & Sasaki (2007)

4.2 Parameters of effects and limitation of coastal forest for tsunami mitigation

Shuto (1987) discussed the effectiveness and limitations of a forest on the reduction of tsunami energy. He classified the degrees of damage to a forest due to tsunamis, i.e., no damage, tilting or turnover due to large scouring, and trunk breakage or uprooting. The classification was based on the characteristics of a forest from the records of five previous huge tsunamis (in 1896, 1933, 1944, 1960, and 1983) that had occurred in Japan and the forest damage at the tsunami events. The relationships between the tsunami inundation depth and tree diameter at breast height and the extent of the forest were statistically analyzed considering the physical aspects of tsunamis. The length of a forest in the streamwise direction, the size and density of the trees, and the undergrowth are important parameters of tsunami energy reduction. In particular, he derived a simple but important parameter for discussing the degree of the forest damage, the '*dn*', where *d* is the tree trunk diameter at breast height (cm), *n* is the number of trees in a forest in the streamwise direction and in unit widths in the cross-stream direction (trees within the area of the forest length in the streamwise direction (m) × 1 m in the cross-stream direction) as a major measure of resistance by a forest, assuming that the drag coefficient is constant with the changing tree density.

After the Indian Ocean tsunami, Tanaka et al. (2007) investigated the effectiveness of Shuto's classification (1987). The damage to *C. equisetifolia* almost satisfied the criteria using the *dn*. However, the *dn* of *Anacardium occidentale* and *Avicennia alba*, broad-leaved trees, and *P. odoratissimus* and *Rhizophora apiculata*, which have aerial roots, were not well classified because the large diameter branches or aerial roots add additional drag that Shuto (1987) did not consider. Thus, to distinguish the effects of tree species, Tanaka et al. (2007) recommended calculating the effective cumulative tree diameter in the stream-wise direction, dN_{all} (cm/(vegetation width in streamwise direction(m) *1 m)), as:

$$dN_{all} = dn \times \frac{1}{h}\int_0^h \alpha(z)\cdot\beta(z)dz = dn \times \alpha\beta$$

Where, h is the tsunami water depth, $\alpha(z)$, $\beta(z)$ are additional coefficients that expresses the effect of cumulative width of a tree in each height on the drag, and the effect of leaves or aerial roots on the drag, respectively, $\alpha\beta$ is about 1.3, 2, 3 and 3.5 for *C. equisetifolia, A. occidentale, R. apiculata* and *P. odoratissimus*, respectively for 3-5 m tsunami.

Mascarenhas & Jayakumar (2008) also reported the length of vegetation damage in the streamwise direction. *Casuarina* plantations remained intact except for the frontal strips, a few trees that faced the ocean, when tsunami run-up levels ranged from 0.7–6.5 m and the length of flooding area from the coast varied from 31–862 m. Although the threshold breaking condition of a single tree has been elucidated (Tanaka & Sasaki, 2007; Tanaka et al., 2008b) and some investigations reported the breaking length of a forest in the tsunami current direction (Tanaka et al., 2007; Mascarenhas & Jayakumar, 2008), the breaking length of a forest has still not been quantitatively evaluated according to the tsunami condition. More study is needed to estimate the length because the breaking affects the dissipation of tsunami energy.

4.3 Limitations of forest

The tsunami force is directly related to the damage of trees, however, the effects of breaking on numerical simulation results were not directly discussed in previous studies except for Thuy et al.(2010a). Thuy et al.(2010a) developed the numerical model for estimating tsunami bending moment on a tree and including tree breakage, and then the damage length of vegetation, reduction of water depth and tsunami force behind the forest are discussed. The study elucidated that: (1) tsunami moment is decreased from the front to the behind due to the decrease in water depth inside the vegetation, and (2) the reduction rate in water depth and tsunami force increases when the incident tsunami water depth exceeds the initial breaking. Fig. 11(a) shows spatial distribution of the maximum tsunami moments (M_{Pmax}) along the forest for two models. According to the result, the damage length of forest by the model including tree breakage (B.M.) is about 77.5m (77.5%) and increased in comparison with the model that does not include tree breaking (N.B.M.) due to the decrease in vegetation resistance in the front area where vegetation is broken. Fig. 11(b) shows the relationship between the tsunami water depth in front of vegetation in the case of no forest and the reduction of water depth $((d_{max0} - d_{max})/d_{max0})$, tsunami force $((F_{*max0} - F_{*max})/F_{*max0})$ behind the forest and survival rate of *P. odoratissimus* (number of remained trees/total tree), where subscript 0 indicates the case of no forest. The results show that trees start breaking at water depth of 4.8m (Fig. 11(b), I.B.). When tsunami water depth reaches to 5.5 m all the trees are broken. The reduction rate in water depth and tsunami force decrease when the tsunami water depth exceeds the initial breaking of water depth.

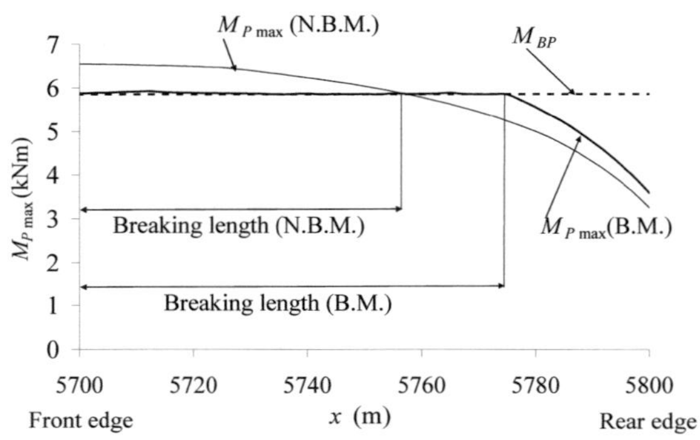

(a)

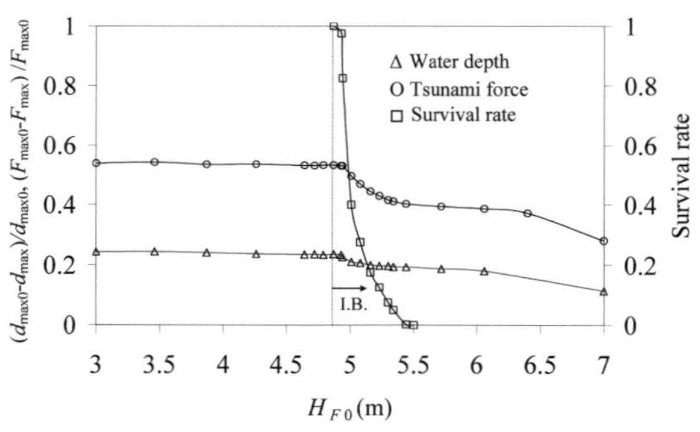

(b)

Fig. 11. (a) Damage length for two numerical model, (b) relationship between tsunami water depth (at the front of vegetation) and reduction of water depth($[d_{max0}-d_{max}]/ d_{max0}$), tsunami force ($[F^*_{max0}-F^*_{max}]/ F^*_{max0}$) and survival rate (number of remained trees /total tree) of *P. odoratissimus*. Subscript 0 indicates the case of no vegetation,(modified from Thuy et al.(2010a))

5. Important aspects of establishment and management of coastal vegetation

5.1 Importance of maintenance as observed in Matara pilot project

A survey was conducted by a team comprising researchers from Saitama University, Japan, and the University of Peradeniya, Sri Lanka (the SU-UoP team) to examine the present situation of coastal vegetation for tsunami mitigation in Sri Lanka. Many government institutions and NGOs have also undertaken such projects along the southern and western coasts in Sri Lanka. An evaluation was carried out to assess whether the coastal vegetation fulfilled the functions of dissipation of wave energy and trapping debris, and offering opportunities to escape and make a soft landing (Tanaka et al., 2007). Currently, 50% of the vegetation barriers are planted in one or two rows due to land availability constraints in areas along main roads or railway tracks. Dissemination of knowledge and coordination with government institutions and research institutions are very important for establishment and management of effective coastal vegetation for tsunami mitigation with community participation, especially at the initial development stage.

The findings of Tanaka et al. (2007) suggested several options for coastal vegetation management that could effectively reduce the impact of tsunamis and other natural disasters in the future. Accordingly, the SU-UoP team proposed a pilot-scale coastal plantation in the Matara Thotamuna area in the southern coastal belt near the mouth of the Nilwala River, Matara City, Sri Lanka (Fig.12(a)), which was severely affected by the 2004 tsunami. *C. equisetifolia* and *P. odoratissimus* were planted in a mixed culture, and their growth was monitored continuously. The average height and girth of *P. odoratissimus* were around 92 cm and 117 cm, respectively, and the average height and trunk diameter of *C. equisetifolia* were 627 cm and 4 cm, respectively, 18 months after planting (Perera et al., 2008). The project was very successful and effective compared to similar projects that suffered from poor maintenance and operation. The project indicated that support from local authorities and communities are vital to make such programs successful, especially in initial development stage. For long-term maintenance of the pilot project, management of the site was handed over to the local temple (Fig.12(b)). This is one of the ideas for sustainable maintenance.

(a)

(b)

Fig. 12. Pilot project in Matara city. (a) growth situation after 18 months, (b) long term maintenance by local temple

5.2 Sustainable maintenance and utilization

De Zoysa (2008) assessed the impact of a *Casuarina* shelterbelt from economic (agricultural crop, fuel supply, timber production, household goods), social (attraction of tourists, prevention of illegal settlements, prevalence of anti-social activity), and environmental (wind speed reduction, sand dune formation, aesthetic value) viewpoints, and concluded that the environmental and social impacts are larger than the economic impacts. Integrated coastal zone management by the residents, city council, and tourist board is recommended in order to increase the benefits of coastal shelter belts.

Considering the Matara project by the SU-UoP team and the research by De Zoysa (2008), the author recommends an integrated coastal vegetation management system (Fig. 13) that maintains the vegetation bioshield effect for a long time. With the growth of trees, the tsunami energy mitigation effect of the vegetation may be reduced, because the tree spacing becomes larger and effective cumulative tree diameter in the stream-wise direction dN_{all} is reduced (Tanaka et al., 2007), and the trees themselves limit the light needed for the undergrowth that prevents soil erosion. To keep the forest dense, proper management is needed by using the forest as a source of firewood or for timber production to thin it, and maintaining various-aged stands of trees by replanting (FAO, 2007). In particular, the front line of the forest close to the sea should be dense vegetation (i.e., a *Pandanus-Casuarina* belt) as Tanaka et al. (2007) proposed. Although *C. nucifera* cannot provide a large drag on the tsunami current, this species is important for economic use. *C. nucifera* has many uses, i.e. fruit production, fuel, timber, and coconut fiber is used as water storage and water purification materials in the substrate of manmade wetlands (Tanaka et al., 2008a). Thus, it is not necessary to remove *C. nucifera* from the coastal region if it is kept at the backside of coastal sand dunes or combined with dense *Pandanus* forest, even though the contribution of *C. nucifera* to increase the overall drag coefficient by itself is not large.

Mangroves in lagoons are quite important and should be maintained for tsunami mitigation because they reduce the tsunami energy overtopping a sand dune or intruding into the land by a river mouth (open gap). In residential areas, broad-leaved trees that have large diameter branches should be maintained close to the houses for people to escape by climbing during a natural disaster event.

Dense coastal belts are sometimes disliked by residents because illegal settlement or anti-social activities are occurred in the place. For keeping the bioshield, it is also quite important to teach people that the coastal vegetation plays a role in preventing strong winds, salt spray, windblown sand, and storm surge propagation, as well as in mitigating tsunami inundation. The local temple or local authorities can utilize the vegetated area for education of the local residents, teaching them to keep the forest dense, control weeds, especially in the developing stage after plantation, and not to cut the forest illegally.

6. Conclusion

Coastal vegetation has been widely recognized as a natural method to reduce the destructive force of tsunami waves, even though the vegetation barrier cannot completely stop a tsunami, and its effectiveness depends on the magnitude of the tsunami as well as the structure of the vegetation. For coastal rehabilitation, optimal planning of natural coastal systems and their maintenance, the capacity of vegetation to reduce the energy of tsunami waves needs to be quantitatively elucidated. The limitations of coastal forests in relation to the magnitude of a tsunami and the maintenance of forests as natural disaster buffer zones have to be understood correctly for effective coastal vegetation planning. Demerits of coastal forests have also been revealed: for example, an open gap in a forest (i.e., a road, river, difference in elevation, etc.) can channel and amplify a strong current by forcing it into the

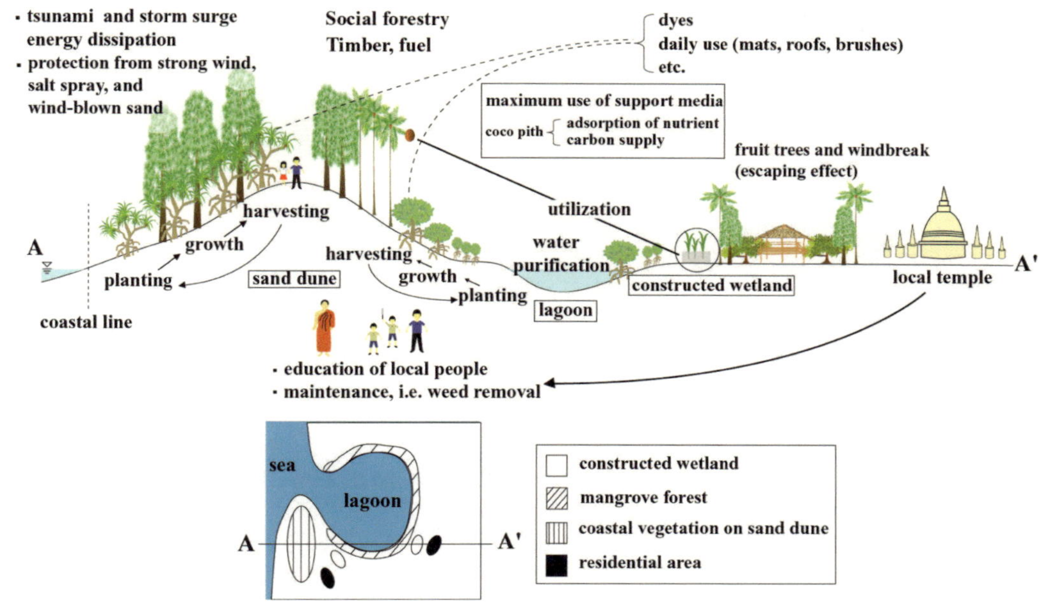

Fig. 13. Integrated coastal vegetation management system (modified from Tanaka et al., 2009)

gap. Floating debris from broken trees also can damage surrounding buildings and hurt people. However, many studies have revealed that these demerits can be overcome with proper planning and management of mangroves and coastal forests, and that coastal vegetation has a significant potential to mitigate damage in constructed areas and save human lives by acting as buffer zones during extreme natural events. The effectiveness of vegetation also changes with the age and structure of the forest (vegetation density, vertical structure and forest shape). This highlights the fact that proper planning and management of vegetation are required to maintain the tsunami buffering function of coastal forests. Although many government and non-government organizations have implemented coastal vegetation projects, many of them have been unsuccessful due to a lack of proper design and maintenance. Many planting project after the Indian Ocean tsunami have dangerous gaps in the coastal forest. A pilot project in Matara City, Sri Lanka, revealed that participation and support from local authorities and communities is essential to make the planting projects successful. An integrated coastal vegetation management system that includes utilization of the materials produced by the forest and a community participation and awareness program are proposed to achieve a sustainable vegetation bioshield.

7. Acknowledgement

Prof. Sasaki, Saitama University, Dr. Mowjood and Dr. Jinadasa, University of Peradeniya, are acknowledged for their useful suggestions during field investigations in Sri Lanka. The author would like to thank Mr. Harsha, Mr. Perera, Mr. Fasley and Mr. Methsiri for their help in field measurements, and Prof. Tanimoto, Mr. Thuy and Mr. Nandasena for their help in numerical studies. This review work is partially funded by the JSPS AA Science Platform Program.

8. References

Cochard, R.; Ranamukhaarachchi, S.L.; Shivakoti, G.P.; Shipin, O.V.; Edwards, P.J. & Seeland, K.T. (2008). The 2004 tsunami in Aceh and Southern Thailand: A review on coastal ecosystems, wave hazards and vulnerability, *Perspectives in Plant Ecology, Evolution and Systematics*, Vol.10, 3-40.

Dahdouh-Guebas, F.; Jayatissa, L.P.; Di Nitto, D.; Bosire, J.O.; Lo Seen, D. & Koedam, N. (2005). How effective were mangroves as a defence against the recent tsunami? *Current Biology*, Vol.15, No.12, 443-447.

Danielsen, F.; Sorensen, M.K.; Olwig, M.F.; Selvam V.; Parish F.; Burgess N.D.; Hiraishi T.; Karunagaran V.M.; Rasmussen M.S.; Hansen L.B.; Quarto A. & Suryadiputra N. (2005). The Asian tsunami: A protective role for coastal vegetation, *Science*, Vol.320 , No.5748, 643.

Dengler, L. & Preuss, J. (2003). Mitigation lessons from the July 17, 1998 Papua New Guinea tsunami, *Pure and Applied Geophysics*, Vol.160, 2001-2031.

De Zoysa, M. (2008). Casuarina coastal forest shelterbelts in Hambantota City, Sri Lanka: assessment of impacts, *Small-scale Forestry*, Vol.7, 17–27.

Fernando, H.J.S.; McCulley, J.L.; Mendis, S.J. & Perera, K. (2005). Coral poaching worsens tsunami destruction in Sri Lanka, *Eos*, Vol.86, No.33, 301-304.

Fernando, H.J.S.; Samarawickrama, S.P.; Balasubramanian, S.; Hettiarachchi, S.S.L. & Voropayev, S. (2008). Effects of porous barriers such as coral reefs on coastal wave propagation, *J. Hydro-environment Research*, Vol.1, 187-194.

Food and Agriculture Organization of the United Nations (FAO) (2007). The Role of Coastal Forests in the Mitigation of Tsunami Impacts – Main report" www.fao.org /forestry /site /coastalprotection /en. ISBN 978-974-13-9321-3, Thammada Press.

Harada, K. & Imamura, F. (2000). Experimental study on the resistance by mangrove under the unsteady flow, *Proceedings of the 1st Congress of the Asian and Pacific Coastal Engineering*, pp.975–984, Dalian, China.

Harada, K. & Imamura, F. (2006). Effects of coastal forest on tsunami hazard mitigation –a preliminary investigation, tsunamis: case studies and recent developments, Advances in Natural and Technological Hazards Research, 279-292.

Hiraishi, T. & Harada, K. (2003). Greenbelt tsunami prevention in South-Pacific region, *Report of the Port and Airport Research Institute*, Vol.42, No.2, 23p.

Honda, S. (1898). Über den Küstenshutzward gegen Springfluten. *Bulletin of the College of Agriculture* 3(4), Imperial University, Komaba, Tokyo, pp.281-298.

Kathiresan, K. & Rajendran, N. (2005). Coastal mangrove forests mitigated tsunami, *Estuarine, Coastal and Shelf Science*, Vol.65, No.3, 601-606.

Kathiresan, K. & Rajendran, N. (2006). Reply to 'Comments of Kerr et al. on ''Coastal mangrove forests mitigated tsunami''' [Estuar. Coast. Shelf Sci. 65 (2005) 601-606], *Estuarine, Coastal and Shelf Science*, Vol.67, 542.

Kerr, A.M.; Baird, A.H. & Campbell, S.J. (2006). Comments on "Coastal mangrove forests mitigated tsunami" by K. Kathiresan and N. Rajendran [Estuar. Coast. Shelf Sci. 65 (2005) 601-606], *Estuarine, Coastal and Shelf Science*, Vol.67, 539-541.

Mascarenhas, A. & Jayakumar, S. (2008). An environmental perspective of the post-tsunami scenario along the coast of Tamil Nadu, India: Role of sand dunes and forests, J. *Environmental Management*, Vol.89, 24–34.

Nandasena, N.A.K.; Tanaka, N. & Takagi, T. (2007). A new derivation of horizontal two dimensional depth averaged momentum equation and continuity equation which include total effect of porosity inside the vegetation, *Annual J. Hydraulic Engineering JSCE*, Vol.51, 163-168.

Nandasena, N.A.K.; Tanaka, N. & Tanimoto, K. (2008a). Perspective of coastal vegetation patches with topography variations for tsunami protection in 2D - numerical modeling, *Annual J. Hydraulic Engineering JSCE*, Vol.52, 133-138.

Nandasena, N.A.K.; Tanaka, N. & Tanimoto, K.(2008b). Tsunami current inundation of ground with coastal vegetation effects; an initial step towards a natural solution for tsunami amelioration, *J. Earthquake and Tsunami*, Vol. 2, No.2, 157–171.

Peltola, H.; Kellomäki, S.; Hassinen, A. & Granander, M. (2000). Mechanical stability of Scots pine, Norway spruce and birch: an analysis of tree-pulling experiments in Finland, *Forest Ecology and Management*, Vol.135, 143-153.

Perera, W.A.R.P.; Mowjood, M.I.M.; Tanaka, N.; Jinadasa, K.B.S.N. & Fasly, M.S.M. (2008). Effective coastal vegetative landscaping for the tsunami protection in Sri Lanka, *Ann. Proc. of SLAAS*.

Sasaki, Y.; Homchuen, S. & Tanaka, N. (2007). The role of coastal vegetation in case of the Indian Ocean tsunami - Coastal area of Thailand and Sri Lanka -, *Mangrove Science*, Vol.4-5, 49-55.

Shuto, N. (1987). The effectiveness and limit of tsunami control forests, *Coastal Engineering in Japan*, Vol.30, No.1, 143-153.

Tanaka, N.; Sasaki, Y.; Mowjood, M.I.M.; Jinadasa, K.B.S.N. & Takemura, T. (2006a). Effective coastal vegetation species and structures with landform, sand dune and lagoon, for tsunami protection at the Indian Ocean tsunami, *Proceedings of 15th APD-IAHR Congress*, pp. 1279-1285, Chennai, India.

Tanaka, N.; Sasaki, Y. & Mowjood, M.I.M. (2006b). Effects of sand dune and vegetation in the coastal area of Sri Lanka at the Indian Ocean tsunami, In: Eds, Namsik Park et al., Advances in Geosciences 6, World Scientific Publishing, Co.

Tanaka, N.; Sasaki, Y.; Mowjood, M.I.M. & Jinadasa, K.B.S.N. (2007). Coastal vegetation structures and their functions in tsunami protection: Experience of the recent Indian Ocean tsunami, *Landscape and Ecological Engineering*, Vol.3, 33-45.

Tanaka, N. & Sasaki, Y. (2007). Limitations of coastal vegetation in the 2004 Indian Ocean tsunami and 2006 Java tsunami, *Proceedings of IAHR 32nd Congress*, (CDROM), Venice, Italy.

Tanaka, N.; Karunarathna, A.K. & Jinadasa, K.B.S.N. (2008a). Effect of coconut coir-pith supplement on nitrogen and phosphate removal in subsurface flow wetland microcosms, *Chemistry and Ecology*, Vol.24, No.1, 1-8.

Tanaka, N.; Nandasena, N.A.K.; Jinadasa, K.B.S.N.; Tanimoto, K.; Sasaki, Y. & Mowjood, M.I.M. (2009). Developing Effective vegetation bioshield for tsunami protection, *Civil Engineering and Environmental Systems*, Vol.26 , No.2, 163 - 180.

Tanaka, N. (2009). Vegetation bioshields for tsunami mitigation: review of effectiveness, limitations, construction, and sustainable management, *Landscape and Ecological Engineering*, Vol. 5, No.1, 71-79.

Thuy, N.B.; Tanimoto, K.; Tanaka, N.; Harada K. & Iimura, K. (2009). Effect of Open Gap in Coastal Forest on Tsunami Run-up - Investigations by Experiment and Numerical Simulation, *Ocean Engineering*, Vol. 36, 1258–1269.

Thuy, N.B.; Tanaka, N. & Tanimoto, K. (2010a). Damage length of vegetation due to tsunami action-Numerical model for tree breaking, *Proceedings of 12th International Summer Sym.* 101-104.

Thuy, N.B.; Tanimoto, K. & Tanaka, N. (2010b). Force due to tsunami runup around a coastal forest with a gap - Experiments and numerical simulations, *Science of Tsunami Hazards,* Vol.29, No.2, 44-69.

10

The 2004 Tsunami in Aceh a Southern Thailand: Coastal Ecosystem Services, Damages and Resilience

Roland Cochard
Natural Resources Management, School of Environment and Resource Development, Asian Institute of Technology, Pathumthani, Thailand

1. Introduction

The expression of a "black swan" became popular during the 2008 economic crisis by a bestselling book by Nicolas Taleb (2007); today a "black swan" is an allegory to describe an event which is deemed highly improbable (unthinkable within experienced reality), yet can cause massive consequences when it actually occurs. Originally, in the Old World, a "black swan" was the quintessential "rare bird on Earth", and only with the discovery of black swans in the 17th century on Australia the conviction that all swans must be white was overturned. In 2010 the allegory, however, reached a new dimension of pictographic realism as the largest-ever oil pest was (and still is) turning thousands of sea-birds into tar-balls in the Gulf of Mexico. This oil pest was caused by the technical limitations to provide a quick fix to a leak from an oil well that is more than a kilometer below sea level (The Economist 2010). The oil pest is now magnitudes larger than anything seen before; it will have a lasting impact on the region's coastal ecosystems for decades to come. It is a stark reminder of today's exposures and fragilities of the coastal zone and the human technical limitations, economic dependencies and short-sighted priorities in dealing with risks.

The great "black swan" of the year 2004 was of natural origin and of an even more sudden catastrophic nature. During only one day (26 December) a huge tsunami killed up to about 300'000 people (two thirds thereof in the Indonesian Province of Aceh), sending physical and psychological shock waves around the planet. In addition to being highly fatal, the event caused massive destruction of housing and infrastructure and the displacement of over 10 million people. Compared to other disasters of similar magnitude, responsiveness of the international community was fast and aid donations were copious (more than 10 billion US$; Wikipedia 2010). In the jargon of some NGO workers and consultants the "second wave" soon came to mean the initial flow of aid money; however, necessities and the efficiency and distribution of aid differed significantly between regions, and the attention of the donor community soon turned away to other priorities.

After the disaster, many NGO's involved in conservation activities were quick to find explanations: according to them the impacts of the giant waves were much increased where coastal ecosystems had been destroyed or damaged, e.g. in areas where mangroves had been replaced by shrimp farms. "Building back better" was a guiding theme for reconstruction; it included the idea of restoring coastal ecosystems - in particular mangroves and beach forests - for use as bio-shields against future tsunamis. While the scientific evidence for a life-protection role of mangroves and other ecosystems still fails to be very convincing (cf. Cochard *et al.* 2008, Feagin *et al.* 2010, Bhalla 2007), mangroves and other coastal ecosystems provide a multitude of socially and economically relevant environmental services; these services may be of particular importance during and after disaster events, contributing to socio-economic robustness and resilience of coastal communities (Adger *et al.* 2005, Alongi 2008, Cochard *et al.* 2008). According to UNEP-WCMC (2006, p. 4) the tsunami was a "wake-up call for the global community, dramatically drawing attention to the vulnerability of coastal ecosystems and the dangers of undermining the services they provide to humankind". This call has been heeded by donors around the world, and since the catastrophe several international and state-funded programs have taken on the task to rehabilitate devastated ecosystems along the affected coastlines (cf. Feagin *et al.* 2010, Stone 2006, Wibisono and Suryadiputra 2006).

How successful have these programs been? Were these programs adequately planned and implemented? Were they at all necessary? In Aceh many attempts made at restoring coastal ecosystems were indeed unnecessarily wasteful, and many expensive schemes to stabilise the coastline were implemented without proper post-disaster surveying. During a visit to Aceh between 26 July and 2 August 2006, I observed many hectares of replanted mangrove seedlings around Banda Aceh and along parts of the tsunami-affected southern coastline. Most of these seedlings were already dead at the time of the visit, and freshly planted seedlings showed signs of rapid weakening with clearly no prospect for survival beyond the seedling stage (Fig. 1a). Similar observations near Banda Aceh were still made in 2008 by B. McAdoo (pers. comm.; Fig. 1e); observations are also confirmed by a consultancy report (Wibisono and Suryadiputra 2006). This contrasts with observations I made in 2006 of vigorous natural growth of the mangrove palm *Nypa fruticans* and the mangrove fern *Acrostichum aureum* in parts of some of the devastated swamps or of nearby periodically flooded areas, e.g. former paddy fields (Fig. 1c). It also contrasts with apparently largely successful mangrove replantation schemes along the affected coastlines in Southern Thailand, as I had observed during a visit to the affected Andaman Sea coastlines between 23 and 26 March 2006.

A "black swan" may be a positive or a negative event; what is most important, however, is the reaction to it (Taleb 2007). Did human communities really "take advantage" from the tsunami (respectively the "the second wave") to "build back better"? What has been learnt from this event? It appears that few restoration programs so far implemented have been accompanied or assessed by independent professionals. There have been some self-critical evaluations by international agencies (e.g. Wibisono and Suryadiputra 2006), but hardly anything is written about the success or failure of restoration programs in the scientific academic literature. This despite the fact that restoration programs would offer major opportunities for innovative scientific studies and progress. At the same time, the efficacy of such aid and development programs could decidedly be improved by implementing recursive learning schemes that strategically use scientific assessments as a means to develop "better" (i.e. more appropriate at any given point in time) management actions (cf. Gross *et al.* 2005).

In a previous article (Cochard *et al.* 2008) we outlined some concepts of risk assessment and management. We concentrated mainly on aspects of vulnerability and the effects of direct fast and catastrophic sea wave hazards, in particular the tsunami. In this article the scope is

Fig. 1. **(a)** Root resprouting in a broken *Rhizophora apiculata* stand at Thale Nok, Thai Andaman coastline. **(b)** *Rhizophora* seedlings planted along the rims of rehabilitated tambak ponds close to Banda Aceh. Most of the seedlings were already dead. No successful seedling establishment was seen around Banda Aceh in July 2006 except for **(c)** vigourous growth (~ 3 m in height) of mangrove palms, *Nypa fruticans*, in an area close to the mouth of the Aceh River. **(d)** In areas where beaches were not completely eroded, natural and aided regrowth of *Casuarina* trees was successful in Aceh as well as in Thailand. Here a regenerating stand on Ko Kho Khao, Thai Andaman coastline. Pictures: R.Cochard, July 2006. **(e)** Dead *Rhizophora* seedlings on a mudflat close to Banda Aceh in January 2008 (picture source: Dr. B. McAdoo).

further extended to aspects of risk robustness and resilience. Attention is drawn to several issues of importance in regard to the development needs of greenbelt restoration schemes. Information is drawn from scientific literature, reports of international organizations and NGO's, and some own experience in the field. Visits to the tsunami-affected coastlines of Aceh Province in Indonesia were conducted between 23 July and 2 August 2006 (Banda Aceh to Meulaboh, including Pulau Weh) and to the Andaman Sea of Thailand between 23 and 26 March 2006 (Thai coastline, Phuket to Ranong). Many of my personal observations, particularly regarding the impacts of the tsunami on coastal vegetation, form thereby one pillar of this article.

The aim of this article was 1.) to provide a short overview over the functions, services and valuation of coastal ecosystems, specifically focusing on resilience aspects in the aftermath of tsunamis, 2.) to review the actual impacts and conditions of coastal ecosystems after the 2004 tsunami, 3.) to review the longer-term anthropogenic impacts on coastal ecosystems, and 4.) to review indirect damages and changes within impacted ecosystems, and 5.) discuss associated problems of ecosystem rehabilitation, respectively natural regeneration.

2. The utility of coastal ecosystems

2.1 The utility of coastal ecosystems as a protection against tsunamis

Following the 2004 tsunami the state of the environment along affected coastlines and the tsunami impacts have been assessed by various international, governmental and private missions (cf. Cochard *et al.* 2008). A part of the concern about the ecosystems was fuelled by the assumption that intact ecosystems had provided significant disaster mitigation services during the tsunami, and that the vulnerability of communities along the affected coastlines was therefore significantly heightened if ecosystems had been destroyed or been degraded during the catastrophe. If this was the case, then efforts at reconstruction would also have been put at increased risk of falling victim to a subsequent tsunami.

Up to this date relatively few informative studies have been published as to what degree intact coastal ecosystems may have provided protection. Primarily based on anecdotal accounts, many reports by aid organisations and environmental NGOs as well as some scientific publications have initially emphasised the positive protection function provided by coastal ecosystems during the disaster, in particular mangroves and coral reefs (see listing in Cochard *et al.* 2008). However later studies, reports and recommendations, based on more analytical assessments, have cautioned against many as yet unsubstantiated claims (see Feagin *et al.* 2010, Cochard *et al.* 2008, Chatenoux & Peduzzi 2007, Kerr *et al.* 2006, Kerr & Baird 2007).

According to an analysis by Chatenoux & Peduzzi (2007), the force and impact of the tsunami mainly depended on the distance to the tsunami source and on coastal topography (particularly near-shore coastal bathymetry). Chatenoux & Peduzzi's (2007) study may, however, contain some weaknesses in that it used multivariate regression statistical methods that did not account for spatial auto-correlation of the data points; this may partly explain why coral reefs appeared to have a positive effect on the tsunami impacts. Regarding the multivariate and spatially interdependent nature of the problem, the relative protection afforded by coral reefs and mangroves is difficult to elaborate and quantify in general terms. In addition, in many cases evaluations appear to contain a certain researcher's bias: remarkably few articles have so far considered that ecosystems can also have the potential to increase tsunami hazards in certain locations, for example by channelling and focussing water flows, or by increasing the impact force by contributing to flow debris (Cochard *et al.* 2008). Evaluating and quantifying tsunami risk at any coastal location requires elaborate multivariate spatial modelling incorporating detailed hydraulic baseline assessments; many of the parameters needed for models are as yet insufficiently understood (Cochard *et al.* 2008).

At the present stage the utility of coastal ecosystems as a tsunami buffer is, therefore, still controversial. It is probably minor as compared to other services provided by these ecosystems; more elaborate research projects would be needed to provide more conclusive scientific answers. Nevertheless, various environmental agencies and NGOs have used the tsunami disaster in order to pressure for greener agendas along the affected coasts, whereby the argument of any tsunami protection may have served for different ends than direct investment in risk management. Whether such observation is testimony to a sort of "machiavellism" by exponents of environmental agencies, as has been stated by some observers (e.g. Baird 2006), is a matter of perspective. Interestingly enough, however, the new TEEB (2010) report still cites Kathiresan and Rajendran (2005) as a source when referring to tsunamis, rather than referring to more substantial and objective, newer literature - and this despite the fact that several authorities on the subject had previously highlighted the scientific shortcomings of that particular paper!

2.2 The role of coastal ecosystems regarding other risk mitigation services

As to what measure investment in environmental reconstruction may influence the risk of future generations living along the coastlines does not merely depend on aspects of direct tsunami hazard mitigation. More holistic approaches to 'risk' also include the concept of "resilience", respectively "robustness" (cf. Adger *et al.* 2005). If conducted in a proper way, coastal restoration projects may not only mitigate the risks of further coastal erosion and decline of livelihood resources; such projects may also play important social and educational functions, implanting resilience to future generations (cf. Pomeroy *et al.* 2006). Considering ecosystem functions there are various valid reasons other than the tsunami *per se* that may justify investment in the rehabilitation and sustainable management of coastal resources.

Firstly, in addition to a potential protection against tsunamis, coastal ecosystems, such as coral reefs, seagrass beds, mangroves and vegetated sand dunes, can provide protection against normal and extreme wind-driven waves (including cyclonic storm surges), coastal erosion, and inland saltwater intrusion (Cochard *et al.* 2008, Das & Vincent 2009). Areas where coastal ecosystems had been destroyed can, therefore, be expected to be exposed to a higher degree to such more common sea-borne hazards. In Aceh in particular, the heavily affected coastlines have been unstable after the tsunami. Coastline changes have been difficult to predict, hampering proper and efficient planning of housing and infrastructure reconstruction in several locations near the coast (O. Nieuwenhuis, pers. comm.). Major reshaping of the coastlines occurred mainly during several days of stormy weather, for example, during a modest rainstorm with high wind speeds (~50 km h^{-1}) that occurred in Banda Aceh on 9 May 2005 (Meilianda *et al.* 2006). Around 15-17 May 2007 disaster struck as waves up to 5 m high (produced by a uncommon constellation) destroyed hundreds of houses and displaced thousands of people along the southern coasts of Indonesia from Bali up to northern Sumatra; in Aceh Barat about 375 houses had to be abandoned by residents in five villages (AlertNet 2010). Losses of mangroves and other vegetation can also lead to significant increases of coastal erosion. According to a study by Thampanya *et al.* (2006) net annual land losses of 0.25 km^2 were measured between 1966 and 1998 along the Andaman coastline; along the western Gulf of Thailand coastline, where about 90% of the former mangrove forests had been destroyed, annual erosion rates were even three times higher (0.91 km^2).

Secondly, coastal ecosystems also provide mitigating functions against land-borne hazards. Forested coastal swamps, for example, can act as overflow areas in flooding periods. Especially peat swamp forests which occur in western coastal Aceh have a great potential to absorb and store floodwaters, and mitigate fast flash flood hazards. Strong rainfall events in catchments (increasingly affected by up-stream deforestation), for example, caused flash floods on 22 December 2006 at Lhokseumawe and in North Aceh, displacing about 170'000 and killing >70 people in 160 villages (ReliefWeb 2010). Swamplands, and to some degree seagrass beds, furthermore act as a filter for sediments and nutrients in domestic wastewater, thereby protecting the marine environment - in particular sensible ecosystems such as coral reefs, which in turn are important for fisheries and coastal protection (Wong *et al.* 1995, Chu *et al.* 1998, Cochard *et al.* 2008).

Thirdly, many other, primarily economic services are provided by coastal ecosystems (Adger *et al.* 2005). Mangrove forests provide shelter and nutrients, and are therefore important feeding, spawning and nursery grounds for fishes, crustaceans and other marine life (Ikejima *et al.* 2003, Chong *et al.* 1990). The richness of fisheries resources along the coastlines of some Asian countries has been attributed to the presence of mangroves (Amarasinghe *et al.* 2002, Islam and Haque 2004, Islam and Wahab 2005). Equally, research has found that fish catch is higher in reefs which are close to mangrove areas, whereby reefs and mangroves are often linked by sea grass beds (Dorenbosch *et al.* 2004, Mumby *et al.* 2004); some economically important species also depend on sea grass beds for some stage of their life cycle (Chou 1994). The value of South East Asia's fisheries in coral reefs alone is estimated to be US$ 2.4 billion per year; in Indonesia economic benefits from reefs including tourism are estimated to be US$ 1.6 billion per year (Burke *et al.* 2002). Commercial annual fish harvests in Indonesian mangroves have been estimated at up to $60'000 km^{-2} (Bann in UNEP-WCMC 2006). Other commercial products from coastal ecosystems include timber, pharmaceuticals, honey, and other plant and animal products; coral reefs in particular are attractive for tourism (UNEP-WCMC 2006). In Aceh, mangrove species are used for roof thatch (the mangrove palm, *Nypa fruticans* and *Pandanus* spp.), fuelwood, charcoal production and building materials for housing and boat construction (*Rhizophora* spp., *Sonneratia* spp., *Avicennia* spp., *Ceriops tagal*), and the production of honey (mainly *Avicennia* spp.). Furthermore, the fruits of *N. fruticans* and *Sonneratia alba* can be consumed for food; the fruit of *N. fruticans* also has potential as a source for biofuel production (Kanagaratnam *et al.* 2006). Estimations from the literature of the use value of mangroves range from less than 3000 US$ ha^{-1} y^{-1} to more than 60'000 US$ ha^{-1} y^{-1} (e.g. Sathirathai and Barbier 2001, Iftekhar 2008, Cabrera *et al.* 1998, Costanza *et al.* 1997).

3. Direct impacts of the 2004 tsunami on coastal ecosystems

Ecosystems weakened by a catastrophe such as a tsunami are likely to be more vulnerable to a subsequent catastrophe, provided there was no time for sufficient recovery. Investments in environmental restoration may, therefore, be a sensible strategy to fasten regeneration processes, thereby enhancing overall robustness and decreasing the vulnerability to subsequent hazards. The question must, however, be raised about what was actually the impact of the 2004 tsunami. If impacts were minor and natural regeneration is fast, such investments may not improve the situation; rather they may represent a waste of resources that could be used for other more tangible efforts to reduce various risks (e.g. early warning systems in the case of tsunamis; cf. Kerr & Baird 2007, Cochard *et al.* 2008).

Cochard *et al.* (2008) summarised and described the major coastal ecosystems of the region. Here an overall account is presented of the damages assessed in Aceh and on the Thai Andaman coast following the 2004 tsunami. References and some data are summarised in Table 1. Despite initial fears of an immense impact, some reports indicated that many ecosystems were surprisingly little affected in many areas, except in coastal areas of Aceh

(Banda Aceh, west coast and some forelying islands), which were directly exposed and close to the epicentre. In Aceh in particular, the database is, however, highly inadequate; most of the figures found in reports were based on rough assumptions.

Coastal ecosystem	Aceh Province coastline			Thai Andaman coast		
	total area or coastline length	affected $^{\alpha}$, heavily affected $^{\eta}$, destroyed $^{\delta}$		total area or coastline length	affected $^{\alpha}$, heavily affected $^{\eta}$, destroyed $^{\delta}$	
		hectares	percent		hectares	percent
Coral reefs	97250 ha ?	~ 19000 ha $^{\alpha}$?	~ 30% $^{\alpha}$?	7900 ha*	~ 1420 ha $^{\eta}$ ~ 4760 ha $^{\alpha}$	~ 18% $^{\eta}$ ~ 60.3% $^{\alpha}$
Sea grass beds	~ 600 ha ?	~ 300 ha $^{\alpha}$??	~ 50% $^{\alpha}$??	7937 ha	120 ha $^{\delta}$ 278 ha $^{\alpha}$	1.5% $^{\delta}$ 3.5% $^{\alpha}$
Mangrove forest	341'000 ha	~ 80'000 ha ?	~ 20% $^{\delta}$??	181400 ha	306 ha $^{\delta}$	0.2% $^{\delta}$
Beach forest	800 km	~ 400 km $^{\delta}$?	~ 50% $^{\delta}$??	4500 ha**	594 ha** $^{\alpha}$	~ 13.2% $^{\alpha}$
'Coastal forest'	?	48'900 ha $^{\alpha}$	?	?	?	?
Agricultural land	1'576'418 ha	64'608 ha $^{\alpha}$	2.4% $^{\alpha}$	287'382 ha*	1438 ha* $^{\alpha}$	0.5% $^{\alpha}$
Plantations	?	28'000 ha $^{\alpha}$	?	6'996'158 ha*	14'123 ha* $^{\alpha}$	0.2% $^{\alpha}$
Aquaculture	47'620 ha	20'429 ha $^{\alpha}$	42.9% $^{\alpha}$	5154 ha*	42.5 ha* $^{\alpha}$	0.8% $^{\alpha}$

* Affected Thai Provinces only (FAO-MOAC 2005); ** Thailand only Takua Pa District (Thanawood *et al.* 2006)

Relevant references

GRCMN 2006, Baird *et al.* 2005; Foster *et al.* 2005, UNEP 2005a, BAPPENAS 2005, BRR 2005a, Thanawood *et al.* 2006, Synergy One 2006, FAO-MOAC 2005, FAO 2005d, EJF 2005, IUCN 2006, NORDECO 2005, FAO-WFP 2005, WIIP 2005, Meilianda *et al.* 2006, FAO 2005e

Table 1. Estimated percentage damages of natural and agricultural ecosystems along the Thai Andaman coast and in Aceh

3.1 Offshore ecosytems

Coral reefs overall seem to have suffered fairly modest impacts from the 2004 tsunami (Pennisi 2005). Along the Thai Andaman coast about 13.2% of corals experienced a heavy impact of >50% damage; 60.4 % of corals were not or just slightly affected (0-10% damage) (GCRMN 2006). At the most exposed Surin Islands corals suffered only about 8% of damage with signs of recovery already eight weeks after impact (EJF 2005). At Ko Phra Tong and Ko Ra on the northern Andaman coast some reefs, however, seem to have been heavily affected (up to 80%; IUCN 2006). In Aceh patterns of impact were variable, but overall damage levels appeared to be similar to Thailand despite the generally much higher wave energy along these coasts (Foster *et al.* 2005). Around Pulau Weh and Aceh reefs were surprisingly little affected; some were intact despite the fact that the tsunami caused severe destruction on adjacent land areas (Baird *et al.* 2005; Foster *et al.* 2005). In places of direct heavy wave impact there was only little evidence of recent coral mortality, e.g. according to Baird *et al.* (2005) an estimated 83% of corals transported on land appeared to have been dead prior to the tsunami (cf. Fig. 6a). They suggested that with an influx of nutrients flushed from the land the tsunami may only have

exacerbated a phase shift from corals to algae which was notable before 2004 as a result of bombing and cyanide fishing in the islands. Overall, direct tsunami impacts were highest in colonies on unconsolidated substratum on the reef slope; damage was mainly influenced by reef aspect, submarine topography and it was highest in bays (Baird *et al.* 2005, Foster *et al.* 2005; cf. discussion of wave build-up in Cochard *et al.* 2008). The most dramatic destruction of coral reefs was, however, due to the effects of the earthquake. On Simeulue, Bangkaru and other islands whole reef flats were lifted above the sea level and were thereby destroyed (Foster *et al.* 2005, GCRMN 2006, Meltzner *et al.* 2006, cf. later discussion).
According to rapid assessments reported in UNEP (2005a) only 1.5% of Thailand's seagrass meadows along the Andaman coast experienced total habitat loss, with about 3.5% being affected by siltation and sand sedimentation. The most impacted seagrass meadows were those of Ko Yao Yai, Phang Nga Province, where an estimated total habitat loss of 10% was recorded. It was suggested that it will take three months for seagrass to recover from siltation (UNEP 2005a). Seagrass meadows are also found in Aceh in the areas of the Aceh Besar Island group (Pulo Aceh), Weh Island group, Simeulue Island group and the Pulau Banyak group. UNEP (2005) put figures of seagrass beds at approximately 600 ha, but according to WIIP (2005), the little information that was available was mostly qualitative. Figures of tsunami impacts on seagrass beds in Aceh are even more in the dark. They were assumed by WIIP (2005) at 50-100% losses based on guesses from location and presumed tsunami impacts. As for coral reefs (cf. Baird *et al.* 2005), the true losses may, however, be considerably lower.

3.2 Mangrove swamps

Along the Thai Andaman coast the extensive mangrove ecosystems were only affected at most exposed locations fronting the sea and along creek systems (cf. Cochard *et al.* 2008), with a total loss of 306 ha or about 0.2% of Thailand's mangroves (UNEP 2005a). In many areas the impacts were selective, uprooting mainly trees of (predominantly young to medium-aged) stilt-rooted *Rhizophora apiculata*, while cable-rooted trees of *Avicennia* spp. and *Sonneratia* spp. remained largely standing and survived (Cochard *et al.* 2008). For Aceh the data which was provided by agencies on mangrove forest cover and impacts were highly inconsistent. The most extensive mangrove forests in the Province of Aceh are found in north-eastern areas in Aceh Timur and Langsa Districts (ca. 297'000 ha; WIIP 2005); these regions were hardly affected by the tsunami. Fewer mangroves (ca. 50'000 ha; WIIP 2005) were found on the heavily affected west coast and about 1000 ha on Simeulue Island before the tsunami. BAPPENAS (2005) put early estimates of mangrove 'area affected' by the tsunami at 25'000 ha. BRR (2005a) presented an 'environmental damage indicator' figure of 2711 ha for mangrove area (no source or explanation). Possibly the most reliable approximations are still some figures listed by WIIP (2005), which estimated a loss of 100% (26'823 ha) for Aceh Besar (north coast, incl. Banda Aceh), 50% (14'000 ha) in Aceh Barat (west coast), 75% (17'000 ha) in Pidie (north coast) and 30% (26'000 ha) in Aceh Utara and Bireun Districts (north coast). According to my own observations, virtually 100% of mangroves were found dead by July 2006 along the west coast (Aceh Jaya and Aceh Barat), and in the surroundings of Banda Aceh probably >98% of pre-tsunami stands had been destroyed. The vast majority of trees had been physically cut or overthrown by the tsunami (Cochard *et al.* 2008); possibly about 50% of the trees that initally withstood the impact later died to other secondary effects (excluding an intact stand on Pulau Weh).

3.3 Beach forests, sand dunes and beaches

Thanawood *et al.* (2006) estimated that 13.2% of beach forests were affected north of Phuket Island in Takua Pa District, however, without any indication to what degree. Their figure of 4500 ha total cover of beach forest for this district furthermore contrasts with the figure of 1465 ha for the entire Andaman coastline stated by UNEP (2005a). On site visits I found that most stands of *Casuarina equisetifolia* between Phuket and Laem Son National Park were affected to some degree, but only in a few places there was complete destruction of *Casuarina* beach forests; in most places only smaller trees were broken (cf. Cochard *et al.* 2008). Following the tsunami there have also been some shifts in tree species composition of other beach forest stands (e.g. the *Barringtonia* formation): typical beach forest trees such as *C. equisetifolia*, *Pandanus* spp., *Cocos nucifera* (coconut palms) and *Terminalia catappa* appeared to be comparatively more resistant to the tsunami impact than other, more generalist species, such as *Eugenia* spp., *Ficus* spp. or leguminous trees, e.g. the exotic *Leucaena leucocephala* (FAO-MOAC 2005, FAO 2005c; Table 2). No detailed data was found in the literature about the impacts on *Casuarina* beach forests and coconut groves dominating the sandy coasts of western Aceh. Along many stretches of coastline these vegetation types, providing the first line of vegetation resistance against the tsunami, experienced complete destruction (own observations; cf. Kerr & Baird 2007). Yet, surprisingly, several large trees also managed to survive on the west coast even at most exposed locations (Fig. 2a,b), much in contrast to most of the mangroves. In many cases, *Casuarina* trees only survived if their roots were protected behind revetments (cf. Meilianda *et al.* 2006). There is no data from the literature as to what degree beach forests actually prevented sand dune erosion during the tsunami. In Aceh the tsunami impact was massive in terms of wave height as well as velocity and the water masses on the beach front have exerted an immense pressure on sand deposits (cf. Gedik *et al.* 2005, Heller *et al.* 2005). In many areas entire beaches and sand spits were washed away (Fig. 2c), some of those beaches were important turtle breeding grounds (NORDECO 2005). A loss of about 50% of beaches was estimated for the west coast (UNEP 2005a, FAO 2005d).

Vegetation / species	Direct physical		Indirect physical		Salinity effects		Waterlogging effects	
	intensity	location	intensity	location	intensity	location	intensity	location
Mangroves	2-3	1	1	1	0	0	0	0
Casuarina	1	2	1	2	0	0	1	1
Ficus	1	1	0	0	1	1	0	0
Eugenia	0	0	0	0	3	1	1	1
Cocos	1	1	0	0	2	2	0	0
*Melaleuca**	1	1	1	1	2-3	2	2	1
Coastal forest*	0	0	0	0	2-3	2	2	1

Table 2. Qualitative indices of tsunami effects on trees as by impact intensity on trees (0 no, 1 mild to 3 intense effect) and overall distribution (0 none, 1 very localised to 3 widespread); assessment by P. Saenger (FAO 2005a) and, on Ko Phra Tong (*), by R. Cochard

Fig. 2. **(a)** A *Casuarina* tree and **(b)** coconut palms that are (still) surviving near the tip of the Meulaboh peninsula. Virtually all the houses of the town that was densily built behind these trees were completely levelled by the tsunami. Post-disaster reconstruction has been prohibited at this very exposed site. **(c, d)** Sand beach erosion and hillside vegetation removal by the tsunami (Ujung Labuhan, South of Lhok Nga), exposing the bare rock surface along the new shoreline and as a trim line along the cliffs. **(e)** Land subsidence and beach erosion (along the coastline between Lho Kruet and Calang) as evidenced by the submersed and killed coconut palms and **(f)** former mangrove stands (see line at horizon, arrow), now exposed to open water wave action and erosion. Pictures: R.Cochard, July 2006.

3.4 Other coastal forestry resources

The effect of salinity on trees is still barely documented and quantified. FAO (2005a) presented some qualitative data as shown in Table 2. The clear indication is that typical beach trees (e.g.

Casuarina) that survived the physical impacts did not have any great problems with salinity while inland trees were much more affected. For example, trees in the affected parts of peat swamp forest and freshwater swamp forests in Aceh and in Southern Thailand were probably mainly killed by salinity. Also a majority of trees in the savanna of Ko Phra Tong (Thai Andaman coastline) were killed by the sea salt rather than being physically affected by the tsunami (>50% killed trees in March 2006, pers. obs.; according to IUCN 2006 only 18% of trees were killed; Fig. 7a). Possibly about 20% of the savanna area on Ko Phra Tong was affected by the tsunami. Similar to observations by Gardner *et al.* (1992), who documented the effects of Hurricane Hugo on a coastal forest in South Carolina, salinity-intolerant tree species were often more affected if standing close to the coastline, in swales or near the edge of marshes. According to BAPPENAS (2005) some 50'000 ha of coastal forests other than mangroves were lost to the tsunami in Aceh; "coastal forest" was not defined in more detail (Table 3). The tsunami also damaged some 28'000 ha of plantations, a major part of the Acehnese economy (BRR 2005a). A collection of recovery indicators (BRR 2005b) provides no environmental data, but a list of possible vegetation, habitat and other environmental indicators is provided under the heading "indicators for which agencies/organisations did not provide adequate data". Along the northwestern shore the trees and thin soils along the rocky hillsides were generally completely removed by the tsunami, giving an indication about the height of incoming water masses ('trim line'; Fig. 2d).

4. Anthropogenic impacts on coastal ecosystems in Thailand and Indonesia

While some beaches and land resources in Aceh were heavily affected by the tsunami, some ecosystems withstood very heavy impacts. Natural stresses, such as tropical storms, earthquakes, volcanoes and tsunamis, as well as disease outbreaks, have shaped coastal ecosystems throughout their evolution and development (cf. Lugo *et al.* 2000, Tomascik *et al.* 1997, Cochard *et al.* 2008). Anthropogenic stresses, on the other hand, are more recent phenomena which in many cases occur in a chronic and cryptic way, their effects often only becoming fully obvious during a catastrophic event (UNEP-WCMC 2006, GCRMN 2006). Different natural and anthropogenic stresses may interact with each other in complex ways. Ecosystems weakened by chronic anthropogenic stresses are likely to be more vulnerable during a catastrophe. Equally, any efforts at post-tsunami restoration of affected ecosystems may be jeopardized, if chronic anthropogenic stresses persist. Such stresses may increase after a catastrophe, as pivotal ecosystem services may have been lost and socio-economic structures may have disintegrated. Coastal resilience, i.e. the capacity to recover from any (catastrophic) disturbance, may be negatively affected in the long term. This section provides an overall account of major anthropogenic stresses and threats to tropical coastal ecosystems; some current figures of impacts for Indonesia and Thailand are provided in Table 3.

4.1 Offshore ecosystems

The actual causes of reef decline are commonly difficult to trace back; reef degradation is normally caused by the interaction of several stressors (Hughes & Connell 1999). Along the coasts of Sumatra the main threats to coral reefs are probably posed by sediment off-flows from land, which are traceable to upland deforestation (GCRMN 2006). According to Abram *et al.* (2003) major coral mortality in the Mentawai Islands west of Sumatra could be traced to iron fertilization from smoke plumes stemming from the 1997 Indonesian wildfires; although this has also been attributed to pollution by wastewater and sediments (Hoeksema & Cleary

2004). On Pulau Weh and Pulau Aceh dynamite fishing had a significant destructive impact (Baird *et al.* 2005, cf. also Pet-Soede *et al.* 1999), yet detailed pre-tsunami data do not exist for Aceh. In contrast, along the Thai Andaman coast coral reef status has been surveyed in relative detail by the Thai Department of Fisheries between 1995-1998, and partially between 2003-2004: 48.8% of the reefs were found to be deteriorated and only 17.5% of reefs were in a 'good' to 'very good' condition (Mather 2005, GCRMN 2006). The relative contribution of different hazards was not known. Threats to some coral reefs along the Andaman coast stem from overfishing, coastal development and tourism (GCRMN 2006). The main threat to reefs does, however, seem to be posed by domestic and industrial pollution, particularly in the Gulf of Thailand. Along the Andaman coast it was estimated in 1986 that 1384 tons of domestic BOD (biochemical oxygen demand) load was discharged into the sea annually (Taranatham in Cheevaporn & Menasveta 2003); today's BOD emissions are probably much higher.

Relatively little is known about the extent and status of seagrass beds in Thailand, and almost nothing in Aceh. According to estimates referred to by UNEP (2004) possibly some 30-40% of seagrass beds in Indonesia have been lost; estimates of losses for Java are as high as 60%. In Thailand, losses were estimated at 20-30% (UNEP 2004), whereby higher losses probably occurred in the more polluted Gulf of Thailand, and the Andaman Sea may be less

Coastal ecosystem	**Percent affected $^{\alpha}$, highly affected $^{\eta}$ and destroyed $^{\delta}$**		**Major threats and impacts**
	Indonesia	**Thailand**	
Coral reefs	86% $^{\alpha}$ 47% $^{\eta}$	77% $^{\alpha}$ 52% $^{\eta}$	overfishing, destructive (e.g. blast or poison) fishing, pollution (by nutrients, poisonous chemicals or land-borne sediments), coastal development, tourism
Sea grass beds	30-40% $^{\delta}$	20-30% $^{\delta}$	siltation, pollution, mechanical damage (e.g. trawling)
Mangrove forest	~ 40% $^{\delta}$	(35-) 55% $^{\delta}$	conversion to aquaculture ponds (tambak), salt ponds and agriculture, coastal development, mining, fuel production (charcoal), hydrological changes, pollution
Beach forest	?	?	replacement with coconut and other plantations, woodchipping, coastal erosion, sand mining
Peatswamp forest	> 53% $^{\alpha}$ > 32% $^{\delta}$	?	logging and conversion to agriculture (e.g. rice fields) and tree crop estates (e.g. oil palm plantations), or abandonment (secondary sedge-fern lands), forest fires

* Thailand only Takua Pa District (Thanawood *et al.* 2006) ** Affected Thai Provinces only (FAO-MOAC 2005)

Relevant references

GRCMN 2004, 2006, Burke *et al.* 2002, Spalding *et al.* 2001, UNEP 2004, Green & Short 2003, Duarte 2002, FAO 2003, Barbier & Sathirathai 2003, Alongi 2002, Soegiarto 2004, Dahdouh-Guebas *et al.* 2005b, Giesen 2004, Sargeant 2001

Table 3. Anthropogenic causes of damage to coastal ecosystems in Indonesia and in Thailand with percentage estimates of impacts affected. In Indonesia and Thailand the main causes are probably traceable to increased siltation rates in estuarine areas and coastal pollution; as with coral reefs the most likely factors are a combination of several impacts (UNEP 2004; Table 3). Other impacts may be due to chronic mechanical damage, such as trawling, dredging, push nets, anchoring and dynamite fishing; though less important in terms of scale, these impacts are normally unambiguous (Duarte 2002).

4.2 Mangrove swamps

Mangrove ecosystems play an important role as a marine pollution control buffer (Mazda *et al.* 2007, Wong *et al.* 1997, 1995). It is obvious that a loss of mangroves commonly not only leads to coastal erosion and increased sedimentation but also to higher domestic pollution discharge into the open sea and onto coral reefs or seagrass beds. Often shrimp farm enterprises that have replaced mangroves contribute highly to marine pollution (Dierberg & Kiattisimkul 1996, Thampanya *et al.* 2006). The destruction of mangrove ecosystems is therefore often directly responsible for the deterioration of offshore ecosystems, whereas the destruction of offshore ecosystems can also affect the function of mangrove environments (e.g. by increased wave impacts; cf. Cochard *et al.* 2008).

According to Aksornkoae & Tokrishna (2003) mangrove forest cover in all of Thailand more than halved between 1961-1997 from 367'900 ha to 167'582 ha; other figures by FAO (2003) suggest, however, that possibly some 240'000 ha are still remaining. Losses of 61'268 ha of mangroves along the Andaman coastline were by 42.5% below the total national level (Aksornkoae & Tokrishna 2003). While there are many reasons for the destruction of mangrove forests, e.g. population pressure, coastal development, mining, conversion to salt ponds and agriculture, the most important single factor in recent years, has been the expansion of aquaculture. Shrimp farming may account for 32.7% of total mangrove losses with a major expansion from 31'906 ha to 66'997 ha between 1983 and 1996 (Aksornkoae & Tokrishna 2003, Barbier & Cox 2002). Other estimates range from 21-64% contribution to destruction (Dierberg & Kiattisimkul 1996). On the Andaman coastline, shrimp farm expansion accounted for only 8.4% according to Aksornkoae & Tokrishna (2003); the most pristine forests could still be found in Ranong and Phang-nga Provinces, i.e. the areas most affected by the tsunami. An analysis using aerial video recording by Erftemeijer (2002), however, indicated an increase of (partially uncontrolled, illegal) shrimp farms of up to 25% of mangrove areas. The analysis did not include Ranong Province, but for Phang-nga an area of 5637 ha of shrimp ponds was estimated instead of the previous estimate of 953 ha. That all these areas were mangrove areas appears, however, unlikely: along the Andaman coastline many shrimp farms are built on higher ground (FAO-MOAC 2005). In Indonesia overall estimates of mangrove losses are from an initial 4250 km^2 to now about 2500 km^2, with most losses recorded along the coasts of Java (~70%), Sulawesi (~49%) and Sumatra (~36%), and the most untouched stands remaining in West Papua (Soegiarto 2004, Chang *et al.* 2006; Table 5). In Indonesia firewood and charcoal production as well as conversion to tambak are major causes for mangrove losses (Soegiarto 2004). There are no detailed data available for Aceh Province, but on the island of Sumatra the Acehnese mangroves were probably the worst affected already before the tsunami. On the vegetation map by Laumonier *et al.* (1997) no primary mangrove forest was indicated in all of Aceh; mangrove destruction has been considerable in Aceh already before 1920 (Luytjes in Laumonier 1997). Large areas of mangrove had been in serious decline, and most of the 36'000 ha of tambak (shrimp and fish farm) ponds in Aceh are thought to have been built in former mangrove areas, including the heavily affected area along the coast of Banda Aceh, where most of the mangroves had been cleared during the last 30 years (GCRMN 2006, NORDECO 2005, cf. Parish 2005; cf. Fig. 2a). According to UNEP (2005a) about 8% of Aceh's total remaining mangroves were considered to be in good condition, 85% in fair condition and 7% in poor condition prior to the tsunami. Brackish water forests adjoining mangrove forests are particularly under pressure from wood uses and have also frequently been converted to agricultural land (Whitten *et al.* 1997).

Cryptic degradation is possibly an underestimated factor leading to higher vulnerability of mangrove forests to coastal wave hazards (cf. Dahdouh-Guebas *et al.* 2005a). According to Dahdouh-Guebas *et al.* (2005b) changes to historic systems of freshwater management in some river catchments in Sri Lanka as well as in other countries have increased dramatically in recent times. As a result of hydrological changes, such as inter-basin transfers via canal systems, ecological and socio-economic coastal equilibria in estuarine lagoon environments have been disrupted. While mangrove areas have increased in some cases due to increased accretion of sediments at river mouths, qualitative degradation of mangrove stands in terms of floristic composition, significance of the species, and biodiversity has often been ongoing (Dahdouh-Guebas *et al.* 2005b). Another possibly underestimated factor of cryptic degradation is chronic pollution. Along the Thai Andaman coast extensive wood chipping for charcoal during the 70ies and 80ies has also lead to major simplification of large areas of mangrove vegetation (FAO-MOAC 2005).

While mangrove destruction may increase the vulnerability of coastal communities overall (cf. Cochard *et al.* 2008), in Thailand the 2004 tsunami mainly hit areas where aquaculture was still protected by mangroves. In total only 42.5 ha of shrimp ponds were affected (FAO-MOAC 2005). In Aceh the situation looks very different: 20'429 ha, or 42.9 % of aquaculture ponds were affected by the tsunami, thereof 1022 ha were completely lost to the sea and 7270 ha were heavily affected (FAO 2005d; Table 3). Given the strength of the waves in Aceh, we may wonder about what difference it had made if mangroves had still been in place; detailed spatial analyses of available impact data may still provide better evaluations about the actual protection benefit of mangroves versus aquaculture. In any case, sound rehabilitation planning should consider the integration of tambak ponds with mangrove replanting schemes, also to account for their fish and shrimp nursery and wastewater cleaning functions (cf. Cochard *et al.* 2008).

4.3 Beach forests, sand dunes and beaches

Causes of beach erosion may be grouped into five categories: increased wave energy, interruption to littoral transport, deprivation of sediment input, human activities and sea level rise (Wong 2003). There do not appear to exist any detailed pre-tsunami data on beach patterns and changes, neither for Aceh nor for the Andaman coastline. According to Meilianda *et al.* (2006) the beaches north of Banda Aceh had, however, been experiencing erosion problems for decades before the tsunami. Whether this is related to the replacement of large mangrove areas by tambak aquaculture and the construction of an artificial river bed for the Aceh River and adjacent rivers is unclear. According to Whitten *et al.* (1997) commercial sand exploitation also had a significant impact on beach vegetation along some coastline stretches near Banda Aceh. The destruction of coral reefs and mangroves may cause a change of sedimentation patterns and wave energy on some nearby beaches (Wong 2003). Changes in wave energy can almost totally destroy beach vegetation (Whitten *et al.* 1997). *Casuarina* beach forests, however, seemed to be in fairly good shape along most of the Thai coastlines visited even after the tsunami (pers. obs.). Anthropogenic impacts on *Barringtonia* beach forests in Sumatra as well as Thailand have certainly been considerably higher already in historic times (Whitten *et al.* 1997).

4.4 Peat swamp forests

A major environmental concern which has so far barely been mentioned in any literature is the past and ongoing large-scale destruction of peat swamp forest in the coastal plains of Aceh and other provinces of Sumatra. This dense forest vegetation type may have mitigated the tsunami impact along some coastal stretches (Cochard *et al.* 2008). Much more important in the longer term is probably that the coastal forest provided a multitude of ecosystem services, thereby enhancing resilience to coastal communities in case of disaster (cf. UNDP 2006). These services are virtually lost by the forests' replacement with oil palm monoculture plantations, which - synonymous to shrimp farm expansion in mangrove areas in the Gulf of Thailand - have expanded at a breath-taking pace during the last two decades (Sargeant 2001, cf. Fig. 5). There are no exact published data, but changes detectible from available satellite images indicate losses of up to 80% of the extensive Tripa forests in Nagan Raya Province south of Meulaboh since about 1994. Also the few remaining forests north of Meulaboh are currently being cut at an alarming rate, followed by fire destruction and conversion into various types of plantations (pers. obs.). Especially repeated impact of fires precludes any potential regeneration of these forests in the longer term (Bowen *et al.* 2000, Giesen 2004). The forests at Singkil are indirectly threatened by water drainage in surrounding plantations. Only the Kluet swamps, which are part of the Leuser National Park, still seem to be in a fairly good shape (Dr. I. Singleton, pers. comm.).

5. Indirect damages and changes within impacted ecosystems

The longer-term consequences of the destructive impacts on coastal environments are in many cases still unknown. Resilience is the capacity of an ecosystem to 'bounce back' to its original state. This capacity, however, depends on many variables (most importantly including the activities of human beings) which may change during and after disaster events. For most places affected by the tsunami (and this applies in particular to the Province of Aceh) there were virtually no published ecological data available from before the impact (cf. UNEP 2005a). Some inferences about biotic and physical changes may still be drawn from studies of post-tsunami environmental dynamics, comparisons of impacted sites to sites that were not impacted and/or the investigation of impact gradients. Post-disaster coastal geomorphological processes were, however, complex and - at least initially - poorly understood (O. Nieuwenhuis, pers. comm.). The same can certainly be said about biological changes and succession patterns, whereby in most cases no systematic surveys were conducted and no detailed and conclusive data exist until today. An understanding of post-disaster ecosystem changes and social dynamics is, however, a pivotal requirement for serious coastal ecosystem restoration projects. This Section provides an overview of some possible indirect and longer-term impacts by the tsunami and the earthquake.

5.1 Damages in estuaries and possible secondary impacts on offshore coastal ecosystems

As virtually all the mangrove areas along the ~250 km western coastline between Banda Aceh and Meulaboh have been destroyed, the ecological long-term implications of mangrove loss in Aceh will be different to Thailand. Firstly, there may be significant negative medium- to long-term effects on fish stocks and marine diversity. Even though in total the mangrove forest areas along the west coast were relatively small, most of the larger river estuaries were forested with some mangroves. These may have been important ecological stepping stones along the coast for various marine and intertidal organisms. Mangrove swamps in estuaries may have been providing safe nursery grounds for economically important fish stocks along these stretches of mostly dynamic, turbid waters that are fringed almost entirely by abiotic shoreline systems. Thereby estuaries were possibly supplying fish stocks to otherwise isolated fringing reefs; coral fish population movements are typically more localized than for pelagic fish (cf. Mumby *et al.* 2004).

The destruction of mangrove vegetation may, furthermore, pose a continuing threat to sensible offshore ecosystems. Many of the inland evergreen forests in Aceh continue to be clearcut for wood production at a fast pace, and sediment loads in the rivers have most likely increased significantly in recent times due to soil erosion in catchments. The filter function of mangroves and related vegetation was possibly essential to mitigate excessive sediment discharge by rivers into the open sea and onto nearby fringing reefs and seagrass beds, particularly also as mangrove swamps can serve as a freshwater sponge, attenuating flooding events after heavy rainfall. The effects on a coral reef after a tsunami and subsequent cyclone were, for example, described in Flores by Tomascik *et al.* (1997). The runoff from the stream was usually directed away from an adjacent fringing coral reef by a small sand bar and carried westward by the longshore current away from the reef. The initial sand bar was destroyed and a different sand bar was formed which then was directing the river run-off towards the fringing reef. After some heavy rains which increased sediment deposition, the reef had been completely smothered (Tomascik *et al.* 1997).

5.2 Restoration of mangrove swamps – what changes need to be considered?

Even if the fast restoration of former mangrove areas may be desirable, such areas may actually no longer be suitable for mangrove replanting. The current regeneration potential of many former mangrove forests seems to be very low, for four main possible reasons: 1.) a new, unfavourable wind and wave environment due to the destruction of protection features; 2.) compression and collapse of completely bare mangrove peat soils; 3.) loss of keystone mangrove infauna during the tsunami; 4.) general land subsidence altering tidal inundation depths and patterns; and 5.) unfavourable change of the rooting substrates by tsunami sediment depositions.

5.2.1 Increased direct wave exposure

Many lagoons and enclosed swamp areas formerly protected by sand spits have been opened up to the sea; for example the tambak area in the former mangrove mudplains west of Banda Aceh could for this reason not be reclaimed (NORDECO 2005, Fig. 3a,b). The presence of mangroves has clearly decreased the overall abrasion of mudflats, as roots of mangroves have provided stability to intertidal flats during the tsunami (Fig. 3a,b). However, the complete demise of coastal forelying sand bars is now directly exposing

Fig. 3. Part of the northwestern shoreline near Banda Aceh at the port of Uleelheue before and after the 2004 tsunami. **(a)** IKONOS satellite image taken on 10 January 2003, showing settlements, the extensive tambak fish ponds in the surrounding former swamp area and some remaining mangrove stands in the river estuary and along the rims of tambak ponds **(b)** IKONOS satellite image taken three days after the tsunami on 29 December 2004. A part of Uleelheue Island, the sand beach to the West as well as most of the tambak areas have been completely scoured by the tsunami waves and form now part of the open sea. As can be seen, intertidal flats still persist in some of the larger former areas of mangroves (arrows). **(c)** A few surviving larger *Rhizophora stylosa* trees along the rims of former tambak ponds, close to the sea front near the mouth of Aceh River, North of Banda Aceh. **(d)** Destroyed

intermediate *Rhizophora* stand close to Uleelheue (see asteriks in 10b). With the complete scour of the forelying sand spit during the tsunami, the mangrove peat is now directly exposed to open sea wave impacts and is slowly eroding, increasingly so, as the binding mangrove root remains are also decaying. No root resprouts were observed, indicating 100% tree death at these sites. Satellite image source: CCI 2006. Pictures: R.Cochard, July 2006.

mangrove mudflats to the physical wave impact of the open seas. As no tree recruitment appears to be occurring and replanting is not successful, the tidal flats are now slowly eroding in the widened bay (Fig. 3d). Also, most mangroves which still had been fringing former tambak ponds have been leveled by the tsunami with the effect that high winds are now freely blowing over the open tambak areas and mudflats. Quite significant waves are thereby produced even within relatively small tambak ponds. These waves probably have a significant impact on unprotected, replanted seedlings (Fig. 4a,b). Unless certain conditions change (e.g. new deposition of sand spits, establishment of wind and wave shields by pioneer plants), there is therefore little prospect of mangrove plantations to be successful in those areas.

Fig. 4. Newly planted mangrove saplings west of Banda Aceh **(a)** during low tide and no wind, and **(b)** during high tide and exposed to wind-driven waves. **(c)** A surviving sapling after such exposure; it can also be seen that there is a layer of rippled sand deposits. Pictures: R.Cochard, July 2006.

5.2.2 Mangrove peat collapse

The former peat soils of the extensive mangrove mudflats may now be largely degraded and unsuitable for mangroves, for several reasons. Peat collapse after catastrophic tree mortality has, for example, been reported in hurricane-affected mangrove stands in the Carribbean. Three years after passage of hurricane Mitch, Cahoon *et al.* (2003) found virtually no mangrove recovery in a former mangrove forest on Guanaja Island, Honduras: neither recolonization by mangrove propagules, nor root growth and also no inorganic sediment deposition. Consequently, the substrate collapsed at a rate of about 11 mm per year in the basin forest as the dead root material decomposed and sediments were compacted. While continued elevation loss of mangrove peat can affect seedling establishment *per se* (via a change in the tidal regime), it could also lead to the development of unfavourable soil conditions, such as low redox potential, increased salinity and sulfides, which may further limit the potential of seedling establishment (Cahoon *et al.* 2003). According to Wanless *et al.* (in Cahoon *et al.* 2003), as per 1992 natural recolonization had not occurred in most of the mudflats created by the 1935 and 1960 hurricanes in South Florida; this despite the availability of plentiful propagules in the immediate vicinity.

5.2.3 Local extinction of mangrove infauna

With the immense impact of the tsunami waves and the associated destruction of the mangrove forests entire communities of crabs, worms and other mangrove infauna may have become locally extinct. Many species, especially burrowing crabs, may be crucially important to maintain conditions essential for post-disaster re-establishment of mangroves. The keystone role of crabs in mangrove forest sediments has, for example, been demonstrated by Smith *et al.* (1991) in a 12-month crab removal experiment in *Rhizophora*-dominated forests in North Queensland. On crab removal plots they found significant increases (in some months) of soil sulfide (up to ×1.4 as compared to control plots) and ammonium (up to × 1.8 increase) concentrations. This affected forest growth and tree reproductive output. Smith *et al.* (1991) primarily noted aeration of the soil by crab burrows as a most important factor to explain their results, but the flushing function of crab burrows for excessive salt removal, as well as nutrient provision to roots is probably of similar importance (Ridd 1996, Stieglitz *et al.* 2000).

5.2.4 Land subsidence or uplift following the 2004 earthquake and aftershocks

Elevation levels of former mangrove mud flats were – respectively still are – not only slowly subsiding because of peat collapse and erosion during and after the tsunami. One of the greatest problems in Aceh for the restoration of coastal and particularly estuarine environments appears to be associated with general land subsidence resulting from the 26 December 2004 earthquake and several later aftershocks. Most of the earthquake-related uplift and subsidence became evident after the first earthquake, which is the second strongest on record globally ($M_w \approx 9.2$; Lay *et al.* 2005). The island of Simeulue to the southwest of the epicentre was mostly uplifted (some southwestern areas of Simeulue slightly subsided) with a maximum uplift of 2.4 m recorded at Salaut Island north of Simeulue (Jaffe *et al.* 2006, Meltzner *et al.* 2006). The coastal areas of Sumatra northeast to the epicentre subsided. Meilianda *et al.* (2006) recorded land subsidence at between 0.75-1.75 m near Uleelheue close to Banda Aceh. This significant subsidence appeared, however, also related to scour by the tsunami (Fig. 3b). At Jantang on the west coast between Banda Aceh

Fig. 5. **(a)** Aerial picture near the Tripa estuary South of Meulaboh, West Aceh coastline, showing the tsunami-buffering by a strip of remaining peat swamp forest. The picture also shows some of the extent of forest destruction by oil palm companies. In this area more forest was cleared than could actually be planted with oil palms and/or the soil conditions were found to be unsuitable. Secondary forest regeneration is clearly visible within the cleared areas. **(b)** An intact stand of peat swamp forest along the coastal road North of Meulaboh. **(c)** Peat swamp forest converted to oil palm plantations in the Tripa region. Aerial picture source: Dr. I. Singleton; other picutures: R. Cochard, July 2006.

and Calang Jaffe *et al.* (2006) estimated minimum subsidence at 0.6 m based on vegetation marks; at a near location they recorded a maximum subsidence of 2 m, but this may also be related by some extent to local scour. Estimates by Meltzner *et al.* (2006) based mostly on analyses of GIS data put minimum subsidence at the Banda Aceh coast and along the west coast at below 20 cm, emphasising however the uncertainty of these estimates which are based on a relatively weak database. In June 2006 subsidence seemed to be obvious all along the western coastline north of Meulaboh (pers. obs.) and may have been at an equal range at Lho Kruet and at Padang Selimeng close to Calang (Fig. 2e,f).

According to Bilham (2005) the 2004 tsunami was triggered by the initial rupture at the plate boundaries which occurred at typical speeds only in the south. At its northern end of the fault line, the surface between the Indian plate and the Andaman archipelago took more than half an hour to slide a distance of 7 to 20 meters. There the slip was too slow to trigger a tsunami - fortunately, considering that tsunamigenic waves triggered in these parts along the faultline would probably have caused considerably more widespread destruction along the coastlines fringing the Andaman Sea and the Sea of Bengal. Most vegetation changes on the Andaman Islands that were described in the 'tsunami' paper by Roy & Krishnan (2005) are therefore actually linked to land subsidence rather than to the tsunami impacts. Additional changes in land elevation levels may have occurred later. According to Lay *et al.* (2005, p. 1129) the most notable aftershock feature was a "swarm of strike-slip and normal faulting events in the Andaman Sea back-arc basin involving more than 150 magnitude 5 and greater earthquakes that occurred from 27 to 30 January 2005." A second large earthquake occurred on 28 March 2005 ($M_w \approx 8.7$), with the epicentre close to Nias Islands, where about 1300 people were killed by the quake. This earthquake also triggered a minor tsunami that destroyed several houses on Nias Island; more significantly it also caused significant topographical changes. The small Sarangbaung Island north of Nias Island subsided by about 1.7 m while reefs on the west coast of Nias were uplifted by about 2.5 m at Afulu and 1-2 m at Asu Island and are now exposed above sea level (USGS 2005).

According to Cochard *et al.* (2008) a sizeable mangrove forest in the Rigaih River floodplain first survived the tsunami, but died later. The exact time of the dieback is not known, but it probably occurred within weeks after the tsunami; its most important cause was likely to be land subsidence and the associated physiological stress due to root asphyxiation and possibly salinity changes. Widespread mortality of *Rhizophora* and other mangrove stands due to land subsidence has so far only been reported from the southern parts of the Andaman Islands and from the Nicobar Islands where some of the Indian territories' most pristine mangrove stands still existed before the tsunami (Roy & Krishnan 2005, Ramachandran *et al.* 2005, Rama Chandra Prasad *et al.* 2009). Roy & Krishnan (2005) speculated on some possible successional vegetation changes that may occur, such as replacement of *Rhizophora* spp. by *Bruguiera* spp. and/or *Avicennia* and *Sonneratia* spp. on elevated, less frequently inundated mudplains, or *vice versa* establishment of *Rhizophora* spp. in slightly subsided areas previously occupied by other mangrove species. Meilianda *et al.* (2006) reported on how after the tsunami typical mangrove crab burrows were found on more elevated locations along the coastline of Banda Aceh, such as in coconut groves. This was also observed in Raya Bay (pers. obs.). From a purely scientific viewpoint, the ecological changes occurring along the coastlines of northern Sumatra and the Andaman Islands may actually provide some unique research opportunities regarding the environmental effects of short-term sea level changes due to seismic activity, possibly with some lessons also for longer-term changes due to climate change. On a more practical side, efforts of mangrove rehabilitation can only meet with success if the relative sea level changes are actually made understood prior to replanting - either by detailed reconnaissance surveys and modelling approaches and/or by trial-and-error recursive learning processes (cf. Gross *et al.* 2005, Gross & Hoffmann-Riem 2005).

5.2.5 Marine sediment deposition in mangrove swamp areas

Another factor which may initially have been causing secondary dieback of mangrove stands may have been excessive sedimentary deposition by the tsunami (Fig. 6b). According to Ellison *et al.* (1998) natural sedimentation rates of mangroves are in a range normally lower than 0.5 cm y^{-1}, with a maximum of about 1 cm y^{-1}, whereby *Rhizophora* spp. with stilt roots appear to be slightly less sensitive than species with pneumatophores, such as *Avicennia* and *Sonneratia* spp. Tree death is also more likely to occur if roots are being smothered by fine sediments, probably because aeration is lower than in rough sediments

Fig. 6. **(a)** Coral fragment tsunami deposits on a rock coast on Weh Island North of Banda Aceh. **(b)** Tsunami mud deposits about 1 km from the coast North of Banda Aceh. Note the crab burrows, possibly indicating land subsidence. **(c)** Former paddy fields near Lamblang, West of Banda Aceh. In July 2006 these fields were swamps dominated by *Typha angustifolia* and sedges. **(d)** Post-tusnami forest regrowth that is mainly dominated by leguminous tree species. Pictures: R. Cochard, July 2006.

such as sand (Ellison *et al.* 1998). No study was found about tsunami deposits in mangroves. It is possible that sediment changes may also explain the success of *Nypa* palms and *Acrostichum* ferns to recolonize mudflats; in a wide range of brackish water conditions these species have probably also a wider tolerance of different sediment substrates (cf. Fig 1c).

5.3 Marine sediment deposition, salinisation and vegetation changes inland

Large tracts of other open land have been affected by marine sedimentary depositions and salinisation, particularly along the coasts of Aceh. Moore *et al.* (2006) provided a detailed description of sedimentary deposits along a 450 m transect south of heavily affected Lhok Nga (west of Banda Aceh): while soils were mostly scoured in the first ~50 m of impact along the transect (cf. Fig. 2c), the sediment layer thickened progressively from 50 to 100 m distance, was then about 10-20 cm thick until at ~375 m distance from the shore, and then abruptly thinned out to the hillside. A very similar pattern of scour (0-50 m) and deposition (50-750 m) has been described by Gelfenbaum & Jaffe (2003) for the 1998 PNG tsunami. Along the western coast tsunami deposits were mainly of sandy texture, whereas the deposits in Banda Aceh were more fine-grained (Jaffe *et al.* 2006, Meilianda *et al.* 2006; cf. Fig. 6b). Typical thickness of deposits was 5-20 cm, with a recorded maximum of 70 cm. Thickest tsunami deposits did not correlate with the deepest tsunami flow depth, but were due to various other factors, including the presence of swales. Deposits were normally composed of multiple layers, representing deposition during multiple waves and/or during uprush and return flows (Jaffe *et al.* 2006). Szczucinski *et al.* (2005) provide data of sedimentary deposits for several sampling sites along the Thai Andaman coast, whereby sediment thickness ranged from 1-20 cm.

According to BRR (2005b) 64'608 ha of productive agricultural land have been affected by the tsunami in Aceh, representing 2.38% of the province's total productive area. Thereof 32.7% have been rehabilitated until December 2005. Land damages due to salinity, and desurfacing of the landscape due to deposition of sand and clay sediments have been mapped for the four most affected districts by Synergy One (2006) using a combination of ground sampling and remote sensing techniques. They estimated that 8118 ha of productive paddy fields or 13.2 % of the total were affected and needed to be rehabilitated. 732 ha of paddy fields were completely lost to scour and to the sea water ("sunk areas"); some of these low-lying areas may now instead be used for aquaculture or mangrove plantations (Slavich 2005; cf. Fig. 3a,b).

The impact of salinity in Aceh and elsewhere was variable. Generally, total salt depositions were highest in areas where sea water had been trapped and logged (Slavich 2005). In these areas production of rice and other crops was often highly diminished. In less affected paddy areas rice production was, however, not affected. In many cases yield even seemed to have increased after the tsunami; although some plant nutrients can be mobilized by salt penetration (cf. Gardner *et al.* 1992), this was, however, probably mainly attributable to improved irrigation schemes following an inflow of agricultural aid (Young 2005, Slavich 2005). It appears that salt generally did not penetrate deep into the coastal clay soil in which rice is usually grown; on the other hand crops growing on sandy soils, such as peanuts, were negatively affected on all the fields hit by the tsunami (Young 2005). It is, however, assumed that negative impacts will not last as salt is washed out by plentiful rainfall in the wet seasons (Young 2005). Even if lands remain potentially productive, major changes in land use along the west coast of Aceh may simply result from the fact that up to 80% of the population were killed by the wave in some of these areas (Slavich 2005).

Apart from agricultural assessments there is so far only very sparse data available from the literature on how salinity or sediments changed the vegetation in the ground layer, particularly in Aceh. Several km² of former paddy fields between Banda Aceh and Lhok Nga were permanently flooded with sea water (Borrero 2005). Much of these areas are now swamps that are dominated by *Typha angustifolia* and sedges (Fig. 6c). *Typha* spp. are known to be very tolerable to various pollution stresses, possibly including elevated levels of salinity (Koottatep *et al.* 2005). In any case, *Typha* spp. absorb pollutants, so that many areas may be rehabilitated naturally and could later again be returned to agricultural use as paddy fields.
As could be expected, many creeping pioneer herb species such as *Ipomea pes-caprae*, *Gloriosa superba* and *Chromolaena odoratum* rapidly increased in the beach forests after the tsunami along the Thai Andaman coast (FAO 2005b). Especially *I. pes-caprae* but also other typical creeping beach species such as *Canavalia rosea* seemed to be stabilizing many of the sand dunes along the Thai and the Aceh coasts which were left bare and unstable after the tsunami. Further inland, secondary regrowth on flat wasteland was observed to be very fast in some areas, exceeding 8 m in height in July 2006; fast growing leguminous trees were dominating (Fig. 6d). In some tsunami-affected coastal areas regrowth of typical pantropical and exotic weeds, such as *Calotropis procera*, *Lantana camara*, *Senna didymobotrya* and *Leucaena leucocephala* was found to be widespread. According to IUCN (2005c) exotic species like *Opuntia dillennii* or *Prosopis juniflora* may pose a problem after the tsunami in Sri Lanka. The extent of any such problem has, however, not been established. In Florida Horvitz *et al.* (1998) found increases in exotic species in hardwood forests after Hurricane Andrew, possibly leading to some long-term shifts in species composition and replacement of native functional groups.

5.4 Natural and human-aided regeneration processes

In Thailand most of the tsunami-affected mangrove areas have been replanted (mostly with *R. apiculata*) and regeneration appeared to be largely successful at most sites which I had visited in March 2006. At some sites natural regeneration by resprouting from root stumps was also observed (Fig. 1b). Overall, it shows that the impacts were more transient, and secondary succession proceeded at a relatively fast rate as ecosystems were to a large degree still intact. In contrast, in Aceh replanting schemes with *Rhizophora*, *Avicennia* and *Bruguiera* seedlings in the mud-flats and along the west coast so far appeared to be largely unsuccessful (Fig. 1a,e), with the exception of a small, only partially affected stand on Pulau Weh, where planted seedlings appeared to be generally surviving. Except in a destroyed *Avicennia* stand in Raya Bay, I observed virtually no natural mangrove regeneration in Aceh from resprouting, indicating mostly complete mortality. Some areas (possibly ~10-20%) were, however, recolonised by mangrove palms, *Nypa fruticans*, such as close to the mouth of the Aceh River just north of Banda Aceh (Fig. 1c). Also mangrove ferns, *Acrostichum aureum*, were a commonly found coloniser species in destroyed mangrove mudflats.
In stands of *Casuarina* beach communities high natural recruitment of seedlings was observed in most of the affected stands of *Casuarina* trees (Fig. 1d). Equally in Aceh, natural regeneration – where some trees were still standing – as well as replanting schemes appeared to be largely successful behind some intact sand dunes. In many areas sand dunes were, however, completely eroded, or continuing sand deposition and shifts of dunes was preventing successful recolonisation for the time being until a new equilibrium would be reached (Fig. 2c; cf. Liew *et al.* 2010).

In a *Melaleuca* savanna on Koh Phra Tong no regeneration was obvious after the tsunami (Fig. 7a), as seedlings of *Melaleuca* and other tree species were observed almost exclusively in non-affected areas. Prolific seedling recruitment was in particular observed in a recently burned area unaffected by the tsunami (Fig. 7b). It is possible that *Melaleuca* seedling establishment was not only facilitated but also triggered by savanna fires. In contrast to statements by IUCN (2006) whereby fires should be prevented, fires could actually represent an intrinsic ecological process in this vegetation type (cf. also Stott 1990, Finlayson 2005) and may potentially serve as a savanna regeneration process following the tsunami.

Fig. 7. **(a)** Tsunami-affected *Melaleuca* savanna on Ko Phra Tong, Thai Andaman coastline. Most of the trees were killed by salinity. No seedling regeneration was evident in the grassland affected by tsunami salinity and sediments, but **(b)** vigorous seedling recruitment was observed in a pristine savanna that had evidently been affected by fire a few weeks earlier. Pictures: R. Cochard, July 2006.

5.5 Future research

There is an urgent need for detailed scientific assessments on the causes of mangrove plantation failures. On the other hand, research into natural post-tsunami mangrove regeneration processes appears to be advisable, as it may provide some useful clues on how to avoid such costly failures of plantation schemes in future. Albeit FAO (2005e) was drawing attention to the negative effects of increased salinity on *Nypa* in some areas, in many other areas *Nypa* and *Acrostichum* possibly seem to act as essential, first pioneer plants initiating an ecological succession. On the other hand, these species may also indicate that some areas may now be more freshwater dominated; according to M. Nasir (pers. comm.) for some reason it also appears that in several areas in the Banda Aceh plains brackish water and freshwater fishes are now more commonly found than before the tsunami.

There may be a lot which can be learnt in terms of natural history of the region - this in turn may actually be helpful for future risk management, including disaster mitigation and ecosystem restoration. For example, by analysing the morphology of coral microatolls and using them as palaeogeodetic 'natural tide gauges' Natawidjaja *et al.* (2004, 2006) reconstructed the historical vertical deformations associated with subduction in western

Sumatra. Their coral records revealed interseismic submergence before and after major earthquake events (e.g. in 1797, 1883, 1935, or also some aseismic slips or "silent events" such as in 1962) in areas of coseismic emergence close to the Sumatran fault line (ie. mainly Mentawai Islands); likewise interseismic emergence occurred in areas of coseismic submergence that are at a greater distance to the fault line (ie. mainland coastal Sumatra). Large seismic events including major mainland subsidence and island uplift, such as in 1833 and 2004/2005, on average seem to occur about every 230 years in the central sections, and probably in other sections of Sumatra as well (Nalbant *et al.* 2005, G. Prasetya, pers. comm.). Hardly any consideration has so far been given in literatures about the possible effects that this observation may have on coastline evolution and ecosystem dynamics in the area. Along some coastlines of Indonesia mangroves may occur in 'drowned bedrock valleys', i.e. geomorpholocial settings that have been sunk by land subsidence. Tomasik *et al.* (1997, p. 941) stated, however, that "it is most likely that this type will be rare in the archipelago, given the tectonic nature of the region, where uplift (ie. regression) is relatively more common than subsidence". Given new recent observations, such statements may soon face revision. In the Rigaih River valley and elsewhere along the west coasts of Sumatra, mangroves possibly would have been replaced by peatswamp forests over time, would there not have been intermittent events of land subsidence (cf. Cochard *et al.* 2008). According to Natawidjaja *et al.* (2006, p. 32), in order to improve the information base on subduction plate deformations, "paleoseismic work in the marshes and estuaries of the mainland coast would likely yield important results". It would certainly also yield important baseline data useful to interpret coastal landscape evolution in the region.

6. Concluding remarks: the costs of failing to invest in human and scientific capital

May it be that post-disaster pressures to reconstruction prevailed over proper planning based on careful analysis of coastline changes, or may it be that the changes were indeed too changeable and complex to interpret in a short time period: several coastal stretches, particularly near the towns of Meulaboh, Calang and Banda Aceh, have been stabilized with hard engineering structures frequently without any proper understanding of their wider impacts or even an assessment of their necessity (cf. Fig. 8). Such concrete structures were not only expensive and unattractive (e.g. to future tourism), they most likely had a significant influence on beach processes. In several cases they were simply misplaced investments buried in sand: as can be seen on Fig. 8a concrete boulders were also placed along coastal stretches where sedimentation rather than erosion was the dominant process. Compared to such concrete investments, lush beach and mangrove forests may appear like an alternate dream.

In Aceh more than 7000 hectares of coastal swamps have been replanted with mangroves in 2005 in just one project - 2550 ha on West coast and 4600 ha on the East coast (cf. Kanagaratnam *et al.* 2006). Planters employed in this project received US$0.04 per planted propagule (on the basis of cash-for-work; Kanagaratnam *et al.* 2006). If we consider a planting density such as is commonly used in Thai mangrove forestry, e.g. about 25'000 propagules planted per hectare (Homcha-Aim 2010), this would amount to a total investment of over 7 million US$. From our observations (cf. Introduction) plantation densities may have been lesser, and plantation implementation more patchy rather than systematically covering large areas. Most of the plantations along the West coast were,

Fig. 8. Protective concrete structures set in place along **(a)** northern and **(b)** southern coasts of the Peninsula of Meulaboh. Note that only the southern coasts were actually eroding; along the northern coasts the beaches were accreting.

however, most likely a complete failure, i.e. a foregone investment which may amount to over 2 million US$ for just this project. There were many other projects in Aceh and elsewhere, in total probably amounting to several million US$, that were invested in a very short time period - too often in the wrong locations and typically in want of a longer-term strategy. Even more disconcerting than the careless use of aid money may eventually be the bitter realization that these projects have actually served as a vehicle to educate local people that mangrove planting and management is essentially a nonsensical activity. If the big aid agencies (employing entire village communities) could not do it, how then should the "locals" be able to do it?

Most of the explanations for the failures of these plantations related to organizational and social aspects of the concerned projects. Wibisono and Suryadiputra (2006), for example, listed various failures, such as mistakes in the selection of planting sites, unsuitable choice of plants, insufficient preparation, inadequate guidance, the low capacity of human resources, poor coordination and information sharing among the stakeholders concerned with the rehabilitation activity, and the very limited amount of community involvement in the rehabilitation activity. They noted (p. iii) that "a mistaken perception among the implementers was that rehabilitation activity ended once the seedlings had been planted in the field. The result as they saw it, therefore, was the number of seedlings planted, not the number that survived after planting." The true problem behind such an observation may, however, be the practice by which aid funds in general are being used by large international organizations: as depicted in Fig. 9 most of the funds for reconstruction had to be spent within only a few years. This is certainly not long enough to foment the foundations for sustainable management of a mangrove forest. But do we actually need to plant mangroves after a disaster? Or is it really something completely different which is needed, e.g. observing, learning and spreading the lessons of the "black swan"?

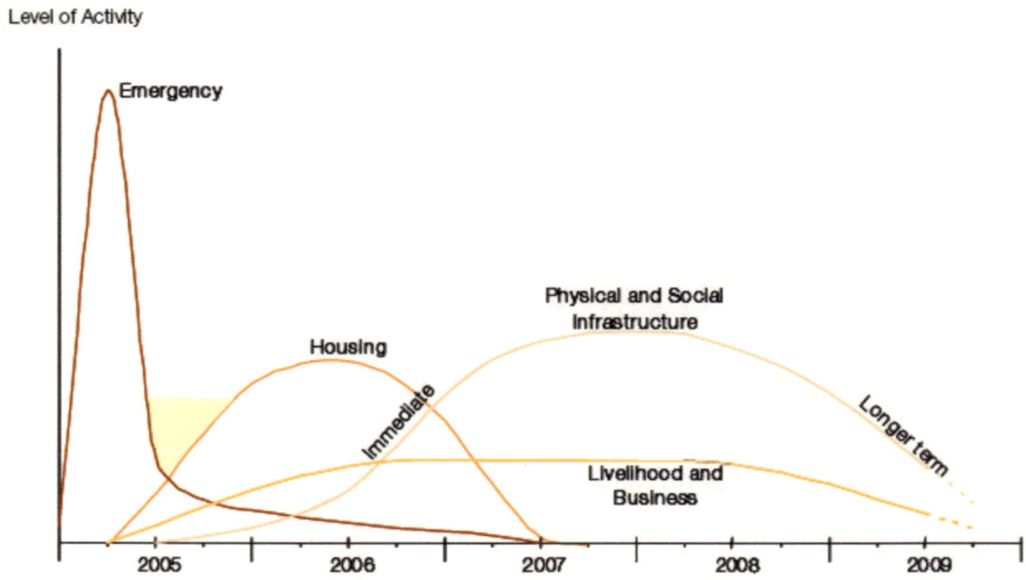

Fig. 9. Sequencing of emergency and recovery efforts after the tsunami in 2004 (adopted from BRR 2005a, p. 19).

The devastation wrought by the tsunami has highlighted a potentially important role of coastal ecosystems for protection of communities from tsunamis and other sea-borne natural hazards. While there exist many anecdotal accounts about the life-saving role of mangroves and coral reefs, truly analytical studies are nonetheless still scarce; the studies so far presented are mostly relatively simple and fail to address the complexity and multi-dimensionality of the issue (cf. Cochard *et al.* 2008). Irrespective of their tsunami risk mitigation functions, coastal ecosystems are in any case an asset in the daily lives of coastal communities, and in particular during and after disaster events. Greenbelts - and this term in a wider sense may include other biotic systems that are not actually green - can significantly contribute to both, decreasing vulnerability as well as to increasing resilience of the coastal resource base and therefore of the communities themselves. As an aide-memoire to emphasize this point the final, commendable paragraph of the publication by Baird *et al.* (2005, pp. 1928-1929) is reprinted here: "Acehnese reefs in the wake of the tsunami provide a profound example of the resilience of marine ecosystems to natural disturbance and also a powerful reminder of the potential for damage from chronic human misuse. The extent of the damage on land and the tragic human cost should not distract attention away from the perennial problems of marine resource management: improving water quality, reducing fishing pressure, and encouraging sustainable coastal development. Neither the conservation priorities nor the risks to reefs have been changed by the tsunami, and it is vitally important that resources are not directed to short-term, small-scale rehabilitation programs, which will not reverse long-term declines in reef condition. The political goodwill and the financial resources the tsunami has generated should rather be used to build economies and societies that provide resilience in both the social and the ecological domain." How better to build such societies than by investing in long-term management-cum-research-cum-education programs? And how better to not waste resources in the first place by starting with the appropriate surveying and research after a disaster - before large plantation schemes are let lose founded on error rather than logically conducted trial? Any such "trials" should be accompanied by the expertise of scientists early in the rehabilitation process (optimally involving an education component). As has been suggested by McAdoo *et al.* (2009) interested scientific experts could be organized in a specialized fast-response team; such a team could serve to consult and accompany NGO's reconstruction efforts, but would operate independently in order to ensure scientific objectivity and credibility.

7. Acknowledgements

This review and the field visits to Southern Thailand and Aceh Province were made possible through the financial support of the Centre for International Agriculture (ZIL Grant No No. 41-3219.4/00256) and the Institute of Integrative Biology, Swiss Federal Institute of Technology (ETHZ). Thanks go to Ms. B.R. Bürgi (ETHZ) for helping to organise the field excursion to the Thai Andaman coast, to Mr. M. Nasir (Syiah Kuala University, Banda Aceh) for spending his time (and car petrol) to show R. Cochard the coastline east of Banda Aceh, to Prof. Dr. Sudarmadji (Jember University) for his company during the trip to Meulaboh and Ms. R. Frey (Paneco Foundation, Switzerland) for leading the way to the remaining Tripa swamp forests. I acknowledge the kind permission to use the image in Fig. 1e from Dr. B. McAdoo, Quickbird and Ikonos satellite images in Fig. 3 from the German Aerospace Centre (Deutsches Zentrum für Luft- und Raumfahrt, DLR, http://www.zki.caf.dlr.de/media/download/applications/2004/indonesia/), and aerial picture in Fig. 5 kindly provided by Dr. I. Singleton.

8. References

Abram N.J., Gagan M.K., McCulloch M.T., Chappell J. & Hantoro W.S. (2003). Coral reef death during the 1997 Indian Ocean dipole linked to Indonesian wildfires. *Science* 301: 952-955.

Adger W.N., Hughes T.P., Folke C., Carpenter S.R. & Rockström J. (2005). Social-ecological resilience to coastal ecosystems. *Science* 309: 1036-1039.

Aksornkoae S. & Tokrishna R. (2003). Overview of shrimp farming and mangrove loss in Thailand. In: *Shrimp Farming and Mangrove Loss in Thailand* (eds E.B. Barbier and S. Sathirathai). Edward Elgar Publishing Ltd., Northampton, USA.

Alertnet (2010). Indonesia tidal waves displace thousands. <http://www.alertnet.org/thenews/fromthefield/222031/118010664443.htm> (accessed in July 2010)

Alongi, D.M. (2002). Present state and future of the world's mangrove forests. *Environmental Conservation* 29(3): 331-349.

Alongi, D.M. (2008). Mangrove forests: Resilience, protection from tsunamis, and responses to global climate change. *Estuarine, Coastal and Shelf Science* 76: 1-13.

Amarasinghe, U.S., Amarasinghe, M.D. and Nissanka, C. (2002). Investigation of the Negombo estuary (Sri Lanka) brush park fishery, with an emphasis on community-based management. *Fisheries Management and Ecology* 9(1): 41-56.

Baird, A. (2006). Myth of green belts. *Samudra Report* 44: 14–19.

Baird A.H., Campbell S.J., Anggoro A.W., Ardiwijaya R.L., Fadli N., Herdiana Y., Kartawijaya T., Mahyiddin D., Mukminin A., Pardede S.T., Pratchett M.S., Rudi E. & Siregar A.M. (2005). Acehnese reefs in the wake of the Asian Tsunami. *Current Biology* 15: 1926-1930.

BAPPENAS (2005). *Indonesia: Preliminary Damage and Loss Assessment. The December 26, 2004 Natural Disaster.* A Technical Report prepared by BAPPENAS and the International Donor Community. Indonesia National Planning and Development Agency/BAPPENAS, Jakarta Indonesia. < http://www.crid.or.cr/digitalizacion/pdf/eng/doc16324/doc16324.htm>

Barbier E. & Cox M. (2002). Economic and demographic factors affecting mangrove loss in the coastal Provinces of Thailand, 1979-1996. *Ambio* 31(4): 351-357.

Barbier E.B. & Sathirathai S. (eds) (2003). *Shrimp Farming and Mangrove Loss in Thailand.* Edward Elgar Publishing Ltd., Northampton, USA.

Bhalla, R.S. (2007). Do bio-shields affect tsunami inundation? *Current Science* 93(6): 831-833.

Bilham R. (2005). A flying start, then a slow slip. *Science* 308: 1126-1127.

Borrero J.C. (2005). Field data and satellite imagery of tsunami effects in Banda Aceh. *Science* 308: 1596

Bowen M.R., Bompard J.M., Andersen I.P., Guizol P. & Gouyon A. (2000). Anthropogenic fires in Indonesia: a view from Sumatra. In: *Forest fires and regional haze in South East Asia* (eds Radojevic and Eaton), New York, USA. <http://www.ssffmp.or.id/ssffmp/file/publication/report/Anthropogenic_Fires_in_Indonesia.PDF >

BRR (2005a). *Aceh and Nias one year after the tsunami. The recovery effort and way forward.* A joint report of the BRR and International Partners, December 2005. Rehabilitation and Reconstruction Agency (Badan Rehabilitasi dan Rekonstruksi), Jakarta. < http://www.e-aceh-nias.org/index.php?id=oneyear_main.html >

BRR (2005b). *Tsunami Recovery Status Report. As of 8 December 2005.* United Nations Information Management Service, Office of the UN Recovery Coordinator for Aceh and Nias in collaboration with the Rehabilitation and Reconstruction Agency (Badan Rehabilitasi dan Rekonstruksi). <http://www.e-aceh-nias.org/index.php?id=indicators_main.html >

Burke L., Selig L. & Spalding M. (2002). *Research Report: Reefs at Risk in Southeast Asia.* World Resources Institute, Washington D.C. < http://marine.wri.org/pubs_description.cfm?PubID=3144 >

Cabreta, M.A., Seijo, J.C., Euano, J. and Perez, E. (1998). Economic values of ecological services from mangrove ecosystem. *International Newsletter of Coastal Management* 33: 1-2.

Cahoon D.R., Hensel P., Rybczyk J., McLee K.L., Proffitt C.E. & Perez B.C. (2003). Mass tree mortality leads to mangrove peat collapse at Bay Islands, Honduras, after Hurricane Mitch. *Journal of Ecology* 91: 1093-1105.

Chatenoux B. & Peduzzi P. (2007). Impacts from the 2004 Indian Ocean Tsunami: analysing the potential protecting role of environmental features. *Natural Hazards* 40(2): 289-304.

Chang S.E., Adams B.J., Alder J., Berke P.R., Chuenpagdee R., Ghosh S. & Wabnitz C. (2006). Coastal ecosystems and tsunami protection after the December 2004 Indian Ocean Tsunami. *Earthquake Spectra* 22(S3): 863-887.

Cheevaporn V. & Menasveta P. (2003). Water pollution and habitat degradation in the Gulf of Thailand. *Marine Pollution Bulletin* 47: 43-51.

Chong H.T., Sasekumar A., Leh M.U.C. & D'Cruz R. (1990). The fish and prawn communities of a Malaysian coastal mangrove system, with comparisons to adjacent mudflats and inshore waters. *Estuarine, Coastal and Shelf Science* 31: 703-722.

Chou L.M. (1994). Marine environmental issues in Southeast Asia: state and development. *Hydrobiologia* 285: 139-150.

Chu H.Y., Chen N.C., Yeung M.C., Tam N.F.Y. & Wong Y.S. (1998). Tide-tank system simulating mangrove wetland for removal of nutrients and heavy metals from wastewater. *Water Science and Technology* 38(1): 361-368.

Cochard R., Ranamukarhachchi S.L., Shivakoti G., Shipin O., Edwards P.J. & Seeland, K.T. (2008). The 2004 tsunami in Aceh and Southern Thailand: a review on coastal ecosystems, wave hazards and vulnerability. *Perspectives in Plant Ecology, Evolution and Systematics* 10: 3-40.

Costanza, R. R., d' Arge, R., de Groot., S., Faber, M., Grasso, B., Hannon, K., Limburg, S., Naeem, R., O'Neill, J., Paruello, R., Raskin, P., Sutton & van den Belt, M. (1997): The value of the Worlds Ecosystem Services and Natural Capital. *Nature* 387: 253-260.

Dahdouh-Guebas F., Jayatissa L.P., di Nitto D., Bosire J.O., lo Seen D. & Koedam N. (2005a). How effective were mangroves as a defence against the recent tsunami? *Current Biology* 15(12): R443-447

Dahdouh-Guebas F., Hettiarachchi S., Lo Seen D., Batelaan O., Sooriyarachchi S., Jayatissa L.P. & Koedam N. (2005b). Transitions in ancient inland freshwater resource management in Sri Lanka affect biota and human populations in and around coastal lagoons. *Current Biology* 15: 579-586. <http://www.citg.tudelft.nl/live/binaries/4de0d195-5207-4e67-84bb-455c5403ae47/doc/2004DeVos.pdf >

Das, S. & Vincent, J.R. (2009). Mangroves protected villages and reduced death toll during Indian super cyclone. *Proceedings of the National Academy of Sciences of the United States of America* 106(18): 7357-7360.

Dierberg F.E. & Kiattisimkul W. (1996). Issues, impacts, and implications of shrimp aquaculture in Thailand. *Environmental Management* 20(5): 649-666.

Dorenbosch M., van Riel M.C., Nagelkerken I. & van der Velde G. (2004). The relationship of reef fish densities to the proximity of mangrove and seagrass nurseries. *Estuarine, Coastal and Shelf Science* 60: 37-48.

Duarte C.M. (2002). The future of seagrass meadows. *Environmental Conservation* 29: 192-206.

EJF (2006). *Mangroves: Nature 's defence against Tsunamis – a report on the impact of mangrove loss and shrimp farm development on coastal defences.* Environmental Justice Foundation, London, UK. < http://www.ejfoundation.org/pdf/tsunami_report.pdf >

Ellison J.C. (1999). Impacts of sediment burial on mangroves. *Marine Pollution Bulletin* 37: 420-426.

Erftemeijer P.L.A. (2002). A new technique for rapid assessment of mangrove degradation: a case study of shrimp farm encroachment in Thailand. *Trees* 16: 204-208.

FAO (2003). *Status and Trends in Mangrove Area Extent Worldwide.* Forest Resources Assessment Working Paper 63, prepared by M.L. Wilkie and S. Fortuna. Forestry Department FAO, Rome.<
http://www.fao.org//docrep/007/j1533e/j1533e00.htm >

FAO (2005a). *In-depth assessment of mangroves and other coastal forests affected by the tsunami in Southern Thailand. THA/05/001.* Consultant Report by P. Saenger, International Consultant, November 2005. FAO, Regional Office for Asia and the Pacific, Bangkok.

FAO (2005b). *In-depth assessment of mangroves and other coastal forest affected by the tsunami in Southern Thailand. THA/05/001/01/12.* Progress Report submitted to the Regional Office for Asia and the Pacific, Food and Agriculture Organisation, Bangkok. Forestry Research Centre, Faculty of Forestry, Kasetsart University, Bangkok.

FAO (2005c). *Back-to-Office Report. Backstopping support for the project: "In-depth assessment of mangroves and other coastal forests affected by the tsunami in Southern Thailand. THA/05/001."* Reporting officer M. Kashio, Forest Resource Officer, November 2005. FAO, Regional Office for Asia and the Pacific, Bangkok.

FAO (2005d). *An assessment of the impacts of the 26th December 2004 earthquake and tsunami on aquaculture in the Provinces of Aceh and North Sumatra, Indonesia.* Report prepared by M. Phillips and A. Budhiman. FAO, Indonesia, March 2005. <http://library.enaca.org/NACA-Publications/Tsunami/indonesian-aquaculture-assessment-report.pdf >

FAO (2005e). *Coastal protection and spatial planning in Indonesia.* Mission Report by M. Pushparajah, May 2005. Food and Agriculture Organisation, Rome.
< http://www.fao.org/forestry/foris/webview/tsunami/index >

FAO-MOAC (2005). *Report of Joint FAO/MOAC Detailed Technical Damages and Needs Assessment Mission in Fisheries and Agriculture Sectors in Tsunami Affected Six Provinces in Thailand, 11-24 January 2005.* Food and Agriculture Organisation of the United Nations (FAO) in cooperation with Ministry of Agriculture and Cooperatives (MOAC), February 2005.
< http://www.apfic.org/apfic_downloads/tsunami/FAO_MOAC_thai.pdf

Feagin, R.A., Mukherjee, N., Shanker, K., Baird, A.H., Cinner, J., Kerr, A.M., Koedam, N., Sridhar, A., Arthur, R., Jayatissa, L.P., Lo Seen, D., Menon, M., Rodriguez, S., Shamsuddoha, M. & Dahdouh-Guebas, F. (2010). Shelter from the storm? Use and misuse of coastal vegetation bioshields for managing natural disasters. *Conservation Letters* 3: 1-11.

Finlayson C.M. (2005). Plant ecology of Australia's tropical floodplain wetlands: a review. *Annals of Botany* 96: 541-555.

Foster R., Hagan A., Perera N., Gunawan C.A., Silaban I., Yaha Y., Manuputty Y., Hazam I. & Hodgson G. (2006). *Tsunami and Earthquake Damage to Coral Reefs of Aceh, Indonesia.* Reef Check Foundation, Pacific Palisades, California.
< http://www.reefcheck.org/PDFs/reefcheck_aceh_jan2006_web.pdf >

Gardner L.R., Michener W.K., Williams T.M., Blood E.R., Kjerve B., Smock L.A., Lipscomb D.J. & Gresham C. (1992). Disturbance effects of Hurricane Hugo on a pristine coastal landscape: North Inlet, South Carolina, USA. *Netherlands Journal of Sea Research* 30: 249-263.

GCRMN (2006). *Status of Coral Reefs in Tsunami Affected Countries: 2005.* Wilkinson C., Souter D. & Goldberg J. (eds). Global Coral Reef Monitoring Network. Australian Institute of Marine Sciences, Townsville. <http://www.aims.gov.au/pages/research/coral-bleaching/scr-tac2005/index.html >

Gedik N., Irtem E. & Kabdasli S. (2005). Laboratory investigation on tsunami run-up. *Ocean Engineering* 32: 513-528.

Gelfenbaum G. & Jaffe B. (2003). Erosion and sedimentation from the 17 July, 1998 Papua New Guinea Tsunami. *Pure and Applied Geophysics* 160: 1969-1999.

Giesen W. (2004). *Causes of Peatswamp Forest Degradation in Berbak NP, Indonesia, and Recommendations for Restoration.* Arcadis Euroconsult. International Agricultural Centre(IAC),WageningenUniversity,Netherlands.<http://www.waterfoodecosystems.nl/docs/AirHitamLaut/Rehabilitation.PDF >

Green E.P. & Short F.T. (eds) (2003). *World Atlas of Seagrasses.* University of California Press, Berkeley.

Gross, M. & Hoffmann-Riem, H. & Krohn, W. (2005). *Realexperimente. Ökologische Gestaltungsprozesse in der Wissensgesellschaft.* Transcript, Bielefeld, Germany.

Gross, M., Hoffmann-Riem, H. (2005). Ecological restoration as a real-world experiment: designing robust implementation strategies in an urban environment. *Public Understanding of Science* 14(3): 269-284.

Heller V., Unger J & Hager W. (2005). Tsunami run-up - a hydraulic perspective. *Journal of Hydraulic Engineering* September: 743-747.

Hoeksema B.W. & Cleary D.F.R. (2004). The sudden death of a coral reef. *Science* 303: 1293.

Homcha-Aim, K. (2010). Carbon stocks in biomass and soil of private mangrove plantations (Rhizophora apiculata Bl.) in Yisan Sub-District, Amphawa District, Samut Songkram Province, Thailand. Masters Thesis, Asian Institute of Technology, Bangkok.

Horvitz C.C., Pascarella J.B., McMann S., Freedman A. & Hofstetter R.H. (1998). Functional roles of invasive non-indigenous plants in hurricane-affected subtropical hardwood forests. *Ecological Applications* 8(4): 947-974.

Hughes T.P. & Connell J.H. (1999). Multiple stressors on coral reefs: a long-term perspective. *Limnology and Oceanography* 44(3): 932-940.

Iftekhar, M.S. (2008). An overview of mangrove management strategies in three South Asian countries: Bangladesh, India and Sri Lanka. *International Forestry Review* 10(1): 38-51.

Ikejima K., Tongnunui P., Medej T. & Taniuchi T. (2003). Juvenile and small fishes in a mangrove estuary in Trang Province, Thailand: seasonal and habitat differences. *Estuarine, Coastal and Shelf Science* 56: 447-457.

Islam, M.S. & Haque, M. (2004). The mangrove-based coastal and near shore fisheries of Bangladesh: ecology, exploitation and management. *Reviews in Fish Biology and Fisheries* 14: 153–180.

Islam, M.S. & Wahab, M.A. (2005): A review on the present status and management of mangrove wetland habitat resources in Bangladesh with emphasis on mangrove fisheries and aquaculture. *Hydrobiologia,* 542, pp: 165–190.

IUCN (2005c). *After the Tsunami: Dealing with Invasive Alien Species.* Information Paper No 2. < http://www.iucn.org/tsunami/docs/ip-ias.pdf >

IUCN (2006). *Guide to Koh Phra Thong Natural Habitats. Preserving the Ecosystems.* IUCN Asia RegionalOffice,Bangkok.DraftReport.<http://www.iucn.org/tsunami/docs/envguidelines_KPT_final.pdf >

Jaffe B.E., Borrero J.C., Prasetya G.S., Peters R., McAdoo B., Gelfenbaum G., Morton R., Ruggiero P., Higman B., Dengler L., Hidayat R., Kingsley E., Kongko W., Lukijanto, Moore A., Titov V. & Yulianto E (2006). Northwest Sumatra and offshore islands field survey after the December 2004 Indian Ocean Tsunami. *Earthquake Spectra* 22(S3): 105-135.

Kanagaratnam, U., Schwarz, A.M., Adhuri, D. & Dey, M.M. (2006). Mangrove rehabilitation in the West Coast of Aceh – Issues and perspectives. *NAGA, Worldfish Center Quarterly* 29(3 & 4): 10-18.

Kerr, A.M., Baird, A.H. (2007). Natural barriers to natural disasters. *BioScience* 57(2): 102.

Kerr A.M., Baird A.H. & Campbell S.J. (2006). Comments on "Coastal mangrove forests mitigated tsunami" by K. Kathiresan & N. Rajendran [Estuar. Coast. Shelf Sci. 65 (2005) 601-606]. *Estuarine, Coastal and Shelf Science* 67: 539-541.

Koottatep T., Surinkul N., Polprasert C., Kamal A.S.M., Koné D., Montangero A., Heinss U. & Strauss M. (2005). Treatment of septage in constructed wetlands in tropical climate: lessons learnt from seven years of operation. *Water Science and Technology* 51(9): 119-126.

Laumonier Y. (1997). *The Vegetation and Physiography of Sumatra.* Kluwer Academic Publishers, Dordrecht, Netherlands.

Lay T., Kanamori H., Ammon C.J., Nettles M., Ward S.N., Aster R.C., Beck S.L., Bilek S.L., Brudzinski M.R., Butler M.R., DeShon H.R., Ekstrom G., Satake K. & Sipkin S. (2005). The great Sumatra-Andaman earthquake of 26 December 2004. *Science* 308: 1127-1133.

Liew, S.C., Gupta, A., Wong, P.P. & Kwoh, L.K. (2010). Recovery from a large tsunami mapped over time: The Aceh coast, Sumatra. *Geomorphology* 114: 520-529.

Lugo A.E., Rogers C.S. & Nixon S.W. (2000). Hurricanes, coral reefs and rainforests: resistance, ruin and recovery in the Carribean. *Ambio* 29(2): 106-114.

Mather R. (2005). Coral Reef Management and Conservation in Andaman Sea Marine Protected Areas, After the Tsunami. Private Report. <http://www.wwfthai.org/thai/resources/download/AndamanCoralImpactsFeature.pdf >

Mazda, Y., Wolanski, E. & Ridd, P.V. (Eds.), 2007. *The Role of Physical Processes in Mangrove Environments. Manual for the Preservation and Utilization of Mangrove Ecosystems.* Terrapub, Tokyo.

McAdoo, B.G., Goff, J.R., Fritz, H.M., Cochard, R. & Kong, L.S. (2009). *The tsunami project: Integrating engineering, natural and social sciences into post-tsunami surveys.* American Geophysical Union, Fall Meeting 2009, abstract #NH34B-01. < http://adsabs.harvard.edu/abs/2009AGUFMNH34B..01M >

Meilianda, E., Wong P.P., Dohmen-Janssen, C. M., (2006). *On the impact of the earthquake and tsunami of 26 December 2004 to the geomorphological settings of the west, north and east coast of Banda Aceh: Report on field work in Banda Aceh,* 22 May - 05 June 2005. *Civil Engineering & Management Research Report 2006W-001/WEM-001.* Waterbeheer *(WEM).* Twente University, Netherlands.
<http://www.wem.ctw.utwente.nl/organisatie/Medewerkers/medewerkers/meilianda/Ella%20Meilianda/ >

Meltzner A.J., Sieh K., Abrams M., Agnew D.C., Hudnut K.W., Avouac J. & Natawidjaja D.H. (2006). Uplift and subsidence associated with the great Aceh-Andaman earthquake of 2004. *Journal of Geophysical Research* 111: B02407, doi:10.1029/2005JB003891.

Moore A., Nishimura Y., Gelfenbaum G., Kamataki T. & Triyono R. (2006). Sedimentary deposits of the 26 December 2004 tsunami on the northwest coast of Aceh, Indonesia. *Earth Planets Space* 58: 253-258.

Mumby P.J., Edwards A.J., Arias-Gonzalez J.E., Lindeman K.C., Blackwell P.G., Gall A., Gorczynska M.I., Harborne A.R., Pescod C.L., Renken H., Wabnitz C.C.C. & Llewellyn G. (2004). Mangroves enhance the biomass of coral reef fish communities in the Carribbean. *Nature* 427(5): 533-536.

Nalbant S.S., Steacy S., Sieh K., Natawidjaja D. & McCloskey J. (2005). Earthquake risk on the Sunda trench. *Nature* 435: 756-757.

Natawidjaja D.H., Sieh K., Chlieh M., Galetzka J., Suwargadi B.W., Cheng H., Edwards R.L., Avouac J. & Ward S.N. (2006). Source parameters of the great Sumatran megathrust earthquakes of 1797 and 1833 inferred from coral microatolls. *Journal of Geophysical Research* 111, B06403, doi:10.1029/2005JB004025.

Natawidjaja D.H., Sieh K., Ward S.N., Cheng H., Edwards R.L., Galetzka J. & Suwargadi B.W. (2004). Paleogeodetic records of seismic and aseismic subduction from central Sumatran microatolls. *Journal of Geophysical Research* 109, B04306, doi:10.1029/2003JB002398.

NORDECO (2005). *Preliminary assessment on impacts of the tsunami on coastal ecosystems and associated livelihoods in Aceh Province, Indonesia.* Global Environment Centre, Wetlands International Indonesia, NORDECO, 10 January 2005.
<http://www.nordeco.dk/nordeco/publications/Preliminary%20Assessment%20on%20Impacts%20of%20Tsunami.pdf >

Parish F. (2005). *Press Release: Mangroves might have helped to protect Banda Aceh.* Global EnvironmentCentre,Malaysia.
< http://www.riverbasin.org/ev_en.php?ID=3794_201&ID2=DO_TOPIC >

Pennisi, E., 2005. Powerful tsunami's impact on coral reefs was hit and miss. *Science* 307: 657.

Pet-Soede C., Cesar H.S.J, Pet J.S. (1999). An economic analysis of blast fishing on Indonesian coral reefs. *Environmental Conservation* 26: 83-93.

Pomeroy, R.S., Ratner, B.D., Hall, S.J., Pimoljinda, J. & Vivekanandan, V. (2006). Coping with disaster: rehabilitating coastal livelihoods and communities. *Marine Policy* 30: 786–793.

Ramachandran S., Anitha S., Balamurugan V., Dharanirajan K., Vendhan K.E., Divien M.I.P., Vel A.S., Hussain I.S. & Udayaraj A. (2005). Ecological impact of tsunami on Nicobar Islands (Camorta, Katchal, Nancowry and Trinkat). *Current Science* 89(1): 195-200.

Rama Chandra Prasad, P., Sudhakar Reddy, C, Sundara Rajan, K., Hazan Raza, S. & Bala Subrahmanya Dutt, C. (2009). Assessment of tsunami and anthropogenic impacts on the forests of the Andaman Islands, India. *International Journal of Remote Sensing* 30(5): 1235-1249.

ReliefWeb (2006a). OCHA Situation Report No. 1. Floods in Aceh, Indonesia. <http://www.reliefweb.int/rw/RWB.NSF/db900SID/ACIO-6WSH99> (accessed July 2010)

Ridd P.V. (1996). Flow through animal burrows in mangrove swamps. *Estuarine, Coastal and Shelf Science* 43: 617–625.

Roy S.D. & Krishnan P. (2005). Mangrove stands of Andamans *vis-à-vis* tsunami. *Current Science* 89(11) : 1800-1804.

Sargeant H.J. (2001). *Vegetation fires in Sumatra, Indonesia. Oil palm agriculture in the wetlands of Sumatra: destruction or development?* European Union Forest Fire Prevention and Control Project with Dinas Kehutanan Propinsi Sumatra Selatan. European Union and Ministry of Forestry, Jakarta, Indonesia. <http://www.fire.uni-freiburg.de/se_asia/projects/ffpcp/FFPCP-16-Oil-Palm.pdf >

Sathirathai, S. & Barbier, E.B. (2001): Valuing mangrove conservation in Southern Thailand. *Contemporary Economic Policy* 19(2): 109–122.

Slavich P. (2006). Washing the salt from Aceh's wounds. *Ecos* 34: 128 < http://www.publish.csiro.au/ecos/index.cfm?sid=10&issue_id=5214 >

Smith T.J., Boto K.G., Frusher S.D. & Giddins R.L. (1991). Keystone species and mangrove forest dynamics: the influence of burrowing by crabs on soil nutrient status and forest productivity. *Estuarine, Coastal and Shelf Science* 33: 419-432.

Soegiarto A. (2004). Research into, and conservation of, mangrove ecosystems in Indonesia. *Mangrove Management and Conservation. Present and Future,* (ed. M. Vanucci), pp. 51-58, United Nations University Press, Tokyo.

Spalding M.D., Balsco F. & Field C.D. (2001). *World Atlas of Coral Reefs.* University of California Press, Berkeley.

Stieglitz T., Ridd P. & Müller P. (2000). Passive irrigation and functional morphology of crustacean burrows in a tropical mangrove swamp. *Hydrobiologia* 421: 69-76.

Stone, R., 2006. A rescue effort for tsunami-ravaged mangrove forests. *Science* 314 (5798): 404.

Stott P. (1990). Stability and stress in the savanna forests of mainland South East Asia. *Journal of Biogeography* 17: 373-383.

Synergy One (2006). *Damages map of tsunami affected agricultural areas in Nanggaroe Aceh Darussalam (NAD) Province. Progress Report of Technical Research Collaboration.* Synergy One Co., Ltd. Thailand and Indonesian Centre for Agricultural Land Resources Research and Development, Ministry of Agriculture.

Szczucinski W., Niedzielski P., Rachlewicz G., Sobczynski T., Ziola A., Kowalski A., Lorenc S. & Siepak J. (2005). Contamination of tsunami sediments in a coastal zone inundated by the 26 December 2004 tsunami in Thailand. *Environmental Geology* 49: 321-331.

Taleb, N.N. (2007). *The Black Swan. The Impact of the Highly Improbable.* Random House, New York.

Thampanya U., Vermaat J.E., Sinsakul S. & Panapitukkul N. (2006). Coastal erosion and mangrove progradation of Southern Thailand. *Estuarine, Coastal and Shelf Science* 68: 75-85.

Thanawood C., Yongchalermchai C. & Densrisereekul O. (2006). Effects of the December 2004 tsunami and disaster management in Southern Thailand. *Science of Tsunami Hazards* 24(3): 206-217.

The Economist (2010). *The gulf oil spill: No end in sight. High tempers and low expectations in the Gulf of Mexico.* The Economist, June 10th 2010.
< http://www.economist.com/node/16322752 >
(accessed August 2010)

Tomascik T., Mah A.J., Nontji A. & Moosa M.K. (1997). *The Ecology of the Indonesian Seas.* Periplus Editions (HK) Ltd., Singapore.

UNDP (2006). *Malaysia's Peat Swamp Forests. Conservation and Sustainable Use.* United Nations Development Programme, Kuala Lumpur, Malaysia.

UNEP (2004). *Seagrasses in the South China Sea.* UNEP/GEF/SCS, Technical Publication No. 3. United Nations Environment Programme, Nairobi.
<http://www.unepscs.org/Publication/Booklets/Seagrass%20booklet%20combine%20version.pdf >

UNEP (2005a). *After the Tsunami: Rapid Environmental Assessment.* United Nations Environment Programme, Nairobi.
< http://www.unep.org/tsunami/reports/TSUNAMI_report_complete.pdf >

UNEP-WCMC (2006). *In the frontline : shoreline protection and other ecosystem services from mangroves and coral reefs.* UNEP-WCMC, Cambridge.
< http://www.unep.org/pdf/infrontline_06.pdf >

USGS (2005). *USGS Scientists in Sumatra Studying Recent Tsunamis.* U.S. Geological Survey. Field Reports by the The West Sumatra International Tsunami Survey Team (ITST).
< http://walrus.wr.usgs.gov/news/reportsleg1.html > (accessed in July 2010)

Whitten T., Damanik S.J., Anwar J. & Hisyam N. (1997). *The Ecology of Sumatra.* Periplus.

Wibisono, I.T.C. & Suryadiputra, I.N.N. (2006). *Study of lessons learned from mangrove/coastal ecosystem restoration efforts in Aceh since the tsunami.* Wetlands International Indonesia Programme. Bogor, Indonesia.

WIIP (2005). Tsunami of Aceh and North Sumatra, 26 December 2004. Report by Wetlands International - Indonesia Programme.
<http://www.wetlands.org/Tsunami/data/TSUNAMI-INDONESIA-WIIP,English.doc >

Wikipedia(2010).
<http://en.wikipedia.org/wiki/Humanitarian_response_to_the_2004_Indian_Ocean_earthquake >
(accessed in July 2010)

Wong P.P. (2003). Where have all the beaches gone? Coastal erosion in the tropics. *Singapore Journal of Tropical Geography* 24(1): 111-132.

Wong Y.S., Lan C.Y., Chen G.Z., Li S.H., Chen X.R. Liu Z.P. & Tam N.F.Y. (1995). Effect of wastewater discharge on nutrient contamination of mangrove soils and plants. *Hydrobiologia* 295: 243-254.

Wong Y.S., Tam N.F.Y. & Lan C.Y. (1997). Mangrove wetlands as wastewater treatment facility: a field trial. *Hydrobiologia* 352: 49-59.

Young E. (2005). Crop revival for Aceh after the tsunami. *New Scientist* 187(2514): 9

Part 2

Observational Records

11

Determination of Tsunami Inundation Model Using Terrestrial Laser Scanner Techniques

Giuseppe Mastronuzzi and Cosimo Pignatelli
University of Bari – II Faculty of Science MM.FF.NN.
Gis Geo-Environmental and Remote Sensing Laboratory (LAGAT-TA)
Taranto
Italy

1. Introduction

The East Java Tsunami occurred on 3 June, 1994 remind us the possibility that these catastrophic events affect coastal area suddenly, without forewarning. On that occasion, human victims were concentrated in the Bali Island and in the nearly areas; the economic loss were limited on very poor regions, surely the most dramatic consequence of tsunami impact. This event has attracted the interest of the scientific community, increasing the number of geophysics and geodynamic studies, mainly concerning the generative mechanisms and the tsunami propagation. At the same time, studies aiming a better understanding of the morphological effects of past-tsunami impact have been performed along the coasts all around the world (Kelletat, 2009).

Recently, the impact of the Indian Ocean Tsunami (IOT) on December 26, 2004 and its charge of about 230,000 victims in Thailand and in other countries facing the Indian Ocean, indigenous and citizen coming from all around the world, underlined the necessity to manage the human activities in coastal areas and to improve of the knowledge of the possible effects of a tsunami impact on the coast. Large coastal sectors were completely inundated, modified or even destroyed by the impact of the IOT (e.g.: Szczuciski et al., 2005; Kelletat et al., 2006, 2007; Lavigne et al., 2006; Richmond et al., 2006; Umitsu et al., 2007; Paris et al., 2007, 2009; Srinivasalu et al., 2007); post event surveys permitted to recognise morphological effects of its impact and, in the same time, to extend all obtained results on coastal sectors where similar evidences were recognised.

Notwithstanding the immense number of data derived by the surveys performed all along the coast hit by the IOT, the debate about the correlation of some landforms/sediments and the extreme event responsible of their genesis/deposit is still open. At present, there are not undisputable signatures of landforms and/or fine sediments that allows discrimination between a past-extreme events - sea storm, hurricane or tsunami deposition. This knowledge gap has given rise to a growing number of papers that examining the source and nature of landforms and sediments (e.g. Frohlich et al., 2009; Goto et al., 2007; 2010a; Kelletat, 2008; Kelletat et al., 2006, 2007; Mastronuzzi et al., 2006; Paris et al., 2009; Scheffers, 2006, 2008; Scheffers & Kelletat, 2005 and others).

New data will come from the more recent events of September 30, 2009 around Samoa Islands and of February 28, 2010 in Chile; this last occurred exactly 50 years after the tsunami that in 1960 hit the coasts of the countries facing the Pacific Ocean (Fig. 1).

Fig. 1. The memorial of the impact along the japanese coast of the 1960 Chile Tsunami in Matsubara Park at Minami - Sanriku, North Japan

All these mentioned tsunami, were not the only occurred in the recent history; many strong tsunami occurred before the Java event in 1994; they killed thousands of people along many coasts of the world and caused a long list of economic, social and natural lost (e.g. Baptista et al., 1993 for Nicaragua; Yeh et al., 1993, Tsuji et al., 1995a, 1995b and Maramai & Tinti, 1997 for Indonesia; Imamura et al., 1995 for the Philippines; Yeh et al., 1995 for the Kuril Islands; Shimamoto et al., 1995 and Shuto & Matsutomi, 1995 for Japan; Pelinovski et al., 1997 for Sulawesi). Tsunami phenomena are rarely taken into account by planners; only in particular high-developed areas such as Japan, Australia, New Zealand and U.S.A. early warning system, modern structures and risk assessment plans seek to mitigate the overall tsunami risk (Narayan et al., 2005, Cochard et al., 2008; Kato et al., 2005). If an important complex early warning system exist in Pacific Ocean, it lack completely in Mediterranean Sea in which the density of coastal settlement is very high. This region is characterised by the presence of all potential mechanism of tsunami generation (Mastronuzzi et al., 2008; Mastronuzzi, 2010 in press). In historical time, it was affected at least by four devastating events: in about 3500 BP connected to the eruption of Santorini Volcano in Greece, in 356 a.D. generated by a strong earthquake in Creta, destroying the area of the Nilo delta, in 1696 along the ionian coasts of Sicily generated by the strong earthquake that destroyed Noto and in 1908 when a strong earthquake and the associated tsunami caused about 100,000 casualties in the Strait of Messina, southern Italy.

In summary, surveys on the field permit to obtain qualitative data about the sediment and the landforms but not a comprehensive view of the hit areas. Luckily, the remote sensing availability (satellite and terrestrial/aerial laser scanner data, aerial photographs, etc.) permit to increase the quantity of information necessary to integrate field surveys like that typical of a sedimentological, morphological and geophysics approach. This typology of studies is finalised in the individuation of the extension of the area hit by extreme event, in the reconstruction the sequence of the tsunami and in the evaluation of the inland penetration in the case of a future depreciable new impact. The elaboration and

interpretation of the data collected could be utilised in order to plan the human activities along the coasts where evidences of past tsunami impact are preserved.
The aim of this paper is suggests as the Terrestrial Laser Scanner (TLS) surveys performed along coastal areas hit by tsunami, can lead to estimate the area inundate by the wave. In particular, the use of the TLS techniques allow to reconstruct 3D model of the boulders and to evaluate coastal roughness; point clouds could be collected during this kind of survey in order to obtain a very accurate measure of the boulder dimensions and of the tsunami inland penetration limit in view to predict the limit of future new flooding applying devoted hydrodynamic mathematical models. All these data are of extreme importance in the definition of the correct integrated coastal zone management (ICZM), in post-tsunami disaster management, and in first aid intervention plans redaction.

2. Morphological effect of past tsunami

Tsunami and exceptional storms are high magnitude/low frequency events that are considered to have played an important role in the evolution of the coastal areas (e.g. Dawson, 1994; Bryant 2001). In fact, whatever the generating mechanism of a tsunami may be - submarine earthquakes, landslides or volcanic eruption, impact of asteroids -, this event can discharge its destructive energy along the coast determining inundations some kilometres extended inland. Not all tsunami are destructive; low energy tsunami cannot flood the coastal areas but in case of flooding, accumulation of a mixture of sediment and debris of marine and continental origin marks the inundated areas whereas waves energy was sufficient to penetrate. Of course, these effects can be produced also by a sea surge generated by exceptional sea storms with very limited difference in extension in the case of flat coastal areas (Scheffers, 2002, 2004; Williams & Hall, 2004; Emanuel, 2005; Fita et al, 2007). On the other hand, as consequence of tsunami flooding the vegetation covers are erased causing a deserted areas along the coast and/or erosive landforms can be shaped directly on the local bedrock (e.g.: Bryant et al., 1996; Bryant, 2001).
Moreover, mega boulders, isolate and sparse or arranged in fields or in berms in intertidal/adlitoral zone, accumulated along the rocky coasts and in coral reef protected areas, represent the most impressive evidence of the impact of extreme waves. These morphological evidence were recognised along many coasts all around the world and have been attributed to the impact of extreme waves occurred in the past or evidenced by post-event survey (e.g. Kawana & Pirazzoli, 1990, Nishimura & Miyaii, 1995 and Goto et al., 2010b for Japan; Goto et al., 2007 for Thailand; Jones & Hunter, 1992 for the Cayman Islands; Bryant et al., 1996, Bryant & Young, 1996 and Young et al., 1996 for the southeastern coast of Australia; Hearty, 1997 for the Bahamas; Mastronuzzi & Sansò, 2000, 2004 and Scicchitano et al., 2007 for Italy; Kelletat & Schellmann, 2002, Scheffers et al., 2008 and Vött et al., 2006, 2007, 2008, 2009a, 2009b for Cyprus and Greece) (Fig. 2).
Unluckily, not always the presence of these unusual landforms find the agreement of the scientist concerning their shaping event; the finding of boulders still pose the question: are they effect of a paleo-storm or a paleo-tsunami impact? On the other hand, the fact that the reply is not absolute, is testified by recent study in which different Authors have evidenced that storms and tsunami can generate phenomena of morphologic convergence in functions of the local coastal features (Mastronuzzi & Sansò, 2004; Williams & Hall, 2004; Hall et al., 2006, Mastronuzzi et al., 2006; Bourrouilh-Le Jan et al., 2007; Scheffers, 2002a, 2002b, 2004).

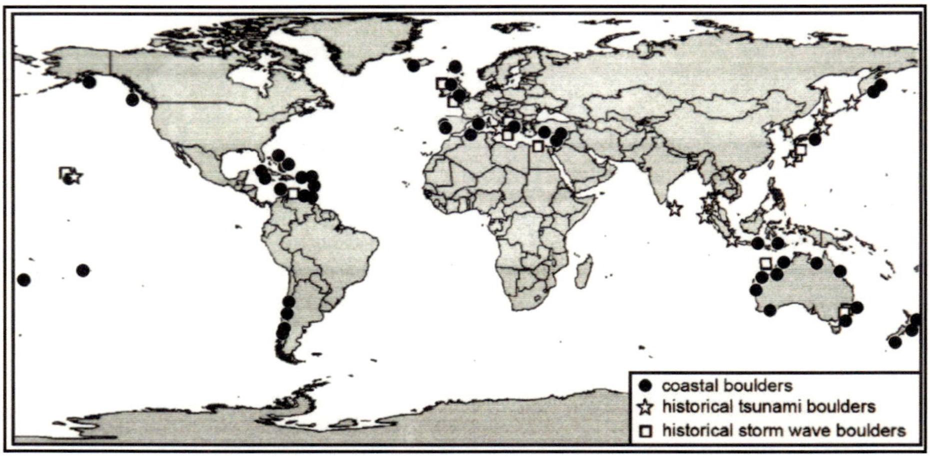

Fig. 2. Megaboulders have been recognised all around the world. Black circles indicate boulder which deposition has been attributed to past event. Stars and boxes indicates accumulations of know origin events occurred in the recent past and testified by direct observation or eyewitness (modified from: Scheffers, 2008; Goto et al., 2010a; Mastronuzzi, 2010).

Fig. 3. A. Coralline megaboulders accumulated along the Ishigaki coasts, South Japan; B. Large boulder accumulated along the eastern coast of Bonaire Island (Netherland Antilles); C. Boulders scattered on the railways station of Reggio Calabria (South Italy) by the December 8, 1908 Messina and Reggio Calabria Tsunami (Archives Alinari/INGV); D. An impressive boulder ridge extends for about 2.5 km between Otranto-S.M Leuca coast (Apulia, Southern Italy) accumulated by February 20, 1743 tsunami.

So storms and tsunami can scatter boulders and mega boulders inland and accumulate them in the emerged coastal areas in function of the dissipation of their energy in function of the coastal bathymetry and topography.
Boulder up to some ten of tons heavy have been recognised in different localities in areas hit by the impact of historical tsunami (Fig. 3A, 3B, 3C, 4D).
Generally they are arranged in fields and are locally embricated; in some case they show a lateral classation, so the biggest and heaviest are close to coastline and the smallest are far inland. In other case boulders are arranged in berms constitute of hundred of them, in chaotic distribution.
The presence of large boulders along the rocky coast can be used to evaluate the inundation limit due to tsunami impact. In particular, the boulders position should indicate the place in which the impacting waves had the energy to transport them but not the limit of its inland penetration.
The integration of morphological and hydrodynamic data allows for an estimation of the water level during the tsunami event responsible for boulder transport. Moreover, it is possible to calculate how far the impacting wave flooded inland; this is a function of the wave parameters and of the degree of roughness of the flooded terrain.
Recent hydrodynamic theories tried to put in relation boulders and coastal features to the wave responsible for their accumulation developing mathematical formulas. Starting from the boulders size and shape, knowing the local waveclimate, it should be possible to discriminate between storm and tsunami inland scattering (Nott, 1997, 2003; Noormets et al., 2002, 2004; Pignatelli et al., 2009; 2010a,b; Imamura et al., 2008; Barbano et al., 2010).

3. Hydrodynamic equations

In order to evaluate the height of the tsunami or severe storm waves able to detaches and move such clasts, Nott (1997) proposed a set of hydrodynamic equations. These equations take into account boulder dimensions (a = major axis, b = medium axis, c = minor axis), boulder shape and the density of rock (bulk density). Nott (2003) improved these equations introducing the pre-transport settings and, to be more precise, the determination of the main morphological features of the deposited boulders: the position prior to the tsunami impact, the size, the shape and rock density. Indeed, the reconstruction of the possible morphodynamic scenario(s) could be extremely important in the prediction of the possible futures ones. Three different scenarios were hypothesized: the first one considers a boulder placed on a cliff edge (joint bounded scenario JBS); in the second one, the boulder is taken up and placed below sea level (submerged scenario SMS); in the last one, the boulder is detached and placed inland (subaerial scenario SAS). Many Authors used Nott's equations to determine if sea storm or tsunami were responsible for boulder displacement (e.g. Scheffers, 2002, 2004; Scicchitano et al., 2007; Maouche et al., 2009); these studies permitted to deduce that storm and tsunami heights calculated by Nott's equations appear overestimated (e.g. Scheffers, 2002, 2004; Paris et al., 2009; Goto et al., 2010b; Bourgeois & MacInnes, 2010).
In order to calculate an accurate value of the wave height able to detaches and/or transport similar boulder, geomorphological surveys have been carried out. In particular, direct surveys performed during strong storms evidenced that boulders have been moved by sliding (e.g.: Mastronuzzi et al., 2004). Frequently large boulder results broken in two ore more pieces placed close one another. Their accumulation has been attributed to the impact

of past tsunami and the most probable transport mechanism is the floating in the water flow and not the sliding on the rocky surface; when wave was not able to transport them they collapsed breaking themselves. Since these considerations and starting from real case studies, Pignatelli et al. (2009) introduced new equations that optimize the Nott's theory. In Pignatelli et al. (2009) hypothesis, boulders come from the emerged - or immediately subemerged - part of the cliff, drawing a scenario that is very similar to the geomorphological situation described by Noormets et al., (2002; 2004); the boulder is placed still in the outcropping rock but loose along joints at all sides. In this condition the a x c face of the boulder - c-axis indicates the thickness – is directly exposed to wave impact (Fig. 4).

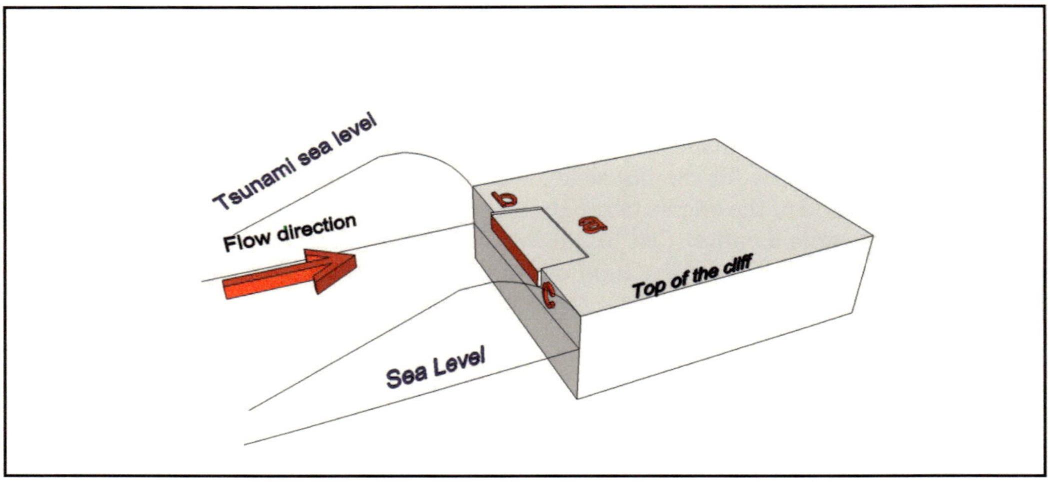

Fig. 4. 3D scheme of the tsunami impact on a rocky coast. The small parallelepiped represents a boulder in joint bounded condition. The a, b, c letter highlighted main axis of the boulder. The red face represents the a x c face directly exposed to wave impact.

The Pignatelli et al. (2009) hydrodynamic equation is:

$$H_T = [0.5 \cdot c \cdot (\rho_b - \rho_w / \rho_w)] \ / \ C_L \tag{1}$$

where H_T is the tsunami wave height breaking point, c axis is the boulder height, ρ_b is the density of boulder, ρ_w is the density of water = 1.02 g/ml; C_L is the coefficient of lift = 0.178 (see Bryant, 2001 and references therein). To apply the appropriate equations is important when aiming to assess the maximum extent of flooding. The hope is that the implementation of mathematical models corresponds realistically to the complexity of processes that occur during coastal inundation. On the other hand, at present mathematical models were used to estimate the run up, the landward flooding limit and the destructive forces that could be produced by future tsunamis.

Many Authors have developed numerical models that simulate the destructive potential of tsunamis (e.g.: Mader, 1974; Tanioka & Satake, 1996; Hills & Mader, 1997; Titov & González, 1997; Titov & Synolakis, 1998; Tinti & Armigliato 2003; Mofjeld et al., 2005; Weiss et al., 2006); they are based on differential equations elaborated with particular boundary conditions that take into account some approximation with respect to local coastal features. Unluckily these models never consider the geomorphological features of the coast expressed by local topography hit by the waves. In fact, tsunami propagation across coastal landscape is influenced by topographic irregularities, slope, the presence of buildings, vegetation textures, etc.

In the scenario hypothesized by Pignatelli et al. (2009) it is possible to calculate the tsunami height at the top of the cliff; the wave heights used to simulate the tsunami flooding are assessed in relation to the boulder scattered inland; if a boulder is detached from cliff edge (joint bounded scenario) the height of the water column flooding, can be obtained from this relationship:

$$H_{FL} = H_T - \text{Altitude of the cliff} \quad (2)$$

In fact, at the edge of the cliff - where the boulders initially were placed - the tsunami wave height had to be the minimum able to transport the biggest block. The tsunami wave must be not smaller then minimum wave able to move the biggest boulder is the height of the water column that cause overland flooding.

Besides, the tsunami inland penetration is influenced by terrain roughness expressed by Manning's coefficient.

In particular, Hills & Mader (1997) have provided an empirical formula that permits to calculate the inland flooding limit X_{FL} of an impacting tsunami wave:

$$X_{FL} = (H_{FL})^{1.33} n^{-2} k \quad (3)$$

where: H_{FL} is tsunami wave height at coastline; n is Manning's number, k = 0.06 is a constant for many tsunamis (see: Bryant, 2001). This formula gives good estimation for coasts characterised by overland flat profile. In the reality, often the coastal sector shows sloping overland profile very complex, conditioned by the local lithostructural features of the outcropping rock. Starting from the cited formula, a factor cos α - where α represents the mean sloping - has been introduced (Pignatelli et al., 2009); this permit to take into account the coastal overland profile:

$$X_{FL} = (H_{FL})^{1.33} n^{-2} k \cos \alpha \quad (4)$$

The Manning number is a coefficient expression of the micro-topography and sinuosity of the surface; it represents the hydraulic roughness of the terrain propriety that causes resistance to the water flow through creating a retarding force (Chow, 1973). Different Authors have estimated Manning coefficient for different terrain typology (Arcement & Schneider, 1989), but the obtained values are often approximated since their absolute value did not derive by strictly quantitative surveys. As consequence, the use of the Hills & Mader (1997) equation adopting different Manning coefficient is not always rigorous. Moreover, an important aspect that should not be neglected is the roughness variability over time. In fact, the Manning coefficient is not constant with time in the same place due to weed and/or urban growth (Asal, 2003). As consequence, in a coastal sector it is possible to obtain an evaluation of the present roughness useful in the elaboration of a near-future scenario, but that is not representative of a past tsunami impact.

According to several studies performed on open channel flows (Smart et al., 2002; Smart, 2004) the roughness characteristics of the channel bottom are very well correlated with the standard deviation of the bottom surface, so it is possible to use *in situ* Terrestrial Laser Scanner (TLS) data to calculate Manning's n for given flow conditions.

Smart (2004) provides an equation linking directly Manning's n to the flow conditions and to the bottom surface characteristics:

$$n = \frac{k\sqrt[6]{R}}{\sqrt{g}} \frac{1-\left(\frac{Z_0}{R}\right)}{\left[\left(\frac{Z_0}{R}\right)-\ln\left(\frac{Z_0}{R}\right)-1\right]} \tag{5}$$

where: k is the Von Karman constant; g is the gravitational constant; R is the flow depth and Z_0 is the standard deviation of the bottom surface.

All mentioned hydrodynamic equations are characterised by different parameters: some of them are directly measurable (e.g.: dimensions of the boulders, density of the rock), other parameters are know in bibliography and/or in technical reports (e.g.: water density, some coefficients, etc.), while there are scheduled parameters (e.g.: coastal roughness) and derived parameters (e.g.: wave length and wave height).

The use of a mathematical approach needs modern surveys aimed to have at one's disposal quantitative data. To have the better prediction of the possible effect of the future tsunami impact, the parameters that describe the morphological features like boulders size and coastal roughness must be surveyed with precision. Recent experiences learned that a traditional manual survey has an approximation respect to the digital one of about 25% of approximation in excess (Marsico et al., 2009). In fact, boulder can be characterised by very irregular shape and by very irregular surfaces; the measurement of a-, b- and c- axis can be not enough for the determination of the volume and of the weight. About this last parameters another difficulty derive by the type of the rock. In case of an homogenous rock like granite or limestone the specific weight is easy to be calculated; the problem is of very difficult solution when the boulder is constitute by coralligenous or algal calcarenite, both characterised by lithological discontinuity (Spitzke et al., 2008). But this problem, at present, cannot be resolved by the remote sensing and by the digital surveys.

4. Remote sensing

Recent TLS technology is based on the reflectorless acquisition of a point clouds of the topography using the time-of-flight distance measurement of laser pulse (Slob & Hack, 2004). The scanner consists of a laser beam generator, a mirror rotating on its horizontal axis and forming a 45° degree angle with the beam direction and a servomotor which makes the instrument rotate around its vertical axis. This setting gives to the scanner a field view of 360° x 270°. The monochromatic and nearly parallel laser pulse is sent out in a precisely known direction. The scanner then records the back-scattered pulse. The time-of-flight of the signal is then converted into the distance between the scanner and the object.; these two values are used to calculate Cartesian coordinates with reference to the centre of the scanner. In the proposed study-case field surveys have been performed using a Leica Scanstation 2. The instrument captures up to a maximum of 50,000 points per second over a maximum range between 200 and 300 m, depending on the reflectivity of the scanned object.

A single scan represents a simple point view, but the TLS kit is supplied with circular/rectangular targets characterised by a very reflective surface (Fig. 5); their position can be acquired for a simple take and distinction to the beam of the scanner. Therefore, to obtain a complete 3D Model, it is necessary to overlap numerous scans using selected targets as reference points.

Fig. 5. Megaboulder scanned on the coastal area of Augusta (Southern Italy). Red circles highlighted the rectangular targets used to co-registrate all the TLS scan.

Fig. 6. Two large boulders recognised along the coast of the Eastern Sicily (Southern Italy): A. San Lorenzo (Siracusa) Area B1; B. Vendicari Area (Siracusa) B2;

The TLS datasets are treated and analyzed using the HDS Cyclone software (Leica Geosystems, 2004). All the raw scans were manually cleaned to remove any outliers such as vegetation.

The co-registration (or alignment) of the scans is achieved first by a manual alignment, which consists in identifying common points (usually the targets) in the different point clouds. An overlap of approximately 20% between the different scans is necessary to ensure good matching between the datasets. At least three targets must be captured to unify the individual point clouds in a unique 3D Model.

TLS surveys generate cloud points that permit a 3D model in order to define shapes and volumes of the boulders present in the studied areas; using a reverse engineering software,

the 3D point clouds was converted to polygonal model which consists of closed surface. Moreover the model showed holes and gaps due both to scans and to surface interpolation. The fill holes operation closes the polygonal structures and makes the layout continuous.
The complete reconstruction of lacking parts, also at the bottom, allows to define the shape of boulder and to gauge linear sizes, surface area and volume. The reverse engineering by mean of Geomagic software, allows to calculate the main axes, the superficial area and the volume when almost of holes were filled. This operation needs a lot of time for a large boulder because of the quantities of holes generated during polygonal conversion and the respective numbers of bytes of the creating file. Therefore, using a performing pc, the resulting 3D model faithfully reproduces the whole real block. In this paragraph we report an example of this operation performed on two large boulder B1 an B2 recognised along the eastern coast of Sicily (Southern Italy) (Fig. 6A, 6B). So, the 3D model reconstruction of the boulders allows to gauge the a, b, c axes more accurately than their measures during classical field survey.
In particular, in order to assess the mean value of each axis several measures can be catch over the whole boulder. Moreover the 3D model provides the correct volume which is essential to estimate the boulder weight, knowing the rock density (Fig. 7).
Applying the (1) hydrodynamic equation, the minimum wave height of tsunami or sea storm able to move the boulders can be obtained. At this moment, it is important to know in detail the coastal topography especially in the coastal sector where the large boulders are recognised. Highly accurate plano-altimetric surveys can be carried out with a Differential Global Position System (DGPS) in Real Time Kinematic (RTK) mode. The new DGPS technology is based on the GNSS Global Navigation Satellite System an advanced solution for the distribution of GNSS data sets from any station via Internet. This permit to obtain the altitude of the cliff and the water column that probably cause inundation using the (2) equation. As consequence, the minimum tsunami landward penetration can be calculated with (3) or (4) equations in function of the coastal sloping.
The TLS and DGPS survey provide to a very large and accurate point clouds representative of the entire coastal area; the point clouds can be converted to .dxf interchange format to be imported in ArcGIS environment where a conversion to "shapefile" (.shp) can be performed; using this software it is possible to extrapolate a Digital Elevation Model (DEM) with a centimetric resolution by means of an Inverse Distance Weighted (IDW) interpolator. Inverse distance weighted methods are based on the assumption that the interpolating surface is mostly influenced by nearby points and less by the more distant points. The interpolating surface is a weighted average of the scattered points and the weight assigned to each scattered point diminishes as the distance from the interpolation point to the scattered point increases. The entire DEM adequately reproduces the surface because of a high density of points, but in some regions the interpolation is totally different from the actual surface.
Since the data needed to be cleaned by the general slope of the area, the average surface was calculated using the trend tool (same cell size) in order to perform the detrending of the DEM; the latter was accomplished using the raster calculator tool subtracting the cell values of the average surface from the real surface ones. The standard deviation of the resulting raster has been used in Smart (2004) equation. The parameters that permit to estimate the Manning number with (5) equation are:

1. the standard deviation Z_0 of the bottom surface obtained directly from laser scanner data elaboration;

Fig. 7. 3D reconstruction of two largeboulders recognised in Sicily, B1 and B2 see figure 6. For boulder B1: A. Scan; B. Scan and photo; C. Surface reconstruction obtained with Geomagic software. For boulder B2: D. Scan; E. Scan and photo; F. Surface reconstruction obtained with Geomagic software.

2. the flow depth R obtained difference between calculated tsunami height H_T and the mean elevation h of the bedrock surface.

At the end, the obtained Manning numbers were used to evaluate the inundation limits using the Hills & Mader (1997) formula and applying a suitable coastal slope. A devoted predisposed GIS (Geographical Information System) allowed for the elaboration of a map of slope angles with an average value for each coastal sector.

5. Conclusions

In the last few years the tsunami field sciences were strongly developed from a descriptive/qualitative approach to a numerical/quantitative one. The awareness that the knowledge of the local coastal reply to a tsunami impact can contribute in the life safeguard drown the scientific community to improve the study devoted to determine the behaviour of the wave running inland. At present, the debate between scientist is concentrated to the possibility to preview the extension of the hit areas, flooded by the possible tsunami using different hydrodynamic theories.

Some features recognisable along the rocky coasts are decisive in these kinds of study: 1 - boulders can permit to recognise areas hit in the past by the impact of extreme wave able to detach and transport them inland; 2 - coastal topography - expressed by the Manning number - is determining in the dissipation of the wave energy and in its capability too run inland. If starting from the presence of boulders is possible to recognise coastal sector hit by extreme waves in the past, so not coupled to eyewitness, at present is still difficult to discriminate between the agent responsible for their accumulation. The experience of the IOT and of the recent Chilean Tsunami evidenced that tsunami wave are able to scatter inland megaboulders. An important degree of uncertainty regards the methodology to investigate them aiming to the definition of the origin of the wave responsible for their deposition. Is the impacting wave height more important of the wave length and of the wave period in the possibility to shape boulders accumulation? Different theories have been proposed in the recent time but the final reply is still far away (Goto et al., 2007; 2009; 2010a; Hansom et al., 2008; Pignatelli et al., 2009; Barbano et al., 2010). The more immediate reply is that if a wave can be described by height, length and period the best way to evaluate its impact on a rocky coast should consider these parameters all together. This could be the way to be went through by the scientists.

Moreover the coastal geomorphology - numerically expressible by the topography - conditions the wave inland propagation. As described in the previous pages, the use of the remote sensing permit to assess different Manning number able to describe numerically the articulation of the coastal landscape (Pignatelli et al., 2010b). The method proposed in the previous pages permit to survey and to schedule different n coefficients able to describe rocky coastal landscape. The final task is to obtain quantitative coastal parameters that permit to assess automatically the tsunami inland penetration knowing the wave features and the approaching direction. First steps have been performed by Pignatelli et al., (2010a) but they are limited to some few case studies. To assess the tsunami risk in areas historically interested by this phenomena, it is necessary to extend the digital survey of the coastal areas and the determination of the Manning numbers all around the world. In this the use of the Terrestrial Laser Scanner has some limits (e.g.: difficult to transport it, number of operators, necessity of long times, extension of coastal areas, instrument range, etc.). The possibility to use its aerial version, the LIDAR, if extensively applied can contribute in the definition of the areas subject to the risk of tsunami flooding and in early warning system development, in the redaction of integrated coastal management plans, in tsunami mitigation plans, and, at last but not least, in first aid plans.

6. References

Arcement, G. & Schneider, V. (1989). Guide for selecting Manning's roughness coefficients for natural channels and flood plains, *U.S. Geol. Surv. Water Supply Pap.*, 2339

Asal, F. F. F. (2003). Airborne remote sensing for landscape modelling, *PhD thesis, The University of Nottingham*, UK, 317 pp.

Baptista, A.M., Priest, G.R. & Murty, T.S. (1993). Field survey of the 1992 Nicaragua Tsunami, *Marine Geodesy*, 16, 169–203.

Barbano, M.S., Pirrotta, C. & Gerardi, F. (2010). Large boulders along the south-eastern Ionian coast of Sicily: Storm or tsunami deposits?, *Marine Geology* doi:10.1016/j.margeo.2010.05.005

Bryant, E.A. (2001). Tsunami. The underrated hazard, *Cambridge University Press*, Cambridge, UK, 320 pp.

Bryant, E.A. & Young, R.W. (1996). Bedrock-sculpturing by tsunami, south coast New South Wales, Australia, *Journal of Geology*, 104, 565–582.

Bourgeois, J. & MacInnes, B. (2010). Tsunami boulder transport and other dramatic effects of the 15 November 2006 central Kuril Islands tsunami on the island of Matua, *Zeitschrift fuer Geomorphologie*, 54, 3, 175–195.

Bourrouilh-Le Jan, F.G., Beck, C. & Gorsline, D.S. (2007). Catastrophic events (hurricanes, tsunami and others) and their sedimentary records: introductory notes and new concepts for shallow water deposits, *Sedimentary Geology*, 199, 1–2, 1–11.

Bryant, E.A., Young, R.W. & Price, D.M. (1996). Tsunami as a major control on coastal evolution, Southeastern Australia, *Journal of Coastal Research*, 12, 831–840.

Chow, V. T. (1973). Open-Channel Hydraulics, *McGraw-Hill*, New York.

Cochard, R., Ranamukhaarachchi, S.L., Shivakoti, G.P., Shipin, O.V., Edwards, P.J. & Seeland, K.T. (2008). The 2004 Tsunami in Aceh and Southern Thailand: a review on coastal ecosystems, wave hazards and vulnerability, *Perspectives in Plant Ecology, Evolution and Systematics*, 10, 1, 3–40.

Dawson, A.G. (1994). Geomorphological effects of tsunami run-up and backwash, *Geomorphology*, 10, 83–94.

Emanuel, K.A. (2005). Increasing destructiveness of tropical cyclones over the past 30 years, *Nature*, 436, 686–688.

Fita, L., Romero, R., Luque, A., Emanuel, K. & Ramis, C. (2007). Analysis of the environments of seven Mediterranean tropical-like storms using an axisymmetric, nonhydrostatic, cloud resolving model, *Natural Hazard Earth Systems Science*, 7, 41–56.

Frohlich, C., Hornbach, M.J., Taylor, F.W., Shen, C.C., Moala, A., Morton, A.E. & Kruger, J. (2009). Huge erratic boulders in Tonga deposited by a prehistoric tsunami, *Geology*, 37 , 2, 131-134.

Goto, K., Chavanich, S.A., Imamura, F., Kunthasap, P., Matsui, T., Minoura, K., Sugawara, D., Yanagisawa, H. (2007). Distribution, origin and transport process of boulders deposited by the 2004 Indian Ocean tsunami at Pakarang Cape, Thailand, *Sedimentary Geology*, 202, 821-837.

Goto, K., Okada, K. & Imamura, F. (2009a) Importance of the initial waveform and coastal profile for the tsunami transport of boulders, *Polish Journal of Environmental Studies*, 18, 53–61.

Goto, K., Okada, K. & Imamura, F. (2009b). Characteristics and hydrodynamics of boulders transported by storm wave at Kudaka Island, Japan, *Marine Geology*, 262, 14–24.

Goto, K., Kawana, T. & Imamura, F. (2010a). Historical and geological evidence of boulders deposited by tsunamis, southern Ryukyu Island, Japan, *Earth-Science Reviews*, 102, 77-99.

Goto, K., Miyagi, K., Kawamata, H. & Imamura, F. (2010b). Discrimination of boulders deposited by tsunamis and storm waves at Ishigaki Island, Japan. *Marine Geology*, 269, 34–45.

Hall, A.M., Hansom, J.D., Williams, D.M. & Jarvis, P. (2006). Distribution, geomorphology and lithofacies of cliff top storm deposits: examples from the high-energy coasts of Scotland and Ireland, *Marine Geology*, 232, 131–155.

Hansom, J.D. & Hall, A.M. (2009). Magnitude and frequency of extra-tropical North Atlantic cyclones: a chronology from cliff-top storm deposits, *Quaternary International*, 195, 42–52.

Hansom, J.D., Barltrop, N.D.P., Hall, A.M., (2008). Modelling the processes of cliff-top erosion and deposition under extreme storm waves, *Marine Geology*, 253, 36–50.

Hearty, J.P. (1997). Boulder deposits from large waves during the last interglaciation on North Eleuthera Island, Bahamas, *Quaternary Research*, 48, 326–338.

Hills J. G. & Mader C. L. (1997). Tsunami producted by the impacts of the small asteroids. *Annals of the New York Accademy of Sciences*, 822, 381-394.

Jones, B. & Hunter, I.G. (1992): Very large boulders on the coast of Grand Cayman, the eff ects of giant waves on rocky shorelines. – J. Coast. Res. 8: 763–774.

Kato, T., Terada, Y., Ito., K., Hattori, R., Abe., T., Miyake T., Koshimura, S., & Nagai, T., (2005). Tsunami due to the 2004 September 5th off the Kii peninsula earthquake, Japan, recorded by a new GPS buoy, *Earth Placet Space*, 57, 297-301.

Kawana, T. & Pirazzoli, P. (1990). Re-examination of the Holocene emerged shorelines in Orabu and Shimoji Islands, the South Ryukyus, Japan, *Quaternary Research*, 28, 419–426.

Kelletat, D. (2009). Bibliography of tsunami research since 2000. *Suppl. Geogr. Rundschau* ,12, (3 parts, 221 references in total), Braunschweig.

Kelletat, D. (2008). Comments to Dawson, A.G. and Stewart, I. (2007). Tsunami deposits in the geological record, *Sedimentary Geology*, 211, 87-91.

Kelletat, D., Scheffers, A. & Scheffers, S. (2007). Field Signatures of the SE-336 Asian Mega-Tsunami along the West Coast of Thailand compared to Holocene Paleo-Tsunami from the Atlantic Region, *Pure and Applied Geophysics*, 164, 2/3, 413-431.

Kelletat, D., Scheffers, S. & Scheffers, A. (2006). Learning from the SE-Asian tsunami: examples from Thailand's west coast, *Geograph. Rundschau Int.*, 1, 4–9, Braunschweig.

Kelletat, D. & Schellmann, G. (2002). Tsunamis on Cyprus - Field Evidences and ^{14}C Dating Results, *ZGeomorph N.F.*, 46, 1, 19-34.

Imamura, F., Goto, K. & Ohkubo, S. (2008). A numerical model for the transport of a boulder by tsunami, *Journal Geophysical Research,* 113, C01008. doi:10.1029/2007JC004170, 2008.

Imamura, F., Synolakis, C.E., Gica, E., Titov, V., Listanco, E. & Lee, H.J. (1995). Field survey of the 1994 Mindoro lsland, Philippines, Tsunami. *Pure and Applied Geophysics,* 144, 3–4, 875–890.

Lavigne, F., Paris, R., Wassmer, P., Gomez, C., Brunstein, D., Grancher, D., Vautier, F., Sartohadi, J., Setiawan, A., Syahnan, T.G., Fachrizal, B.W., Mardiatno, D., Widagdo,

A., Cahyadi, R., Lespinasse, N. & Mahieu, L. (2006). Learning from a Major Disaster (Banda Aceh, December 26th, 2009): A Methodology to Calibrate Simulation Codes for Tsunami Inundation Models, *ZGeomorph. N.F.*, Suppl.-Bd. 146, 253-265.

Leica Geosystems (2004). HDS Cyclone Training Manual. Milton Keynes, UK, 231 pp.

Mader, C.L. (1974). Numerical simulation of tsunamis, *Journal of Physical Oceanography*, 4, 1, 74-82.

Mader, C.L. (1999). Modeling the 1958 Lituya Bay Mega-Tsunami, *Science of Tsunami Hazards*, 17, 57-67.

Marsico, A., Pignatelli, C., Piscitelli, A., Mastronuzzi, G. & Pennetta, L. (2009). Ricostruzione digitale di blocchi accumulati da eventi estremi in Italia meridionale. Atti 13a Conferenza Nazionale ASITA, 1 - 4 dicembre 2009, 1377-1385.

Maramai, A. & Tinti, S. (1997). The 3 June 1994 Java Tsunami: a post-event survey of the coastal effects, *Natural Hazards*, 15, 31–49.

Mastronuzzi, G. (2010). Tsunami in Mediterranean Sea, *The Egyptian Journal of Environmental Change*, 2,1,1-12.

Mastronuzzi, G. & Sansò, P. (2000). Morphological effects of catastrophic waves along the lonian coast of Apulia (southern ltaly), *Marine Geology*, 170, 93–103.

Mastronuzzi, G. & Sansò, P. (2004). Large boulder accumulations by extreme waves along the Adriatic coast of Southern Apulia (Italy), *Quaternary International*, 120, 173–184.

Mastronuzzi, G., Pignatelli, C. & Sansò, P. (2004). Assessment of catastrophic wave impact in Apulia region (Southern Italy). In: Brebbia C.A. (ed) Risk Analysis IV, Wessex Istitute of Technology Press, 681-689.

Mastronuzzi, G., Pignatelli, C. & Sansó, P. (2006). Boulder fields: a valuable morphological indicator of paleotsunami in the Mediterranean Sea, *ZGeomorph. N.F.*, Suppl.-Bd. 146, 173–194.

Mastronuzzi, G., Pignatelli, C., Sansò, P. & Selleri, G. (2007). Boulder accumulations produced by the 20th February 1743 tsunami along the coast of southeastern Salento (Apulia region, Italy), *Marine Geology*, 242, 191-205.

Mastronuzzi, G., Sansò, P., Brückner, H., Vött, A., Pignatelli, C., Caputo, R., Coppola, D., Di Bucci, D., Fracassi, U., May, S.M., Milella, M. & Selleri, G. (2008). 2nd International Tsunami Field Symposium. Ostuni (Puglia, Italy) - Lefkas (Ionian Islands, Greece), 21-27 September 2008, *GI²S Coast Coast – Gruppo Informale di Studi Costieri, Research Publication*, 6, 202 pp., Bari.

Maouche, S., Morhange, C. & Meghraoui, M. (2009). Large boulder accumulation on the Algerian coast evidence tsunami events in the western Mediterranean, *Marine Geology*, 262, 96–104.

Mofjeld H.O., Gonzalez F.I., Titov V.V., Venturato A.J. & Newman J.C. (2005). Effects of tides on maximum tsunami wave heights: Probability distributions, *J Atmos Ocean Tech.*, 24 ,1 , 117–123.

Narayan, J.P., Sharma, M.L. & Maheshwar, B.K. (2005). Run-up and inundation pattern developed during the Indian Ocean Tsunami of December 26, 2004 along the coast of Tamilnadu (India), *Gondwana Research*, 8, 4, 611–616.

Nishimura, Y. & Miyaji, N. (1995). Tsunami deposits from the 1993 Southwest Hokkaido earthquake and the 1640 Hokkaido Komagatake eruption, northern Japan. - In: Satake K. & Imamura, F. (eds): Tsunamis: 1992–1994, their generation, dynamics, and hazard, *Pure and Applied Geophysics*, 144, 3–4, 719–733.

Noormets, R, Felton, E.A. & Crook K.A.W. (2002). Sedimentology of rocky shorelines: 2. Shoreline megaclasts on the north shore of Oahu, Hawaii - origins and history, *Sedimentary Geology*, 150, 31-45.

Noormets, R, Crook, K.A.W. & Felton E.A. (2004). Sedimentology of rocky shorelines: 3. Hydrodynamics of megaclast emplacement and transport on a shore platform, Oahu, Hawaii, *Sedimentary Geology*, 172, 41-65.

Paris, R., Wassmer, P., Sartohadi, J., Lavigne, F., Barthomeuf, B., Desgages, E., Grancher, D. , Baumert, P., Vaultier, F., Brunstein, D. & Gomez, C. (2009). Tsunamis as geomorphic crisis: lessons from the December 26, 2004 tsunami in Lhok Nga, west Banda Aceh (Sumatra, Indonesia), *Geomorphology*, 104, 1-2, 59-72.

Pignatelli, C., Ferilli, S., Capolongo, D., Marsico, A., Micella, M., Pennetta, L., Piscitelli, A. & Mastronuzzi, G. (2010a). Evidenze morfologiche, rilievo digitale ed applicazioni informatiche al fine della valutazione del limite di inondazione da tsunami, *Italian Journal of Remote Sensing*, 42, 2, 129-142.

Pignatelli, C., Piscitelli, A., Damato, B., Mastronuzzi, G. (2010b). Estimation of the value of Manning's coefficient using Terrestrial Laser Scanner techniques for the assessment of extreme waves flooding, *Zeitschrift für Geomorphologie*, 54, 3, 317-336.

Pignatelli, C., Sansò, P. & Mastronuzzi, G. (2009). Evaluation of tsunami flooding using geomorphologic evidence, *Marine Geology*, 260, 6-18.

Pelinovsky, E., Yuliadi, D., Prasetya, G. & Hmayat, R. (1997). The 1996 Sulawesi Tsunami, *Natural Hazards*, 16, 29–38.

Richmond, B.M., Jaffe, B.E., Gelfenbaum, G. & Morton, R.A. (2006). Geologic impacts of the 2004 Indian Ocean Tsunami on Indonesia, Sri Lanka, and the Maldives, *ZGeomorph. N.F. Suppl.*, 146, 235–251.

Scheffers, A. (2002a). Paleo-tsunamis in the Caribbean. Field evidences and datings from Aruba, Curaçao and Bonaire, *Essener Geographique Arbeiten*, 33, 186 pp.

Scheffers, A. (2002b). Paleo-tsunami evidences of tsunami from boulder deposits on Aruba, Curaçao and Bonaire, *Science of Tsunami Hazard*, 20, 1, 26–37.

Scheffers, A. (2004). Tsunami imprints on the Leeward Netherlands Antilles (Aruba, Curaçao and Bonaire) and their relation to other coastal problems, *Quaternary International*, 120, 1, 163-172.

Scheffers, A. (2006). Sedimentary Impacts of Holocene Tsunami Events from the Intra Americas Seas and Southern Europe – A Review, *ZGeomorph N.F.*, Suppl. Bd. 146, 7-37.

Scheffers, A. (2008). Tsunami Boulder Deposits.- in: Shiki, T., Tsuji, Y., Yamazaki, T. & Minoura, K. (eds.). *Tsunamiites. Feature and Implications*, 299-318, Elsevier.

Scheffers, A. & Kelletat, D. (2005). Tsunami Relics in the Coastal Landscape West of Lisbon, Portugal, *Science of Tsunami Hazards*, 23, 1, 3-16.

Scheffers, A., Kelletat, D., Vött, A., May, S.M. & Scheffers, S. (2008). Late Holocene tsunami traces on the western and southern coastlines of the Peloponnesus (Greece), *Earth Planetary Science Letters*, 269, 271–279.

Scicchitano, G., Monaco, C. & Tortorici, L. (2007). Large boulder deposits by tsunami waves along the Ionian coast of south-eastern Sicily (Italy), Marine Geology, 238, 1–4, 75–91.

Shimamoto, T., Tsutsumi, A., Kawamoto, E., Miyawaki, M. & Sato, H. (1995). Field survey report on tsunami disasters caused by the 1993 Southwest Hokkaido Earthquake, *Pure and Applied Geophysics*, 144, 3–4, 665–691.

Shuto, N. & Matsutomi, H. (1995). Field survey of the 1993 Hokkaido Nansei-Oki Earthquake Tsunami, *Pure and Applied Geophysics*, 144, 3–4, 649–663.

Slob, S. & Hack, R. (2004). 3-D Terrestrial Laser Scanning as a New Field Measurement and Monitoring Technique, In: Engineering Geology for Infrastructure Planning in Europe. A European Perspective, Lecture Notes in Earth Sciences, edited by: Hack, R., Azzam, R., & Charlier, R., 104, Springer, Berlin/Heidelberg, Germany, 179–190.

Smart, G. M. (2004). An Improved Flow Resistance Formula. In: Greco M., Carravetta A., Della Morte R. (Eds) River Flow 2004. Proceedings of the second International Conference on Fluvial Hydraulics, 23-25 June, Napoli, Italy, Taylor & Francis, 1, 1455 pp.

Smart, G. M., Duncan, M. J. & Walsh, J. (2002). Relatively rough flow resistance equations, *Journal of Hydraulic Engineering*, ASCE, 128(6), 568-578.

Spiske, M., Böröcz, Z. & Bahlburg, H. (2008). The role of porosity in discriminating between tsunami and hurricane emplacement of boulders — A case study from the Lesser Antilles, southern Caribbean, *Earth and Planetary Science Letters*, 268, 3-4, 384-396.

Srinivasalu S., Thangadurai N., Switzer A.D., Ram Mohan, V. & Ayyamperumal, T. (2007). Erosion and sedimentation in Kalpakkam (N Tamil Nadu, India) from the 26th December 2004 Tsunami, *Marine Geology*, 240, (1–4), 65–75.

Szczuciski, W., Niedzielski, P., Rachlewicz, G., Sobczyski, T., Ziola, A., Kowalski, A., Lorenc, S. & Siepak, J. (2005). Contamination of tsunami sediments in a coastal zone inundated by the 26 December 2004 tsunami in Thailand, *Environmental Geology*, 49, 2, (month and year of the edition), 321–331, ISSN

Tanioka, Y. & Satake, K. (1996). Tsunami Generation by Horizontal Displacement of Ocean Bottom, *Geophysical Research Letters*, 23, 8, 861-864.

Tinti, S. & Armigliato, A. (2003). The use of scenarios to evaluate tsunami impact in South Italy, *Marine Geology*, 199, 3-4, 221-243.

Titov, V. & González, F.I. (1997). Implementation and testing of the Method of Splitting Tsunami (MOST) model, *NOAA Tech. Memo. ERL PMEL-112 (PB98-122773), NOAA/Pacific Marine Environmental Laboratory*, Seattle, WA, 11 pp.

Titov, V. & Synolakis, C.E. (1998). Numerical modelling of tidal wave runup, *Journal of Waterway Port Coastal Ocean Engineering*, 124, 4, 157-171.

Weiss, R., Wünnemann, K & Bahlburg, H. (2006). Numerical modelling of generation, propagation and run-up of tsunamis caused by oceanic impacts: model strategy and technical solutions, *Geophysical Journal International*, 167, 1, 77-88.

Tsuji, Y., Matsutomi, H., Imamura, F., Takeo, M., Kawata, Y., Matsuyama, M., Takahashi, T., Sunarj, O. & Haraji, P. (1995a). Damage to coastal villages due to the 1992 Flores Island Earthquake Tsunami, *Pure and Applied Geophysics*, 144, 3–4, 481–524.

Tsuji, Y., Imamura, F., Matsutomi, H., Synolakis, C.E., Nanang, P.T., Jumadi Harada, S., Han, S.S., Arai, K. & Cook, B. (1995b). Field survey of the East Java Earthquake and Tsunami of June 3, 1994, *Pure and Applied Geophysics*, 144, 3–4, 839–854.

Umitsu, M., Tanavud, C. & Patanakanog, B. (2007). Effects of landforms on tsunami flow in the plains of Banda Aceh, Indonesia, and Nam Khem, Thailand, *Marine Geology*, 242, 1–3, 141–153.

Vött, A., Brückner, H., Brockmüller, S., Handl, M., May, S.M., Gaki-Papanastassiou, K., Herd, R., Lang, F., Maroukian, H., Nelle, O. & Papanastassiou, D. (2009a). Traces of Holocene tsunamis across the Sound of Lefkada, NW Greece, *Global Planetary Change*, 66, 112–128.

Vött, A., Brückner, H., May, S.M., Lang, F. & Brockmüller, S. (2007). Late Holocene tsunami imprint on Actio headland at the entrance to the Ambrakian Gulf, *Méditerranée*, 108, 43–57.

Vött, A., Brückner, H., May, S.M., Lang, F., Herd, R. & Brockmüller, S. (2008). Strong tsunami impact on the Bay of Aghios Nikolaos and its environs (NW Greece) during Classical-Hellenistic times, *Quaternary International*, 181, 105–122.

Vött, A., Brückner, H., May, S.M., Sakellariou, D., Nelle, O., Lang, F., Kapsimalis, V., Jahns, S., Herd, R., Handl, M. & Fountoulis, I. (2009b). The Lake Voulkaria (Akarnania, NW Greece) palaeoenvironmental archive - a sediment trap for multiple tsunami impact since the mid-Holocene, *ZGeomorph. Suppl.*, 53, 1–37.

Vött, A., May, S.M., Brückner, H. & Brockmüller, S. (2006). Sedimentary evidence of late Holocene tsunami events near Lefkada Island (NW Greece), *ZGeomorph. Suppl.*, 146, 139–172.

Williams, D. M. & Hall A. M. (2004). Cliff-top megaclast deposits of Ireland, a record of extreme waves in the North Atlantic—storms or tsunamis?, *Marine Geology*, 206, 101-117.

Yeh, H., Imamura, F., Synolakis, C., Tsuji, Y., Liu, P. & Shi, S. (1993). The Flores lsland tsunamis, EOS, *Trans. American Geophys. Union*, 74, 33 , 369–373.

Yeh, H., Titov, V., Gusiakov, V., Pelinovsky, E., Khramushin, V. & Kai-Strenko, V. (1995). The 1994 Shitokan earthquake tsunamis, *Pure and Applied Geophysics*, 144, 855–874.

Young, R.W., Bryant, E.A. & Price, D.M. (1996). Catastrophic wave (tsunami?) transport of boulders in southern New South Wales, Australia, *ZGeomorph. N.F.*, 40, 2, 191–207.

12

Sedimentary Characteristics of the Holocene Tsunamigenic Deposits in the Coastal Systems of the Cadiz Gulf (Spain)

Juan A. Morales[1], José M. Gutiérrez Mas[2], José Borrego[1] and Antonio Rodríguez-Ramírez[3]
[1]*Departamento de Geología. Facultad de Ciencias Experimentales. Universidad de Huelva. Campus Universitario del Carmen, s/n. 21007 Huelva.*
[2]*Departamento de Ciencias de la Tierra. Facultad de Ciencias del Mar y Ambientales. Universidad de Cádiz. Campus Universitario Río San Pedro, s/n. Puerto Real (Cádiz).*
[3]*Departamento de Geodinámica y Paleontología. Facultad de Ciencias Experimentales. Universidad de Huelva. Campus Universitario del Carmen, s/n. 21007 Huelva. Spain*

1. Introduction

From a depositional point of view, tsunami deposits can be defined as high-energy event sequences, generated by highly energetic dynamic processes involving a high degree of erosion and transport of sediments. Tsunamis are waves induced either by plate tectonics or by the impact of meteorites or large underwater landslides. They are characterized by a wavelength of dozens of kilometers and by their high speed. Interest in the study of tsunamigenic deposits has increased in recent times, with numerous works published in the last few years on tsunamites generated by documented events, such as the Hokkaido-Nansei-Oki earthquake in 1993 (Nanayama et al., 2000; Sawai, 2002), the Okoropunga tsunami in New Zealand in the 15th century (Goff et al., 2004) or the one associated with the famous volcanic eruption of the Krakatoa in 1883 (Van den Bergh et al., 2003).

In coastlines protected from swell, such as bays, estuaries or coastal lagoons, tsunamis usually generate complex but easily recognizable deposits (tsunamites) clearly differentiated from storm deposits, as storms cannot reach these environments. In the inner areas of these protected coastal systems the arrival of the tsunami wave takes place after dispersion of the initial energy, which occurs through friction with the bottom when the wave crosses a shallow tidal system. The innermost limit of the tsunamigenic layer is simply evidenced as a fine but continuous level of plant fragments that may also contain a variable amount of small soft boulders (Bondevik et al., 1997; Dawson and Smith, 2000). In the most open areas of these systems, other surface deposits characteristic of high-energy events and linked to geomorphological features characteristic of storms, such as cheniers or washover fans, develop. These features can also be generated during tsunamis, and in this case it is difficult to distinguish the depositional mechanisms of both phenomena.

Criteria in favor of the action of tsunamis in this coastal sector are the tectonic setting and the rich historical record of earthquakes and tsunamis. Specifically, the zone is near an active tectonic area, located SW of Cape Saint Vincent (Fig. 1), the Gorringe Bank, where recent geophysics studies show the existence of active faults (Udias et al., 1976; Ribeiro, 1995, Baptista et al., 1996, 1998; Terrinha et al., 2003; Gutsher, 2005 and 2006), which could be the focus of several well-documented earthquakes and tsunamis (Galbis, 1940; Udias *et al.*, 1976; Martínez Solares et al., 1979; Levret, 1991; Ribeiro, 1995; Dawson *et al.*, 1996; Luque *et al.*, 2001; 2002; Silva, 2002; Silva *et al.*, 2005; Gutierrez-Mas *et al.*, 2009a and b). The best known is the tsunami that followed the Lisbon earthquake (1st November 1755 AD), which produced catastrophic effects on the coast of the Algarve and West Andalusia, as well as significant morphological changes in the coastal systems (Udias *et al.*, 1976; Campos, 1992; Ribeiro, 1995; Luque *et al.*, 2001 and 2002). The magnitude of this earthquake was 8.3 and its intensity at the epicenter was XI-XII. The earthquake generated large tsunami waves, which razed the coasts of Portugal, Atlantic Andalusia and Morocco. Another documented tsunami in the area occurred on 26th January 1531, and was caused by an earthquake of magnitude 7.7 and intensity IX. Earlier, on 10th June 881 another earthquake caused a large tsunami in the area, while still earlier, in 395, an earthquake followed by a tsunami destroyed the ancient Roman town of Baelo Claudia, on the Coast of Cadiz (Bermúdez and Peinado, 2005).

Evidence of deposits generated by the tsunami that followed the Lisbon earthquake in 1755, similar to those found in the South of Portugal (Andrade et al., 1994; Dawson et al., 1995; Hindson and Andrade, 1999), in the Bay of Cadiz (Dabrio et al., 1998; Luque et al, 2002), or the accumulations of large blocks on the coastal platform of Cape Trafalgar (Whelan and Kelletat, 2005; Gracia et al., 2006), have been found in the study area. Some washover generated by the 395 tsunami has also been documented in the Coast of Cadiz (Luque et al., 2001) and a sedimentary record of all historic tsunamis has been characterized in the Odiel and Tinto Estuaries (Morales et al., 2008) and the Bay of Cadiz (Gutierrez Mas et al, 2009 a and b). A record of older tsunamis has been documented at the mouth of the Guadalquivir River (Ruiz et al., 2005).

Previous works conducted in the Huelva estuaries, Guadalquivir estuary and the Bay of Cadiz have described the presence of coarse grain-sized surface layers and macroforms generated by high-energy events (Dabrio et al., 1998; Luque et al, 2002; Ruiz et al., 2005; Gracia et al., 2006; Morales et al., 2008, Gutiérrez Mas et al., 2009 a and b). Some of these works have focused on proving that some of these layers correspond to the action of tsunamis and, using radiometric dating techniques, they associate the layers with corresponding well-documented events which occurred in the historical period. Once this fact has been proved, there is evidence of variety in the typology of surface deposits and forms which have not been properly characterized. Thus, a methodology has been established to achieve a sedimentological description and characterization of these deposits, as well as to determine the factors responsible for their spatial distribution.

2. Lithofacies characteristic of tsunami deposits

Composition, grain size and area of tsunamigenic deposits are as variable as the height and intensity of incident waves, the configuration of the coast and the bottom, the type of sedimentary substrate in the bottom, the availability of sediments, and the path of the tsunami wave. Nevertheless, there are common features which characterize tsunamites and which have been described by numerous researchers. These are: 1) erosive base, 2) coarser sediments than in the overlying and underlying beds, with a fabric characterized by a remarkably bad internal arrangement, 3) presence of exotic sedimentary particles from environments external to those where they are deposited, and 4) abundance of alien marine organisms such as foraminifera, diatoms, and even mollusk shells.

Fujiwara et al. (2000) distinguish three main types of tsunamites:

1. Massive accumulation of shells and shell fragments with sandy or muddy matrix. These deposits usually have an erosive base and present a common characteristic: a specific high diversity with a mixture of open marine species and others typical of protected low-energy environments. These shell deposits can also contain a variable percentage of alien lithoclasts, which can be associated with the most energetic zones of restricted environments, such as the bottom of estuarine or deltaic channels.
2. Fining-upwards successions which start with shell accumulation on an erosive base, and a muddy sandy or sandy gravel ceiling with muddy matrix. Shell accumulation usually includes a mixture of whole organisms (2 valves), disarticulated valves and shell fragments. This deposit also contains a large number of species, mixing shells from different environments. The upper sandy part of the succession usually shows bioturbation and may frequently present convolute bedding. These tsunamigenic sequences are indicative of a lower energy than the previous deposit and are usually associated to restricted environments such as tidal plains and channel margins or deeper channel zones. Other times they appear in the same environments as the deposit described above, although they are generated during lower energy events.
3. Sand layers with erosive bases. They usually appear in the form of decreasing grain-size sequences with shell and plant fragments and soft boulders. Very often they present internal arrangement structures characteristic of bidirectional currents like herringbone cross-stratifications and convolute bedding. Marine microorganisms are also present, especially diatoms (Sawai, 2002) and ostracods (Smoot et al., 2000).

Nichol et al. (2003) describe the tsunamite in the Whangapoua Bay (New Zealand), as being made up of coarse sands and gravels with sub-rounded pebbles from local lithologies of nearshore rock outcrops. The minor remains of marine organisms generally appear among lithoclasts. Deposits of this kind are smaller in size and exclusive to barrier-island environments, especially in the zone of the backshore, although they can also appear in cliffs. With regard to these deposits, other authors (Whelan y Kelletat, 2005; Gracia et al., 2006) have described large block deposits on cliffs or dune systems located behind abrasion platforms. In these cases the blocks also correspond to the lithologies present in the submerged rock outcrops in front of the coastal system.

Hindson and Andrade (1999), Goff et al. (2000) and Nanayama et al., (2000) have described facies variations in the sedimentary record along coastal systems originating in known events. All of them agree on a decreasing grain-size distribution of facies landwards, where coarse sand and gravels are located around closure barriers, while shell accumulations are distributed along the channels and intertidal zones of estuaries, especially in marsh areas and supratidal deltaic plains.

Goff et al. (2000) observed enrichment in heavy metals such as Pb, Cu, Ni, Fe and Cr in these deposits duet to the accumulation of particles selected by high-energy currents following extraction from adjacent formations in coastal areas. Van der Bergh et al. (2003) also observed this event in the tsunamite of the Krakatoa eruption. In this case, the energetic layer is also accompanied by an increase in magnetic susceptibility.

The only agents able to generate the studied deposits are great storms or tsunami waves (Dawson *et al.*, 1988; Dott, 1996; Nott, 1997, 2003; Bryant and Nott, 2001; Goff *et al.*, 2001;

Kennedy *et al.*, 2007; Kortekaas and Dawson, 2007). Due to the similarity of the facies, both deposit types are sometimes confused (Bryant and Nott, 2001; Shanmugam, 2006). Nevertheless, some differences can be established (Nanayama *et al.*, 2000; Dawson *et al.*, 1988, 1996; Dawson, 2005; Kortekaas and Dawson, 2007). Tsunamites extend much further inland than storm deposits, have an erosive base, are poorly classified and often contain soil and vegetable remains (Paskoff, 1991; Dawson et al., 1988, 1996; Nanayama et al., 2000; Dawson, 2004).

Nanayama et al. (2000) and Goff et al. (2004) use distance, shoreline, topographic height of deposits and clast size as differentiation criteria. Three main features may be considered as essential distinctive characteristics: 1) Tsunamites almost always reveal an erosive base, while storm layers sometimes present net, though depositional, limits; 2) grain size in storm layers is usually coarser than in tsunamites, while the latter present a lower internal arrangement; 3) tsunamites extend inland or towards estuarine or deltaic areas much more than storm layers.

Goff et al. (2004) analyze the differences between storm and tsunamigenic deposits, comparing storm deposits on Easter Island in 2002 with those of the Okorocumba tsunami which occurred in the 15th century. The differences are evident in the area, thickness, and grain size. The tsunamigenic deposit diminishes sharply at the margins and inland, is better selected, coarser and does not present an erosive base with vegetation and buried soil. In addition, the storm deposit extends landwards up to 40 m, as opposed to the tsunamigenic deposit, which extends up to 200 m.

However, sometimes those features are not sufficient to establish the depositional mechanism. Other criteria may be necessary to distinguish the dynamic agent that generated the deposits (Cita and Aloisi, 2000; Shanmugam, 2006).

3. The study area

The Gulf of Cadiz forms a coastal sector in the SW of the Iberian Peninsula which extends from Cape Saint Vincent in Portugal to the Strait of Gibraltar. From a physiographic point of view, the coast of the Gulf of Cadiz can be divided into several sectors according to its topographical configuration and its erosive or depositional nature. As a result, we can distinguish low sandy sections of coast which extend along the Portuguese Algarve coast in the form of barrier-island systems, and along the Huelva coastline in the form of long beaches developed on the basis of a quaternary paleo-cliff system (Fig. 1).

To the South, in the coast of Cadiz and the western part of the Algarve, rocky coast and bottoms prevail, with good examples of cliff systems. The rectilinear configuration of the shoreline is broken at the mouths of the Guadiana, Piedras, Tinto-Odiel, Guadalquivir and Guadalete rivers, which are protected from the direct action of swell and in whose interior tidal sedimentation prevails.

The zone is affected by a mesotidal regime, where the mean tidal range is 2.2 m. In spring tides the highest range reaches 3.7 m and in neap tides the lowest range is 0.65 m (Benavente et al., 2000). Tidal range varies along the coast, reaching maximum values at the mouth of the Tinto and Odiel rivers, whereas it decreases in the coasts of Portugal and Cadiz. The speed of the tidal current is highly variable, with the highest values reached inside the Bay of Cadiz, where the highest speeds have been determined along the Strait of Puntales (Alvarez et al., 1999), and in the Punta Umbría channel (Odiel-Tinto Estuary), reaching values over 1.5 m/s during spring tide ebbs.

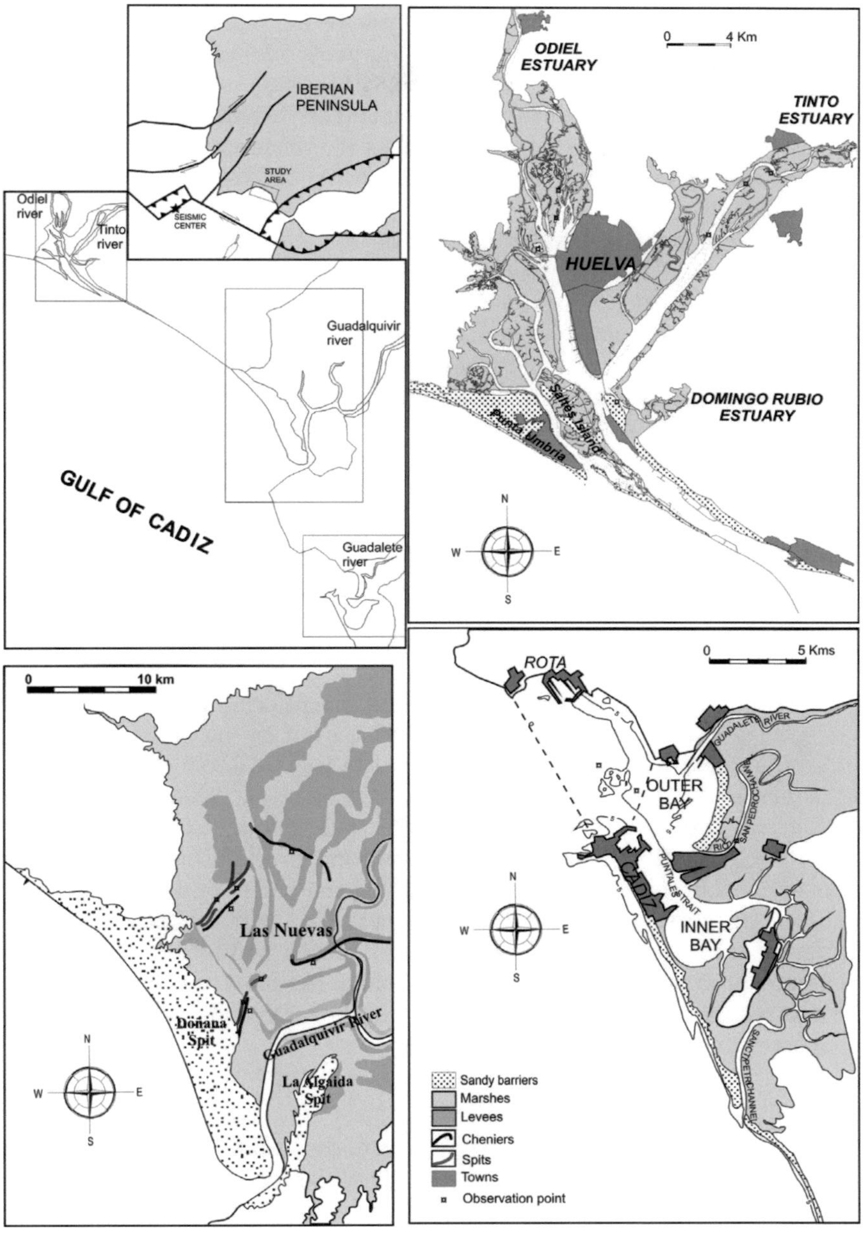

Fig. 1. Location of the seismic center and the study areas.

The prevailing swell is from the west. The data from the point WANA 1051048, in the WANA network, show that the waves from the SSW, SW and WSW represent 49.19% of the time for the period from the end of 2005 to the end of 2007. Mean significant wave height ($H_{1/3}$) is 0.69 m, and the period is 5.04 seconds.

Storm waves are related to southwest Atlantic storms. Mean values of $H_{1/3}$ reached during storms are 2.20 m. with periods of 6.24 s (Rodríguez-Ramírez et al., 2003).

The infilling in the subtidal zones of the inner systems are made up of estuarine accretion bodies (Frey and Howard, 1986). In the central and inner domains of estuaries in this coast, as well as in the Bay of Cadiz, where swell influence is scarce, sedimentary bodies generally present muddy lithology with scarce sand content and are usually highly bioturbated by the activity of annelids (*Arenicolides ecaudata*), bivalves (*Scrobicularia plana, Crassostrea angulata* and *Mytilus galloprovincialis*), and gastropods (*Cymbium olla, Murex brandaris* and *Cerithium rupestre*).

In estuarine accretion bodies within the marine domain of the above-mentioned tidal systems, muddy sand lithologies prevail, also with a high percentage of bioturbation by annelids, (*Nereis diversicolor* and *Arenicolides ecaudata*), bivalves (*Cerastoderma edule, Cerastoderma glaucum, Ruditapes decussatus, Solen marginatus, Crassostrea angulata* and *Mytilus galloprovincialis*) and gastropods (*Murex brandaris* and *Cerithium rupestre*).

The sedimentary infilling of the linear coast exposed to swell presents noticeable differences between the Huelva sector and the Cadiz sector. The subtidal zone of the Coast of Huelva under the influence of waves presents a sandy and sandy-muddy nature, with the rocky substrate being covered by several meters of sediments in the zone closest to the coast. On the other hand, the equivalent zone in the coast of Cadiz presents large rock outcrops in the form of rock slabs or beach rock composed of quartzitic-bioclastic conglomerates only partially covered by a decimetric layer of sandy sediments in the deepest zones of rock outcrops.

4. Approach and methodology

The presence of coarse-grained surface layers and macroforms generated by high-energy events has been described in previous works carried out in the Odiel-Tinto Estuaries, in Guadalquivir river mouth and in the Bay of Cadiz (Rodríguez Ramírez et al., 1996; Dabrio et al., 1998; Luque et al, 2002; Ruiz et al., 2005; Gracia et al., 2006; Morales et al., 2008; Rodríguez Ramírez and Yañez 2008; Gutiérrez Mas et al., 2009 a and b). Some of these works have focused on proving that these layers correspond to the action of tsunamis and, by using radiometric dating techniques, on associating the layers with corresponding events which occurred in well-documented historical periods. Once this fact has been proved, variety is clearly seen in the typology of surface deposits and forms, which has not been properly characterized yet. As a result, a methodology has been established with the aim of describing and characterizing the sedimentology of these deposits, and also of determining the factors responsible for its spatial distribution.

4.1 Field work

This involved observation of outcrops and trenches. Both in the Odiel-tinto and Guadalquivir estuaries and in the Bay of Cadiz, there are zones where the lateral erosion of tidal channels has excavated previous deposits, generating taluses of over one-and-a-half meters in height, uncovering some of the most recent tsunamites and allowing direct observation of the facies and the morphology of sedimentary bodies. The campaigns involved direct observation and exploration of the terrain, uplifting representative facies sequences, trenches excavations and drilling to extract sediment samples. Guide measurements of shell levels present in the sandy deposit were conducted to determine the relationship of these shell accumulations with flow directions. Several lacquer peel tests were performed to determine the sedimentary structures present in the deposits. Some sea samplings were also carried out to determine the existence of mollusk species similar to

those present in the analyzed deposits. Orientation and position data were carried out using GPS and a compass.
Surface geoforms, attributable to the action of high-energy events (cheniers and spits), were also described in the Odiel and Guadalquivir estuaries. The facies of these events were also studied by making trenches about one meter deep using a shovel.

4.2 Core logging

A study of the tsunamigenic facies in the sedimentary record was conducted in order to characterize the oldest tsunamites that do not crop out in the trenches. This study was carried out by obtaining sediment cores using different techniques.
A total of 73 vibracores (VH, VT, VB and VR) obtained using the method described by Lanesky et al. (1979) and with a maximum length of 6.40 m were analyzed in the Odiel-Tinto Estuary. This same method was used to a lesser extent in the Bay of Cadiz, where only 4 of these cores were studied, although another type of vibracore (Asther-2) was recovered in the subtidal zone, obtaining a total of 35 cores with a maximum length of 4 m.
Elsewhere, manual logging was conducted using an Eijkelkamp Beeker-type hand core sampler, which allows other sediment sequences with a maximum length of 2 m. to be obtained. A total of 57 cores were obtained using this method in the Ria of Huelva and 20 in the Bay of Cadiz.

4.3 Texture analysis

Grain-size analyses were conducted at different levels of the tsunamigenic sequences using a Malvern Mastersizer 2000 laser diffraction particle analyzer which allows precise grain-size distribution between 2-mm and 2-μm sizes to be determined, so that each of the types of sediments differentiated by the Udden-Wentworth scale (1972) is in turn divided into 6 types, with measurements obtained for each one. A conventional sieving method was used for sizes bigger than 2 mm. The results were represented on grain-size distribution histograms in order to characterize the different types of sediments.
For grain-size distribution analysis and classification, we used the scale presented in Tucker (1988), based on data by Folk (1974) and Folk and Ward (1957), along with data on high-energy littoral deposits by Visher (1969), Dawson (2005), Dawson et al. (1988; 1996), Long et al. (1989a and b), Nanayama et al. (2000), Nichol et al. (2003) and Paskoff (1991).

4.4 Facies analysis

This allowed us to establish depositional features distinguishable among relict and functional sedimentary environments present in the study zone, evidenced through small variations in the textural and compositional characteristics of the deposits. These small variations, unnoticeable by routine observation, were very useful when distinguishing the deposits corresponding to environments which present similar sedimentological characteristics, but which have been generated by different depositional processes, such as, for example, sediments in environments or subenvironments affected by reworking, limited transport and almost immediate resedimentation. As part of the facies analysis, a microscopic analysis was conducted, involving determination of the sand-fraction components by first separating the different size fractions present in the sediment by sieving them mechanically and then observing them with a binocular magnifying glass. Determination of microfossils and other components was made by counting 1500 grains per

fraction. Subsequently, the percentage content of each component and its proportions in the whole of the sample were determined.
The different facies and sequences considered tsunamigenic in nature in the study zone were described and distinguished. The facies description was performed according to the criteria suggested by Frey and Howard (1986) for facies identification in tidal systems. In each facies deposit the following were described: texture, mean grain size, composition, color, physical sedimentary structure, bioturbation type and degree. The macrofauna species present were also identified in order to establish possible source areas for the sediment.

5. Results

Different sedimentary sequences, whose characteristics vary according to the acting depositional mechanisms and the energy of the processes, were characterized in the sedimentary record observed:

5.1 Bioclastic sand and microgravel sequences

The finest grain-size sequence among those observed is that made up of sandy sediment, ocher in color, with a large amount of coarser material of a different typology, dominated by bioclastic fragments of sea mollusk shells, although quartz boulders, large shell-marble fragments and soft boulders are also frequent. The most abundant identifiable species in these deposits are the bivalves *Cerastoderma glaucum, Cerastoderma edule, Crassostrea angulata, Solen marginatus, Chamelea gallina, Ruditapes decussatus, Glycimeris variabilis* and *Chlamys sp.* Some gastropod individuals, *Turritella communis* and *Cerithium rupestre,* were also observed. Only a few specimens of *Cerastoderma glaucum* were found complete with their valves disarticulated. In the estuaries of the Coast of Huelva the shells are accompanied by quartzitic particles rounded to sub-rounded clasts with sizes ranging from a few mm to 3 cm in diameter and corresponding to hydrothermal quartz. Shell-marble fragments are the biggest clasts, reaching 10 cm in diameter and are made up of the whole remains of immature bivalve individuals belonging to the *Cerastoderma glaucum* and *Crassostrea angulata* species. In Huelva, shell-marble fragments are bound together by goethite cement (Fig. 2A), while it is carbonate cement in Cadiz.
This sequence was observed in the Odiel estuary on high marsh bodies, above the level of extreme equinoctial spring tide. The deposit is organized in successive grain-fining-grain-coarsening sequences, always starting from erosive bases. Each sequence is organized showing sedimentary structures which go from parallel lamination to cross stratification, with sets of decimeters in size and herringbone beddings, although those inclined towards the mouth prevail (Fig. 2B). The deposit total power is over 1.5 m.
This kind of sequence was also observed alternating with subtidal deposits in the external sector of the Bay of Cadiz. In this case, the thickness of the sequences is much smaller, hardly reaching 20 cm, and apart from quartzic clasts, Pleistone-old oyster rock clasts are also observed.
Strong subsurface levels mainly composed of sands with shell fragments appear in the supratidal areas of the Bay of Cadiz (Fig. 2C). These levels present sand content higher than 97%, which vary slightly in the vertical sense. Gravel content is always lower than 1% and mud content is between 1 and 3%. Mean grain size is 0.22 mm (fine sand), and sorting is 0.63 (moderately well classified). A characteristic feature of these levels is their relatively high mean content in foraminifera found in current sand deposits in different environments of

the Bay of Cadiz. Among foraminifera, the high content of miliolida is remarkable. This high content in foraminifera reaches maximum values of up to 14.34%, contrasting with the results obtained in other nearby coastal environments such as current dunes (with a foraminifera content of 2.8%) or nearby beaches (where total foraminifera content is 3.8%).

Fig. 2. Sand and bioclastic microgravel sequence (A). Photograph of the sequence studied on the Odiel Estuary (B) and photograph of the facies studied on the Bay of Cadiz (C).

Fig. 3. Sand and bioclastic gravels studied on the Guadalquivir River mouth.

On top of the Guadalquivir estuarine marsh muddy deposits (Fig. 3) we find a series of sandy surface formations basically composed of the abundant remains of molluscs, which have recently been characterized as cheniers (Rodríguez Ramírez and Yañez 2008). This deposit is characterized by sandy and shelly material, which runs roughly parallel to the paleocoast and is isolated from the shore by mudflats. They form a sharp contrast to the silty or clayey lower deposits. The sediment's succession starts with a slight erosional unconformity with the underlying deposit. The internal structure of the sedimentary build-up is characterized by a gentle landward-dipping lamination, interrupted by large-scale cross-bedded sets with some mud debris in the base. In general the macrofauna is strongly dominated by the disarticulated valves of *Cerastoderma edule* and *Crassostrea angulata* accompanied by other marine bivalves (*Tellina sp., Glycymeris glycimeris, Chlamys sp., etc.)* which evidences the varied origin of the contributions.

5.2 Shell and sand fining-upwards sequences

The most frequently observed sedimentary sequence consists, in broad terms, of a mixture of disarticulated valves, shell fragments and whole bivalves in a dark gray sandy-muddy matrix (Fig. 4A). The shells and fragments include typical marine species, but also species typical of the inner estuary (Fig. 4B), although these species are different in the different layers that could be identified. On these layers of shells, the sequence ends with a very bioturbated decimetric muddy sand package, on top of which a number of dispersed valves and shells in life position accumulate (Fig. 4C).

In more detail, the base of the sequence is usually composed of a first level of shells 5-8 cm thick, with an erosive base and made up mainly of whole bivalve shells from *Cerastoderma edule, Cerastoderma glaucum Mytilus galloprovincialis, Solen marginatus and Scrobicularia plana* and the gastropod *Murex brandaris,* accompanied by the disarticulated valves of mature specimens of *Cerastoderma edule, Crassostrea angulata* and *Cerastoderma glaucum*. In the Odiel estuary, the erosive base of the layer is covered by a goethite crust and the surface of the shells also appears covered in a patina of Fe oxides (Fig. 4D).

On this shell level there is another muddy sand level, 10 to 15 cm thick and black to dark gray in color, with progressive grain-size fining. This is usually a bioturbated level, so that in the internal zones of estuaries it usually presents vertical galleries of *Scrobicularia plana,* which start from the lower shell level and finish with the shell itself in life position. In other zones, muddy sands are bioturbated only by a dense network of galleries of the annelid *Arenicolides ecaudata*. The upper part of the sequence is composed of a fine layer of mud with parallel lamination.

Layers of this kind were observed in erosive trenches along the tidal channel margins of the Tinto and Odiel river estuaries, more than 15 km away from the open coast. Continuity in these layers is usually very high towards the interior of the continent in low zones, keeping the same topographic level for distances from several hundred meters to several kilometres, but in the coast longitudinal direction continuity is lower and the layer adapts to the already existing topography.

5.3 Massive shell strata

These are shell deposits with non-articulated whole valves and muddy, sandy or microgravel matrix, which can even be absent or subsequently entered the pores after percolation. Sometimes the sandy-gravel matrix is made up of smaller bioclastic fragments

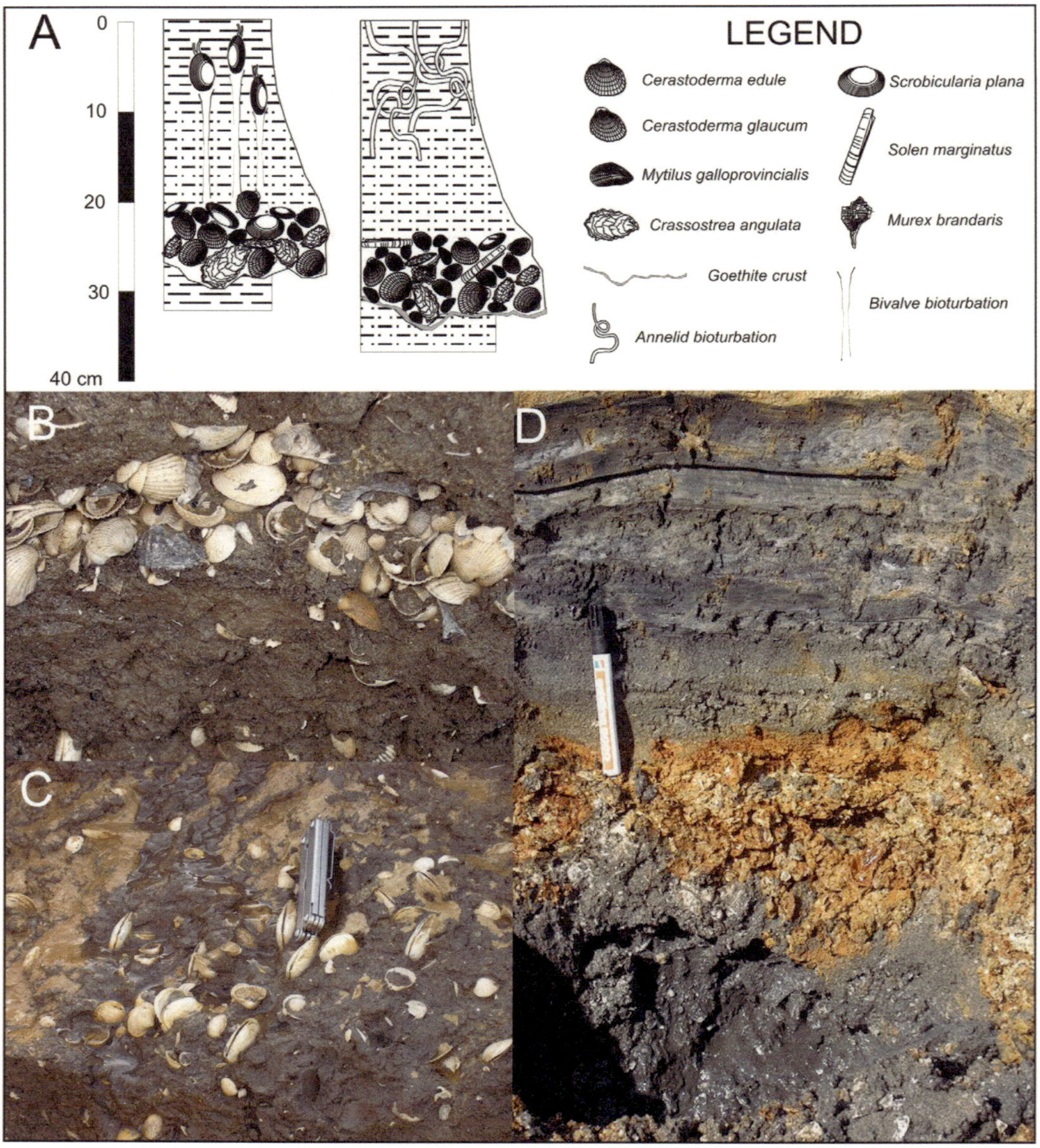

Fig. 4. Shelly sands fining-upwards sequence (A). Detail of the erosive character and species variability constituting the base of the sequence (B). Detail of the living position estuarine molluscs in the top of the sequence (C). Complete ferruginized sequence in the Odiel Estuary (D).

(Fig. 5A). The shells and clasts present strong imbrication in some deposits, as in the Bay of Cadiz (Fig. 5B), while in others, internal arrangement is observed or both horizontal and cross lamination can be slightly seen. Thickness in the deposits varies from 0.2 to 0.5 m. In many points the deposits are intersected by the present relief, while shells appear disperse forming shell pavements (Fig. 5C). In most cases, the shells present in these accumulations are of a monospecific nature and the species present in these packages are different from one place to another and from one topographic location of the deposit to another. In the Tinto and Odiel river estuaries the most abundant species are *Crassostrea angulata*, accompanied on occasions by immature individuals of *Mytilus galloprovincialis* (Fig. 5A), while in the Coast of Cadiz layers are more usually made up of *Glycymeris variabilis*. (Figs. 5B and C).

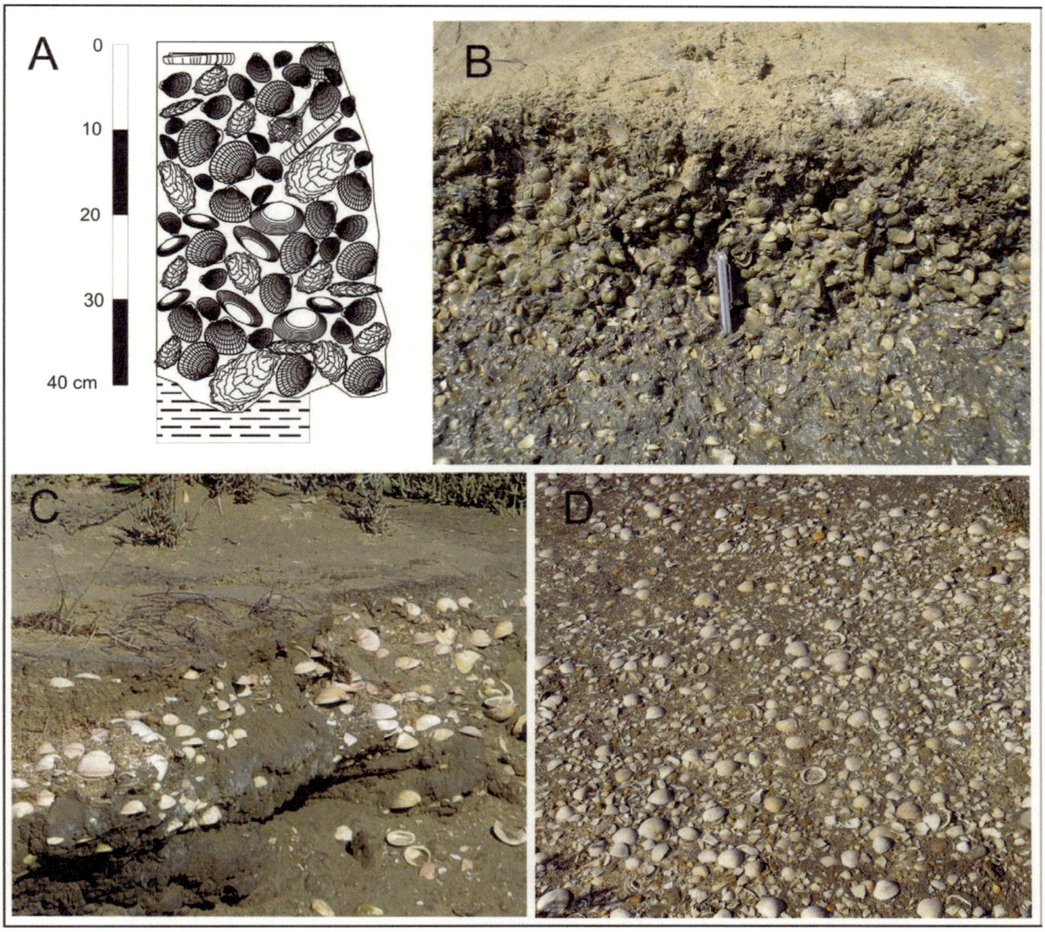

Fig. 5. Massive Shell sequence (A). Detail of the facies present in the Tinto Estuary (B). Massive strata of *Glicymerys* with muddy matrix in the Bay of Cadiz (C). *Glycymerys* pavement on a marsh in the Bay of Cadiz (D).

In the Tinto river mouth, massive ostreid packages were always observed in drillings and located under the Mean Spring Tide level, while in the Coast of Cadiz *Glycymeris* packages appear both in the intertidal and supratidal zones. These last deposits are usually accompanied by gastropod shells such as *Murex* and *Ocenebra*. *Glycymeris* valves are from 3 to 5 cm in size, although some of them reach up to 7 cm. They do not show defined orientations, although they generally appear in a subhorizontal position and concavity downwards. In some levels, valves appear in an oblique position and show a significant degree of imbrication.

5.4 Layers of mixed shells-and-pebbles

In some cases, shell accumulations appear mixed in variables proportions with quartzitic clasts which may even reach 5 cm in diameter (Fig. 6A). In this case, species are usually also of highly resistant shells, the most abundant being *Glycimeris variabilis*, *Crassostrea angulata*, *Chamelea gallina* and *Ruditapes decussatus* (Fig. 6B). These facies were observed in the lower levels of the cores in the central zone of Odiel and Tinto Estuaries.

In the Bay of Cadiz, in some outcrops, well-rounded clasts of quartz, quartzite, calcite and other rocks, such as sandstones and calcarenites, appear along with mollusk shells. The deposits are made up of siliciclastic sediments, mainly quartz, mica, quartzite pebbles, limestone and rock fragments, with abundant mollusk shells, especially bivalves of the *Glycimeris* species and some gastropods which appear complete and fragmented. There are also fragments of wood or trunk pieces of plants swept from the continent. In these cases, the deposits usually display a high degree of imbrication of the whole clasts and mollusk remains. Facies of this kind appear in the lowest intertidal and subtidal levels. Subtidal levels make up the basis of the emerged sections of the San Pedro creek.

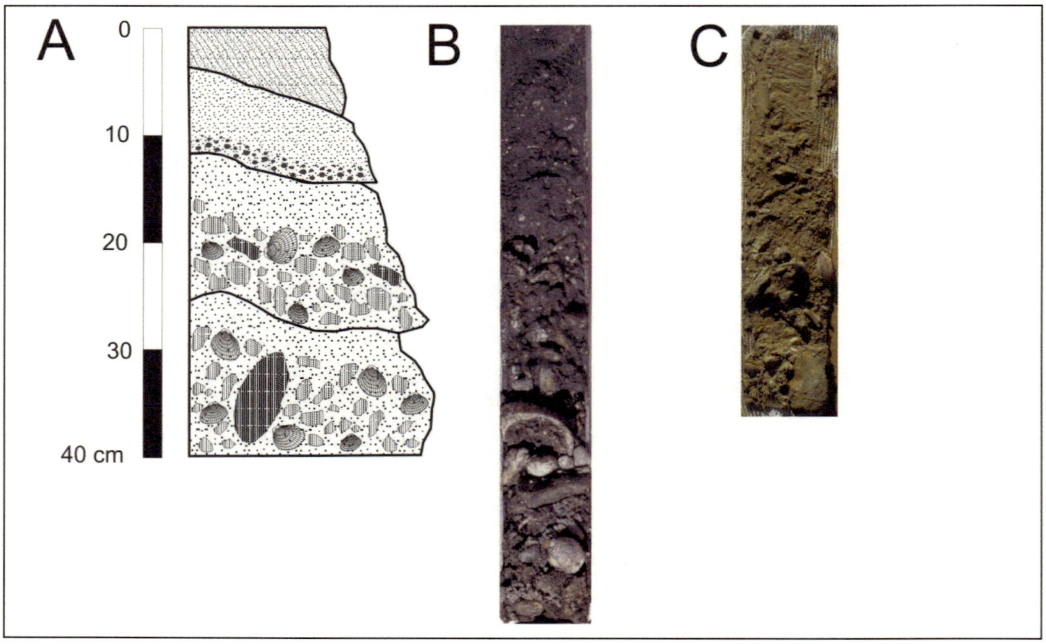

Fig. 6. Mixed shell-and-lithoclast sequence (A). Detail of the sequence in a core of the Tinto Estuary (B). Detail of a core of the subtidal area of the Bay of Cadiz (C).

The sections are grain-fining in nature, with frequent internal erosion surfaces. On the base there is a level of gravels made up of rounded blocks of rock fragments, mollusk remains, especially whole and fragmented valves of *Glycymeris* and quartzite pebbles, both with a large degree of imbrication and heterometric grain size and low classification. The sand fraction has mean contents over 78 %, gravel 13.4%, and fines 13.4%, while sorting value is 1.6, corresponding to a poorly classified sediment.

5.5 Pebble layers

The presence of quartz, quartzite and oyster rock included in deposits where the majority coarse elements are shells, has been described in the levels above. However, there is a type of high-energy deposit made up almost exclusively of pebbles of the above mentioned lithologies (Fig. 7A), clasts variable in size, but exceeding 1 cm in diameter. These facies were observed in cores in the central zone of the Odiel Estuary (Fig. 7B and C), over 20 meter deep and could be easily confused with fluvial facies were it not for the presence of some disperse fragments of seashells. The power of these packages does not exceed 30 cm and by correlation between different cores, these packages are known to display a certain lateral continuity and a morphology adapted to the bottom of tidal channels, extending landwards and severar kilometers into the interior of the estuaries.

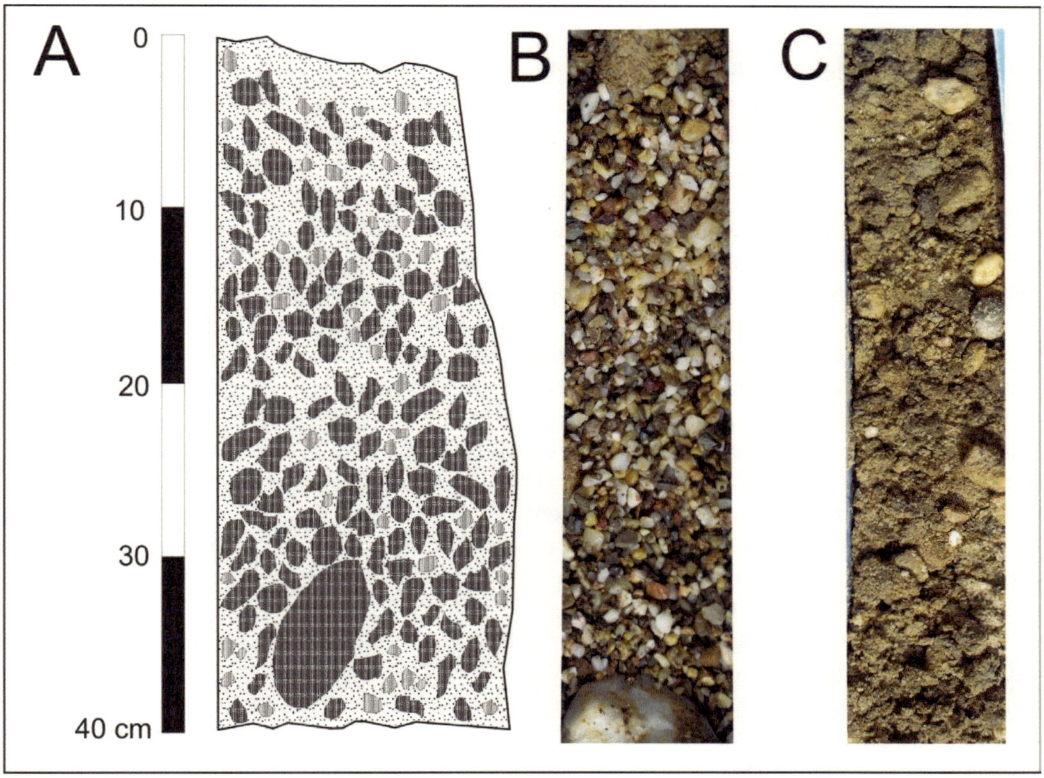

Fig. 7. Lithoclastic sequence (A). Detail of a core showing clean clast supported gravel (B). Detail of a core showing gravel with sandy matrix (C).

6. Discussion

In present coastal environments it is relatively common to find old littoral deposits such as aeolian dunes, beach ridges or washover fans, formed in the not-too-distant past by wind, waves or tidal action, in a climatic and hydrodynamic regime that was somewhat different from the present. These deposits, which at present display a relict character and are found fossilized among the estuary sediments or on the surface, anchored under the vegetation, appear at a different height and distance from the line of the present coast and, due to their morphology and facies, it generally proves easy to recognize the agent and environment in which they were deposited.

In the case of the estuary of the Odiel and Tinto rivers the succession of five sequences of bioclastic sands and microgravels separated by erosive surfaces with a total thickness of approximately 2 meters is clearly tsunamigenic (Morales et al., 2008). The sequences of bioclastic sands and microgravels are similar to the third type of tsunamites described by Fujiwara et al. (2000) and can be interpreted in different ways according to their location.

These sequences would represent the action of successive run-up and run-down currents (ebbs and flows) in the upper part of the estuary. The orientation of the internal cross-stratification is predominantly inclined towards the sea, which would indicate the action of an erosional flood followed by a depositional ebb. This body of sand was interpreted by Morales et al. (2008) as the sedimentary result of the dissipation of the tsunami wave over the high mud flats following the Lisbon earthquake of 1755, based on 14C data, and excluding other possibilities since the topographic height is greater than that reached by extreme tides and the 16-kilometer distance to the coast prevents storms reaching this zone.

Very similar facies also appear in the surface formations in the Bay of Cadiz (the Algaida sand sheet). The deposit extends along the bank of the tidal channel of the San Pedro Creek, reaching a height of roughly 8 m above the average level of the present low tides. It is composed of fine and very fine sand, with insertions, at different heights, of various shell levels of variable thickness. This sand sheet had been interpreted differently by previous authors: its granulometry, morphology and orientation parallel to the coast made Fernández-Palacios et al. (1988) think that it was formerly a coastal aeolian sheet, while Zazo et al. (1994) considered it a former spit bar, abandoned at present, which formed part of the barrier-island system located to the south of the mouth of the Guadalete river. Given the character of the deposit, fine and very fine sand, and its position on a barrier-island system, thinking that it is a former sand dune deposit is one of the easiest hypotheses to accept. These sediments present very similar granulometry and texture to coastal aeolian deposits. However, in spite of the similarity in the facies, there are very significant differences. The main difference is the foraminifera content of these sands and the planktonic/benthic indices, both of which are much higher than those of present eolian and intertidal formations, as they present similar values to sediments which are characteristic of deeper zones away from the coast (Gutiérrez Mas et al., 2009b).

Very similar sequences, although of coarser granulometry, have also been identified in subtidal zones in the Bay of Cadiz. In this case the high-energy facies appear in a sub-recent (terminal Holocene) littoral deposit which is relict in character, however, due to their location in zones which are easily reached by storms, it cannot be stated categorically that these facies were deposited by tsunamis.

The cheniers described in Doñana National Park (Guadalquivir river mouth) are also composed of sequences of this type, but in this case their genesis is related to the moments following high-energy events, which caused the breakthrough of large littoral spits and the interruption of the progradation of muddy formations. These events carry sand and shells towards the interior of the estuary. Subsequent wave-winnowing produced by the dynamics of the estuary concentrates the sand and shells forming a longshore bar or ridge (chenier) which is driven shoreward with intense landward migration. Below these facies are found other coarser facies, which would indeed correspond to higher-energy events that are possibly tsunamigenic in origin.

In the last two cases the dating coincides more or less with the date of certain historical tsunamis, however, these could have broken the sandy formations which closed off the tidal systems, allowing storms to enter inner sectors of these and making possible the subsequent deposition of cheniers.

The fining sequences of shells and sands display clear similarities to the tsunamigenic layers described in the literature (Fujiwara et al., 2000): 1) erosive base, 2) accumulation of a mixture of articulated and disarticulated valves and shell fragments, 3) the mixture of a large number of species from different domains in the tidal systems, 4) the inclusion of many immature bivalve and gastropod individuals, 5) the large area of the deposit without any significant modification of the topographic position but with adaptation to a previous topography. These points of agreement allow us to state that the sequences of shells studied directly in the outcrops in banks in the interior of the estuary are tsunamigenic in origin, with a similar sedimentary record to the second type of tsunamites described by Fujiwara et al. (2000). The presence of disarticulated valves is explained by the uplifting of already dead bivalves from the bottom, while complete organisms with articulated valves were dragged up by the flow while they were still alive on the bottom. Some of the organisms with finer, lighter shells, such as *Scrobicularia plana* survived the transport, producing vertical escape tunnels to continue living buried close to the bottom.

The massive sequences of shells observed in the sediment sample cores from the Odiel-Tinto estuaries or in the trenches in the supratidal zones of the Bay of Cadiz are also very similar to the first type of tsunamites described by Fujiwara et al. (2000) and can also be interpreted as the result of tsunami action. In this case, the massive size of the deposit may be interpreted as the result of low efficiency in the selective capacity of the currents, either because the high energy of the process is maintained at all times, or else due to the unavailability of finer material. With regard to the massive levels of *Glicymerys* observed in the Bay of Cadiz, there is no evidence of these in previous studies, except in those carried out recently by Gutierrez-Mas et al. (2009 a and b). These levels may be related to storm overwash processes, however no washover-fan-type morphologies are observed in the zone, rather the shells appear as continuous or scattered paving in varying degrees. In this case the sequences were simply interpreted as the result of high-energy events; it was not possible to state categorically that they were generated by tsunamis, due to their location in zones which are easily reached by extreme storms.

These massive layers of shells and pebbles are very similar to those described by Nichol et al. (2003) in Whangapoua Bay (New Zealand), although in this case gravel with sub-rounded edges come from the reworking of fluvial deposits from the Upper Pleistocene which are widespread in the area, both in terrestrial outcrops and under the sea. As in the case described by Nichol, deposits of this type are small in area and are exclusive to the zone behind sandy formations which close estuaries.

The sedimentary record studied in the logs of the vibracores in the Huelva estuaries shows wide distribution of each tsunamigenic layer under the muddy estuarine accretion bodies. The tsunamigenic deposits of the deeper zones consist of massive coarse shell accumulations with a sandy matrix and are indicative of higher energy than the deposits of shallower zones, which typically consist of fining-upward sequences with finer shells. Each layer can be observed in different topographic positions along the estuaries, because the deposit is adapted to the existing topography.

This fact may be interpreted as the result of the estuaries' sedimentary infilling. Thus, the first tsunamis arrived in a deeper, much more open estuary with less energy dissipation when entering it, while the last waves arrived in an evolved estuary with a well-developed tidal channel network and large salt-marsh bodies. In this surface environment, the tsunami wave is dissipated, hindering propagation to more internal zones of the estuaries.

In the case of the Bay of Cadiz, on the other hand, the coarser layers are found at the highest levels. This fact can be interpreted as the action of tsunamis in a more open coast oriented in the direction of the origin of the tsunami waves, so that energy dissipation is lower and the wave can reach higher levels more efficiently.

7. Conclusions

This work presents a summary of the different high-energy facies studied in previous works in different protected systems (estuaries and bays) located along the Gulf of Cadiz. These works carried out in the Odiel, Tinto and Guadalquivir estuaries and the Bay of Cadiz described the presence of layers and superficial macroforms with coarse grain size, generated by high-energy events. Some of these works have emphasized the existence of a relationship between these layers and the action of historically documented tsunamis. These papers used radiometric dating techniques to demonstrate that the age of each layer corresponds to a well-documented event which occurred in the historical period. Once this fact is demonstrated, a variety of facies and surface forms are evidenced. The nature of these different deposits still stays uncharacterized. Among these sequences the following have been identified: a) sequences of sands and bioclastic microgravels, b) fining-upwards sequences of shells and sands, c) massive levels of shells, d) levels of shells with pebbles and e) levels of pebbles. Each of these sedimentary sequences presents different features, which vary according to: a) the energy of the event, b) the distance to the open coast and c) the depth of the sedimentary record.

All the sequences described are similar to the tsunamigenic sequences defined by other authors in coasts with a high frequency of tsunamis. The position of these high-energy sequences in the interior part of estuarine systems dozens of kilometers away from the action of storms or in topographical levels where storms cannot reach eliminates the possibility of these being storm deposits. However, in other cases such as subtidal facies in the Bay of Cadiz, the possibility that these facies were formed by very large storms may not be discarded. In the case of the Doñana cheniers, in the Guadalquivir River estuary, the high-energy levels found in the record are related to lower-energy surface formation. In this case, it is interpreted that the action of the tsunamis broke the estuary closure barrier on different occasions, and that this tsunami was responsible for the depositing of layers which have appeared in the record, but not for the surface formations, which would be formed by the action of storms, once the system remained open after the tsunami action.

8. Acknowledgements

This work has been funded by the PB93-1205, AMB94-0160-CO4-04 and 1FD97-0900-C02-01 projects (R+D National Plan, D.G.I.C.Y.T.), the CTM2006-06090/MAR project of the Science and Technology Interministerial Commission (CICYT) of the Spanish Government, and by the Andalusian Government project "Environmental and Health Diagnosis of the Ria of Huelva".

9. References

Alvarez, O., Izquierdo, A., Tejedor, B., Mañanes, R., Tejedor, L. and Kagan, B. A. (1999). The Influence of Sediment Load on Tidal Dynamics, a Case Study: Cádiz Bay. *Estuarine, Coastal and Shelf Science,* 48: 439-450.

Andrade, C.; Hindson, R.; Freitas and Dawson, A. (1994). Sedimentary evidence of tsunami flooding in Algarve coastal lowlands. *Proceedings of the Symposium of the Littoral '94,* 1: 1035-1036.

Baptista M.A, Miranda P.M.A, Miranda J.M, Mendes L.V. (1996). Ruptura extent of the 1755 Lisbon earthquake inferred from numerical modelling of tsunami data. *Physics and Chemistry of the Earth,* 21: 65-70.

Baptista, M.A.; Miranda, P.M.A. and Mendes V. L. (1998). Constraints on the source of the 1755 Lisbon tsunami inferred from numerical modelling of historical data. *Journal of Geodynamics,* 25: 159–174.

Benavente, J., Gracia, F. J. and López-Aguayo, F. (2000). Empirical model of morphodynamic beachface behaviour for low-energy mesotidal environments. *Marine Geology,* 167: 375-390

Bermudez, J.L. and Peinado, M.D. (2005). El Riesgo de Tsunami en Andalucía. *Spin Cero,* 9: 3-8.

Bondevik, S.; Svendsen, J.I. and Mangerud, J. (1997). Tsunami sedimentary facies deposited by the Storegga tsunami in shallow marine basins and coastal lakes, western Norway. *Sedimentology,* 44: 1115-1131.

Bryant E.A. and Nott J. (2001). Geological indicators of large tsunami in Australia. *Natural Hazards,* 24: 231–249.

Campos, M.L. (1992). El riesgo de tsunamis en España. Análisis y valoración geográfica. *IGN, Monografías,* 9, 204 pp.

Cita, M.B., Aloisi, G. (2000). Deep-sea tsunami deposits triggered by the explosion of Santorini (3000 y BP), eastern Mediterranean. *Sedimentary Geology,* 135: 181-203.

Dabrio, C. J.; Goy, J. L and Zazo, C. (1998). The record of the tsunami caused by 1755 Lisbon earthquake in Valdelagrana spit (Gulf of Cadiz, Southwestern Spain). *Geogaceta,* 23: 31-44.

Dawson, A.G. (2005). Tsunami hazards in Europe. *Environment Internacional,* 30: 577-585.

Dawson, A.G. and Smith, D.E. (2000). The sedimentology of Middle Holocene tsunami facies in northern Sutherland, Scotland, UK. *Marine Geology,* 170: 69-79.

Dawson, A.G., Long, D. and Smith, D. (1988). The Storegga slides: evidence from eastern Scotland for a possible tsunami. *Marine geology,* 82: 271-276.

Dawson, A.G.; Foster, I.D.L.; Shi, S.; Smith, D.E. and Long, D. (1991). The identification of tsunami deposits in coastal sediment sequences. *Science of Tsunami Hazards,* 9: 73-82.

Dawson, A.G.; Hindson, R.; Andrade, C.; Freitas, C.; Parish, R. and Bateman, M. (1995). Tsunami sedimentation associated with the Lisbon earthquake of 1 November AD 1755: Boca do Rio, Algarve, Portugal. *The Holocene*, 5: 209-215.

Dawson, A.G., Shi, S., Dawson, S., Takahashi, T. and Shuto, N. (1996). Coastal sedimentation associated with the June 2nd and 3rd, 1994 tsunami in Rajegwesi, Java. *Quaternary Science Reviews*, 15: 901-912.

Dott Jr R.H. (1996). Episodic event deposits versus stratigraphic sequences shall the twain never meet? *Sedimentary Geology*, 199: 1–11.

Fernández-Palacios Carmona, A., Fernández-Palacios Carmona, J., Gil Gómez, B.J. (1988). *Guias Naturalistas de la Provincia de Cádiz. I. El Litoral.* Libros de la Diputación de Cádiz, 352 pp.

Folk, R.L. (1974). *Petrology of Sedimentary Rocks.* Ed. Hemphill. Austin, 170 pp.

Folk, R.L and Ward, W.C. (1957). Brazos River bar: a study in the significance of grain size parameters. *Journal Sediment, Petrol.*, 27: 3-26.

Frey, R.W. and Howard, J.D. (1986). Mesotidal estuarine sequences: A perspective from the Georgia Bight. *Journ. Sed. Petrol.*, 56: 911-924.

Fujiwara, O.; Masuda, F.; Sakai, T.; Irizuki, T. and Fuse, K. (2000). Tsunami deposits in Holocene bay mud in southern Kanto region, Pacific coast of central Japan. *Sedimentary Geology, 135: 219-23.*

Galbis, R.J. (1940). *Catálogo sísmico de la zona comprendida entre los meridianos 5º E y 20º W de Greenwich y los paralelos 45º y 25º N.* Dirección General del Instituto Geográfico y Catastral. Madrid. 277 pp.

Goff, J. R.; Rouse, H. L.; Jones, S. L.; Hayward, B. W.; Cochran, U.; McLea, W.; Dickinson, W. W. and Morley, M. S. (2000). Evidence for an earthquake and tsunami about 3100–3400 yr ago, and other catastrophic saltwater inundations recorded in a coastal lagoon, New Zealand. *Marine Geology*, 170: 231-24.

Goff J.R, Chague-Goff C and Nichol S. (2001). Palaeotsunami deposits: a New Zealand perspective. *Sedimentary Geology*, 143: 1–6.

Goff , J.; McFadgen, B. G. and Chagué-Goff, C. (2004). Sedimentary differences between the 2002 Easter storm and the 15th-century Okoropunga tsunami, southeastern North Island, New Zealand. *Marine Geology*, 204: 235-250.

Gracia, F.J, Alonso, C, Benavente, J, Anfuso, G and Del Río, L, (2006). The different coastal records of the 1755 tsunami waves along the South Atlantic Spanish coast. *Zeitschrift für Geomorphologie, Supplementbände*, 146: 195–220.

Gutierrez-Mas J.M, Juan C and Morales J.A. (2009a). Evidence of high-energy events in shelly layers interbedded in coastal Holocene sands in Cadiz Bay (south-west Spain). *Earth Surface Processes and Landforms*, 34: 810-823.

Gutierrez-Mas , J.M, López-Arroyo, J. and Morales, J.A. (2009b). Recent marine lithofacies in the Cadiz Bay (SW Spain). Sequences, Processes and control factors. *Sedimentary Geology*, 218: 31–47.

Gutster, M. R. (2005). Active seafloor deformation in the Gulf of Cadiz: New cruise results and the possible link to the source of 1755 Lisbon earthquake and tsunami. Proceedings *of the Iberian Margin Paleoseismology, Active Tectonics and Sedimentology (IMPACTS Workshop)*. OUEM, Plouzané and Brest (France).

Gutscher M.A, Baptista M.A and Miranda J.M. (2006). The Gibraltar Arc seismogenic zone (part 2): Constraints on a shallow east dipping fault plane source for the 1755

Lisbon earthquake provided by tsunami modeling and seismic intensity. *Tectonophysics,* 426: 153–166.Hindson, R.A. and Andrade, C. (1999). Sedimentation and hydrodynamic processes associated with the tsunami generated by the 1755 Lisbon earthquake. *Quaternary International,* 56: 27-38.

Kennedy D.M, Tannock KL, Crozier M.J. and Rieser U. (2007). Boulders of MIS 5 age deposited by a tsunami on the coast of Otago, New Zealand. *Sedimentary Geology,* 200: 222-231.

Kortekaas, S., Dawson, A.G. (2007). Distinguishing tsunami and storm deposits: an example fromMartinhal, SW Portugal. *Sedimentary Geology,* 200: 208-221.

Lanesky, D.E.; Logan, B.W.; Brown, R.G. and Rine, A.C. (1979). A new approach to portable vibracoring underwater and on land. *Journ. Sed. Petrol.,* 48: 654-657.

Levret, A. (1991). The effects of the November 1, 1755 "Lisbon" earthquake in Morocco. *Tectonophysics,* 193: 83-94.

Long, D., Dawson, A.G. and Smith, D.E., (1989a). Tsunami risk in northwestern Europe: a Holocene example. *Terra Nova,* 1: 532–537.

Long, D., Smith, D.E. and Dawson, A.G. (1989b). A Holocene tsunami deposit in eastern Scotland. *Journal of Quaternary Science,* 4: 61-66.

Luque, L.; Goy, J. L.; Zazo, C.; Dabrio, C. J.; Silva, P. G. and Lario, J. (2001). Tsunami deposits as paleoseismic indicators : examples from the Spanish coast. *Acta geológica hispánica,* 36: 197-212.

Luque, L., Lario, J., Civis, P.G., Zazo, C., Goy, J.L., Dabrio, J.C. and Silva, P.G. (2002). Tsunami deposits as paleosismic indicators: examples from the spanish coast. *Acta Geológica Hispánica,* 36, N° 3-4: 197-211.

Martínez Solares, J.M.; López Arroyo, A. and Mezcua, J. (1979). Isoseismal map of the 1755 Lisbon earthquake obtained from Spanish data, *Tectonophysics,* 53: 301–313.

Morales, J.A., Borrego, J., San Miguel, E.G., Lopez-Gonzalez, N. and Carro, B.M. (2008). Sedimentary record of recent tsunamis in the Huelva Estuary (Southwest Spain). *Quaternary Science Reviews,* 27: 734-746.

Nanayama, F.; Shigeno, K.; Satake, K.; Shimokawa, K.; Koitabashi, S.; Miyasaka, S. and Ishii, M. (2000). Sedimentary differences between the 1993 Hokkaido-Nansei-Oki tsunami and the 1959 Miyakojima typhoon at Taisei, southwestern Hokkaido, northern Japan. *Sedimentary Geology, 135: 255-26.*

Nichol, S. L.; Lian, O B. and Carter, C. H. (2003). Sheet-gravel evidence for a late Holocene tsunami run-up on beach dunes, Great Barrier Island, New Zealand. *Sedimentary Geology,* 155: 129-145.

Nott, J. (1997). Extremely high-energy wave deposits inside the Great Barrier Reef, Australia: determining the cause-tsunami or tropical cyclone. Marine Geology, 141: 193–207

Nott, J. (2003). Waves, coastal boulder deposits and the importance of the pre-transport setting. Earth and Planetary Science Letters, 210: 269– 276.

Paskoff, R. (1991). Likely ocurrence of a megatsunami near Coquimbo, Chile. *Revista Geológica de Chile,* 18: 87-91.

Ribeiro, A. (1995). Deformable plate tectonics of the Azorez-Gibraltar boundary-where the next 1755 earthquake Hill strike again? *Actas del 1er Simpósio sobre a margen continental Ibérica Atlântica,* Lisboa: 46-47.

Rodríguez-Ramírez A., Rodríguez Vidal, J., Cáceres, L., Clemente, L., Belluomini, G., Manfra, L., Improta, S. & de Andres, J.R. (1996) Recent coastal evolution of the

Doñana National Park (S.Spain). *Quatern. Sci. Reviews.* 15, 803-809.Rodríguez-Ramírez, A. (1996). Geomorfología continental y submarina del Golfo de Cádiz (Guadiana-Guadalquivir). Ph.D. Thesis, University of Huelva, Huelva. 370 pp.

Rodríguez-Ramírez, A., Ruiz, F., Cáceres, L. M., Rodríguez Vidal, J., Pino, R. and Muñoz, J. M. (2003). Analysis of the recent storm record in the southwestern Spanish coast: implications for littoral management. *The Science of The Total Environment,* 303: 189-201.

Rodríguez-Ramírez A. & Yáñez-Camacho C.M. (2008). Formation of Chenier Plain of the Doñana marshland (SW Spain): Observations and geomorphic model. *Marine Geology* 254: 187-196.

Ruiz, F.; Rodríguez-Ramírez, A.; Cáceres, L.M.; Vidal, J.R.; Carretero, M.I.; Abad, M., Olías, M. and Pozo, M. (2005). Evidence of high-energy events in the geological record: Mid-holocene evolution of the southwestern Doñana National Park (SW Spain). *Palaeogeography, Palaeoclimatology, Palaeoecology,* 229: 212-229.

Shanmugam, G. (2006). The Tsunamite Problem. *Journal of Sedimentary Research,* 76: 718-730.

Silva, P. G. (2002). *Paleoseismic record at the ancient Roman city of Baelo Claudia (Cadiz, South Spain). Environmental Catastrophes and Recoveries in the Holocene.* 1p. Dpto. Geography & Earth Sciences, Brunel University, Uxbridge, UK., 78 pp.

Silva, P.G., Borja , F., Zazo, C. J.L., Bardají, T., Luque, L., Lario, J. and Dabrio, C. J. (2005). Archaeoseismic record at the ancient Roman City of Baelo Claudia (cadiz, South Spain). *Tectonophysics,* 408: 129-146.

Smoot, J.P., Litwin, R.J., Bischoff, J.L. and Lund, S.J. (2000). Sedimentary record of the 1872 earthquake and "Tsunami" at Owens Lake, southeast California. *Sedimentary Geology,* 135: 241-254.

Terrinha, P.; Pinheiro, L. M.; Henriet, J.P.; Matias, L. ; Ivanov, M. K. ; Monteiro, J. H.; Akhmetzhanov, A.; Volkonskaya, A.; Cunha, T.; Shaskin, P. and Rovere, M. (2003). Tsunamigenic-seismogenic structures, neotectonics, sedimentary processes and slope instability on the southwest Portuguese Margin. *Marine Geology, 195: 55-73.*

Tucker, M.E. (1986). *Sedimentary Petrology. An Introduction.* Blackwell Scientific Publications. 252 pp.

Udias, A., López Arroyo, A. and Mezcua, J. (1976). Seismotectonic of the Azorez-Alboran region. *Tectonophysics,* 31: 259-289.

Van den Bergh, G. D. ; Boer, W.; de Haas, H.; van Weering, Tj. C. E. and van Wijhe, R. (2003). Shallow marine tsunami deposits in Teluk Banten (NW Java, Indonesia), generated by the 1883 Krakatau eruption. *Marine Geology, 197: 13-34.*

Visher, G.S. (1969). Grain size distribution and depositional processes. *Journal of Sedimentary Petrology.* 39. N° 3: 1074-1106.

Whelan, F. and Kelletat, D. (2005). Boulder deposits on the southern Spanish Atlantic coast: possible evidence for the 1755 AD Lisbon tsunami? Science of Tsunami Hazards, 23 (3): 25-38.

Zazo, C.; Goy, J.L.; Somoza, L.; Dabrio, C.J.; Belluomini, G.; Improta, S.; Lario, J.; Bardají, T. and Silva, P.G. (1994). Holocene sequence of sea-level fluctuations in relation to climatic trends in the Atlantic-Mediterranean linkage coast. *Journ. Coastal Res.,* 10: 933-945.

13

Grain-Size and Thin Section Characteristics of Tsunami Sediments from Thai-Andaman Coast, Thailand

Wagner Jean-Frank[1] and Srisutam Chanchai[2]
[1]*Department of Geology, University of Trier, 54286 Trier,*
[2]*Soil Investigation Section, Office of Topographical and Geotechnical Surveys, Royal Irrigation Dept., 10300 Bangkok,*
[1]*Germany*
[2]*Thailand*

1. Introduction

The 2004 Indian Ocean tsunami killed more than 300,000 people and damaged the infrastructures, properties and environments of many coastal and island nations of the Indian Ocean, principally Indonesia, Sri Lanka, Thailand, India, The Maldives and the African Coast. In Thailand, the coastal provinces of Ranong, Phang Nga, Phuket, Krabi, Trang and Satun on the Andaman coast at latitude 5° to 11°N, longitude 98° to 102°E (Fig. 1) suffered 1–10 m run-up heights (Japanese Survey Team 2006). Phang Nga was the worst affected province attacked by 4–10 m high run-ups. Tsunamis cause seawater to flood onto land, creating distinctive horizontally widespread sandy deposits. Tsunami sediments from Japan are the typical sediments discussed for the historical tsunami events and the geological evidence (e.g., Minoura et al., 1996; Fujiwara et al., 2000; Sawai, 2002; Fujino et al., 2006).

Tsunami waves transport sandy sediments from offshore areas and the beach and deposit sediments on the inundated land surface. Tsunami deposits are very characteristic for tsunami affected areas. A local tsunami chronology can be determined by using geological methods to identify and characterize recent and ancient tsunami events in coastal sediment sequences (Dawson et al., 1996; Dawson, 2007; Dawson & Stewart, 2007). Sediment characteristics from general coastal processes, storms, ancient and recent tsunamis have been reported from many parts of the world for distinguishing deposit types, geomorphological effects and sedimentation processes (e.g. Dawson et al., 1991; Dawson, 1994; Nanayama et al., 2000; Shi & Smith, 2003; Gelfenbaum & Jaffe, 2003; Anthony & Hequette, 2007; Morton et al., 2007; Kortekaas & Dawson, 2007). Tsunami deposits are provisionally distinguished in the field on the basis of anomalous sand horizontals, fining-upward and fining-landward, coupled with organic-rich, fragmented backwash sediment (e.g. Dawson et al., 1991; Shi et al., 1995; Gelfenbaum & Jaffe, 2003; Smith et al., 2004; Srisutam & Wagner, 2009; Srisutam & Wagner, 2010a). Typically, tsunami sediments comprise medium to fine sand with small amounts of fine silt and clay (e.g. Dawson & Shi,

2000; Smith et al., 2004) . Detailed analysis of the grain-size distributions of tsunami deposits in vertical and horizontal directions reveal both landward and upward fining sedimentation sheets (Srisutam & Wagner, 2009; Srisutam & Wagner 2010a).

Grain-size and various parameters derived from the statistical analysis of the grain-size distribution, such as sorting, reflect the energy gradients responsible for transportation and deposition of sand bodies and are therefore important in reconstructing the geological environment (Friedman, 1962). Smith et al. (2007), Srisutam & Wagner (2009) and Srisutam & Wagner (2010a) identified the sedimentation sequences by grain-size distribution curve analyses. Hindson & Andrade (1999) proposed that the break points in a plot of standard deviation with depth of the tsunami deposit mark a break in turbulence associated with a transition to a lower or higher Reynolds number run-up.

Although a sedimentary signature for onshore tsunami deposits has been defined (e.g. Nelson et al., 1996; Dawson & Shi, 2000; Whelan & Kelletat, 2003), the particular hydrodynamic behavior of the long-period surge associated with tsunamigenic flooding is still under debate. In many tsunami studies, scientists have used data on tsunami run-up (e.g. number of tsunami run-ups, run-up height) to calibrate numerical models for undersea fault-generated and submarine landslide-generated tsunami (Dawson et al., 1991). In almost all cases, the number of tsunami run-ups have been based on eyewitness reports of the tsunami flood.

This chapter presents the composition and physical properties of tsunami sediments from Thai-Andaman Coast, Thailand.

2. Study area

Tsunami-affected areas at the Thai-Andaman coast (Fig. 1, western coast of southern Thailand) were investigated in 7 transects at 5 different locations.

Study locations are:

1. Ao Kheuy beach (Phang Nga province), the reference coordinate is 9.30°N, 98.38°E. Two transects have been examined.
2. Khuk Khak beach (Phang Nga province), the reference coordinate is 8.69°N, 98.23°E. Two transects have been examined.
3. Bang Thao beach (Phuket province), the reference coordinate is 7.99°N, 98.29°E. One transect has been examined.
4. Kamala beach (Phuket province), the reference coordinate is 7.96°N, 98.26°E. One transect has been examined.
5. Ao Krabi (Krabi province), the reference coordinate is 7.96°N, 98.95°E. One transect has been examined.

The geomorphological features of the study areas are described as open coastal zone except Ao Krabi (estuary zone). The pre-existing sediments are coastal deposits and agricultural soils.

The tsunami inundation distance (200-2000m) and the run-up height (0.80-10m) are different for the 5 areas due to a different geomorphology of the coastal area (steep or mild slope), bathymetry of the seawater body, orientation of approaching waves, interference with tides, etc. (Table 1).

The selected locations are also different in the shape of coastline. For Ao Kheuy beach, Khuk Khak beach and Ao Krabi the shape is a long straight coastline and for Bang Thao beach and Kamala beach a circular bay. The selected locations were affected by direct and indirect (diffracted and/or dissipated) tsunami waves as well.

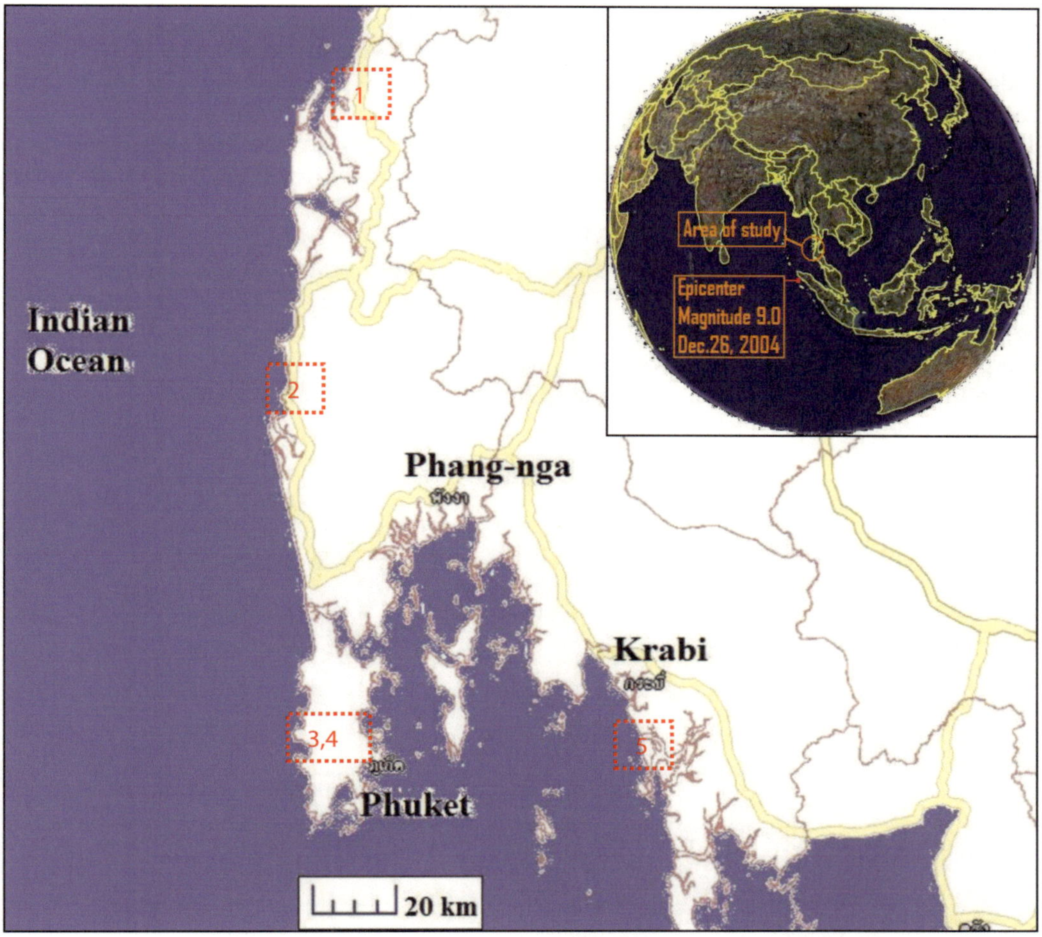

Fig. 1. Location map of southern Thailand. Areas in red dashed line rectangles are the study areas: (1) Ao Kheuy beach; (2) Khuk Khak beach; (3) Bang Thao beach; (4) Kamala beach; (5) Ao Krabi.

3. Methods

The depositional characteristics of tsunami sediments were evaluated from grain-size distribution curves, analyzed by wet-sieving and from micro-morphological features from thin sections, respectively.

Topographical profile investigations, tsunami deposit thickness measurements and sediment sample collections were made in the study areas from August to October 2007, more than two and a half year after the December 2004 Indian Ocean tsunami event. Topographical profiles were established by the combination of an automatic level measurement and Thai topographical maps.

Sediment samples were collected from four sub-locations which are offshore (S1), swash zone (S2), berm/dune (S3) and tsunami deposits plus underlying soil/sediment (S4). S1 samples have been drilled from a boat about 300 to 500 m from coastline (Fig. 2; only at

Bang Thao beach and Ao Krabi), S2 samples (swash zone, Fig. 3a) have been sampled at coastline. S3 samples (dune, Fig. 3b) and S4 samples (tsunami deposit, Fig. 4) have been collected in test pits at a distance between 20 to 40m and in trenches 50 to 300 m from coastline, respectively.

The S1, S2 and S3 samples have been analyzed in 1- to 5-cm intervals. The S4 samples were separated into a tsunami layer (S4d) and the pre-existing soil (S4p) (Fig. 4) and analyzed in 1-cm intervals.

In total 41 samples from offshore (S1), 33 from swash zone (S2), 54 from berm/dune (S3) and 123 samples from tsunami sediments resp. soil (S4) have been collected and analyzed.

In addition undisturbed samples of tsunami sediment from Khuk Khak beach were collected in 3 inches diameter PVC tubes in December 2006 for studying the micro-morphological features in thin sections (Fig.5).

The number of tsunami run-ups and run-up heights were obtained from eyewitness reports and local people's observations.

Province	Location	Latitude (°N)	Longitude (°E)	Run-up height (m)	Direction of approach (degrees)	Max. inundation distance (m)	Offshore slope
Phang-Nga	Ao Kheuy beach	9.30	98.38	4-6	80 **	800	1:100
	Khuk Khak beach	8.69	98.23	8-10	85 **	2,000	1:600
	Khao Lak *	8.68*	98.24*	8.59*	85 **	1,800 ***	1:600
Phuket	Bang Thao beach *	8.00*	98.30 *	3.76 *	85 **	1,500 ***	1:50
	Bang Thao beach	7.99	98.29	4-6	85 **	1,500 ***	1:50
	Kamala beach	7.96	98.26	4-6	85 **	1,000	1:75
	Kamala Beach *	7.95*	98.28*	4.85 *	85 **	1,000	1:75
	Kamala Beach *	7.95*	98.28*	5.29 *	85 **	1,000	1:75
	Leam Him	7.94*	98.40*	0.72 *	-	-	-
Krabi	Ao Krabi	7.96	98.95	0.8-1.2	40 **	200	1:100

* HARADA, K. (http://www.drs.dpri.kyoto-u.ac.jp/sumatra/thailand/phuket_survey_e.html)

** Estimated from Titov et al.; 2005

*** Chulalongkorn Tsunami Research Team, 2005

Table 1. Survey results of the 2004 Indian Ocean tsunami at the Thai Andaman coast.

Fig. 2. Offshore sampling

Fig. 3. Coastal deposits at Khuk Khak beach, transect 1; a) Test Pit P0 (swash zone); b) Test Pit P5 (dune)

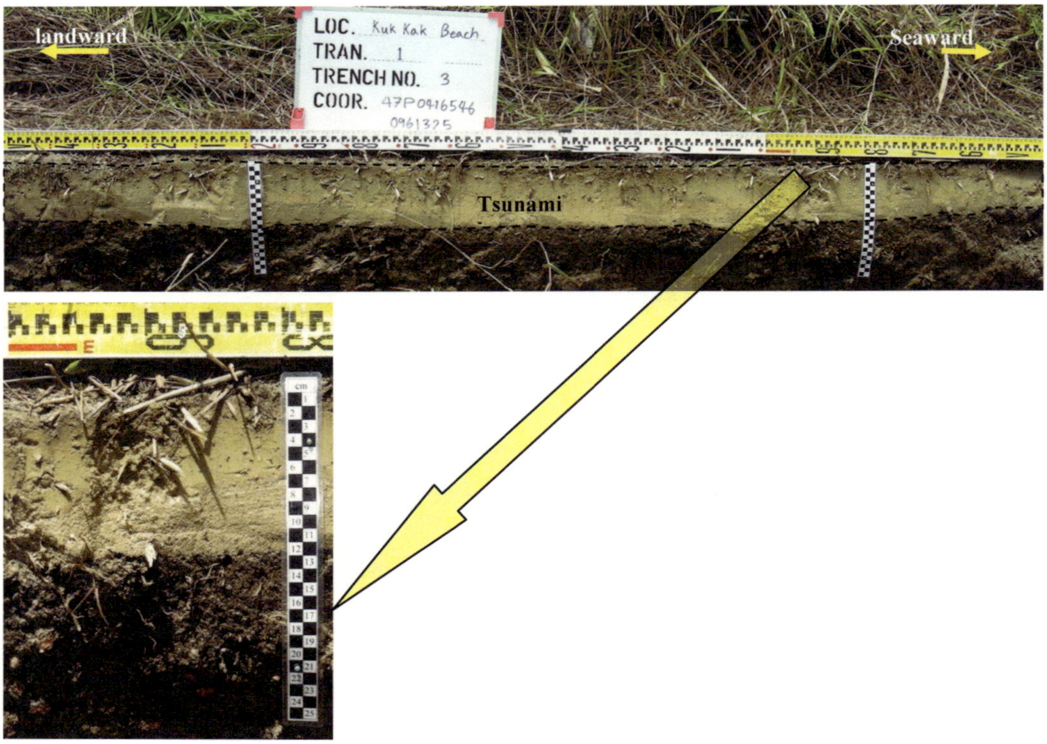

Fig. 4. Tsunami sediments in a sampling trench at Khuk Khak beach

(a) (b)

Fig. 5. (a) Tsunami deposit rests on the pre-existing soil; (b) an opened drilling core sample for the preparation of thin sections (from Srisutam & Wagner, 2010c).

3.1 Grain size analysis

Collected samples were analyzed for grain-size distribution by the wet-sieve method. Mean grain-size, standard deviation, skewness and kurtosis were calculated on the basis of the percentile statistics of Folk & Ward (1957). The characteristics of coastal non-tsunami deposits and tsunami sediments could be studied and compared by using grain-size analysis results. The tsunami sediment and tsunami run-up characteristics were evaluated from (1) grain-size distribution curves, (2) percent content of sediment fractions, (3) plotting of mean grain-size with standard deviation and (4) plotting of mean grain-size, standard deviation, skewness and kurtosis with depth.

The different run-up sequences in the tsunami deposit are evaluated from particle size and plot of standard deviation with depth. The sediment transportation pattern is analyzed by a sediment trend analysis (STA) following McLaren et al. (2007).

3.2 Sample preparation for thin sections

Samples for thin-sectioning are collected by coredrilling with a PVC tube. As the samples are cohesionless sands, the pushing of thing-walled aluminum boxes into the core of soft sediment similar to that made by Francus (1998) could not be used. Therefore the whole section of sample core is left to dry at air first. Then it is impregnated by a low-viscosity epoxy resin (Fig. 6). When the core has hardened, slices for the preparation of thin sections are cut from the depth 0-16 cm, providing an undisturbed vertical slice of the tsunami sediment and the sediment/soil interface.

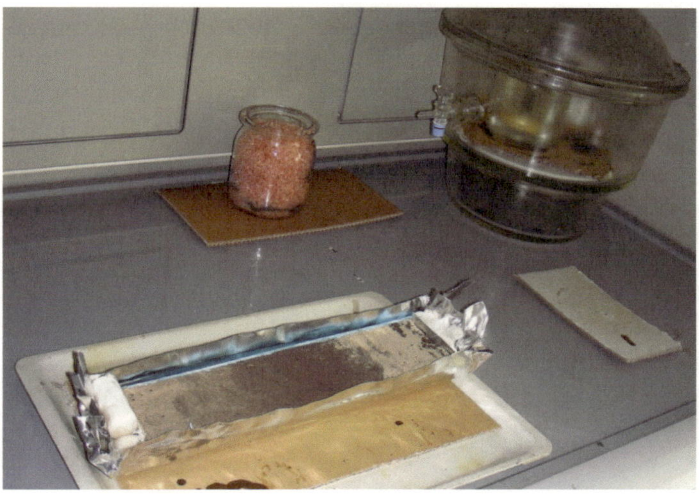

Fig. 6. Impregnation of the soft core sediment to get a hardened sample for the preparation of a thin section (from Srisutam & Wagner, 2010c)

4. Results

4.1 Grain size analyses

4.1.1 Tsunami sediment layer

Tsunami deposits are not observed at swash zone, berm/dune and the ending of water inundation. The locations where the recent tsunami sediment layer can be found is about

50–200 m inland from coastline. We observe three zones within the tsunami affected areas: (1) a zone of erosion extending from the shore inland, (2) a broad zone of tsunami deposition landward of the erosion zone and (3) a narrow zone with neither deposition nor erosion near the limit of inundation (Fig.7). Tsunami sediments consist of a distinct sediment layer of gray sand with shell fragments resting on a pre-existing soil with a sharp, often erosional contact including eroded clasts. These observations resemble to the results of recent tsunami field investigations from literature (Gelfenbaum & Jaffe, 2003, Jaffe et al., 2003, Jaffe & Gelfenbaum, 2007). Fig. 8 displays example photos of the 2004 Indian Ocean tsunami sediments overlying the pre-existing soil at the Thai Andaman coast. Table 2 gives tsunami sediment thicknesses at study areas which vary from 2 to 22 cm. The thickness variations may depend on coastal morphology and local topography as discussed by Gelfenbaum & Jaffe (2003).

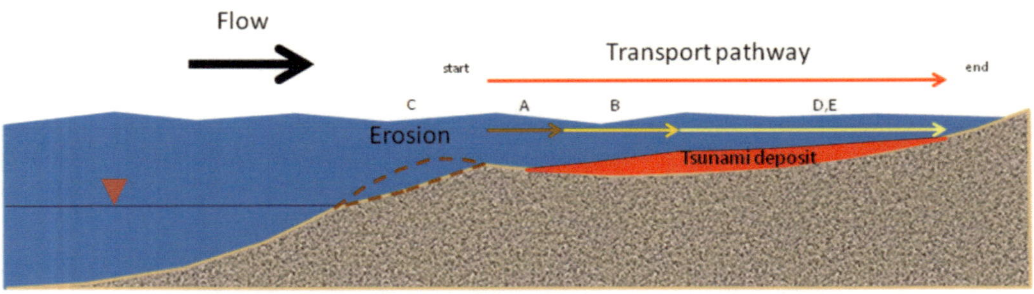

Fig. 7. Schematic of tsunami sediment transportation pattern during the tsunami run-up (following McLaren et al., 2007). A detailed description is given by Srisutam & Wagner (2010a)

Fig. 8. Tsunami deposits overlying a pre-existing agricultural soil with a bottom layer of coarse sand containing rip-up soil clasts and an upper layer of fine sand (from Srisutam & Wagner, 2010c)

4.1.2 Comparison of tsunami sediment characteristics with coastal non-tsunami sediment characteristics

The grain-size distribution curves from the four sub-environment deposits which were a) offshore deposits, b) swash zone deposits, c) berm/dune deposits and d) tsunami deposits did vary concerning average grain size, sorting and modality respectively. Representative grain-size distribution curves of the deposits from the different littoral environments are shown in Fig. 9. A low degree of variability in the shape of the distribution curves was observed for the offshore sediments. The fine sandy deposits were mainly uni-modal and well to moderately sorted. The swash zone samples did consist of coarse to fine sand, poorly to moderately sorted with a bi-modal character. Also coarse to fine sand but better sorted (well to moderately) and with a uni-modal distribution were found in the berm/dune deposits. The characteristics of the particle size distribution in tsunami sediments change progressively from a poorly-sorted multi-modal distribution in the bottom layer to a better sorted distribution within the upper sequences.

Deposit mean grain-sizes from all study areas of the Thai Andaman coast are plotted against their standard deviations (sorting values) in Fig. 10. The plot shows the trend of scatter for each sub-location. The scatter in Fig. 10 demonstrates that the S1 deposit is a fine to very fine sand, poorly to moderately well sorted; the S2 deposit is a coarse to fine sand, poorly to well sorted; the S3 deposit is a medium to fine sand, moderately to well sorted; the S4d deposit is coarse to very fine sand, poorly to moderately well sorted; and the S4p deposit is coarse to fine sand, poorly to moderately sorted. The plot of the mean grain-size of tsunami sediment versus sorting indicates that tsunami deposits (S4d) are a mixture of offshore sediments, swash zone deposits and berm/dune deposits as well. Generally tsunami sediments are not characterized by a typical grain size distribution and their grain size distribution depends strongly on the coastal morphology. E.g. for a coastal area where the offshore slope is mild such as at Khuk Khak beach, the major origin of tsunami sediment is the offshore area. But also large quantities of beach sediments were eroded as Fig.11 does show.

Locations	Tsunami sediment's thickness (cm)	Coastal morphology
1. Ao Kheuy beach	4-10	~1:100 offshore slope, ~1:150 coastal slope, short inundation zone (~250 m inland) which is limited by hill.
2. Khuk Khak beach	2-15	~1:600 offshore slope, coastal plain, long inundation zone (~2 km inland), there are natural and manmade structures (pond and road) in direction of run-up flow.
3. Bang Thao beach	7-20	~1:75 offshore slope, coastal plain, long inundation zone (~1 km inland), there are manmade structures (road) in direction of run-up flow.
4. Kamala beach	2-22	
5. Ao Krabi	3-9	~1:100 offshore slope, coastal plain, short inundation zone (~200 m inland) which is limited by road.

Table 2. Tsunami sediment thickness from field surveys at the Thai-Andaman coast

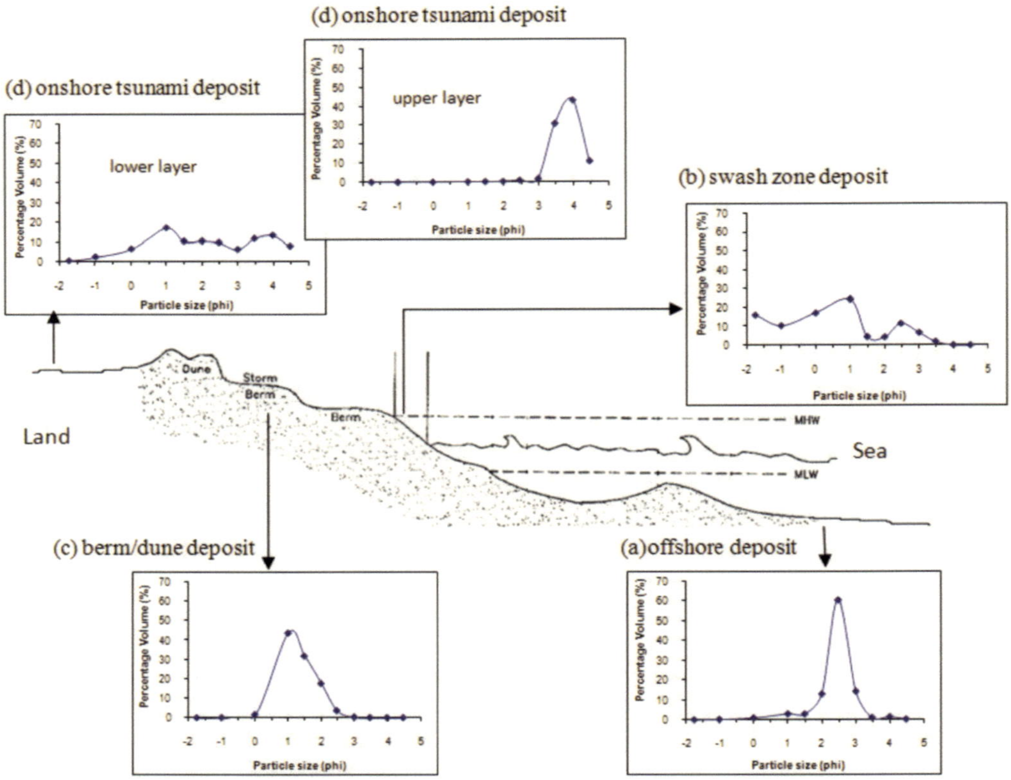

Fig. 9. Representative grain-size distribution curves of littoral deposits from the four sub-environments at the Thai Andaman coast (from Srisutam & Wagner, 2010a)

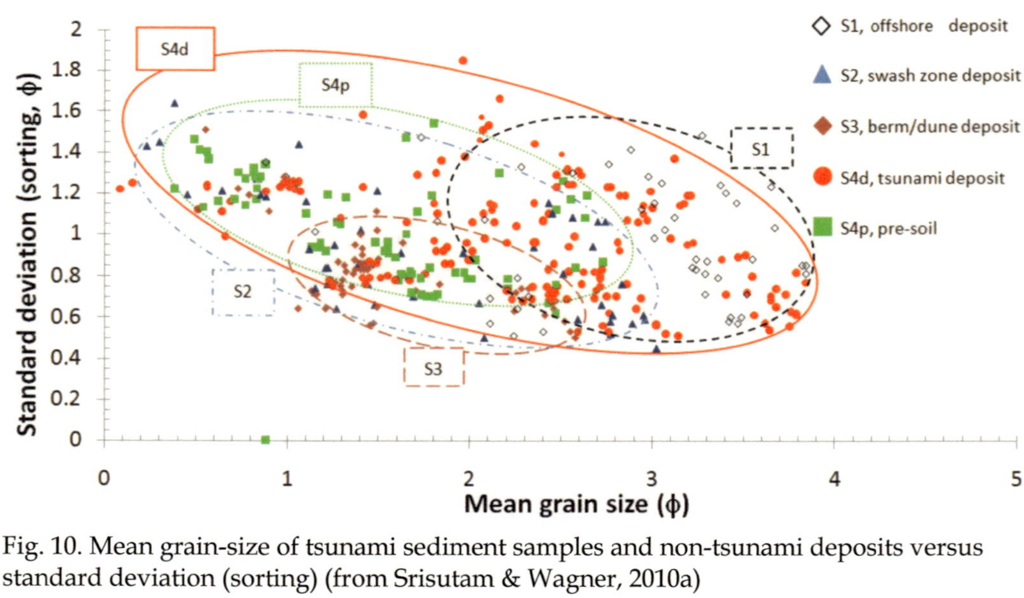

Fig. 10. Mean grain-size of tsunami sediment samples and non-tsunami deposits versus standard deviation (sorting) (from Srisutam & Wagner, 2010a)

Fig. 11. Coastal erosion caused by the 2004 Indian Ocean Tsunami (Pakarang Cape, about 4 km north of Khuk Khak beach)

4.1.3 Vertical grain size variations in the tsunami layer

The vertical development of tsunami sediment textures are shown in Fig.12 and Fig.13.
At Ao Kheuy and Khuk Khak beach mean grain-size fines generally upward and landward. There appears high vertical variability of mean grain-size at the beginning of the deposition zone. Shi et al. (1995) did propose that during periods when tsunamis inundate the coastal zone, provided an adequate sediment supply, sedimentation rates are so high that tsunami sediment is frequently composed of several populations of particles in different size ranges. As a result, the interaction of turbulence, rapid sedimentation and the characteristics of the transported material play a large part in dictating the characteristics of the particular tsunami deposit (Dawson et al., 1996). The set of individual fining upward sequences in the tsunami sediment suggests that tsunami waves inundated the study areas more than once. The coarse-grained sediments are supplied during a tsunami run-up wave. The overlying fine grained sediments with lower settling velocities probably have accumulated when run-up flows decelerate or stand still before the subsequent backflows (Hori et al., 2007).
Fig. 14 shows an example of some more textural parameters of the tsunami sediment at Khuk Khak beach. Sorting values may be used to determine the wave period. The sediment main layer with a higher variation of sorted value suggests that the sediment may be created by a shorter wave period. The value of skewness exhibits a large magnitude change over relatively short vertical intervals. The high value of skewness corresponds to a grain-size distribution with a modal peak corresponding to medium-grain sands and the corresponding tail to a distribution composed of finer-grain sediments. By contrast, low skewness values correspond to grain size distributions in which there is a significant coarse-tail element (Dawson et al., 1996). The kurtosis value generally refers to the degree in which the grain size distribution exhibits marked peakedness. High kurtosis values correspond to

grain size distributions which are characteristically unimodal and which possess a peaked distribution. Conversely, low kurtosis values correspond to flatter grain-size distributions. From these data we conclude that at the Thai Andaman coast, tsunamis are associated with rapid changes in the energy regime so that periods of high water turbulence (tsunami run-up at the beginning of the inundation) are followed by a still-water condition (pre-backwash phase) and, in turn followed by increased turbulence (backwash). The grain-size distributions of examples shown in Fig.12-14 demonstrates a distinctive change from a multi-modal sediment distribution to a uni-modal one as it was observed in previous publications (e.g., Flores event, Shi et al., 1995; Grand banks tsunami, Moore et al., 2007).

4.1.4 Identification of tsunami wave run-ups

The changing of slope in the plots of standard deviation with depth can be used to identify the boundaries of tsunami sediment layers (number of run-up sequences). The break points of slope in the plots of standard deviation with depth mark a break in turbulence associated with a transition to a lower or higher Reynolds number run-up. This may be due to a number of factors, including a deceleration or reflection of the wave bore, backwash from a

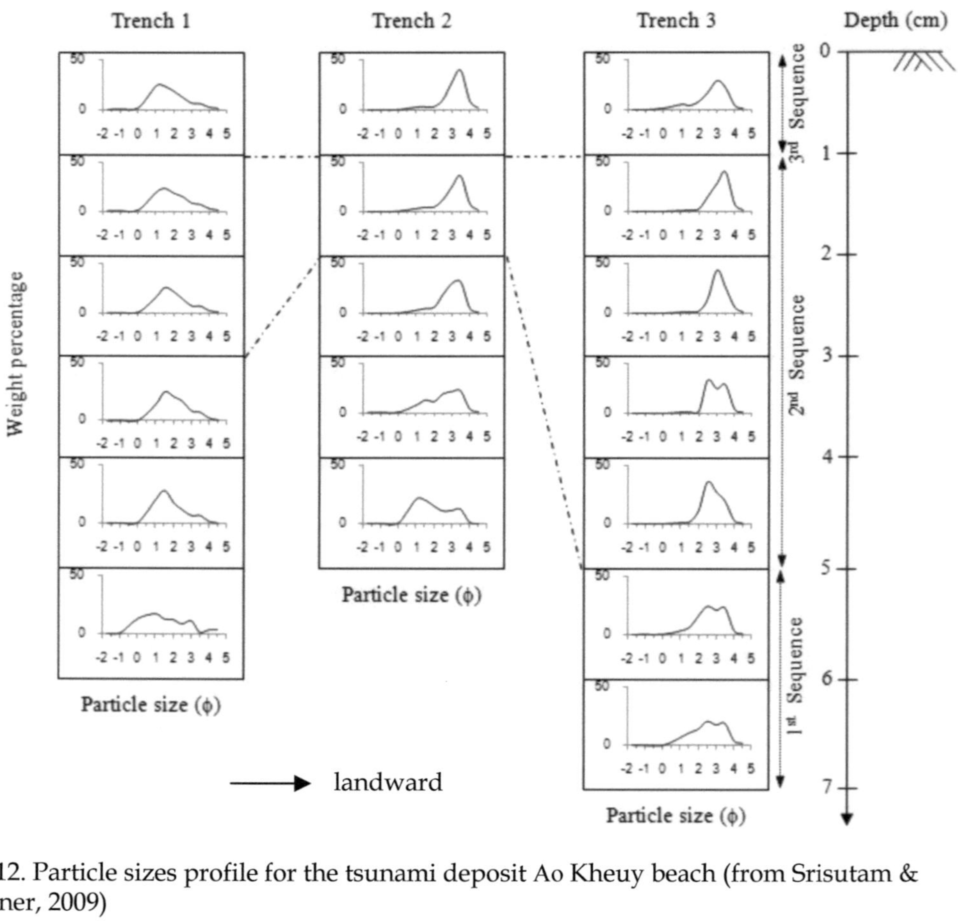

Fig. 12. Particle sizes profile for the tsunami deposit Ao Kheuy beach (from Srisutam & Wagner, 2009)

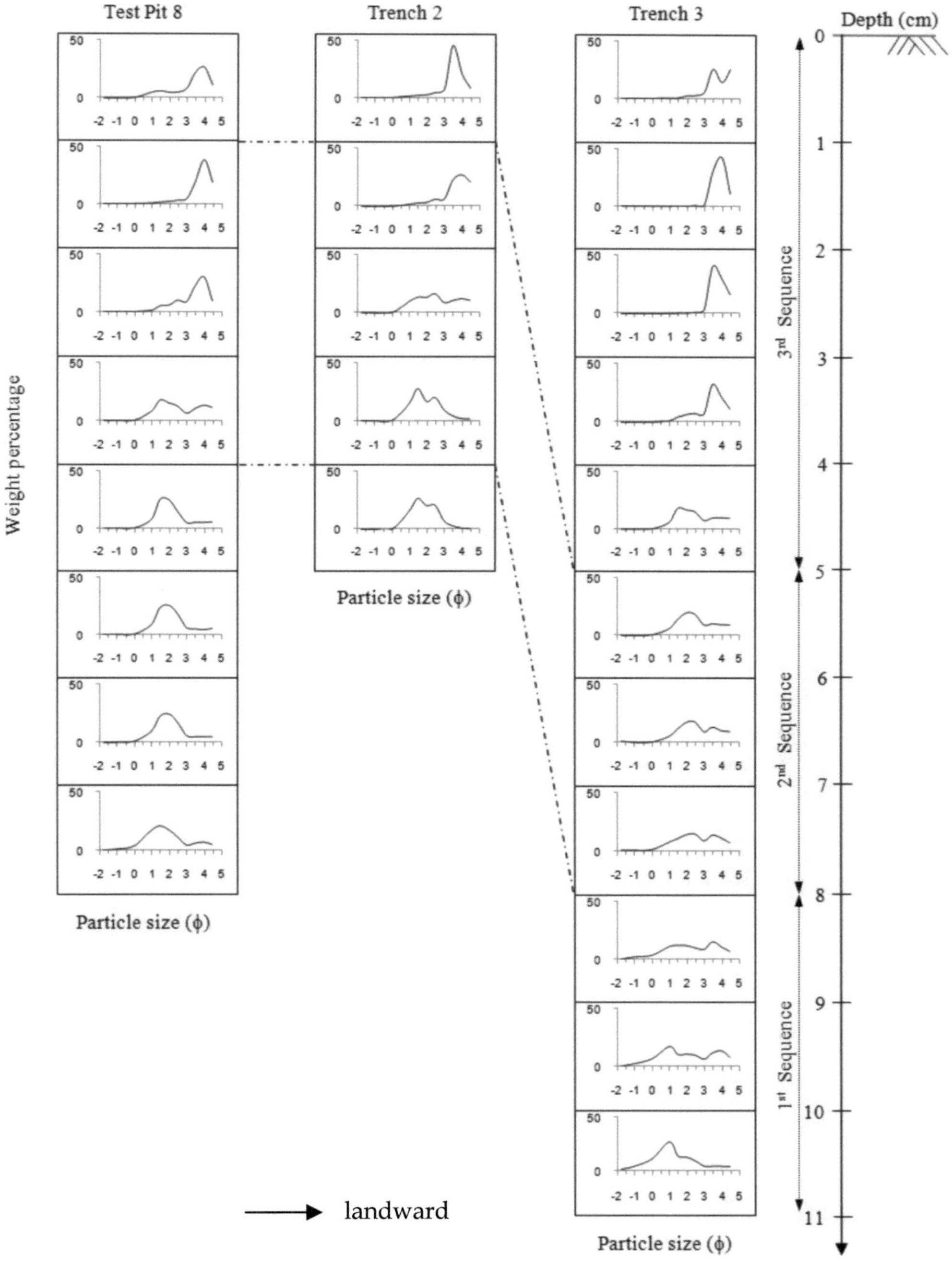

Fig. 13. Particle sizes profile for the tsunami deposit deposit at Khuk Khak beach (from Srisutam & Wagner, 2009)

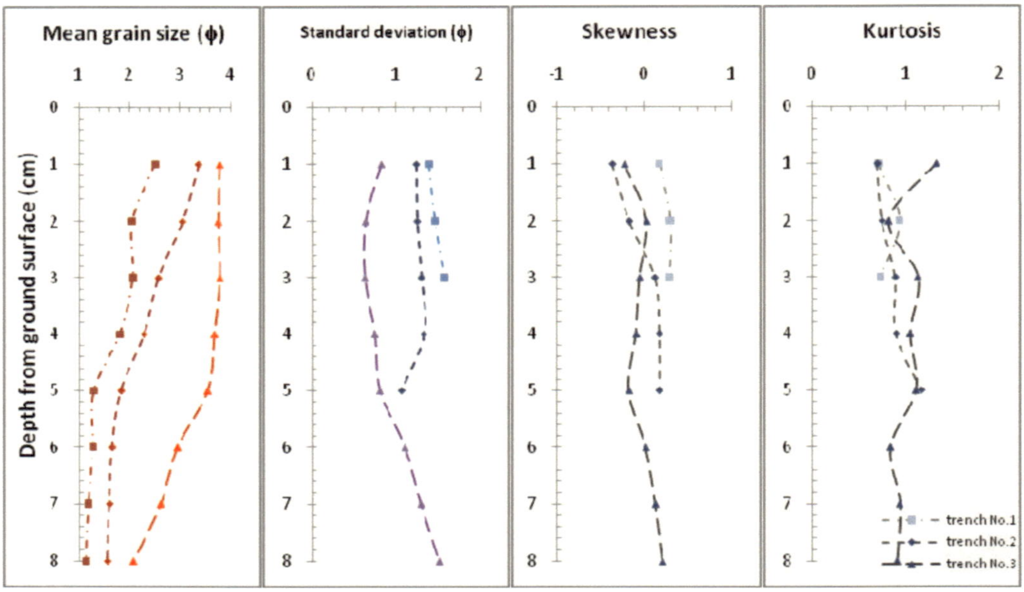

Fig. 14. Plots of tsunami deposits mean grain-size, standard deviation (sorting), skewness and kurtosis versus depth (Khuk Khak beach, Transect 2)

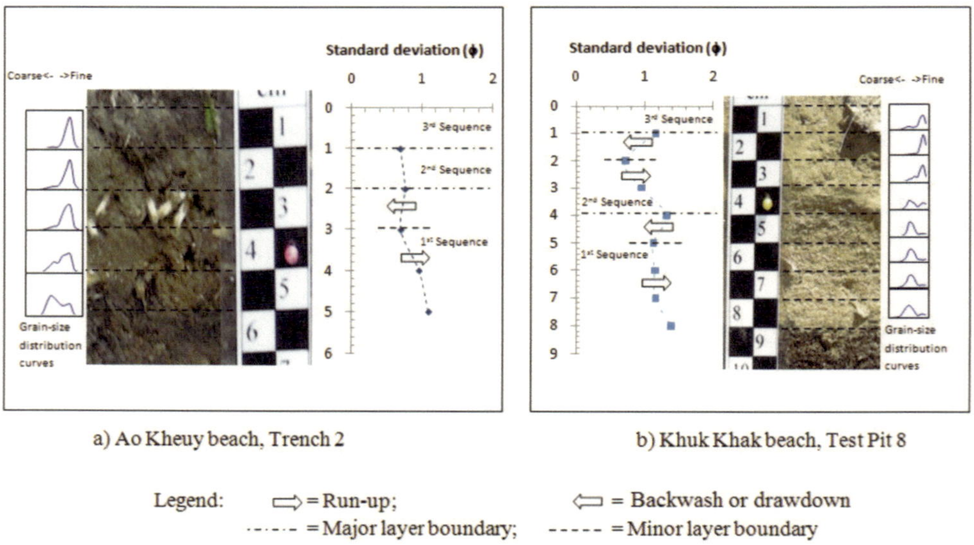

Fig. 15. Plot of standard deviation versus depth for identification of run-up sequences (from Srisutam & Wagner, 2009)

previous wave or another incoming wave running over an already flooded surface (Hindson & Andrade, 1999).

To identify the tsunami run-up layers at Thai Andaman coast, plots of standard deviation with depth and grain-size distribution curve for the sediment layer in 1-cm intervals are evaluated (Fig. 15). Three main layers of tsunami sediment (or three tsunami run-ups) could be identified

from the grain size profiles at Thai Andaman coast. It corresponds to the observed run-up numbers from local eyewitnesses. The tsunami flow consists of a run-up and a backwash (return flow or drawdown) during the inundation event (e.g., Morton et al., 2007). In each tsunami sediment layer, there is a break point of standard deviation. It shows that tsunami sediment layer is composed of a deposit from run-up and backwash. Generally, the thickness of the deposit from run-up is higher than from backwash at the study areas.
Fig.16 shows a decrease of the grain-size from the first to the third run-up. A flattening of the curve can also be observed. The distribution curve is much flatter for the first run-up than the following, i.e. an increase in sorting occurs with every new run-up.

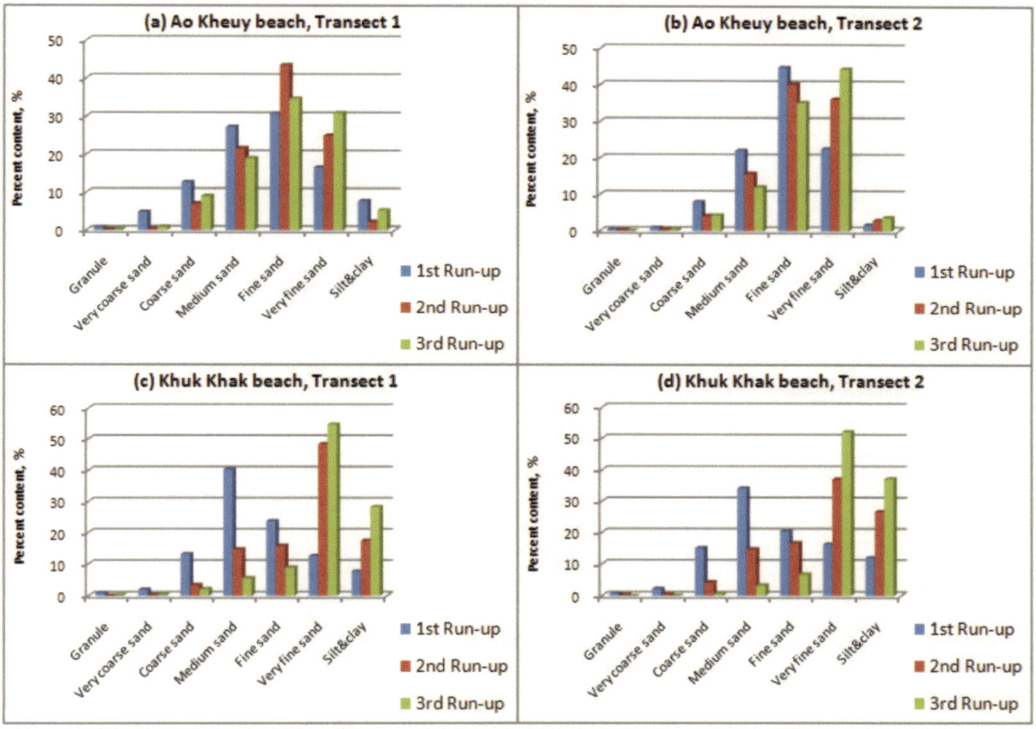

Fig. 16. Mean grain-size classes of run-up sequences (from Srisutam & Wagner, 2010a)

4.1.5 Influence of coastal morphology on tsunami deposits

The offshore slope influences tsunami waves in term of wave build-up (wave amplitude) and wave breaking. A wave breaks closer to the coastline at a steep offshore slope. When the wave breaks, the energy is dissipated, the near shore currents are induced and the mean water level is increased. The onshore slope however influences the velocity and inundation length of tsunami run-ups. At a steep onshore slope, the run-up velocity and inundation length are low and short, respectively. In contrast, the run-up velocity and inundation length are high and long at a mild onshore slope.
Two different offshore slopes are selected to discuss thickness and mean grain-size of the tsunami deposits at different slope inclination. Ao Kheuy beach, transect 1 (Morphology A) and Khuk Khak beach, transect 1 (Morphology B) represent a steep and a mild offshore

slope, respectively. The Wentworth size classes for the mean grain-size of the run-up layers are plotted in Fig.17 and show that for a steep slope the grain-size characteristics are very similar for every run-up sequence. On the contrary, a mild slope leads to a distinctly coarser deposit from the first run-up.

Further tsunami deposit characteristics for different slope inclinations are described in detail in Table 3.

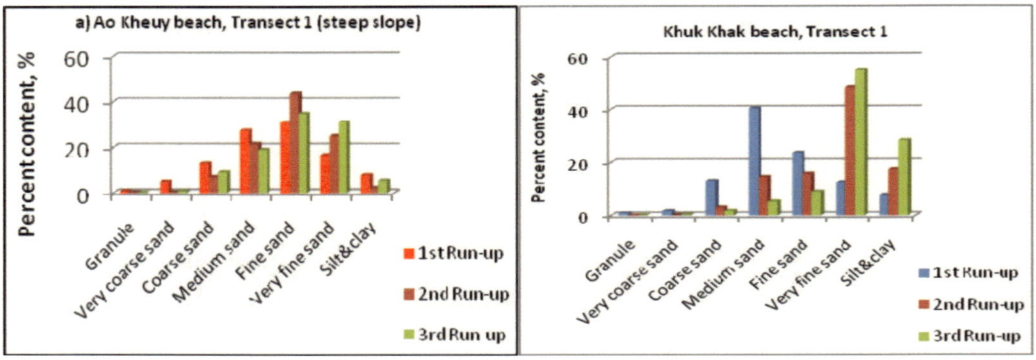

Fig. 17. Mean grain-size distribution charts for the deposits created by the tsunami run-ups at a steep (Ao Kheuy beach) and a mild (Khuk Khak beach) offshore slope

Tsunami deposit characteristics	Offshore	
	Steep slope (Ao Kheuy beach)	Mild slope (Khuk Khak beach)
Started point of deposit	About 40 m inland from coastline	About 70 m inland from coastline
Thickness	4-10 cm (inundation ends at hill slope)	2-12 cm (longer inundation and more deposits)
Mean grain-size	Medium to fine sand, the modal of deposit in each major run-up is same (fine sand)	Fine to very fine sand, the 1st run-up deposited coarser sediments (medium sand), the followed run-up deposited finer sediments (very fine sand)
On-land run-up energy	Lower energy, more run-up energy is dissipated by offshore slope	Higher energy, less run-up energy is dissipated by offshore slope

Table 3. Details of tsunami deposit characteristics at the steep and mild offshore slope

4.1.6 Influence of direct or indirect attacking tsunami waves

Two coastal areas (Khuk Khak beach and Ao Krabi) are considered for tsunami deposit characteristics created by direct and indirect tsunami waves. Khuk Khak beach is located at the upper part of the Thai Andaman coast. It was attacked by direct tsunami waves. Ao

Krabi was attacked by indirect tsunami waves which were diffracted and dissipated by Phuket Island. It is located at the lower part of the Thai Andaman coast. Plots of standard deviation with depth and plots of mean grain-size with standard deviation are used to define the characteristics of tsunami deposits created by direct and indirect tsunami waves, which are described in Table 4.

The tsunami deposit characteristics show that the energy at the coastline and onshore of the area attacked by the direct tsunami waves is higher and more variable than at the area attacked by the indirect tsunami waves (higher scatter in the plot of mean grain-size and standard deviation). A wide range of different source sediments is found at the area attacked by direct tsunami waves. The lower layer is mainly built up by nearshore and onshore sediments, the upper deposits comes from offshore sediments. At the area attacked by indirect tsunami waves, nearshore sediments are less found in the lower part of the deposits. More onshore sediments are found in the upper part of the tsunami deposit. These suggest that the tsunami deposits at the area attacked by indirect tsunami waves are mainly transported and deposited by the backwash flow.

Tsunami deposit characteristics	Affected coastal areas	
	Direct tsunami waves (Khuk Khak beach)	Indirect tsunami waves (Ao Krabi)
Mean grain-size	Medium to very fine sand, fining upward	Fine sand, particle size is uniform
Grain sorted	Moderately well sorted to poorly sorted	Well sorted to moderately sorted
Base	Major deposit base are easy to notice	Major deposit base are not easy to notice.
Source of deposits	Offshore, nearshore, and onshore	Nearshore and onshore
Attacked wave	High energy (~8-10 m run-up height)	Low energy (~1 m run-up height)

Table 4. Tsunami deposit characteristics at areas attacked by direct and indirect tsunami waves.

4.2 Thin sections

More details about the tsunami sediment composition can be studied in thin sections (see also Srisutam & Wagner, 2010c). The tsunami sediments (light colored sediments in Fig. 18) are medium to very fine grained. Sub-angular to sub-rounded grains are widely dispersed. There is evidence for grass roots which are the brown materials at the top of the tsunami deposit (0-2 cm). A lot of organic matter (dark colored materials) is also found in the sediment at the bottom of the tsunami deposit (9-12 cm). The tsunami deposit fines upward with three distinguishable sedimentation layers (Fig. 18). The base of each major sediment layer is identified by the changing of grain-size from fine grained to coarse grained upwards. The interfaces of the major sediment sequences are identified at the depths of 3-4 cm (the base of the upper major sediment layer, Fig. 18(d)) and 7-8 cm (the base of the 2nd sequence, Fig. 18(g)). The basis of the tsunami deposit is not seen by a changing in grain

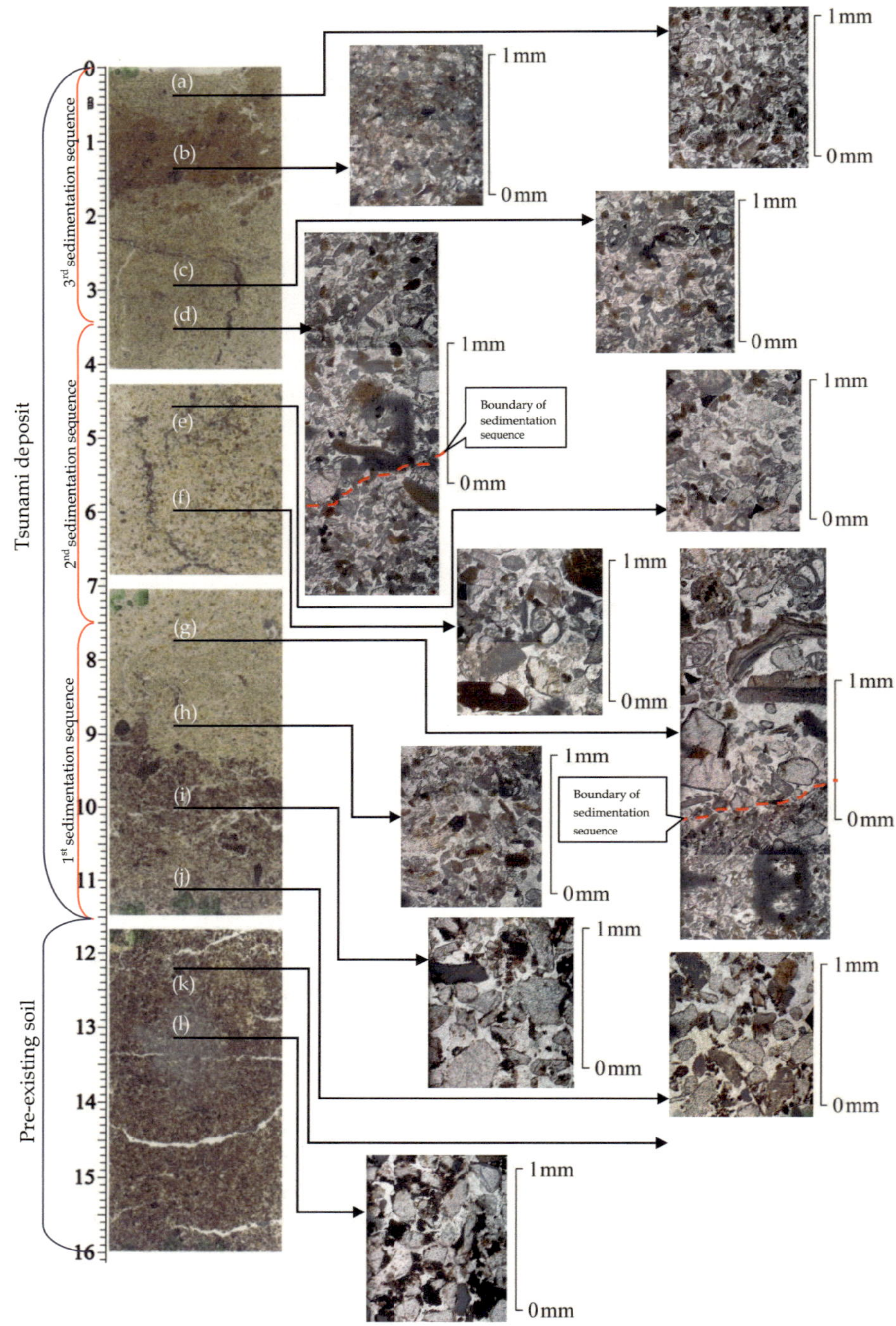

Fig. 18. The pictures of thin sections show details of the sediments in the tsunami deposit. The upper parts (light colored sediments, at the depth of 0-11.5 cm) are the tsunami deposit. The lower part (dark colored sediments) is the pre-existing soil. The sediments fine upward with three run-up sequences (from Srisutam & Wagner, 2010c)

size. It is identified by differences in the roundness of the grains and the content of organic matter. The pre-existing soil is composed of rounded grains and the percentage of organic material in the soil is higher. Therefore, the depth which separates the tsunami deposit from the pre-existing soil is at a depth between 11-12 cm (the base of the 1st major sediment layer, see Fig. 18(j) and Fig. 18(k)). There are three sediment layers or three sedimentation sequences in the tsunami deposit in total. The 1st, 2nd and 3rd sedimentation sequences are at the depth from 7.5-11.5 cm, 3.5-7.5 cm and 0-3.5 cm, respectively.

Shell fragments and shallow water micro-organisms are commonly found in the tsunami deposit as shown in Fig. 19(a). It is possible to observe micro-soil rip-up clasts or micro-sand pockets (a cluster of sediments which are coarser or finer than the surrounding sediment) in each sedimentation sequence (Fig. 19(b)). The most abundant sediment grains are quartz, some muscovite mica and some organic material (Fig. 20).

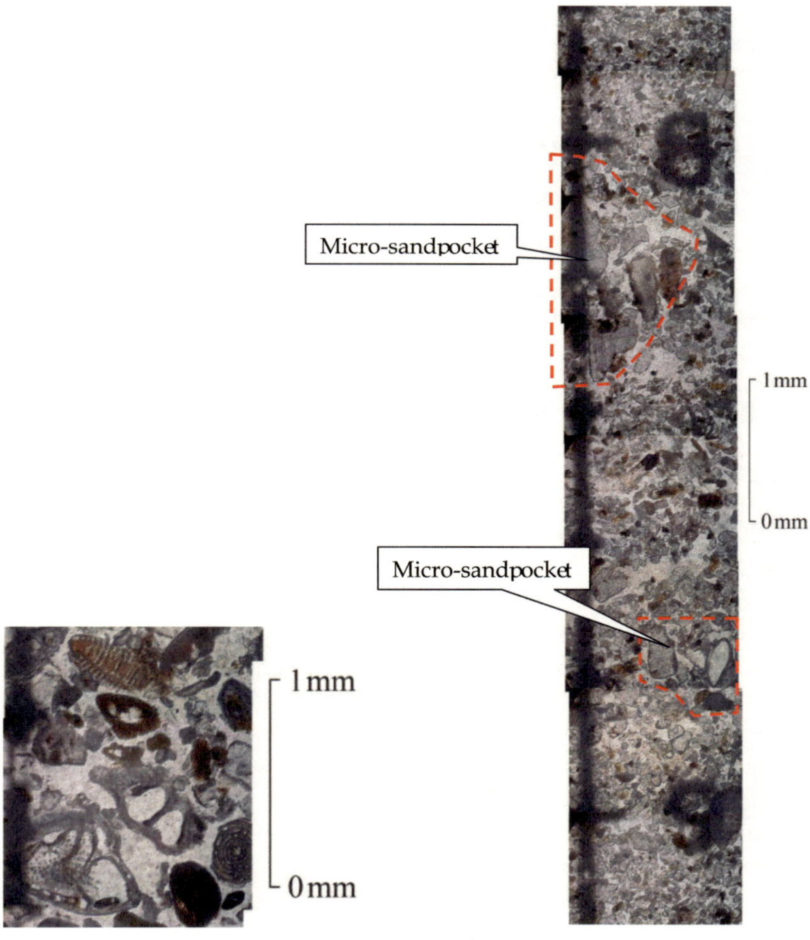

Fig. 19. (a) Shell fragment and shallow water micro-organisms in the tsunami deposit (an example from the depth of 6.3-6.4 cm) and (b) example of the micro- sand pockets from the depth 8.5-9.0 cm (from Srisutam & Wagner, 2010c)

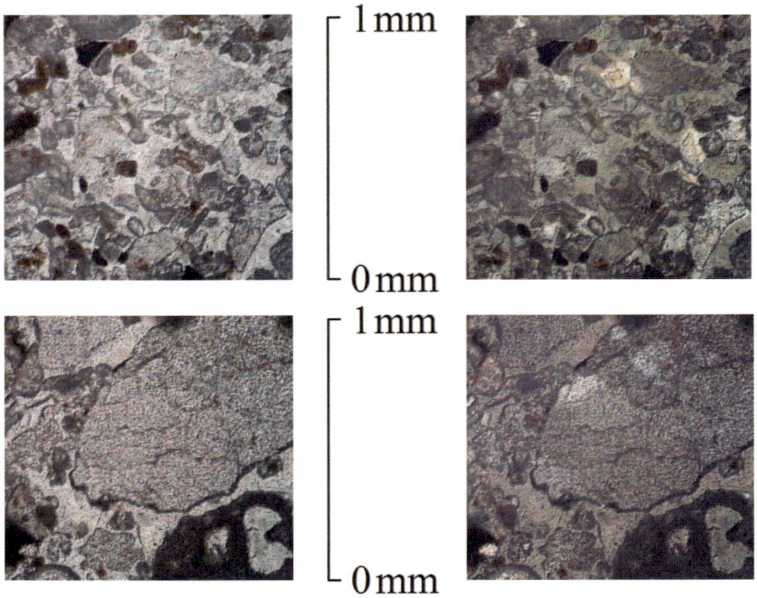

(a) Plane-polarized photomicrograph (b) Cross-polarized photomicrograph

Fig. 20. Polarizing microscope pictures of the tsunami deposit thin sections. The cross-polarized picture better displays the mineralogy of the sediment. The quartz grains are the large grey grains, which are displaying corroded margins. The brightly colored mineral is muscovite mica, in typical lath shaped appearance. The dark colored spots are organic components in the tsunami deposit (from Srisutam & Wagner, 2010c)

5. Discussion

The results of the authors demonstrate that the wet-sieve grain-size analysis as well as thin section analyses can be used to illustrate the characteristics of a tsunami deposit. Both techniques show best the unique characteristics of tsunami sediments which are very similar for both methods. The tsunami deposit is generally composed of multiple fining upward sediment layers. The vertical variations of grain-size parameters from the wet-sieve suggest that there are three major sediment layers in the tsunami deposit which can be interpreted as three run-up/backwash sequences. The same observation was made from the thin sections. Here the interfaces of the major sediment layers were at depths of 3-4 cm, 7-8 cm and 11-12 cm. In the vertical grain-size profiles they were identified at the break points in plot of standard deviation with depth and at the locations of grain-size changing from fine grained to coarse grained. The similar characteristics from both methods can be explained by the previous works of Friedman (1958, 1962) and Johnson (1994). They presented a linear relationship existence for grain-size analysis between the quartile measures of sieving and thin section data. There is very little scatter on either side of the regression line for this relationship and the coefficient of correlation is close to 1.0. The finding that three run-ups or three tsunami waves affected the study area is in accordance with eyewitnesses presented by Szczuciński et al. (2006) and Hori et al. (2007).

In this study, minor details which could not be observed in wet-sieve analysis such as the exact contact between tsunami deposit and the pre-existing soil (the base of the tsunami deposit), the micro-soil rip-up clasts or micro-sand pockets in the major sediment layer and the mineral content could be identified in thin sections. Generally, the contact between tsunami deposit and pre-existing soil can be identified in the field at the boundary where the color of sediments is changing (normally dark to light color of sediments). The thin section shows that the contact of tsunami deposit and pre-existing soil is located lower than the boundary where the color of the sediment changes from dark to light. The sediment lying immediately above the soil/ tsunami sediment contact has a darker color than the sediments in the upper part. Sediments at the base of a tsunami deposit are composed of coastal sand (light color), onshore sediments and the dark color may be due to organic matter or resedimentated soil clasts (see Fig. 18(j)). This is a further proof that tsunamis erode and transport pre-existing sediments from coastal and onshore areas to be deposited further inland. This phenomenon does not only occur in the 1st run-up, but also in the 2nd and the 3rd. It is however more easy to be observed in the 1st run-up layer where more micro-soil rip-up clasts or micro-sand pockets can be observed. The clasts and sand pockets may be the result of turbulence associated with a transition to a lower or higher Reynolds flow number.

The mineral content of a tsunami deposit can be evaluated under the polarizing microscope. At the Thai-Andaman Coast the tsunami sediment grains are mainly quartz and some muscovite mica. These minerals are the major minerals found in sand and silt deposits of the coastal area (e.g. Smith et al., 1976; Harris et al., 1989).

6. Conclusions

The following depositional characteristics of tsunami sediments were obtained from field observations, grain-size distributions (Srisutam and Wagner, 2009, 2010a), analyzed by wet-sieving and from micro-morphological features from thin sections (Srisutam & Wagner, 2010c), respectively

- Tsunami sediments overlay the pre-existing soil with a sharp basal erosion surface.
- The tsunami deposit is a multiple-layered sediment. It generally fines upwards and landwards.
- The material of the tsunami layer is eroded sediment from offshore, coastal and onshore areas transported by the tsunami to inland areas.
- The tsunami sediment in this study is very often a moderately to poorly sorted, fine to very fine sand with coarser particles like e.g. shell fragments. Generally tsunami sediments are not characterized by a typical grain size distribution.
- A plot of the mean grain-size of tsunami sediment versus sorting indicates that tsunami deposits are a mixture of offshore sediments, swash zone deposits and berm/dune deposits as well.
- The grain size distribution depends strongly on the coastal morphology.
- At Thai-Andaman Coast three sedimentation sequences could be distinguished in the tsunami deposit which reflects three run-ups or three tsunami waves affecting this area. The interface of each sedimentation sequence can be located at break points in a plot of grain-size standard deviation with depth. The break points are the changing points of grain-sizes from fine to coarse grains upwards and can be observed in the thin sections as well.

- The 1st run-up transported and deposited coarser particles than the following run-ups.
- Shallow water microorganisms, shell fragments and micro soil rip-up clasts or sand pockets can be observed in almost all sediment sequences.
- Micro-morphological features suggest also that the tsunami sediments are deposited by a run-up flow and a drawdown (backwash) flow as well. Mean grain-sizes of the sediments deposited by both flows are not significantly different.
- The wet-sieving and thin section analyses for the characterization of tsunami deposits give significantly similar results concerning grain size distribution. However, the reconstruction of the sedimentological environment from tsunami sediment characteristics is more convenient from the grain size distribution by wet sieving than from thin sections. Thin sections give additional information about micro-morphological features and composition.

The characteristics of tsunami deposits can be used to calculate tsunami run-up height and velocity. From these data risk assessment and coastal development programs can be deduced for tsunami affected areas (Srisutam and Wagner, 2010b).

7. References

Anthony, E. J. & Hequette, A.(2007). The grain-size characterization of coastal sand from the Somme estuary to Belguim: Sediment sorting processes and mixing in a tide- and storm-dominated setting. *Sedimentary Geology,* 202, 369-382.

Chulalongkorn Tsunami Research Team, Department of Geology, Faculty of Science, Chulalongkorn University (2005). Physical marine factor analyses. http://www.geo.sc.chula.ac.th/Geology/Thai/Knowledge/html/Tsunami/

Dawson, A.G. (1994). Geomorphological effects of tsunami run-up and backwash. *Geomorphology,* 10, 83-94.

Dawson, S. (2007). Diatom biostratigraphy of tsunami deposits: Examples from the 1998 Papua New Guinea tsunami. *Sedimentary Geology,* 200, 328-335.

Dawson, A.G., Foster, I.D.L., Shi, S., Smith, D.E. & Long, D. (1991). The identification of tsunami deposits in coastal sediment sequences. *Science of tsunami hazards,* 9, 73-82.

Dawson, A. G., & Shi, S. (2000). Tsunami Deposits. *Pure and Applied Geophysics,* 157, 875-897.

Dawson, A.G., Shi, S., Dawson, S., Takahashi, T. & Shuto, N. (1996). Coastal sedimentation associated with the June 2nd and 3rd, 1994 Tsunami in Rajegwesi, Java. *Quaternary Science Reviews,* 15, 901-912.

Dawson, S., Smith, D. E., Ruffman, A., & Shi, S. (1996). The diatom biostratigraphy of tsunami sediments: Examples from recent and middle Holocene events. *Phys. Chem. Earth,* 21, 87-92.

Dawson, A. G. & Stewart, I. (2007). Tsunami deposits in the geological record. *Sedimentary Geology,* 200, 166-183.

Francus, P. 81998). An image-analysis technique to measure grain-size variation in thin sections of soft clastic sediments. *Sedimentary Geology,* 121, 289-298.

Friedman, G.M. (1958). Determination of sieve-size distribution from thin-section data for sedimentary petrological studies. *J. Geol.,* 66, 394-416.

Friedman, G.M. (1962). Comparison of moment measures for sieving and thin-section data in sedimentary petrological studies. *Journal of Sedimentary Petrology,* 32, 15-25.

Folk, R.L. & Ward, W.C. (1957). Brazos River bar: A study in the significant of grain-size parameter. *Jour. Sed. Petrology,* 27, 3-26.

Fujino, S., Masuda, F., Tagomori, S., & Matsumoto, D. (2006). Structure and depositional processes of a gravelly tsunami deposit in a shallow marine setting: Lower Cretaceous Miyako Group, Japan. *Sedimentary Geology,* 187, 127-138.

Fujiwara, O., Masuda, F., Sakai, T., Irizuki, T., & Fuse, K. (2000). Tsunami deposits in Holocene bay mud in southern Kanto region, Pacific coast of central Japan. *Sedimentary Geology,* 135, 219-230.

Gelfenbaum, G., & Jaffe, B. (2003). Erosion and Sedimentation from the 17 July, 1998 Papua New Guinea Tsunami. *Pure Appl. Geophys.,* 160, 1969-1999.

Harada, K. The December 26, 2004 Sumatra Earthquake Tsunami, Tsunami field Survey around Phuket, Thailand.
http://www.drs.dpri.kyoto-u.ac.jp/sumatra/thailand/phuket_survey_e.html

Harris, W.G., Hollien, K. A. & Carlisle, V. W. (1989). Pedon Distribution of Minerals in Coastal Plain Paleudults. *Soil Sci. Soc. Am. J.,* 53, 1901-1906.

Hindson, R.A. & Andrade, C. (1999). Sedimentation and hydrodynamic processes associated with the tsunami generated by the 1755 Lisbon earthquake. *Quaternary International,* 56, 27-38.

Hori, K., Kuzumoto, R., Hirouchi, D., Umitsu, M., Janjirawuttikul, N. & Patanakanog, B. (2007). Horizontal and vertical variation of 2004 Indian tsunami deposit: An example of two transects along the western coast of Thailand. *Marine Geology,* 239, 163-172.

Jaffe, B. E., & Gelfenbuam, G. (2007). A simple model for calculating tsunami flow speed from tsunami deposits. *Sedimentary Geology,* 200, 347-361.

Jaffe, B., Gelfenbaum, G., Rubin, D., Peters. R., Swensson, M., Olcese, D., Bernales, L., Gomez, J., & Reiga, P. (2003). Identification and interpretation of tsunami deposit from the June 23, 2001 Peru Tsunami. *Proceedings of the international conference on coastal sediments 2003,* 13 pp.

Jimenez, J. A., & Madsen, O. S. (2003). A Simple Formula to Estimate Settling Velocity of Natural Sediments. *Journal of Waterway, Port, Coastal and Ocean Engineering,* pp. 70-78 ASCE /March/April 2003.

Johnson, M. R. (1994). Thin section grain size analysis revised. *Sedimentology,* 41, 985-999.

Kortekaas, S. (2002). Tsunami, storm and earthquake: distinguishing coastal flooding events. Unpublished thesis, University of Coventry, Coventry, England. 179 pp.

Kortekaas, S., & Dawson, A. G. (2007). Distinguishing tsunami and storm deposits: An example from Martinhal, SW Portugal. *Sedimentary Geology,* 200, 208-221.

McLaren, P., Hill, S. H., & Bowles, D. (2007). Deriving transport pathways in a sediment trend analysis (STA). *Sedimentary Geology,* 202, 469-498.

Minoura, K., Gusiakov, V.G., Kurbatov, A., Takeuti, S., Svendson, J.I., Bondevik, S., & Oda, T. (1996). Tsunami sedimentation associated with the 1923 Kamchatka earthquake. *Sedimentary Geology,* 106, 145-154.

Moore, A. L., McAdoo, B. G., & Ruffman, A. (2007). Landward fining from multiple sources in a sand sheet deposited by the 1929 Grand Banks tsunami, Newfoundland. *Sedimentary Geology,* 200, 336-346.

Morton, R.A., Gelfenbaum, Guy, & Jaffe, B.E. (2007). Physical criteria for distinguishing sandy tsunami and storm deposits using modern examples. *Sedimentary Geology,* 200, 184-207.

Nanayama, F., Shigeno, K., Satake, K., Shimokawa, K., Koittabashi, S., Miyasaka, S., & Ishii, M. (2000). Sedimentary differences between the 1993 Hokkaido-nansei-oki tsunami and the 1959 Miyakojima typhoon at Taisei, southwestern Hokkaido, northern Japan. *Sedimentary Geology,* 135, 255-264.

Nelson, A.R., I Shennan, & Lonh A.J. (1996). Identifying coastal subsidence in tidal wetland stratigraphic sequences at Cascadia subduction zone of western North America. *J. Geophys. Res.*, 101, 6115-6135.

Sawai, Y. (2002). Evidence for 17th-century tsunami generated on the Kuril-Kamchatka subduction zone, Lake Tokotan, Hokkaido, Japan. *Journal of Asian Earth Sciences* 20, 903-911.

Scheffers, A. & Kelletat, D. (2003). Sedimentologic and geomorphologic tsunami imprints worldwide - a review. *Earth-Science Review*, 63, 83-92.

Shi, S., Dawson, A.G. & Smith, D.E. (1995). Coastal sedimentation associated with the December 12th, 1992 tsunami in Flores, Indonesia. *Pure Appl. Geophys.*, 144, 525-536.

Shi, S. & Smith, D.E. (2003). Coastal tsunami geomorphological impacts and sedimentation processes: case studies of modern and prehistorical events. *International conference on estuaries and coast*, November 9-11, 2003, Hangzhou, China, 189-198.

Smith, B. R., Granger, M. A. & Boul, S.W. (1976). Sand and Coarse Silt Mineralogy of Selected Soils on the Lower Coastal Plain of North Carolina. *Soil Sci. Soc. Am. J.*, 40, 928-932.

Smith, D. E., Foster, I.D.L., Long, D. & Shi, S. (2007). Reconstructing the pattern and depth of flow onshore in a palaeotsunami from associated deposits. *Sedimentary Geology*, 200, 362-371.

Smith, D. E., Shi, S., Cullingford, R. A., Dawson, A. G., Dawson, S., Firth, C. R., Foster, I. D. L., Fretwell, P. T., Haggart, B. A., Holloway, L. K. & Long, D. (2004). The Holocene Storegga Slide tsunami in the United Kingdom. *Quaternary Science Reviews*, 23, 2291-2321.

Srisutam, C. & Wagner, J.F. (2009) Multiple layer identification and transportation pattern analysis for onshore tsunami deposit as the extending tsunami data- A case study from the Thai Andaman coast. *Science of tsunami hazards*, 28, 205-217.

Srisutam, C. & Wagner, J.F. 2010(a). Tsunami Sediment Characteristics at the Thai Andaman Coast. *Pure Appl. Geophys.*, 167, 3, 215-232.

Srisutam, C. & Wagner, J.F. 2010(b). Reconstructing tsunami run-up from the characteristics of tsunami deposits on the Thai Andaman Coast. *Coastal Engineering*, Vol. 57, 5, May 2010, 493-499.

Srisutam, C. & Wagner J.F. 2010(c). Grain-size and thin section characteristics of tsunami sediments from Khao Lak, Thailand. *Proceedings of the 9th U.S. National & 10th Canadian Conference on Earthquake Engineering,* paper No 1842, Toronto, Ontario, Canada, July 2010

Switzer, A.D. (2005). Depositional characteristics of recent and late Holocene overwash sandsheets in coastal embayments form southeast Australia. A doctoral thesis, School of Earth and Environment Science, University of Wollongong.

Szczuciński, W., Chaimanee, N., Niedzielski, P., Rachlewicz, G., Saisuttichai, D., Tepsuwan, T., Lorenc, S. & Siepak, J. (2006). Environmental and Geological Impacts of the 26 December 2004 Tsunami in Coastal Zone of Thailand-Overview of Short and Long-Term Effects. *Polish J. of Environ. Stud.* 15, 793-810.

Titov, V.V., Rabinovich, A.B., Mofjeld, H.O., Thomson, R.E. & González, F.I. (2005). The Global Reach of the 26 December 2004 Sumatra Tsunami. *Science* 23, 2045-2048.

Whelan, F. & Kelletat, D. (2003). Analysis of tsunami deposits at Trafalgar, Spain, using GIS and GPS technology. *Essener Geographische Arbeiten* 35, 25.

14

Geological Study of Unusual Tsunami Deposits in the Kuril Subduction Zone for Mitigation of Tsunami Disasters

Futoshi Nanayama[1], Kiyuyuki Shigeno[2],
Yorinao Shitaoka[3] and Ryuta Furukawa[1]
[1]*Geological Survey of Japan, AIST*
[2]*Graduate School of Science and Engineering, Ibaraki University*
[3]*Institute for Geothermal Sciences, Kyoto University*
Japan

1. Introduction

On 26 December 2004 a magnitude M 9.3 earthquake deformed the ocean floor 160 km off the coast of Sumatra, generating the Indian Ocean tsunami and thus causing large sediment transfers due to tsunami run-up in coastal lowlands around the Indian Ocean (e.g., Goff et al., 2006; Moore et al., 2006; Hori et al., 2007; Hawkes et al., 2007; Choowong et al., 2007, 2010). Sediment transfers of this scale are rare events historically. Only when an unusual tsunami strikes coastal lowlands does a large-scale sediment transfer occur, leaving a sedimentary record, that is, tsunami deposits, in the geological strata on shore (Dawson & Stewart, 2007). In this chapter, we seek to understand the run-up process of past unusual tsunamis by examining a series of tsunami deposits on the Pacific coast of eastern Hokkaido, northern Japan, and we estimate the average recurrence interval of such tsunamis from the geological record.

Large earthquakes with M > ~8 in the Kuril subduction zone have historically generated tsunamis that caused damage in eastern Hokkaido between Nemuro and the Tokachi coast (Satake et al., 2005; Fig. 1). Most recently, the 1952 Tokachi-oki, the 1960 Chilean, the 1973 Nemuro-oki, and the 2003 Tokachi-oki tsunamis caused considerable damage and great loss of life in this district. Therefore, it is very important to estimate the likely timing and size of the next large, earthquake-generated tsunami. Information about historical earthquakes in the Kuril subduction zone is limited, however, and no documents from before the 19th century that might refer to tsunami events are available. The earliest written records from eastern Hokkaido are the "Nikkanki" series of documents from Kokutai-ji Temple, which was built by the Edo government at Akkeshi in 1805 (Soeda et al., 2004; Fig. 1).

In the hope of finding traces of past giant tsunamis to use to evaluate the frequency and extent of past tsunami inundation in east Hokkaido, late Holocene coastal sediments such as peat beds and lagoon sediments have been studied since 1998 by our research group and other researchers (e.g., Hirakawa et al., 2000; Nishimura et al., 2000; Sawai, 2002; Nanayama et al., 2003; Soeda et al., 2004). Nanayama et al. (2003, 2007) and Sawai et al. (2009) have reported the general stratigraphy of unusual tsunami deposits due to "500-year earthquake" along survey lines in Kiritappu marsh and adjoined lagoons, and have helped establish the regional stratigraphy of unusual tsunami deposits from the Tokachi coast to Kiritappu marsh (Fig. 1). The stratigraphy of unusual tsunami deposits in the Nemuro coastal area, however, has yet to be fully explained (Nanayama et al., 2008, 2009).

Since 2000, we have carried out a geological study of unusual tsunami deposits in the Nemuro lowland (Fig. 1). In this chapter, (1) we describe the typical sedimentary occurrence of tsunami deposits; (2) we discuss the dating of the deposits by tephrochronology and by accelerator mass spectrometry (AMS) ^{14}C and optically stimulated luminescence (OSL) methods; and then (3), on the basis of these data, we estimate the average recurrence interval of unusual tsunamis in the Nemuro lowland. We also describe the use of our results for mitigation of tsunami disasters.

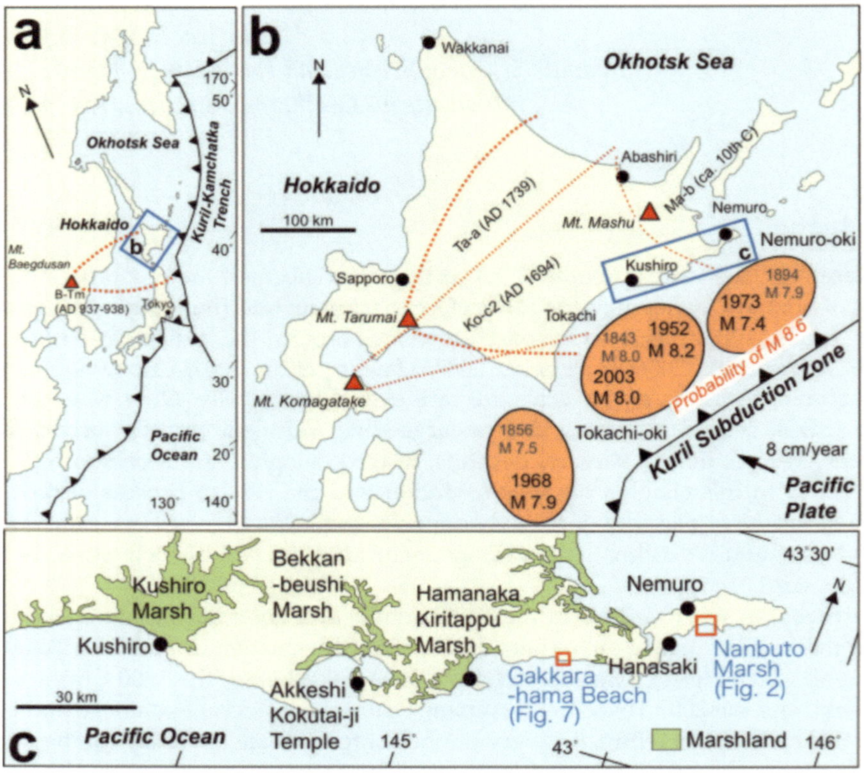

Fig. 1. Tectonic map of the ocean off Hokkaido showing earthquake sources along the Kuril subduction zone (a); index map of the Kushiro-Nemuro coastal area, eastern Hokkaido (b); and a detailed map of the area between Nemuro and Kushiro showing the location of Nanbuto marsh and Gakkara-hama beach (c). The locations of Mt. Baegdusan (a) and Mts. Komagatake, Tarumai, and Mashu (b), sources of tephra layers, are also shown.

2. Tectonic setting of the Pacific coast of eastern Hokkaido

Eastern Hokkaido is situated on a continental plate, the Okhotsk plate, under which the Pacific plate is being subducted at the rate of 8 cm/year, and many earthquakes with M > ~8 have occurred in the Kuril subduction area (Satake et al., 2005, 2008; Fig. 1). The most recent, the 2003 Tokachi-oki earthquake (M 8.0), produced a tsunami with a height of less than 3–4 m (Tanioka et al., 2004). This region has been steadily subsiding at a rate of 1 cm/year since the 19th century (Atwater et al., 2004), but previously it may have been uplifted, either about 0.5–1 m (Atwater et al., 2004) or 1–2 m (Sawai et al., 2004), by repeated great earthquakes (probably M 8.6) as a result of multi-segment interplate ruptures linking the Tokachi-oki and Nemuro-oki segments (Fig. 1; Satake et al., 2005, 2008), with an average recurrence interval of 400–500 years (Nanayama et al., 2003, 2007). The last great earthquake tsunami occurred in this area in the 17th century and left widespread tsunami deposits (Nanayama et al., 2003). Because of the complex history of seismic uplift and interseismic subsidence, sea-level changes during the late Holocene are not well understood in this study area.

3. Geomorphic setting of Nanbuto marsh and Gakkara-hama beach

The Nemuro coastal lowland is on the Nemuro Peninsula, the easternmost part of eastern Hokkaido (Fig. 1). The southern Kuril area, including the Nemuro Peninsula, is an active seismic area. The population of Nemuro City, which is the second largest city along this coast, is about 30000. In 1973, the Nemuro-oki earthquake tsunami (M 7.9) struck Hanasaki Port; its measured tsunami height was 2–3 m, and it caused heavy damage. Earlier earthquake tsunamis, the 1960 Chilean (M 9.5), the 1952 Tokachi-oki (M 8.2), and the 1894 Nemuro-oki (M 7.4) tsunamis, also struck Hanasaki Port (Satake et al., 2005). In addition, a Tokachi-oki earthquake tsunami (M 8.0) occurring in 1843 is described in the "Nikkanki," the earliest written records from this area (Soeda et al., 2004). We investigated tsunami deposits at two sites along this coast: Nanbuto marsh and Gakkara-hama beach.

Nanbuto is a small marsh along the coast near the Nemuro urban district (Fig. 2). The marsh is on a flat coastal plain ranging from 1 to 4 m in elevation with an area of about 8 km^2. Its maximum extent is about 2 km from north to south and about 4 km from east to west. The plain is surrounded on the north, west, and east by marine terraces, 60–80 m in elevation, of Pleistocene age, formed during marine oxygen isotopic stage (MIS) 9 (ca. 300 ka; Okumura, 1996) (Fig. 2). No large streams are present that might bring sandy sediments to the Nemuro lowland.

Nanbuto marsh is one of a group of marshes developed on low-lying coastal plains and valley floors of eastern Hokkaido that were inundated by seawater during the Jomon transgression since 10000 years BP, as shown by the presence of an abandoned sea cliff on the north side of the marsh that is estimated to date to 6000–5500 years BP (Nanayama et al., 2003; Sawai et al., 2009)(Fig. 2). Aerial photographs show that Nanbuto marsh is a typical strand plain, which probably formed as the sea retreated gradually from the plain, leaving up to three beach ridges along its southern edge (Fig. 2). These ridges may have formed during forced regressions associated with seismic uplift or subsidence (Atwater et al., 2004). Thus, these marshes developed and peat deposition began only after the sea retreated from the area, after 5500–6000 years BP.

At Hanasaki Port (Fig. 1), the spring-tide range is 1.2 m, and the neap-tide range is 0.9 m. The corresponding ranges are 1.3 and 1.0 m at Kushiro (Fig. 1), where the extreme tidal range, between the highest and lowest astronomical tides, is 1.7 m. We thus estimated the corresponding ranges at Nanbuto to be 1.2–1.3 m and 0.9–1.0 m (Fig. 3). During the last 200 years, typhoons and heavy storms have rarely struck this coast; thus, no large seawater flooding events or sand movements associated with huge storms or typhoons have occurred. We estimated the landward limit of the annual storm run-up from the distribution of flotsam washed up on the present beach (Fig. 2). Hanasaki Port suffered damage from the 1973 Nemuro-oki earthquake tsunami, which had a wave height of 2–3 m. Figures 2 and 3 show the area inundated by the 1973 tsunami and the flooding elevation.

Our other study site, Gakkara-hama beach, is situated on the western margin of Nemuro City, 30 km west of Nanbuto marsh (Fig. 1), at the edge of a marine terrace formed during MIS 9 (Okumura, 1996). There is private ranch and no colony. Behind the rocky Gakkara-hama beach, there is a small marsh, 2 to 6 m in elevation. The marsh deposits are exposed in the sea cliff that extends along the shoreline.

Fig. 2. Aerial photograph showing the topography around Nanbuto marsh, and the locations of three of the trenches excavated along survey line NB. Circles show sites where 17th century tsunami deposits (layer NS1) were observed. The blue shading shows the area inundated by the 1973 Nemuro-oki tsunami. Dark gray shading shows MIS 9 marine terrace surfaces, and light gray shading shows the area inundated by the 17th century tsunami (estimated from the distribution of NS1).

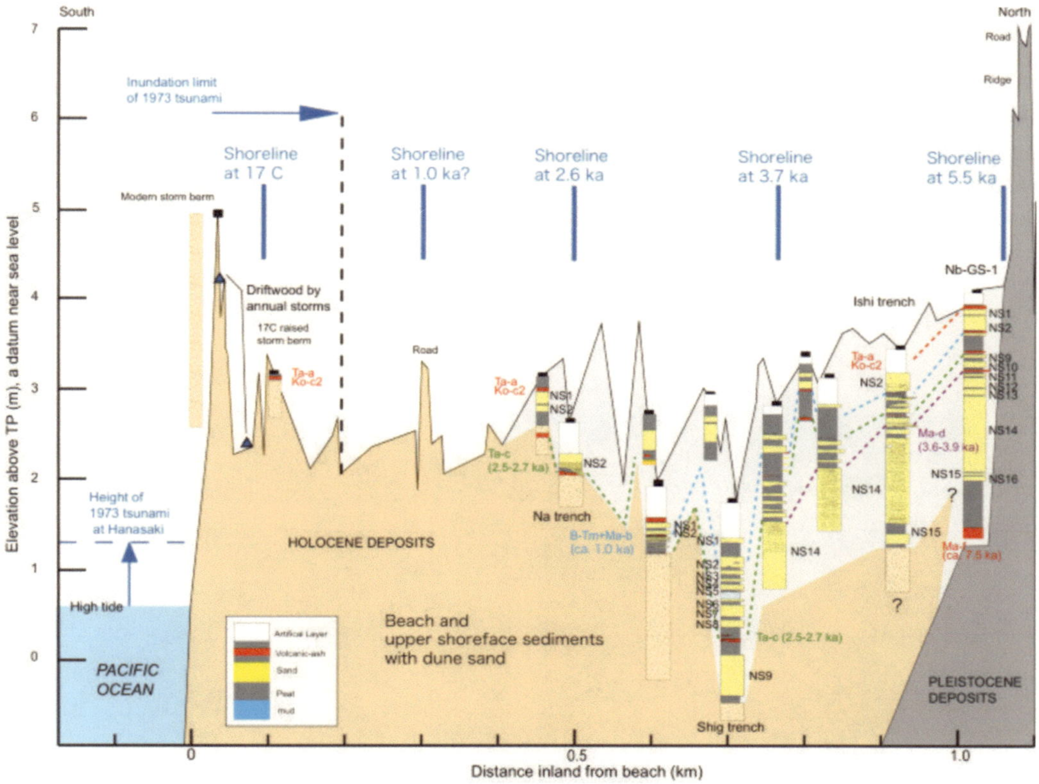

Fig. 3. Geomorphologic profile along line NB. Seven tephra layers were used to correlate the sedimentary stratigraphy among trenches in Nanbuto marsh (colored dashed lines).

4. Methods

4.1 Field survey and sampling

Line NB is 1000 m long and traverses the western part of Nanbuto marsh along the eastern side of Lake Nanbuto (Fig. 2); it is approximately perpendicular to the present shoreline. We measured elevation and distance from the present shoreline with a tape measure, a leveling instrument, and a Handy GPS system (Leica System 1200 GNSS). Most of the marsh is part of a private ranch; therefore, it is partially cultivated and also used by domestic animals.

We looked for regional tephra layers and sand beds of possible tsunami origin within the peat beds and other marsh deposits and traced every sand bed that we found, taking samples with a scoop or a peat sampler at 10-m intervals along the survey line (Fig. 2). In addition, we used heavy equipment to dig seven trenches, up to 2.7 m wide, 14.4 m long and 2.7 m deep, and described the trench walls. Finally, we obtained a wide, oriented sample, 300 cm long by 30 cm thick, with a Geo-slicer soil sampler (Atwater et al., 2001) at Nb-GS-1 (Fig. 3).

4.2 Sedimentary description

We made three-dimensional oriented peels of the trench walls and described major sedimentary structures in the field. We also used a large plastic box (21 cm × 30 cm) to take oriented samples of sand beds for radiographic observation from each trench wall and from the Geo-slicer sample.

In our laboratory at the Geological Survey of Japan, National Institute of Advanced Industrial Science and Technology (AIST), we took radiographs of typical sand beds in this area to infer their depositional processes. We also described the sedimentary structures of each sand bed, including bed form, grain size, and current directions, and we recorded the color of each bed with a digital soil color reader (Minolta SPAD-503). Finally, we examined each sand sample under a binocular microscope for marine components such as rounded beach sand grains and marine microfossils. We then integrated all this information and used it to identify tsunami deposits in the Nemuro lowland.

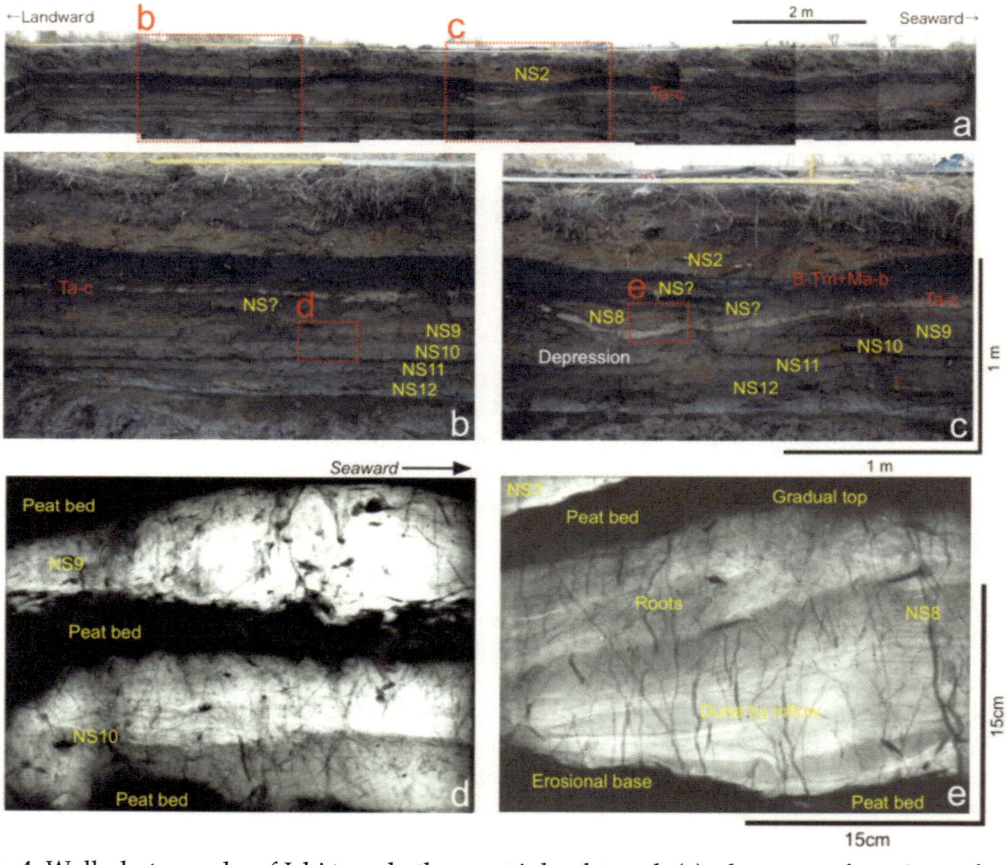

Fig. 4. Wall photographs of Ishi trench, the most inland trench (a); close-ups of portions of the wall (b, c); and radiographs of typical tsunami deposits (d, e) in this trench.

4.3 Tephra study

We described the thickness, color, and grain size of tephra layers and collected samples in the field. In our laboratory, the tephra samples were mounted on slides with resin and double-polished and their petrographic features, including glass shard morphology and phenocryst assemblage, were examined under a stereoscope (polarizing microscope).

Chemical analyses were performed with a JEOL JXA-8900R electron probe microanalyzer at the Geological Survey of Japan. Nine major elements (SiO_2, TiO_2, Al_2O_3, FeO, MnO, MgO, CaO, Na_2O, and K_2O) were analyzed with an accelerating voltage of 15 kV and a beam current of 12 nA. The narrow beam scanned within a 10-μm grid, with counting times of 20 and 10 s for peak and background, respectively.

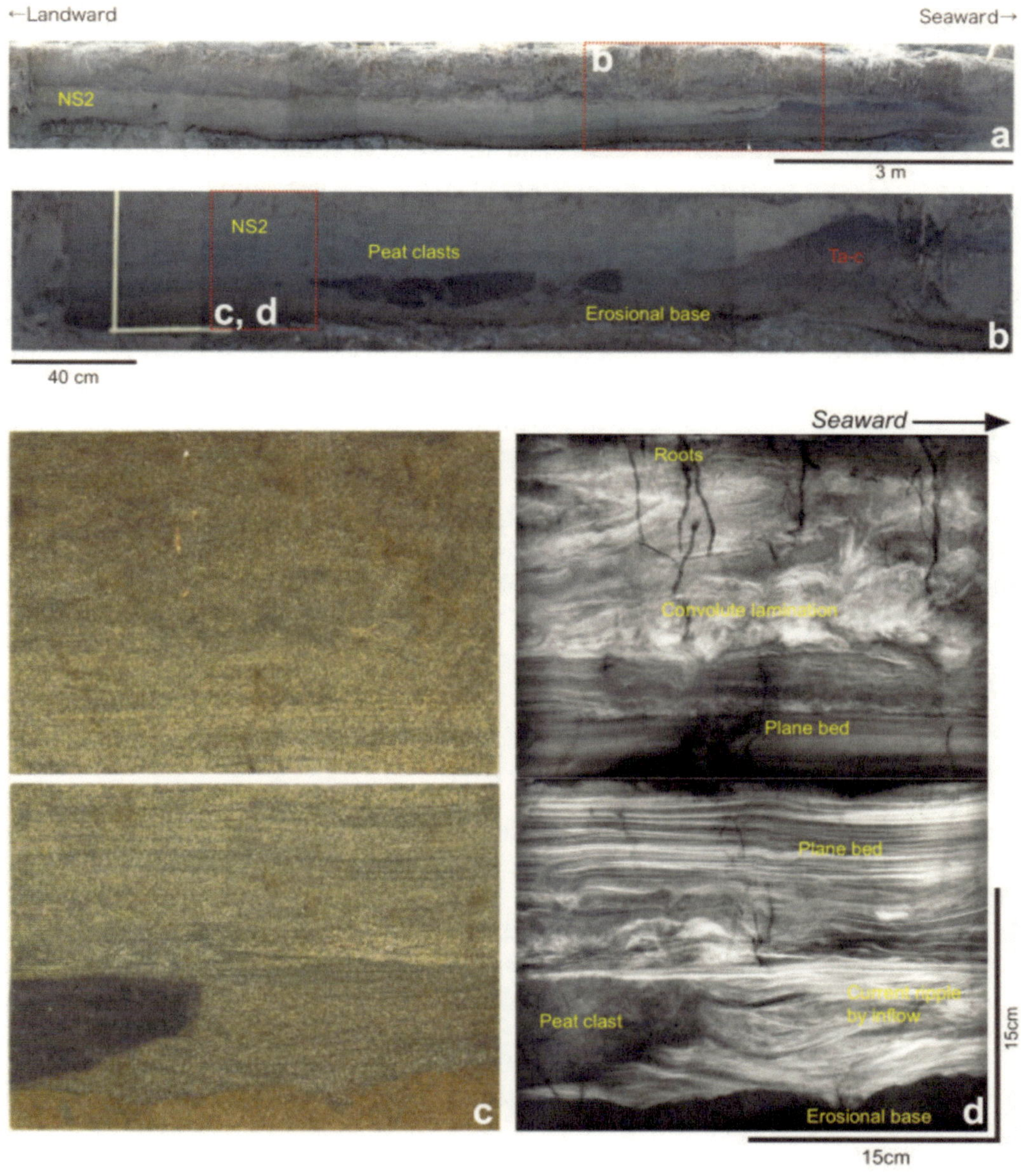

Fig. 5. Na trench, at the seaward location of the survey line, in Nanbuto marsh. Trench wall photographs (a, b); photograph of the oriented plastic-box sample of NS2; (c) radiograph of the same sample with sedimentary interpretation (d).

At least 30 volcanic glass shards from each tephra sample were analyzed to identify the origin of the volcanic ash (Furuta et al., 1986). We compared our results with the known chemical compositions of regional tephra layers as reported by Furukawa et al. (1997), Furukawa & Nanayama (2006), and Yamamoto et al. (2010).

4.4 Radiocarbon dating

Plant materials and one bone sample were selected for ^{14}C dating from each peat bed and basal beach sediment. In all, 20 samples were dated at the AMS facilities at Geo-Science Laboratory Co. in Nagoya, Japan and the Institute of Accelerator Analysis Ltd. in Kawasaki, Japan. The dates were converted to calendar years before present (cal. yBP) and calendar thousand years before present (cal. ka BP) with the calibration program IntCal04 (Reimer et al., 2004), where BP is relative to the year AD 1950. We refer only to the 2-sigma range of calendar years in this chapter.

4.5 OSL dating

Dose estimation in the OSL dating method has improved greatly over the last 10 years, with the result that OSL has been increasingly used to date late Quaternary sediments. However, the age range to which OSL dating of quartz, feldspar and other minerals can be applied, which depends on both the saturation dose and the dose rate, is limited (Tsukamoto & Iwata, 2005). Nevertheless, it is a very useful method because it can be used to obtain age data directly from tsunami deposits.

We collected 14 samples from 12 sand beds interpreted as tsunami deposits (NS1–NS12) for OSL dating by the radiation laboratory of Nara University of Education. These samples, which contained both quartz and feldspar grains with other mineral grains, were dated by the infrared stimulated luminescence method (Nanayama et al., 2009).

5. Results and discussion

5.1 General stratigraphy

In the Nanbuto marsh, the most seaward trench (Na) was dug about 490 m from the shoreline, and the most inland trench (Ishi) was approximately 970 m from the shoreline. The survey found the formation of about 2.2-m-thick peat layer in its deepest zone within the subject area. We identified 7 regional tephra layers and 16 tsunami deposits (NS1–NS16) (Figs. 4 and 5).

On the basis of macroscopic examinations and the known stratigraphy of volcanic ashes in the region (Furukawa & Nanayama, 2006; Yamamoto et al., 2010; Fig. 1), we identified, in descending order, ashes Ta-a (erupted from Mt. Tarumai in 1739), Ko-c2 (Mt. Komagatake, 1694), Ma-b (Mt. Mashu, 10th century), B-Tm (Mt. Baitousan, about AD 937–938), Ta-c (Mt. Tarumai, 2.5–2.7 ka), Ma-d (Mt. Mashu, 3.6–3.9 ka), and Ma-f (Mt. Mashu, ca. 7.5 ka).

In addition, we obtained AMS ^{14}C dates on plant materials from peat beds (Table 1). Of 18 samples dated by OSL from NS1 to NS12 (Fig. 5), we considered six to yield reasonable dates for the tsunami deposits. The other luminescence dates were apparently too old, by comparison with the tephrochronology and AMS ^{14}C data, as discussed below (Table 1).

On the basis of the tephra age of the deepest peat horizon immediately overlying beach sediment at each trench site, we estimated past shoreline positions at the 17th century, 1.0

ka?, 2.6 ka, 3.7 ka, and 5.5 ka (Fig. 3), which we used to estimate the run-up distance between the observation site of each tsunami deposit and the corresponding shoreline. The estimated run-up distances exceeded several hundred meters, which was an important consideration in our association of these deposits with unusual tsunamis.

We also described 12 tsunami deposits, NS1–NS12 (Nanayama et al., 2009; Fig. 7), in the sea cliff scarp at Gakkara-hama beach. The stratigraphic sequence of the tephra layers and tsunami deposits were the same as at Nanbuto, indicating that these unusual tsunami deposits are distributed regionally in the Nemuro lowland.

5.2 Sedimentary structures in unusual tsunami sand beds

When the 1973 Nemuro and 2003 Tokachi-oki tsunamis struck the Nemuro lowland, no large sediment transfer occurred, nor was there much coastal erosion. These tsunamis were too small to leave tsunami deposits in this environment. Giant tsunamis on the scale of the 2004 Indian Ocean tsunami very likely struck the coast in the past, generating large-scale sediment transfers that formed the tsunami deposits that we observed in the marsh environment of the Nemuro lowland.

The major component of each of the 16 tsunami deposits (NS1 to NS 16) is very well sorted fine sand. These sands were mainly scoured from beach and dune sand after the tsunami hit the coast. In marsh environments, tsunami deposits are usually interbedded with peat (Dawson & Stewart, 2007). No shell fragments or carbonate microfossils were observed, presumably because they were dissolved by submersion. At Shig trench, some sand beds were covered by mud layers (Fig. 3). We interpreted this as mud deposited in a trough between two sand ridges, possibly during seawater flooding. The thickness of the tsunami sand layers is important information because it indicates the magnitude of local topographic depressions in the marsh environment. For example, NS2 ranges in thickness from a few centimeters to tens of centimeters; where it reaches maximum thickness of 95 cm, it clearly displays parallel lamination and resembles beach sediment (Fig. 5). NS2 also shows clear landward thinning.

Although we did not observe any clear graded bedding, the sand beds included internal sedimentary structures such as plane beds, dunes, and current ripples, suggesting bedload transport (Figs. 5 and 6). Moreover, within each bed, dune forms and current ripples indicate two flow directions, thus recording both the tsunami inflow and its outflow (Nanayama & Shigeno, 2006).

The gradual upper boundary and the erosional base of each sand bed are characteristic features of tsunami sedimentation. We inferred that the tsunami run-up eroded the underlying stratum, and that the gradual upper boundary reflects the regrowth of marsh vegetation in the years following the tsunami sand deposition. The erosional lower bounds and associated peat blocks or clasts that characterize the tsunami sand layers in this area constitute important evidence of past tsunami deposits in a marsh environment (Bondevik et al., 2003; Gelfenbaum and Jaffe, 2003). Without the application of large stress, it is difficult to detach peat clasts from a peat bed, because of the fibrous nature of peat. According to our radiograph observations, the deposits contained accretion structures generated by flowing water such as plane beds and current ripples. We also observed convolute lamination, reflecting rapid sedimentation and water drainage (Fig. 5), both of which occur during tsunami run-up.

Fig. 6. Collection of samples for OSL dating at Shig trench (left). Photograph of the trench wall showing the OSL sampling horizons and dating results (right). AMS ^{14}C dating results, tephra layers and dates, and sand beds NS1 to NS9 are indicated to the right of the photograph.

Fig. 7. View of Gakkara-hama beach looking southwest (a), and photograph of the sampled outcrop (b). Photograph of a large peel sample (c) showing the locations of tsunami sand beds NS1–NS12, tephra layers and dates, OSL sampling horizons and dating results, and AMS ^{14}C dating results.

5.3 Estimation of the recurrence interval of unusual tsunamis

We identified 16 tsunami sands (NS1 to NS16) within peat beds in the Nemuro lowland, and ascertained the chronology of their deposition using tephrochronology, AMS ^{14}C dating, and OSL dating. The date of each peat bed can be measured by AMS ^{14}C dating to within a 2-sigma range of several hundred years. Because the base of each sand layer is usually erosional, it is not possible to estimate the exact age of each tsunami deposit from the AMS ^{14}C ages of the peat horizons. We thus inferred the average recurrence interval of unusual tsunamis to within about 100 years by using the regional tephra ages along with the AMS ^{14}C ages from certain important peat horizons as follows (Table 1).

1. Ko-c2 (AD 1694) overlies tsunami deposits NS1 and NS2, which overlie both B-Tm (AD 937–938; Fukusawa et al., 1998) and Ma-b (10th century). We correlated NS1 with the 17th century tsunami and NS2 with a 13th century tsunami. They are estimated to be separated by 379 years, obtained by dividing the interval between the two time markers by two.
2. Underlying B-Tm and Ma-b, and overlying Ta-c (2.5–2.7 ka), are six tsunami deposits (NS3 to NS8). Their estimated recurrence interval is thus 250–283 years.
3. Underlying Ta-c and overlying Ma-d (3.6–3.9 ka), there are three tsunami deposits (NS9 to NS11), and the estimated recurrence interval is 200–367 years.
4. Underlying Ma-d and overlying the lowest peat horizon (4.8–5.0 ka) are five tsunami deposits (NS11 to NS16), for an estimated recurrence interval of 220–320 years.

Therefore, the estimated average recurrence interval of unusual tsunami events in the Nemuro coastal area is 200–379 years (Table 1). However, the recurrence interval was not estimated by using dates obtained directly from the tsunami sands, so this value should be understood as a maximum. The number of tsunami deposits in the Nemuro area between the regional tephra layers Ko-c2, B-Tm, and Ta-c is greater than the number in the Tokachi-Kiritappu area, suggesting a possible tsunami source off Habomai, Shikotan, Kunashiri, and Etorofu Islands (southern Kuril Islands), in addition to tsunami-generating multisegment interplate earthquakes along the Tokachi-oki and Nemuro-oki ruptures (Fig. 1).

We also attempted to date the tsunami deposits directly by using the OSL dating technique (Fig. 6; Table 1). We hope that it will be possible to obtain the formation ages of individual tsunami deposits more exactly by future advances in OSL dating technology. However, our luminescence results yielded numerous, erroneously old ages compared with ages ascertained by tephrochronology and AMS ^{14}C dating. These erroneous ages may be attributable mainly to (1) insufficient zeroing or (2) the mixing of deposits of different ages by scouring during the tsunami run-up, or both. We plan to investigate the sources of the tsunami sand deposits by using the OSL technique, and to conduct basic research on erosional and depositional processes during tsunami run-up in detail in a new study area.

5.4 Open trench demonstration and donation of large peel samples

Unusual tsunami deposits can be traced as high as 18 m above the current sea level and as far as 1–4 km inland from the Pacific shoreline of eastern Hokkaido, and such unusual tsunamis have recurred at intervals of several hundred years, with the most recent event in the 17th century (Stake et al., 2008). The results of this study has thus improved the unusual tsunami hazard map, produce in accordance with government guidelines, municipalities of eastern Hokkaido, including Nemuro City.

Because Nemuro is an active seismic area, it is important for citizens to be informed with regard to tsunami hazards. Therefore, on 15 October 2005, we conducted an open trench demonstration for the people of Nemuro at Ishi trench. The Geological Survey of Japan, Hokkaido University, Nemuro City Museum of History and Nature, and the Historical Museum of Hokkaido cosponsored this outreach event. About 200 residents of Nemuro City and eastern Hokkaido participated, giving us a good opportunity to explain the importance of our research results directly to local citizens.

After our investigation, we donated some large peel samples to the Nemuro City Museum of History and Nature and to the Historical Museum of Hokkaido, Hokkaido University, to use as educational materials in regard to tsunami disaster mitigation (Fig. 8).

Stratigraphy	OSL dating (ka)	AMS ^{14}C dating (2-sigma cal. ka)	Recurrence interval
Ta-a (AD1739) / Ko-c2 (AD1694)			
NS1 (17C)	2.9±0.2		379 years
NS2 (13C)	0.8 ± 0.3		
B-Tm (AD937-938) +Ma-b (ca. 1ka)			
NS3	1.7 ± 0.2, 1.5 ± 0.1		250-283 years
NS4	1.4 ± 0.2		
NS5	5.1 ± 0.8		
NS6	10.3 ± 1.9		
NS7	5.3 ± 0.6		
		2.1-2.3 (wood)	
NS8	5.6 ± 0.7, 9.3 ± 1.0		
Ta-c (2.5-2.7 ka)			
		2.5-2.8 (wood)	
NS9	2.7 ± 0.4, 8.3 ± 0.6		200-367 years
NS10	9.5 ± 0.8		
		3.2-3.3 (wood)	
NS11	3.0 ± 0.4		
		3.4-3.6 (wood)	
Ma-d (3.6-3.9 ka)			
		3.4-3.7 (wood)	
NS12	5.1 ± 0.6		220-320 years
		3.9-4.1 (wood)	
NS13			
		4.5-4.8 (wood)	
NS14			
		4.8-4.9 (wood)	
NS15			
		4.9-5.0 (wood)	
NS16			
		4.8-5.0 (wood)	
Beach sand		*5.5-5.7 (bone)*	

Table 1. Tephrochronology and OSL and AMS ^{14}C dating results and estimated recurrence intervals of unusual tsunamis (NS1–NS16) during the last 5000 years. OSL dates in black type are considered reasonable ages by comparison with AMS ^{14}C dating and tephrochronology results, and those in purple type are considered to be erroneous (too old).

Fig. 8. Open trench demonstration held for Nemuro residents on 15 October 2005 at Ishi trench (left). A large peel sample that was donated for tsunami education (right).

6. Conclusion

1. We conducted a geological study of unusual tsunami deposits in the Nemuro lowland, along the Pacific coast of eastern Hokkaido, to obtain information useful for planning for tsunami disaster mitigation.
2. On the basis of sedimentary descriptions, tephrochronology, and AMS ^{14}C and OSL dating results, we identified 16 tsunami deposits in Nanbuto marsh and estimated the unusual tsunami recurrence interval to be generally 220–379 years during the last 5000 years.
3. We described internal sedimentary structures such as plane beds, dunes, and current ripples that indicate bedload transport. Within each bed, dune forms and current ripples show both the inflow and outflow directions of the tsunami inundation.
4. The geological information obtained by this study is valuable for mitigation of tsunami disasters. We show two examples. An open trench demonstration afforded a good opportunity to present our results to local residents. In addition, we donated large peel samples to local museums for effective tsunami education.

7. Acknowledgements

We thank Prof. Minoru Kasahara (Hokkaido University), Prof. Kenji Satake (The University of Tokyo), Prof. Tsuneto Nagatomo (Nara University of Education), Mr. Masayuki Ishii (Meiji Consultant Co.), Dr. Mitsuru Nakagawa (Geological Survey of Hokkaido, AIST), Mr. Yuji Soeda (Historical Museum of Hokkaido), Mr. Shigeto Inokuma (Nemuro City Museum) provided valuable comments about this study. This research was supported in part by Grants-in-Aid for Scientific Research from the Ministry of Education, Science and Culture of Japan (project leader: Futoshi Nanayama, Nos. 16540423 and 20540450).

8. References

Atwater, B.F., Baker, D., Barnhardt, W.A., Burrell, K.S., Haraguchi, T., Higman, B., Kayen, R.E., Minasian, D., Nakata, T., Satake, K., Shimokawa, K., Takada, K. & Cisternas, M. (2001). Grouted sediment slices show signs of earthquake shaking, *EOS, Trans. Am. Geophys. Union* Vol. 82: 603-608.

Atwater, B.F., Furukawa, R., Hemphill-Haley, E., Ikeda, Y., Kashima, K., Kawase, K., Kelsey., H.M., Moore, A.L., Nanayama, F., Nishimura, Y., Odagiri, S., Ota, Y., Park, S.C., Satake, K., Sawai, Y. & Shimokawa, K. (2004). Seventeenth-century uplift in eastern Hokkaido, Japan, *Holocene* Vol. 14: 487-501.

Bondevik, S., Mangerud, J., Dawson, S., Dawson, A. & Lohne, O. (2003). Record-breaking height for 8000-year-old tsunami in the North Atlantic, *EOS, Trans. Am. Geophys. Union* Vol. 84: 289-300.

Choowong, M., Murakoshi, N., Hisada, K., Charusiri, P., Daorerk, V., Charoentitirat, T., Chutakositkanon, V., Jankaew, K. & Kanjanapayont, P. (2007). Erosion and Deposition by the 2004 Indian Ocean Tsunami in Phuket and Phang-nga Provinces, Thailand, *Journal of Coastal Research* Vol. 23: 1270-1276.

Choowong, M., Murakoshi, N., Hisada, K., Charoentitirat, T., Charusiri, P., Phantuwongraj, S., Wongkok, P., Choowong, A., Subsayjun, R., Chutakositkanon, V., Jankaew, K. & Kanjanapayont, P. (2010). Flow conditions of the 2004 Indian Ocean tsunami in Thailand, inferred from capping bedforms and sedimentary structures, *Terra Nova* Vol. 20: 141-149.

Dawson, A. G. & Stewart, I. (2007). Tsunami deposits in the geological record, *Sedimentary Geology* Vol. 200: 166-183.

Fukusawa, H., Tsukamoto, S., Tsukamoto, H., Ikeda, M., Okamura. M. & Matsuoka, H. (1998). Falling age of Baegdusan-Tomakomai tephra (B-Tm) estimated by using non-glacial varves, *LAGUNA* Vol. 5: 55-62.

Furukawa, K. & Nanayama, F. (2006). "Holocene pyroclastic fall deposits along the Pacific coastal region of eastern Hokkaido," *Bull. Volcanol. Soc. Jpn.* Vol. 51: 351-371.

Furukawa, K., Yoshimoto, M., Yamagata, K., Wada, K. & Ui, T. (1997). Did Hokkaido Komagatake Volcano erupt in 1694? Reappraisal of the eruptive ages of 17–18th centuries in Hokkaido, *Bull. Volcanol. Soc. Jpn.* Vol. 42: 269-279.

Furuta, T., Fujioka, K. & Arai, F. (1986). Widespread submarine tephras around Japan—petrographic and chemical properties, *Marine Geology* Vol. 72: 125-142.

Gelfenbaum, G., & Jaffe, B. (2003). Erosion and sedimentation from the 17 July 1998 Papua New Guinea tsunami, *Pure Appl. Geophys.* Vol. 160: 1969-1999.

Goff, J., Liu, P. L-F., Higman, B., Morton, R., Jaffe, B. E., Fernando, H., Lynett, P., Fritz, H., & Synolakis (2006). Sri Lanka Field Survey after the December 2004 Indian Ocean Tsunami, *Earthquake Spectra* Vol. 22: 155–172.

Hawkes, A.D., Bird, M., Cowie, S., Grundy-Warr, C., Horton, B.P., Hwai, A.T.S., Lisa Law, Macgregor, C., Nott, J., Ong, J.E., Rigg, J., Robinson, R., Tan-Mullins, M., Sa, T.T., Yasin, Z., Aik, L.W. (2007). Sediments deposited by the 2004 Indian Ocean Tsunami along the Malaysia–Thailand Peninsula, *Marine Geology* Vol. 242: 169–190.

Hori, K., Kuzumoto, R., Hirouchi, D., Umitsu, M., Janjirawuttikuld, N. & Patanakanogd, B. (2007). Horizontal and vertical variation of 2004 Indian tsunami deposits: An example of two transects along the western coast of Thailand, *Marine Geology* Vol. 239: 163-172.

Hirakawa, K., Nakamura, Y. & Echigo, T. (2000). Giant tsunami along the Pacific coast of the Tokachi region, *Gekkan Chikyu, supplement* No. 31: 92–98.

Moore, A., Nishimura, Y., Gelfenbaum, G., Kamataki, T. & Triyono, R. (2006). Sedimentary deposits of the 26 December 2004 tsunami on the northwest coast of Aceh, Indonesia, *Earth Planets Space* Vol. 58, 253–258.

Nanayama, F. & Shigeno, K. (2006). Inflow and outflow facies from the 1993 tsunami in southwest Hokkaido, Sedimentary Geology 187, 139-158.

Nanayama, F., Satake, K., Furukawa, R., Shimokawa, K., Atwater, B.F., Shigeno, K. & Yamaki, S. (2003). Unusually large earthquakes inferred from tsunami deposits along the Kuril trench, Nature 424, 660-663.

Nanayama, F., Furukawa, R., Shigeno, K., Makino, A., Soeda, Y. & Igarashi, Y. (2007). Unusually nine large tsunami deposits from the past 4000 years at Kiritappu marsh along the southern Kuril Trench, *Sedimentary Geology* Vol. 200: 275-294.

Nanayama, F., Inokuma, S., Furukawa, R., Shigeno, K., Kitazawa, T. & Nakagawa, M., (2008). Stratigraphy of large tsunami traces in Nemuro coastal area along the Kuril subduction zone, in Wallendorf, L. et al. (ed.), *Solutions to Coastal Disasters 2008 tsunamis*, Published by the American Society of Civil Engineers, ISBN 978-0-7844-0978-7: 224-234.

Nanayama, F., Nagatomo, T., Shitaoka, Y., Suzaki, K., Furukawa, R, Shigeno, K., Ishii, M., Inokuma, S., Kitazawa, T. & Nakagawa, M. (2009). The experiment on dating of the tsunami deposits using the luminescence method: A case study of samples from Nemuro coastal lowland, eastern Hokkaido, Jour. Geol. Soc. Japan 115, 249-260.

Nishimura, Y., Miyaji, N., Yoshida, M., Murata, T. & Nakagawa, M. (2000). The 1843 tsunami deposits found in the peat deposit at Kiritappu marsh, eastern Hokkaido, Japan, *Daiyonki Kenkyu* Vol. 39: 451–460.

Okumura, K. (1996). Tephrochronology, correlation, and deformation of marine terraces in eastern Hokkaido, Japan, *Geogr. Rep. Tokyo Metropol. Univ.* Vol. 31: 19-26.

Reimer P.J., Baillie, MGL., Bard, E., Bayliss, A., Beck, J.W., Bertrand, C., Blackwell, P.G., Buck, C.E., Burr, G., Cutler, K.B, Damon, P.E., Edwards, R.L., Fairbanks, R.G., Friedrich, M., Guilderson, T.P., Hughen, K.A., Kromer, B., McCormac, F.G., Manning, S., Ramsey, C.B., Reimer, R.W., Remmele, S., Southon, J.R., Stuiver, M., Talamo, S., Taylor, F.W., van der Plicht, J., & Weyhenmeyer, C.E. (2004). IntCal04 terrestrial radiocarbon age calibration, 0-26 cal kyr BP, *Radiocarbon* Vol. 46: 1029-1058.

Satake, K., Nanayama, F., Yamaki, S., Tanioka, Y. & Hirata, K. (2005). Earthquakes and tsunamis along the Kuril trench in the 17th through 21st centuries, Tsunamis: their sources, monitoring and coastal hazard, in Satake, K. (ed.), *Tsunami Research: Case Studies and Recent Developments*, Springer: 157-170.

Satake, K., Nanayama, F. & Yamaki, S. (2008). Fault models of unusual tsunami in the 17th century along the Kuril trench, *Earth Planets Space* Vol. 60: 925–935.

Sawai, Y. (2002). Evidence for 17th-century tsunamis generated on the Kuril-Kamchatka subduction zone, Lake Tokotan, Hokkaido, Japan, *Journal of Asian Earth Science* Vol. 20: 903–911.

Sawai, Y., Satake, K., Kamataki, T., Nasu, H., Shishikura, M., Atwater, B.F., Horton, B.P., Kelsey, H.M., Nagumo, T. & Yamaguchi, M. (2004). Transient uplift after a 17th-century earthquake along the Kuril subduction zone, Science 306, 1918-1920.

Sawai, Y., Kamataki, T., Shishikura, M., Nasu, H., Okamura, Y., Satake, K., Thomson, K.H., Matsumoto, D., Fujii, Y., Komatsubara, J. & Aung, T.T. (2009). Aperiodic recurrence of geologically recorded tsunamis during the past 5500 years in eastern Hokkaido, Japan, *Journal of Geophysical Research* Vol. 114, B01319, doi:10.1029/2007JB005503.

Soeda, Y., Nanayama, F., Shigeno, K., Furukawa, R., Kumasaki, N. & Ishii, M. (2004). Large prehistorical tsunami traces at the historical site of Kokutaiji Temple and Shiomigawa lowland, eastern Hokkaido: significance of sedimentological and diatom analyses for recognition of past tsunami deposits, *Mem. Geol. Soc. Jpn.*, No. 58: 63-75.

Tanioka, Y., Nishimura, Y., Hirakawa, K., Imamura, F., Abe, I., Abe, Y., Shindou, K., Matsutomi, H., Takahashi, T., Imai, K., Harada, K., Namegawa, Y., Hasegawa, Y., Hayashi, Y., Nanayama, F., Kamataki, T., Kawata, Y., Fukasawa, Y., Koshimura, S., Hada, Y., Azumai, Y., Hirata, K., Tamikawa, A., Yoshikawa, A., Shiga, T. & Masaka, S. (2004). Tsunami run-up heights of the 2003 Tokachi-oki earthquake, *Earth Planet Space* Vol. 56: 359-365.

Tsukamoto, S. & Iwata, S. (2005). Recent development in luminescence dating; on the extending the age range and the behavior of different OSL components from quartz, *Jour. Geol. Soc. Japan* Vol. 111, 643-653.

Yamamoto, T., Itho, J., Nakagawa, M., Hasegawa, K. & Kishimoto, H. (2010). 14C ages for the ejecta from Kutcharo and Mashu calderas, eastern Hokkaido, Japan, *Bull. Geol. Survey. Japan* Vol. 61, 161-170.

15

Sand Sheets on a Beach-Ridge Plain in Thailand: Identification and Dating of Tsunami Deposits in a Far-Field Tropical Setting

Kruawun Jankaew[1], Maria E. Martin[2],
Yuki Sawai[3] and Amy L. Prendergast[4]
[1]*Department of Geology, Faculty of Science, Chulalongkorn University, Bangkok 10330*
[2]*Department of Earth and Space Sciences, University of Washington, Seattle, Washington 98195-1310*
[3]*Geological Survey of Japan, National Institute of Advanced Industrial Science and Technology (AIST), Tsukuba 305-8567*
[4]*Department of Archeology, University of Cambridge, Downing Street, Cambridge CB2 3DZ*
[1]*Thailand*
[2]*USA*
[3]*Japan*
[4]*UK*

1. Introduction

Surveys of recent tsunamis, especially the 2004 Indian Ocean tsunami, demonstrate that tsunami deposits are different from place to place and that tsunami geology is very complex. This complexity has increased the challenges of identifying and interpreting of past tsunamis in many parts of the world as no set of criteria are suitable in all locations. In this chapter we present a case study of tsunami deposits, both modern and ancient, from Thailand. We discuss further complications and challenges associated with identifying and dating paleotsunami events in this tropical setting that include post-depositional change, the absence of marine microfossils and lack of appropriate materials for dating. Preservation potential of 2004 tsunami deposit in Thailand will also be considered.
2004 Sumatra-Andaman earthquake and associated tsunami seemed unprecedented as there was no written or instrumental record of great earthquakes (M > 8) in this region. No event comparable to the 2004 tsunami is recorded in known historical sources from the Indian Ocean region (Dominey-Howes et al., 2007). Surviving historical accounts in Thailand, mainly from European visitors and Indian and Chinese traders to the Siam Kingdom (Thailand) via Takola- or later known as Takau Pa, a major trading southern coastal town, also lack identified accounts of a large tsunami affecting the Thailand Andaman Coast. Accordingly, prior to the 2004 tsunami, Thailand had almost no awareness of tsunami hazards, let alone any warning system or mitigation measures in place. This situation contributed to the loss of life in Thailand during the 2004 tsunami.

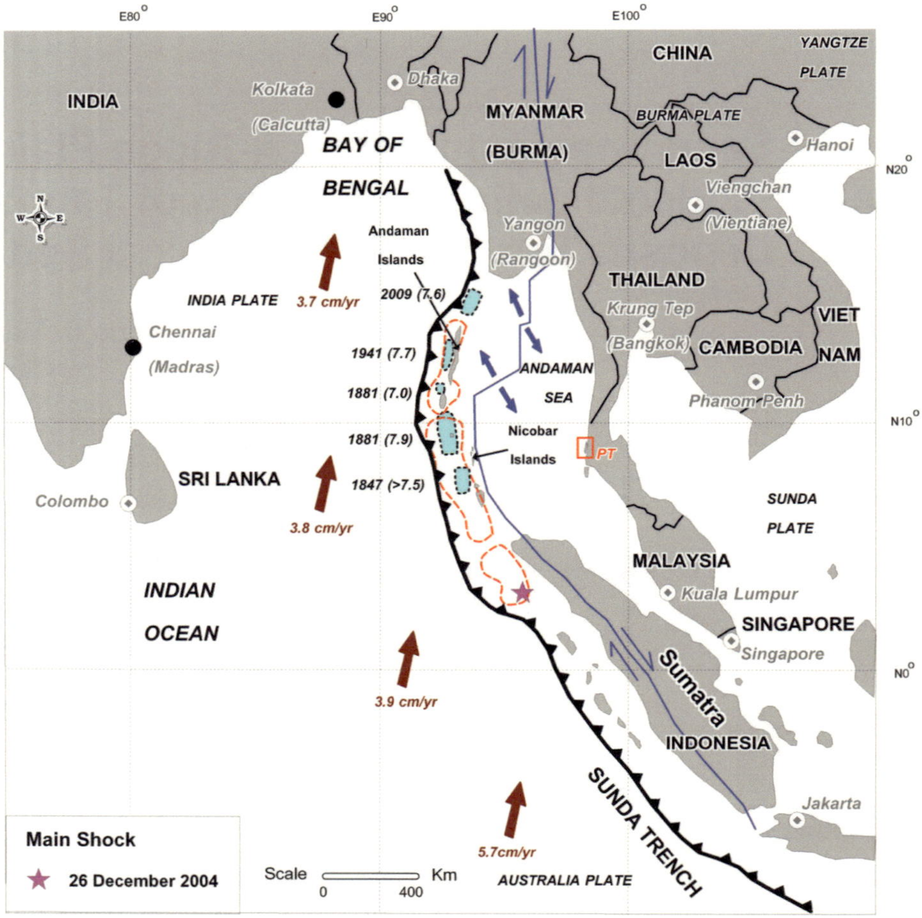

Fig. 1. Setting and rupture areas of large and giant earthquakes in the Sumatra-Andaman area. Star is 26 December 2004 earthquake epicenter. Red dash defines the 2004 rupture area (from Chlieh et al., 2007). Plate convergence rates are from Chlieh et al. (2007). The areas in light blue are rupture areas of historical earthquakes (from Ortiz & Bilham, 2003; Bilham et al. 2005). PT is Phra Thong Island.

It had been thought that subduction zones capable of producing great earthquakes are those with young subducting plates and fast subduction rates (Ruff & Kanomari, 1980). But after the 2004 Sumatra-Andaman earthquake, which occurred where old plates have modest subduction rates, Stein & Okal (2007) have shown that plate ages and subduction rate do not fully predict earthquake magnitudes on subduction zones.

Earthquake magnitude and recurrence interval are important information needed for hazard mitigation. Estimation of earthquake size and tsunami recurrence intervals can be obtained from geophysical studies alone or in combination with paleoseismological studies. Current geophysical estimates of great earthquake recurrence intervals along the Sunda Trench are 140-420 yr (Chlieh et al., 2007) and 200-1000 yr (Stein & Okal, 2007). The large variation in the range is due to the complex and poorly understood tectonic setting of this subduction zone (Stein & Okal, 2007).

A history of past earthquakes and resultant tsunamis can be extended into the prehistoric past by means of geology. Paleoseismological records can be derived from ages and heights of uplifted terraces (e.g. Aung et al. 2008; Rajendran et al., 2008) and ages and level of vertical movement recorded in coral microatolls (e.g. Sieh et al., 2008; Meltzner et al., 2010) above or near the active subduction zone. Paleotsunami deposits are sediments, typically sand-sized but grain size can range up to boulders, moved and left behind by tsunamis (Dawson and Shi, 2000; Bourgeois, 2009). Tsunami deposits can be found both near and far from the tsunami source (Dawson et al., 1988). Paleoseismological and paleotsunami records, if correctly identified and dated, can extend the chronology of earthquake and tsunami of the area for several thousand of years beyond the historic time (e.g. Kelsey et al., 2005).

To improve understanding of the risk from large Indian Ocean tsunamis, there is a need to study tsunami deposits in order to extend the tsunami record in this region. Thailand is situated about 600 km east of the northern Sunda subduction zone-the main and probably the only tsunami source for Thailand. We aim to understand the characteristics of the 2004 tsunami deposits as well as searching for paleotsunami deposits in Thailand. Once identified with confidence, the ages of tsunami deposits is hoped to extend the chronology of large earthquake and associated tsunamis in Thailand. Satake & Atwater (2007) were able to show--using examples of written history in combination with geological study from Chile, Cascadia and Japan-- that earthquake sizes and repeat time along the Pacific Realm subduction zone are varied (non predictable, non characteristic). However, at present, data available on the history and behavior of the Sunda subduction zone is rather scarce. Ages of tsunami deposits from Thailand will help characterize the long-term behavior of the Sunda subduction zone and provide recurrence estimates that can be used to support future tsunami mitigation and risk assessment of the country.

Importance of paleotsunami study of this region

The historical record of large earthquakes and tsunamis in the Indian Ocean is rather brief, as written history of Banda Aceh only extends back to c.1530 AD (Reid, 2005). Long-term behavior of this subduction zone is still obscure; hence risk from tsunami of this region is poorly understood. Earthquake source characteristics, size and recurrence are also not comprehend. Prior to 2004, historical records of large earthquakes in the northern Sunda subduction zone contained no earthquakes bigger than Mw 8 (Fig.1); M 7.5-7.9 (1847, north of Little Nicobar), 7.0 and 7.9 (1881, Car Nicobar) and 7.7 (1941, Middle Andaman) (Ortiz & Bilham 2003). On 10 August 2009 there was an earthquake of magnitude 7.6 with an epicenter reported to be just north of 1941 rupture area.

2. Previous works

Background research in paleotsunami study

Paleotsunami studies involve identification, mapping, and correlation and dating of tsunami deposits (Bourgeois & Minoura, 1997). Paleotsunami deposits are often identified in low-energy coastal environments, such as tidal marsh, back-barrier marshes and coastal lakes.

Most publications on paleotsunami studies during the past two decades are from temperate zones such as the north Atlantic (Bondevik et al., 1997; Bondevik et al., 2003; Dawson et al., 1988), the circum-Pacific shores of the Americas and Asia (Atwater & Moore, 1992; Atwater et al., 2005; Bourgeois et al., 2006; Cisternas et al., 2005; Fiedorowicz & Peterson, 2002; Kelsey et al., 2005; Minoura & Nakata, 1994; Nanayama et al., 2003; Tuttle et al., 2004) and New

Zealand (Chague-Goff et al.,2000; Goff et al., 2001; Goff et al., 1998; Grauert et al., 2001). In many cases tsunami studies occur in conjunction with other evidence for large earthquakes such as the study of coastal geological evidence of past tsunamis and associated post seismic uplift in Japan by Sawai et al. (2009b).

Publications describing historic tsunami deposits from the tropics include the 1883 Krakatau tsunami (Simkin & Fiske, 1983), 1992 East Java tsunami (Shi et al., 1995), 1998 Papua New Guinea tsunami (McSaveney et al., 2000; Gelfenbaum & Jaffe, 2003), and 2006 West Java tsunami (Shi et al., 1995).

2004 Sumatra-Andaman tsunami deposits were examined around the Indian Ocean coasts by many groups of scientist (e.g. Goff et al., 2006; Jaffe et al., 2006; Moore et al., 2006; Paris et al., 2010; Singarasubramanian et al., 2006, etc.). In Thailand, 2004 tsunami deposits were investigated by Choowong et al. (2007); (2008a); (2008b); Fujino et al. (2008); Hawke et al. (2007); Hori et al., (2007) and Sawai et al. (2009a) among others. Preliminary results of paleotsunami study in Thailand are reported by Fujino et al. (2008; 2009) and Jankaew et al. (2008). The 2004 tsunami height at several locations along Thailand Andaman coast was surveyed by Tsuji et al. (2006).

Following the 2004 Sumatra-Andaman tsunami there were a few publications detailing paleotsunami study in this area, including Fujino et al., (2009); Jankaew et al., (2008); Monecke et al. (2008); Rajendran et al., (2006). Jankaew et al., (2008) reported at least three sand layers, which are likely to have been deposited by the predecessors of the 2004 tsunami. Using radiocarbon dating, the youngest sand layer (sand layer B) postdates AD 1300-1450. This age overlaps with the age of the tsunami sand layer from Meulaboh, Sumatra by Monecke et al. (2008) and may be correlative to the age of a shell bed deposit of 500 years found in a cave near Ao Nang, southeast of Phuket reported by Harper (2005). The ages of the two lower paleo-sand layers (sand layers C and D in Jankaew et al., 2008) are not well constrained but ages of non-abraded shells in tidal flat sediments underlying these sand layers suggested that both layers were deposited after 2,500-2,800 years ago (Jankaew et al., 2008).

3. Methods

We have been searching for evidence of past tsunamis in the Thai coastal areas inundated by the 2004 tsunami. Sites investigated include Kamala Beach and Le Pang Bay of Phuket Province, Thap Lamu, Pakarang Cape, Ban Nam Khem tidal inlet, Kho Khao Island and Phra Thong Island of Phang Nga Province (Fig. 2). We cored and dug small pits and trenches on the coastal low-lying area including tidal flat, back beach lagoon and mangrove fringed tidal inlet.

Moist environments such as beach ridge swales and coastal lakes preserve tsunami deposits in other environments (e.g. Pinegina et al., 2003; Kelsey et al., 2005) and in Thailand the equivalent environments are beach ridge swales and mangrove swamps.

In our fieldwork we learned that the 2004 tsunami deposits are not well preserved in mangrove environment, mainly because of burrowing by crabs. The 2004 tsunami sands are already being muddled up by these organisms causing low or no tendency of these sands being preserved as a paleotsunami deposit for future observation. Preservation potential of previous tsunamis along the Andaman coast of Thailand is further jeopardized by the extensive placer tin mining, coastal developments and other human disturbance of the area. Thus far, only on nearly undisturbed beach-ridge plains at Phra Thong Island has it been possible to find numerous sites for preserving geological records of pre-historic tsunamis in Thailand.

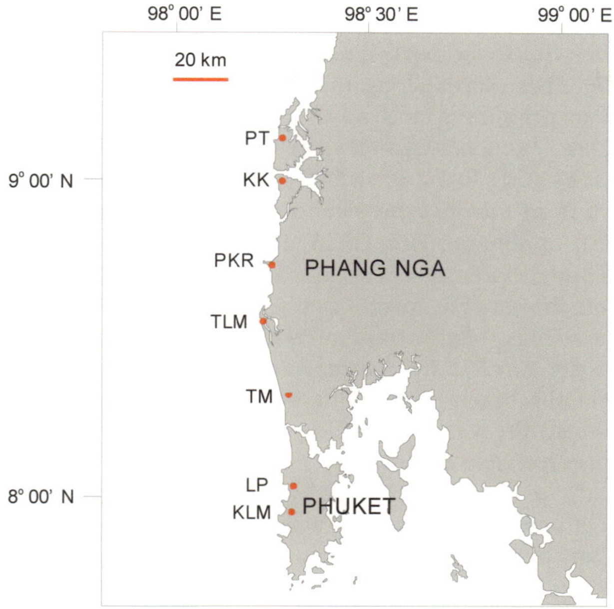

Fig. 2. Map shows locations along the Andaman Coast of Thailand that we have investigated. KLM= Kamala Beach, LP = Le Pang Bay, TM = Tai Maung, TLM = Tap Lamu, PKR = Pakarang Cape, KK = Kho Khao Island, PT = Phra Thong Island

4. Setting of Phra Thong Island

Phra Thong Island is located about 125 km north of Phuket. The island is about 10 km wide and 15 km long, facing the Andaman Sea without any obstruction. It is completely cut off from the mainland by tidal channels. There is no running river on the island except some small tidal channels. Due to its location of just north of the Equator, Phra Thong's exposure to tropical cyclones is probably limited to typhoons that have been weakened by crossing the Thai-Malay Peninsula (Murty & Flather, 1994) for example the case of Typhoon Gay in 1989.. Indian Ocean cyclones that originate near the Equator typically moved west- and northwestward towards India, Bangladesh and Myanmar.

Phra Thong is subjected to inundation by tsunamis due to its relatively flat terrain and its exposure to transoceanic tsunamis from the Northern Sunda Trench. The western half of the island comprises series of beach ridges and swales, extending over 2 km inland. The beach-ridge series is believed to have been formed by prograding beach as a result of the receding sea level following a early-mid Holocene sea level highstand. However, currently there is little information on the numerical ages of these ridges. The eastern half of the island is covered with mangroves which extend from the tidal channels. Tidal range average in this area is about 1.8 meters, with a maximum of 2.7 meters.

Phra Thong History

The name Koh Phra Thong (Phra Thong Island) appeared for the first time in Chotmaihet (account or report) of a visit to Malaya peninsula by King Rama V during 16-22 April 1890 but there was no mention if there was settlement on Phra Thong during that time. It was believed that the first settlement on Phra Thong was by sea gypsies. During 1797-1853, Chinese workers migrated to Phra Thong due to the expansion of tin mining in Southern Thailand.

Phra Thong, as with other places along the Andaman Coast of Thailand but to a lesser degree, was subject to tin-mining and other human disturbance. The settlement and tin mining activity on Phra Thong has existed since before 1909 as described in the Chotmaihet of Southern Thai visit by King Rama VI in 1909. The earliest period of tin mining is on Koh Chad, eastern part of Phra Thong, where concession was granted by the government. The three main sites of tin mining are Koh Chad, Tung Oon Nua and Koh Ra. On the beach-ridge plain of Phra Thong, evidence left by small-scale tin-mining can still be seen (Figs. 3-4) and easily spotted on the satellite images and on the ground by the steep-walled ponds, typically dug into the swales, next to large piles of spoil sand. Tin mining in the beach-ridge plains could only be done in the dry season as during the rainy season (April-November) the swales of the island's beach-ridge plain are submerged. After 1972, mining suffered because the easily mined tin was diminishing and more importantly, the price of tin was dropping making it unprofitable to mine.

At Phra Thong Island, we studied stratigraphy and sediment characteristics from shallow pits and trenches, avoiding the tin mining areas. Maximum depth of the pits dug at a given location is often defined by the ground water level and the stratigraphy. Ground water level at Phra Thong is relatively high even in the dry season, so most of the pits are less than 1m deep.

Fig. 3. Pond dug by tin miners to use as a source of water for placer mining. It has steep slopes compared to a natural swale. The pond is also accompanied by spoil piles of sand.

Fig. 4. Pile of sand, about 1.5 meter high, left behind after separation from tin. The less vegetation covering the tops of these spoil piles made them vulnerable to transport by the 2004 tsunami. Next to the sand pile there is a man-made pond.

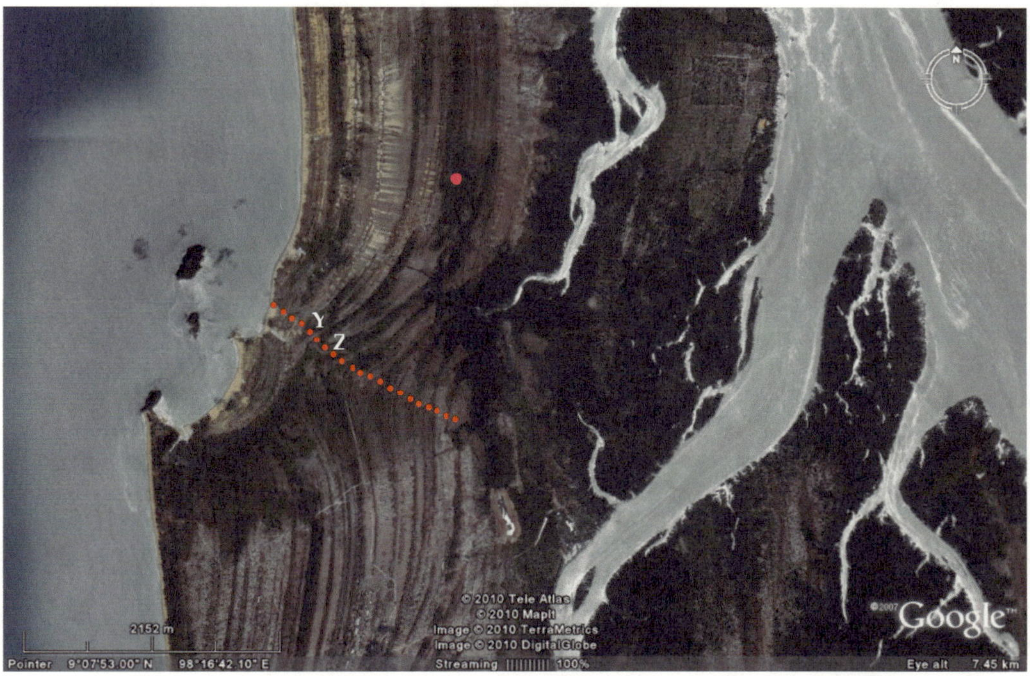

Fig. 5. Google Earth image showing locations of pits (small dots) and trench (large dot) investigated for 2004 and paleotsunami sand sheets as described in this chapter.

Tsunami sediments were collected from pits and trench walls for grain size analysis, at 0.5 cm intervals. Grain size analysis was conducted using 1.5-metre long settling tube, and grain size distribution data at 0.1 phi interval was obtained. Organic materials at or near the contact between tsunami deposits and underlying soils, where present, were collected for 14C dating. Non-abraded shells contained in the tidal flat sand at the base of the swale, where present, were also collected for 14C dating. More recently, we have been collecting OSL samples of the sand sheets and bounding beach ridges. The OSL beach-ridge ages will be used to test their geomorphic superposition and provide age bracketing for the tsunami sand sheet. Fig.5 shows locations of shallow pits along the 2-km transect from the shoreline (NW) to the end of 2004 sedimentation (SE). We studied sedimentary characteristics of 2004 and paleotsunami sand layers along this transect. Moreover, we also described a section further north of a transect in Fig.12.

5. Results

Thickness of deposits of the 2004 tsunami and its predecessors varies across the beach-ridge plain and is primarily controlled by local topography. Thickness tends to be greatest in a low-lying swale (as much as 30 cm), whereas in ridges and higher ground it tends to be thinner (few cm, 10 cm at most). Thicker, and sometimes coarser, deposits in topographic lows is a response of the flow to the deeper water depth, which slows the flow and drive deposition (Apotsos et al., 2009). In the case of Phra Thong, thickness is increased if the swale is a path for back flow, which is often channelized and can eroded sediments from landward location as shown in Fig 6. 2004 deposit in these locations are almost 30 cm thick. The deposit contains up to 4 fining-upward sequences. The lowest fining-upward sequence was probably deposited out of the first wave. On top of this sequence there is a continuous dark gray fine silt-clay layer which was probably deposited when the tsunami flow was still and between the two main waves. Above this silty clay layer there are 2-3 fining upward sequences, which were possibly deposited from back flow of the first or second waves. Sets of cross bedding in these upper sequences indicate a flow direction towards the sea (Fig. 6).
On high ground, with less vegetation cover, the 2004 tsunami deposit lies above the sand ridge sediment with faint soil in between (Fig. 7). It is still possible to identify the 2004 tsunami deposits in the high ground 5 years after the event, but allowing more time differentiating it from the sand ridge below will be problematic. At places, distinguishing the 2004 tsunami laid sand and the underlying ridge sand was already difficult due to the lack of organic soil formation on top of the ridge sand. On the high ground right next to the swale, after tsunami dumped majority of sediments in the swale, 2004 sediments compose mainly of silt which fell out of suspension. The tsunami flow then picked up more sediments from the ridge area along its flow path. As a result, grain size of 2004 tsunami on high ground is composed mainly of silt in contrast to sand and silt in the swale deposit.
Topographic profile across the transect in Fig 5. is shown in Fig. 12. The thicknesses of the 2004 deposit at different locations are also shown.
Among more than 200 hand-dug pits, at least 4 paleotsunamis sand layers have been found on Phra Thong Island. These deposits are in the fresh-water swales between beach ridges. The marshy swales of Phra Thong Island serve as an excellent recorder for the tsunami sand layers because they contain peaty soil that gives contrast in colour and grain size to the tsunami sediments, and which protects tsunami deposits by building up on top of them. The fresh water of these swales excludes crabs that would otherwise destroy the sand layers by mixing them with the peat.

Fig. 6. Thick 2004 deposit of almost 30cm is found at the landward location of swale Z in Fig.5. This location is a landward side of the swale, which is in the main direction of the tsunami back flow. Pre-2004 sand layer is also present at this location, though when compared with the 2004 it is very thin (2-3cm) and discontinuous.

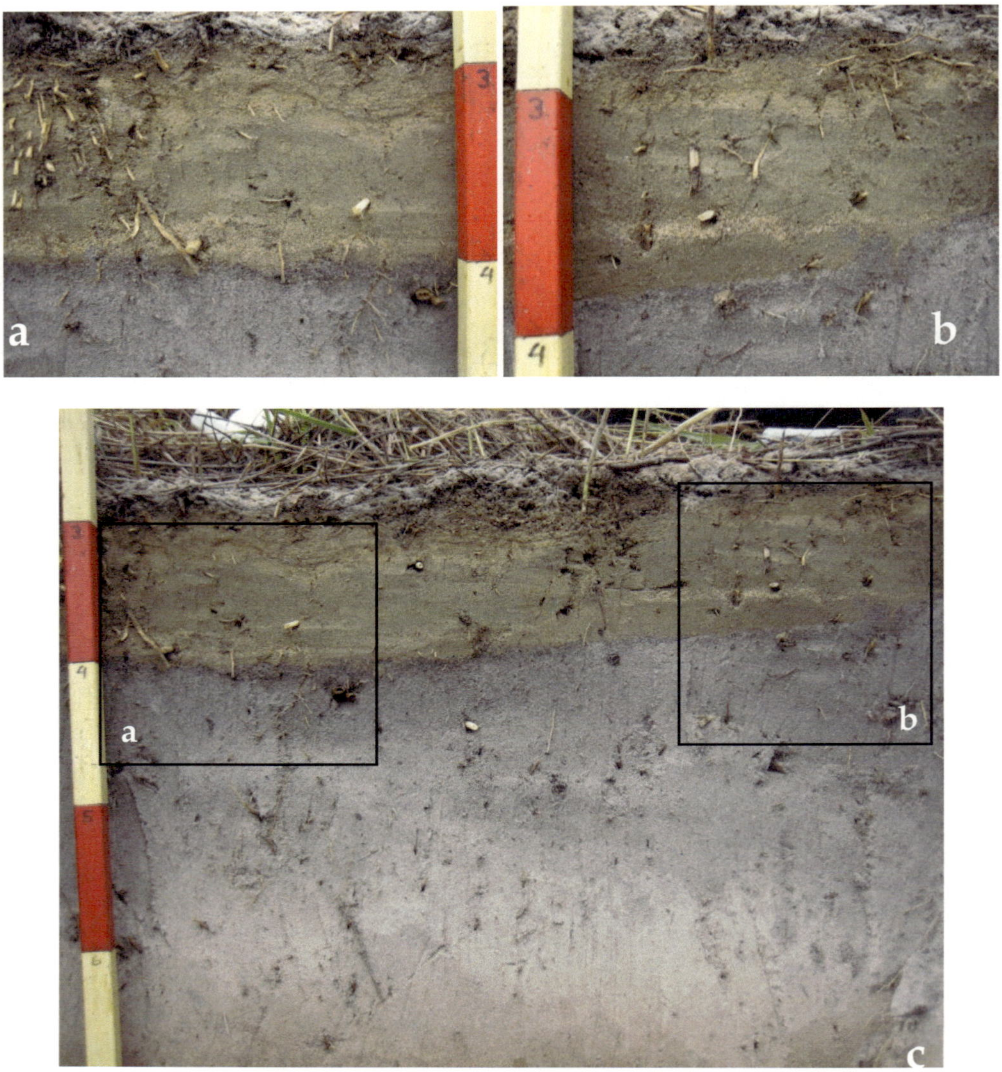

Fig. 7. Thick deposit of 2004 tsunami sediments on the sand ridge with only thin faint organic soil between them.

5.1 Sedimentary characteristics of Phra Thong Tsunami deposit

Apart from variation in thickness of the deposit, internal characteristics of 2004 deposit on Phra Thong also varies greatly from one place to another. Often the deposits show no internal structure layering. Typically the deposits appear as a massive bed or faintly normally graded. In locations close to the sea, some deposits coarsened upward. Examples of sediment grain size and composition described below are from swale Y (see Fig. 5).

a. 2004 Tsunami Deposit

2004 tsunami sediments range in grain size from coarse sand to coarse silt, with distinctive bimodal distribution with first mode in fine-sized range (2 phi), and second and highest mode in very fine-size range (phi 3.3) (Fig. 8A). The coarser sediments are composed of coarser sands from beach berm and possibly from offshore area, and angular shell fragments and large forams and other macrofossils (Fig 9, top row), whereas the finer sediments are possibly from the subtidal zone. 2004 sediment composes about 85% of clear to white quartz grains, 8% of shell fragments and micro fossils, 4% of muscovite and 3% of heavy minerals, mostly small grains of tins. Although typically 2004 tsunami deposits at Phra Thong appear as a massive bed or faintly normally graded, detailed grain size analysis shows that they can contain 2-3 fining-upward sequences (Fig. 10). Mean grain size of the 2004 tsunami is slightly bigger than that of paleotsunami sand (sand B).

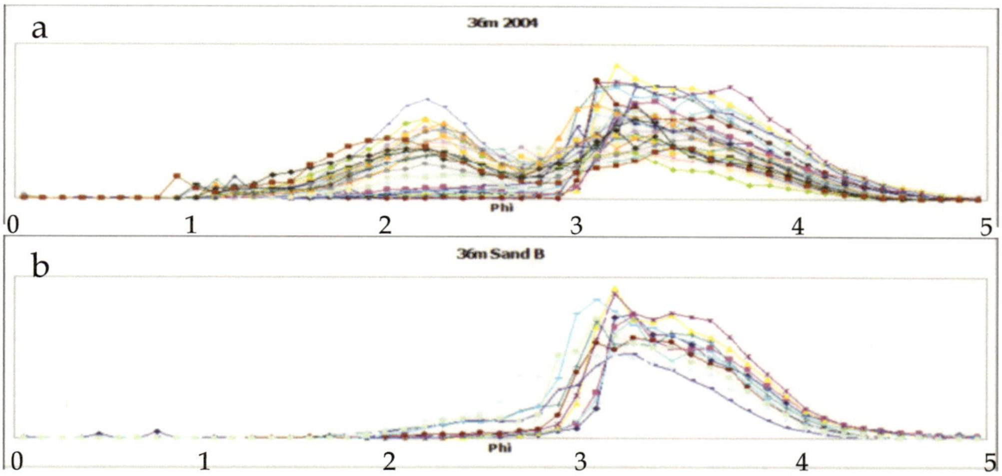

Fig. 8. Sediment grain size distributions of (a) 2004 tsunami deposit compared with (b) sand layer B. Each line represent sample of 0.5 cm interval.

Fig. 9. Photograph of sediments of 2004 tsunami (top row) compare with paleotsunami sand layer B (bottom row). The scale bar, bottom right, is 1 mm.

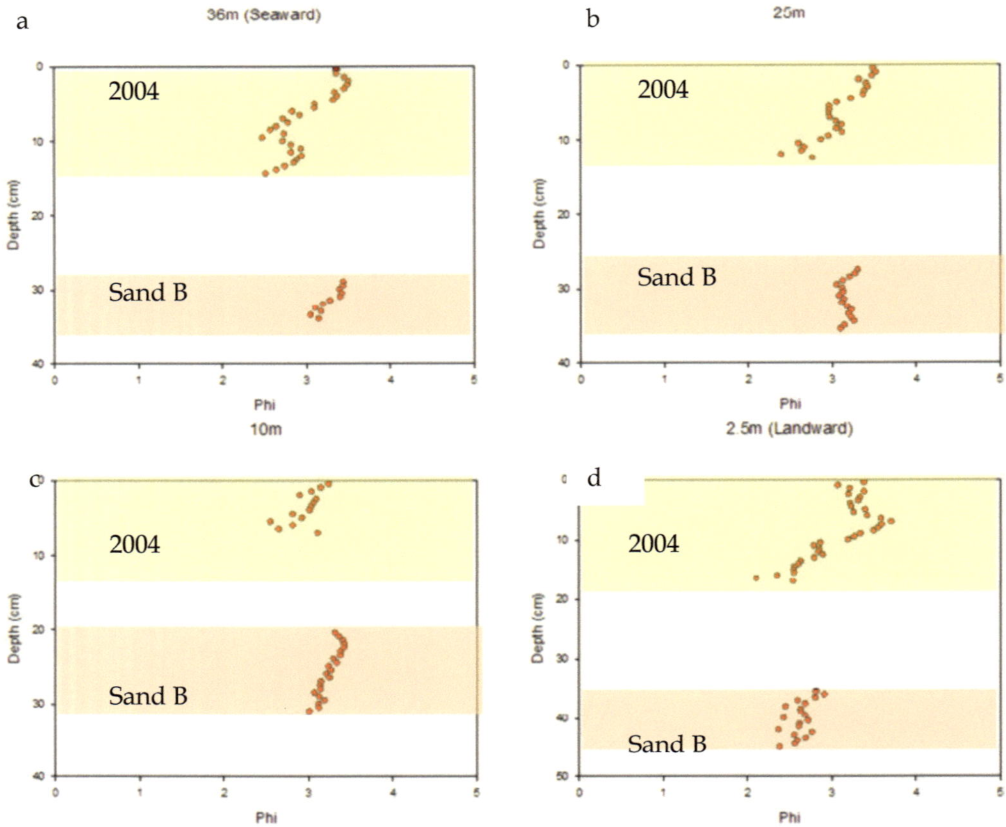

Fig. 10. Plots of mean grain size of 2004 tsunami deposit compared with paleotsunami deposit (sand B) at different locations along the same swale. Each data point represents a sample at 0.5 cm intervals. Thicknesses of 2004 and sand B are varied along the swale's width. 2004 tsunami layer contains 2-3 fining-upward sequences while sand B deposit contains 2 fining-upward sequences. Mean grain size of the 2004 tsunami is slightly larger than that of sand B.

b. Paleotsunami Sand Layers

Paleotsunami deposits at Phra Thong Island are composed of grains ranging in size from coarse sand to coarse silt, with bimodal distribution with small first mode in fine-sized range (2.3 phi), and second and highest mode in very fine-size range (phi 3.2) (Fig. 8B). Sediment of sand B composes about 91% of clear to white quartz grains, 7% of muscovite and 2% of heavy minerals, mostly small grains of tin. The shell fragments and microfossils are absent in the paleosand layer. The cause of the absence will be discussed further below. Phra Thong paleotsunami deposits appear as a massive bed or faintly normally graded, more so than the 2004 deposits, but mean grain size at 0.5 cm thickness intervals show that they contain up to two fining-upward sequences (Fig. 10).

5.2 Complications and challenges in studying Phra Thong Paleotsunami deposits

5.2.1 Identification

Challenges in studying paleotsunami deposits world-wide include positive identification, and dating and correlation of events (Bourgeois & Minoura, 1997). Tsunami deposits are different from place to place depending on the sediment source, nearshore bathymetry, coastal topography and tsunami intensity. Tsunami deposits are only found in locations inundated by the tsunami that were favorable to their deposition and preservation, and coastal conditions at the time of deposition may be different than current conditions. At Phra Thong, paleotsunami deposits are found in fresh water swales between beach ridges. Identification of these deposits is based on their sedimentary character and stratigraphic context.

Marine and brackish diatom and foraminifera contained in tsunami deposits help in identifying the marine origin of the sediments, and in many cases can discriminate between tsunami and storm deposits. They are especially helpful in paleoseismology and paleotsunami studies in temperate climates.

Both diatom and foraminifera are abundantly reported in 2004 tsunami deposits (e.g. Hawke et al., 2007, Sawai et al., 2009a). However, diatoms and foraminifera are absent in the paleotsunami deposits at Phra Thong (Jankaew et al., 2008). High underground water temperature in the tropical setting may have caused dissolution of their carbonate and opaline tests. The absence of diatoms and foraminifera in old tsunami deposits in Thailand deprives us of a simple way to confirm that an inferred tsunami deposit was in fact deposited by sea water.

Non-occurrence or non-recognition of tsunami deposits at a given locality does not mean that there was no tsunami inundation in the past (Bourgeois & Minoura, 1997). This is also the case of Phra Thong tsunami deposits as best shown by example in Fig. 11. At this location, the large dot in Fig. 5, there are at least three paleotsunami sand layers present. However, tracing these sand layers across the trench wall of less than 10-metre distance prove that correlating these sand layers are indeed challenging. The second paleo-sand layer is absent abruptly at various locations along the trench wall. The local disturbance of sand layers after deposition, by animals and other means, causes discontinuities in the deposit. This demonstrates the ambiguity of interpreting and correlating paleotsunami deposits based on core data only. The stratigraphy exposed at this trench also attests to the challenge in correlating paleotsunami deposits in the far-field tropical setting without reliable ages.

5.2.2 Dating

14C dating

Sampling of appropriate materials for radiocarbon dating, a more acceptable dating method, can be problematic (Bourgeois & Minoura, 1997). Organic materials that are alive and then killed by the event, as opposed to transported of older or younger materials, are not always present. Radiocarbon dates of organic materials within or directly underneath the tsunami deposits, thus, represent only a maximum possible age for the tsunami event because they may have been reworked by the tsunami. Limiting maximum ages of the tsunami sand deposits in eastern Hokkaido, Japan, were obtained by dating plant materials (e.g. fruits, seeds and leaves) using accelerated mass spectrometry (AMS) radiocarbon method (Sawai et al., 2008; 2009b). Cisternas et al. (2005) also dated entombed stems of herbaceous plants, using AMS radiocarbon methods, to infer the age of tsunami sand layers.

Precise and accurate dates are often difficult to determine using radiocarbon dating (Peters et al., 2007). In Cascadia, for example, a typical error for tsunami deposit ages using radiocarbon techniques is in the order of ±100–200 years, which does not allow discrimination of tsunamis closely spaced in time (Peters et al., 2007). Numerical dating of pre-historic tsunami deposits is also difficult if a time-scale of decades separate between events (Bourgeois & Minoura, 1997).

Tephra layers found in the stratigraphic sections aid in correlation of tsunami sand layers and identifying the relative ages of tsunamis in eastern Hokkaido, Japan (Nanayama et al. 2003; 2007; Sawai et al., 2008; 2009b) and Kamchatka, Russia (Pinegina et al., 2003).

The plant materials and plant rhizome tend to survive better in Quaternary sediments of temperate climates than in those of the tropics. In Thailand, materials suitable for C14 dating are not easy to find due to high bioturbation and biodegradation. From more than 200 pits and trenches that we dug, only a few of the locations contained organic materials suitable for dating. We searched for seeds and fruits in peaty soils between the sand layers, but typically without success.

Our experience with dating detrital leaves formed as lamina within the tsunami sand layer showed that leaf fragments are less reliable (Jankaew et al., 2008). The leaf ages conflict one another and exceed the barks, collected from the base of the sand layer, age. The ages of the paleotsunami sand layers are still unknown, or not well constrained, except the youngest paleo-sand layer, which was inferred to be older than 550-700 years (Jankaew et al., 2008). The challenge in dating these sand layers exists because the organic materials suitable for radiocarbon dating are not always present in the highly-weathered tropical settings.

OSL Dating

Where organic material is not present for radiocarbon dating, optical dating techniques such as optically stimulated luminescence (OSL) are used to date the tsunami deposits, such as in Cascadia by Ollerhead et al. (2001) and Huntley & Clague (1996).

From our preliminary result of OSL dating, both two 2004 tsunami samples exhibited relatively symmetric D_e distributions, suggesting adequate bleaching. Results from multiple grain single-aliquot and single grain samples from the 2004 sand sheet were compared. The multiple grain single-aliquot samples overestimated the age of the 2004 sand sheet by around 100 years (80 ± 10 years and 110 ± 20 years). Single grain analyses showed between 70 to 76 % of near-zero grains with some partially-bleached grains, which, when averaged in the multiple-grain aliquots, generated the older central ages (Prendergast et al., in prep). Examination of the individual equivalent dose estimates in multiple-grain single-aliquots is not an accurate technique for identifying partial bleaching because each aliquot could contain equal proportions of partially bleached quartz (Wallinga, 2002). Using aliquots of single grains overcomes this problem.

In situations where incomplete bleaching is suspected, Olley et al. (1999; 2004a, b) suggest that the lowest equivalent dose will provide the best estimate of burial age. We therefore applied a minimum age model (Galbraith et al., 1999) to calculate the ages of the sand sheets at Phra Thong Island. In well-bleached samples and in those where bioturbation has caused the intrusion of younger and older grains, a central age model (Galbraith et al., 1999) may provide a more accurate estimate of burial age. This model has been applied to the beach ridges samples.

Preliminary results of optically stimulated luminescence (OSL) age dating of the sand sheets and the beach ridges are presented below (Prendergast et al., in prep).

In addition to tsunami sand sheets, bounding beach ridges; west, middle and east, were also dated. The OSL ages of the beach-ridge sequence will help bracketing the age of the tsunami sand layers deposited in the swales and helping correlating tsunami sand layers across the island. Preliminary results of OSL ages of the ridges prove that the beach ridges young seawards, consistent with their interpretation as a prograding sequence; the east ridge (2560

Fig. 11. Trench wall, about 9 meters long, exposing 2004 tsunami deposit (top light colour layer) and up to three paleotsunami layers. The top two paleotsunami sand layers (light colour) are discontinuous, especially the lower one which disappear laterally. Bioturbation and disturbance by animals are thought to be responsible for the discontinuity of this sand layer. The third paleotsunami sand layer is under the water level and only the top edge can be seen in this fig. Stripe on stick is 10 cm.

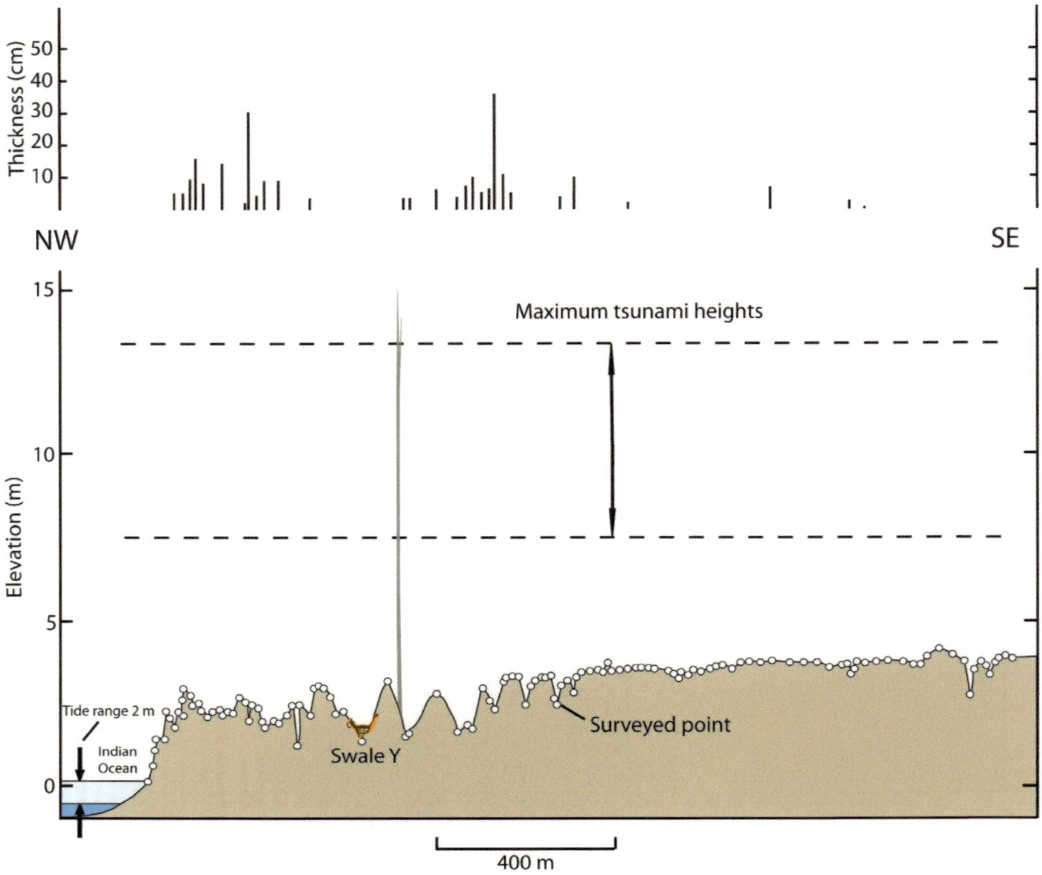

Fig. 12. Topographic profile across the beach-ridge plain (the same transect in Fig. 5). Vertical lines are 2004 tsunami thickness (in centimeter), at different locations along this transect.

± 350 years ago) is older than the middle (2160 ± 280 and 2100 ± 270 years ago) and west ridges (1600± 210 years ago). The nearly identical paired OSL ages from the middle ridge (2160 ± 280 and 2100 ± 270 years) affords confidence in the use of OSL in this environment. Based on the age of the most landward ridge (2560 ± 350 years ago) and its distance from the modern shoreline (500-700 m), the beach ridges prograde at 0.19 - 0.27 m per year by average.

Three tsunami sand layers (B, C and D) have OSL ages of 350 ± 50 - 380 ± 50; 990 ± 130 - 1410 ± 190, and 2100 ± 260 years ago (at 1-sigma precision), while tidal-flat deposit under these sand layers has an OSL age of 2540 ± 340 years ago. The ages are concordant and give the impression that at least three large tsunamis impacted the coast during the Late Holocene, yielding a tsunami recurrence interval of between ~350-1000 years. However, when comparing the OSL ages from beach-ridges, sand sheets and tidal-flat deposit the picture is not clear and the data seem to be conflicting with one another. The problem persists when comparing OSL data of the beach-ridges and sand sheets from this study with 14C ages from previous studies (Jankaew et al., 2008; Monecke et al., 2008; Rajendran et al., 2006,

2008) and uranium-series age (Meltzner et al., 2010). The OSL ages are consistently younger than 14C, uranium-series ages and written records. The complication of OSL ages may lie with the variation of the environmental dose through time.

Other Complications and Challenges

Relative sea level change along the Andaman coast of Thailand in the Holocene is poorly defined. Sea level highstands can enhance the ability of a tsunami to inundate a coast and lowstands can decrease the ability of a tsunami to inundate. Interpretation of tsunami deposits may also be complicated by post-depositional changes in shoreline position. Phra Thong coastline is believed to be prograded after the mid-Holocene highstand. In this case the distance between the present shoreline and the limit of inundation is greater than the inundation distance at the time of the tsunami.

5.2.3 Correlation

On Phra Thong Island it is impossible to trace tsunami deposits from one swale to the next because the sand layers are identifiable only in the low-lying areas, separated by higher topography beach ridges. Furthermore, preservation potential of the sand layers may differ from place to place with physical (micro topography) and biological (bioturbation, human disturbance) factors. An insufficient number of event dated thus far also limits the correlation of these deposits. There are no tephra layers that may help in correlating the sand sheets. Confidence in correlating these deposits relied solely on the deposit ages. Ambiguities s and error in dating also exist.

6. Post depositional change and preservation potential in tropical environment

Phra Thong Island is an excellent site to observe how the 2004 tsunami was preserved. Five years after the Indian Ocean tsunami, black organic rich soil as thick as 5 centimeters has already developed on top of the 2004 tsunami sand layer in the swales (Fig.13), which are submerged under water for most of the year. High precipitation causes a high level of weathering of inorganic material and also aids in the decomposition of organic material that makes up this black organic rich soil. This quick development of soil in areas such as swales and other well protected areas aid in the preservation of 2004 deposits. In comparison, deposit of 2004 tsunami at Pakarang Cape has only a thin veneer of soil developed on top of the 2004 deposits in Fig. 14 which was taken on 11 July 2006. At this location, less than 200m from shore, the 2004 tsunami deposit contain two layers separated by layer of mud and fine-silt about 2.5 centimeters thick. The two layers represent two main waves of the tsunami wave train. Mud and fine silt layer deposited out of suspension from standing water between waves.

On the beach ridges, where there is no or little vegetation cover, top few centimetres of the deposits which is mostly composed of coarse-fine silt were reworked by aeolian and animal and human activities. In the future, should the 2004 deposit at these locations survive and preserve it will lack the very top of the deposit making interpretation of flow property of past tsunami using deposit thickness unsound.

In many places, thickness of 2004 sand layer is much thicker than the paleotsunami sand layer found at the same location (such as in Fig. 6). This could be due to bioturbation or erosion of part of the paleosand layer. Distance of the paleo-shoreline to the site may also be varied through time. The slip distribution of the paleoearthquake might have also generated smaller waves on Phra Thong.

Fig. 13. Thick organic rich soils already developed on top of the 2004 tsunami layer in (a), (b) and (c) (March 2008). One and two paleotsunami sand layers are present in (b) and (c), respectively. Stripe on stick is 10cm.

In certain areas where plant and animal diversity are high, such as the mangroves swamp, the 2004 tsunami deposits suffer from intense bioturbation that has already obliterated much of the deposit (Atwater, 2007). Mangrove crabs eat the tsunami sand and rework the sediment and deposit it at the surface as balls (Fig 15, left). Mud-lobsters, commonly found in tidal fringe mangrove at Phra Thong and other locations such as Tap Lamu, mixed the mangrove sediments with the tsunami sands.

Fig. 14. 27-centemetre thick 2004 tsunami deposit at Blue Village Resort, Pakarang Cape (July 2006). Bluish gray layer consists of mud and silt deposited out of suspension between tsunami wave trains.

Fig. 15. Balls of tsunami sands were created by mangrove crab (a). Mud lobster capable of mixing up sediments, including tsunami sediments, and can build mounds as high as 1 metre (b).

7. Conclusion

Factors such as identifying deposit mechanism and bioturbation make paleoseismology in Thailand difficult. The lack of co-seismic deformation, to indicate an earthquake origin of deposits along the Thai coast caused difficulties in distinguishing and identifying paleotsunami deposits. High levels of vegetation and human disturbance along the western coasts of Thailand further hinder the possibility of finding the deposits of past tsunamis. We found the mangrove environment to be unfavourable for preserving tsunami records. Marshy swales of beach ridge plains, however, can serve as a suitable recorders for tsunami events, as shown on Phra Thong Island.

The 2004 tsunami deposit on Phra Thong varies in thickness, structure and preservation potential. Thickness of the 2004 tsunami deposit at Phra Thong is primarily controlled by local topography and distance inland, with thick deposits in topographic lows and closer to the shore. Internal structure of the 2004 deposits at Phra Thong is different from one location to another. In most cases the deposit shows no internal structure and contains one fining-upward sequence. In thicker deposits, especially in the swales, the deposit is composed of 2-4 fining upward sequences. Preservation potential of the deposit in these swales is higher than other locations because quick soil development offers protection. 2004 tsunami sediments range in grain size from coarse sand to coarse silt. They show prominent bimodal distribution with the larger fraction consisting of beach sand and shell fragments and foraminifera tests.

The fresh-water swales in the beach ridge plain at Phra Thong offer an excellent recorder for tsunami deposits. Up to four paleotsunami sand layers were found on the island thus far. Pre-2004 deposits display up to 2 upward-sequences, but often show just one. Range of grain sizes for paleotsunami sand layers though similar to the 2004 sand, generally they are smaller and do not contain shell fragments and forams. Principal component of both 2004 and paleotsunami sediments are clear quartz, which is ubiquitously present along the Andaman coast of Thailand.

Challenges in studying paleotsunami deposit in tropical settings include identifying the source of the layer, bioturbation, dating and correlation. As foraminifera, shell fragments, and diatom in deposits are dissolved, identification of the marine origin of sand layers becomes more difficult. High bioturbation and disturbance of sand layer after deposition is also a main problem in this environment, leading to difficulty in correlating these sand layers across the area. Dating of these deposits by means of 14C and OSL is also problematic as the high rate of biodegradation in this environment destroyed most, if not all, suitable materials for 14C dating. OSL dating of the tsunami sediments is also doubtful, and is not concordant with ages obtained by other means. OSL ages are found to be younger, when compare with 14C ages. This concordance may be caused by high bioturbation rates or variation of the environmental dose rate through time in this environment.

8. Future research

Priorities in the future study of paleotsunami deposits at Phra Thong is establishing chronology of tsunami sand layers and correlation of sand layer across the island using ^{14}C, OSL and other methods. Combining several methods of dating may help overcome the problem of dating in the tropical environment. But careful interpretation of these ages is of utmost important in inferring the tsunami event age. Future work on other Indian Ocean shores will further provide a more detailed regional picture of the recurrence of great and giant subduction zone earthquakes and tsunami. The magnitude of the tsunamis responsible for the tsunami sand layers at Phra Thong can be assessed using numerical models with help from tsunami deposits to evaluate model results. Because of its relatively flat terrain, Phra Thong offers a suitable site to study velocity and depth of flow from properties of the deposits. Future research also includes establishing evolution of beach ridge-swale series at Phra Thong Island and with more ages of the sand ridges we hope to construct the Holocene relative sea-level curve for the Andaman Coast of Thailand. Hydraulic interpretation of 2004 and past tsunamis from deposits will also be carried out. Data from tsunami deposits-thickness and grain size distribution- can be used in sediment transport models (Jaffe &

Gelfenbaum, 2002; Jaffe & Gelfenbaum, 2007) to derive estimates of flow velocity and water depth (Peters et al., 2007). Estimates of flow velocity and tsunami flow depth are important to building design and in planning evacuation route.

9. Acknowledgement

The authors would like to thank AIST for allowing KJ to use settling tube and lab facility for grian size analysis. Thanks are also extended to Dr. Brian Atwater, Associate Prof. Montri Choowong, Assistant Prof. Thasinee Chareontirat, Dr. Vichai Chootakositkanon, Dr. Santi Pailoplee, Dr. Terry Machado, and graduate and undergraduate students from Chulalongkorn University for their help during fieldworks in 2007-2008. Review of manuscript by Dr. Brian Atwater is also greatly appreciated.

10. References

Apotsos, A.; Jaffe, B.; Gelfenbaum, G & Elias, E. (2009). Modeling time-varying tsunami sediment deposition. *Proceedings of Coastal Dynamics 2009*, pp. 1-15

Atwater, B.F. (2007). Hunting for ancient tsunamis in the tropics. *Eos Trans. AGU*, Vol. 88

Atwater, B.F. & Moore, A.L. (1992). A tsunami about 1000 years ago in Puget Sound, Washington. *Science*, Vol. 258, 1614-1617

Atwater, B.F.; Musumi-Rokkaku, S.; Satake, K.; Tsuji, Y.; Ueda, K. & Yamaguchi, D.K. (2005). The orphan tsunami of 1700: Japanese clues to a parent earthquake in North America, 133 p.

Aung, T.T.; Satake, K.; Okamura, Y.; Shishikura, M.; Swe, W.; Saw, H.; Swe, T.L.; Tun, S.T. & Aung, T. (2008). Geologic evidence for three great earthquakes in the past 3400 years off Myanmar. *Journal of Earthquake and Tsunami*, Vol. 2, 259-265

Bilham, R.; Engdahl, R.; Feldl, N. & Satyabala, S. P. (2005). Partial and complete rupture of the Indo-Andaman Plate boundary 1847-2004. *Seismological Research Letters*, Vol. 76, 299-311

Bondevik, S.; Mangerud, J.D.S. & Dawson, A.L.O. (2003). Record-breaking height for 8000-year-old tsunami in the North Atlantic. *Eos Trans. AGU*, Vol. 84, 289-293

Bondevik, S.; Svendsen, J.I.; Johnsen, G.; Mangerud, J. & Kaland, P.E. (1997). The Storegga Tsunami along the Norwegian Coast, its age and runup. *Boreas*, Vol. 26, 29-53

Bourgeois, J. (2009). Chapter 3. Geologic effects and records of tsunamis, In: *The Sea, Vol. 15: Tsunamis*, Robinson, A. R. & Barnard, E. N., (Ed.), 53-91, Harvard University Press

Bourgeois, J. & Minoura, K. (1997). Paleotsunami studies-contribution to mitigation and risk assessment. In: *Tsunami Mitigation and Risk Assessment*, Report of the International Workshop, Gusiakov, V.K. (Ed.), 1-4, Computing Center, Siberian Branch, Russian Academy of Sciences

Bourgeois, J.; Pinegina, T.; Ponomareva, V. & Zaretskaia, N. (2006). Holocene tsunamis in the southwestern Bering Sea, Russian Far East, and their tectonic implications. *Geological Society of America Bulletin*, Vol. 118, 449-463

Chague-Goff, C.; Nichol, S.L.; Jenkinson, A.V. & Heijnis, H. (2000). Signatures of natural catastrophic events and anthropogenic impact in an estuarine environment, New Zealand. *Marine Geology*, Vol. 167, 285-301

Chlieh, M.; Avouac, J.; Hjorleifsdottir, V.; Song, T. A.; Ji, C.; Sieh, K.; Sladen, A.; Hebert, H.; Prawirodirdjo, L.; Bock, Y. & Galetzka, J. (2007). Coseismic slip and afterslip of the great M_w 9.15 Sumatra-Andaman earthquake of 2004. *Bulletin of the Seismological Society of America*, Vol. 97, No. 1A, January 2007, S152-S173, doi: 10.1785/0120050631

Choowong, M.; Murakoshi, N.; Hisada, K.; Charusiri, P.; Charoentitirat, T.; Chutakositkanon, V.; Jankaew, K. & Kanjanapayont, P. (2007). Erosion and deposition by the 2004 Indian ocean tsunami in Phuket and Phang-nga Provinces, Thailand, *Journal of Coastal Research*, Vol. 23, No. 5, 1270-1276

Choowong, M.; Murakoshi, N.; Hisada, K.; Charusiri, P.; Charoentitirat, T.; Phantuwongraj, S.; Wongkok, P.; Choowong, A.; Subsayjun, R.; Chutakositkanon, V.; Jankaew, K. & Kanjanapayont, P. (2008a). Flow conditions of the 2004 Indian Ocean tsunami in Thailand, inferred from capping bedforms and sedimentary structures, *Terra Nova*, 20, 141-149.

Choowong, M.; Murakoshi, N.; Hisada, K.; Charusiri, P.; Charoentitirat, T.; Chutakositkanon, V.; Jankaew, K.; Kanjanapayont, P. & Phantuwongraj, S. (2008b). 2004 Indian Ocean Tsunami Inflow and Outflow at, Phuket, Thailand, *Marine Geology*, Vol. 248, No. 3-4, 179-192

Chotmaihet of Malaya peninsula visit by King Rama V during 16-22 April 1890 (1924). available online at http://dspace.kids-d.org/handle/123456789/2659 (in Thai)

Cisternas, M.; Atwater, B.F.; Torrejón, F.; Sawai, Y.; Machuca, G.; Lagos, M.; Eipert, A.; Youlton, C.; Salgado, I.; Kamataki, T.; Shishikura, M.; Rajendran, C.P.; Malik, J.K.; Rizal, Y. & Husni, M. (2005). Predecessors to the giant 1960 Chile earthquake. *Nature*, Vol. 437, 404-407

Dawson, A.G.; Long, D. & Smith, D.E. (1988). The Storegga slides; evidence from eastern Scotland for a possible tsunami. *Marine Geology*, Vol. 82, 271-276

Dawson, A. G. and Shi, S. (2000) Tsunami deposits. Pure and Applied Geophysics, Vol. 157, 875-897

Dominey-Howes, D.; Cummins, P. & Burbidge, B. (2007). Historic records of teletsunami in the Indian Ocean and insights from numerical modeling. *Natural Hazards*, Vol. 42, 1-17

Fiedorowicz, B.K. & Peterson, C.D. (2002). Tsunami deposit mapping at Seaside, Oregon, USA. In: *Geoenvironmental Mapping; Methods, Theory, and Practice*, Bobrowsky, P.T. (Ed.), 629–648, A.A. Balkema Publishers, Lisse, Netherlands

Fujino, S.; Naruse, H.; Matsumoto, D.; Jarupongsakul, T.; Suphawajruksakul, A. & Sakakura, N. (2009). Stratigraphic evidence for pre-2004 tsunamis in southwestern Thailand. *Marine Geology*, Vol. 262, 25-28

Fujino, S.; Naruse, H.; Suphawajruksakul, A.; Jarupongsakul, T.; Murayama, M. & Ichihara, T. (2008). Thickness and grain-size distribution of Indian Ocean tsunami deposits at Khao Lak and Phra Thong Island, south-western Thailand. In: *Tsunamites: Features and Implications*, Shiki, T.; Tsuji, Y.; Minoura, K. & Yamazaki, T. (Ed.), 123-132, Elsevier, Amsterdam

Galbraith, R. F.; Roberts, R. G; Laslett, G. M; Yoshida, H. & Olley, J. M. (1999). Optical dating of single and multiple grains from Jimnium rock shelter, northern Australia: Part 1, experimental design and statistical methods. *Archaeometry*, Vol. 41, 339-364

Gelfenbaum, G. & Jaffe, B. E (2003). Erosion and sedimentation from the 17 July, 1998 Papua New Guinea tsunami. *Pure and Applied Geophysics*, Vol. 160, 1969-1999

Goff, J.; Chague-Goff, C. & Nichol, S. (2001). Palaeotsunami deposits: A New Zealand perspective. *Sedimentary Geology*, Vol. 143, 1-6

Goff, J.R.; Crozier, M.; Sutherland, V.; Cochran, U. & Shane, P. (1998). Possible tsunami deposits from the 1855 earthquake, North Island, New Zealand. *Coastal tectonics*, Vol. 146, 353-374

Goff, J.; Liu, P.L.F.; Higman, B.; Morton, R.; Jaffe, B.E.; Fernando, H.J.S.; Lynett, P.; Fritz, H.; Synolakis, C. & Fernando, S. (2006). Sri Lanka Field Survey after the December 2004 Indian Ocean Tsunami. *Earthquake Spectra*, Vol. 22, S155-S172

Grauert, M.; Bjorck, S. & Bondevik, S. (2001). Storegga tsunami deposits in a coastal lake on Suduroy, the Faroe Islands. *Boreas*, Vol. 30, 263-271

Harper, M., 2005, Bedded shell deposit at Ao Nang, Krabi Province, southern Thailand: a record of a prehistoric tsunami event or extreme storm event or neither. *Geological Society of America Abstracts with Programs*, Vol. 37, 75

Hawkes, A.D.; Bird, M; Cowie, S; Grundy-Warr, C; Horton, B.P; Tan Shau Hwai A.; Law, L; Macgregor, C; Nott, J.; Eong Ong, J.; Rigg, J.; Robinson, R.; Tan-Mullins, M.; Tiong Sa, T. & Zulfigar, Y. (2007). The sediments deposited by the 2004 Indian Ocean Tsunami along the Malaysia-Thailand Peninsula. *Marine Geology* , Vol. 242, 169-190

Hori, K.; Kuzumoto, R.; Hirouchi, D.; Umitsu, M.; Janjirawuttikul, N. & Patanakanog, B. (2007). Horizontal and vertical variation of 2004 Indian tsunami deposits: An example of two transects along the western coast of Thailand. *Marine Geology*, Vol. 239, 163–172

Horton, B.P.; Gibbard, P.L.; Milne, G.M.; Morley, R.J.; Purintavaragul, C. & Stargardt, J.M. (2005). Holocene sea levels and palaeoenvironments, Malay-Thai Peninsula, southeast Asia. *The Holocene*, Vol. 15, 1199-1213

Huntley, D. J. & Clague, J. J. (1996). Optical dating of tsunami-laid sands. *Quaternary Research*, Vol. 46, 127-140

Jaffe, B.E.; Borrero, J.C.; Prasetya, G.S.; Peters, R.; McAdoo, B.; Gelfenbaum, G.; Morton, R.; Ruggiero, P.; Higman, B.; Dengler, L.; Hidayat, R.; Kingsley, E.; Kongko, W.; Lukijanto; Moore, A.L.; Titov, V.V. & Yulianto, E. (2006). Northwest Sumatra and offshore islands field survey after the December 2004 Indian Ocean tsunami. *Earthquake Spectra*, Vol. 22, S105–S135

Jaffe, B. E. & Gelfenbaum, G. (2002). Using tsunami deposits to improve assessment of tsunami risk. *Solutions to Coastal Disasters'02, Conference Proceedings, ASCE*, pp. 836-847

Jaffe, B. E. & Gelfenbaum, G. (2007). A simple model for calculating tsunami flow speed from tsunami deposits, *Sedimentary Geology*, Vol. 200, 347-361

Jaffe, B. E.; Gelfenbaum, G.; Rubin, D.; Peters, R.; Anima, R.; Swensson, M.; Olcese, D.; Bernales, L.; Gomez, J. & Riega, P. (2003). Tsunami deposits: Identification and interpretation of tsunami deposits from the June 23, 2001 Peru tsunami. *Proceedings of the International Conference on Coastal Sediments 2003*, 13 pp., ISBN: 981-238-422-7, Publish by World Scientific Publishing Corp and East Meets West Productions, Corpus Christi, TX, USA

Jankaew, K.; Atwater, B.F.; Sawai, Y.; Choowong, M.; Charoentitirat, T.; Martin, M.E. & Prendergast, A. (2008). Medieval forewarning of the 2004 Indian Ocean tsunami in Thailand. *Nature*, Vol. 455, October 2008, 1228-1231

Kelsey, H.M.; Nelson, A.R.; Hemphill-Haley, E. & Witter, R.C. (2005). Tsunami history of an Oregon coastal lake reveals a 4600 yr record of great earthquakes on the Cascadia subduction zone. *Geological Society of America Bulletin.* Vol. 117, 1009-1032

McSaveney, M.J.; Goff, J.R.; Darby, D.J.; Goldsmith, P.; Barnett, A.; Elliott, S. & Nongkas, M. (2000). The 17 July 1998 tsunami, Papua New Guinea; evidence and initial interpretation. In: *International Geological Correlation Program Project 367 "Records of rapid change in the late Quaternary" Vol. 170*, Stiros, S.C. & Scott, D.B. (Ed.), 81-92, Elsevier

Metlzner, A.J; Sieh, K.; Chiang, H.-W.; Shen, C.-C.; Suwargadi, B.W.; Natawidjaja, D.H.; Philibosian, B.E.; Briggs, R.W. & Galetzka, J. (2010). Coral evidence for earthquake recurrence and an AD 1390-1455 cluster at the south end of the 2004 Aceh-Andaman rupture.

Minoura, K. & Nakata, T. (1994). Discovery of an ancient tsunami deposit in coastal sequences of Southwest Japan; verification of a large historic tsunami. *The Island Arc*, Vol. 3, 66-72

Monecke, K.; Finger, W.; Klarer, D.; Kongko, W.; McAdoo, B.G.; Moore, A.L. & Sudrajat, S.U. (2008). A 1,000-year sediment record of tsunami recurrence in northern Sumatra. *Nature*, Vol.455, October 2008, 1232-1234

Moore, A.; Nishimura, Y.; Gelfenbaum, G.; Kamataki, T. & Triyono, R. (2006). Sedimentary deposits of the 26 December 2004 tsunami on the northwest coast of Aceh, Indonesia. *Earth, Planets and Space*, Vol. 58, 253-258

Murty, T. S. & Flather, R. A. (1994). Impact of storm surges in the Bay of Bengal. *Journal of Coastal Research*, Vol. 12, 149-161

Nanayama, F.; Furukawa, R.; Shigeno, K.; Makino, A.; Soeda, Y. & Igarashi, Y. (2007). Nine unusually large tsunami deposits from the past 4000 years at Kiritappu marsh along the southern Kuril Trench. *Sedimentary Geology*, Vol. 200, 275-294

Nanayama, F.; Satake, K.; Furukawa, R.; Shimokawa, K., Atwater, B.F.; Shigeno, K. & Yamaki, S. (2003). Unusually large earthquakes inferred from tsunami deposits along the Kuril Trench. *Nature*, Vol. 424, 660-663

Olley, J.M.; Caitcheon, G.G. & Roberts, R.G. (1999). The origin of dose distributions in fluvial sediments, and the prospect of dating single grains from fluvial deposits using optically stimulated luminescence. *Radiation Measurements*, Vol. 30, 207-217

Olley, J.M.; De Deckker, P.; Roberts, R.G.; Fifield, L.K.; Yoshida, H. & Hancock, G. (2004a). Optical dating of deep-sea sediments using single grains of quartz: A comparison with radiocarbon. *Sedimentary Geology*, Vol. 169, 175-189

Olley, J.M.; Pietsch, T. & Roberts, R.G. (2004b). Optical dating of Holocene sediments from a variety of geomorphic settings using single grains of quartz. *Geomorphology*, Vol. 60, 337-358

Ollerhead, J.; Huntley, D. J.; Nelson, A. R. & Kelsey, H. M. (2001). Optical dating of tsunami-laid sand from an Oregon coastal lake. *Quaternary Science Reviews*, Vol. 20, 1915-1926

Oritz, M. & Bilham, R. (2003). Source area and rupture parameters of the 31 December 1881 Mw=7.9 Car Nicobar earthquake estimated from tsunamis recorded in the Bay of Bengal. *Journal of Geophysical Research*, Vol. 108, No. B4, ESE 1-1-ESE11-16, doi:2002JB001941RR 2003

Paris, R.; Fournier, J.; Poizot, E.; Etienne, S.; Morin, J.; Lavigne, F. & Wassmer, P. (2010). Boulder and fine sediment transport and deposition by the 2004 tsunami in Lhok Nga (western Banda Aceh, Sumatra, Indonesia): A coupled offshore-onshore model. *Marine Geology*, Vol. 268, 43-54

Peters, R.; Jaffe, B. & Gelfenbaum, G. (2007). Distribution and sedimentary characteristics of tsunami deposits along the Cascadia margin of western North America, *Sedimentary Geology*, Vol. 200, 372-386

Pinegina, T.K.; Bourgeois, J.; Bazanova,I.V. & Braitseva, O.A. (2003). A millennial-scale record of Holocene tsunamis on the Kronotskiy Bay coast, Kamchatka, Russia. *Quaternary Research*, Vol. 59, 36-47

Rajendran, C. P.; Rajendran, K.; Machado, T.; Satyamurthy, T.; Aravazhi, P. & Jaiswal, M. (2006). Evidence of ancient sea surges at the Mamallapuram coast of India and implications for previous Indian Ocean Tsunami events. *Current Science*, Vol. 91. 1242-1247

Rajendran, K.; Rajendran, C. P.; Earnest, A.; Prasad, G. V. R.; Dutta, K.; Ray, D. K. & Anu, R. (2008). Age estimates of coastal terraces in the Andaman and Nicobar Islands and their tectonic implications. *Tectonophysics*, Vol. 455, 53-60

Reid, A. (2005). An Indonesian Frontier, Achenese and Other Histories of Sumatra. University of Hawaii Press. Honolulu

Ruff, L. & Kanamori, H. (1980). Seismicity and the subduction process. *Physics of the Earth and Planetary Interiors*, Vol. 23, 240-252

Satake, K & Atwater, B. F. (2007). Long-term perspectives on giant earthquakes and tsunamis at subduction zones, *Annual Review of Earth and Planertary Sciences*, Vol. 35, 349-374

Shi, S.Z.; Dawson, A.G. & Smith, D.E. (1995). Coastal sedimentation associated with the December 12th, 1992 tsunami in Flores, Indonesia. *Pure and Applied Geophysics*, Vol. 144, 525-536

Simkin, T. & Fiske, R.S. (1983). *Krakatau 1883; the volcanic eruption and its effects*, Smithson. Inst. Press, Washington, DC, United States (USA)

Singarasubramanian, S.R.; Mukesh, M.V.; Manoharan, K.; Murugan, S.; Bakkiaraj, D. & Peter, A.J. (2006). Sediment characteristics of the M 9 tsunami event between Rameswaram and Thoothukudi, Gulf of Mannar, southeast coast of India. *Science of Tsunami Hazards*, Vol. 25, 160-172

Sawai, Y.; Fujii, Y.; Fujiwara, O.; Kamataki, T.; Komatsubara, J.; Okamura, Y.; Satake, K. & Shishikura, M. (2008). Marine incursions of the past 1500 years and evidence of tsunamis at Suijin-numa, a coastal lake facing the Japan Trench. *The Holocene*, Vol. 18, No. 4, 517-528

Sawai, Y.; Jankaew, K.; Martin, M.E.; Prendergast, A.; Choowong, M. & Charoentitirat, T. (2009a). Diatom assemblages in tsunami deposits associated with the 2004 Indian Ocean tsunami at Phra Thong Island, Thailand. *Marine Micropaleontology*, Vol. 73, 70-79

Sawai, Y.; Kamataki, T.; Shishikura, M.; Nasu, H.; Okamura, Y.; Satake, K.; Thomson, K.H; Matsumoto, D.; Fujii, Y.; Komatsubara, J. & Aung, T.T. (2009b). Aperiodic recurrence of geologically recorded tsunamis during the past 5500 years in eastern Hokkaido, Japan. *Journal of Geophysical Research*, Vol. 114, B01319, doi:10.1029/2007JB005503

Sieh, K.; Natawidjaja, D.H.; Meltzner, A.J.; Shen, C.-C.; Cheng, H.; Li, K.-S.; Suwargadi, B.W.; Galetzka, J.; Philibosian, B. & Edwards, R. L. (2008). Earthquake supercycles inferred from sea-level changes recorded in the corals of West Sumatra, *Science*, Vol. 322, December 2008, 1674-1678

Stein, S. & Okal, E. A. (2007). Ultra-long period seismic study of the December 2004 Indian Ocean earthquake and implications for regional tectonics and the subduction process. *Bulletin of the Seismological Society of America*, Vol. 97, S279-S295

Tsuji, Y.; Namegaya, Y.; Matsumoto, H.; Iwasaki, S.; Kanbua, W.; Sriwichai, M. & Meesuk, V. (2006). The 2004 Indian tsunami in Thailand: Surveyed runup heights and tide gauge records. *Earth Planets Space*, Vol. 58, 223-232

Tuttle, M.P.; Ruffman, A.A.T. & Jeter, H. (2004). Distinguishing tsunami from storm deposits in Eastern North America: The 1929 Grand Banks tsunami versus the 1991 Halloween storm. *Seismological Research Letters*, Vol. 75, No. 1, 117-131

Wallinga, J., 2002, On the detection of OSL age overestimation using single-aliquot techniques. *Geochronometria*, Vol. 21, 17-26

Witter, R.C.; Kelsey, H.M. & Hemphill-Haley, E. (2000). Stratigraphic evidence for twelve Cascadia earthquakes in the last 6600 years from the Coquille River estuary, Oregon. In: *Penrose conference "Great Cascadia earthquake tricentennial"*, Clague, J.J.; Atwater, B.F.; Wang, K.; Wang, Y. & Wong, I.G. (Ed.), 140, Oregon Department of Geology and Mineral Industries

16

The 2004 Indian Ocean Tsunami: Impact on and Rehabilitation of Fisheries and Aquaculture in Thailand

Suwat Tanyaros[1] and David Crookall[2]
[1]Rajamanagala University of Technology Srivijaya, Trang
[2]Université de Nice Sophia Antipolis
[1]Thailand
[2]France

This chapter is dedicated to the memory of Isabelle Attalli, a wonderful friend who, with her two children, was taken by the waters in Sri Lanka on that fateful day. Search for "Isabelle Attalli" in Google.

"My boat is everything. My life totally depends on it. When the tsunami came, it destroyed our boats and put an end to everything for all of us. But now we have the boats back and our livelihood as well. We can go fishing again." (A fisher on Koh Lanta.)

1. Introduction

Southern Thailand, also known as Peninsular Thailand, lies between latitudes 50 and 110 N, and longitudes 980 and 1020 E. It covers an area of 7,153,917 ha and has over 2,705 km of shoreline, with the western coastline facing the Andaman Sea and the eastern coastline facing the Gulf of Thailand - see Figure 2. The result of the phenomenal growth of tourism and fishing is that much of the mangrove and beach forest along the coastal shores have been replaced by human-built infrastructures, such as aquaculture industries and tourist resorts (UNEP, 2005). As a consequence, the coastal areas of southern Thailand have become particularly vulnerable to natural disasters (Thanawood et al., 2006).

The earthquake that created the tsunami was the world's fifth largest, with a magnitude of 9.15 to 9.3 on the Richter scale. It occurred at 00:58:53 (GMT) on Sunday, 26 December, 2004, with the epicenter, at a depth of 30km, just off the west coast of North Sumatera, Indonesia (Harinarayana & Hirata, 2005); see Figure 1. The sudden vertical rise of the seabed by several meters during the quake displaced massive volumes of water, resulting in a devastating tsunami. This seismic sea wave traveled thousands of kilometres across the Indian Ocean, and ravaged the Andaman coast of southern Thailand at 9.30 am local time.

The earthquake was the largest since the 1964 Alaska quake, and the fourth biggest this century. The tsunami that ensued was among the five largest recorded, as measured in tsunami magnitude - see Table 1. The waves that hit the Andaman coast were, in some places, as high as 20 metres. The Indian Ocean tsunami was the first in living memory to strike the shorelines of southern Thailand.

The tsunami caused extensive damage to life and property in six southern coastal provinces along the Andaman coast of Thailand (Figure 2). The people suffered a total of 5,395 deaths, 8,457 injured and 3,144 missing. The severely affected areas included 292 villages in 78 sub-districts of 24 districts. In these areas, 20,537 households with a total population of 91,638 are considered to have been directly affected through loss of, or injury to, a family member (FAO-MOAC, 2005). Detailed tables for damage and effect on people, for number of deaths, injured and missing, and property damage (in THB), are provided by Srivichai et al. (2007).

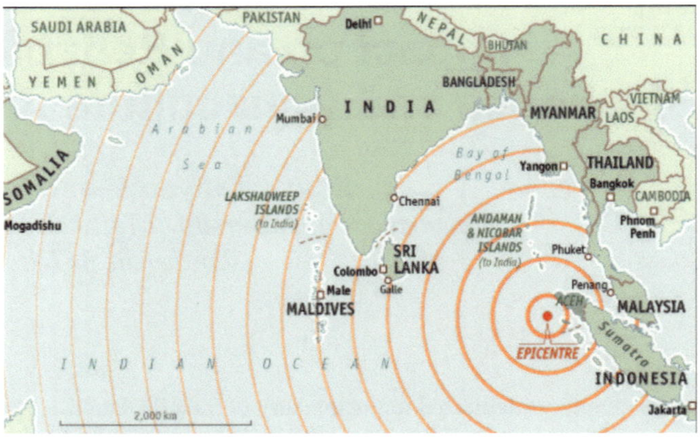

Fig. 1. Indian Ocean earthquake propagation

Year	Location	Mt *
1837	Valdivia, Chile	9.3
1846	Aleutians	9.3
1960	Chile	9.4
1964	Alaska	9.1
2004	Sumatra, Indonesia	9.1

Table 1. The five greatest recorded tsunamis (Data source: Kohl et al., 2005.)
* Mt = tsunami magnitude

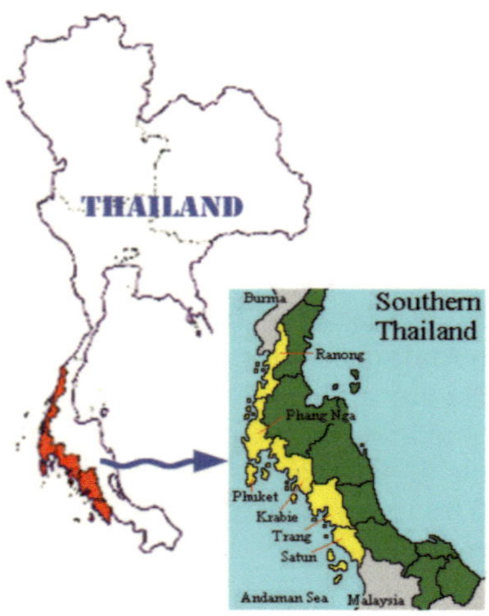

Fig. 2. Provinces affected by the tsunami

The most seriously affected sectors of the economy were tourism and coastal fisheries. Prior to the 2004 tsunami, the tourism and fishing industries provided most of the livelihoods in the affected areas along the Thai Andaman coast. The immediate result of the tsunami was the near destruction of the local economy. The economic disaster was such that it led to a 0.8% reduction in the Thai national gross domestic product (GDP) growth rate for the year 2005, compared to the rate that would have existed without the tsunami (Israngkura, 2005). We shall concern ourselves in this chapter to fishing, although a certain indirect interdependence existed (and exists) between tourism and fishing.

> The catastrophe left hotels, resorts, shops, businesses, private houses, boats and vehicles in ruins, but the greatest number of victims were people living in 400 affected fishing villages. The rehabilitation happened faster in Thailand than other countries, partly because the waves hit one of the most important tourist areas, which rakes in a third of the country's total tourism income, so there was a commercial incentive to get everything back to normal as fast as possible. Also, because a third of the 8,562 people who died were foreign tourists …, the crisis got a lot of attention … from the first day. (Kerr, 2005.)

The tsunami caused major losses in the fishing industry and coastal aquaculture in terms of fishing boats and gear, culture ponds, cages and shrimp hatcheries. Seventy-four affected sub-districts and 386 villages reported losses for fisheries and aquaculture (DoF, 2005). The estimated damage, as reported by the Fisheries Rescue Coordination Centre on 10 January 2005, is summarized in Table 2.

Small fishing boats	3,714	Fish & shellfish cage farms	6,063
Large fishing boats	1,199	Shrimp farms	42
Ecotourism boats	554	Hatcheries	573
Public harbours & piers	83	Shellfish concession plots (819 rai)	17
Sets of fishing gear	47,063		

Table 2. Damage (in units) caused by the tsunami (FAO-MOAC, 2005)

Damage to private jetties and piers for private boats or fishery-associated businesses, for example, ice plants, gas stations, fish landings, markets and others, has still to be assessed. The tsunami caused severe damage not only to fisher folk and aquaculturists, but also to fisher folk earning additional income from ecotourism. The total damage to fisheries and aquaculture is estimated at 343 M USD (FAO, 2005).

Fishing and aquaculture, although present, were marginal activities some 30 years ago. Fishing developed steadily and tourism grew exponentially in the years leading up to the tsunami. Of course, fishing (particularly artisanal) in Thailand has traditionally brought in much less money than tourism; the per capita and total household incomes of fishers are lower than the national averages. Within the fisheries sector, artisanal fishers as a group earn the lowest income (Silvestre et al. 2003). The livelihoods of many coastal fishery communities in Thailand were completely or partially destroyed by the tsunami. Economies at the community level were severely affected, which led to 19,968 already poor fishers to fall into even deeper poverty (Paton et al., 2007).

Note on units. Throughout this chapter a number of different units are used, and vary according to the source cited. The two monetary units are the Thai baht, written as THB and the US dollar, written as USD. They have not been converted one to the other, as the exchange rate rates varied greatly over the period of publication the various sources. For current and historical exchange rates, see Onanda, XE, National Bank of Thailand and other websites. Other units used are: ha (hectare), equivalent to 6.25 rai in Thailand.

Fishing is one of the oldest industries in Thailand and has contributed significantly to the development of the country - see Table 3 for a number of fishing industry indicators in Thailand and the Andaman provinces.

Fish production of Thailand (both Andaman Sea and Gulf of Thailand)	4 M mt	2005 population in the 6 Andaman Sea coast provinces	2.03 M
N° of Andaman fishery establishments	20,703	Proportion of fishers in Thai population (FAO, 2005)	1%
Total marine catch taken in the Andaman Sea (% of total national catch, in weight)	31.7%	N° of fishers (in Thailand) during peak season (Nov to March), of which:	47,000+
Thai GDP in 2005 (World Bank, 2006) - USD	1,766,000 M	- family members - employees	30,000 18,000
Proportion of fisheries in GDP in 2005	2.5%	GPP of the 6 Andaman Provinces - USD	4,754.8 M
GPP for Andaman fisheries - USD	482.5 M	Per capita GPP for Andaman (Panjarat, 2008) - USD	2,335

Table 3. Indicators relating to Thai fisheries (GPP = Gross Provincial Product)

Fig. 3. Damaged fishing boats and trawlers

2. Impact on fisheries

In contrast to the small impact on commercial fisheries, the Andaman coast tsunami significantly impacted small-scale and artisanal fisheries. These are generally subsistence fishermen. The main characteristics of such fishing households (FAO, 2004) are that they:

- use relatively small amounts of capital and energy;
- use small fishing vessels;
- undertake short fishing trips close to the shore;
- target a wide range of fish species;
- use multiple types of fishing gear;
- use a variety of fishing strategies to suit the natural availability of fish stocks or the seasonality of fish; and
- catch mainly for local consumption.

After the tsunami, these small-scale and artisanal fisheries suffered major losses in terms of damage to or loss of fishing boats and gear, and fishery resources. We summarize each of these here.

2.1 Damage to or loss of fishing boats

Some 5,431 fishing boats were either damaged or totally wrecked, of which 93% were small-scale and artisanal fishing craft and 7% commercial fishing ships.
Estimates indicate that 2,923 fisheries households were affected. Estimated damages to fisheries alone amounted to 16.6 M USD (DoF, 2005). See Table 4 and Figure 4.

Province	Large boats damaged	Large boats sunk	Small boats damaged	Small boats sunk	USD value of all boat losses
Krabi	147	1	808	54	19,269
Phang-nga	322	124	754	46	915,546
Phuket	490	157	642	41	1,884,618
Ranong	204	13	414	27	12,331
Satun	35	6	552	49	20,520
Trang	1	-	648	-	-
Total	1,199	301	3,714	217	2,852,284

Table 4. N° of affected fishing boats (DoF, 2005)

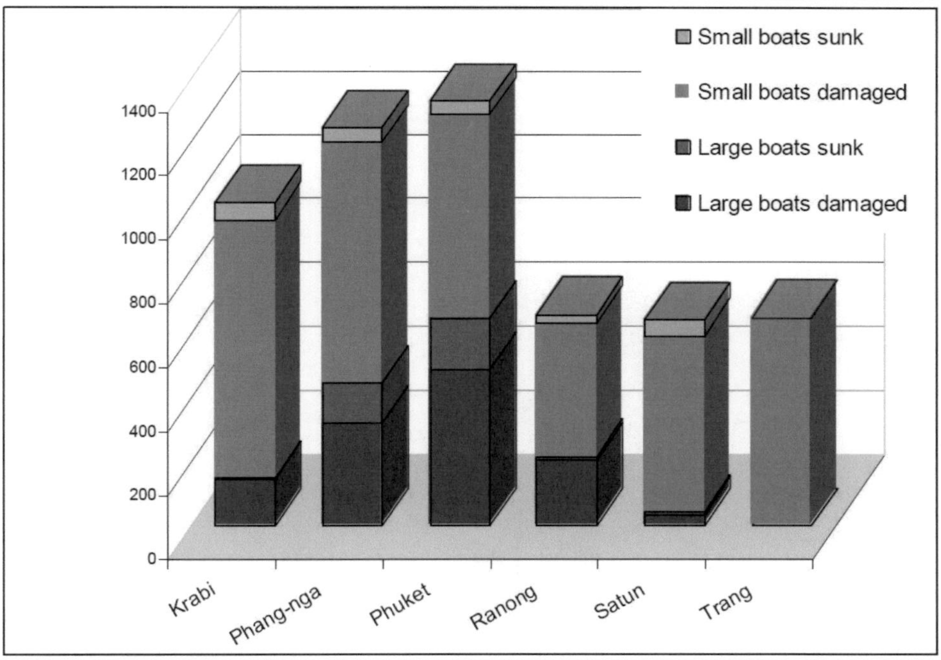

Fig. 4. N° of affected fishing boats (based on Table 4)

We took the figures provided by the DoF (2005), and calculated percentages, allowing a more balanced appreciation of the variations among the six provinces. These are given in

Table 5. Phuket and Phang-nga suffered the greatest losses for large boats, while small boat losses were largest in Krabi and Phang-nga.

Large boats						
Province	**N° damaged**	*% damaged*	**N° sunk**	*% sunk*	***damaged** as % of all boats*	***sunk** as % of all boats*
Phuket	*490*	*41%*	*157*	*52%*	***33%***	*10%*
Phang-nga	*322*	*27%*	*124*	*41%*	***21%***	*8%*
Ranong	204	*17%*	13	*4%*	*14%*	*1%*
Krabi	147	*12%*	1	*0%*	*10%*	*0%*
Satun	35	*3%*	6	*2%*	*2%*	*0%*
Trang	1	*0%*		*0%*	*0%*	*0%*
Total	1,199	*100%*	301	*100%*	***80%***	*20%*
Small boats						
Province	N° **damaged**	*% damaged*	N° **sunk**	*% sunk*	**damaged** as % of all boats	**sunk** as % of all boats
Phuket	642	*17%*	41	*19%*	*16%*	*1%*
Phang-nga	754	*20%*	46	*21%*	***19%***	*1%*
Ranong	414	*11%*	27	*12%*	*11%*	*1%*
Krabi	808	*22%*	54	*25%*	***21%***	*1%*
Satun	552	*15%*	49	*23%*	*14%*	*1%*
Trang	648	*17%*		*0%*	*16%*	*0%*
Total	3714	*103%*	217	*100%*	*97%*	*6%*

Table 5. Percentages (rounded) of large and small boats damaged and sunk (based on Table 4).

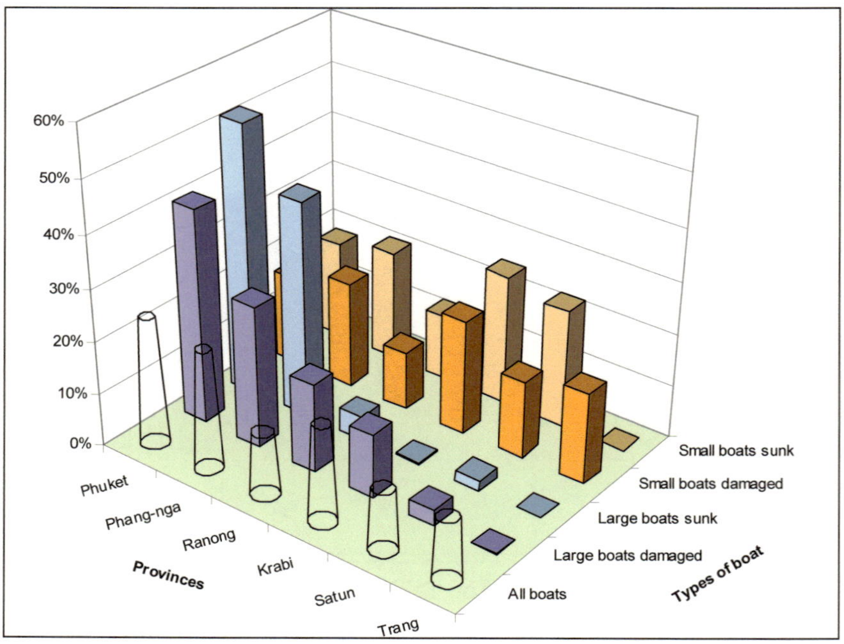

Fig. 5. Percentages of boats damaged and sunk (based on Table 5).

2.2 Damage to or loss of fishing gear

Loss of fishing gear is extremely difficult to assess. After the tsunami, for the purpose of assessing monitory loss, the vessel and the gear are were assumed to have been lost together. Gear replacement is a lower-cost item, but compensation of which can help return fishing folk to their livelihood as fishers. It assumes that they are able to repair damaged vessels themselves. In such circumstances, gear provision is probably significantly less costly than expenses for boat repair.

2.3 Impact on fishery resources

From 1976 to 2003, the DoF monitored the implementation of measures to conserve marine resources during seasons of fish spawning and nursing young fish along the Andaman coast; the project covered 2,400 k^2 along the Andaman coast. Their aim was to identify study areas and periods in order to implement conservation measures for spawning and breeding of mature economic shallow-water fishes and marine shrimps. Surveys conducted by the DoF after implementation reported an abundance of marine fishes and shrimps in several study sites along the Andaman coast (DoF, 2005). Preliminary assessment, in early January 2005, of fishery resources of the Andaman coast indicated that those resources had, in some areas, declined by half after the tsunami (Bueno, 2005). It was also found that the price of marine animals dropped within three months after the tsunami, and then gradually climbed back to normal levels.

The DoF reported that the density of marine resources had decreased during the: month following the tsunami. However, if only economic fish species were counted, the density around Phang-nga Bay and adjacent areas had decreased compared with a similar pre-tsunami period. In the two provinces of Trang and Satun, which were less severely hit by the tsunami, the catch increased (Office of Natural Resources and Environmental Policy and Planning, 2006). The marine fish catch off the west coast of Phang-nga and Phuket Provinces decreased after the tsunami struck (before 72.69 kg/hr; after 34.92 kg/hr). The juvenile fish catch increased after the tsunami struck; similar results were found in a fish larvae survey (Nootmorn, 2006).

An explanation for these differentials can be found in the physical effect of water movement. The tsunami waves caused a huge undercurrent with massive movement of water, disturbing sediments at the sea bottom and stirring a water mass rich in nutrients at the pycnocline level in the deep sea, bringing them up towards the continental shelf, a part of which was the Andaman coast. In this way, the tsunami moved great amounts of nutrients from the Sundra Trench and water mass from the deep sea up to fishery areas along the Andaman coast in Thailand. The food chain was thus impacted, with links from water movement through plankton and young marine animals to increased juvenile fish catch. Whanpetch et al. (2010) have shown that the patterns of temporal change in abundance and diversity of macrofaunal assemblages before and after the tsunami varied greatly from site to site, and that the degree of temporal changes in assemblage structure was not solely related to the magnitude of the tsunami disturbance. More importantly, the presence or absence of seagrass vegetation altered the patterns of temporal change in macrofaunal assemblages and recovery processes after a tsunami disturbance.

3. Coastal habitats

The coastal habitat and environment were impacted in various ways. In some coastal areas, coral reefs and seagrass beds were damaged or destroyed, greatly degrading fisheries resources and thus fishers' livelihoods, both directly and indirectly. Both coral reefs and seagrass have been able to recover to a certain extent.

From a rapid assessment three weeks after the event, only about 13% of the coral reefs in the Andaman Sea were found to be severely damaged (>50% of corals destroyed), while almost 40% showed no measurable impact by the tsunami (DMCR, 2005; Brown, 2005; Wilkinson et al. 2006). A study on coral recruitment and recovery after the 2004 tsunami by Sawall et al. (2009) near the Phi Phi Islands (Krabi Province) and Phuket on the Andaman Sea found rapid recovery to be the norm, suggesting that the duration of disturbance, degree of sorting and hence stability of coral rubble is a key determinant of recruitment success. The tsunami-impacted coral reefs have now revived to a certain degree as a result of rehabilitation activities instituted by government agencies, private groups, communities and NGOs.

Seagrass shows a similar picture. Only some 5% of seagrass beds and 321.6 ha of mangrove forests were damaged or destroyed (Office of Natural Resources and Environmental Policy and Planning, 2006). A comparative analysis (Nakaoka et al., 2007) of seagrass biomass and coverage before and after the tsunami revealed that seagrass beds were severely affected by the tsunami. A broad-scale coastal census after the tsunami showed that the effects on seagrass beds were spatially variable; some seagrass beds disappeared completely, whereas others were only negligibly impacted (Department of Marine and Coastal Resources of Thailand, 2005). The Department of Marine and Coastal Resources studied and monitored seagrass beds destroyed by the tsunami, and revealed that the ecology of seagrass beds in all areas hit by the tsunami were able to survive and regain previous levels within a year without replanting.

From monitoring by the Office of Natural Resources and Environmental Policy and Planning (2006), it appears that local people were satisfied with the level of resilience of natural resources, such as mangrove forests, beach forests, seagrass beds and coral reefs. Local perception, gathered from personal interaction, is that the rehabilitation of coral reefs had been successful to a certain extent. People thus requested continued efforts to restore the natural resources back to the original state of abundance.

4. Impact on aquaculture

The Andaman coast of Thailand has significant amounts of coastal aquaculture based in and around mangrove areas, especially in the creeks and delta mouths. Several types of aquaculture were affected by the tsunami. We will review fish cages, shrimp ponds, hatcheries and shellfish.

4.1 Cage culture

Cage culture was one of the primary occupations for those living in the coastal communities devastated by the December 2004 tsunami. The typically fragile construction of cages made them particularly vulnerable to the tsunami attack, which resulted in the breakup of cages and escape of the stocks - see Figure 6. Some 5,568 cage culture farms, covering a total cage area of some 1,123,176 m^2, were reportedly affected by the tsunami impact, with a total of 15,802 cages damaged. The government estimated losses from aquaculture cages to be 20 M USD (DoF, 2005). Grouper and sea bass suffered the most, and Krabi and Phang-gna suffered the widest losses, in terms of variety of species; losses are summarized in Table 7.

Province	Fish cage farms	Shrimp ponds (ha)	Hatchery farms	Shellfish (ha)
Ranong	677	1.61	-	3.41
Phang-nga	3,008	16.88	180	64
Phuket	315	5.84	209	58.02
Krabi	389	18.24	-	4.86
Trang	243	-	144	0.84
Satun	966	-	40	-
Total	**5,568**	**42.57**	**573**	**131.15**

Table 6. Damage to aquaculture (source: DoF, 2005)

Fig. 6. Fish cages - Smashed and washed ashore

Cultured species	%age lost	Distribution among provinces					
		Ranong	Phang-gna	Phuket	Krabi	Trang	Satun
Grouper	37.7	x	x	x	x	x	x
Sea Bass	35.0	x	x	x	x	x	x
Red Snapper	11.8	x	x	x	x	x	x
Cobia	10.1		x		x		
Green mussel	4.4	x	x			x	
Oysters	0.7				x		
Lobster	0.3						x
Total	100.0						

Table 7. Cage aquaculture loss (source: Tanyaros et al., 2010)

4.2 Shrimp farms
Many shrimp farms exist on the Andaman coast, but little damage to shrimp farm operations have been reported, due to the smaller number of farms located near the coast, as compared with the Gulf of Thailand coast. The DoF (2005) reported that only 42.56 ha of shrimp ponds in four provinces were totally damaged by the tsunami. The most damaged farms were located in the immediate vicinity of the shore of low-lying land.

4.3 Fish hatcheries
Only one private grouper hatchery was reported to have been damaged, while the government stations were reported not to have incurred any significant damage.

4.4 Shrimp hatcheries
Destruction of shrimp hatcheries considerably reduced production. The six affected provinces are the main areas in Thailand for marine shrimp fry production. The 573 hatcheries damaged accounted for a 30% loss in seed production, which translates into 70,000 metric tons of cultured shrimp for only one crop (Bueno, 2005). It was estimated that it would take at least six months to get most of the hatcheries back into operation. The industry speculated that this would be an additional lost opportunity of more than 28 M USD and would inevitably lead to shrimp seed scarcity and higher prices. We have not been able to ascertain whether reality matched speculation.

Province	Fish cage farms	Shrimp ponds	Hatcheries	Shellfish farms	Total
Ranong	4,405,403	-	-	93,434	4,498,837
Phang-nga	3,994,861	391,744	3,215,205	N/A	7,601,810
Phuket	3,083,026	12,385	513,551	820,319	4,429,281
Krabi	3,187,173	176,641	-	-	3,363,814
Trang	10,760,56	-	N/A	8,462	1,084,518
Satun	4,604,375	-	N/A	385,898	4,990,273
Total (USD)	**20,350,894**	**580,770**	**3,728,756**	**1,308,113**	**25,968,533**

Table 8. Financial loss to coastal aquaculture (source: DoF, 2005)

4.5 Shellfish concessions
Shellfish consist principally of cockle beds, green mussels, oyster and land-based abalone operations, all of which suffered some damage.

5. Inequitable compensation distribution

Careful scrutiny of some of the figures from the DoF reveals wide discrepancies among the six provinces. As a general rule, compensation for farmers in high-loss provinces tended to be higher than for formers in low-loss provinces. In other words, for similar type of loss, initially more well off fishers received more compensation than poorer fishers. Two examples are cage farms and boats.

5.1 Differential declared cage loss
From Table 9, we can see that some cage farms were estimated at a much greater USD loss in some provinces than in others. These differentials are illustrated in Figure 7.

Province	N° of fish cage farms	USD loss	Mean USD loss per farm
Ranong	677	4,405,403	6507
Phang-nga	3008	3,994,861	**1328**
Phuket	315	3,083,026	**9787**
Krabi	389	3,187,173	8193
Trang	243	10,760,56	4428
Satun	966	4,604,375	4766
Total	5568	20,350,894	3655

Table 9. Mean loss for fish cage farms.

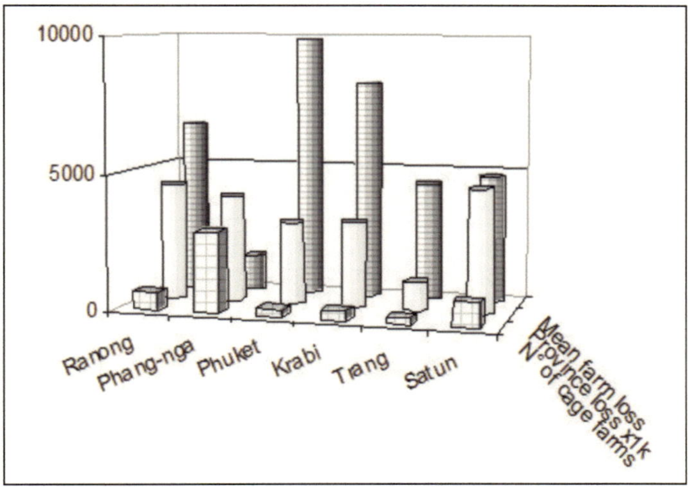

Fig. 7. Differences in declared loss for cage farms

Several factors may account for these differences. For example, farm size varies greatly across the provinces, and larger farms have more cages than smaller ones. Personal contacts with locals indicated too that many non-registered farms filed opportunistic claims for compensation. Additionally, this also illustrates the world-wide phenomenon of wealthier people having easier access to money. In the present case, Phuket fishers are generally better off than fishers from Phang-nga; thus they are better able to make claims, to make most use of the system, and to influence authorities (both legally and surreptitiously).

5.2 Differential declared boat loss

A similar situation existed for boats, although the great variation in boat sizes and types will account for some of the cross-province differences. It is clear from Table 10 and Figure 8 that, compared to the actual loss or damage (as a % of boats lost), the USD loss for all boats was estimated to be far greater proportionately in one province (Phuket) than in four other provinces (Ranong, Krabi, Satun, Trang). It is unclear what factors account for this huge discrepancy. It was probably a mix both of objective elements (such as boat size, Phuket may have had a greater proportion of larger, and thus more highly compensated, boats) and of human elements of the kind discussed above for cage farm loss. Even if Phuket did have a greater proportion of larger vessels, this does also reflect the relatively better off Phuket fisher.

Province	Total N° of boats	*% of boats lost*	USD loss for all boats	USD boat[2] value	*% of USD loss*
Phuket	1330	24%	1,884,618	1417	66%
Phang-nga	1246	23%	915,546	735	32%
Ranong	658	12%	12,331	19	0%
Krabi	1010	19%	19,269	19	1%
Satun	642	12%	20,520	32	1%
Trang	649	12%	0	0	0%
Total	5,535	102%	2,852,284	515	100%

1. Includes all categories of boat lost
2. USD value per boat = total value / N° of boats

Table 10. Differences in boat value and loss

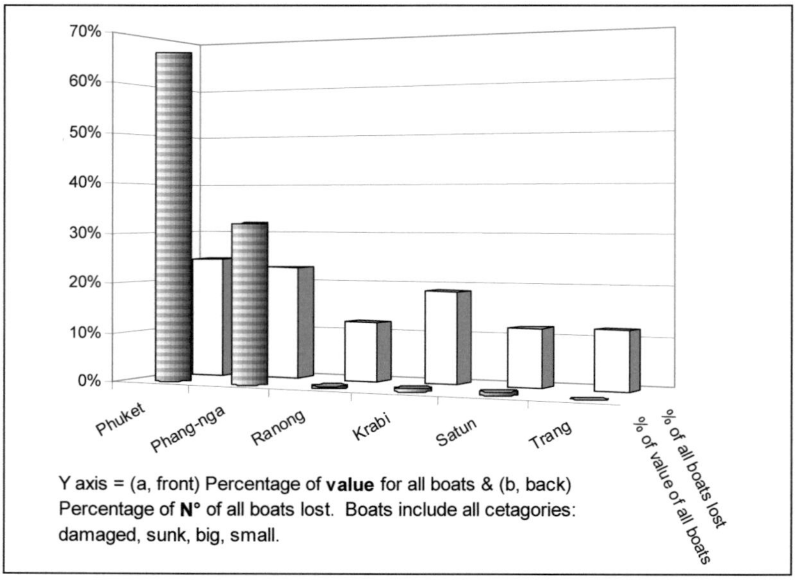

Fig. 8. Differences in boat value and loss, based on Table 10

These are examples of abstract data pointing up real-world enigmas. Compensation claims varied, from 'inaccurate' (at best) and fraudulent (at worst). This type of reporting of damage and loss was inevitable, given the greed element in human nature, and given the devastation inflicted at the time, with the ensuing social and economic disarray. Such data variation and the vagaries of such situations give an idea of the immense work undertaken to assess the loss and to calculate compensation. It can easily be argued that it is better for many to receive some form of quick compensation, even with fraud and corruption, than for fewer people to get delayed compensation, in the hope that less money will get into the wrong hands, but with inevitable hardship imposed on the really needy.

6. Post-rehabilitation

After the tsunami, several Thai government agencies, the private sector, NGOs and international organizations set up a wide range of projects for emergency assistance and long term rehabilitation, in both fisheries and aquaculture sectors, with the aim of enhancing the quality of life for all victims. We have already seen that this effort was not equitably distributed, and below we note that the unequal sharing of compensation led to friction within and among communities.

6.1 Fisheries sector

A first step in the assistance plan was swiftly implemented by the DoF; its primary goal was to established a Fisheries Rescue Centre (FRC). The centre coordinated and continues to coordinate the collection of damage and loss information from the six affected provinces for the Department of Fisheries. The detailed information is listed by village, and it was therefore possible to identify at sub-district and even village level where significant losses had been reported. This data set can therefore be used as an indicator to assist in identifiying the most affected areas.

The second stage of the DoF emergency plan was to compensate the victims in cash for their losses or damaged boats, fishing gear and aquafarms. The aim of this DoF compensation programme was to assist victims to start up their fishing and aquafarming activities and to restore their livelihoods. The full Cabinet of the Thai Government approved a tsunami response budget of 5,252 M THB. Of this, 1,343 M THB was allocated to the DoF to assist 27,828 fishermen. This included the repair of 3,426 small fishing boats (under 10 metres in length) and 544 larger fishing boats (DoF, 2005).

Loss or damage claims that might be eligible for government compensation were registered at the Provincial Fishery Office, and in some cases at the district level if a fishery officer was present. Boat owners (as listed on boat registration records) were required to register the loss with the registering district or province within 15 days in the area where the vessel was affected or the registered home place of the vessel owner. The Provincial Fisheries Office (or District Fisheries Office) had to collate the documents and check for accuracy before sending to the Department of Fisheries within 120 days. Typical documentary evidence required for compensation was the vessel registration or permission to fish document (with date). Since the majority of vessels that were damaged or lost were in the small-scale category, very few actually had registration documents. In this case, alternative evidence of ownership was considered as supporting evidence. These included:

- Statement of ownership by the Provincial Fisheries Office (stating the boat is repairable or beyond repair);
- Document of guarantee from a fishers' association, farmers' group or fishery cooperative stating that the owner is a member;
- Sub-district headman's statement;
- Guarantee document from the village headman;
- Record in the fishery survey;

We will now survey the compensation according to type of article damaged. We include fishing vessels and fishing gear.

6.1.1 Compensation for fishing vessels

Several categories of loss were eligible for compensation:

- Support for the recovery of a vessel (i.e., re-floatation, or movement of the vessel, as many had been swept some distance inland above the high tide mark).
- Support for repairs to a damaged vessel.

- Compensation for loss or damage beyond repair.

Fishing vessel damage and loss was assisted by the government compensation package as shown in Table 11.

Type	Compensation per vessel	
	<10m	>10mm
Retrieval	10,000	25,000
Repair	20,000	70,000
Vessel lost	66,000	200,000

Table 11. Compensation (THB) package for fishing vessels, damaged and lost (Source: DoF, 2005)

6.1.2 Compensation for fishing gear loss

Fishing gear loss is extremely difficult to assess, unless the gear is assumed to have been associated with the vessel. In this case, the vessel and the gear are assumed to have been lost together. Small-scale artisanal gear may be reconstructed (such as fish and crab traps). Larger gear such as nets, require purchase. Loss of fishing gear was compensated at a relatively low rate (about 3,000 THB per case). Table 12 shows the government compensation for the loss of fishing gear.

Type	Unit	Total USD budget*	Max. USD compensation / victim		
			Retrieval	Repair	Replacement
Small fishing boat	3,426 boats	6,676,308	256	512	1,692
Large fishing boat	1,222 boats	7,050,000	641	1795	5,128
Bamboo trap (legal)	421 fishermen	107,949	-	-	256
Other traps	13,690 fishermen	3,510,256	-	-	256
Nets	1,871 fishermen	479,734	-	-	256
Total		17,824,247			

Table 12. DoF budget for emergency assistance for the fishery sector (DoF, 2005)
* Cost (or calculated loss). Actual compensation payment was less than this.

6.2 Aquaculture sector

Loss or damaged of aquaculture facility was compensated by the Royal Thai Government to assist the farmers to re-start their farming activities. Two broad targets for compensation were aquaculture holdings and fish cage culture operations.

6.2.1 Compensation for aquaculture holdings

Information regarding loss from aquaculture holdings does not appear to have been announced, however the document made available to the mission containing the dates of compensation also included rates for aquaculture. Compensation was payable in the case of:

- Loss of fish stocks 1,400 baht per rai (for a total area not exceeding 5 rai).
- Loss of shrimp and crabs 3,800 baht per rai (for a total area not exceeding 5 rai).

6.2.2 Compensation for fish cage culture operations

Compensation was payable in the cases of culture of fish in a cage, cement tanks or other (e.g., aquarium fish, frogs, soft-shelled turtle). The rate was 150 baht per square metre of production up to an area not exceeding 80 square metres. The Thai government provided funding through the DoF of over 111.6 M THB to restore the cage farms on the coast of the Andaman Sea after the tsunami (DoF 2005). Excluded here are funds from the private sector, NGOs and philanthropic foundations.

Several types of assistance were provided from those organizations for the cage farmers, including provision of cash funds, materials for cage re-construction and fish seed. The money used to buy materials for cage re-construction and repair accounted for 28.1% of the funds provided; the purchase of seed, feed and chemicals for culture operation made up 15.8%. It was surprising to find that a substantial sum, 56.1% of the funds, was used for other purposes (Tanyaros et al., 2010). The loss of aquaculture operation systems was supported through emergency assistance from the government as illustrated in Table 13.

Type	N°of farmers	Total USD budget*	Max. repair & seed for restocking / farmer
Cage	27,828	1,4270,769	513 (cage & seed)
Shrimp pond	42	21,538	513 (pond & post larva)
Hatchery	573	293,846	513
Shellfish farm	80	40,770	-
Total	28,523	14,626,923	

Table 13. DoF budget for emergency assistance for the aquaculture sector (source: DoF, 2005). * Cost (or calculated loss). Actual compensation payment was less than this.

Anantasuk et al. (2008) reported that 69.7% of the main financial sources supporting the farmers in the effected areas came from government organizations, as listed in Table 14. Non-governmental organizations, at the Thai national and international levels, also played an important role in assisting the affected farmers. About 31.3% of the farmers received assistance, and each of those obtained 30,000 baht from these organizations. The DoF was the main body supporting aid to the farmers; more than half of the farmers received materials for cage re-construction, seeds, fishing gear, and boats. NGOs were also important bodies that furnished materials (Tanyaros et al., 2010). Table 15 gives the percentages of farmers who obtained materials from the DoF and various NGOs.

Level	Organisation	per farm
National	Department of Fisheries (DoF)	20,000
Local	Tambon Administrative Office (TAO)	2,000
Local	Education institutes (some farms only)	15,000

Table 14. Funding (THB) sources fish cage culture compensation

Target	Source %ages DoF	NGOs
Material for cage construction	55.6	47.6
Seed	11.1	18.3
Fishing gear	31.1	13.4
Fishing boats	2.2	20.7
Totals	100%	100%
Proportion from each source	57.1%	42.9%

Table 15. Source of materials support (source : Anantasuk et al., 2008)

6.3 Public satisfaction regarding post-rehabilitation

Since the tsunami on 26 December 2004, government agencies, private organizations and civic groups have undertaken a large number of projects to rehabilitate the economy and improve the well-being of people in the six tsunami-affected provinces. The actions varied from short- to long-term measures.

Using Weight Average Index (WAI), Tanyaros et al. (2010) reported on the level of public satisfaction regarding post-tsunami rehabilitation. The results show that the mean satisfaction in regard to quality of housing and land tenure was at a high level (3.71; on a scale of 1 to 5). Other measures of satisfaction were at the middle level. However, the mean quality of all parameters for the post-rehabilitation period (three years post-tsunami) was significantly higher in comparison to that of the pre-rehabilitation period (one month post-tsunami).

Parameters related to aquaculture farms follow a similar pattern. During the post-rehabilitation period the mean level of satisfaction with equipment quality and seed and feed invested in aquaculture farms increased. Satisfaction with farm areas, growth rates, amount of product and revenue from farms also improved. These results indicate that the immediate and mid-term rehabilitation actions have generally been fairly successful. In comparison to the pre-rehabilitation period, most post-rehabilitation parameters related to livelihood, farm and culture operations improved - see Table 16. Farmer satisfaction with the rehabilitation programs was overall fairly high. However, the farmers need more effective support, both financial and non-financial. Building the capacity of farmers to help themselves is one of the best practices for sustainable support.

	Parameters	Rehab WAI		t
		Pre-	Post-	
Livelihood	house and land tenure	3.40	3.71	**5.598****
	household consumer goods (basic necessities)	3.39	3.62	4.195**
	electricity & water	3.28	3.51	3.693**
	health care	3.39	3.58	3.427**
	education	3.39	3.64	4.933**
Farm	farm area	3.40	3.58	2.303*
	culture equipment	3.23	3.75	**6.501****
	seed	3.24	3.65	4.606**
	feed	3.22	3.46	2.701**
	growth rate	3.27	3.41	1.995*
	amount of product	3.25	3.43	2.179*
	marketing	3.21	3.26	0.444ns
	price of cultured species	3.18	3.17	0.069ns
	revenue	3.23	3.42	0.266*

** = highly significant level ($p<0.01$); * = significant level ($p<0.05$); ns = not significant; ($p>0.05$)

Table 16. Farmer satisfaction related to livelihood and farms (Source: Tanyaros et al. 2010)

6.4 Constraints for rehabilitation in the fisheries sector

So far the picture appears to be a rather rosy one, with compensation appearing, at least in monitory terms, to have had much potential for good. However, on the ground, the results of money compensation were far from desirable - due to both internal and external factors. Internal factors include greed and family structure; external include distribution methods. Here we outline each set of constraints and make suggestions as to how they may be alleviated, for the benefit of all and in such a manner as to avoid hardship on those who must give way to better, even best, practices.

7. Intra and inter-communal conflicts

The way in which compensation was given and monitored directly or indirectly caused serious problems, which in turn led to serious conflicts within communities and among communities in tsunami-affected areas.

One set of problems was related to the donation of fishing gear for tsunami victims by the government and private sectors; this encouraged an increase in the number of fishermen, sometimes beyond resource sustainable levels. This in turn generated conflicts between new and old fishermen; the old fishermen had well-tried fishing traditions and devices. In addition, this assistance was unequally distributed. For instance, some tsunami victims received a house, compensation packages, scholarships and children's allowances, fishing boats and gear, while others received little or none despite having lost their occupations and properties. Even worse, some tsunami victims who had never worked as fishermen were, incredulously, given fishing boats, which they then sold at great personal profit.

Another set of problems revolved round the migration of outside workers into the area, giving rise to social problems as well as improper use of the local natural resources. For example, outside workers caught protected aquatic animals. Thus, in addition to conflicts within and among local communities, conflict emerged between migrant workers and local communities (Department of Marine and Coastal Resources, 2005).

7.1 Illegal fishing

Fishing with trawlers and push nets still exists in several areas. Illegal fishing causes damage to marine life and benthos and also to coastal ecology. Before the great Indian Ocean tsunami, the status of fishery resources in the Andaman Sea was over-exploited. Many rehabilitation projects were introduced with the aim of enhancing the fishery resources of this area. However, illegal fishing is still practiced and constitutes a major obstacle to sustainable management of fishery resources. The government of Thailand has banned commercial operations in a near-shore (5km) zone in an attempt (a) to provide a nursing ground for aquatic animal, and (b) to reduce conflict with artisanal fisheries. In such cases, conflicts continue when trawlers encroach into these demarcated areas.

The DoF must play a significant pro-active role in controlling the use of non-destructive fishing gear by strict law enforcement. Allowing unfair and unsustainable competition for limited resources between artisanal and commercial sectors inevitably spawns human disputes and reduces fish spawning.

7.2 Management democratization

Fisheries experts now recognize that resource conflicts can be diminished and resources better managed when fishers and other resource stakeholders are more directly involved in management, and access rights are distributed more effectively and equitably. This is part of management democratization. The Thai government is increasingly committed to policies and programmes of decentralization and community-based management and co-management (Pomeroy, 1995).

The DoF has formulated a fisheries development plan for small-scale fisheries, with a view to improving both livelihood and quality of life in the sector. The short-term objectives aim to upgrade the socio-economic status of the small-scale fisher, develop the coastal fishing ground to be more productive, and promote and develop the fishing occupation. The long-term objectives seek to allocate the existing resources to achieve most benefit, and lessen conflict between the small-scale and commercial fishers.

Several small-scale fisheries development projects have been implemented to meet the short-term objectives. These include construction of artificial reefs and provision of basic infra-structure for fishing and community activities, such as fish processing and fish landing, as well as institutional support, such as credit and fishery information. However, we are compelled to warm that government development programmes alone cannot achieve the long-term objectives as long as the fishing grounds are left as an open-access resource and the enforcement of fisheries regulations is ineffective.

7.3 Constraints and needs for continuing rehabilitation

Before the December 2006 tsunami, many aquaculturists faced an array of problems in aquaculture operation. After the three-year rehabilitation period, observations (Office of Natural Resources and Environmental Policy and Planning, 2006) indicated that most aquaculture farm became or were on the way to becoming viable - commercially and environmentally. However, a number obstacles still needed to be overcome three years after the tsunami, and many still remain today (five plus years later). We outline some of the obstacles here, as well as some possible ways through.

7.4 Culture operation

The main operational constraints for cage farmers lie in the high cost and shortage of seeds. The current seed supply is problematic because the trend of seed catching in the wild has gradually diminished over recent years. As a consequence, the price of grouper seed has increased three- to four-fold over the past few years. The grouper and snapper hatchery technique was poorly developed, while the production of sea bass fry from the DoF hatcheries met the demand because the technique of nursing the fry was well developed.

One constraint is the high cost of trash fish, used as feed for the grouper, red snapper and cobia cage. Cost of trash fish is the single most expensive item, comprising more than 40% of the total production cost. Typically, trash fish from fishing operations are used as feed source, although in times of low catch, trash fish may be purchased for use as feed. Pellet feed is available only for sea bass, but not for grouper, red snapper and cobia cage culture.

Fish disease and mortality were the most common reported problems associated with cage culture. The problem of disease was experienced by all cage culturists.

7.5 Marketing

Three years after the tsunami, cage farmers in general were faced with fluctuating market situations. With an increasing number of cage culturists and rising demand for live fish, an excess in supply over demand developed. Marketing of live sea bass and grouper was also seriously affected. The local farmers found it difficult to compete with the prices set by the middlemen. The marketing problem appeared in Cobia culture when fish reached marketable size.

7.6 Technical support

The need for future support for sustainability in rehabilitated aquaculture farms on the coast of the Andaman Sea can be divided into three categories:

Providing seed for grow-out. As mentioned earlier, the problem of seed shortage had a major impact on the cage farms on the Andaman coast of Thailand, and occurred before the impact of the 2004 tsunami, especially in connection with the grouper seed. High demand for seed and high seed cost originated in the decline in wild seed. All cage farms in this area used the grouper and red snapper seed collected in the wild, whereas the sea bass seed was mainly bought from the DoF hatcheries. Therefore, the DoF government sector should pay attention to research and development of hatchery seed production techniques. As seedling resources are developed, fish seed holding centers will be needed. Each center should have the dual function of supplying low-cost fish seeds to cage farms, and undertaking nursery and quarantine work to ensure optimal survival rate of the fish seed.

Formulated feed development. In response to the limitations and problems associated with trash fish use for grouper and sea bass aquaculture, new and better practices are being developed. Foremost among these is the use of formulated feeds instead of trash fish. The economic benefit of using formulated feeds is that it is more cost-effective than trash fish. Because formulated feeds provide a nutritionally complete diet, fish grow faster and are healthier in comparison with fish that are fed trash fish. The Research Institute for Mariculture, Gondol, Indonesia, showed that grouper grew 75% faster when fed formulated pelleted feed, compared with grouper-fed trash fish (Sim et al. 2005). Formulated feeds also generate less pollution, since water stability is better and less of the feed is wasted. In turn, this provides a better environment for the fish, leading to a reduction in disease problems.

Disease control and prevention. From a study by Tanyaros et al (2010), it became clear that disease was a major factor affecting loss of production in cage farms. In order to diagnose the disease, fish samples needed to go to a laboratory. Fisher farmers would contact the DoF laboratory technician to take their specimen to the closest DoF fishery station, but some DoF technicians were slow to respond. Sometimes also the fish died before a DoF technician arrived. Subsequently, most cage farmers were not consistently helped by the government sector and turned to local wisdom from past experience rather than waiting for the government sector to respond. Today, cage farmers still need the public sector fish disease monitoring programs, so that measures can be implemented promptly if disease breaks out.

8. Lessons learned and opportunities for the future

Probably more time and detailed data are needed to provide sufficient distance to evaluate fully the whole rehabilitation effort. However, people cannot wait for that; they must live now. Thus, five years on, several promising opportunities present themselves. We provide a patchwork of lessons so far and opportunities to be seized now.

8.1 Creating supplementary income

The Community Development Office in affected provinces should provide fishers with additional skills to work in their communities in areas such as batik painting, home-stay tourism services and repairing fishing gear.

8.2 Training programmes

To facilitate the initiation of cage culture development, the creation of a core group of functional personnel should be an activity of high priority. This core group should provide the initial personnel capacity in handling farm management as well as the whole range of tasks of a production farm. It should also be capable of participating in the planning and the subsequent implementation of production and marketing programs.

8.3 Development of a marketing system

The marketing system and the basic essential supporting infrastructure both play a key role in the well-being of any fishery. The experience gained within the limited time spent in the field has clearly indicated that these aspects of the fishery industry need urgent review. In cases of failure and success, research should be conducted to establish and understand the underlying reasons. Thus, patterns will emerge and help to make future endeavours more effective. This should be carried out now as an ongoing activity.

8.4 Marketing statistics

A representative inventory of the basic information on the biological and economic aspects of cage farms should be kept to serve as a reliable source of data on which future development actions can be reliably based.

8.5 New marketing techniques – local branding

Techniques of marketing that have worked elsewhere should be implemented for Andaman fisheries. For example, local branding has proven extremely effective in many countries and localities attempting to sell their produce elsewhere and abroad. Such marketing techniques may be new to Andaman fishers and their cooperatives, but with care, they can be implemented relatively easily and inexpensively. One can easily imagine labels like "Phang-nga Grouper" or "Trang Oysters". Local branding for fish can also take advantage of local branding for tourism in the Andaman provinces. Secondary effects of local branding include greater pride and thus care by local fishers, greater sense of long-tem goals, greater inter-village solidarity. One province has already begun to look seriously at local branding as a way of marketing its fish -- we invite you to guess to which province this quote refers:

> The human geography of the province consists mainly of hundreds of settlements built around the coves and inlets that stretch along our coasts. ... fishing has been the foundation of the local economy and fishing continues to be one of our primary industries. Moreover, those picturesque fishing villages are also the main lure of the tourist trade - another primary industry. Our cultural geography is, likewise, constructed around fishing and the sea; ... fishing serves as a cultural self-identifier. (Bousquet, 2009.)

We provide the anwser in Appendix A.

8.6 Financial support

Micro-credit arrangements/revolving funds. It is important to provide fish farmers with a low interest grant for revolving funds that they can borrow for their small business.
Improved compensation package. As mentioned earlier, the compensation package for cage farms is rather low compared to the loss, especially for the poorer farmers in some provinces. Compensation levels to replace essential livelihood inputs must be realistic. If people cannot replace their equipment, then assistance just is not effective

8.7 Greater coordination and communication

It is critical to establish coordination among all players. A lack of effective and comprehensive coordination and communication mechanisms has been the single biggest constraint on assistance. It is clear that assistance could have been faster and more effective with better coordination, and that greater input and participation from the fisher community could have facilitate more realistic and responsible decisions.

The benefits are increased when donors collaborate in complementary activities. Assistance could complement government policies and activities to rehabilitate the fisheries sector. Donors need guidance on what kind and how much assistance to provide, and who needs it. Sometimes, external donors have trouble engaging with local communities. Some kinds of assistance or distribution may not match actual need, or they may duplicate effort. Inappropriate assistance can be detrimental, and even cause community conflict. Communities should be fully involved in identifying priorities for assistance and in the selection of beneficiaries. Focused assistance is required to strengthen internal community coordination.

8.8 Need for a strategic approach

Mid-term and longer-term assistance to rehabilitate livelihoods should be carefully planned, based on assessments of actual need, not assumed needs. This means creating channels so that the views of the people affected are heard, and can be acted upon. It cannot be top-down planning; grass-roots initiative, involvement and coordination are essential.

8.9 Focus on self-help

Building the capacity of communities to help themselves is the only way to sustain support. Revolving funds and micro-credit arrangements are useful tools and an alternative to donations. Most farmers depend on loans, economic activities cannot generate enough income and the farmers need the revolving funds to subsidize their livelihood and maintain their farm. Therefore, good implementation and management of credit programs should be promoted.

8.10 Conflict management,

Communities can be strengthened by conflict management mechanisms that involve them directly. These might include the following:

- Independent local ombudsman can be appointed for small conflicts, as long as this respected local cultural norms.
- Local committees can be set up to look into larger conflicts, with possible input from research institutes.

8.11 Participatory methods

Innovative methods can be introduced, such as Companion Modeling, used successfully in other parts of Thailand (see Barnaud, 2007, for a remarkable example). The closely related technique of computerized role-play simulation has been effective with subsitence fishers (see Cleland et al., forthcoming). Methods such as these also provide greater knowledge about and more enlightened involvement in fishery dynamics and sustainability and thus lead to greater harmony among stakeholders.

Methods such as SWOT (Strengths, Weaknesses, Opportunities and Threats) might be applied to specific actions, and provide increased probability that assistance will have the desired effect.

8.12 Research

Of course, research is also a key to rehabilitation development. Rigg et al. (2005) put it well:

> In addition to the spectacular, tragic and obvious destruction, the tsunami event has produced numerous socially transformative challenges and opportunities that necessitate longer-term research engagement.

Any rehabilitation action, short- or long-term, needs to be backed up with rigorous studies by independent researchers.

Local institutions and authorities are the best placed to understand and assist local communities. Enhancing local capacity to manage rehabilitation programs is essential to sustain them. Long-term support is needed beyond the funding horizons of most donors.

The question that arises in complex situations such as these is: How far will people, groups and governments act on the knowledge that is available and that is generated through research? The connection between knowledge and action is an uneasy one - for a discussion see Crookall and Thorngate (2009).

9. Conclusions

Finally, we need to heed these words.

> ... the havoc ostensibly wreaked by nature is often a reflection of the ways that humankind makes itself vulnerable to natural phenomena. And it is humankind that may have the best solution to these problems. ...
>
> Yet two measures of critical importance, "vulnerability mapping" - identifying potential humanitarian crisis threats - and information sharing, are still fraught with conflicting interests, hampering the dialogue needed between scientists and policymakers. To make a real difference, natural and social scientists, policymakers and planners must work more closely to identify threats, ... Social anthropology and psychology have as much to contribute as engineering, agriculture, economics and medical assessments.
>
> As was evident in the aftermath of Hurricane Katrina, the knowledge of social scientists about ways to engage local communities is often bypassed by the "experts." ... Most governments say they "use" scientists; that they are in the room when it comes to policymaking. They may be in the room, but they are not at the table.
>
> As the prospect that more complex and devastating humanitarian crises are on the rise, the dialogue between scientists and policymakers is needed more than ever. (Kent, 2010.)

10. References

Anantasuk R., Songrak, A., Tanyaros, S. & Sangchan, S. (2008). *Socio-economic status of the post-tsunami cage aquaculture farmers along Andaman Sea coast, Thailand*. The International Institute of Fisheries Economics & Trade (IIFET 2008), Nha Trang, Vietnam, July 22-25, 2008.

Barnaud C., Promburom, T., Trébuil, G., Bousquet, F. (2007). An evolving simulation/gaming process to facilitate adaptive watershed management in northern mountainous Thailand. *Simulation & Gaming: An Interdisciplinary Journal*, Vol. 38 No. 3, September 2007 398-420.

Bueno, P.B. (2005). *Impacts of Tsunami on Fisheries, Coastal Resources and Human Environment in Thailand*. 4th Regional Network of Local Governments Forum, 27 April 2005, Bali, Indonesia.

Brown, B. E. (2005). The fate of coral reefs in the Andaman Sea, eastern Indian Ocean following the Sumatran earthquake and tsunami, 26 December 2004. *The Geographical Journal*, Vol. 171, No. 4, December 2005, pp. 369–386.

Cleland, D., Anne Dray, A., Perez, P., Annabelle Cruz-Trinidad, A., & Geronimo, R. (forthcoming). Simulating the Dynamics of Subsistence Fishing Communities: REEFGAME as a Learning and Data-Gathering Computer-Assisted Role-Play Game. *Simulation & Gaming: An Interdisciplinary Journal.* doi:10.1177/ 1046878110380890

Crookall, D. & Thorngate, W. (2009). Acting, Knowing, Learning, Simulating, Gaming. *Simulation & Gaming: An Interdisciplinary Journal*, 40; 8.

DoF. (2005). *The compensation package for cage farms* [Cited 10 June, 2007) Available from http://www.cffp.th.com/tsunami.

Department of Marine and Coastal Resources. (2005). *Rapid assessment of the tsunami impact on marine resources in the Andaman Sea, Thailand.* Phuket Marine Biological Center, Phuket, 76 pp.

FAO. (2004). *Governance of small scale fisheries.* Fisheries global information system. FAO, Rome. http://www.fao.org/figis/servlet/topic.

FAO. (2005). *Thailand: Fishers and fish farmers.* FAO data base.

FAO. (2005). *Impacts of the Tsunami on Fisheries, Aquaculture and Coastal Livelihoods.* http://ftp.fao.org/FI/DOCUMENT/tsunamis_05/jan_07_2005/tsunami_7_1_05.ht m

FAO-MOAC. (2005). *Report of joint FAO/MOAC detailed technical damages and needs assessment mission in fisheries and agriculture sectors in tsunami affected six provinces in Thailand.* Food and Agriculture Organization of the United Nations (FAO) in cooperation with Ministry of Agriculture and Cooperatives (MOAC), 126pp.

Harinarayana, T and N. Hirata. (2005). Destructive earthquake and disastrous tsunami in the Indian Ocean, What next? *Gondwana Research*, 8: 246-257.

Israngkura, A. (2005). *Economic impact of tsunami on Thailand.* Natural Resource and Environment Program. Thailand Development Research Institute (TDRI). 27 pp.

Kent, R. (2010). It's good to talk: A new dialogue is needed between policymakers and scientists to tackle natural crises. *Prospect: Good Writing about the Things that Matter*, October 2010.

Kerr, T. (ed) (2005). *How Asia's precarious coastal communities are coping after the tsunami.* Special issue of *Housing by People in Asia*, N° 16, August 2005. Asian Coalition for Housing Rights

Kohl, P A., O'Rourke, A. P., Schmidman, D. L., Dopkin, W. A., Birnbaum, M. L. (2005). The Sumatra-Andaman earthquake and tsunami of 2004: The hazards, events, and damage. *Prehospital and Disaster Medicine*, Vol. 20, No. 6. November - December 2005

Nakaoka, M., Y.Tanaka, H. Mukai, T. Suzuki, C. Aryuthaka. (2007). Impacts of Tsunami on biodiversity of seagrass ecosystems along the Andaman Sea Coast of Thailand: (1) Seagrass abundance and diversity. In: Rigby, P.R., Shirayama, Y. (Eds.), Selected Papers of the NaGISA World Congress 2006, pp. 49–56.

Nootmorn, P. (2006). Impact of Tsunami aftermath with marine fish resources and fisheries along Andaman Coast, Andaman Sea Fisheries Research and Development Center, Phuket, Thailand

Office of Natural Resources and Environmental Policy and Planning. (2006). *Two years after tsunami: restoration of Thailand's natural resources and environment.* Ministry of Natural Resources and Environment. Bangkok, Thailand. 150 pp.

Panjarat, S. (2008). *Sustainable fisheries in the Andaman Sea coast of Thailand.* Division for Ocean Affairs and the Law of the Sea, Office of Legal Affairs, The United Nations, New York, USA.

Paton, D., Gregg, C.E., Houghton, B.F., Lachman, R., Lachman, J., Johnston, D.M & Wongbusarakum, S. (2007). The impact of the 2004 tsunami on coastal Thai communities: assessing adaptive capacity. *Journal compilation.* doi:10.1111/j.0361-3666.2007.01029.x

Pomeroy, R.S. (1995). Community-based and co-management institutions for sustainable coastal fisheries management in Southeast Asia. *Ocean & Coastal Management.* 27:143-162.

Rigg, J., Law, L., Tan-Mullins, M. & Grundy-Warr, C.. (2005)The Indian Ocean tsunami: Socio-economic impacts in Thailand. *The Geographical Journal,* Vol. 171, No. 4, December 2005, pp. 369–386.

Sawall, Y., Phongsuwan, N. & Richter, C. (2009). Coral recruitment and recovery after the 2004 Tsunami around the Phi Phi Islands (Krabi Province) and Phuket, Andaman Sea, Thailand. *Helgol Marine Research.* DOI 10.1007/s10152-010-0192-5

Silvestre, G.T., Garces, L.R., Stobutzki, I., Ahmed, M., Santos, R.AV., Luna, C.Z. & Zhou W. (2003). South and South-East Asian coastal fisheries: Their status and directions for improved management: conference synopsis and recommendations, p. 1 - 40. In Silvestre, G. Garces, L. Stobutzki, I. Ahmed, M., Valmonte-Santos, R.A., Luna, C., Lachica-Aliño, L., Munro, P., Christensen V. & Pauly D. (eds.) *Assessment, management and future directions for coastal fisheries in Asian countries.* World Fish Center Conference Proceedings 67, 1120 p.

Sim, Y.S., Rimmer, M.A., Williams, K., Toledo, J.D., Sugama, K., Rumengan I. & Phillips, M.J. (2005). *A practical guide to feeds and feed management for cultured groupers.* Publication No. 2005–02 of the Asia-Pacific Marine Finfish Aquaculture Network. 18 p.

Siripong, A., Yumuang, S. & Swangphol, N. (Undated). *The assessment of the damage from Indian Ocean tsunami on the coastal morphology of the Andaman seacoast of Thailand by remote sensing.* Available at http://pdfcast.org/pdf/the-assessment-of-the-damage-from-indian-ocean-tsunami-on-the-coastal-morphology-of-the-andaman-seacoast-of-thailand-by-remote-sensing.

Srivichai, M., Supharatid, S. & Imamura, F.. (2007). Recovery process in Thailand after the 2004 Indian Ocean Tsunami. *Journal of Natural Disaster Science,* Volume 29, Number 1, 2007, pp3-12

Tanyaros, S., Songrak, A., Anantasuk, R., Sangchan, S. & Templer, B. (2010). Post-tsunami rehabilitation of cage farms on the Andaman coast of Thailand. *Aquaculture International,* 8 (4) 693-703.

Thanawood, C., Yongchalermchai C. & Densrisereekul, O. (2006). Effects of the December 2004 Tsunami and disaster management in southern Thailand. *Science of Tsunami Hazards*, 24 (3) 206-217.

Whanpetch, N., Nakaoka, M., Mukai, H., Suzuki, T., Nojima, S., Kawai T. & Aryuthaka C. (2010). Temporal changes in benthic communities of seagrass beds impacted by a tsunami in the Andaman Sea, Thailand. *Estuarine, Coastal and Shelf Science*, 87: 246–252.

World Bank. (2006). *Total GDP 2005, World development indicators database*. 1-1 pp.

Wilkinson, C., Souter, D. & Goldberg, J. (eds) (2006). Status of coral reefs in tsunami affected coiuntris: 2005. GCRMN (Global Coral Reef Monitoring Network) & Australian Institute of Marine Science.

UNEP. (2006). *Tsunami Thailand.: One year later*. UN, Bangkok. 118 pp.

Appendix A: Local branding for Andaman fish

This appendix contains quotes about local branding, either in fisheries or in poor islands. Local branding for Andaman fish could give a sustainable boost to fisher livelihoods, on condition, of course, that the sustainability of fish is managed correctly.

> On the Caribbean island of Tobago, a small group of local women have formed a loose **affiliation** to produce fruit preserves, jams and jellies, pepper sauce, herb seasonings, etc. under the **label** Taste of Tobago. They were assisted by the Caribbean Agriculture Research and Development Institute (CARDI) in food processing techniques and other support was provided by the Marketing Department of the Tobago House of Assembly. Their products have become **known** for their high quality, both locally and abroad, and recently won a European award.
> Abdool and Carey 2004: Making All-Inclusives More Inclusive: A research project on the economic impact of the all-inclusive hotel sector in Tobago for the Travel Foundation.
> **Citizen groups** are particularly demanding and active. Under this scenario, **fish farmers** become **actors** acknowledged for their efficient management of ecosystems. Its success is based on ... **local branding**. Fish farming benefits from **local** development plans, cultivates regional roots but must take considerable account of scrutiny ... of its practices and production conditions. Available at http://www.international.inra.fr/research/some_examples/possible_scenarios_for_french_fish_farming_2021

Local catch key to fisheries by Tim Bousquet

Our quote from Bousquet (2009) refers to a very different part of the world, but one where fishers have similar concerns. It is about Nova Scotia, Canada.

> We can save Nova Scotia's fishing communities and the fish they depend on, by bringing the local food movement to the sea. On the one hand, Nova Scotia pretty much *is* fish. The human geography of the province consists mainly of hundreds of settlements built around the coves and inlets that stretch along our coasts. ... fishing has been the foundation of the local economy and fishing continues to be one of our primary industries. Moreover, those picturesque fishing villages are also the main lure of the tourist trade---another primary industry.
> Our cultural geography is, likewise, constructed around fishing and the sea; from the ballads and songs about Nova Scotia to the decor of the student bars downtown, fishing serves as a cultural self-identifier.
> February 26, 2009. From http://www.thecoast.ca

Appendix B: Fishers' stories

This appendix contains extensive extracts from a wonderful journal (Kerr, 2005), published about six months after the tsunami. They are provided here because they convey and illustrate the down-to-earth, real, **human side** of our rather abstract, statistical, factual, generalizations in our main article. (Our emphasis in bold.)

> After the tsunami, a lot of NGOs, activists and professionals who'd cut their teeth on problems of urban poverty and housing suddenly found themselves plunging into the unfamiliar work of emergency relief and coastal village rehabilitation. The scale of the calamity meant that everyone's help was needed and **no time to be fussy about qualifications**. But as they got deeper into the work, a lot of these city folks found that the underlying forces at the seaside are not so different than those in the city after all. These fisher folk might have had better incomes and living environments than urban slum-dwellers, but **when the tsunami swept away their homes, families, friends, communities, boats and jobs, it left them poorer than anybody**. And like their urban cousins, **these battered fishing communities suddenly found themselves facing some of the same big, structural issues of land commercialization which push people out of the places they need to be to survive** - whether it be in the city or near the sea.
>
> For these urban groups, the tsunami has been a development challenge on a gigantic scale. And some of the old techniques and organizing principles which have helped urban slum dwellers find solutions to serious problems of land tenure, housing and livelihood have come in quite handy in this dire, new seaside setting: techniques like network building, surveying, making space for people to come together and plan together, negotiating in blocks with government, model house exhibitions, using inaugurations ... to create alliances and showcase people's ideas.
>
> Of course there's much need for help with rescue, medical aid, food, shelter and funds. But the **tsunami has shown that even badly traumatized disaster survivors can begin taking care of things** very early on, organizing themselves, coming together to talk, plan and work. Getting busy like this sets a healthy ball rolling which makes the long-term process of rehabilitation go much faster and better. Activity can be an excellent post-trauma therapy and one of the best antidotes to the helpless victim mind-set, which reduces formerly active people to passive recipients of what everyone else thinks they need.
>
> The notion that disaster survivors need to be centrally involved in decisions about their own rehabilitation is not a radical new concept, but a lot of governments and aid organizations still don't get it. Disasters bring out a daddy-knows-best attitude in many of the best-intended state agencies and aid institutions.
>
> There's no point waiting for the formal systems, with their bureaucracies and professionals, to solve the problems - the tsunami is too big, too sudden, too complex, too off-the-map for any of the conventional systems of governance to understand, much less deal with. So instead of waiting for the system to do what it can't do, it's possible for affected people to initiate things, find ways of resolving these huge problems in their own ways. The tsunami is proving that when this happens, people can show their societies ways of dealing with calamities and these **people-driven solutions can become part of the system**. Plus, **if the victims themselves, who know best what they need, are centrally involved in all aspects of relief and rehabilitation, there will be fewer aid mismatches, fewer conflicts, less waste**.
>
> Natural disasters hit everyone in their path, rich or poor. But disaster relief and rehabilitation are seldom so democratic. The crisis in many disasters is that those with the right credentials get the aid coming through official channels, while those without don't: you need proof of residence to get a place in the relief camp, ..., title deeds to get housing compensation, registration certificates to get new boats, etc. In these ways, disaster aid can become an extension of the unbalanced power

structures which enriched some and impoverished and marginalized others in the first place. So another lesson in the Asian tsunami has been that the relief process must also target the untitled, the unregistered, the unlisted, ... and the undocumented, so they don't fall through the cracks. It may comfort some to think that land-grabbers are going to rot in hell for their sins against the poor, tsunami-battered fisher folk they are trying to dispossess. But in the here-and-now, imperiled communities need more than divine retribution to defend their turf and to convince a sometimes-greedy system that in fact, their villages, their livelihoods and their ways of life can co-exist quite nicely with development. They can only do that by developing pragmatic, workable solutions. And this involves gathering accurate information, preparing alternatives and using all the tools of networking, negotiation, good design, ecological sensitivity and persuasiveness to battle the powerful and unscrupulous commercial forces that are only too ready to use such catastrophes to eliminate obstacles in the way of their plans.

In Thailand, there was a lot of hand-wringing about overdevelopment and environmental destruction in tsunami-hit areas, and calls for more stringent shoreline planning regulations, ... But in southern Thailand, where tourism is law, **the greater threat** to tsunami-hit coastal communities comes not from environmental controls but **from economic forces**. In April, the government tabled a new bill that would create "special economic zones"... Then in June, the government's Designated Area for Sustainable Tourism Administration (DASTA) began drafting tourism development "blueprints" for several tsunami-hit areas, drawing more lines and readjusting land-use patterns **without consulting any local communities**, many of which were already exhausted from six months battling to keep their old land.

The alternative

Thailand's **embattled coastal fishing communities** with unclear tenure rights have also employed the strategy of reoccupying their former land and rebuilding as soon as possible, to **assert their rights** to the land, and to negotiate from a position of occupation. Especially in cases where powerful **business interests are making spurious ownership claims** to the villagers' land. And so far, in most of the cases, the reoccupation strategy has paid off, and the communities have been able to negotiate some kind of secure land rights, either through long- term lease, or land-purchase, land-sharing or relocation to nearby land.

The fishing villages along Thailand's Andaman coast have a long **history of being pushed around** in a titan's **game of money and power**. A lot of this land, especially in the worst affected province Phangnga, used to be public land. Despite being already occupied by long-established fishing villages, huge tracts of this land have been concessioned out over the past century for commercial exploitation, first to tin mining companies and later to shrimp farming interests. Now tourism is upping the pressure to chuck these perpetually vulnerable communities off their ancestral land. Many of the resorts where so many foreign tourists perished were built on land previously occupied by fishing communities. Those which survive have faced increasing threats of eviction. The **land status of most of the tsunami-hit communities is extremely precarious**. Some are on public land (under the control of many different ministries and government agencies and subject to many different policies), some are on national park land while others are on land being fought over by two owners or claimed by private businessmen. Though people have lived here for decades - even centuries - many have no title deeds or lease contracts, and therefore considered by some to be illegal squatters. Even within these villages, the tenure situation is a messy patchwork of murky tenure rights, conflicting claims of ownership, spurious land titles and criss-crossing land disputes.

What the tsunami has done is to tear open and aggravate all these already difficult issues of land: who determines how it's supposed to be used and who has the right to use it.

For the affected communities and their supporters, the biggest and most difficult post-tsunami project has now become finding ways to resolve the overlapping "traditional" land rights of these fishing villages, and the so- called "legal" land rights of the speculators, developers and politicians. **Post-tsunami land grab resumes in earnest just a few weeks after the waves.** For a while after the tsunami, there was a lot of hand-wringing in the press about the ravaged coastal ecosystems even before the waves, vanishing mangroves, illegal resort building, banished indigenous peoples and unrestrained capitalism. This rhetoric wasn't much help for Andaman villagers, however, who found that after a very brief lull, the assault on their land rights was resumed with even greater energy and viciousness than before. Of the 47 villages destroyed by the tsunami, at least 32 are now embroiled in serious land conflicts - about half of these in the province of Phangnga. In the village of Ban Nam Khem alone, more than 80 court cases over disputed land have been filed since the tsunami, mostly by wealthy capitalists.

In recent months, variations on the same story have played themselves out in 30 or 40 villages, as armed thugs, policemen, officials and perspiring lawyers try to prevent villagers from returning to their land. And it's not only private land-owners. In other cases, local administrative bodies have conjured up bogus civic projects or newfangled zoning plans as a pretext for preventing villagers from rebuilding houses on the public land they have occupied for ages. Because both local and national politicians have been partners in - or beneficiaries of - schemes to commercialize the Andaman coastline, the **government's role in managing these public lands is deeply compromised by conflicts of interest**. To these powerful interests, the tsunami has been like a prayer answered, since it literally wiped the coast clean of the last communities which stood in the way of their plans for resorts, hotels, golf courses and shrimp farms. As far as they're concerned, these ruined villages are now open land! Senator Chirmsak Pinthong, on a recent tour of tsunami-hit areas to investigate land rights, put it this way: "**The developers have tried before to chase people away. Now the tsunami has done the job for them**." Under Thai law, squatters can apply for legal title to a plot of land after 10 years of continuous occupation. In practice, few succeed and millions of people around the country continue to live on what is technically public land in a kind of legal limbo, without papers, without clear rights. Speculators exploit this ambiguity by using various "informal" means to get land purchase records back-dated or documents issued in their names, and then accusing villagers of encroaching. **Battles over land title are common, particularly where tourist dollars are at stake**.

But the **voices of the fisher folk who want to go back to their land have become very strong now**. Newspapers are filled with stories of small fishing communities fighting the fat cats to reclaim their land and rebuild their lives. Their land problems have also come out in a series of well-publicized meetings and large seminars organized by different ministries and civic groups in the aftermath of the tsunami. Behind the scenes, people's groups and prominent figures have also been lobbying government advisors and officials, on these people's behalf. In these ways, **the issue of land for these traditional coastal communities has become much more open**. There is now more information, more discussion, more awareness of the needs of these fishing communities among all the groups involved in tsunami rehabilitation. All this public discussion has helped to slow down the land grabbing (and the eviction of traditional fishing villages) considerably.

"Our most urgent need is to repair our boats and replace our lost fishing gear so we can earn a living again." (Ahlee Charnnam, a fisher from Koh Poo, Krabi). ... The government's compensation scheme promised fishermen [money] who lost registered boats ... . Thousands of the lost boats were not registered, however. In these cases, Deputy Prime Minister Suwat Liptapanlop announced magnanimously that "fishermen whose unregistered boats or fishing equipment were destroyed ... will get compensation equivalent to 70% of the actual damage, even though they have no right to claim compensation."

Either way, it was officials deciding who got compensation and who didn't, leaving plenty of room for **manipulation** and **palm-greasing**. By February, only a fraction of fishermen with registered boats had got the compensation, while others were still going through endless red-tape to verify their claims. Anyway, all the fishermen agreed the compensation was too little to buy new boats or even to repair damaged ones.

It didn't take long for fishermen in many different areas to come to the same conclusion: that there was **no point waiting around passively for state assistance**. Their clearest need was to get their boats fixed (or built new) as soon as possible so they could get start earning their living again. But how?

As Hat Thip, the community leader in Sang Kha-U village says, "All of us fishermen used to know how to repair and even build our boats. But in many communities nowadays, we just buy our boats, so our skills have gotten rusty." That's why the Community Planning Network, the Save the Andaman Network (SAN) and CODI decided to support several of these communities by securing donation funding and hiring some local skilled boat-builders to work with the communities, and teach them how to build boats. These networks encouraged fishing communities to identify their needs, draw up their own restoration plans and manage them as a group. Boat-building and boat-repair was item number one on most villages' list. Promoting activities like boat building was not only a strategy for reviving the spirit of self reliance, but also a means of building villagers' collective strength to tackle other problems in the future.

The first community boat yards were set up at Sang Kha-U (on Koh Lanta Island in Krabi) and at Batuputeh (on Koh Libong Island in Trang). Word spread quickly, and by early March, about 20 boat yards were in operation in tsunami-hit villages. The number grew after the first boat-building exchange was organized in March, in which a big group of fisher folk from 10 coastal villages in Krabi and Phangnga came to see the boat-yard at the Sang Kha-U community and to learn about how to repair boats.

All these **boat-yards are managed by committees of local fishermen**, who make their own rules about how to use the grant funds and set criteria for who gets new and repaired boats first. In Ban Nam Khem, the town which lost the most boats, the boat yard they set up at the Bang Muang relief camp began with only ten fishermen and a seed grant of 100,000 Baht (US$ 2,500) from CODI. This wasn't enough for a full boat (which costs 130,000 Baht), but it allowed them to start building right in the middle of this big, busy, much-visited relief camp. Two boat-builders from Krabi were hired, whom the local fishermen assisted and learned from. Once visitors could see this clear, concrete self-help activity, the Bang Muang boat yard became an **attractive target for donations** (especially from Toyota and Cement Thai) which allowed the boat-building fund to grow rapidly. Like many other groups, the Bang Muang fishermen adopted a system where the fishermen repay half the cost of their new boat as soon as they start fishing again. This money goes back into the fund to finance more boats.

In the first of March, the fishermen at Baan Nam Khem launched the first batch of 15 new fishing boats they had built themselves (out of a target of 300 new boats!). ... Before putting out to sea, the men gathered at the boatyard to pray and have their boats blessed with colorful pieces of cloth tied around the prow for good fortune. By mid-May, 34 community boat yards had been set up in the six affected provinces and 700 boats had been repaired or built, with another 1,300 damaged boats yet to finish. There is still a long way to go, and many are still in dire need, but it's a considerable dent in the problem.

The boat yards have given fishing communities devastated by the tsunami another common point for building their **networks for mutual support**. Buoyed by the renewed **sense of community**, many villages are extending their work with community boat yards to start providing boat repair services to other villages and setting up new collective projects such as **community fish markets**, to cut out the middlemen. Many of the villages have also learned from each other that setting up community savings groups can help them with low-interest loans - and wean them from extortionate loan sharks as well.

"My boat is everything. My life totally depends on it. When the tsunami came, it destroyed our boats and put an end to everything for all of us. But now we have the boats back and our livelihood as well. We can go fishing again."
(A fisherman from Ban Hualam Village on Koh Lanta Island, whose community boat-building group has almost finished rebuilding the 70 boats that were destroyed by the tsunami.)

17

Tide Gauge Observations of the Indian Ocean Tsunami December 26, 2004, at the Río de la Plata Estuary, Argentina

Walter Dragani, Enrique D'Onofrio, Dernis Mediavilla
Walter Grismeyer and Mónica Fiore
Servicio de Hidrografía Naval and UBA/CONICET
Argentina

1. Introduction

The Río de la Plata (RDP), located on the eastern coast of southern South America at approximately 35°S (Fig. 1), is one of the largest estuaries of the world (Shiklomanov, 1998). It has a northwest to southeast oriented funnel shape approximately 300 km long that narrows from 220 km at its mouth to 40 km at its upper end (Balay, 1961). The estuarine area is 35,000 km^2 and the fluvial drainage area is 3.1 x 10^6 km^2. The system drains the waters of the Paraná and Uruguay rivers, which constitutes the second largest basin of South America. Therefore, it has a large discharge with a mean of around 25,000 m^3 s^{-1}, and maximum values as high as 50,000 m^3 s^{-1} under extreme conditions (Jaime et al., 2002). The RDP can be divided into three regions: upper, with an averaged depth of less than 3-5 m, intermediate, 5-8 m deep, characterized by the presence of several shallow sand banks and an outer region with depths ranging from 10 to 20 m (Dragani & Romero, 2004). Throughout its system of dredged channels the RDP estuary constitutes the main maritime access to Argentina and Uruguay.

Water level stations located along the estuary constitute a tide gauge network with the main purpose of recording water level heights associated not only with tides but also with the atmospheric forcing which produces storm surges (D'Onofrio et al., 1999). Tides in the RDP present a mixed, primarily-semidiurnal regime. Tides have a spring range of 1.58 m at Santa Teresita (Argentina) and 0.38 m at Punta del Este (Uruguay) located on the Atlantic coast, at the south-western and north-eastern side of the RDP mouth, respectively. The tidal range increases north-westward: 0.72 m at Punta Indio Channel, 1.01 m at La Plata and 1.10 m at Buenos Aires (Fig. 2), along the RDP southern coast. On the other hand, along the Uruguayan coast, the tidal range varies: 0.68 m at Montevideo, 0.66 m at Colonia del Sacramento (known as Colonia) and 0.76 m at Martín García Island (SHN, 2010). The coincidence of large or even moderate high tides and large meteorologically induced surges has historically caused catastrophic floods in many coastal areas of the Buenos Aires Province (D'Onofrio et al., 1999).

Sea level oscillations in the frequency band from a few minutes to almost two hours have been frequently observed at different tide stations along the Buenos Aires coast (Balay, 1955; Inman et al., 1962; Dragani et al., 2002; Dragani et al., 2009). Dragani (1988) studied a

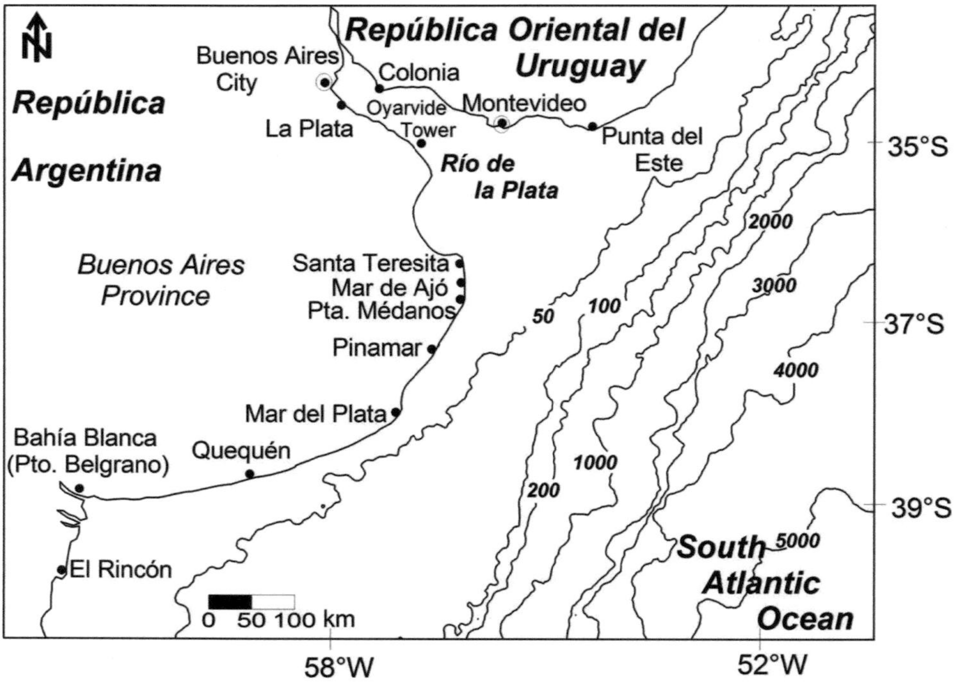

Fig. 1. Buenos Aires Province and Río de la Plata estuary. Bathymetry contours are in meters. Source: nautical charts SHN (1992, 1993)

possible relationship between oceanic seismic activity and energetic sea level oscillation events detected at Pinamar (Fig.1) but, due to the low correlation between both phenomena, it was suggested that a different mechanism would be required to generate these sea level oscillations. Dragani (1997) and Dragani et al. (2002) showed that, during high sea level activity, spectral peaks covered almost the whole frequency band between 1.1 and 4.7 cph (cycles per hour). Significant coherence values estimated between Mar de Ajó and Mar del Plata (172-km apart, Fig. 1) have clearly shown that this is a regional phenomenon. Based on the occurrence of simultaneous atmospheric gravity waves and long ocean wave events, the similarities of the spectral structures of both waving phenomena (Dragani et al., 2002) and the high effectiveness in the atmospheric-ocean energetic transference (proved by numerical simulations), it was possible to conclude that atmospheric gravity waves are the most probable forcing mechanism able to generate long ocean waves on the Buenos Aires continental shelf (Dragani, 2007). Dragani (1997) described the typical synoptic situation during sea level oscillation events in the coastal waters of the Buenos Aires province. Low level atmospheric cyclonic circulation and the passage of atmospheric fronts were always present prior to and during those events. Upper-air soundings obtained at Bahía Blanca (Fig. 1) meteorological station showed a lower pronounced tropospheric inversion that depicts an example of the state of the atmosphere when a frontal surface lies overhead. This tropospheric inversion constitutes an optimal interface for the propagation of high amplitude atmospheric gravity waves (Nuñez et al., 1998).

At 00:59 UTC on 26 December 2004, a moment magnitude (MW) 9.3 megathrust earthquake occurred along 1300 km of the oceanic subduction zone located 100 km west of Sumatra and

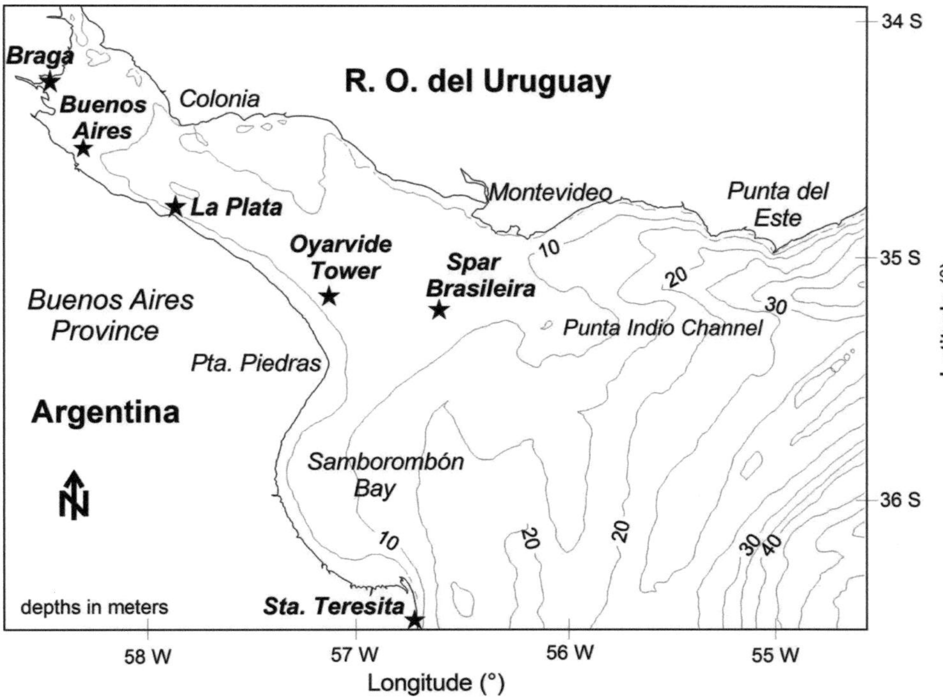

Fig. 2. Río de la Plata estuary. 5 m depth contours are depicted

the Nicobar and Andaman Island in the eastern Indian Ocean. The waves recorded around the world revealed unprecedented, truly global reach of the waves generated on December 26 (Titov et al., 2005). This tsunami is the first for which there are high-quality worldwide tide-gauge measurements. Tsunami amplitudes from many gauges along the Indian Ocean coast-line reached several meters and in some cases the waves were large enough to destroy the tide gauge recording equipment (Merrifield et al., 2005). All these records, together with those from other countries, demonstrate that this particular tsunami was a truly global event (Woodworth et al., 2005).

Filtered sea level oscillations recorded at Santa Teresita, Mar del Plata and Puerto Belgrano (Fig. 1) were studied by Dragani et al. (2006). At Mar del Plata, Santa Teresita and Puerto Belgrano, the Indian Ocean tsunami first arrived as a trough and the perturbations persisted approximately two days after the initial arrival. Initial waves were not the largest ones in the group. At Mar del Plata and Santa Teresita the largest waves were not observed until 7.8 and 17.8 hours after the first arrival, respectively. The first arrival detected was at Mar del Plata on December 27, 2004, at 00:15 UTC. At Santa Teresita and Puerto Belgrano, the tsunami reached the coast 33 minutes and 4.5 hours, respectively, after the arrival at Mar del Plata. Maximum wave height determined at Santa Teresita, Mar del Plata and Puerto Belgrano stations were 0.27, 0.15 and 0.20 m, respectively. Amplitudes and periods (ranging from 20 to 120 minutes) presented noticeable temporal variability while the oscillations persist. Oscillations lasted 40 hours at Mar del Plata and the perturbation was longer at Santa Teresita (54 hours). The mean (December 25 to 28 2004) sea level pressure field (NOAA-CIRES/Climate Diagnostics Center, www.cdc.noaa.gov) showed a low-pressure cell located east of the RDP estuary and a high-pressure cell located south of the low cell, at

50°S. During December 25 to 28 easterlies prevailed in the lower troposphere over the whole coastal area of Buenos Aires province. This weather pattern is completely different to the ones associated with frontal passages over the area. The tropospheric inversion necessary for the development of atmospheric gravity waves (capable of generating sea level oscillations in the region with periods similar to the tsunami waves) was not present in the RDP or in the adjacent continental shelf on December 25 to 28, 2004.

At the present, water level perturbations associated with the Indian oceanic tsunami have never been reported in the Río de la Plata estuary. Both tide gauges operating in the Río de la Plata estuary (at Buenos Aires and Oyarvide Tower stations), maintained by the Servicio de Hidrografía Naval (SHN) of Argentina, have a sampling interval of 60 minutes which is too long to resolve the signals corresponding to the tsunami. Recently, Jan de Nul (Hidrovía S.A. at Argentina), an international dredging company, provided several temporal high resolution water level data series positioned at different locations (Fig. 2) of the RDP. Consequently, it was possible to present the first description of the water level oscillations in the upper RDP associated to the tsunami generated as a response to the Sumatra earthquake. Even though these perturbations reached only 0.05 - 0.10 m, it was practically unthinkable that a tsunami wave would reach and propagate along the RDP before the Sumatra earthquake. In addition, a numerical experiment was performed to determine the vulnerability of the coast of the Buenos Aires Province to a hypothetical severe tsunami generated in the Mid Atlantic Ridge of the South Atlantic Ocean. The validation of this numerical model was carried out by comparing the maximum amplitudes simulated and the ones recorded at the selected tide stations located in the RDP , after the occurrence of the earthquake and subsequent tsunami in Sumatra (December 2004).

2. The Indian ocean tsunami, 2004, at the RDP. Observational study

Sea level records from a tide gauge located in the open sea (Santa Teresita) and from five tide gauges located along the RDP were analyzed. Pressure sensors were installed at Santa Teresita (ST: 36° 32' S, 56° 40' W), Spar Brasileira (SB: 35° 10′ S, 56° 37′ W), Oyarvide Tower (OT: 35° 06′ S, 57° 08′ W), La Plata (LP: 34° 50′ S, 57° 53′ W), Buenos Aires (BA: 34° 34′ S, 58° 23′ W) and Braga (BR: 34° 20′ S, 58° 29′ W) (Fig. 2). Out of all tide stations, ST is the only one exposed to the open sea, located about 20 km south of the RDP mouth (Figs. 1 and 2). SB is approximately in the middle of the imaginary line connecting Montevideo and Punta Piedras, which is the limit between the outer and intermediate RDP. OT is located in the intermediate RDP and LP, BA and BR stations were emplaced in the upper RDP. Except ST, the five tidal stations are disposed approximately 50-80 km apart, along the navigation channel access to the Buenos Aires Port (Argentina).

Water levels were measured by pressure transducers 312P TECMES (http://www.tecmes.com) which are commonly used to measure hydrostatic column height variations. Thermally compensated, the sensor is housed in a stainless steel capsule highly resistant to impacts. The system includes a vent tube for automatic compensation of the variations of the atmospheric pressure. Pressure sensors were fixed approximately 2 m below the tidal datum. The accuracy given by the pressure sensor is approximately ±0.1% of the instrument depth (±0.002 m). At SB, OT, LP, BA and BR these sensors were set with a data acquisition sampling interval of 20 min but, at ST, a lower sampling interval (6 min) was adopted. Sea level oscillations from December 26 12:00 to 29 00:00, 2004, (h = +3) recorded at ST, SB, OT, LP, BA and BR are shown in Fig. 3. It can be noted that the tide

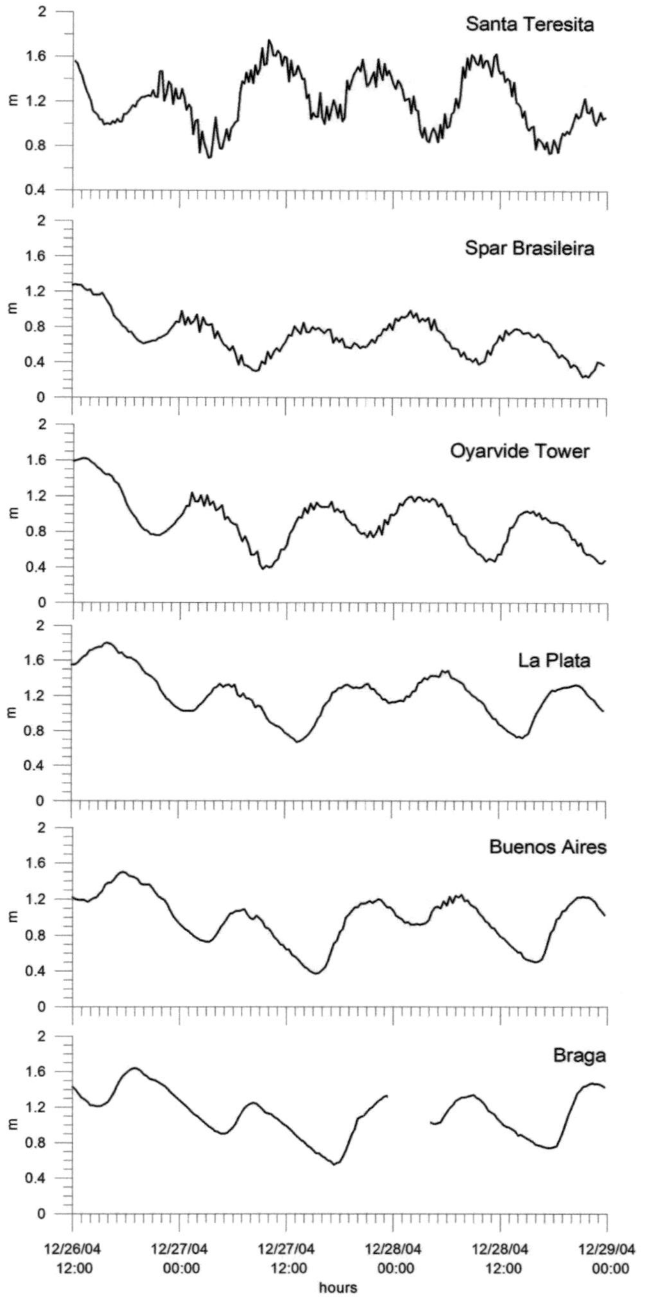

Fig. 3. Sea level records (m) at Santa Teresita, Spar Brasileira, Oyarvide Tower, La Plata, Buenos Aires and Braga. Time in hours from December 26, 2004, 12:00 (h = +3)

gauge at Braga did not work between 12/27 23:00 and 12/28 03:00. Sea level data contain diurnal and semidiurnal tides and higher-frequency oscillations ranging from a few minutes to two hours. Water level data were convoluted by means of Kaiser-Bessel bandpass filters

(201 elements), which select periods of approximately 12 - 180 minutes, and provide an attenuation factor of 100 dB outside that range (Hamming, 1977; Harris, 1978).
Filtered sea level oscillations at selected stations (Fig. 2) are shown in Fig. 4. At ST and LP the Indian Ocean tsunami first arrived as a trough and at SB, OT, BS and BR it arrived as a crest. The perturbations persisted approximately two days after the initial arrival. Initial waves were not the largest ones in the group. At ST, SB, OT, LP, BA and BR the largest waves were not observed until 7.8, 2.5, 7.6, 1.6, 23.3 and 23.0 h after the first arrival, respectively (Table 1). Arrival times were estimated by the first measured increase or decrease before the crest or trough. The first arrival detected was at ST on December 27, 2004, at 21:48. At SB and OT, the tsunami reached the gauge at 1.86 and 3.52 h, respectively, after the arrival at ST. The wave reached the locations of LP, BA and BR, in the upper RDP, at 6.52, 7.85 and 9.18 h, respectively, after the arrival at Santa Teresita. Consequently, the tsunami wave lasted approximately 9 h to propagate from the RDP mouth (ST) to the upper RDP (BR). Maximum wave height (distance from trough to crest or vice versa) determined at ST, SB, OT, LP, BA and BR were 0.27, 0.20, 0.13, 0.09, 0.08 and 0.05 m, respectively (Table 1). It can be clearly appreciated that maximum wave height decreases noticeably from the outer to the upper RDP. A possible explanation of this feature could be related to the shallowness of the RDP estuary. In the middle RDP, near to the Argentine coast, the mean depth is less than 5 m and, in the upper part of the estuary, the mean depth is less than 3 m (Fig.2). Consequently, tsunami waves could be significantly attenuated due to friction dissipation and refractive effects.
Amplitudes and periods presented noticeable temporal variability while the oscillations persisted. It was considered that the activity persisted as long as amplitudes in the filtered records were higher than two root mean square of the water level before the beginning of the activity. Then, it was estimated that the tsunami wave activity lasted approximately 40-50 h in the RDP estuary. The main characteristics of the recorded tsunami waves at ST, SB, OT, LP, BA and BR (arrival time, maximum wave height and approximate duration of the sea level activity) are summarized in Table 1. Lags in the time of first arrivals between consecutive tide stations were also calculated considering the tsunami wave speed along the imaginary line connecting gauges in the RDP. Distances between consecutive gauges from SB to BR (Fig. 2) are 50 km (SB – OT), 80 km (OT – LP), 60 km (LP – BA) and 35 km (BA – BR). Mean depth through each consecutive pair of stations were computed from nautical charts SHN (1999a; b).

Location	Arrival time (h= +3)	Max. height (m)	Duration (h)
ST	12/26 21:48	0.27	54.00
SB	12/26 23:40	0.20	44.33
OT	12/27 01:20	0.13	41.33
LP	12/27 04:20	0.09	59.33
BA	12/27 05:40	0.08	58.33
BR	12/27 07:00	0.05	36.00

Table 1. Arrival time (h = +3), maximum wave height (m) and duration of the activity (h) obtained from water level records gathered at Santa Teresita (ST), Spar Brasileira (SB), Oyarvide Tower (OT), La Plata (LP), Buenos Aires (BA) and Braga (BR)

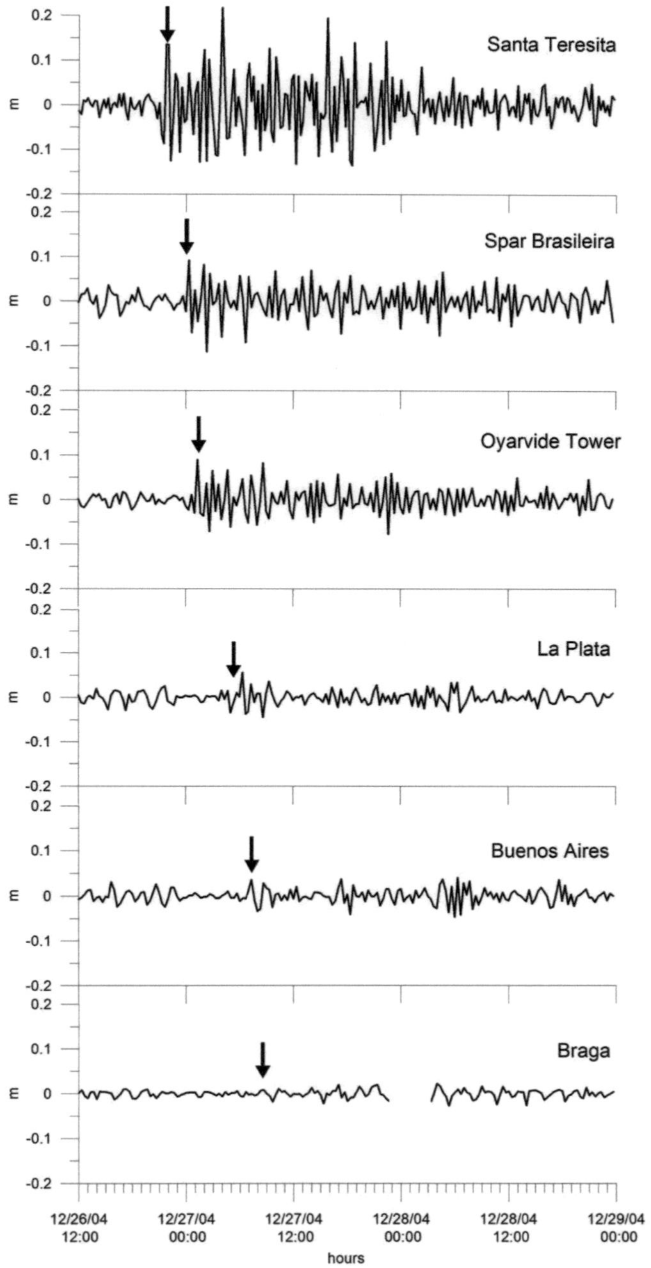

Fig. 4. Passband filtered sea level records (m) at Santa Teresita, Spar Brasileira, Oyarvide Tower, La Plata, Buenos Aires and Braga. Time in hours from December 26, 2004, 12:00 (h = +3). Arrows point out the beginning of the activity

The wave speed, c, of long ocean waves could be reasonably approximated by $(g\ h)^{1/2}$ (e.g., Dean & Dalrymple, 1984) where g is the acceleration due to gravity and h the local depth. According to the expression for c and using the corresponding distances and depths, the computed time lags between SB – OT, OT – LP, LP – BA and BA – BR resulted in 1.55, 2.65, 1.97 and 1.53 h, respectively, which match reasonably well with the observed times (1.66, 3.00, 1.33 and 1.33 h, respectively). It must be highlighted that depth variations produced by tide have not been included in this rough estimation of times. Computed and observed travel time, from SP to BR, resulted 7.70 and 7.32 h, respectively.

3. The Río de la Plata estuary: vulnerability to regional tsunamis

The southeastern coast of South America, and specially the RDP, has been traditionally considered to have low hazard probability of being affected by tsunamis. The main reason is the emplacement in a stable, tectonically inactive and passive continental margin (Mouzo, 1982; Urien & Zambrano, 1996). However, some geological, topographical and oceanographic aspects of the surrounding regions were reviewed as they could favor the occurrence of destructive events or contribute to increase damage. The Scotia Arc, located between the southernmost extreme of South America and the Antarctic Peninsula, is a highly dynamic convergent margin with frequent occurrence of earthquakes and volcanic eruptions (USGS-NEIC, 2008). The geographical positions of the strongest earthquakes in this region are pointed out in Fig. 5. Five relatively stronger earthquakes (magnitudes ≥ 7.5) have been recorded between 45°S - 65°S and 15°W - 75°W, nine earthquakes, with magnitudes ranging from 7.0 to 7.4, and ninety-two earthquakes, with magnitudes ranging from 6.0 to 6.9, have been recorded in this region. These earthquakes present the highest spatial density of hypocenters around the South Sandwich Islands, located at the eastern extreme of the Scotia Arc.

Considering the regional geotectonic setting, the configuration and the relative geographical position of Argentine coasts, the vulnerability to long ocean waves originated by submarine events like earthquakes, volcanic eruptions and slides in the Scotia Arc and surroundings was numerically studied by Dragani et al. (2008). The smooth and gently sloping continental shelf (Parker et al., 1997) is a feature that can behave as a "ramp" for ocean waves approaching the coast driven by either climatic or geological processes. The variable shelf width (decreasing from 600 to 170 km from South to North) and the morphology of the adjacent continental slope were considered as probable factors affecting the propagation of ocean waves. Regarding coastal morphology, there is an alternation of areas with different susceptibility to damaging effects coming from processes and events originated in the ocean. Results obtained from numerical simulations (Dragani et al., 2008) showed that bathymetric refraction is highly significant on the continental shelf producing a considerable divergence of energy which attenuates the long ocean wave amplitudes on the Patagonian coast (between 45° and 55°S). Maximum amplification factors were obtained at Southern Malvinas and Burdwood bank because the wave propagates almost without refracting and the transformation by shoaling is highly significant. Based on these preliminary numerical simulations, it was concluded that the Argentinean coast, and specially the RDP estuary, present a relatively low vulnerability to the tsunami wave generated by earthquakes whose epicenters are located at Scotia Arc. On the other hand, it is evident that tectonism, magmatism, volcanism and submarine slides at different settings of the Mid Atlantic Ridge in the South Atlantic Ocean must be seriously taken into account when evaluating the possibility of generation of ocean waves affecting the RDP. But, no strong submarine earthquakes, between 20°S and 50°S, have been reported in this region. Furthermore, no evidence of paleotsunamis have been mentioned for the Argentine coasts and marine

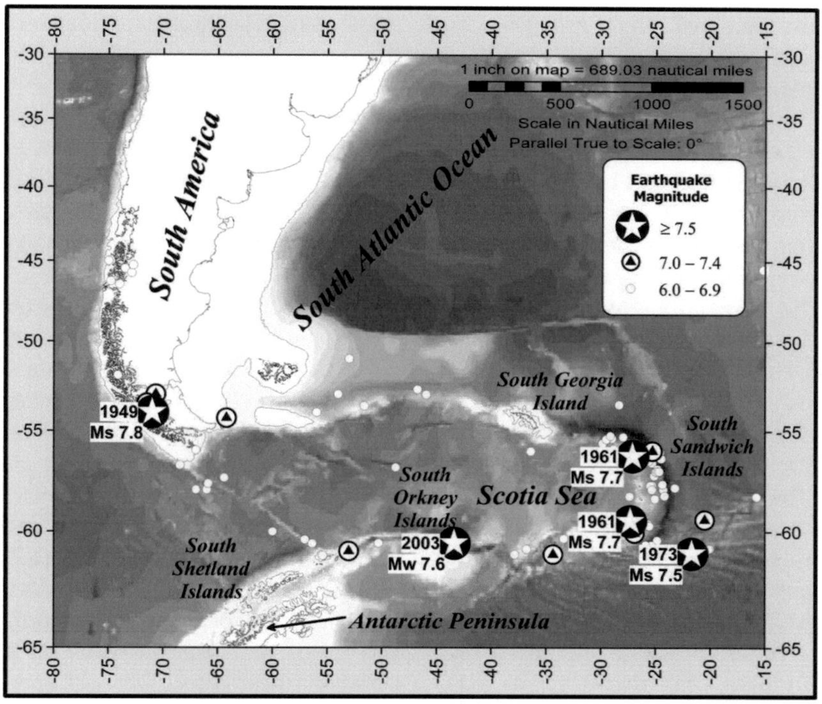

Fig. 5. Southwestern Atlantic Ocean and Antarctic Peninsula region. Locations of earthquakes which occurred in this area are pointed out with white stars, black triangles or white circles for earthquakes whose magnitudes were equal or higher than 7.5, ranged from 7.0 to 7.4 or from 6.0 to 6.9, respectively. Source: USGS-NEIC (2008)

geological records. However, the potential hazard of the RDP region should not be disregarded. Consequently, the amplification and propagation of a tsunami wave produced by a hypothetical strong earthquake in the South Atlantic Ocean will be analyzed by means of a validated numerical model.

4. Numerical simulations and numerical experiments

The model used for the numerical investigations is WQMap version 5.0 (Water Quality Mapping), developed by Applied Science Associates Inc. (ASA, 2004). The mathematical description of sea levels and currents requires the simultaneous solution of the dynamic equations of motion and the continuity equation. It is assumed that vertical accelerations are negligible, pressures are hydrostatic over depth and fluid density is homogeneous. The bottom stress is parameterized by means of a quadratic law in terms of the depth averaged current velocity and a constant value equal to 0.002 was selected for the non dimensional drag or bottom friction coefficient. Transient floor perturbations vertically forcing the ocean water column, like those produced during submarine earthquakes can not be included as forcing in WQMap model. In other words, this model is not able to generate surface sea level oscillation produced by earthquakes whose epicenters are located in the computational domain. The tsunami wave (when it is far enough from the epicenter) can be considered as a long wave (wavelength is longer than hundreds of kilometers) propagating in the ocean like

a shallow water wave (Dean and Dalrymple, 1984). WQMap model resulted to be a very efficient tool (Dragani et al., 2008) to simulate the propagation of this kind of waves on the Buenos Aires and Patagonian continental shelf waters (Fig. 5).
A realistic tsunami wave perturbation was imposed as boundary condition in the eastern border of the implemented computational domain (31°S to 45°S and 44°W to 66°W), between the RDP and the hypothetic epicenter of the earthquake. Sea level perturbations along selected grid points of the eastern open boundary were defined by means of a simple sinusoidal expression. Sea level oscillated only during three periods at every grid point of the open boundary, representing the incoming of the tsunami wave to the computational domain. The original bathymetry was obtained from an assemblage of nautical charts published by the Servicio de Hidrografía Naval of Argentina (SHN, 1992 and 1993), for the continental shelf (depths lower than 200 m) and a 1'x1' resolution depth data set coming from GEBCO (2003) for the continental slope and deep ocean. A square inverse distance method was used to create a regularly spaced grid from irregularly spaced bathymetry data. The resulting bathymetry was taken at 6.2 km x 5 km intervals, thus there are 281 grid points in the east-west and 312 grid points in the north-south directions. This grid represents a rectangular domain with a distance of 1,924 km in the east-west and 1,560 km in the north-south directions. The resulting grid has the shallowest depths (1.2 m) in the upper estuary and the greatest depths (deeper than 5,000 m) at the south-eastern corner of the domain. The adopted time step was 5 minutes. A set of experiments (Dragani, 2008) was designed to simulate the propagation characteristics of the tsunami wave from the deep ocean (boundary, 44°W meridian) to the coast and the best result was obtained for wave amplitude equal to 0.03 m and period 1 h.
Observed bandpass filtered water level (Fig. 4) and simulated water level oscillations (Fig. 6) are compared. Simulated maximum wave height at ST, SB, OT, LP and BA resulted 0.22, 0.13, 0.09, 0.07 and 0.05 m, respectively, and, although they are a little underestimated, they are in reasonable good agreement with the measurements (Table 2). Simulated lags between initial activity times corresponding to consecutives locations were also computed and presented in Table 3. Simulated water level at Braga was discarded from these comparisons because the modeled maximum wave height resulted extremely low, under 0.02 m. Similarly to the observed filtered records (Fig. 4) the largest simulated heights were observed after the first arrival of the wave. Perhaps, the main difference between observed and simulated water levels is the larger activity period corresponding to the observations. A possible explanation of this difference could be that the spatial resolution of the model is not enough to represent the complex geometry of the natural sand banks and channels in the RDP.

Location	Observed max. height (m)	Simulated max. height (m)
ST	0.27	0.22
SB	0.20	0.13
OT	0.13	0.09
LP	0.09	0.07
BA	0.08	0.05
BR	0.05	< 0.02

Table 2. Observed and simulated maximum wave height (m) at Santa Teresita (ST), Spar Brasileira (SB), Oyarvide Tower (OT), La Plata (LP), Buenos Aires (BA) and Braga (BR)

Consecutive stations	Observed lag (h)	Simulated lag (h)	Theoretic lag (h)
SB - OT	1.66	1.50	1.55
OT - LP	3.00	2.66	2.65
LP - BA	1.33	1.50	1.97

Table 3. Observed, simulated and theoretical lags (h) between consecutive tidal stations, from Spar Brasileira to Braga

It was shown that maximum amplitudes recorded at the RDP (after the occurrence of earthquake and subsequent tsunami occurred on December 26, 2004 in Sumatra) were realistically simulated using WQMap model. This validated model was used to study the potential vulnerability in the RDP to a hypothetical severe tsunami occurred in the Mid Atlantic Ridge of the South Atlantic Ocean.

The afore described numerical experiment was replayed, but a simple sinusoidal wave of 0.5 m amplitude (instead of 0.03 m) was imposed as forcing at the eastern boundary of the computational domain. Simulated water levels at ST, SB, OT and BA are presented in Fig. 7. LP and BR are not included in this figure because the simulated water levels resulted very similar to the ones of BA.

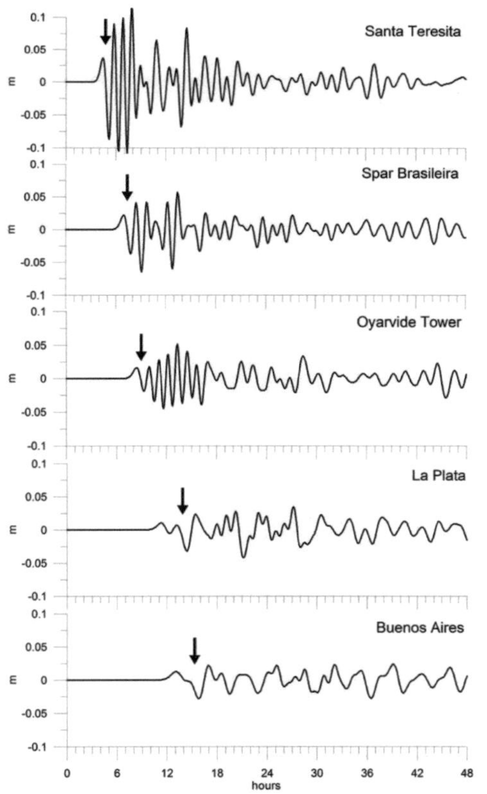

Fig. 6. Simulated sea level oscillation (m) at Santa Teresita, Spar Brasileira, Oyarvide Tower, La Plata and Buenos Aires, forced with wave amplitude equal to 0.03 m at the eastern boundary. Arrows point out the beginning of the activity

Maximum wave height determined at ST, SB, OT and BA were 3.25, 1.10, 0.70 and 0.25 m, respectively. It can be clearly appreciate that maximum wave height decreases drastically from ST to BA. This simulation shows that shallow water effects (especially refraction and bottom friction) could play a very significant roll in the propagation and transformation of the tsunami wave from the outer to the upper RDP. Maximum wave height decreased 13 times from the RDP mouth (ST: 3.25 m) to the upper RDP (BA: 0.25 m) and 3-4 times from the RDP mouth to the outer and intermediate RDP. Therefore, it can be concluded that Buenos Aires City and the coast at the upper RDP, where millions of inhabitants are settled, present very low vulnerability to tsunamis generated by earthquakes with epicenters located at the Mid Atlantic Ridge of the South Atlantic Ocean.

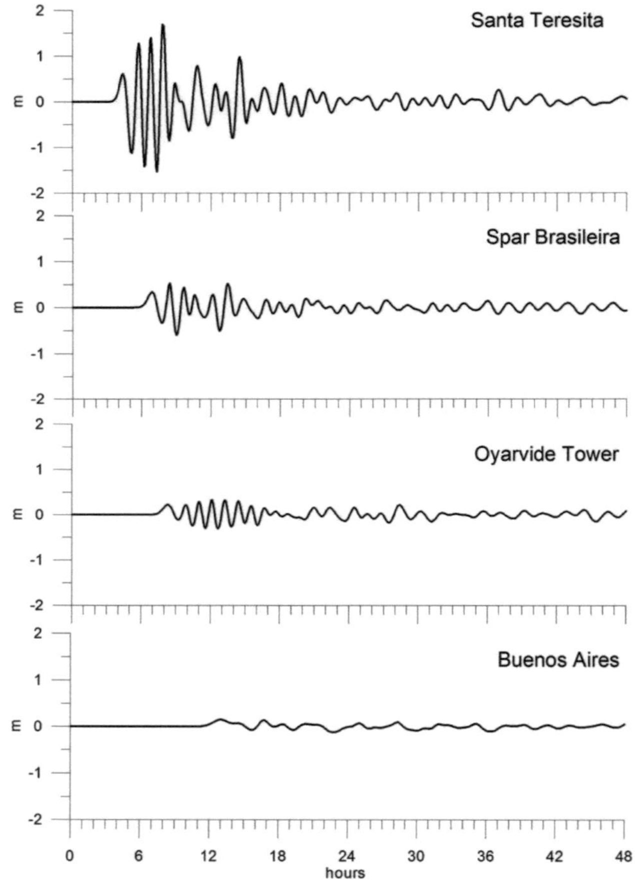

Fig. 7. Simulated sea level oscillation (m) at Santa Teresita, Spar Brasileira, Oyarvide Tower and Buenos Aires, forced with wave amplitude equal to 0.50 m at the eastern boundary

On the contrary, Samborombón bay presents relatively high vulnerability to this kind of natural hazards. This bay is located 160 km southeast of the city of Buenos Aires (Fig. 2) and receives the Samborombón River from the northwest and the Salado River from the west, as well as several canals draining the bay's low, irrigated hinterland, which is part of the humid Pampa, an extensive area of fertile plains. Samborombón bay was established as

Ramsar site 885 (01/24/1997) in response to Article 2.1 of the Convention on Wetlands (Ramsar, Iran, 1971, www.ramsar.org). A very high amount of organisms of the intertidal zone of the southernmost part of the bay are predated every summer by a high number of migratory birds that use Bahía Samborombón as a refuel station. The whole bay is characterized by soft bottoms of fine sediments with high percentage of mud and organic matter and very low slope beach. Maximum wave heights ranged from 3.25 m (at ST, 20 km southward the southern limit of the bay) to 1.10 - 0.70 m (at SB and OT, respectively, located at the northernmost part of the bay) could flood a very significant coastal area, especially, if a severe tsunami wave reached these low coasts during spring high water and/or during storm surge conditions.

5. Conclusions

The south-eastern coast of South America, and specially the RDP, has been traditionally considered to have low hazard probabilities of being affected by tsunamis. At 00:59 UTC on 26 December 2004, a moment magnitude (MW) 9.3 megathrust earthquake occurred in the eastern Indian Ocean generating a wave tsunami measured in almost all the World Ocean demonstrating that this particular wave was a truly global event (Woodworth et al., 2005). Even though this perturbation reached only a few centimeters high at the upper RDP, before the occurrence of the Sumatra earthquake, the propagation of a tsunami wave along the RDP was unthinkable. Water level records from a tide gauge located at the open sea and from five tide gauges located along the RDP were analyzed and discussed in this work. Water level oscillations at selected stations persisted approximately two days after the initial arrival of the tsunami. Initial waves were not the largest ones in the group. Maximum wave height determined at Santa Teresita, Spar Brasileira, Oyarvide Tower, La Plata, Buenos Aires and Braga (Fig. 2) were 0.27, 0.20, 0.13, 0.09, 0.08 and 0.05 m, respectively showing that maximum wave height decreased noticeably from the outer to the upper RDP. A possible explanation of this feature could be related to the shallowness of the RDP estuary, where the tsunami wave heights could be significantly attenuated due to friction dissipation and refractive effects.

Even though, no strong submarine earthquakes, between 20°S and 50°S, have been reported in this region of the ocean, the vulnerability of the RDP to a tsunami wave produced by a hypothetic strong earthquake in the South Atlantic Ocean was numerically studied by means of a validated model. The model used for this investigation was WQMap (version 5.0). Simulated water level oscillations were slightly underestimated but they were in good agreement with the observations (Fig. 4, Fig. 6 and Table 2). Similarly to the observed filtered records (Fig. 4) the largest simulated heights were produced after the first arrival of the wave. The main difference between observed and simulated water levels was the longer duration of the observed events. This difference could be attributed to the relatively low spatial resolution of the model domain, which might have not been enough to represent the complex geometry of the natural sand banks and channels at the RDP. This validated model was used to carry out a numerical experiment in which a hypothetical strong tsunami propagated from the Mid Atlantic Ridge in the South Atlantic Ocean to the upper RDP. The numerical experiment, using a simple sinusoidal wave of 0.5 m amplitude, showed that the maximum wave height decreased 13 times from the RDP mouth (ST: 3.25 m) to the upper RDP (BA: 0.25 m) indicating that shallow water transformation (especially refraction and bottom friction) could play a very significant role in the propagation of the tsunami wave from the outer to the upper RDP. Consequently, it could be preliminarily concluded that the Buenos Aires City and the coast along the upper RDP, would present very low vulnerability to tsunamis generated by earthquakes whose epicenters were located in the South Atlantic Ocean. On the contrary, Samborombón bay (Fig. 2) shows a relatively higher vulnerability to this kind of natural hazards because it has a very low slope beach and the simulations showed that the maximum wave height ranged from 3.25 m to 1.10 - 0.70 m, between the southernmost and northernmost extremes of the bay. It is evident that this hazardousness could be greater if a severe tsunami wave reached this low coast during spring high water and/or during a storm surge conditions.

6. References

ASA (2004). Applied Science Associates. WQMap User Manual, Version 5.0. Applied Science Associates Inc, Narraganset, RI, USA.

Balay, M. (1955). La determinación del nivel medio del Mar Argentino, influencias de las oscilaciones del mar no causadas por la marea, Dir. Gral. de Nav. Hidrog., Min. de Marina, pp. 46.

Balay, M.A. (1961). El Río de la Plata entre la atmósfera y el mar, Publ. H-621, 153 pp., Serv. de Hidrogr. Nav. Armada Argentina, Buenos Aires.

Dean, R.G. & Dalrymple, R.A. (1984). Water wave mechanics for Engineers and scientistics, Prentice-Hall, Englewood Cliffs, New Jersey, pp. 353.

D'Onofrio, E. E.; Fiore, M. M. E. & Romero, S. I. (1999). Return periods of extreme water levels estimated for some vunerable areas of Buenos Aires, Continental Shelf Research, 19, 1681-1693.

Dragani, W.C. (1988). Análisis del proceso físico generador de ondas largas en la costa bonaerense argentina, Thesis, Inst. Tec. Bs. As., Buenos Aires, pp. 65.

Dragani, W.C. (1997). Una explicación del proceso físico generador de ondas de largo período en la costa bonaerense argentina, Doctoral Thesis, Facultad de Ciencias Exactas y Naturales, Universidad de Buenos Aires, pp. 222.

Dragani, W.C.; Mazio, C. A. & Nuñez, M. N. (2002). Sea level oscillations in coastal waters of the Buenos Aires Province, Argentina, Continental Shelf Research, 22, 779-790.

Dragani, W.C. & Romero, S. I. (2004). Impact of a possible local wind change on the wave climate in the upper Río de la Plata. International Journal of Climatology, 24(9), 1149-1157.

Dragani, W. C.; D'Onofrio, E.E.; Grismeyer W. & Fiore M.E. (2006). Tide gauge observations of the Indian Ocean tsunami, December 26, 2004, in Buenos Aires coastal waters, Argentina. Continental Shelf Research 26, 1543-1550.

Dragani, W. C. (2007). Numerical experiments of the generation of long ocean waves in coastal waters of the Buenos Aires Province, Argentina, Continental Shelf Research. doi:10.1016/j.csr.2006.11.009.

Dragani, W.C.; D'Onofrio, E.E.; Grismeyer, W.; Fiore, M.E.; Violante, R. & Rovere, E. (2008). Vulnerability of the Atlantic Patagonian coast to tsunamis generated by submarine earthquakes located in the Scotia Arc region. Some numerical experiments. Natural Hazards, doi:10.1007/s11069-008-9289-4.

Dragani, W.C.; D'Onofrio, E.E.; Grismeyer, W.; Fiore, M. & Campos, M. I. (2009). Atmospherically-induced water oscillations detected in the Port of Quequén, Buenos Aires, Argentina. Journal of Physics and Chemistry of the Earth 34 998–1008.

GEBCO (2003). User guide to the centenary edition of the GEBCO Digital Atlas and its data sets. Ed. M. T. Jones, Natural Environment Research Council.

Hamming, R. W. (1977). Digital filters, Prentice Hall, Signal Processing Series, pp. 221.

Harris, F. J. (1978). Of the use of windows for harmonics analysis with the discrete Fourier Transform, Proc. IEEE, 66, 51-83.

Inman, D.; Munk, W. & Balay, M. (1962). Spectra of low frequency ocean waves along the Argentine shelf, Deep-Sea Res., 8, 155-164.

Jaime, P.; Menéndez, A.; Uriburu Quirno, M. & Torchio, J. (2002). Análisis del régimen hidrológico de los ríos Paraná y Uruguay, Informe LHA 05-216-02, 140 pp., Inst. Nac. del Agua, Buenos Aires.

Merrifield, M. A.; Firing, Y. L.; Aarup, T.; Agricole, W.; Brundrit, G.; Chang-Seng, D.; Farre, R.; Kilonsky, B.; Knight, W.; Kong, L.; Magori, C.; Manurung, P.; McCreery, C.; Mitchell, W.; Pillay, S.; Schindele, F.; Shillington, F.; Testut, L.; Wijeratne, E.M.S.; Caldwell, P.; Jardin, J.; Nakahara, S.; Porter, F.Y. & Turetsky, N. (2005). Tide gauge observations of the Indian Ocean tsunami, December 26, 2004, Geophysical Research Letter, 32, LO9603, doi:10.1029/2005GL022610.

Mouzo, F.H. (1982). Geología Marítima y Fluvial. In: Historia Marítima Argentina, Armada Argentina. Cuántica Editora, Buenos Aires, Tomo I, Secc. 2: 117.

Nuñez, M.N.; Mazio, C.A. & Dragani, W. C. (1998). Estudio espectral de un lapso de intensa actividad de ondas de gravedad atmosféricas registradas en la costa bonaerense argentina, Meteorológica, 23, (1 and 2), 47-54.

Parker, G.; Paterlini, C.M. & Violante, R.A. (1997). El Fondo Marino. In: El Mar Argentino y sus Recursos Pesqueros (E.E. Boschi, Ed.), INIDEP, Mar del Plata, Argentina. 1: 65-87.

SHN (1992). Nautical Chart H-1: Acceso al Río de la Plata, Serv. Hidrog. Nav., Buenos Aires.

SHN (1993). Nautical Chart H-2: El Rincón, Golfos San Matías y Nuevo, Serv. Hidrog. Nav., Buenos Aires.

SHN (1999a). Nautical Chart H-113: Rio de la Plata Exterior, Serv. Hidrog. Nav., Buenos Aires.

SHN (1999b). Nautical Chart H-116: Rio de la Plata Médio y Superior, Serv. Hidrog. Nav., Buenos Aires.

SHN (2010). Tablas de Marea, H-610, Serv. Hidrog. Nav., Buenos Aires.

Shiklomanov, I. A. (1998). A summary of the monograph world water resources: A new appraisal and assessment for the 21st Century, report, U. N. Environ. Programme, Nairobi.

Titov, V.; Rabinovich, A.B.; Mofjeld, H.O.; Thomson, R.E. & González, F.I. (2005). The global reach of the 26 December 2004 Sumatra tsunami, Science, 309, 2045-2048.

Urien, C.M. & Zambrano, J.J. (1996). Estructura de la Plataforma continental, in "Geología y Recursos Naturales de la Plataforma Continental Argentina" (V.A. Ramos and M.A. Turic, eds.). Relatorio XIII Congreso Geológico Argentino y III Congreso de

Exploración de Hidrocarburos, Asociación Geológica Argentina-Instituto Argentino del Petróleo, Buenos Aires: 29-65.

USGS-NEIC (2008). United States Geological Survey - National Earthquake Information Center, http://earthquake.usgs.gov/regional/neic/

Woodworth, P.L.; Blackman, D.L.; Foden, P.; Holgate, S.; Horsburgh, K.; Knight, P.J.; Smith, D.E.; Macleod, E.A. & Bradshaw, E. (2005). Evidence for the Indonesian Tsunami in British tidal records, Weather, 60, 9, 263-267.

18

Traces of Tsunami Events in off- and on-Shore Environments. Case Studies in the Maldives Scotland and Sweden

Mörner, Nils-Axel[1] and Dawson, Sue[2]
[1]*Paleogeophysics & Geodynamics*
[2]*University of St. Andrews*
[1]*Sweden*
[2]*Scotland*

1. Introduction

A tsunami wave is characteristic by its large diameter (up to a couple of kilometres) and high speed of displacement (in the order of 700 km/h). Tsunami waves are generated by submarine earthquakes, by submarine slides and, occasionally, even by meteor impacts. When the wave starts to trim the seabed off a coast, it may rise to considerable heights and break in over the coast with disastrous force. The 1755 Lisbon earthquake and its disastrous tsunami effects are classical. The Boxing Day tsunami in 2004 in the Indian Ocean woke up the world in realizing what terrible effects a tsunami event may cause (NOAA, 2010). The death toll was around 230,000 persons. Suddenly, the word *"tsunami"* became known to a broad audience.

The tsunami hazard of a region can only be assessed in a meaningful way if we have a reasonable record of the past events in that region. Consequently, there is an urgent need of establishing such records; i.e. a database of the paleotsunami events of the region in question (Mörner, 2009a, 2010). This, in its turn, calls for a methodology of how to record past tsunami events. This implies a careful search of imprints left in morphology and stratigraph, and means of interpreting those imprints in terms of a past tsunami event.

This paper is devoted to the traces and imprints left in nature from former tsunami events. We explore such signals in the off-shore sediments as well as in the on-shore deposits left in three different regions from where we have primary studies of our own. Hence, we may be talking about three case studies: the Maldives, Scotland and Sweden. As other tsunami studies are lacking in those areas, this paper tends to become a compilation of our previous work with recent additions.

2. The Maldives

The Maldives are located in the middle of the Indian Ocean. They consist of some 1200 islands arranged in 20 larger atolls. The entire island archipelago rises steeply out of the deep ocean, and is directly surrounded by depths in the order of 2500 m. This implies quite specific conditions at the passing of a tsunami wave; instead of actually breaking in over land, it may rather passes over the low-lying islands (Fig. 1). This was even the case at the 2004 mega-tsunami.

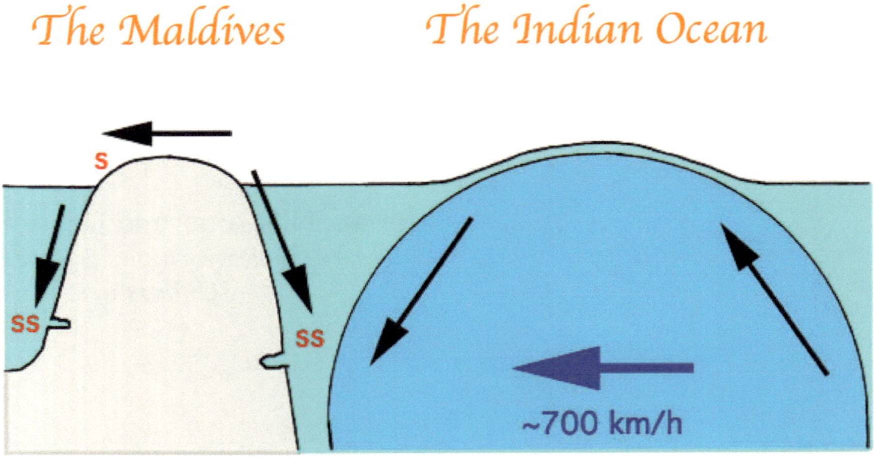

Fig. 1. A tsunami wave is characterized by its exceptional vertical diameter and by its high lateral speed. The December 26, 2004, tsunami began with a withdrawal of coastal water before the main disastrous breaking wave hit the coasts. In the Maldives, rising steeply from the abyssal plain, the wave rather washed over the islands than broke in over them (hence only 19 persons were killed). In the submarine environment, divers reposted the occurrence of "a submarine sandstorm". This gave us the idea that our dated submarine cave-fillings might be remains of submarine sandstorms (ss) from former tsunami events (Mörner et al., 2008)

We worked in the Maldives for several years in association with an international sea level project (Mörner et al., 2004; Mörner, 2007). During those studies we came across imprints of former tsunami events both as a part of our methodical studies (Mörner, 2007) but also as a total surprise (Mörner et al., 2008a).

We visited the Maldives in February 2005 in order to record the traces left after the 2004 event. The washing from east to west over the individual islands was quite clear in morphology (Fig. 2), damages of houses and other constructions, and in the erosion/deposition of sediments (Fig. 3). On the atoll of Goidhoo, we had in two cored swamps (located on opposite sides of the island) recorded a sand layer in the middle of peat deposits. In 2005, we recorded identical signals on top of both swamps. I combination with C14-dates and micropaleontological studies of the cores retrieved, we were now quite sure that we were dealing with a paleo-tsunami record. The top of the peat below the tsunami sand was dated by two C14-dates overlapping at AD 1730-1735.

In order to date former submarine sea level positions, we had sampled submarine cave-fillings and subtracted shallow-water corals and gastropods for C14-dating. This study failed completely with respect to the dating of formed sea level positions. At the 2004 tsunami, divers had reported the observation of what they called *"submarine sandstorms"*. Suddenly our cave dates made sense; instead of referring to former sea level positions, they recorded former tsunami events creating "submarine sandstorms" (Mörner et al., 2008). In the Addu Atoll, there is a cave at -27 m where two C14-dates of gastropods and corals gave ages overlapping at 1735 ±25.

Fig. 2. In lee of the tree, there is a clear ridge left after the 2004 tsunami wave that crossed straight over the island. The material eroded was deposited on the west side of the island (Fig. 3)

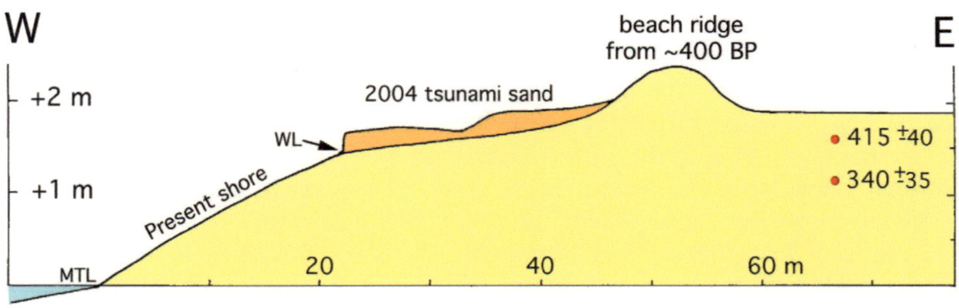

Fig. 3. Profile across the western shore of the island of Gan in the Laamu Atoll. The 2004 tsunami completely over-washed the island, eroding littoral sand from the eastern shore and depositing a 20-30 cm sand layer on the western side graded into two terrace surfaces 1.95 and 1.65 cm above MTL. The present shore has a clear washing limit (WL), an erosion notch, 1.45 cm above mean-tide level (MTL). A sub-recent beach ridge, now over-grown by trees and vegetation, post-dates two radiocarbon dates with a mean of 375 BP or AD 1575. It represents a sea level by about 0.5-0.6 m above the present one

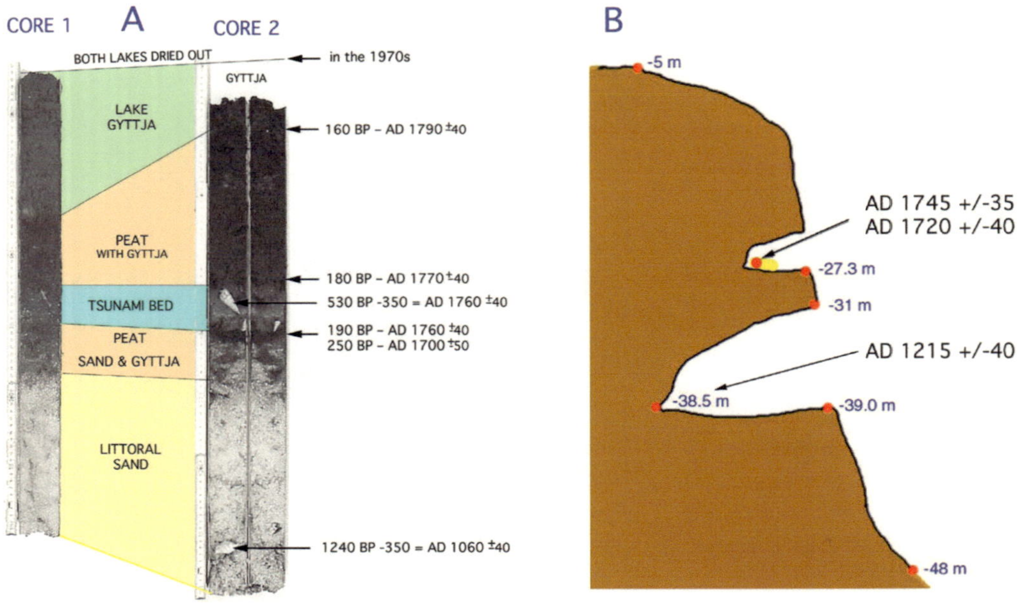

Fig. 4. Evidence of the AD 1733 tsunami. A: the 2 swamp cores on the island of Goidhoo, both recording an intra-peat layer of sand. The overlapping C14-dates give an age of about 1730-1735. B: the caves off Addu with sand fillings containing shallow-water corals and gastropods, which were used for C14-dating. The -27.3 m cave infill gives and overlapping C14-age of about 1735 ±25. The time coincidence between the two sites is striking, and fits perfectly well with the historical notation of a major flooding event in the year 1733. The -38.5 m cave infill gives a C14 age that fits with other records of a tsunami event

Finally, in the chronicle of the Maldives by Bell (1940), there is a record of a disastrous flooding event in AD 1733. The time coincidence with the intra-peat sand on Goidhoo and the submarine sandstorm filling the -27 cave at Addu is striking (Mörner, 2007; Mörner et al., 2008). Hence, we are confident that they are all expressions of one and the same tsunami event occurring in 1733.

Besides the dates, correlations and interpretation in terms of a tsunami event, we established a new methodology of recording past tsunami events; i.e. the recording of "submarine sandstorms" being deposited at great depths in submarine caves (Mörner et al., 2008).

In the Maldives, we have established 5 events of past "submarine sandstorms", viz (1) of modern age (probably 1988 or 1991) in a cave at -21 m, (2) at AD 1733 in the cave at -27 m, (3) at around AD 1550 in a cave at -20 m (with a possible simultaneous on-shore beach ridge in Fig. 3), (4) at around AD 1215 in a cave at -38 m (Fig. 4) probably also recorded by beach ridges on the islands of Viligili and Lhosfushi (where it covered the skeleton of "the reef woman", Mörner, 2007), and (5) at around 6300 BP in a cave at -26 m.

There is also a number of cases of a sudden beach ridge deposition that might - but need not - be understood in terms of a tsunami event (Mörner, 2009a). They occur at 1950 ±10 years, 1115, 850, 530 and 200 ±50 AD. They are still only considered as possible paleotsunami events. At any rate, we have the 2004 event, 5 events of "submarine sandstorms" and related effect, and 5 possible events of a sudden beach ridge formation; i.e. all together 11 events.

This is a good beginning of a long-term tsunami record. The Maldives seem to offer ideal conditions for amplified work and the establishment of a detailed long-term record of tsunami events in the Indian Ocean, which could serve as a base for improved hazard assessments.

3. Scotland

In Scotland, there are clear records of the tsunami event from the huge Storegga submarine slide dated at around 7000 C14-years BP (Dawson et al., 1988; Dawson, 1999). This event is seen as an extensive sand layer in the estuarine mud and clay beds. It is dated by several C14-dates. The characteristics of this tsunami sand layer are its content of planctonic microfossils in contrary to littoral deposits that would have a bentic, shallow-water micro fauna and flora. We use this technique (Dawson, 1999; Dawson & Smith, 2000) in the analyses of the Swedish tsunami beds.

4. Sweden

In recent years (e.g. Mörner 2003, 2004, 2005, 2008a), it has become evident that Sweden, at the time of deglaciation, was a high-seismic region, in magnitudes as well as in frequency. The driving force seems predominantly to have been the extremely high rate of glacial isostatic uplift. Event in the order of M7 have also been recorded in the Late Holocene, however (e.g. Mörner, 2009a), the last one of which occurred at 900 BP, displacing the Viking shoreline by 1 m or more.

Thanks to the Swedish varve chronology (De Geer, 1940), most of the paleoseismic events could be assigned an age with a resolution of one single year; in one case even the season of a year (Mörner, 2003).

In association with those earthquakes, tsunami events have also been recorded. Up to now, a total of 17 events (Table 1) have been documented and described (Mörner, 1996, 1998, 2003, 2008b, 2009a; Mörner et al., 2000). They are traced both in on-shore environment as intra-clay sand-layers containing a planctonic microfossil fauna and flora, and in off-shore environment as extensive turbidites (followed over distances as large as 320 km). A few of those events will be highlighted with respect to their occurrence and characteristics. We have selected the 10,430 BP, 9663 BP, 2900 BP and 2000 BP events for a more extensive analysis.

4.1 The 10,430 BP paleoseismic event

In varve-year 10,740 BP, the Baltic Ice Lake drained (Brunnberg, 1995) and the Baltic became in level with the Atlantic. As ice retreated, a straight opened across southern Sweden (Fig. 5), known as the Närke Straight. The straight remained blocked by pack-ice and icebergs, however, and the conditions in the Baltic remained lacustrine (Mörner, 1995). In the autumn of varve 10,430 BP, the situation changed totally, and the Baltic became brackish-marine within one single year. Something exceptional and revolutionary must have happened. We now know that it was a very large earthquake of a magnitude well above M8 (Mörner, 1996, 2003). The location of the epicentre was in the Stockholm region (Fig. 5) along an old fault that crosses over southern Sweden from the west to the east and probably continues in the Bay of Finland and northeastwards into Russia. Bedrock fracturing is recorded over a zone of 50x200 km. Liquefaction is recorded over a huge area of 200x320 km. Magnetic grain

age *in* BP	*area affected*	*Earthquake magnitude*	*observed tsunami record*
10,400	Kattegatt	at least 8	very high wavy, strong coastal effects
11,600	Kattegatt	at least 7	high wave with significant costal effect
11,200	Kattegatt	about 7	high wave with coastal effects
10,430	Mälardalen Valley	well above 8	very high wave, very strong effect, extensive turbidite
9663	Hälsingland	above 8	at least 15 m high wave, extensive turbidite
9428	Umeå area	at least 7	height unknown
9291	Umeå area	7–8	at least some metres run-up, coastal effects
8600	Södermanland	6–7	probably some 5–10 m wave
7800	Stockholm region	at least above 6	maybe 13 m run-up, extensive turbidite
6100	Hälsingland	around 8.5	at least 10–15 m wave on-shore and -25 m off-shore
4000	Umeå area	6–7	height unknown
4000	Södermanland	around 6	uncertain height, distinct run-up
3200	Södermanland	around 7	local lake tsunami, turbidite over 5.5 km
2900	Northern Uppland	unknown	6 to 20 m run-up, -20 m off-shore
2000	Hälsingland	explosive gas venting	at least 20 m wave height
1600	South Kattegatt	unknown	some metres run-up
900	South Kattegatt	around 7	height unknown, sudden silting over of 2 Viking ships

Table 1. Tsunami events recorded in Sweden, all in association with paleoseismic events

rotation is recorded over an immense area of 500x600 km (Mörner & Sun, 2008). A turbidite (seismite) was deposited in the autumn of varve 10,430 BP over a distance of 320 km. No doubts, this must have been a very large earthquake (Mörner, 2003, 2011).

Obviously, a major tsunami was set up by this earthquake (Mörner, 1995, 1996, 1998, 2003, 2008a, 2011). The tsunami broke into the Närke Straight and washed it free of ice-bergs and pack-ice so that the Atlantic water could enter the Baltic turning it into a marine basin; the Yoldia Sea *sensu structo* (Mörner, 1995). The tsunami wave also broke in over land and invaded separate lake basins. Nine sites are recorded spread over a distance from the epicentre of 320 km to the W and 400 km to the SW (Mörner, 2003, 2008b). The earthquake and tsunami occurred in the autumn of varve 10,430 BP, as evidenced from three independent varved clay sections located 76 km apart (Mörner, 2003, 2011, Fig. 14).

The 10,430 BP varve constitutes a true "marker varve" that stands out in the stratigraphical records. This was first noted by De Geer (e.g. 1940), and he let it be the marker of the sudden drainage of the Baltic Ice Lake and the onset of the Yoldia Sea stage. We now know that this drainage occurred some 300 years earlier, however. This left the sudden change from freshwater to marine-brackish environment unexplained. Mörner (1980) noted that the

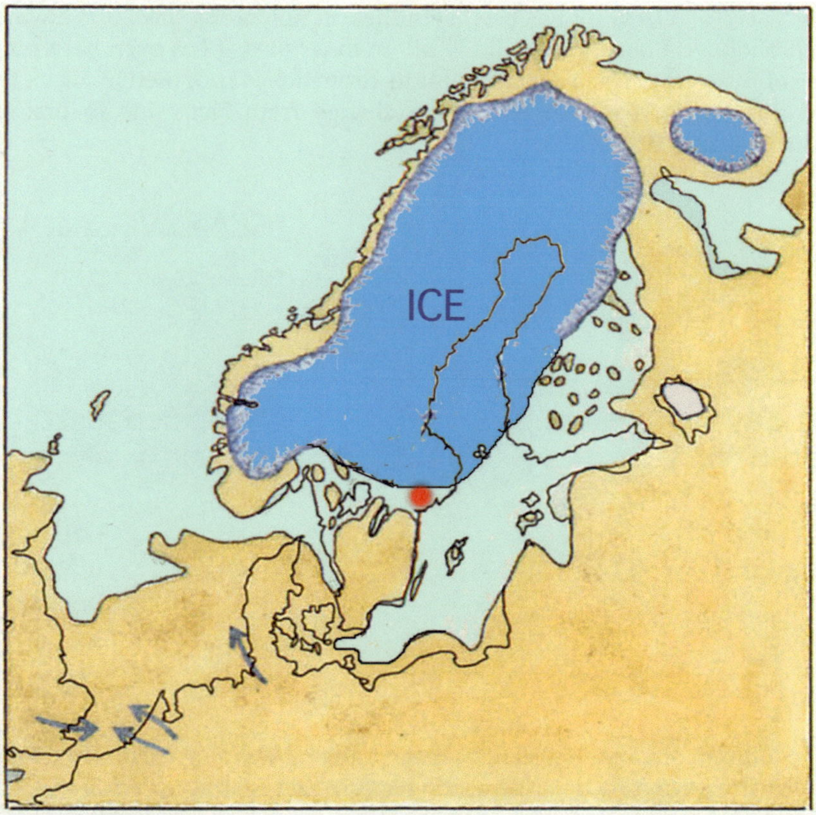

Fig. 5. Paleogeography of NW Europe at the time of the huge 10,430 BP paleoseismic event: ice cap (blue), sea (light blue), land (brown) and earthquake epicentre (red dot). The Baltic was on a level with the Atlantic via the Närke Straight, which remained blocked by pack-ice and icebergs up to the earthquake event. The 10,430 BP tsunami washed the straight free of ice allowing marine water to enter the Baltic and turn it, within 1 varve, into a brackish-marine basin; the Yoldia Sea *sensu stricto* (Mörner, 1995)

10,430 BP varve had special characteristics suggesting that it, in fact, was a seismite; i.e. formed as a consequence of an earthquake. In 1995, we found conclusive field evidence to assign it a paleoseismic origin and an age of "the autumn of varve 10,430 BP" (Mörner, 1995, 1996, 1998, 2003, 2008a, 2011; Mörner & Tröften, 1993; Tröften, 1997; Tröften & Mörner, 1997). At the time of the earthquake, the water depth was in the order of 150 m (above the assumed epicentre) implying that we may be dealing with a tsunami wave with a diameter of about 150 m. When this wave, at a high speed, moved laterally and the water-depth became shallower, it started to trim the sediment surface of the seabed (Fig. 6). This bottom erosion generated local slides, extensive turbiditic bed-loads and huge "clouds" of suspended matter (clay and silt). At many sites, we record anomalously thick varves of silt, sand and gravel that sometimes even include eroded "clay pebbles". They are easily identifiable as they stick out of the normal records (Fig. 7). They are identified over an area of 200x320 km. At other sites, we record exceptionally thick clay varves (from the setting of suspended clay particles).

The marker varve of 10,430 BP has been identified in numerous sites. It is always found at the same chronological level; varve 10,430 BP. In three sites, it has even been pinpointed at the autumn of this varve. The mechanism for its formation is proposed in Fig. 6. It should be noted that this varve year also marks the change from lacustrine to brackish-marine environment (Mörner, 1995, 2003).

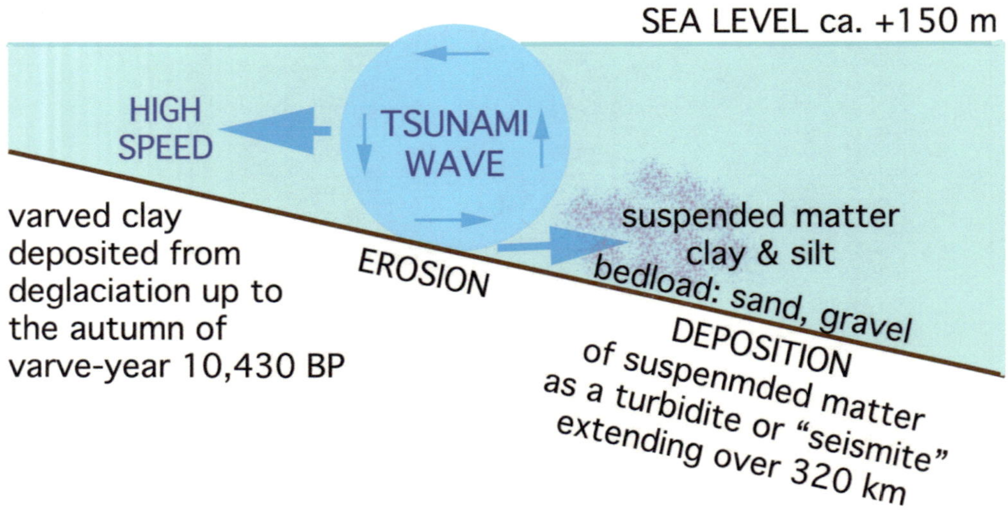

Fig. 6. In the autumn of varve 10,430 BP, a major tsunami wave rapidly moved laterally from the epicentre (primarily documented in western and southwestern directions). The diameter may have been in the order of 150 m. When the rotating wave hit the seabed, it generated erosion that set up both a bed-load of silty-sandy-gravely grains and a suspension of clay and fine silt. Sediments were also set in motion by the ground shaking. All together, it generated the deposition of a turbidite (seismite) over a very large area, viz. 200x320 km

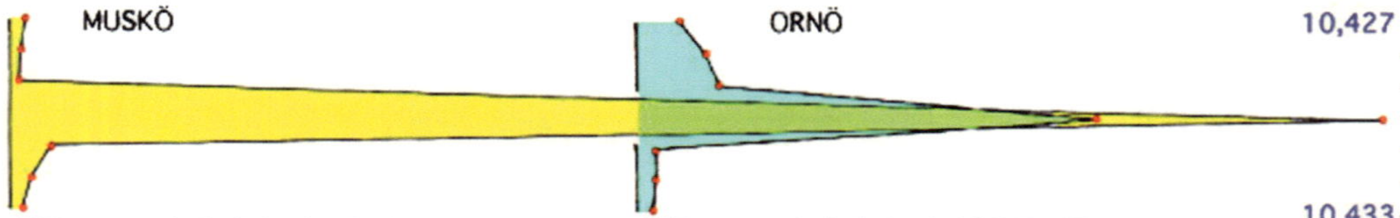

Fig. 7. The turbidite of the 10,430 vBP event as recorded by two sites south of Stockholm. After 274 and 321 normal varves, respectively, there is a very thick layer in the same varve in both sites located 20 km apart; viz. thick sandy-silty layers with incorporated pieces of contorted clay and clay pebbles, i.e. evidence of a massive turbidite flow. This turbidite (marker varve) is found over a wide area of 200x320 km

4.2 The 9663 BP paleoseismic event

In the Hudiksvall area of central Sweden, there occurred a very large earthquake in the varve-year 9663 BP (Mörner et al., 2000; Mörner, 1998, 2003, 2008a, 2011). The paleogeography of this event is very well known (Fig. 8; Mörner, 2003). Sea level was in the order of +231-236 m, and only minor islands stack out of the sea in front of the ice margin. The primary fault was along a sheer zone. A considerable up-thrusting is recorded. The water depth at the epicentre is likely to have been in the order of 250 m (allowing for a tsunami wave with the same diameter). Bedrock fracturing is recorded in some 100 sites over an area of 50x50 km. Liquefaction is recorded at 12 separate sites covering an area of 80x40 km. The liquefaction event is directly tied to varve 9663 BP.

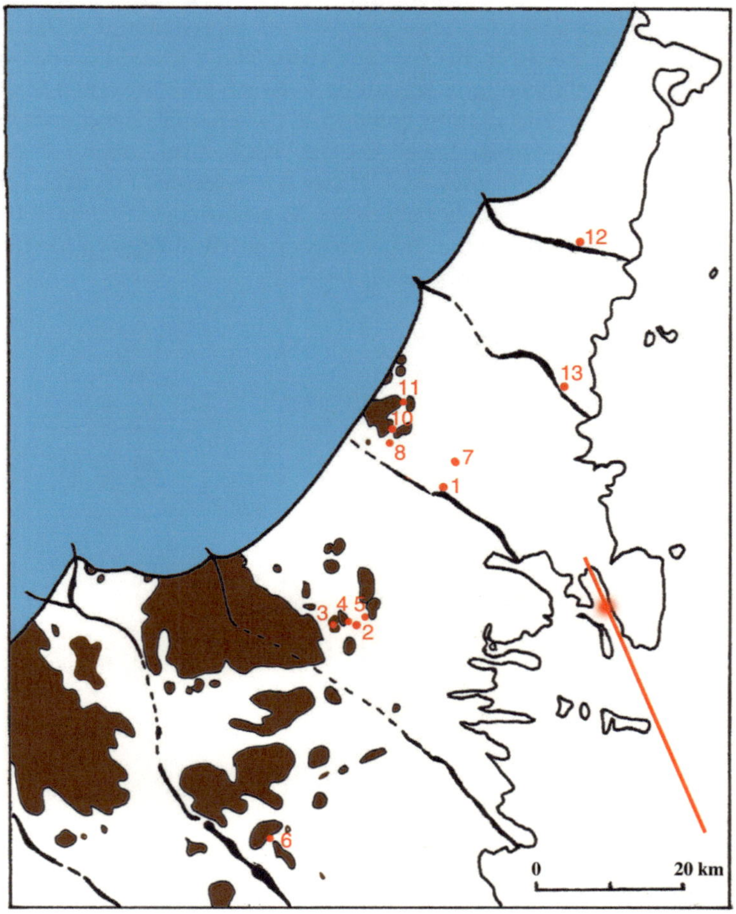

Fig. 8. Paleogeography of the Hudiksvall region at the time of the 9663 BP paleoseismic event: ice cover (blue), land (brown), sea (white) and primary fault with epicentre (red). Sea level was in the order of +231-236 m (with a NW-SE tilt). Tsunami beds were recorded at 13 sites (red dots and numbers). A related turbidite layer is recorded and dated in 27 additional sites within the region, and in numerous sites all the way down to Uppsala (Fig. 11)

The earthquake, which must have been in the order of M8 (or even above), also set up a tsunami wave. It is recorded at 13 different sites, including 9 lakes where a total of 44 cores were taken. It is dated both by varves (at 9663 varve-years BP) and by radiocarbon (at about 9150 C14-years BP). In the lakes, the tsunami event is recorded as a sand layer of graded bedding (fining upwards). The microfossil content of those sand layers is characterized by a planctonic Lake Ancylus fauna and flora (i.e. the same criteria as used in Scotland by Dawson, 1999). The investigation is closely described in Mörner (2003). We will here discuss two sites of methodological interest.

Lake Svartsjön has a present elevation of +223.4 m. The highest Baltic level (BL) is closely determined at +231.3 m and the level at the tsunami event (TL) is determined at +223.5 m. This means that Lake Svartsjön was close to sea level at the time of the tsunami (that is the western side). To the east, facing the open Baltic, there was a 10 m high sill, however. We have 15 cores from this lake basin. In varve-year 9663 BP (and at about 9150 C14-years BP), a tsunami wave broke over the sill in the east and deposited a graded on-swash bed (fading from east to west). It was followed by a graded back-swash bed deposited from west to east. This is illustrated in Fig. 9. The diatom content of the tsunami sand contains planctonic deep-water species from the open Lake Ancylus basin (*Aulacosira islandica, Aulacosira ambigua, Alacosira italica, Aulacosira arenaria, Gyrosigma attenuatum, Fragilaria construens*, etc.). This implies the same tsunami bed characteristics as recorded in Scotland (Dawson, 1999; Dawson & Smith, 2000). We believe that this is a good methodology of identifying tsunami beds, and discriminate them from littoral beds.

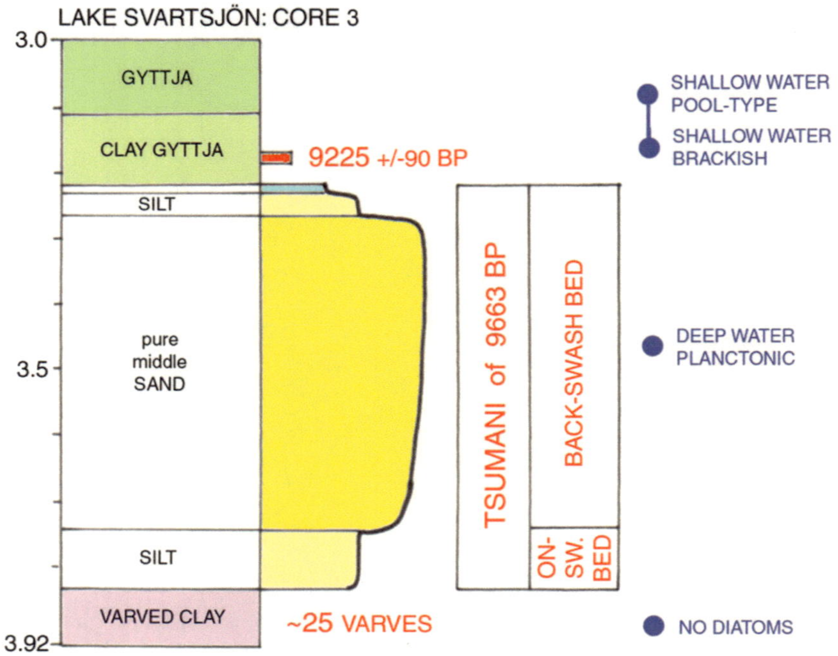

Fig. 9. Lake Svartsjön is densely cored. In 9663 BP, a tsunami wave broke over the 10 m sill in the east and deposited down-washed material (wedging out from east to west). It was followed by a back-swash wave that deposited a graded sand-bed wedging out from west to east. The tsunami bed is characterized by a planctonic deep-water diatom flora from the open Baltic basin (the Lake Ancylus stage)

Lake Källsjön is located above the highest coastline and has had no open connection with the Baltic. It has a present level of +232 m but a former damming sill seems to have had an elevation of about +236 m. The highest Baltic level (BL) is determined at +231,3 m. The shoreline at the tsunami event (TL) was at +223.5 m. The situation is illustrated in Fig. 10. The lakebed was cored at 3 sites (5 cores), and a graded tsunami sand-bed was recorded in all cores. Its fining-upward bedding is consisting with a tsunami wave ingression. The very evidence of a tsunami comes from the microfossil content of the sand-bed, which records a planctonic diatom flora of the Baltic Lake Ancylus stage with species like *Aulacosira islandica, Gyrosigma attenuatum, Aulacosira ambigua, Aulacosira italica, Aulacosira distans* and *Gomphonema angustatum*. To enter the lake, the tsunami wave must have crossed a 700 m long land area at a height of 12.5 m above the tsunami shore level (Mörner, 2003, 2011). In this lake (contrary to other lakes in the area), there occur a small fish *Osmerus eperlanus* (smelt), which is likely to be a relict from the Ancylus Lake. It can only have arrived into this lake basin via the 9663 BP tsunami event. This tsunami may even be recorded in sites 300 km to the south (Mörner, 2003).

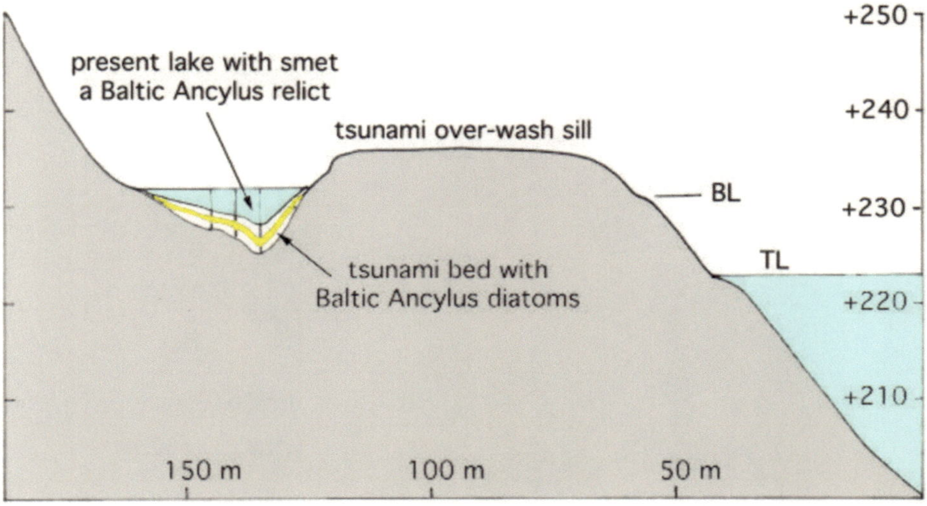

Fig. 10. The situation at Lake Källsjön (+232 m). This lake was located above the highest Baltic level (BL at +131,3 m). The sill in-between had an original level of about +236 m. At the time of the 9663 BP tsunami, the Baltic level was at +223,5 m (TL). In the lakebed, there is a sandy tsunami bed (in fining-upward sequence) that contains a Baltic Lake Ancylus diatom flora. In order to enter the lake basin, the tsunami wave must have been more the 12,5 m high and to have over-wasted a sill at +236 m for about 700 m. Therefore, the tsunami wave is likely to have been, at least, 15 m high. In today's lake, a small fish, smelt, is living, which is likely to be a relict from the Lake Ancylus water, washed into the lake by the tsunami wave

The 9663 BP tsunami left extensive records in the offshore environment, too. It is recorded (and dated) as a turbidite (seismite) within 27 varve records in near-field area. In total, it is recorded in numerous varved clay records extending from Sundsvall in the north to Uppsala in the south; that is over 320 km along the coast (Fig. 11). This layer cannot be a strict density-driven turbidite. Obviously, we are dealing with a seismite strongly affected by the bottom effect of the tsunami wave as illustrated in Fig. 6 (above).
A secondary effect recorded and dated thanks to the varve chronology, is the seepage of methane gas through the clay up to varve 9663 BP where it ends (Mörner, 2003). This is

understood as an earthquake effect triggering methane ice stored in the bedrock to transfer to methane gas that tried to seep through the seabed at the time of the earthquake (i.e. varve 9663 BP). This methane gas venting was, at some sites, explosive, which must have added to the bed-load and turbidite formation (Fig. 11).

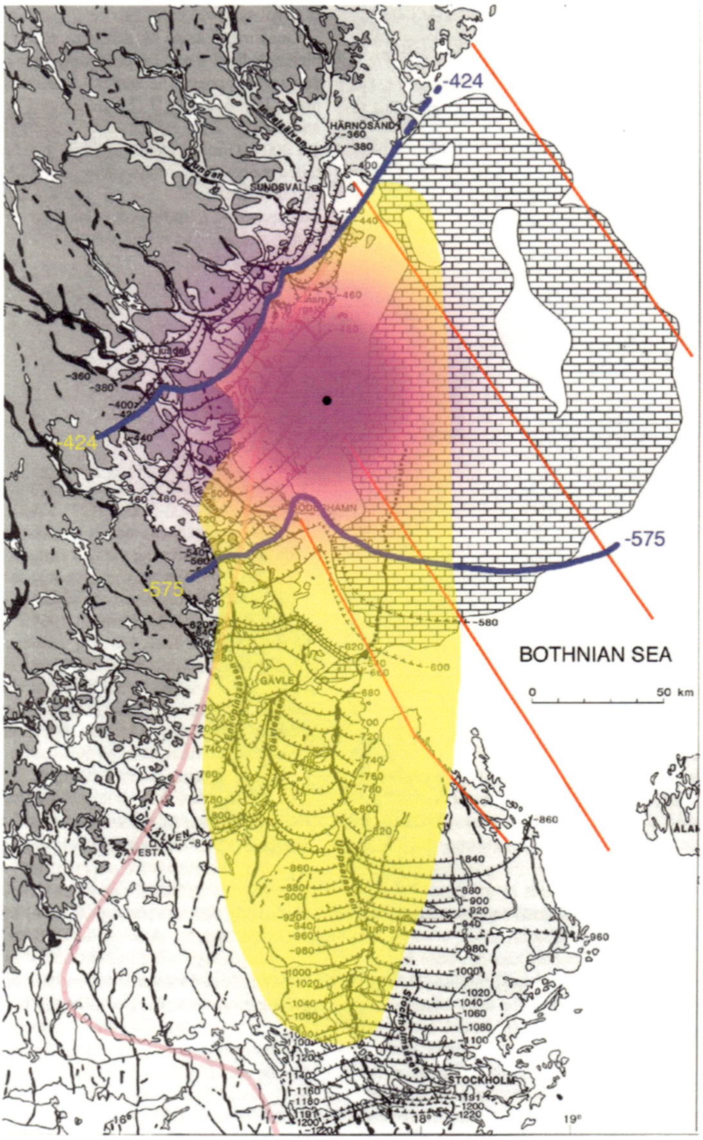

Fig. 11. At the 9663 BP earthquake, a turbidite (seismite) was spread over the seabed for a distance of at least 320x100 km (yellow zone). The ice margin is marked by a blue line with figure –424. The epicentre is marked in purple along the fault zone in red. This extensive turbidite must have been formed by interacting forces setting sediments in motion along the seabed; viz. ground shaking, tsunami wave trimming (Fig. 6) and gas venting

4.3 The 2900 BP tsunami event

The coastal area of northern Uppland is traversed by the Singö Fault zone. This zone seems to have been reactivated during the deglaciation phase some 10,000 years ago (Mörner, 2003, 2004). A Late Holocene tsunami event has been described from the same area (Mörner, 2008b, 2009b). The age is 2900 BP. The event is only recorded in lakes and bogs. No simultaneous faulting is recorded. We suspect, however, that it is another reactivation event of the Singö Fault zone.

A tsunami bed was recorded in offshore sediments, in shore-zone sediments and in lake and bog sediments; all together ranging in elevations from 20 m below to 20 m above (or, at least, 6 m above) the corresponding sea level (Fig. 12). We followed the tsunami beds from offshore basins (15 to 35 cm sand and gravel in graded bedding), via lagoonal basins (with 70 cm sandy beds at the clay/gyttja interface) up into lake basins above the corresponding shore (40–50 cm sandy-gravelly beds erosively deposited between the marine clay and lacustrine lake gyttja). Six C14-dates provide a close age of the offshore and lagoonal sites and a strong erosive effect in the lake basins at least up to a level 5 m above the corresponding shore as illustrated in Fig. 12. The data record a vertical spread of the tsunami beds from –20 m to +6 m. The lake and bog coring suggests that the tsunami may have had a run-up of 20 m. This is not yet supported by dates; only a 6 m run-up (Fig. 12).

The recording of tsunami beds in offshore environment is important. Ordinary waves have been shown only to leave sedimentary signals down to a depth of maximum 10 m. All the 3 beds recorded in offshore environment have identical C14-ages that differ significantly from the ages of the isolations of the lake basins (Fig. 12). This fits perfectly well with an interpretation in terms of a tsunami event, but is totally inconsistent with a normal uplift history. The on-shore run-up might be in the order of 20 m, judging from field observations. This has not yet been confirmed by C14-dates, however.

4.4 The 2000 BP explosive gas venting event

An event of violent methane venting was recorded north of Hudiksvall (Mörner, 2003, 2009b). It occurred at about 2000 C14-years BP when local sea level was at about +18 m. It set up a tsunami wave of significant height. Five bogs ranging from +8 m to +38 m were investigated by coring (Fig. 13). The +38 m record is especially interesting. The basin was isolated from the Baltic (due to uplift) some 4000 BP. Above the isolation level and between a freshwater gyttja and a covering peat, there is a 2.65 m thick bed of gravel with numerous shells of Baltic brackish-water origin. This implies a tsunami wave that reached, at least, from +18 m to +38 m; i.e. having a run-up of 20 m or more. Consequently, this was a very strong event. Tsunami beds were also recorded in basins located at +23, +18 and +14 m. A core taken at +8 m recorded, at the 2000 BP level, a black layer of FeS-rich clay (i.e. strongly reducing environment interpreted as a result of the methane venting), which rested on older deposits with a significant erosion contact (i.e. a hiatus of about 7000 years).

Lake Dellen is a large lake located 25 km to the west. Today it has a level of +37 m. A peat layer found 3 m below the present water level indicates that the lake must once have had a level, at least below +34 m. Because the peat is C14-dated at about 2000 BP, the rise in water level seems coincidental with the tsunami event recorded further to the east. Therefore, it seems likely that also the Lake Dellen pounding was an effect of this tsunami event (Fig. 13). The pounding of Lake Dellen was in the order of 3–4 m.

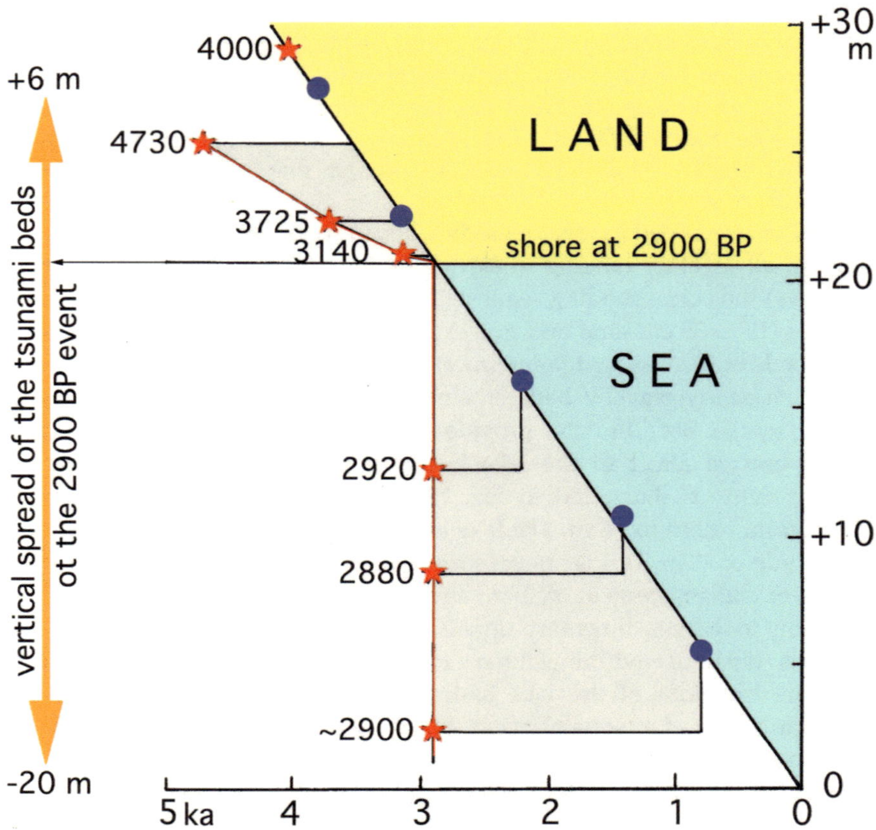

Fig. 12. The 2900 C14-years BP tsunami event in northern Uppland with respect to the rate of land uplift and shore displacement over the last 4000 years (the oblique line of ~7 m uplift per millennia passing through dated anchor points marked by black dots). At 2900 C14-years BP, the shore was at +20.7 m with land above (yellow) and sea (blue) below as marked on the right side of the diagram. The stars mark tsunami beds recorded and dated in off-shore sediments (all falling sharply at the 2900 BP level), in coastal deposits and in lakes and bogs on land where the tsunami beds have eroded down into the older sediments. The supposed tsunami bed in the +29 m basin has an age that coincides with the time of isolation. Therefore, we can in this case not discriminate between a normal shore sand from the isolation and a subsequent tsunami bed. Consequently, the graph gives evidence of a tsunami event that deposited typical tsunami beds over a vertical range from -20 m to +6 m. The tsunami run-up might have reached even 20 m above the shore level, judging from lake and bog coring at higher altitudes, however

5. Perspectives

The examples studied shed light, we hope, on the general processes of tsunami deposits and their characteristics to be used in order to be able to identify the deposits as traces of past tsunami events. Long-term records of past tsunami events offers means of assessing the tsunami risk for a given region.

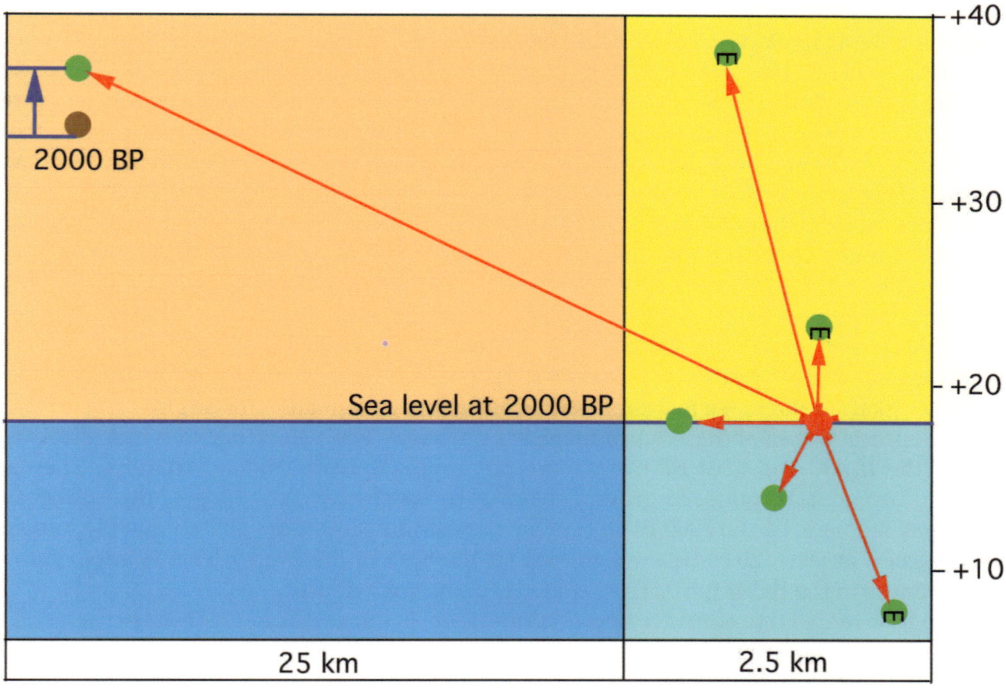

Fig. 13. Vertical and horizontal spread (red arrows) of the 2000 BP tsunami generated by an explosive venting of methane gas (at red dot). Green dots mark bogs invaded by the tsunami. Lying E-signs mark erosion below tsunami bed

In the Maldives, for example, we have traced 11 potential events; out of which 1 is the 2004 event, 5 refer to submarine sandstorms (and additional facts) and 5 refer to occasional beach ridge events of possible tsunami origin.

The Swedish records include 17 events (Table 1). This is not only of academic interest but also a serious fact to consider when it concerns super-long future safety predictions in the case of nuclear waste handling, which calls for a safe deposition for, al least, 100,000 years.

6. Conclusions

We need to have long records of past tsunami event in order to be able to present meaningful hazard assessments for the future. This calls for the use of a reliable methodology. We have tried to give some case studies where the methodology plays an important role.

In the Maldives, we have introduced the concept of "submarine sandstorms" in association with tsunami events, and we have shown that such "sandstorms" may reach considerable depths where they are even depositing significant sand wedges in the submarine caves (Mörner et al., 2008). Radiocarbon dating of shallow-water coral and gastropod species provide ages that can be compared with land records (swamp stratigraphy, coastal stratigraphy and morphology) and historical descriptions.

In Scotland, there are excellent records of the Storegga tsunami. The identification of planctonic deep-water species (not shallow-water bentic or lagoonal species) in the tsunami sand layers offers means of discriminating its origin as to a tsunami wave deposit and not a shallow sea deposit (Dawson, 1999; Dawson et al, 2000).

In Sweden, we have identified a very high number - 17 - of past tsunami events. The reason is not so strange. Because of the postglacial land uplift, present land was previously covered by water. As most of the paleoseismic events seem to have been normal fault events, those events in submarine position are likely also to generate tsunamis. Finally, we have undertaken quite extensive studies of those paleoseismic events (Mörner, 2003). We have cored numerous lakes and basins, and identified potential tsunami beds. Their graded character (fining upwards) and their sudden appearance in otherwise unbroken deposits (e.g. marker varves of turbidite character) provide sedimentological characteristics suggesting a tsunami origin. The micropaleontological analyses revealing planctonic deep-water faunas and floras provide more conclusive indications of a tsunami origin. The data are integrated with the other paleoseismic data. Only by applying this integrated, multi-parameter methodology, we feel confident in assigning a paleotsunami origin.

The evidence in the offshore environment is important. The very extensive turbidites (e.g. those of the 10,430 and 9663 BP events) are not simple density driven turbidites. They are seismites formed in response to ground shaking, tsunami wave trimming of the seabed and violent gas seepage. At the 2900 BP event, the deposition of a sandy-gravely-pebbly tsunami seismite goes, at least, 20 m below sea level (at the time of the event). This is important, as normal waves in the Baltic never have effect reaching such depths.

We learn that we need to consider observational facts in the on-shore environment as well as in the off-shore environment. Besides this, we should consider historical observations. Sometimes, even the core of some myths has a message to tell.

7. Acknowledgements

Other persons taking part in the Maldives project were: Jaques Laborel, Michael Tooley, William Allison, Jaques Collina, Francois Laborel, Caroline Rufin, Shaidul Islam, Bjarne Lembke, David Dominey-Howes, Alex Dupuch, Michael Banfield and our local friends from the National Centre of Linguistic and Historical Research (Naseema Mohamed and co-workers), from Ecocare (Mohamed Zhair, Moosa Manik, Hassan Maniku), from Sea Explorer (Shaahina and co-workers) and Whale Submarine (O. Naseer).

Other persons taking part in the Swedish project (known as the Boda project) were: Franch Audemard (Venezuela), Douglas Grant (Canada), Andrej Nikonov (Russia), Dimitri Zykov (Russia) and from the P&G unit at Stockholm University; Rabbe Sjöberg, Per Einar Tröften, Christian Bronge, Ole Kvamsdal, Guangyu Sun, Hans Wigren and the local experts Alf Sidén and Lars Strandh; besides numerous international experts taking part in the 1999 and 2008 excursions.

We express our gratitude to all of them.

For further communications use: morner@pog.nu

8. References

Bell, H.C.P. (1940). The Maldive Islands. Monograph on the history, archaeology and epigraphy. Ceylon Government Press, Colombo, 204 pp.

Brunnberg, L. (1995). Clay-varve chronology and deglaciation during the Younger Dryas and Preboreal in the easternmost part of the Middle Swedish Ice Marginal Zone. Quaternaria, Ser. A, No. 2, 1-94.

Dawson, A.G., Long, D. & Smith, D.E. (1988). The Storegga slides: evidence from eastern Scotland for a possible tsunami. Marine Geology, Vol. 82, 271-276.

Dawson, S. (1999). Flandrian relative sea level changes in northern Scotland. Ph.D.-thesis, Coventry University, 370 pp.

Dawson, S. & Smith, D.E. (2000). The sedimentology of Middle Holocene tsunami facies in northern Sutherland, Scotland. Marine Geology, Vol. 170, 69-79

De Geer, G. (1940). Geologia Suecia Principles. Kungl. Sv. Vet. Akad. Handl., 3rd, No. 18:6, 1-360.

Mörner, N.-A. (1980). A 10,700 year's paleotemperature record from Gotland and Pleistocene/Holocene boundary events. Boreas, Vol. 9, 283-287.

Mörner, N.-A. (1995). The Baltic Ice Lake - Yoldia Sea transition. Quatern. International, Vol. 27, 95-98.

Mörner, N.-A. (1996). Liquefaction and varve disturbance as evidence of paleoseismic events and tsunamis: the autumn 10,430 BP event in Sweden. Quaternary Science Revue, Vol. 15, 939-948.

Mörner, N.-A. (2003). Paleoseismicity of Sweden - a novel paradigm. A contribution to INQUA from its Sub-commission of Paleoseismology, Reno 2003, ISBN-91-631-4072-1, 320 pp, Stockholm.

Mörner, N.-A. (2004). Active faults and paleoseismicity in Fennoscandia, especially Sweden. Primary structures and secondary effects. Tectonophysics, Vol. 380, 139-157.

Mörner, N.-A. (2005). An investigation and catalogue of paleoseismology in Sweden. Tectonophysics, Vol. 408, 265-307.

Mörner, N.A. (2007). Sea Level Changes and Tsunamis. Environmental Stress and Migration over the Seas. Internationales Asienforum, Vol. 38, 353-374.

Mörner, N.-A. (2008a). Paleoseismicity and Uplift of Sweden. Guidebook, Excursion 11 at 33rd IGC, Oslo 2008, 107 pp, www.33IGC.org.

Mörner, N.-A. (2008b). Tsunami events within the Baltic. Polish Geological Institute, Special Papers, No. 23, 71-76.

Mörner, N.-A. (2009a). Paleo-tsunami events in the Maldives and surrounding areas. Proceedings GeoDev 2009, Daka, Bangladesh.

Mörner, N.-A. (2009b). Late Holocene earthquake geology in Sweden. Geological Society in London, Spec. Publ., Vol. 316, 179-188.

Mörner, N.-A. (2010). Natural, man-made and imagined disaster. Disaster Advances, Vol. 3, No. 2, 3-5.

Mörner, N.-A. (2011). Multiple parameter interaction in paleoseismology. Submitted.

Mörner, N.-A., Laborel, J. & Dawson, S. (2008). Submarine "sandstorms" and tsunami events in the Indian Ocean Journal of Coastal Research, Vol. 24, 1608-1611.

Mörner, N.-A. & Sun, G. (2008). Paleoearthquake deformations recorded by magnetic variables. Earth Planetary Science Letters, Vol. 267, 495-502.

Mörner, N.-A., Tooley, M. & Possnert, G. (2004). New perspectives for the future of the Maldives. Global Planetary Change, Vol. 40, 177-182.

Mörner, N.-A. and Tröften P.E. (1993). Paleoseismotectonics in glaciated cratonal Sweden. Zeitschrift für Geomorphologie N.F., Vol. 94, 107-117.

Mörner, N.-A., Tröften, P.E., Sjöberg, R., Grant, D., Dawson, S., Bronge, C., Kvamsdal, O. & Sidén, A. (2000). Deglacial paleoseismicity in Sweden: the 9663 BP Iggesund event. Quaternary Science Revue, Vol. 19, 1461-1468.)

NOAA, (2010). NOAA Centre for Tsunami Research. December 26, 2004 Indonesian Sumatra earthquake and tsunami Web link compilation and data. http://nctr.pmel.noaa.gov/sumatra20041226.html

Tröften, P.E., (1997). Neotectonics and paleoseismicity in southern Sweden with emphasis on sedimentological criteria. Stockholm University, Ph.D.-thesis, P&G No. 8, 1-124.

Tröften, P.E. and Mörner, N.-A. (1997). Varved clay chronology as a means of recording paleoseismic events in southern Sweden. Journal of Geodynamics, Vol. 24, 249-258.

19

Tsunami Deposit Research: Fidelity of the Tsunami Record, Ephemeral Nature, Tsunami Deposits Characteristics, Remobilization of Sediment by Later Waves, and Boulder Movements

Barbara H., Keating[1], Charles E. Helsley[2], Matt Wanink[3], and Daniel Walker[4]
University of Hawaii, SOEST, Honolulu, HI, USA

1. Introduction

Historically, much that has been reported about tsunamis has been based upon mathematical modeling. However, modeling tells little about the impact of tsunami on land and people, inundation and run-up, drain back, movement of sediments and boulders, the natures of deposition and erosion, the carrying capacity of waves, etc. In order to study these topics, the question can be asked, what do tsunami deposits tell us about tsunamis?

Since the Hawaiian Islands have experienced numerous tsunamis it is an ideal place for field studies. Observations and analysis have been carried out on the North Shore of Oahu. The objectives were: to observe tsunami vs. storm activity, document the impact of the 1946 Aleutian and 2006 Kuril Tsunami, observe the nature and distribution of sediments associated with these events, and quantify the nature of transport of mega-boulders, and calculate the transport factor.

The results indicate: (1) There was little difference in the sedimentary record and inundation record between the November 2006 Kuril Islands Tsunami and the December 2006 Winter Storm. (2) The 2006 tsunami and storm deposits were ephemeral, i.e., the tsunami record was removed by the winter storm. (3) The sedimentary record is of low fidelity in discriminating and preserving tsunami and storm events. From the analysis of the 1946 Aleutian Tsunami, the real-time aerial photography proved invaluable, since the patterns of inundation and drawback become obvious. A sand sheet, with only isolated and rare boulders, covered the inundation zone. The drain back utilized pre-existing drainages in a limited way. A stratigraphy is found in the archaeological sampling showing that cultural items were relocated by the tsunami and some vegetation was buried in place.

The study of mega-boulders suggests that giant waves are capable of moving rock boulders around on the reef platform, but did not significantly displace them. The waves surging over the platform plucked rocks from the back or leeward side of the mega-boulders. The fragments of boulders struck the karstic platform and broke off additional rock. The transport paths of the fragments could be followed intermittently across several meters length on the shoreward side of the platform. No large boulders were added or removed from the platform by the storm. Transportation calculations indicate Hawaiian giant surf and tsunamis occupy the lower range of transport values, when compared globally. There is remarkable similarity between the boulders of Hawaii and Morocco that are presumed to be of tsunami origin. This similarity likely reflects the changing sea level within a shallow carbonate beach and back beach setting.

One of the most valuable products of the tsunami deposit studies is their potential for revealing the periodicity of tsunami, and thus allows risk assessments. The Kahuku study provides evidence that a high-resolution geologic record of tsunami activity is unlikely, and thus the possibility of establishing reliable recurrence records is low. The study also indicates that the Kahuku beach, dune field, and back beach coastal facies are poor choices for obtaining a complete record of tsunami recurrence. Future research should evaluate the submarine setting as a favorable environment.

2. Geologic setting

The island of Oahu is famous for its scenic coastal plain, fringing reef, and surf. The coastal plain extends around the periphery of the island and is interrupted by drainages and becomes very narrow at the eastward and western tips of the island, coincident with the ancient volcanic rift zones (Stearns, 1974). The sediments of the coastal plain generally consist of lithified sand dunes, reef deposits and talus, and in places stream deposits and soils. Wells drilled around Oahu show that the carbonate sediments underlying the coastal plain extend to -400 m depth, with the thickest sediments in valleys and the outer edges of the exposed coastal plain (Stearns and Vaksvik, 1935). The coastal plain rocks display several terrace levels and erosional unconformities marking shifts in global sea level.

Stearns (1970) suggested that the island of Oahu shows important evidence for sea level changes, including: (1.) The presence of sea cliffs separated from the sea by emergent marine deposits, (2.) Evidence that there are older marine deposits at elevations of up to 30 m, and (3.) The base of coastal sand dunes and the lithified beach rock occurs several meters below sea level (mbsl) indicating that sea level was at this (modern) position or slightly lower during dune accumulation. Modern studies of changes in sea level indicate that global sea level has not been significantly higher than present during the last several hundred thousand years.

Because Oahu is situated near the northern fringe of the live Pacific Ocean coral distribution, isolated from the source of Indo-Pacific coral larvae (which do not remain viable over such great distances) and separated by the barriers of sub sea mountain chains and circulatory gyres, the coral reef growth is limited. The reefs of Hawaii favor small numbers of coral species associated with more abundant lime-secreting algae mat. Where a modern fringing reef is present, it is narrow, extending only 0.1-0.3 km from the shore.

The coastal zone from Turtle Bay to Kahuku Point consists of modern coastal deposits superimposed on lithified Pleistocene calcareous reef rock and lithified dune ridges. The modern beach consists of unconsolidated calcareous sands, adjacent to exposures of lithified beach rock and platform reef rock of Pleistocene age. The lithified beach rock platform is approximately the same width as the unconsolidated modern beach.

The dune ridges extend 4.5–6 m above sea level (masl) and are for the most part stabilized by vegetation. At Kahuku Town, southeast of Kahuku Point, several patches of older dunes reach a maximum height of 15.2 m, approximately 0.8 km inland (Stearns, 1970). The Kahuku Town dune sands show good cross bedding and poor consolidation. The deposits are readily differentiated from modern beach sediments by their smaller grain size. Radiometric dating of samples from the Kahuku dunes yield ages of 114,000-122,000 yrs (Szabo et al., 1994), and an age of 121,000 yrs (Ku et al., 1994). The lithified beach rock along the shoreline at Kahuku Point is cut by numerous fractures and the upper surface displays deeply pitted karst micro topography as well as pitting from bio-erosion. Slabs of lithified beach rock have been detached from the uppermost rock platform and transported landward (vertical change of 2 m and horizontal distance of 6 m observed in several instances). Many white scars suggest that more slabs have been quarried and transported however there is an insufficient inventory of boulders in the near vicinity. A likely scenario is that ongoing mechanical degradation has reduced the slabs to loose sand within the wave zone.

3. Fidelity of the tsunami record

The exposed northern coastline of the island of Oahu, Hawaii, is vulnerable to tsunami inundation and extreme winter storm swells. Because this shoreline has yet to be altered by development, it is an excellent location for studies of inundation and sedimentary processes related to tsunamis, giant surf, and sea level changes. The results of field studies on the North Shore (see map, Fig. 1) are presented here, including: observations of a Nov. 15, 2006 tsunami and a winter storm event during December 2006, observations of the 1946 tsunami, and an quantitative analysis of boulder movements. (North Shore is a geographic name, thus capitalized.)

The 2006 Kuril Earthquake, a large magnitude (M=8.3) event, with relatively shallow focal depth and long duration, was tsunamigenic and provided an opportunity to observe a tsunami and evaluate the environmental parameters, sedimentary processes and the extent of tsunami deposits left in the environment. The observation of modern tsunamis is fundamental to understanding sedimentary processes, establishing recurrence rates, for risk assessments involving flooding in coastal zones. Studies of tsunami recurrence rates generally focus on obtaining records of marine inundations (paleo-seismic or paleo-tsunami records). The fidelity of these records is problematic since it is yet to be determined if tsunami deposits and storm deposits can be differentiated.

3.1 Earthquake and tsunami

The 2006 Kuril Tsunami arrived in Hawaii at 7:16 a.m., during the transition from low to high tide (projected tidal range for November 15 was 0.3 m). Real-time reports to the Tsunami Bulletin Board (ITIC, 2006) indicated that the initial run-up at tide gauges in the Kuril Islands near field was less than 1 m. Gusiakov (2006) provided an initial model of tsunami propagation using a low-angle thrust mechanism typical of the Kuril region and a 4-m movement on a fault plane and dimensions of the source area of 200 by 60 km. The analysis suggested that the main energy of the tsunami propagated into the NW Pacific and central part of the Okhotsk Sea. The maximum peak-to-trough tide gauge measurements, Pacific-wide, did not exceed 1 m (Pacific Tsunami Warning Center, 2006).

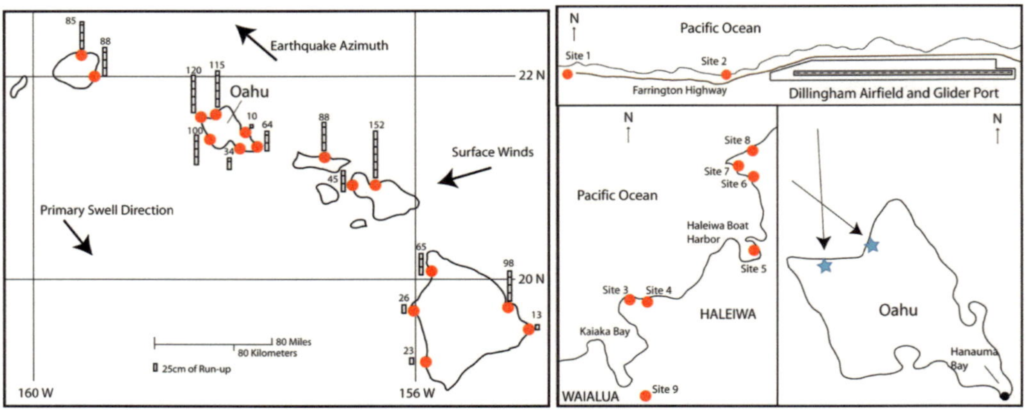

Fig. 1. A map of the Hawaiian Island chain is shown at left. The observed run-up from this study and the maximum trough-to-peak heights recorded by tide gauges are indicated by columns on the map at left (bars on the vertical axis are at 25 cm intervals, and with the recorded data values listed above) based on the NOAA Pacific Tsunami Warning Center, 2006 (Tsunami Observations Kuril Tsunami by Tide Gauges website). The map of Oahu is at the lower right. Wind and wave directions shown by arrows on map were obtained from the Preliminary Local Storm Report (Wroe and Caldwell, 2006). The azimuth to the earthquake source was obtained from the report of Titov (2006). The maps (to the right) show the locations of run-up observations (GPS) Latitude (N); Longitude (W): Site 1 Trail Head to Kaena Point 21 ° 34.6, 158 ° 14; Site 2 W. Dillingham Air Field Mokuleia Army Beach 21 ° 34.6, 158 ° 13; Site 3, West end of site 4 Kaiaka State Recreation Area 21 ° 35.4, 158 ° 07; Site 4 North side of Kaiaka State Recreation Area 21 ° 35.4, 158 ° 07; Site 5 Pier at Haleiwa Small Boat Harbor 21°35.6, 158 ° 07; Site 6 Surfing beach, eastern side Haleiwa beach park 21 ° 36.1, 158 ° 06; Site 7 Near Puaena Pt. (west side) 21 ° 36.1, 158 ° 06; Site 8 North side of old Haleiwa airfield (east of Puaena Pt.) 21 ° 36.2, 158 ° 06; Site 9 Road bridge 21 ° 34.7, 158 ° 07; Puuiki Beach (West of Haleiwa) showed no changes 21 °35.1, 158 ° 08. Locations and Place Names from Bryan's Sectional Maps Oahu, 2007 Edition, EMIC Graphics, Waipuhu, HI. The maximum runup for each site is listed in Table 1.

3.2 2006 Tsunami impact

Dr. Dan Walker (University of Hawaii seismologist and resident of the North Shore of Oahu) made direct observations of tsunami activity from the Highway 930 Bridge at Haleiwa Road over Paukauila Stream (Location 9 in Fig. 1). The first indication of the 2006 tsunami was the rapidly increasing seaward movement of leaves floating on the stream. The water then rose and flooded the nearby sand bar; traveling faster than a person could run. Turbulent flow began in the stream as the tsunami flooded inland and overcame the outgoing stream flow. Black sediment boiled up from the stream bottom as the first wave receded. Four significant tsunami waves covered the sand bar at the mouth of the river. Oscillations associated with the tsunami continued for at least 10 hours after the onset of the tsunami. In attempting to measure run-ups along the banks of the stream just prior to sunset Walker was surprised to be caught in additional significant high water oscillations that prevented any valid run-up measurements at that time. On the next day, the maximum measured run-up on the banks of the otherwise quiet stream was 1 m.

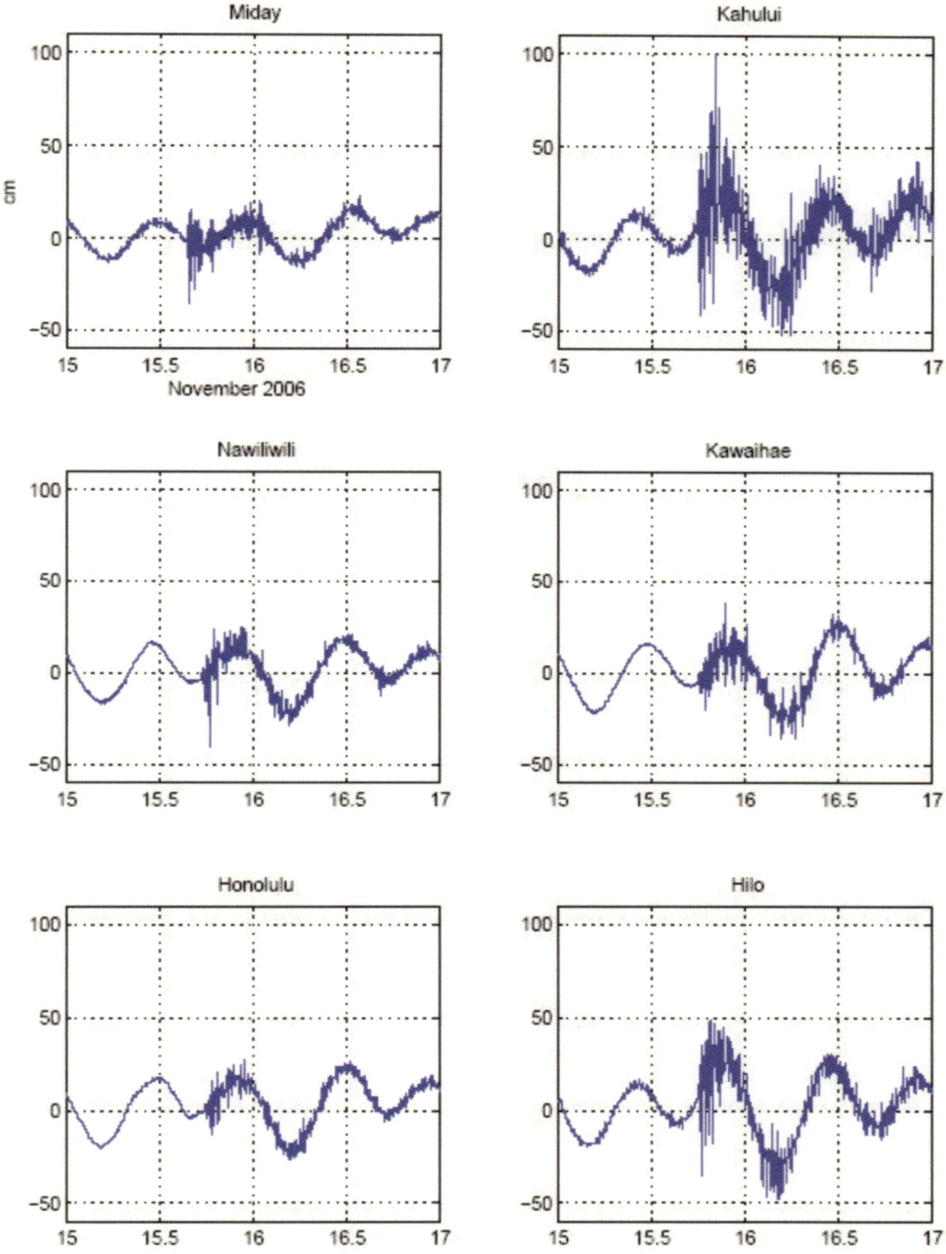

Fig. 2. Tide gauge records of the 15 November 2006 tsunami for various Hawaii ports (courtesy of Hawaii Sea Level Data Center). Note the extended duration of the event as energy from the earthquake was reflected from bathymetric features and the coastlines of the Pacific basin.

Hanauma Bay, the underwater park on the SE shore of Oahu, was closed for 2 days. The strong currents within the bay raised concerns that inexperienced swimmers might be carried out of the sheltered bay into deep waters. The closure of the park is unprecedented, and was a consequence of the chaotic nature of wave and current activity that continued well after the tsunami passed (KGMB News, 2006). The tide gauge records on the east and south shore of the Hawaiian Islands generally showed tsunami wave amplitudes of about 20 cm.

A Coast Guard video of tsunami activity on Kauai (next island to west) also shows strong water movements associated with inundation and flow-back (U.S. Coast Guard, 2006). The tide records (Fig. 2) show that the series of oscillations associated with the tsunami continued for a day and a half after the tsunami passed the Hawaiian Islands. Newspaper and television reports indicate that there were also strong currents on the north sides of the islands of Kauai and Oahu associated with the tsunami.

3.3 Coastal traces

The 2006 tsunami inundation of NW Oahu left a record, observed 3 days after the event, even though these beaches have high usage by local surfers, tourists and residents. The tsunami left a small but significant debris line 1.2 m above the sea level (tsunami plus tide) at several localities (see the list of locations and paired post-tsunami and post-storm photographs). A photograph (from television news) shows a man standing in knee-deep tsunami waters on the submerged pier walkway at the Haleiwa Small Boat Harbor during the tsunami (Location 5, Figure 3a and 3b). The largest of the tsunami waves flooded to the grass line, just above the embankment in these pictures.

Fig. 3. a & b. The photograph on the left was taken during the tsunami by KGMB News (2006) Channel 9 television and appeared on their web site. The photograph shows a man standing in water to his mid-leg on the flooded dock and walkway next to the Shark Encounters boat and sign (published with the kind permission of the photographer). Fig. 3b. The photograph (at right) shows a person at the same location on the pier after the tsunami (Site 5, Fig.1). (Please note that both the boat and the sign in the photograph on the left are absent in the photograph at right, and the photograph on the left was taken using a fish-eye lens.) The tsunami inundation exceeded the high tide mark evident on the pier and the concrete embankment by more than a half meter. Notice that in the photograph at left, that there is a heavy sediment load suspended during the tsunami.

Fig. 4. a & b. At Location 6, lines of coral and basalt clasts were observed lying on an exposed reddish-brown soil horizon. The tsunami inundated the sandy beaches at the east end of Haleiwa Beach Park and Puaena Point (Locations 7 and 8 on Fig. 1). Fig. 4a (left) shows the bay after the tsunami and Fig. 4b (right) shows the shoreline after the winter storm.

Fig. 5. a & b. These images were taken on the North Shore of Oahu, at Haleiwa, Site 8, shown in Fig.1. The scenes have in common, a fragment of a WW II concrete bunker (straight arrows) and the stump of a palm tree (bent arrows). The stump was situated on the exposed coral platform was moved onto the beach sand by the tsunami. A line of vegetation was washed away by the tsunami (above, right). The white lines on the images mark the maximum run-up.

At the Haleiwa surfing beach (Location 6 on Fig. 1), lines of coral and basalt clasts were observed lying on an exposed reddish-brown soil horizon (Fig. 4a and 4b). The tsunami inundated an elevated fossil reef and sandy beach at the east end of Haleiwa Beach Park and Puaena Point (Locations 7 and 8 on Fig. 1). The tsunami and superimposed surf left numerous small sedimentary features and desultory clastic debris traces on the scoured and eroded reef platform, and transported debris westward, depositing it near and in the eastern margin of Waialua Bay.

Only at the beach location 0.5 km west of the glider port (Location 2 on Fig. 1) did the debris line extend above the beach into the dune or back beach to an elevation of 2 m. At this location, the incoming waves can be observed to wrap around an offshore feature and focus into one small area of the beach, where a locally double run-up and inundation were measured.

3.4 Winter swell and strong trade winds

Waves from a North Pacific winter storm reached high levels on the North Shore of Oahu on December 6 and by the evening, they were in the extra-large category, with peaks of 6.6 to 8.5 m on December 7-8. Concurrently, a weak front with strong E-NE trade winds passed over the western Hawaiian Islands on Dec. 7-8. Sustained winds at (25mph) 45 km/h with gusts as high as (50mph) 80 km/h resulted in reports of roof damage, windows shattering, fallen trees, telephone, and power poles (Wroe and Caldwell, 2006).

Following the tsunami, winter swell, and wind event, each of the coastal sites were reexamined and photographed. The winter surf, and strong winds left debris lines, composed predominantly of decaying vegetation, situated only a few cm above the previously photographed tsunami debris lines. At one site, on the exposed coast NE of Haleiwa Bay (Location 7 on Fig 1), the winter swell, augmented by the strong winds, scoured the fossil reef platform of the small and isolated accumulations of sand, shell, and coral rubble left by the tsunami. In some areas, the fossil reef surface was scoured to a depth sufficient to observe a change the color of the rock from gray to white, indicating an apparent beach retreat of roughly 1 m (Fig. 7). In addition, at the eastern end of Haleiwa Beach Park, (Location 6, Fig. 1) the distribution of clasts on the beach was altered by the winter swell and there was visible infilling of sand around the coral clasts.

A sample of the coarse winter swell deposit at Location 7 contained coral fragments, algal nodules, beach rock, twigs, bi-valve fragments, whole micro-mollusk, portions of worm tubes, gastropods, mono-valve shells, and echinoderm spines, with a mean grain size of 0.6 cm with rare clasts 2 - 4 cm across. The coral-algal mixture included rounded, angular, and fragmented clasts; preservation of both whole shells and broken fragments was good (the colors of the sea shells were still intact). A debris line of mostly dead and dry vegetation, with only occasional sparse collections of shell and coral hash debris was also present.

Fig. 6. The 2006 tsunami inundated an elevated fossil reef and sandy beach at the east end of Haleiwa Beach Park (left) and Puaena Point (right); Locations 7 and 8 on Fig. 1.

The high surf event of December 7-8 was more intense and of longer duration than the tsunami and effectively removed all visible traces left by the Kuril tsunami by eroding and remobilizing traces of the tsunami into a higher storm event debris line. Moreover, the seasonally high surf left an event debris line that included floating debris and in most places minor concentrations of coarse sand and coral gravels that were not associated with the previous tsunami trace.

3.5 Ephemeral high-energy record

The 2006 tsunami produced debris lines on several beaches, but at other intervening locations, no obvious traces were observed. The storm scouring of the fossil reef platform and erosion of clastic sediments from Puaena Point (Locations 7 and 8; 7a-7b) exceeded the

Fig. 7. At Haleiwa Bay (Location 3) the beach rock was scoured to a depth sufficient to observe a change of the color of the rock from gray to white, suggesting a beach retreat of roughly 1 m.

deposition and erosion caused by the tsunami. By comparing post-tsunami and post-storm photographs, it was observed that the sedimentary trace of the winter swell was consistently higher than that of the tsunami; with the storm traces situated a few centimeters up the beach slope. Because normal coastal processes easily remove the traces of tsunami and large swell events, smaller events such as those described here are unlikely to be preserved in the geologic record. The absence of these events will result in a low fidelity paleo-tsunami record.

4. The 1946 Tsunami at Kahuku, NE Oahu, Hawaii

The 1946 Aleutian Tsunami was analyzed in the second phase of this study in order to evaluate the tsunami damage, patterns of erosion and deposition, areas of flooding and inundation, and pene-comtemporaneous drain back patterns.

The Hawaiian Islands are prone to flooding damage from far-field tsunamis generated around the Pacific Rim, and the near-field sources within Hawaii. Walker (1994) found that 11 tsunamis, with runup greater than or equal to 1 meter, have occurred on Oahu since 1837. The far-field tsunamis are more numerous and have had significant impact within the Hawaiian Islands. Walker (2004) generated maps of the Hawaiian Islands showing the runup values for large, Pacific-Wide 20th Century Tsunamis. These maps show that Kahuku experienced tsunamis having maximum runup of 7.2 m (1946 Aleutian Islands Earthquake), 1.2 m (1952 Kamchatka Peninsula Earthquake), 6.9 m (1957 Aleutian Earthquake) and 1.2 m (1960 Chilean Earthquake). Shepard et al. (1950) and MacDonald et al. (1947) published post-tsunami surveys; offshore bathymetry was described by Pararas-Carayannis (1965); and known run-up levels for all islands were published (Walker, 2004). The runup values for sites along the North Shore of Oahu are summarized here. The sites are distributed from the northeast point of the island, Kahuku Pt., to the northwest point, Kaena Pt., shown in Fig. 8.

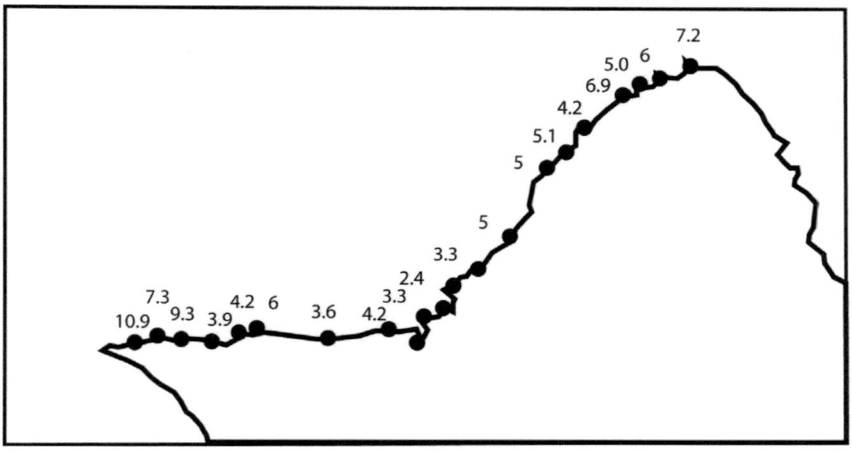

Fig. 8. Map of the North Shore of Hawaii showing the locations of the published 1946 tsunami runup measurements (in meters) between Kahuku Point (at right) and Kaena Point (left), also see Table 1. These values indicate that the maximum tsunami runup (in meters) occurred at the NE and NW tips of the island. Adams (1977) suggests that the bathymetry around both points allows refraction and focusing of the incoming tsunami waves, leading to increased run-up.

Kahuku Pt. 7.2, Kawela Bay 6, Waialee 5.7, Sunset Beach 6.9, Waimea Bay 4.2, Kawailoa Beach NE 4.2, Kawailoa Beach 5.1, Kawailoa Beach 5.7, Puaana Pt. 3.0, Waialua Bay 3.3., Kaiaka Pt.3.3, Kaiaka Bay 2.4, West side of Kaiaka Bay 2.4, Mokuleia Beach 3.6, West of Dillingham Airfield 4.2, Kaena Pt. Trail Head 3.9, Kaena Pt. Trail (west of railhead) 6.0, Kaena Pt. (Intermediate site along trail) 9.3, Kaena Pt. coastal plain 7.2, Kaena Pt. (old light house) 10.2 meters.

Table 1. List of 1946 Runup Values (in meters) by Site Name, (East to West along North Shore, Walker, 2004)

4.1 Historical background

The history of tsunami in the Hawaiian Islands (Daniel Walker, 1994) indicates that tsunamis with runup greater than 1 m have occurred four times during the last century. At Kahuku, the maximum 1946 tsunami runup was reported to be 7.0 m with a nearby measurement at 8.5 m (Shepard et al., 1950). Jagger (1946) reported that 20 waves from the 1946 Aleutian Tsunami were recorded on the Honolulu tide gauge. Some of the tsunami waves were sufficient to over-wash the sand dune ridge along the coast. A second tsunami in 1957 was reported to have reached 7.3 m high (Walker, 1994). A photograph of Kahuku Point coast, taken in 1968, shows that the Kahuku beach was nearly devoid of unconsolidated beach sands, at that time.

The 1946 Aleutian Tsunami took place 9 months after the end of World War II. The military command in Hawaii sent airplanes aloft to photograph the damage to the airfields in the coastal zone around the island of Oahu. The resulting photographs have proven to be extremely valuable resource information and they provide insight into the processes of tsunami inundation and drain-back. The aerial photographs were found in the Hawaii State Archives and scanned for this study. The goal was to analyze the images, delineate the tsunami inundation, observe tsunami-generated features, and to employ reconnaissance-level geologic fieldwork to examine the modern day distribution of the tsunami deposits. Aerial photographs of the vacant land prior to building the Kahuku air base were taken in 1940. The airfields and facilities were constructed largely with wooden buildings constructed between 1941 and 1942 (Allan, 1950). The aerial photographs of the Kahuku Air Base discussed here were taken during the 1946 tsunami and compared to aerial photography from 1951.

4.2 Methods of investigation

The Hawaii State Archives holds the Governor Stainback Photographic Collection, consisting of a folder of oblique aerial pictures of the 1946 tsunami damage. The collection contains numerous photographs of Queens Beach (S. Makapu'u Pt., SE Oahu), Kahuku Field, and a few other airfields. The locations of the photographs can generally be identified using known coastal features since no time code is available. It proved difficult to relocate all of the photographs without a base map. Many roads present in the photographs taken during 1946 have disappeared and are no longer on modern maps; also roads names have changed. Finally, a golf course has been built on top of the NE-SW runway. In an attempt to remedy this limitation, archaeologist Nancy Farwell, was able to provide copies of two schematic drawings of portions of the air base electrical systems. A 1943 Army Corp of Engineers map was also located (University of Hawaii map collection). Bath et al. (1984) included 3 pages of blueprints and aerial photos (taken in 1951) in an unpublished report. Prior to analyzing the photographs, a field expedition (Fig. 9) was organized to examine the study area. No unambiguous evidence of a past tsunami was obvious on the surface, largely due to overgrowth of vegetation.

The first iteration of the analysis focused on establishing the sequence of the photographs using the frame numbers present on most, but not all photographs. This examination revealed that there are two sets of images- wide-angle photographs and close-up pictures. The close-ups are mixed in the same numbering sequence as the wide-angle photographs. The second iteration of the analysis involved trying to relocate the photographs (in an over-lapping sequence) using common features, e.g., runways, control tower, a circular distribution of buildings, or large warehouses. From this study, three different areas of the air base became obvious: the two long runways, the NE-SW runway (closest to Kahuku Pt. and 1920 m long) and the eastern runway (here called Kahuku Runway, 1500 m long) and the control tower located at the junction of the 2 intersecting runways. The air base east of the runways contained numerous white canvas tents and large wooden barracks. The warehouse district, furthest inland and at a seemingly higher elevation was not damaged.

Fig. 9. Elongate slabs (left) were situated at the modern shoreline. Several slabs of rock had been overturned leaving the dark (weathered) side down and the light side on top. In addition, a large coral-algal boulder (approximately 1 m across) was found in a back-beach setting.

4.3 Observations

The aerial photographs show that the airfield was extensively flooded. The flooding and deposits of white sand extends to the lower left corner of the photograph (Fig. 10), and they show sand transported a maximum of 1.6 km (1.4 miles) from the coastline and deposited along the landmark road (Brooks Drive). The tsunami waves overtopped the sand dunes and transported the sand (from beach and dunes) across the coastal zone, forming a large sheet of sand (Fig. 11 & 12). The NE-SW runway (Fig. 11) was also partially buried by sand transported landward by the tsunami flooding. Several of the buildings were washed off their foundations by the tsunami. The drain back of the tsunami waves carried the buildings near the coastline seaward.

At Kahuku, several military buildings seaward of the runways were moved off their foundations, into the entrance of one of the c-shaped ground works situated in the lee of the sand dunes, other buildings were transported and left standing in the middle and seaward edge of the runway (Fig. 11). The seaward flow directions are evident in the area where coastal vegetation was flattened (lower left in photo) and where plant debris was carried onto the beach (lower right of photograph).

Along the seashore, large ground-works were constructed to protected aircraft, with interior concrete walls. The tsunami stripped much of their vegetation from the seaward side of the structures and the ground-works were subsequently partially buried under a layer of sand. In addition, in the landward view (Fig. 13), it is clear that the sand sheet left after tsunami drain-back was extensive, extending across much of the air base. There is limited evidence that mega-boulders (i.e., boulder larger than 1 m) were moved by the tsunami (lower half of image). One mega-boulder was observed on the aircraft runway seen in close-up views. This mega-boulder is also present in photographs taken in a post-tsunami survey by Shepard et al. (1950). Wooden military buildings situated landward of the runways were washed off their foundations in a landward direction (by incoming waves). When several buildings were moved, they often collided so that they were literally stacked one upon one another (Fig. 13).

In the upper and center portion of Fig. 14, gray areas of ground are visible on either side of an area that was scoured and the sand transported seaward during the tsunami drain back phase. In this gray ground, there appears to be an exposure of the gray hard-ground (weathered coral and algal reef) and cemented beach-rock that can be seen today exposed at the beach. Between the stripped surfaces, an area of white rock and sand is exposed. At this site, evidence can be seen that the erosion due to drain back has been localized. In addition, the drain back was focused in low ground and pre-existing drainages. This localized drain back produced a desultory geologic record of tsunami wave activity. In Hawaii, residential buildings of the 1940's were traditionally built with single wall construction. During tsunami inundation, these homes often have the walls collapse and the roof fall intact to the ground. Few of the military buildings however displayed this problem, indicating that different building standards were used. Some of the wooden buildings were built on concrete foundations, and the vacant foundations can easily be identified in photographs. Other barracks buildings that were situated further inland, appear to have been built on the ground, leaving slightly different traces.

The aerial photographs show little debris scattered on the ground. Reports in the *Honolulu Advertiser* (1946) indicate that World War II surplus supplies and equipment were being sold or shipped out of Hawaii when the tsunami took place. The majority of the airfield buildings were probably empty by April 1946. Abundant material from the air base was incorporated into the sediments (Bath et al., 1984).

Observations from Aerial Photographs: Buildings can be carried either seaward or landward, Collisions of building against building obvious, Debris occurs locally near damaged buildings, Foundation are exposed, Sand sheet covers inundation zone, Sand Dunes were lowered or removed and the sand was transported elsewhere, One giant boulder was observed, Tsunami Flooding left some buildings flooded to the eves of the roofs. Walls washed away, Glass broken by tsunami flooding, Items having buoyancy, e.g., wharfs, lawn furniture, etc. were washed away, Vegetation was killed, damaged or bent over parallel to the drain back direction, Salt waters penetrated the ground, making agriculture ineffective until the rains washed out salt from the seawater.
Geologic Impact: Deposition of a large sand sheet, Scouring of the reef surface, and erosion of tens of feet of dune sand (measured laterally from the beach, Adams, 1977), Deposition on reef surface, Minimum inundation was 1.6 km inland.
Human Impact: One person died within the Marconi or RCA Building at the Kahuku Air Base (p. Comm., Dan Walker), 150 people dead and 163 badly injured or missing in the Hawaiian Islands and $25 million dollars of property damage (Shepard et al., 1950).

Table 2. Tsunami observations and impact

Fig. 10. This oblique aerial photograph (Frame 71 close-up photograph, taken after a tsunami wave drain-back) shows the junction of the NW-SE Runway and the Kahuku south runway. The Flight Control Tower can be seen near the center and left in the bottom quarter of this photograph. The runways are largely covered by a sheet of sand derived from the beach and coastal dunes. The large square in the photograph with white debris and small trees near the center (bottom) of the image appears to be a garden (?) with walls toppled by the tsunami. A rectangular, wooden building lies at the middle of the sand-covered runway.

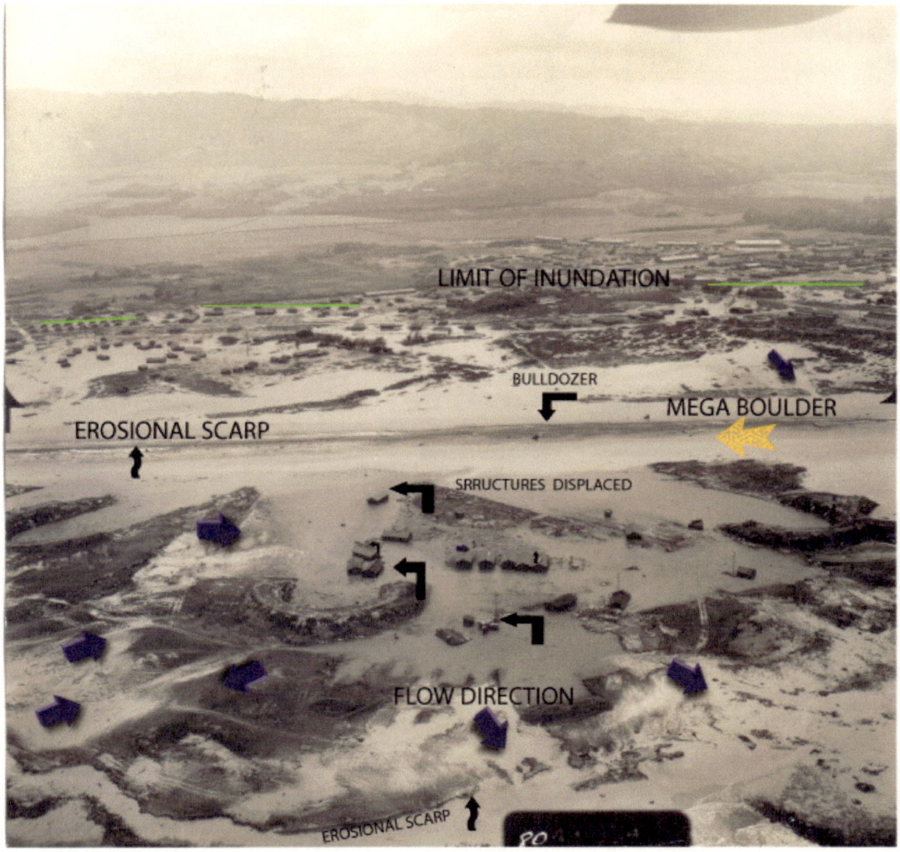

Fig. 11. Looking landward in this oblique aerial photograph (Frame 80), it is clear that the sand sheet left after tsunami drain-back was extensive, extending across much of the air base. The sand sheet completely buries the NW-SE runway. A line on the image marks the limit of inundation. There is a mega-boulder that is comparable in size to a bulldozer sitting nearby on the runway (right-center). An erosional scarp can be seen at the landward edge of the runway. The bent arrows, at the bottom of the picture, mark the displaced buildings. Broad arrows, at the lower edge of the picture, mark the seaward (drawback) flow directions. In the former dune ridge, hard ground having a marked seaward erosional scarp is evident as dark areas in the bottom of the photograph. The drain-back direction can be identified (lower right of photograph) where plant debris has been carried on to the beach (marked by a broad arrow).

Fig. 12. In this oblique aerial photograph (Frame 72, seaward-looking direction) the flight control tower can be seen at the right and center edge of the photograph. Water is standing on the ground near the buildings. The NE-SW Runway lies buried seaward of the buildings (upper half of the image). On the left, bottom corner of the photograph, the foundations of three building can be seen (bright white areas). In the upper and center portion of the photograph, a gray area is visible. This shore appears to be the underlain by gray hard-ground (weathered coral and algal reef, and beach-rock) that was scoured and exposed during tsunami drain back. On either side of this area, the sand sheet remains intact.

Fig. 13. The aerial photograph (Frame 63, taken during a drain back phase) shows a barracks area (in the lower portion of the picture) and a tent barracks area (upper portion of the photograph). At the bottom center, several buildings have collided with one another. (The precise location of this photograph is unknown. The frame number is appropriate for the Kahuku area and the layout matches a fragment of a map (also from Kahuku). The arrow at the bottom of the image indicates the inundation direction.

Fig. 14. The oblique aerial photograph (Frame 78, was taken with a seaward-looking view, during the drain back phase) shows evidence of extensive flooding. The flooding and depositions of white sand extends beyond Brook Drive (lower left corner). The inundation measured from Brooks Drive to the coastline is 1.6 km (1.0 to 1.4 miles). Compare this image to Fig. 12 (Frame 72) to see the extent of the water standing on the ground.

Following the tsunami, the vegetation was killed or damaged by the physical effects of the flooding or by damage done to the flora by salt in the water. The seawaters penetrated the ground, making agriculture nonviable until the rains wash away the salt. Effects of the tsunami on vegetation can best be seen in a photograph published in the *Honolulu Advertiser* (04-03-1946, p. 11).

Since the tsunami took place 9 months after the end of WW II, the extensive damage to the base lead to the facility being abandoned, eventually the land reverted to the State of Hawaii and the Campbell Estate. Some of this land has since been converted to the James Campbell Wildlife Refuge, while the Campbell Estate continues to push for permits to build 4 or 5 Condominium buildings on the site most vulnerable to tsunami damage.

Based upon the study of the aerial photographs a model is suggested to describe the processes associated with tsunami activity. Five phases of activity are outlined in Fig. 15, however, since there are multiple waves in a tsunami wave train, thus the full sequence is repeated. During Phase 1, the leading wave is a negative, meaning the sea recedes (N-wave). The first wave, of course, can be either positive or negative (thus this phase may be absent). During Phase 2a the tsunami inundates the inter-tidal zone and beach face. During Phase 2b, the tsunami inundation reaches the back-beach. Phase 3, the stand still phase, is when the water movement reaches a near-zero velocity, and any sediment suspended in the water is deposited. After the stand still, the drain down phase, Phase 4, takes place. As the tsunami waters drain back to the sea, they channelize and erode the underlying deposits. Phase 5, represents a phase that may or may not be present, that is when an incoming wave overrides the outgoing wave. The profile below illustrates the additive nature of the tsunami wave height. This entire sequence of activities will be repeated multiple times.

4.4 Archaeological studies

The archaeological survey of the Kuilima Resort Project by Bath et al. (1984) is an extremely valuable source of information and provides insight into the processes involved in tsunami inundation. The report, written in 1984, predates modern geological studies of tsunami, thus the observations were strictly interpreted in archaeological terms, e.g., if a layer contained human bones or modern or ancient artifacts it was designated a cultural layer. Despite this, evidence of modification was obvious and restated repeatedly through the report (available in the Hawaiian Collection, University of Hawaii Hamilton Library). Bath et al. (1984) carried out a subsurface archaeological survey, for the Campbell Estate Resort Expansion Project at Turtle Bay (western portion of the airbase), which documents the nature and distribution of sediments in the study area. The field study involved subsurface sampling of several sites as well as a general subsurface reconnaissance. The majority of the subsurface sampling was done with augur sampling. In areas associated with surface drainages, the field crew cut the surrounding banks back to obtain an exposure of the underlying rock.

4.5 Observations

At Kewala Bay, modern aeolian sand or mixed deposits were found overlying the cultural horizon (layer containing artifacts of occupation). Furthermore, the thinness of deposits and their position at the surface indicated some truncation had occurred. A representative profile for this area consisted of (top to bottom): 30 cm of fine silty calcareous sand, with roots common and a smooth but abrupt boundary above 30-150 cm of calcareous sand with common coral fragments. A Soil Conservation Service stratigraphic profile lists three C-horizons. In some profiles, a layer 10 cm thick, contained coral fragments, a bone fishhook, several artifacts, a fragment of human bone, and charcoal.

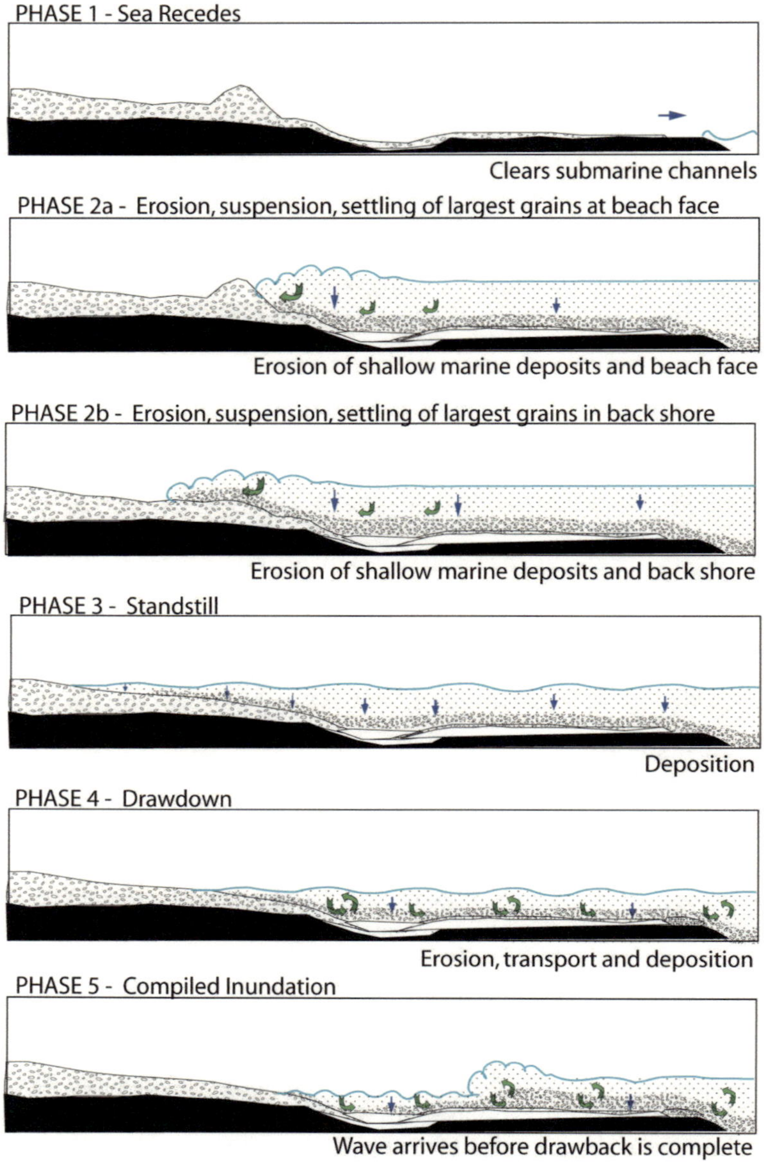

Fig. 15. Above: A series of five illustrations are used to describe the processes associated with tsunami activity. Phase 1, (top panel) the leading wave is a negative, meaning the sea recedes (N-wave). The first wave, of course, can be either positive or negative (thus this phase may be absent), Phase 2a (next panel; top to bottom) shows the inundation of the inter-tidal zone and beach face. Phase 2b, or the inundation in the back-beach. Phase 3, the stand still phase in which the water movement reaches a near-zero velocity. Phase 4, the drain down phase when the tsunami waters drain back to the sea locally in channels. Phase 5, an incoming wave overrides the outgoing wave. This profile illustrates the additive nature of the tsunami wave height.

Along the Kawela Stream Alignment, the B-horizon clay deposit contained charcoal and other flecks of organic material. Three distinctly different stratigraphic profiles were discovered. Inland from the beach, there is substantial alluvial sedimentation. Two units containing angular blocky structure were found from 1-80 and 80-140 cm. It was suggested that at some time in the past the drainage channel had been scoured. Flecks of organic material were thought to represent material from former agricultural activities or natural processes, rather than occupation.

At the Turtle Bay Beach situated behind the berm, the sampling found a remnant pocket of an older deposit. The archaeologists interpreted the stratigraphy as implying that the beach berm had been reworked by wave action in the possibly recent past.

On the east side of Kulima Point, sampling along the berm revealed, "lens of darker sand" at various depths. Between 50-265 cm, the sand layer contained shoe leather, modern glass, and cement fragments. Along the coastal stream alignment, inland of the berm, a buried horizon contained ironwood cones and needles (Ironwood trees were introduced to Hawaii during the 19th century). Bath et al. report that the beach berm itself does not appear to be a natural deposit (Kraft, pers. comm.). In other profiles, a typically fine-grained calcareous beach sand became progressively coarser with depth.

West of Kahuku Point, a minimum of 2 and maximum of 3 layers were found containing cultural material, buried beneath sand. Unit 3, a calcareous sand with dark charcoal, was found from 35-47 cm, and Unit 4, a fine to crumb sand with charcoal, at 47-60 cm, contained aluminum pop-top tabs. The archaeological report suggests these layers may well be

A: Unit: Historic event: 1946 Aleutian Earthquake and submarine Landslide? **B: Author:** Keating (2008) and this paper **C: Age:** 1946, **D: Place:** Kahuku, NE Oahu, HI, **E: Deposition:** sand sheet, **F: Sorting:** Normal, **G: Soil** A, B, C, and Layers, **K. Fine grain component:** beach and dune sand, **L. Coarse grain component:** "crumb" or pebble, and blocky structure, **M: Boulders:** limited, **N: Rip up clasts:** "some clay films on some ped faces" by Archaeologists Bath et al (1984), **O: Marine Conglomerate:** called midden remains by Archaeologists, Bath et al (1984); **P: Upper Boundary:** desultory, **Q: Couplet:** Sand and coral fragments on clay or peat, **R: Lower Boundary:** abrupt smooth boundary, **S: Nature of basin**: open coastal plain, **T: Environment:** beach and sand dunes, and back beach, **U: Matrix:** Sand, **V: Geomorphic Change:** offshore reef alternately covered with sand then scoured free of sand, **W: Layers:** 3-5, **X: Thickness:** roughly 20 to 200 cm, **Y: Run-up:** 6 m or 24 ft., **Z: Inundation:** 1.6 km or 1 mile, **AA: Publication:** Conference Paper and Field Trip Guide, **AB: Dating:** Observed 1946 Tsunami Event, and Carbon dated organic material 3,865 BC. to 1630 A.D. Bath et al (1984), **AC: Archaeological Site?** Yes, and human burials, **AD: Location:** between Kahuku town and Turtle Bay, **AE: Latitude** 21° 41′ N; **Longitude:** 157 ° 59′ W, **AF: Erosion Features:** Scoured ground surface, scoured river drainage, abrupt boundaries between units, **AG: Vegetation:** Damaged by salt water and flooding, **AH: Wave Height:** 7.5 m (Adams, 1977), **Number of waves:** 20 in Honolulu (tide gauge), **AI: Sampling:** Reconnaissance Archaeology by Bath et al (1984) over 100 auger holes and cut or trench sections, **AJ: TDDB #** 1001, **AK: Notes:** The study area (Kahuku Field) was a large Air Base during World War II.

Table 3. Characteristics of 1946 Tsunami Deposits on NE Oahu (Tsunami Deposit Data Base Format per NOAA Geophysical Tsunami Data Base)

intrusive. Charcoal and midden material was retrieved from deeper layers. The report states that the Turtle Bay area and Kulima Point area both provide evidence that these areas were recently modified by storm activity [here interpreted as disruptions that resulting from the 1946 tsunami].

At the west side of Kahuku Point, calcareous sands were found with sparse charcoal and human skeletal material (not in-situ) and abrupt but smooth boundaries, at 50-70 cm the remains of a fire pit (of blackened stones) were found. A single flake of volcanic glass and a rusty metal fragment were found at 0-12 cm, a shell casing and a carbon rod were collected at the surface. Other military artifacts were found on the beach. At a proposed water hazard on the current golf course, the surface layer contained broken glass, leather, a wooden post, glass bottles (3 ink bottles, 3 pickle/preserve bottles, 2 medicine bottles), a nail, a single piece of shell and a ceramic insulator. The surface layer consisted of humus with a crumb structure, a sandy loam with fine angular to sub-angular blocky structure, underlain by sandy clay, all with very abrupt but smooth boundaries. A stratigraphic section along the railroad right of way contained inverted stratigraphy, and assigned a construction related origin.

Four sites were described as coastal midden remains, and contained: gastropods, bivalves, echinoderms, crustaceans, bird fish and mammal bones and land snails, kukui nuts, and charcoal. Thus shallow water, reef, land animals, and vegetation were recovered. Dating studies suggest transport of alluvium from the uplands down to the coastal plain. Additional sites to the south of the runways are not described here for brevity sake. They describe peat and clay units, underlying sand, near the Punahoolapa Marsh.

4.6 Discussion

Modern studies of tsunami sediments (Keating et al., 2008) now allow us to characterize tsunami deposits. The deposits described by Bath et al. (1984) share characteristics with other known tsunami deposits. Bath et al. (1984) interpreted the thin layers (10-20 cm thick) at the surface as indication that some truncation had occurred. Similarly, roots were common in deposits as well as flecks of organic matter. The lower boundaries of the sedimentary units were abrupt but smooth. Sands contained a mix of common coral fragments; artifacts, human bones, shell or broken coral that were described as not being in-situ. Remnant pockets and lens of older (clay) sediments were found in stratigraphic profiles. The beach berm was reworked. Layers that contain cones, needles, seeds and tree trunk were found. Sand units were described with normal sorting (fining upward). Sedimentary layers contained crumb or irregular angular and blocky structures. Some sedimentary units contained mixed assemblages of organisms from different environments (shallow water, reef and land). The program also found a stratigraphy of sand overlying clay, and/or peat. All of these characteristics are consistent with reworking associated with tsunamis.

5. Boulders deposits

The North Shore of Oahu has routinely been exposed to tsunamis as well as large seasonal storms. Both events are capable of moving boulders. When compared to other published reports, these Hawaiian storms and tsunami have transport parameters in the lower range of published values. Noormets et al. (2002, 2004) published studies of large boulders at Shark's Cove on the North Shore of Oahu. The field studies were augmented by the laboratory study of a set of aerial photographs. The observations from Kahuku Point, thus allow us to extend the initial Shark's Cove studies by Noormets et al. and provide insight into the nature and carrying power of tsunami and storm waves. By comparing the geomorphology, rock units, and types, numbers, and distribution of boulders on the North Shore to deposits elsewhere, it was observed that both Hawaii and Rabat, Morocco, display similar geologic features, that reflect the sea level change along a coastline dominated by calcareous rocks.

The winter swell of 2006 provided an opportunity to test whether giant winter surf could move boulders and mega-boulders. The wave climate in Hawaii includes swell heights that can reach a maximum of ca 14 m or face heights of about 24 m (Noormets, et al. (2004). During the winter of 2006, the mega-boulders at Shark's Cove were photographed. In Dec 2006, the high surf associated with a winter swell produced waves that reached 14 m high at Shark's Cove. The same boulders were again photographed one month after the storm. During the storm, heavy surf broke at the outer margin of the reef platform, and flooded over the platform and around the mega-boulders. The flooding of the reef platform (between 1-2 m deep) rose to roughly half the vertical dimension of the largest mega-boulder at Shark's Cove (Fig 16).

Fig. 16. A photographic image of the surf breaking on the edge of the reef at Shark's Cove, during the 2006 giant swell.

The mega-boulders at Shark's Cove are porous, i.e. contain many cavities filled with air, and consist of low-density rocks. By comparing "before" and "after" photographs, no clear or significant transport of mega-boulders could be identified. However, the largest mega-boulder shows evidence of limited movement. Six large nails (ca 20 cm long) had been driven into the reef near the bottom of the largest mega-boulder. After the winter storm, five out of the six nails were gone. While the winter surf did not substantially transport the mega-boulder, it appears that the mega-boulder did jostle around on the reef surface, to a degree sufficient to remove the nails. The largest of the mega-boulders sits in a shallow depression, it appears to have remained roughly in the same position but ground down the shallow depression. The roughness of the platform (which is significant) obviously plays a part in boulder transport or absence thereof.

Fragments of rock were plucked from the leeward side of the mega-boulders, exposing a white colored interior. The surficial scars occurred on the lee side of megaclasts, at their base, midsection, and upper edge. In addition, many more (cobble size) rock fragments were present in the micro karst pits on the shoreward side of the platform after the storm. The extreme swell did erode the boulders and platform but only in a limited fashion.

At Kahuku a mega-boulder moved by the 1946 tsunami was seen in the aerials photographs, roughly the size of a bulldozer. Observations of the tsunami damage associated with the 2004 Indian Ocean Tsunami in the Maldive Atolls shows that tsunamis can deeply scour poorly consolidated sediments around dug pits such as toilet trenches, water collection trenches, etc. Thus, it is possible that the mega-boulder observed on the runway at Kahuku was plucked from the poorly consolidated sediments around the edges of the runway during runup and was not transported a great distance. Thus, the transport of this boulder may be a "special case" related to tsunami scour of a poorly compacted man-made feature.

Boulders on the order of 1m in diameter were also observed on the airfield runway. In addition, 0.5 m size boulders were observed in another photograph and another phase of the tsunami. The latter appears to be a relic deposit of isolated boulders where the sand size portion of the feature was winnowed away from the boulders. The deposit is widest inland and narrowest at the beach where the margins are marked by vegetation debris with drain back of a tsunami wave. The 1.5 m cores collected from the Kahuku-Turtle Bay area for archaeological evaluation (Bath et al., 1984) show desultory sand units bounded by erosional conformities and containing: metal roofing material, bone fragments (from ancient Hawaiian burials) and vegetation (including trunks still standing).

5.1 Comparison of geologic features

Mhammdi et al. (2008) describe mega-boulders situated along the Rabat Coast of Morocco and suggests that the emplacement of the boulders was associated with the Nov. 1st 1755 A.D. Lisbon Tsunami. The emergent deposits exposed along the Kahuku shore (Fig. 17) are very similar to the descriptions of the lithified dunes of Morocco, in terms of the geomorphology, rock units, and types and numbers of boulders. We compare the Rabat observations to that of the North Shore, where the coastline is exposed to tsunamis as well as large seasonal winter storms. The comparison shows that the sites have similar geologic settings, distribution of large boulders upon platforms, and similar types and groupings of large boulders.

When the schematic cross-sections are compared (Fig. 20) to that of the North Shore of Oahu, both coastlines are characterized (seaward to landward) by: 1) An offshore "Plateforme a vasques", 2) A notch at modern sea level, (3) An ancient rock unit with a karst surface, (end of the Ouljian in Morocco and Pleistocene Waimana high stand in Hawaii), 4.) Red soils containing land snails in Hawaii and Morocco (Soltanian red clays in Morocco), 5.) a thin lithified rock unit with karst surface, 6.) Boulders or slabs on top of the wave-cut platform, 7.) Modern unconsolidated sand beaches with poorly lithified beach rock in Hawaii, generally containing microfossils as old as 5 kya in Hawaii, 8.) The modern sand dune ridge, (Flandrian/Mellahian, 5,000-6,000 yrs in Morocco), and 9) A dune belt (dated at 114-122 kya at Kahuku, Oahu).

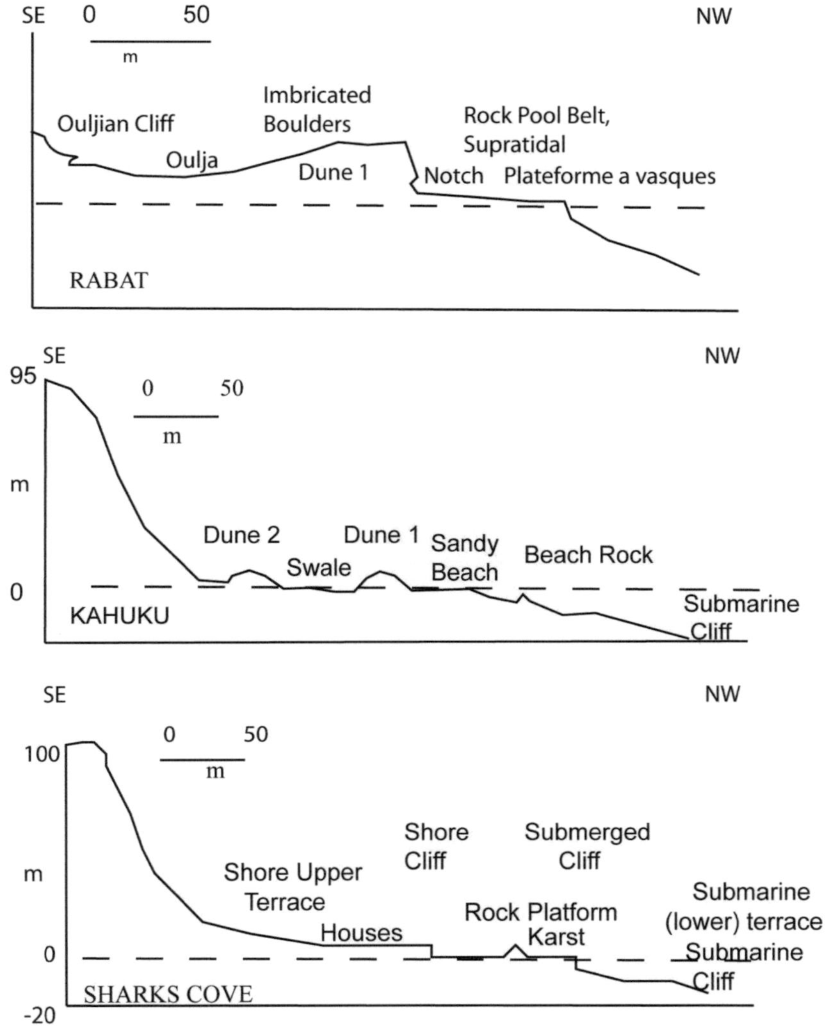

Fig. 17. Comparison of the schematic cross sections of Rabat Coast (Morocco, at top) with that of Kahuku Point of Oahu (middle) and Shark's Cove (bottom). The dashed line marks sealevel.

At Rabat, Morocco, the coast consists of relatively small sand beaches separated by rocky cliffs and lapiezed platforms. The area displays great biodiversity with species that modify the development of the landscape by their construction and bio-erosional activities. In Morocco, the morphology of the coastal dune system has been divided into two types: a generally flat "dissolution-driven type" and a mechanically driven type. Mhammdi et al. (2008) point out that the "dissolution type" can be broken down into several zones. They include: 1) lapiez and pools, 2. "Plateforme a vasques", and 3) the area of break-up of the Plateforme a vasques. The salient points of the Mhammdi et al. (2008) description are compared to similar morphological features observed on the north shore of Oahu in Table 4. Mhammdi et al. (2008) describe large platy boulders (slabs) that sit upon the coastal outcrops. The weight of the boulders ranges from a few tons to 100 tons. Several types of arrangements were observed: single boulders, imbricate boulder trains, large chaotic clusters, or more rarely trains or ridges, with the amount and arrangement of the boulders quite variable from one site to another. The dimensions range from large blocks 3 to 6 m in width and a maximum length of 8.4 m. Noormets et al. (2002) and Whelan and Keating (2004) observed: solitary mega-boulders (Fig. 18), as well as a large chaotic cluster, and imbricate clusters. At Kahuku Point, we see a chaotic cluster of slabs one to two meters long with largest dimension of one to two meters. In addition, the boulders in Morocco occur above the shore cliff. In Hawaii, the mega-boulders were described by Noormets et al. (2002, 2004) and lay on the rock platform adjacent to or within meters of the modern sea level rather than on cliffs.

Fig. 18. This photograph was taken after heavy winter surf (8 m wave face). The extreme waves broke at the edge of the carbonate platform (at left) and then flooded on the platform to roughly half the height of the mega-boulders without significant displacement of the boulders. A branch of a tree has been washed in and trapped at the base of the rock. A walking stick can be seen in the lower right. Six large nails (several cm long) had been driven into the reef along the bottom of the "mega-boulder" prior to the winter storm (source unknown). After the winter storm that took place less than one month later, five of the six nails were gone.

Like Rabat, the boulders at Kahuku Point are derived from the coastal rocks; their source is the fractured rock pool belt, the lithified sand dunes and fragments of platy beach rock. The boulders and outcrops show vermetids, carvings of sea urchins, seashells, encrusting corals and thick algal layers. The exposed fractures within the rocks at Kahuku Point are numerous and many incipient boulders appear perched along fracture surfaces, with the narrow pedestals of the rock between them sometimes intact.

The boulders at Kahuku Point are significantly smaller than those at Shark's Cove. The boulders at Shark's Cove are a mixture of mega-boulders 3-4 m in diameter and weighing 1.5 to 96 tons and boulders of lesser size. At Kahuku and Shark's Cove (15.3 km away) we find the amount, arrangement, and sizes are also quite variable. Mhammdi et al. (2008) suggested the greatest number of boulders occur where the platform is widest. A similar relationship is observed at Shark's Cove and Kahuku Point, Hawaii.

5.2 Quantitative assessment

Mhammdi et al. (2008) applied the general formula of Noormets et al. (2004) in order to assess the role of breaking waves in the displacement of boulders. In the Rabat case, the scenario used was that of a joint bounded block (as described by Noormets et al., 2004). Mhammdi et al. (2008) report that the lift required in order to overturn a ten-ton boulder would require a storm wave of at least 18.5 m. They concluded that because the weights of mega-boulders in South Rabat were in the 20-100 ton range, that storm waves could be excluded. These authors believe that only tsunami waves would be capable of overturning boulders due to their higher flow velocity.

Mhammdi et al. (2008) applied the formula to estimate the height of waves on a platform. They found the tsunami waves were capable of moving the blocks much further than storm waves, largely related to the longer period of the tsunami wave. They also applied the formula for quantifying the energy of displacement of a boulder. The resulting value is called the Transport figure (Tf). They reported a Tf of 3,000, for a twenty-ton block transported 150 m (taking into account the height reached). They also report a Tf value of 1,500-6,000 for a 30 to 60 ton block moved 10-20 m inland to an elevation of +5m.

Noormets et al. (2004) examined the forces necessary to dislodge and transport a mega-boulder at Sharks Cove, a distance of 20 m and concluded that "balancing the overturning and resisting moments, show that breaking swell waves as well as turbulent bores are capable of quarrying large clasts from the edge of the cliff given that sufficient fracturing is present." Noormets et al. (2004) suggests, "the large number of clasts on the sea floor suggests that waves of the breaking swell were capable of quarrying the megaclasts but are seldom capable of emplacing them onto the platform… However the historical records exclude a tsunami as a viable transporting event for the most recent 30 m movement of the megaclast to its current location."

Parameters affecting the transport of a boulder include: weight, shape, density, porosity, water velocity, water depth, duration, roughness of platform, steepness of beach, direction and rate of flow, etc. Scheffers and Kelletat (2005) studied the tsunami relics in the coastal landscape of Portugal. They found: single large isolated boulders, imbricate boulders, boulder ridges, pebbles and shells high above the modern storm level, etc. They conclude, "There is only one force which can move boulders of this size high above the surf line, namely tsunami waves" (reference is to boulders weighing 10-20 tons at ca 29 masl).

Studies suggest that the shape of the boulder is critical both to quarrying and to transport. In Hawaii, e.g., at Shark's Cove and Kahe Beach (SW Oahu), many boulders and mega-boulders break from pre-existing fractures and eventually fall from cliffs without being transported. Since many of the boulders near the shore have fallen from outcrops, rather than being actively pried off, or quarried away from the reef by wave energy, many of the assumptions made by Noormets et al. (2004) would not apply.
It is possible to examine additional information from Hawaii based upon several examples (see Table 4): (1) 0.45 m diameters round coral, entrained in the surf at Waimea Bay, Oahu. 2) A 1 m diameter round coral and conglomerate at Kahuku Point, 3) A 2 m slab of sandstone moved 3.5 m inland, and 2 m upward by storm waves, Kahuku Point. 4) 0.5m round boulders moved 15 m inland, and 4 m vertical, during a winter storm (the boulders were thrown up from the ocean and filled the lower story of a condominium).
The use of a "Transport figure" (Tf) is a method of quantifying the energy of displacement of a boulder as has been shown by the Scheffers and Kelletat (2005) equation:

$$Tf = W \times D \times V \quad (1)$$

where W is weight (tons); D is distance moved (m), and V is the vertical distance (m). The value of 2,000 is considered the upper limit of storm wave transport energy. To compare the observations in Hawaii with those reported elsewhere in the world, boulder volumes were estimated and appropriate densities between 2 gm/cc and 3 gm/cc were assigned to each boulder depending on rock type, lithification, and amount of porosity observed on the surface faces. Our results, along with two of Mhammdi's examples, are shown in Table 4.

Mhammdi 1	W= 20 tons	D=159	V=	Tf = 3000	Petit Val d'Or
Mhammdi 2	W= 30-60	D= 10-20	V 5m	Tf = 1,500-6,000	Average
Winter Storm	W= 0.3 tons	D= 15 m	V=2 m	Tf =5.0	Waimea Bay
Tsunami ?	W= 2.68 tons	D= 6m	V=2.1	Tf = 33.8	Kahuku Point
Winter Storm	W= 28.8 tons	D= 6m	V= 2m	Tf =28.8	Kahuku Point
Winter storm	W= 90 tons	D= 20m	V= 0.5	Tf = 900	Shark's Cove
Winter Storm	W= 0.3 tons	D=15 m	V=4m	Tf =39.4	Kona, HI
Winter Storm	W= 1.6 Tons	D=13	V=4m	Tf =83	Kona, HI

Table 4. Comparison of mega-boulder Transport Figures from Morocco and Hawaii

With the exception of one large boulder, the sizes moved during storms were restricted, ca. 1 m or less, and the transport figures were small varying from 5.0 to 83. The Transport figure, Tf, for the Hawaii data set (Table 4) was combined with those of Scheffers and Kelletat (2005). Table 4 shows that the Hawaii storm data clustered at low values; while the data associated with tsunami have higher Tf values and larger block sizes. While, the Tf numbers are small and seemingly insignificant the observations are important since they are associated with a known (rather than inferred) storm origin.

6. Discussion

A considerable amount of discussion can be found in the literature regarding the emplacement of boulders by tsunami vs. storm. Among those examining boulders and comparing tsunami vs. storm origins are: Nott (2004) from Western Australia; Scicchitano et al. (2007) from SE Sicily (Italy); Susmilch (1912) from E. Australia; Goff et al. (2004) from the North Island, New Zealand; Hindson et al. (1996) from the Algarve Coast, Portugal; Whelan and Kelletat (2005) from Cabo de Trafalgar, Portugal, and others.

Were the boulders at Kahuku Point transported by tsunami or storm waves? Clearly, the observation of movement during storm events documents that storms can move boulders up to 100 tons in size across the relatively flat surface of a platform. But they probably could not have been responsible for their initial placement on the platform. The clastic slabs (beach rock) are less resistant to mechanical abrasion than the limestone. The more-resistant reef blocks are clearly reduced in size by bio-erosion, mechanical erosion and dissolution and probably have a longer shelf life. The smaller boulders and tabular slabs of beach rock can easily be moved by storm waves. At Kahuku, boulders 1-3 m (a-axis) were transported onto the runways, based upon the study of the 1946 aerial photographs (Keating, 2008). Intense storm activity at Sharks Cove shows that erosion of boulders can take place without significant transport. The boulders at Shark's Cove appear to be sufficiently far from the reef edge that they are not necessarily moved by waves that break at the outer reef edge and then flood the platform. These observations provide insight into the nature and carrying power of tsunami and storm waves.

These observations are not unique to Hawaii; clearly similar observations are likely to vary at different sites around the world that have similar settings. Likewise, the observations described here are not unique in terms of tsunami and storm activity within Hawaii. Both storms and tsunami are variable in nature. If there is no source material of mega-boulder size to be transported by storm or tsunami, the carrying power of the waves will not be reflected by the deposits. Likewise, the source, intensity, and direction of movement of storms and tsunami are variable. The use of the Tf parameter may be misleading for materials having undergone large displacement for it may be that the large displacements are the result of an accumulation of many small displacements, such as has been observed for the largest megaclast on the platform at Sharks Cove. Each displacement episode for the largest block seems to be in the 10 to 30 m range (Noormets, et al., 2002). The cumulative displacement is substantially larger, thus it's Tf value is smaller than would be calculated if the total displacement from a presumed source location were used.

The Hawaii and Rabat examples discussed here represent a geologic environment in which reef and dune sediments are preserved. The stratigraphy at both sites reflects the oscillation of sea level during the Pleistocene. In both places, the mega-boulders/boulders are most abundant in a setting with a broad shallow submarine platform. During times of glaciations, when waters are locked up in glaciers, the sea level drops and the beaches shift seaward; a regressive sequence is formed capping the former reef platform (Stearns, 1978). Later in time, but still during the low stand, aeolian sands cap the platform and karst topography is locally developed. We now see sand dunes exposed at Kahuku Point that were formed when the sea was lower than the modern sea level. The base of the deposits occurs offshore, and below modern-day sea level. Upon rise of sea level, these lithified dunes and the margins of the platform become sources for blocks of material that can be transported by both storms and tsunami. Since the sea level changes take place globally, it is reasonable that coastal carbonate sequences around the world would display a similar patterns of mega-boulder/boulder distribution (Table 5) e.g., Western Australia (Nott, 2004); Caribbean (Scheffers and Kelletat, 2004); Cyprus (Kelletat and Schellmann, 2002); Curacao, Netherlands Antilles (Schnellmann et al., 2004). The positive correlation appears to be in large part controlled by a similar geologic setting of the sites and reflect the impact of changes in climate and sea level.

Based upon studies in Cascadia, Nelson et al. (2006) suggested that there are thresholds for creating and preserving evidence of earthquakes and tsunami. Clearly, the Kuril Islands Tsunami with a 1 m tsunami runup and a limited inundation zone did not pass the preservation threshold in the Hawaiian high-energy beach sites. Nelson et al. (2006) further suggest that depositional sites must favor preservation for a tsunami event to leave a geologically identifiable trace. Clearly, the Hawaiian beach facies studied does not constitute a positive preservation regime. Sediment availability, erosion, deposition, steep seasonal beaches, and meteorological conditions that include high winds, large swells, heavy rain, etc. combined together to quickly remove the visible traces of the November 2006 Tsunami. Our observations suggest that within the seasonally high-energy coastal environment such as Hawaii, tsunami traces can be easily and quickly erased.

The modern run-up data shows that the greatest number of occurrences of run-up values occur within the 0-1 m range (Keating et al., 2008). This study shows that tsunami recurrence rates, based on the absence of sedimentary data, are likely to be improperly estimated even if only M8+ events are used, e.g. the 2006 Kuril Tsunami. Furthermore, the inability to differentiate storm from tsunami deposits suggests that the paleo-tsunami record at least in Hawaii has low fidelity.

Ref	Place	Rock Type	Karst	Max. Size (m)	Wt. tons	Im-bricate	Storm or Tsunami	Source
1.	Sicilly	Calcareous	Yes Yes	8,8,0.7	182	Yes	Tsunami	Edge of platform
2.	W. Aus-tralia	Eolianite Sandstone	No Yes	Several m.	100	Yes	Tsunami	Edge of platform
3.	S. Spain	Eolianite Sandstone	No Yes	8,5,1	75	Yes	Tsunami	Intertidal platform
4	Bahamas	Eolianite Limestone	Yes Yes	13,11.5, 6.5	2300	No	Tsunami	Adjacent cliffs
5	Antilles	Limestone Beachrock	Yes	0.2-2	195	Yes	Tsunami	Adjacent cliff or subtital

Notes- References: 1. Scicchitano et al. (2007), 2. Nott (2004), 3. Whelan & Kelletat (2005), 4. Hearty (1997), 5. Scheffers and Kelletat (2003)

Table 5. Summary of mega-boulder deposits and their geologic characteristics

7. Conclusions

The major findings of this study regarding the 1946 Tsunami includes: widespread inundation (flooding from the sea) was sufficient to destroy the wooden barracks buildings, even if they were well constructed. Wooden buildings literally became sandwiched together by collision (which would have resulted in disastrous consequences to occupants, had the Kahuku barracks been occupied); construction glass and other materials become incorporated into the tsunami deposits during the turbulent runup of the tsunami and added to the hazards. The sand that was transported in the tsunami flood deprived the beach of protective cover. The sand moved by the tsunami covered everything inland as a sand sheet. At Kahuku the inundation zone was 1.6 km wide (a broad inundation zone) and the flood of salt water mechanically and chemically damaged the vegetation.

Eyewitness observations at the only non-beach site (Paukauila Stream) revealed that the Kuril tsunami did inundate the bay and the stream that drains the coastal wetlands. Although floating debris, e.g. plastic items, especially food wrappers, that were deposited to heights of 1 m along the stream provided evidence of an approximately 1 m run-up, the only sediments observed in transport in those backwaters was mud in water moving seaward. Conceivably, the long periods of a larger tsunami could have carried sands derived from beach into the quiet waters of the wetland. However, the 2006 Kuril Tsunami did not overflow stream banks where preservation may be possible, and any sand in the channels was likely to have been removed by the receding part of the tsunami wave, by tidal action, or the next rainstorm.

The wave action associated with the 1946 Aleutian tsunami repeatedly eroded sand during inundation, deposited the sand during the standstill phase and then selectively scoured the deposits (from the inundation and stand-still phase) during the drain back. The tsunami also eroded and widened drainages, leaving the ground surface covered by a giant sand sheet. During the course of the tsunami activity, large pools of water ponded in low areas after individual tsunami waves drained away.

The archaeological studies (by Bath et al., 1984) revealed desultory distribution of deposits, changing stratigraphy over a small area in discontinuous layers or lens, coral-rich units containing marine organisms from the shallow offshore, charcoal-rich units showing organic matter including trees, needles, seeds, etc derived from land and mixed with marine organisms as the tsunami drain back remobilizes sediment. These layers are generally thin (10-20 cm), have abrupt lower boundaries associated with tsunami waves scouring the back beach sediments and drainages, evidence of beach berm or dune being modified and showing "crumb" or blocky internal structures, and "midden" material consisting of shells, fish bones, animal bones, etc., often mixed with sand, coral and cobbles that may also be called a marine conglomerate. These characteristics of the 1946 tsunami are shared with other tsunami deposits worldwide. [The end]

8. References

Adams, W. M., 1977, Tsunami Effects and Risk of Kahuku Point, Oahu, Hawaii, *Geology Case Histories J.*, 8, 63-70.

Allen, G., *Hawaii War Years, 1941-1945*, University of Hawaii Press, Honolulu, 226-227.

Allen, G. (1950). *Hawaii's War Years, 1941-1945,* University of Hawaii Press, Honolulu.

Bath, J. E.; Rosendahl, M. L. K. & Rosendahl, P. H. (1984). Subsurface archaeological reconnaissance survey, Kulima Resort Expansion Project…island of Oahu, *Report 137-100784* for Kuilima Development Corporation, 62 pp.

Goff, J.; MaFadgen, B. G., & Chague-Goff, C. (2004). Sedimentary differences between the 2002 Easter Storm and the 15th century Okoropunga tsunami, southeastern North Island, New Zealand, *Mar. Geology,* 204, 235-250.

Gusiakov, V. K. (ITIC, 2006) *The Tsunami Bulletin Board* is a web subscriber-based notice, for more information contact itic.tsunami@noaa.gov

Hawaii Sea Level Data Center and Pacific Warning Center, Tide Gauge Records, University of Hawaii

Hearty, P. J., 1997, Boulder deposits from Large Waves during the Last Interglaciation on North Elethera Island, Bahamas, *Quat. Res.,* 48, 326-338.

Hindson, R. A.; Andrade, C., & Dawson, A. G. (1996). Sedimentary Processes Associated with the Tsunami Generated by the 1755 Lisbon Earthquake on the Algarve Coast, Portugal. *Phys. Chem. Earth,* 2, 12, 57-63.

Honolulu Advertiser (Issue 04-03-1946, p. 11, Issue 04-04-1946, p. 2; Issue 04-05-1946, p. 5; Issue 04-06-1946, p. 6; Issue 04-08-1946, p. 1) Ceased publication in 2011.

KGMB News (2006) Channel 9 television. KGMB News (2006) website: (Haleiwa, Oahu) http://kgmb.com/kgmb/print.cfm?sid=1183&storyID=9683 (Hanauma Bay, Oahu).

Keating, B. H.; Wanink, M. & Helsley, C. E., (2008). Introduction To A Tsunami Deposits Data Base, In *Tsunamiites,* (T. Shiki, Y. Tsuji, T. Yamazaki and K. Minoua, Eds.), Elseveir Publ., New York, p. 359-361.

Keating, B. H.; Helsley, C. E., Waheed, Z. & Dominey-Howes, D., (2005). 2004 Indian Ocean Tsunami on the Maldive Islands: Initial Observations. *Sci. of Tsunami Haz.,* 23, 2, 19-70.

Keating, B.H. (2008). The 1946 Tsunami at Kahuku, NE Oahu, Hawaii. In: Solutions to Coastal Disasters 2008. (L. Wallendorf, C. Jones, L. Ewing and B. Jaffe, Eds) *Am. Society of Civil Engineers,* Reston, Va., p. 157-168.

Kelletat, D., & Schnellman, G. (2002). Tsunamis on Cyprus: Field evidences 14C dating results, *Zeitschrift fur Geomorphologie,* 46, 1, 19-34.

Ku, T.-L., Kimmel, M.A., Easton, W.H. and O'Neil, T.J., 1994. Eustatic Sea-Level 120,000 Years Ago on Oahu, Hawaii. *Science,* 183: 959-962.

Macdonald, G.A., F.P. Shepard, and D.C. Cox. 1947. The Tsunami of April 1, 1946 in the Hawaiian Islands, *Pacific Science,* 1: 21-37.

Mhammdi, F.M.; Kelletat, D. & M'Fedal Ahmamou, L.A. (2008). Large Boulder along the Rabat Coast (Morocco): Possible Emplacement by the November 1st, 1755 A. D. Tsunami. *Sci. of Tsunami Haz.,* 27, 1, 17-30.

Nelson, A. N.; Kelsey, H. M., & Witter, R. C. (2006). Great earthquakes of variable magnitude at the Cascadia subduction zone, *Quaternary Research,* 65, 354-365.

Noormets, R.; Felton, E.A. & Crook, K.A.W. (2002). Sedimentology of rocky shorelines: 2; Shoreline megaclasts on the north shore of Oahu, Hawaii--origins and history. *Sedimentary Geology*, 150, 1-2, 31-45.

Noormets, R.; Felton, E.A. & Crook, K.A.W. (2004). Sedimentology of rocky shorelines: 3. Hydrodynamics of megaclast emplacement and transport on a shore platform, Oahu, Hawaii. *Sedimentary Geology*, 172, 41-65.

Nott , J. (2004). The tsunami hypothesis- comparisons of the field evidence against the effects, on Western Australian coast, of some of the most powerful storms on Earth, *Mar. Geology*, 208, 1-12.

Pacific Tsunami Warning Center (2006*). Tsunami Observations Kuril Islands, Magnitude 8.3,* November 15, 2006. Website: www.prh.noaa.gov/ptwc/

Scheffers, A. & Kelletat, D. (2003). Sedimentologic and geomorphic tsunami imprints worldwide - a review, *Earth-Science Rev.*, 63, 83-92.

Scheffers, A. & Kelletat, D., (2004). Bimodal tsunami deposits - a neglected feature in paleo-tsunami research, *Coastline Reports*, 1, 67-75. Geographie der Meere & Kusten.

Schellmann, G.; Radtke, U., Scheffers, A., Whelan, F., & Kelletat, D. (2004). ESR Dating of Coral Reef Terraces on Curacau (Netherlands Antilles) with Estimates of Younger Pleistocene Sea Level Elevations, *J. Coastal Research*, 20, 4, 947-957.

Scicchitano, G., Monaco, C., and Tortorui, L. (2007). Large Boulder Deposits by Tsunami waves along the Ionian coast of south-eastern Sicily (Italy). v. 208, 1-12.

Shepard, F.P.; MacDonald, G.A. & Cox, D.C. (1950). The tsunami of April 1, 1946. *Bull. Scripps Inst. Oceanography*, 5, 6, 391-528.

Stearns, H.T. (1935). Shore benches on the island of Oahu, Hawaii Islands. *Geol. Soc. Am. Bull.*, 46: 1467-1482.

Stearns, H.T. and Vaksvik, K.N. (1935). Geology and Ground-water Resources of the island of Oahu, Hawaii. *Ground Water Bulletin 1*, Territory of Hawaii, Honolulu, 168 pp.

Stearns, H. T. (1935). Shore benches on the island of Oahu, Hawaii Islands. *Geol. Soc. Am. Bull.*, 46: 1467-1482.

Stearns, H.T. (1970). Ages of Dunes on Oahu, Hawaiian Islands. *Bernice P. Bishop Museum Bulletin*, XXIV (May 8, 1950, #4): 50-71.

Stearns, H.T. (1974). Submerged Shorelines and Shelves in the Hawaiian Islands and a Revision of some of the Eustatic Emerged Shorelines. *Geol. Soc. Am. Bull.*, 85: 795-804.

Stearns, H.T. (1978). Quaternary Shorelines in the Hawaiian Islands. *Bernice P. Bishop Museum Bulletin*, 237, pp 1-57.

Sussmilch, F. G. S. (1912). Note on some Recent Marine Erosion at Bondi, *Roy. Soc. Of New S. Wales, Journal And Proc.*, v. 46.

Szabo, B.J. and Ludwig, K.R., Muhs, D. R., Simmons, R., 1994. Thorium-230 Ages of Corals and duration of the Last Interglacial Sea-Level Stand on Oahu, Hawaii. *Science*, 266(5182): 93-94.

U. S. Coast Guard (2206) United States Coast Guard, *2006 Website:* http:/www.uscghawaii.com/go/doctypearchive/800/73848 and select video: Coast Guard Storm Surge Station Kauai, November 15, 2006

Walker, D. (1994). *Tsunami Facts,* University of Hawaii, Honolulu, pp 1-93.

Walker, D. (2004). Map of Run ups of Large Pacific-wide 20th century Tsunamis, Oahu Map. Laminated Map of Oahu Runup Values, Tsunami Memorial Foundation, Honolulu, HI.

Whelan, F. & Keating, B.H. (2004). Tsunami Deposits on the island of Oahu. *Geographie der Meere und Kushen, Coastline Report,* 1, 77-82.

Whelan, F., & Kelletat, D. (2005). Boulder Deposits on the Southern Spanish Atlantic Coast: Possible Evidence for the 1755 AD Lisbon Tsunami? *Sci. Tsunami Haz.,* 23, 3, 25-38.

Wroe and Caldwell P., 2006, *Preliminary Local Storm Report,* National Weather Service, Honolulu, 12/8/2006. Website: http://www.prh.noaa.gov/data/HFO

20

The Atlantic Tsunami on November 1st, 1755: World Range and Amplitude According to Primary Documentary Sources

Paul-Louis Blanc
Institut de Radioprotection et de Sûreté Nucléaire,
Fontenay aux roses,
France

1. Introduction

On Saturday, November 1st, 1755, at about 9:40, Lisbon underwent the strongest tsunamigenic earthquake ever reported by witnesses in Western Europe. Many people, attending the religious offices of All Hallows, were killed by the collapse of churches. Some took refuge on the quays of the Tagus, attempting to escape the fall of debris from collapsing buildings and to flee on boats: but a tsunami, 5 to 10 m high according to various accounts, struck downtown Lisbon. Fire outbreaks, lit by houses collapsing on their kitchen fire, set the town to flames, which raged for almost one week.

The event triggered a competition between Scientific Institutions and between Gazettes to provide their members or readers with original accounts: a reaction similar to what we witnessed on the WEB after the December 26th, 2004 earthquake in Sumatra. In that case, thanks to the speed at which the information now travels, and constraint provided by photography and video-movies, the result has been the gathering of a rich array of accurate observations and data. At variance, in 1755-56, under the escalation of the competition between institutions or journals, with a small number of qualified observers and slow travel of letters conveying the accounts, the exaggerations of compilers were let loose. The reliability of the testimonies suffered in the process: many are contradictory, both in terms of lack of internal consistency and of contradiction between reports.

The Andaman-Sumatra earthquake and tsunami have awoken a public awareness of the risk of tsunami. The November 1st, 1755 tsunami caused by the Great Lisbon Earthquake, remains the only destructive event of the kind on the European Atlantic shores to have been described in some detail in the past, partly supplemented by the tsunami of March 31st, 1761. The primary accounts of the 1755 event are in no danger to loose their documentary importance : the simulations now attempted to evaluate the tsunami risk to human lives and coastal settlements must be constrained by "real data, deducted from historical reports" (Mendes-Victor *et al.*, 2005). But the data gathered from such reports are so dispersed that they cannot be applied without a critical analysis.

In the Ibero-Moroccan gulf, the recent study of a restricted selection of primary documents made it possible to unravel some relationships between classical reports on the tsunami in Cadiz, and to show that it was not as high as the classical estimates had made it : the conclusion was to accept a tsunami wave at shoreline 2.5 m high above the expected

astronomical tide and to disregard a previously accepted height of 19.5 m, already downgraded to 18.3 m with the abandonment of the old French "Pied de Roy" units.
A review of early Europeans documents reporting on the same event in Morocco led to similar conclusions, but the analysis was made more difficult : [1] by the confusion between the effects of the November 1st earthquake and tsunami and those of the November 18th-19th Rifian earthquake (purportedly also dated November 29th by Arabic sources) ; [2] by the general loose knowledge, in Europe, at the time, of Moroccan geography and toponymy ; [3] by the scarcity of European observers and the ill-qualification of some of them for the task.
The idea that in Tangier the tsunami wave reached 14 to 16 m at shoreline must be abandoned, as well as the notion of a 24.5 m wave in the Portuguese stronghold of Mazagão in El Jadida.
This article extends these results to the whole Atlantic basin, by reviewing the extent, amplitude or other available parameters of the November 1st, 1755, tsunami, considering the earliest accounts (Primary Documentary Sources) recorded in European archives and journals. This implies that later documents (1757 to present) can only be of use if we trust that the actual writing was contemporaneous, or as far as they quote older documents, if the original character of the cited document resists critical analysis.

2. British Isles

2.1 Ireland

2.1.1 Cork

Two letters to the Royal Society of London (Nicola, 1756 ; Hen. Tom. 1756) and a dispatch in the Gentleman's Magazine (Gent's Mag., 1756b) mention that the shock of earthquake was felt in Cork. The Royal Society letters then undermine the reliability of the information: Hen. Tom. allocates it the same time as the tsunami in Kinsale: but seismic waves travel much faster in the mantle and crust than tsunami waves on the ocean, they should only be simultaneous at the epicentre, which was definitely not South-East Ireland; Nicola declares: "I have not met any persons of credit who pretend to say they felt it" ; but a lack of so-called creditable witness is no proof that it did not happen.
The second part of the dispatch in the Gentleman's Magazine carries valuable details. The time of the event, 9:30 AM, may be correct within an error margin of some minutes, as Cork is under a more easterly longitude than Lisbon. Limits are assigned to the places affected in town and the surroundings:
"the limits of the places affected were, southward, *Watergate-lane, Christ-church-lane* and *Playhouse-street*; Northward, *Broad-lane, Coal-kay* and *Draw-bridge*. I do not find that it was felt more easterly than alderman *Bradshaw's*, not more westerly than *Hammond's Marsh*. The limits were so narrow, I // think that the great torrent of waters that ran down the river Lee has found a passage through the middle part of our city."
This cannot be a description of an earthquake in the common acceptance of the word, they do not show sharp limits to the zones affected. The place-name mentioned are related to low areas and water : *Water-gate lane, Coal-kay, Draw-bridge* and *Hammond's Marsh* speak by themselves. *Christ-church* was not far from the ancient Water-gate. These limits are described as narrow : they follow the banks of the river Lee. The account ends up with a torrent of water running down the river Lee, finding a passage through the middle part of the town.
The explanation reconciling these elements, plus the doubts of Nicola, is that the earthquake itself was not felt, but the bay of Cork, a wide ria reinvaded by marine waters in post-glacial times, was travelled up by a bore. Cork being at the narrow end of the ria, the bore invaded

part of the town, flooding the river quays and a few streets, and a swampy area beyond town, between the main arms of the river. This bore, almost simultaneous to the quake itself, was caused by the seismic low-frequency surface ground-waves. If a torrent of water ran down the river Lee, it means that a bore had first to run it up.

No ground movement was felt in Cork, but the almost simultaneous consequence was: the water effects of the event were reported all over northern Europe at the time, as far as Norway and Finland. It is just a matter of definition: the Hydrosphere is an integral part to the Earth, but it is unusual to consider water movements as earth movements, let alone quakes.

This interpretation is reinforced by the description given six years later by the London Chronicle, and re-circulated to the Royal Society by Borlase, of the earthquake of March 31st, 1761, again in Cork:

"*Extract of a letter from Cork, March 31.* A quarter after twelve this day a shock of an earthquake was felt here in the Exchange, in the Merchants Coffee-house, and from the beginning of the Red-House Walk to the East Marsh, between the gates only, just as it was in November 1755 ; but allowed to have been more violent by all that felt it. It did not continue above one minute, undulating from East to West, and *vice versa*." (Lond. Chron., 1761 A).

An East to West undulation, reflected at the end of its course, can only describe the propagation of a wave at the surface of a liquid medium. Again, the time difference with Kinsale, only reached by the tsunami six hours later on the same day, is decisive (Lond. Chron., 1761 B).

2.1.2 Kinsale

Several papers and useful comments report the tsunami in Kinsale, in the Gentleman's Magazine (Gent's Mag., 1756b), and in the Philosophical Transactions of the Royal Society (Hen. Tom., 1756; Nicola, 1756; Huxham, 1756; Borlase, 1756). These are indeed observations of the tsunami: the time is tolerably acknowledged: between 2 and 3 PM for Nicola; "not till 3 or 4 afternoon" for Huxham; Borlase adds "not a full hour after us, but above two hours sooner than at Swansea", making sense because his observations in Cornwall were made little after 2 PM. We can accept 3 PM as the arrival time in south-east Ireland, 5 ½ hours later than the surge in Cork.

Kinsale is also an estuary harbour, re-invaded by marine waters during the last glacial retreat, and as such penetrating far inland. The progressively diminishing depth and decreasing width explain that the run-up remained high. Nicola attributes to one Rev^{d} Mr Keef, a 1.68 m measurement for the final run up at the Market quay, against an estimate of 1.83 to 2.13 m by the Gentleman's Magazine. Nicola states that the Market-quay was overflowed, and workers "ran off, but could not do it with expedition enough to prevent their being overtaken, and up to the knees." This is contradictory with the purported height of 1.7 to 2.1 m, which has to be discarded as exaggeration: only the Gentleman's Magazine reports that some ships were not only tossed to and fro, but also lifted and left aground: According to Hen. Tom, they were just left on shore, not quite grounded, and the Market-quay, the dead end of the cul-de-sac of the ria, was inundated: but only to a level of 0.5 m.

Despite havoc among the ships, the event remained rather mild for a tsunami, but its nature is also demonstrated by its periodic character: The period is given as "seldom more than a quarter of an hour between each return" by the Gentleman's Magazine. and the oscillations were perceived from 3 PM till 10 PM or, for Nicola, till 3 AM on November 2nd. This is the northernmost record which we know of: if it reached further than Kinsale, there is no record.

2.2 Wales: Swansea

The northernmost indication of this tsunami on the main island is in Swansea (S-W coast of Wales). There are two references, in the Philosophical Transactions (Philos. Trans., 1756a), and in the Gentleman's Magazine (Gent's Mag., 1756b) also published early in 1756. A third one, the earliest to mention the tsunami there, cannot be considered a true primary document: it was written in Cornwall and only discusses the time of occurrence (Borlase, 1756). All three concur on the time of the event, 6:45 PM. A single bore "rushed up with a great noise, floated two large vessels, the least above two hundred tons (one whereof was almost dry before), broke their moorings, and hove them, across the river, and had like to overset them, by throwing them on the banks. The whole did not last ten minutes; the rise and fall; and // it was not felt in any other part of the river". A ferry on the same estuary was unaffected. The location where this was observed is clear: the ruins of the White Rock Copper Works, first established 1736, were designated an industrial archaeology park during the 1980's. The river there is subject to tides. Neither amplitude estimate, nor indication of multiple waves, are given : but the late time of the observation, nine hours later than the earthquake, shows that it was the true tsunami, not a seiche caused by the transmission of low frequency surface waves in the solid crust.

2.3 Cornwall

2.3.1 St-Ives & Hayle

Borlase (1756) engaged in investigations extending not only to the St-Michael' Mount's Bay, but also to the harbours of St-Ives and Hayle, facing St George's Channel and the Irish Sea, 5 km to the North. He did not witness the facts which he records, but his position as the Church of England Pastor for Ludgwan and St Just allowed him to meet people of various condition and gather a wide array of data. The maximum crest to trough amplitude is surprisingly high, from 2.44 to 2.74 m at St-Ives, but it is also observed that the wave did not break: it had lost impetus. The reason given by Borlase, that the swell had to skirt a circuit of land around Land's End, from Mount's Bay to St-Ives and Hayle, is questionable: the Irish sea is a direction quite as open to a wave coming from the south as the entrance of the Channel. The wave was even weaker in Hayle, secluded from the open sea by the Porth Kidney sands, where Borlase ascribes a 2.13 m crest to trough amplitude to the highest oscillation, the others being given for only 0.61 m: but the wording "rose and fell but two feet only" might well mean the simple, positive amplitude: rising two feet plus felling two feet would do for a 1.22 m crest to trough wave train. The waves only sent vagrant a few ships or boats: "…at St *Ive*'s, a few miles distant, a vessel that was near dry, floated off some distance from the place." (Gent's Mag., 1756b).

2.3.2 Mount's-Bay: St-Michael's Mount, Penzance, Newlyn, Mousehole

The observations of the Rev[d] William Borlase (Borlase, 1756) are among the most detailed and the less anecdotal, as he recorded the time, the amplitude and the direction of the water movements, rather than the feelings, impressions and frights of the witnesses.

The phenomenon occurred "a little after two o'clock in the afternoon, // about half an hour after ebb" in the sense of high tide. At the Mount, the apparent flow was "from the eastward // running to the west, and south-west". At the Mount's pier, the measurement of the crest - to-trough amplitude was 1.83 m :

"It continued to swell and rise for the space of ten minutes; it then began to retire for about ten minutes, till the water was six feet lower than when it began to retire. The sea then began to return, and in ten minutes it was at the before-mentioned extraordinary height; in ten minutes more it was sunk as before, and so it continued alternately to rise and fall between five and six feet, in the same space of time."

Borlase thus also indicates that the period was 20 min. The maximum amplitude of the wave was reached during the third and fourth periods, and the tsunami was perceived for almost six hours.

In Penzance, the amplitude was stronger: 2.44 m and a direction seeming from South-East, stronger again in Newlyn, where it broke into a bore more than 3.05 m high, travelling in a northerly direction. In Mousehole the wave is only mentioned. Borlase suggests an explanation of these differences, and demonstrates a spontaneous perception of the influence of coastal physiography on variability of Run-up.

Borlase explains the wave direction by a reflection on Cape Lizard: his analysis may not be fully reliable, as at the Mount, the eastern direction reaches the coast after about only one kilometre, and at Newlyn the true south direction should cut across the Cornish coast. In fact it must have come from the south-east and south-south-east direction. Borlase also mentions an attempt to measure the speed of the flow in Gwavas lake (the name applies to the roadstead of Newlyn) as a Ketch of War threw its log and measured a speed of seven knots: 12.96 kph.

No damages were done to the fishers' boats, neither does Borlase mention any inundation of the shores: it appears that the tide was low enough to prevent the run-up to reach higher than the high tide mark. or to overflow the embankments : this is reconfirmed by the description of the boats being "hurried back again through the gap or mouth of the pier with incredible velocity: when they had gone off as far as the reflux determined, they were carried in, and out again". The true safeguard of the Cornish harbours may well have been that the tide was already low enough after two hour of ebbing.

2.4 Devonshire

2.4.1 Plymouth

John Huxham (Huxham, 1756) reports the event at the limit between Cornwall and Devon, on the sites of two ferries crossing the rias on which the town and harbour are established :

At the Oreston Ferry, located east of Plymouth, upstream from the harbour of the time, an initial drawdown of 1.22 to 1.52 m occurred, leaving the ferry-boat lying on the mud of the bottom. It was followed by a swell of 1.83 m, giving the wave a run-down to run-up amplitude of 3 m : The wave did not break, but from the given time, about 4 PM, it was indeed the tsunami. And the phenomenon was indeed repeated a few times, the period being maybe a quarter of an hour.

The second ferry was at Cremyl passage, on a bottleneck of the river Tamar, west of Plymouth, and the border between Devonshire and the Duchy of Cornwall. The wave did not break there either. In Stonehouse lake, one of the branches of the ria, close to Cremyl, the wave developed into a true breaking bore. The tidal currents in Cremyl were deemed dangerous, and Huxham's words "tearing up of mud and sand bank in a very alarming manner" were cited by Dawson *et al.* (Dawson *et al.*, 2000), accounting for the deposit of sedimentary bodies by such tsunami.

Huxham wrote this letter before he knew of the earthquake in Lisbon, and owns up to his ignorance of the cause. He was aware of observations done elsewhere by some correspondents or published in journals, and emphasizes the problem of the time of the events reported:

"at Portsmouth, Holland, in Ireland, Germany, &c. the waters were so strongly agitated on the very same day, it roused my attention greatly, though I am quite lost in conjecture of the cause. // it happened not here till about 4 p. m., at Portsmouth about 11 a.m., in Holland about 11 a.m., at Kinsale, &c. in Ireland not till 3 or 4 a.m."
Paradoxically, morning in the eastern Channel and North Sea, against afternoon to late afternoon to the west of the British Isles (Cornwall, Devon, Ireland). The discrepancies were not yet understood as differences in the medium of transmission (solid crust against water) of waves caused by the same event.

2.4.2 Dartmouth

Dartmouth is in South-Devon, about 40 km east of Plymouth. The report (Holdsworth, 1756) was compiled five month later than those already discussed. The evidences presented are difficult to analyse :
"...there was a surprising agitation in the waters about nine in the morning //there was a great and sudden swell // the boats, riding near the mouth of the river, tumbled and tossed as if they would have leaped into each other ; and two of them broke loose from their moorings. // though it was four hour ebb, the water rose as high, or higher than they usually do on the highest spring tide. This violent motion lasted about three quarters of an hour, and then the waters fell ever since without any perceivable alteration..."
The time ascribed to the water agitation, about 9 AM, is earlier than the earthquake in Lisbon, which took place between 9:30 and 9:45. The longitude difference between Lisbon and Dartmouth does not warrant such a difference in local (solar) time: and the time should be later in Dartmouth than in Lisbon. If the water agitation in Dartmouth happened in the morning, this report can only refer to a water effect, in the estuary, despite the reference to the tide level : But the inquiries were conducted four or five months after the facts, and the memory of the witnesses must be questioned.

3. France: Brest

The claim that the earthquake was felt on the coasts of Brittany is not convincing, for want or particulars. Observations published in French gazettes or kept in Archives only deal with the agitation of waters caused by the crustal long waves. The only one worth examination (Abeille, 1756) deals with the harbour of Brest, in Brittany:
"The earthquake has been felt, more so on the Coasts than in Brest: it was on the 1st of November, at nine in the morning, it was very slight, as it was only felt by a small number of people : but on the same day, and at the same hour, waters suddenly rose to a three feet height in a new Basin which they dig at present // at the foot of Saint-Magdelaine's mount. these Waters disappeared on the instant. Five minutes later they reappeared at the same height, and disappeared in the same way as the first time"
The wording so convoluted as to be almost impossible to interpret. The basin under excavation became the back-end of basin 3 of the arsenal, in the area of Pontaniou. Upon reading, one gets the impression that 0.98 m of water suddenly exuded from the bottom of the basin, and were reabsorbed quite as quickly! This is of course impossible: the water came from the river Penfeld, breaking through or overflowing any kind of cofferdam isolating the basin. This could only be the tsunami if the reported time was wrong, but this report describes the hydraulic perturbation brought about by crustal low frequency surface waves. Obviously, the north-west corner of the Galician coasts efficiently shielded France from the tsunami wave.

4. Iberian Peninsula

4.1 Galicia: A Coruña, Corcubión

Abbot Frishman's consular letter from Coruña is not widely known, due to its diplomatic status: few copies were distributed among the Ministries in Paris: A dispatch in the JHMT (JHMT, 1756b) must derive from Abbot Frishman's letter.

"At ten in the morning // the Earth Shook // The tide, which was at twelve, was higher than it should have been // From one PM to one and a half, the sea rose and fell seven times. The same thing was observed during the whole ebbing of the tide; The water rising and falling as much as six feet of height." (Frishman, 1755)

The tsunami reached Coruña about three hours after the earthquake happened. It also provides us with an estimate of the amplitude, 1.95 m. There may be a run-up component in this height, as Coruña harbour is situated on a rather jagged, indented coast. The frequency of the fluxes and refluxes (seven in half an hour) shows that the tsunami wave was already quite perturbed by diffraction and multiple reflections along the coasts when it reached the area. The anomalous wave ended only on the morning of November 2nd. The JHMT introduces some confusion by overrating the swelling of the sea during the quake itself, while Frishman, at that moment, mainly mentions the whirling of the waters.

The tsunami is also mentioned at Corcubión, about 75 km south-west of Coruña, with a crest to trough amplitude slightly higher than the tidal range (Lemaur, 1755, in Pérez-Gómez., 2009). The travel time may have been only 1 & ½ hour.

4.2 Portugal

4.2.1 Oporto

The November 1st, 1755, tsunami in Portugal has been re-discussed by Baptista & Miranda (Baptista & Miranda, 2009), and we can add little to their assessment, as the available primary sources are the same. Most documents on November 1st in Porto originate from the British community, and were printed in British journals. Wolsall (Wolsall 1756) states that this town "quite escaped" the earthquake and tsunami: perhaps only by comparison to the major destructions undergone by Lisbon. The estuary of the Douro river has an east-west orientation, which did not favour the funnelling of the tsunami wave towards Porto. The accounts provide us with an estimated range for the wave height of 1.22 to 1.83 m. (Philos. Trans., 1756b; Philos. Trans., 1756d; Gent's Mag., 1756a): also, three of these original documentary sources (Gent's Mag., 1756a ; Philos. Trans., 1756c ; Muysson, 1756) state that two ships, which were going out of the harbour, were forced in again, above the bar which crossed the estuary. This may require the maximum water height cited by the authors, but it implies that this estimate describes the run-up: the simple amplitude at shoreline may have been slightly less. The Philosophical Transactions letter LIX-VIII (Philos. Trans., 1756b) gives a wave period of 15 min, while Letter LIX-XI (Philos. Trans., 1756d), giving only one minute, must be erroneous on that notion.

4.2.2 Setúbal, Portimão, Faro

Most classical documents on the tsunami in Setúbal post-date some of the French diplomatic letters (Partyet 1755a). These diplomatic letters are not evidences from a true witness, as he resided in Madrid. Later, these indications found their way into the French Journals (JHMT, 1755 ; Merc. Fr., 1755), and maybe to the British Journals, as it took the British Special Envoy two weeks to send news to his country.

The annotations of the Priest of Saint-Julian Church, (in Gardi *et al.* 2009), clearly states that the tsunami waves inundated the village three times, but the amplitude of the waves is not mentioned.
Tedim Pedrosa & Gonçalves (2008) mention that in Ferragudo and Portimão, Forte São João do Arade was completely washed out while on the opposite bank of the estuary, Fortaleza Sta Catarina was also damaged ; and the Convento de São Francisco was completely destroyed. The sources given are the *Dicionário Geográphico,* a survey registered in 1758, and a book (J.B. Lopes, 1841?) both unknown to us. The evidences that the run-up reached 20 m above average sea level (a.s.l.) are not reliable: damages to fortresses could have been the collapse of pieces of the walls, which might have been triggered by the tsunami if the rock or lower levels of masonry were already weathered by salt-water alteration and undermined by the waves : but it would not demonstrate any level of run-up.
Actually, Fortaleza Sta Catarina is still in existence, if no longer in any military use, and Forte São João was sometimes a residence of the poet Joaquim José Coelho de Carvalho (1852-1934). The buildings of the Convento de São Francisco are indeed in a state of advanced ruination, as the result of fire in 1884, but the original mannerist period walls are still standing. Any 18th century repairs or re-building cannot be estimated, but damages could have been caused by the earthquake itself, as much as by the tsunami. However it appears that the monastic community may have been dissolved as a consequence of the earthquake and tsunami.
Most documentary sources assign to Faro the same fate as Setubal (Partyet, 1755a; Merc. Fr., 1755; Lond. Gaz., 1755): but the evidences are not clear and mostly hearsay: they provide no description, nor even the name of a building having existed in the allegedly destroyed town, and are unable to attribute the damages specifically to the earthquake or to the tsunami. In a recent article, Baptista & Miranda (Baptista & Miranda, 2009), citing a manuscript by Castro (1786), advocate that the city and harbour of Faro underwent nothing worse than a strong spring tide… though at a moment when they expected half-tide, of course. The tsunami was felt, but it was not that destructive: no one would have dared negate the destructions if it had been common local knowledge that Faro had indeed been washed away. When Castro wrote, thirty-one years after the event, there was no place left for hearsay, as real travellers had revisited the place.

4.3 Andalusia

4.3.1 Cadiz

The primary record of the tsunami in Cadiz was extensively discussed in Blanc (2008), so that the present purpose is only to consider how additional sources can now affect our previous conclusions.
A London Gazette (Lond. Gaz., 1755) dispatch dated 4th November, published on November 29th 1755, dealing with Cadiz may derive from a letter written by the French Consul in Cadiz (Desvarennes, 1755) on the same day. Both may originate from an official statement by the Governor of Cadiz, which he must have issued after consultation with Godin.
An anonymous 16-pages brochure in French, titled "*Relation du Tremblement de terre arrivé a Cadix le premier de Novembre 1755*", now remains in the library of the University of Göttingen founded by the Elector Georg August, Duke of Hanover and King of Great Britain as George II. Bearing the eighteenth-century library stamps on verso of first leaf, "EX BIBLIOTHECA REGIA ACAD. GEORGIAE AUG", registered as 8 MIN III, 8655, it entered the library in 1773 as part of the private collection of Gottfried Achenwall (1719-1772), an historian and statistician in the University of Göttingen (personal communication from Frau Cornelia Pfordt, Librarian in the *Niedersächsische Staats- und Universitätsbibliothek Göttingen*). This text

is obviously contemporaneous with the earthquake. The photocopy now in my institution was probably obtained there during the 1980/ 90's, by the late Jean Vogt, then very active in the research of original documents on historical seismicity.
This brochure (Anonyme "de Göttingen", 1756-1772) reports that the initial wave hit the town-walls a little below the string-course of the Parapet: it validates the testimonies of Godin (Godin, 1755) and Ulloa (Ulloa, 1756b) by wording differently the same observation (Figure 1) : the water-height of the tsunami at shoreline was 5.85 m above the marine zero datum, 3.85 m above mean sea-level. and 2.5 m above the expected high tide at noon on that day, November 1st 1755 (Blanc, 2008).
At variance, such a statement as "considering the height of the sea when in its natural state, at mid tide, the increase caused by the earthquake must have been of 36 feet" (11.69 m) remains puzzling. The same estimate is found in Ulloa (1756b), but it is referred to the High Equinoctial Spring Tide level, locally 2 m above average sea level (natural state, at mid tide). No justification is given of this height estimate, contradictory to facts: the town was not submerged. The sentence "The parapet // was broken in flush with the course string, in 6 different places, which the waters used as so many inlets" clearly shows that the water did not overflow the whole circumvallation, but only breaches in the Parapet, so described : "One // was of 85 ells, & another of 55, the 4 others // smaller" (the French word for "ell" was the usual translation for the Spanish *Vara de Burgos*, the Spanish standard for a somewhat shorter equivalent to the Yard). A run-up height of 11.7 m would not allow for the frantic run of the populace from one town gate to the other, described by Governor Antonio de Azlor (1755, *in* Blanco-Moyano, 2005): "It is not possible to express the clamour, the tears and the prayers which were heard in the town, during the three hours that lasted the unusual motion and the threat of the sea". As all the streets would have been immediately drowned, and the houses foundered, that noise would have been soon put to end.
The account of the arrival and timing of the tsunami is corrupt: the time of arrival of the first positive wave is correct, but the detail of the different assaults by the successive wave has been badly transcribed:

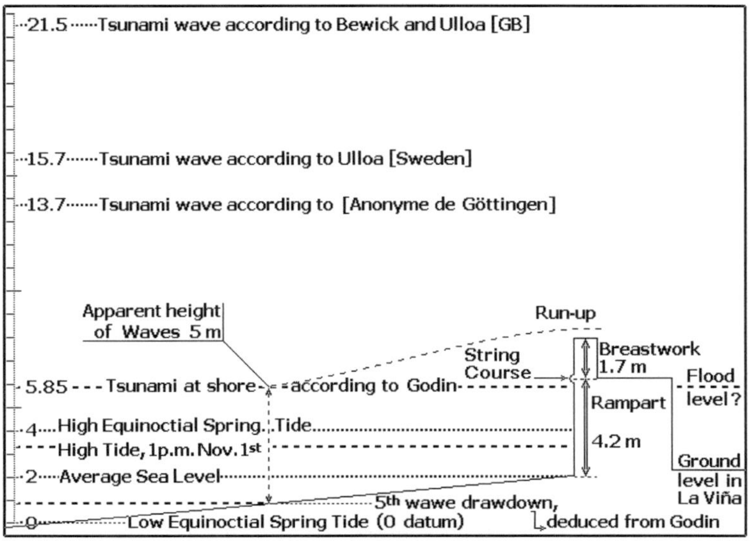

Fig. 1. Comparison of the normal scale of hydrographical levels in Cadiz (0–4 m) and wave setting proposed by contemporary authors. (Modified from Blanc, 2008).

"The 1st at 11 hours 10 minutes // The 2nd at 11 hours 30 minutes // The 3rd at 11 hours 33 minutes // The 5th at 1 hour 10 minutes // The 6th at 1 hour 50 minutes." Any mention of the fourth wave is lacking, and the intercalated comments and explanations. We need not alter the lag between the 1st and 2nd wave, nor between the 5th and 6th : they correspond to the only data really clear in the timing by Ulloa (1756a), showing that the period increased from 20 to 40 minutes in six phases. The arrival of the third wave can have taken place at 11 hours plus something, but not three minutes only after the second. It is therefore between "11 hours" and "33 minutes" that the restoration which these data require must be placed. Admitting that the indication "33 minutes" is correct, it must refer to the fourth assault, but it should follow "12 hours". And 12h 33 for this fourth wave is acceptable, the table by Godin (1755) gives 12h 35. One may then transfer a similar difference for the third assault, which would have taken place one or two minutes before the 12 hours of the table of Godin, be it 11h 58 or 11h 59. This restoration bears on only one line (or two half-lines) in table 1, and gives an ideal damping curve. The difference with the timing by Godin regularly increases, from 0 to 5 min for the five first assaults, but only from 0 to 3 min if we consider the duration of periods : just slight differences in the rounding of the number of seconds ! This also allows us to understand how the damping tables diverged in Ulloa's apocryphal communication to the Royal Society (Ulloa , 1756a), which is inaccurate because of a poor restoration, and in the Acts of the Royal Academy of Sciences of Sweden (Ulloa, 1756b), where they did not detail the periods, for want of a consistent table (Blanc, 2008).

Author	Godin, 1755	Ulloa, 1756a	Ulloa, 1756b	Anonyme de Göttingen 1756/1772	proposed restoration (this chapter)
1st wave	11 h 10 Δ	11 h 10 Δ	11 h 10	11 h 10 Δ	11 h 10 Δ
2nd wave	11 h 30 20 min	11 h 30 20 min	'til two hours after noon, the sea rose six times…	11 h 30 20 min	11 h 30 20 min
3rd wave	12 h 00 30 min	11 h 50 20 min		11 h ***33*** ***3 min***	11 h ***58*** ***29 min***
4th wave	12 h 35 35 min	12 h 30 40 min		--------- -------	***12 h*** 33 34 min
5th wave	13 h 15 40 min	13 h 10 40 min		13 h 10 -------	13 h 10 37 min
6th wave	about 14 h 00	13 h 50 40 min	about 14 h 00	13 h 50 40 min	13 h 50 40 min

Table 1. Periodicity of the tsunami waves in Cadiz on November 1st, 1755.

But it corroborates our recognition of the brochure as being closest to the source of the letter of Ulloa which was summarized and printed in the Swedish text in 1756. Not a copy, it is an earlier state of the text, most probably the same that Bouguer read at the Royal Academy of Sciences of Paris on December 3rd, 1755. Whoever edited it was honest, and did not alter the description or introduce uncontrolled ideas. It may have been printed up to fifteen years after the event, but the data carried by the accounts mentioned (Godin 1755; Godin *fide* Martínez Solares *et al* 2004; Ulloa 1756a; Ulloa 1756b; Anonyme de Göttingen 1756/1772) complete each others and demonstrate their common origin. They compose a consistent corpus of what happened in Cadiz on November 1st, 1755.

4.3.2 Gulf of Cadiz

The coast of the gulf of Cadiz, on both side of the Portuguese-Spanish border, is a low coastal plain, some miles wide, in front of higher ground. There are no clear descriptions of tsunami damages to the towns: The description by Ramos Dávila (in Blanco, 2005) shows that the towns stood on firmer ground that the marshlands, only mentioning the inundations of the first streets in Huelva: "Three quarters of an hour after the earthquake ended, the sea was furiously agitated, and the waters pushed towards the coast // the river came out of its course, sending waves to the first streets, inundating the plains, and with much damage to the boats". But the fishermen's cabins and salting sheds, on the shore, were destroyed, and people were drowned: from three hundred (Anonyme de Göttingen, 1756/1772) to one thousand (Ramos Dávila). This cannot be considered as a proof that the wave was so very high, since the speed of the tsunami (a few tens of km per hour) and a ventral segment several km long in the direction of the wave movement, account for the destruction of all but the strongest buildings within the reach of the wave. Lima *et al.* (Lima *et al.*, 2010) mention a possible run-in of 11 km from the coast line to Huelva (town centre): it must be pointed out that this occurred through the estuary and valleys of the rivers Odiel and Tinto, rather than on the flats themselves. In the 18th century map reproduced in their article, the salt-marshes are deemed liable to flooding: only more so to tsunami-flooding ! But from the map itself, the run-in on the salt-marshes appears to have been little more than one league.

Both the London Gazette and the Gentleman's Magazine mention the periodical nature of the phenomenon: "...at Port St Maria's, the Sea had rose eight several Times, and frightened all the Inhabitants out of the Town"; (Lond. Gaz., 1755; Gent's Mag., 1755). Unfortunately they forget to mention the unit applied to the measurement of the amplitude at shore: "eight" according to the London Gazette, "8" (Arabic numerals) in the Gentleman's Magazine. Admitting it was the foot, the conversion may be from 2.23 to 2.60 m, considering the range from the Spanish to the French units: This does not differ much from what Godin (1755) reported in Cadiz, 2.51 m. As Puerto de Sta Maria is on the mainland, people were able to flee to higher land, while the Gaditans could not.

Many French documents describe Conil as having been ruined: this may be true, but no real evidence is supplied, and none of these are true primary documents.

French dispatches have no value on Gibraltar, a British harbour, on which only English accounts can be relevant: the falling of the Rock on the town was just wishful thinking from enemies (JHMT, 1756a). The indication of water swell and recess of 1.83 m (Gent's Mag., 1756a) must be a crest to trough figure: the Strait is close to 900 m deep, just South and East of Gibraltar; the wave, travelling from west to east, was loosing its amplitude : and no damage or loss are properly described.

5. Africa

5.1 Morocco

The contemporaneous accounts on the tsunami of November 1st, 1755, have been re-examined recently by Kaabouben *et al.* (2009) and by Blanc (2009), in a critical analysis of documents, closest to primary sources, having reached France and England within about two month of the event. Almost everywhere in Morocco the amplitude of the tsunami has been vastly overestimated : In Tangiers and Mazagão the exaggeration may have been six to ten-fold. In Tangier a conservative estimate is that it was the same as the measurement given by Louis Godin (1755) for Cadiz, *i.e.* 2.5 m above the calculated astronomical tide or a crest-to-trough double amplitude of 5 m. Kaabouben (Kaabouben *et al.* 2009) cite a height of 2.72 m in Assilah from the Gentleman's Magazine.

A parallel drawn between the event in the two countries, showing that there were no scale differences, justifies this conclusion: [1] in Morocco, Mazagão was surrounded by the tsunami waters, and the outer defences were swept away ; in Andalusia, Cadiz is naturally surrounded by the Atlantic Ocean and the Puntales Basin, but the causeway, on the tombolo leading to the isle of León, was dismantled. [2] also close to Cadiz, on the same causeway, travellers going to or leaving the town were carried away from the road and drowned; Near Salé, a caravan following a track, either on the shore or along the Bou Regreg estuary, was engulfed by the tsunami.

As in the case of Cadiz, we only comment on reports not taken into consideration previously.

5.1.1 Ceuta (Spanish enclave)

For Ceuta we find the same account both in the JHMT and in the Gazette de Cologne: "... The sea swelled to the height of seven feet, & a quarter of an hour later, in came down so much, that great quantity of boats and fish remained dry on the sand. These fluxes and re-fluxes followed alternately, until the next morning: but at two after-noon they began to decrease progressively." (JHMT, 1756b). These accounts are the only one for this place in French Journals, as they are copies of each other. The water height is about the same as described in Gibraltar, but for possible differences in units. Applying the Spanish official units of Burgos, the measurement given is equivalent to 1.95 m. As in Gibraltar, this must be a crest to trough amplitude: the Strait deepens from west to east, and reaches about 900 m just North of Ceuta, so that the amplitude of the waves decreased : accordingly, no damage or loss are described.

5.1.2 Tangiers, Asilah, Larache, La Mamora

Kaabouben *et al.* (2009) recently cited the very short mention of Assilah in the Gentleman's Magazine for Dec. 1755: "At Arzilla in Barbary the water rose nine feet, and a settee of the town foundered..." (Gent's Mag., 1756a). This despatch, purportedly written on November 7th in Gibraltar, may have been the very first hint of the occurrence of a tsunami in Morocco to reach Europe, and it did not even acknowledge the earthquake.

This estimate of the water height at shoreline (2.73 m), compares well with the figure given by Godin in Cadiz, within a 10 % error: we cannot expect better, due to the lack of consistency of the systems of measure used. It also confirms our opinion that in the London gazette and Gentleman's Magazine reports (Lond. Gaz., 1755; Gent's Mag., 1755) on Puerto de Sta Maria, the intended unit after "eight" or "8" was indeed the foot (in that case with a 12.5 % error). Both these estimates endorse our conclusion that the tsunami was the same in Morocco as on the south-west coasts of Spain (Blanc, 2009).

We do not understand what is the foundering of a *settee*: the present meaning of the word makes no sense in this context.

Concerning Tangier, the legend of sea-water desalinated by a tsunami (Gent's Mag., 1756a; JHMT, 1756b), already commented on (Blanc 2009), was spreading as early as November 20th. It must be discarded as impossible in term of Physics or Chemistry. The estimate of a wave height of 15 m and overflowing the ramparts is not better justified. The Gazette d'Amsterdam (Gaz. Amsterdam, 1756) conveys no news not found elsewhere, except that the wording implies that the tsunami occurred on November 18th, but the printing of this Gazette is too late to give it any authority. Damages are often mentioned in Larache and La Mamora (Kenitra), without any detail liable to analysis.

5.1.3 Salé

The tsunami in Salé has been discussed both in Kaabouben *et al.* (2009) and in Blanc (2009). The testimony of the Gentleman's Magazine, "A caravan with 200 persons going from Sallee to Morocco along the coast, was destroyed by the sea, which rose to a prodigious height almost in an instant; and another caravan, with yet a greater number of persons, going from Sallee to Fez, was destroy'd by the sudden rise of the inland rivers..." (Gent.'s Mag. XXV, Supp. 1755, p.587), appears to double the number of caravans carried away by the tsunami. We do not exactly know if the caravan was washed away on a truly coastal tract or along the estuary of the Bou Regreg.

5.1.4 Safi, Sta Cruz de Agadir

The documents considered in Blanc (2009) all report heavy damages in Safi and in Sta Cruz (Agadir), ranging from several houses destroyed to many houses and buildings overturned, but the image of the destruction remains extremely blurred. The French documents did not even grant a separate paragraph to these harbours (Blanc, 2009), while the British reports did not consider Sta Cruz at all.

Some Dutch primary sources show a very different image: In a letter written on May 5th, 1992 to A. Levret, the late J. Vogt mentioned that he had found in the *Leidse Courant* (Leidse Courant, 1756b; a Dutch newspaper issued thrice a week or so) that the earthquake did no more damages in Sta Cruz de Agadir than in Tetuan; that it was already known that it had caused no damages in Safi and that the tsunami only damaged a few boats there. The Friday, January 9th issue of this newspaper reports that nothing happened: "By a ship, which arrived in 22 days from Sta-Cruz in Barbary in England, we have news of Dec. 4th, "according to which not the slightest damage has been felt from the earthquake there, and also that no one knew of a catastrophe to have happened on those shores..." (Leidse Courant, 1756a) The Dutch captains certainly had the scientific rigour to deliver faithful testimonies. The position of these two harbours is not quite in direct line from the likely positions considered for the epicentre and origin of the tsunami, so that it should not have been stronger than in Coruña or even the Cornish Mount's bay.

5.2 West Africa: Coast of Guinea

An anonymous letter to the Académie Royale des Sciences de Paris, dated Dec. 15th, read by Duhamel on March 24th, 1756, mentions coastal freshwater lagoons, on the West-African coast, where the Slavers from Nantes used to take in fresh water before crossing the Atlantic towards the Caribbean : "The same event occurred on the same day on the coast of Guinea, according to what told us Captain Charlery, who is from your shire, who commands a Slaver from Nantes, and who has been happy enough to escape in the Channel from 5 English ships cruising there. All the lagoons at Cape *****, were they used to take water, happened to be filled by the sea." (Anonyme "lu par Duhamel", 1756). Captain Charlery reported only the fact that the water had become brackish. This was due to the tsunami, as agitation by crustal low frequency waves could have only caused a loss of freshwater from the lagoons, not an input of saltwater. The name of the Cape marking the position of the lakes is omitted, due to the naval war between the French and English : the French captains tried to avoid giving to the English information which could have allowed them to lay in ambush, had the Martinique mail been caught : a witness to this is the fact that this same boat had had to escape five English men-of-war cruising in the channel of St Lucy. These lakes were anywhere on the coasts from Senegal to Liberia, so as to shorten as much as possible the duration of the crossing of the Atlantic, during which they would live on that water. The vague location prevents any speculation on the height of the tsunami waves : but this account is the most southerly one we know of, on the coasts of the Old World.

6. Macaronesia

6.1 Azores: Terceira

A short dispatch published in the Leidse Courant (Leidse Courant, 1756a) mentions that in the Azores "only slight shocks of earthquake had been felt, but that the Ebb had been strong at that moment in the harbour 200 that it was almost possible to see the Anchors of the ships on the bottom". In the description of the drawdown, it is difficult to understand why the number "200" (Arabic numerals) in mentioned after the word "*haven*", unless the typographer forgot to mention a unit, likely the foot : in which case it could be an horizontal distance of drawdown of 61 m.

6.2 Madeira: Funchal, Machico, Porto da Cruz

Two letters to the Royal Society (Heberden, 1756 ; Chambers, 1756), dealing with the tsunamigenic earthquake in Funchal and the island of Madeira, complete one another. The earthquake was felt at 9:18 AM according to the timing by Chambers, in accordance with the time-record in the Astronomical Observatory in Cadiz, 9:52, and with the 10°37' difference in longitude. It was shorter in Madeira than in Lisbon. The tsunami took place 1 ½ hour later : 11:00 (Heberden) or at 10:48 (earthquake time from Chambers plus travel-time from Heberden).

The first wave allegedly reached an amplitude of 5.64 m in Funchal (from half-ebb level to 15 feet above high water mark), though the harbour faces to the South, while the wave came from the North-North-East. A contradiction is that both accounts mention that the inundation was worse to the North and East of the island, causing damage and loss of property, these coasts being in the general direction of origin. The losses described in Porto da Cruz are loss of goods (grain, pipes of wine) or equipment (distillation stills) : the Madeira wine production being mainly exported, these pipes of wine were stored on the quays, or in sheds with direct access, and the stills must have been installed there too, unless they had just been unloaded from some ship and waited for transportation further inland. Heberden insists that the subsequent waves only oscillated between high and low tide levels, and this observation must be reliable : witnesses must have been relieved that no re-inundation happened. : this means that the amplitude of the second and subsequent waves was about 1.07 m, or 2.13 m peak to trough, but it implies that the fist wave in Funchal was heavily overestimated by Heberden : it is doubtful that it could have been five time higher than the subsequent ones : the damages reported do not warrant as strong a wave as in Portugal or in Spain. Two metres may be an proper estimate of the amplitude at shore.

6.3 Canary Islands: Sta Cruz de Tenerife

"In a letter from Sta-Cruz de Teneriffe on Nov. 12, no unfortunate tidings is ever mentioned about the earthquake…" (Leidse Courant, 1756a) : The wording of this letter implies just that : there was no damage in Sta Cruz de Tenerife, and no one was endangered. It does not mean that the tsunami was not observed, but it is strictly impossible to extract anything more accurate.

7. Caribbean Islands

7.1 Barbadoes

News from Barbadoes arrived rather quickly in Europe : the first mention of the place by the Gentleman's Magazine (Gent's Mag., 1756b) shows that the dispatch arrived probably early in December. Those early news contain very little quantifiable details.

Overall, we come up with three wave amplitude evaluations :

The JHMT (JHMT, 1756c) puts forward an estimate of five British feet (1.52 m) : British, because no Frenchman could have made measurements and reported to a French journal, from an English colony, while the two countries were at war : this must originate from an English dispatch passed to France through neutral ground (the Netherlands or a German harbour).

Captain Affleck (Affleck, 1756) gives an estimate of 3.66 m : "at Antigua, Nov. 1, 1755 // The tide rose here twelve feet perpendicular several times, and returned almost immediately : the same at Barbadoes." showing that he was not in Barbadoes at the time. ; then, with a second, more reliable, testimony, he re-assesses the amplitude in Antigua to have been 1.52 m : thus, "same in Barbadoes" meant nothing more than that the tsunami was observed.

The report by William Hillary is somewhat convoluted : "At 20 minutes after 2 o'Clock after Noon, above an Hour after it was High-water there, the Sea suddenly flowed and rose more than two Feet higher than it does in the highest Spring Tides, and in three Minutes time it ebbed so as to be as much lower than the usual lowest Ebb" (Hillary, 1756). The total crest-to-trough height which we can compute from this indication is two feet of positive run-up plus the usual tide range plus a draw-down similar to the run-up. As the Atlantic tidal range on the Caribbean Islands is about 30 cm, this figure turns out to fit the JHMT estimate of 1.52 m.

	Gent's Mag., 1756b, 589-590	JHMT, 1756c, 463-464	Affleck, 1756	Hillary, 1756
Time of arrival :	2 PM	2 PM		2:20 PM
Peak to trough amplitude :	over wharfs into houses	1.52 m	3.66 m	1.52 m
Duration :		8 hours, 64 periods.		every 5 min

Table 2. Comparison of available tsunami parameters in Barbadoes.

A tsunami wave causing a run-up little more than 61 cm high and a crest to trough amplitude of 1.52 m may be enough to make a strong impression on witnesses used to a tidal range of 30 cm. The considerations on the period of the wave train are less clear, such a sentence as "the Times between its Ebbing and Flowing decreased, so as to be a little longer" being somewhat contradictory ! It only appears that the period increased from 5 to 20 min between 2:20 PM and 7 PM.

7.2 Martinique

True primary documents for this tsunami in Martinique are not numerous, nor widely known. Two excerpts of letters to the *Académie Royale des Sciences de Paris,* read there on January 28th and March 24th , 1756, were recorded in the *Procès-Verbaux* (Lettée, 1756 ; Anonyme lu par Duhamel, 1756) : no doubt is possible as to their contemporaneous character. A third contemporaneous document, published by the Royal Society (Affleck, 1756), mentions Martinique in two lines : but it cannot be considered as a true primary testimony : the author was certainly not on the spot at the time of the event.

We also consider three local documents printed in the middle of the 19th century, the authenticity of which relies on the archaic style and spelling. The first was published in Saint-Pierre in 1846 by a Martinican historian, Sidney Daney de Marcillac. In the pages concerning the earthquake and tsunami (Daney, 1846), the form of the imperfect tense is modern (*-ait, -aient,* not the archaic *-oit, -oient*), but some word are spelled so as to drop the mute "T" at the end of nouns or adjectives such as "*savans*", "*environnans*" and "*événemens*", an obsolete spelling in the 19th century, showing that Daney borrowed the ancient text itself, not just from it. The same applies to an extract of a book by the sometimes mayor of Saint-Pierre, Dr Etienne Rufz de Lavison (Rufz, 1850), in which a similar archaic spelling of some words is observed : "*instans*" without a mute "T" ; "*isle*" instead of "*île*", absolutely obsolete in French in the 19th century ; or "*appercevoir*" with two "P". In this case also the forme of the imperfect tense was up-dated. Measurements are given in the old French units rather than in Metric units. This 18th century text published by Rufz was later reproduced with the spelling of words fully corrected to more recent usage, but not the units (Ballet, 1896 ; Cottrel, 2003)

The letters recorded in the *Procès-Verbaux* clearly state that it was in the *Cul-de-Sac de la Trinité* that this tsunami was felt, and Rufz also insists that it was ignored everywhere else : "this event was little perceptible in Galion and not at all at Robert or at Ste-Marie, though la Trinité is only one league from Ste-Marie and Galion, and four from Robert." The bay of La Trinité in Martinique is funnel-shaped, oriented toward the North-North-East, between the North-East coast of the island and the North-West coast of the Caravelle peninsula, projecting at right angle from the main coast.

The time of arrival of the wave is not perfectly established : two of the witnesses suggest 2 PM and only the earliest one, 4 PM, local time, of course. Daney (1846) ascribes a four hours leg to the tsunami wave (towards 10 AM at Lisbon, 2 PM in Martinique) but he omits the time lag between the European coasts and the lesser Antilles, which is at least 3 ¼ hours. This cannot be part of a true primary report anyway.

The amplitude and drawdown show a similar discrepancy : the two late texts concur on the distance of withdrawal and on the amplitude of the fist wave : drawdown to half the distance of anchoring, *i.e.* 200 paces or 140 m (the French *Pas* being the length of a single step) and the amplitude 0.65 m ; while the documents contemporaneous to the event do differ by an order of magnitude, with a run-up of 0.975 m (Lettée 1756) to 9.75 m (Anonyme lu par Duhamel, 1756).

Had the amplitude been 9.75 m, the descriptions would not just lament the loss of stored coffee, as reported by both Lettée (1756) : "all the houses in the town were inundated of more than three feet of water, all the stores were filled with coffee, which was soaked, as a consequence it is spoilt coffee. "and Rufz (1850) : "this third instance, it broke open the doors of the houses, though well secured from inside ; the coffee which happened to be there were lost, and some people just escaped being submerged ;".

Waves 9.75 m high would have only left the destruction of lives and village to mourn : towns hit by such waves in Sumatra on Dec. 26th, 2004 were merely reduced to parcelling-lots, drawn on the ground.

	LETTÉE, 1755	ANONYME lu par Duhamel 1756	DANEY 1846	RUFZ de Lavison 1850
Time	4 PM		2 PM	
Travel time			"about four hours" local-time lag omitted	
Drawdown distance			½ anchoring distance, 125/ 140 m ?	
Drawdown	6.5 m			
Amplitude			0.65 m	1st wave, 0.65 m 2nd wave, 0.49 m 3rd wave, 1.30 m
Run up	0.975 m in houses	9.74 m a. s. l.		3.9 m above high tides
Run-in	65 m			
Period	½ hour ?		¼ hour	

Table 3. Comparison of available tsunami parameters in Martinique.

The last paragraph dedicated by J. Ballet to this tsunamigenic earthquake "In December we learned that a dreadful earthquake had thrown down on the same day the town of Lisbon, and many towns in Morocco, and had extended its devastations in the whole of Europe, in Africa, in Asia, in America, where the shock was felt as far as Boston" (Ballet, 1896) typifies the confusion prevailing from the mid-18th to the mid-20th centuries, between the November 1st, 1755, tsunamigenic earthquake of Lisbon, the Rifian earthquake of November 18th & 19th, 1755 and the Cape Ann earthquake, in North America, also on November 18th, 1755. And Ballet adds Asia to make a good measure…

7.3 Guadeloupe

One reference to the tsunami in Guadeloupe island was only published 140 years later. This was in Sainte-Anne, the main harbour in the 18th century :

"Guadeloupe had no event to record during the year 1755 // Only, on November 1st, // On several point of the coast, there was a considerable withdrawal of the sea. In Saint-Anne, it ebbed down to the line of the keys which surround the roadstead, leaving only two channels, and, rising back with violence, it invaded the ground. In the borough, then considerable, of that parish, the waves came and broke against the porch of the church…" (Ballet , 1896)

The sea receded to a ring of tabular reefs surrounding the bay, the keys : their horizontal character is mostly determined by weak tides and agitation of the waters. In a recent paper (Roger *et al.,* 2010) a height of 3.2 m a.s.l. is ascribed to the sill of the porch of the present chuch in Sainte-Anne, but it was built by the Architect Ali Tur in 1928 after the passage of a strong hurricane over Guadeloupe. The assumption that the elevation of the church did not change must be discarted, as banking-up of the ground must have taken place before erecting the new concrete Modern-Art style building : That the inundation allegedly reached the porch of the 18th century church offers no run-up calibration.

7.4 Antigua

The date to which captain Affleck (Affleck, 1756) refers this agitation of the water in the harbour of Antigua is indeed the 1st of Nov., 1755. It has to be noted that there are essentially two different measurements of the amplitude of the wave in his testimony : in the first paragraph the crest-to-trough amplitude is given as 3.66 m. But in the second paragraph he insists on the trustworthy character of a new witness and supplies us with the time of the tsunami, 3:30 PM, the period, 5 min, the significant duration of the phenomenon, at least 2 ½ hours, but the crest-to-trough amplitude is reduced to 1.52 m, the surface of the water quiet. And as indicated earlier, "the same in Barbadoes" only means that the wave was observed there. A short article from the Gentleman's Magazine (Gent's Mag., 1756b)do not add anything significant to the account of Capt. Affleck. A JHMT dispatch must have been picked from the British papers (JHMT, 1756b).

7.5 Saint-Martin/ Sint Maarten; Saba

The report of captain Affleck (Affleck, 1756) expresses nothing convincing in these islands : In Sabia, a measurement given is 6.4 m, without even saying what the figure applies to. It might be a crest-to-trough, double, amplitude, but not without a very high degree of exaggeration. At St-Martin's, a ship anchored "in fifteen feet of water was laid dry on her broad side" : it did drag on its anchor under the heave of the tsunami, to upper ground, and was abandoned there on the water retreat. No indication of amplitude or run-up can be obtained from such fact.

The case of St Martin is complicated by the fact that Professor Winthrop (Winthrop, 1757) attributes the tsunami to the November 18th, 1755 "Cape Ann" earthquake. Reid (1914) clearly states that the November 1st Lisbon earthquake was only known in New England when a ship arrived from Cadiz on December 15th, and the news appeared in the local Gazettes on December 22nd : Winthrop learned of the Lesser Antilles tsunami after the Cape Ann earthquake had taken place, and he attached the accounts of the sea movements to the earthquake he knew of, the North-American one. As he did not know of any other, he corrected the date from November 1st to November 18th. The time of the tsunami in St. Martin's is given as 2:00 PM, but the Cape Ann earthquake was felt at 4:30 AM in Boston : If it had been on the same day, as the time-lag, in local time, between Boston and St Martin is about 35 min, this would allow for a travel time close to 9 hours. A teletsunami could not have been that slow.

The data provided are not very reliable either : according to Winthrop, the initial drawdown was 5.5 to 7.3 m. The positive height was only 1.83 m, and this should be a run-up rather than amplitude at seashore. The negative wave may have preceded the positive one, but nothing accounts for the dissymmetry in the phenomenon : where did the water, removed from the shore by such a negative amplitude, go ? The initial drawdown of three or four fathoms is over-evaluated. If the Cape Ann earthquake had been strong enough to cause such waves, the coasts of New England should have been devastated :

One entry in the JHMT (JHMT, 1756b) attributes a tsunami to the November 18th Cape Ann earthquake : "On November 18th, the town of Boston suffered several shocks, & and a great number of chimneys & roofs of houses fell. During this quake, the waters raised everywhere suddenly of twenty feet. We also learned that on November 18th there had been strong shocks of earthquake at Philadelphia and in New York, as well as in other places on the coasts of northern America, but that they had caused no notable damage. This is the only allusion we know of, to a November 18th tsunami on the coasts of New England : None of the British journals we consulted refers to anything of the kind, though their correspondents should have been on the front line to observe it. Furthermore, no French journal was in a

position to obtain first-hand information about the coasts of North America, as there was a war between the two colonial empires, in which the French would lose their North-American possessions. The notion of twenty feet high waves tsunami hitting Boston is nothing but a figment of the imagination of a loose pen-pusher.
Professor Winthrop (Winthrop, 1757), even as he ascribes the tsunami in St Martin's to the Cape Ann earthquake, only acknowledges that the ships in the harbours were shaken and that some fish were floated belly up. Some other accounts of this earthquake (Philos. Trans., 1756e ; Colden, 1756 ; Philos. Trans., 1756f), published by the Royal Society earlier than Winthrop's, do not mention any sea-shore damages. Thus in Saint-Martin, as well as in all the other Caribbean islands, all accounts pertains to the tsunami caused by the Great Lisbon Earthquake : a conclusion which R. Rothman (Rothman, 1968) reached years ago.

8. Conclusions

8.1 Methodology

This critical review of the earliest accounts available on the tsunami of November 1st, 1755, on both sides of the Atlantic basin, allows us to reach beyond what each of these descriptions conveys. At each scale, local, regional or global, the analysis must not be just a juxtaposition of observations brought by all the various accounts. But the descriptions are not always complete : we often have to borrow one parameter from one document, another from a second one, and so on.
When the data reconstructed from several sources remains consistent, there is a good chance that these different sources actually originate from one single initial report, but that, in their effort to make the article they were editing appear as original, the second-hand compilers did not quite understand the fundamental unity of the phenomena they were summarizing.
It should not be a surprise to find common descriptions in papers attributed to a single author (Ulloa 1756a & b) but part of the data set were corrupt before they arrived to one of the journals. Finding that the letter of L. Godin to the French King and registered by the *Académie Royale des Sciences de Paris* (Godin, 1755) had a common origin with the letters by Ulloa was only possible through the comparison of the data reported in the account, and the same comparison again led us to observe that the Anonyme de Göttingen (1756/1772 ?) also had the same origin. The frame of these reports is also the same, however different the length of the various parts in each of them. On an identical basis of similarity of data and phrasing, the origins of the two reports on this tsunami published in the middle of the 19th century in Martinique (Daney, 1846 ; Rufz, 1850), appear to be the same.
In most cases, the tsunami was correctly evaluated by the original describers, as long as they only described events which they actually witnessed, or at least only transcribed descriptions from true witnesses, in places which they personally knew. Trouble began when the authors tried to widen the scope of their account, gathering exogenous pieces of information.
When we have to face contradiction instead of redundant information, the discrepancies between the estimated parameters cannot be solved in any way other than blunt choice. The principles to solve such cases is to confront them with simple physiographic data : simple, because 18th century geographic documents are generally not accurate enough to allow any better, and some change in the layout may have taken place since the event ; and then to judge which of the contradictory statements is to be trusted, according to the laws of Physics and to common sense, which should concur in this.
An example is the height of run-up in La Trinité, in Martinique : 3 feet (0.975 m) according to Lettée (1756) as against 30 feet (9.75 m) according to a letter read by Duhamel on March

24th (Anonyme lu par Duhamel, 1756). This is an easy case : no one would have survived a 9.75 m wave to regret the wasted coffee. The discrepancy must result from the carelessness of a copyist, who wrote 30 (numerals) rather than 3 : or, under the impression of other extraordinary estimates already published for other places, he believed he was rectifying a mistake !

8.2 Results

Admittedly, Lisbon remains the populated place hardest hit both by the earthquake and by the tsunami. A new-built quay on which people had taken refuge against the fall of debris, was said to have been swallowed by the sea, to a depth of more than 100 fathoms (Lyell 1830). Reid (Reid, 1914) pointed out that the harbour of Lisbon was situated on the Tagus estuary, which was nowhere deeper than 5 fathoms : so that the masonry of the quay could only have been shattered and scattered on the bed of the Tagus' mouth. This is one of the few destruction of stonework attributed to the tsunami alone, apart from some length of the parapet of the Town walls of Cadiz : The alleged destructions of two fortresses in Ferragudo and Portimão (Tedim Pedrosa & Gonçalves 2008) are spurious : the forts are still standing, if no longer in use.

On the Spanish coasts of the gulf of Cadiz, the quality of the reports is due to the presence of some outstanding individuals :

Louis Godin, a former member of the *Académie Royale des Sciences de Paris*, had been the originator and head of the geodetic expedition to Peru, in charge of measuring the length of one angular degree of the meridian arc, at the equator. He was the head of the *Real Academia de Guardias Marinas* (School of Naval Cadets) in Cadiz, in charge of the scientific courses of junior officers.

Antonio de Ulloa, an Officer in the Spanish Navy, had also been a member of the expedition to Peru, together with Jorge Juan y Santacilia : both had been rewarded with membership of several European Academies and Scientific Societies.

In 1755-56, besides his official reports to the Spanish authorities, the observations of Godin, intertwined with those of Ulloa, were circulated to the learned societies of Europe : *Académie Royale des Sciences de Paris* (both a letter from Godin and one from Ulloa, the anonymous text in French now in the Göttingen *Georgia Augusta* Library) ; *Kongliga Svenska Vetenskaps Academiens*.

Unfortunately, editorial conditions worked in such a way that these accounts were quickly forgotten : [1] the Royal Society of London did not receive a proper copy of Ulloa's letter, but they published under his name an adulterated abstract obtained from the Netherlands : due to the high status of the Philosophical Transactions, the best known published account now conveys the enormous tall-story of 60 feet high waves in Cadiz ; [2] a bad choice by the French Royal Academy of Paris was to reserve the subject of Earthquakes and Tsunami until Ph. Buache, the King's Geographer and a member of the Academy, would be ready to deliver a synthesis on the subject : he died sixteen years later, leaving only a series of ill-assorted reports received at the Academy, at best registered, but never circulated : the French Royal Academy of Paris did not publish anything of value, despite the fact that they had received the two best reports on the subject. How their un-recorded letter from Ulloa came to be published as the brochure now found in the *Georgia Augusta* Library in Göttingen is not clear : it was printed any time between 1756 and 1772, and it is strange that it does not bear any editorial indication or bookshop where to obtain it. Its distribution being limited, it

made no strong impression at the time, and was ignored by later generations. The official Spanish report, still extant (Martínez Solares & López Arroyo, 2004), remained confidential.
The quality of the reports deriving from observations by Godin and Ulloa is such that they must be established as the yardstick by which measure the other accounts and the likelihood of the physical parameters implied by or deduced from the descriptions, after the necessary critical examination of the physical setting of the events described. These reports provide a moderate figure for the amplitude of the tsunami : 2.51 m.
When this earthquake occurred, the fact that tsunami result from earthquakes was only understood in areas where both are frequent, if not always catastrophic (Windward Islands). "If the ground had shaken, it would be said that it was the reason" (Lettée (1756). The notion that tsunami could strike totally unaware areas, across a whole oceanic basin, was not understood at all. The confusion of Winthrop was due to this un-established notion of travel beyond the distance of earthquake perception.
The result of the study is that the best known testimonies, constantly referred to by the scientific community for 2 & ½ centuries, are the most striking ones, those which describe the worse apocalyptic waves… if a superlative can apply to the Apocalypse. The point is that, except in Lisbon and in Cadiz, reliable descriptions of structural damages to the harbours and towns stricken, are generally lacking. More reasonable witnesses reported moderate measurements or estimates, almost everywhere : but when they attempted to extend their words beyond what they had actually seen by themselves, most overindulged in sensationalism. The amplitude at shoreline of the November 1st tsunami may have been 5 m in Lisbon, but it decreased to 2.5 m in the Gulf of Cadiz and in northern Morocco, to less than 2 m in Galicia and in southern Morocco. It must have been about 1 m in the Atlantic islands (Madeira) and decreased further to 0.7 m in the Caribbean Islands.
Can such result be favourable to the safety of people, coastal establishments and activities ?
It can give us confidence in the modelling experiments conducted to estimate the local risks : it has now been a long time since M.A. Baptista concluded that the mathematical models could not simulate wave higher than 10 m, in the Ibero-Moroccan Gulf, when investigating plausible seismic sources (Baptista *et al.*, 1998).
Does it mean that moderate estimates of the tsunami amplitude decrease the risks ? The mechanical parameters were sometimes widely exaggerated, but the sufferings of the coastal population were quite real. Risks which were believed attached to tsunami amplitudes of 10 to 15 m are in fact attached to amplitudes at shore between 2 and 3 m.
The need for efficient warning systems can only be made more urgent by such a conclusion.

9. Acknowledgements

The author gratefully acknowledges the contribution of Frau Cornelia Pfordt, Librarian in the *Niedersächsische Staats- und Universitätsbibliothek Göttingen*, in tracing the origin of the brochure referenced here as the Anonyme de Göttingen. The author's thanks also go to Madame Rose-Marie Benoistel-Roebook for the translation from Dutch to French of the texts extracted from the Leidse Courant (1756). Thanks are also due to Dr Ana Maria Baptista for informally reviewing an early version of this chapter.

10. References

(Most Anonymous reports are referenced by the name of the journal which published them or the Institution which detains the books recording them.)

Abeille M[r], 1756. Relation de ce qui s'est passé en Bretagne pendant le cours du mois de novembre 1755 au sujet des secousses de tremblement de Terre. *P. V. Séances Acad. Sci. Paris,* 75, 1-3, 7 janvier 1756.

Affleck Capt., 1756. CIII. An Account of the Agitation of the Sea at Antigua, Nov. 1, 1755. *Philos. Trans. Roy. Soc.* XLIX (II) 668-670.

Anonyme "de Göttingen", 1756/1772. Relation du tremblement de terre arrivé à Cadix le premier de novembre 1755. *Booklet,* 16 pp.

Anonyme "lu par Duhamel", 1756. Extrait... d'une lettre de la Martinique, du 15 Décembre dernier. *P. V. Séances Acad. Sci. Paris,* 75, 145, 24 mars 1756.

Ballet J., 1896, La Guadeloupe : Renseignements sur l'Histoire, la Flore, la Faune, la Géologie, la Minéralogie, l'Agriculture, le Commerce, l'Industrie, la Législation, L'administration. *Imprimerie du Gouvernement, Basse-Terre.* T. 2, pp. 286-287.

Baptista M.A. & J.M. Miranda, 2009. Revision of the Portuguese catalog of Tsunamis. *Nat. Hazards Earth Syst. Sci.* 9, 25-42.

Baptista M.A., P.M.A. Miranda, J.M. Miranda, & L. Mendes Victor, 1998. Constraints on the source of the 1755 Lisbon tsunami inferred from numerical modelling of historical data on the source of the 1755 Lisbon tsunami. *J. Geodynamics,* 25 (2) 159-174.

Bewick B., 1756. LIX. LETTER XIV. An account of the Earthquake at Cadiz, Novem. 1, 1755. *Philos. Trans. Roy. Soc.* XLIX (I) 424-427.

Blanc P. -L., 2008. The tsunami in Cadiz on 1 November 1755: a critical analysis of reports by Antonio de Ulloa and by Louis Godin, *C.R. Geoscience,* 340 (4), 251-261.

Blanc P. -L., 2009. Earthquakes and tsunami in November 1755 in Morocco: a different reading of contemporaneous documentary sources, *Nat. Hazards Earth Syst. Sci.,* 9, 725-738.

Blanco-Moyano B., 2005. 250 Aniversario del Tsunami que Barrió Andalucía Occidental. *Marina Civil,* (79) 59-64 (Oct.-Dec. 2005). (Excerpts of L. Godin, A de Azlor and B. Ramos da Vila).

Borlase W., 1756. LVIII. LETTER XV. In Cornwall : At Mount-Bay & Penzance. *Philos. Trans. Roy. Soc.* XLIX (I) 373-378.

Castro, D. A. L. F. : Historia Geral de Portugal e suas conquistas, *Database Fundo Antigo, Academia de Ciências de Lisboa, Livro 1, Capitulo IV, Manuscript 691,* 21–36, 1786 (in Portuguese ; *fide* Baptista & Miranda 2009).

Chambers Ch., 1756. LIX. LETTER XVIII. Another account of the same Earthquake at Madeira, Novem. 1, 1755. *Philos. Trans. Roy. Soc.* XLIX (I) 435-436.

Colden C., 1756. LIX. LETTER XXI*[B]* An account of the Earthquake felt in New York, Novem. 18, 1755. *Philos. Trans. Roy. Soc.* XLIX (I) 443.

Cottrel Ph. 2003. Journal d'un vieil habitant de Sainte-Marie (1745-1765). *Malesherbes.*

Daney S., 1846. Histoire de la Martinique depuis la colonisation jusqu'en 1815. *Fort-Royal, E. Ruelle, Imprimeur du Gouvernement. 1846.* T. III, pp. 237-238.

Dawson A.G., R.M.W. Musson, I.D.L. Foster & D. Brunsden 2000. Abnormal historic sea-surface fluctuations, SW England. *Marine Geology* 170, 59-68.

Desvarennes M[r], 1755. Nouvelles de Cadix. Du 4 novem[e] 1755. *Archives Nationales, Fond Marine B7/396. 3 p.* Annexée à Partyet, 1755

Fowke Th., 1756. LIX. LETTER XVI. An account of the Earthquake, that happened in Barbary. *Philos. Trans. Roy. Soc.* XLIX (I) 428-432.

Frishman Abbé de, 1755. Relation d'un tremblement de terre arrivé à La Corogne capitale de la Galice et dans les environs. Lettre de M. l'Abbé de Frishman du 7 9bre 1755. *Archives Nationales de France, Fond Affaires Etrangères, B1/255.*

Gardi A., A. Silva, M. Koscielny, E. Alparslan, R. Guillande, B. Ranguelov, N. Valencia, A. Atillah, S. Scheer, St. Tinti & Ph. Buckle, 2009. Scenarios for Hazard-induced Emergencies Management : State of the art. *GSC report D1.1*, 90 pp. , February, 16 2009 (Excerpt of : Annotations made by the Priest of Saint Julian Church).

Gazette d'Amsterdam (Anonyme) 1756. *Gaz. Amsterdam*, VII, 23 Janvier 1756.

Gent's Mag., 1755 (Anonymous). *Gentleman's Magazine*, XXV, p. 521, November 1755.

Gent's Mag., 1756a (Anonymous). *Gentleman's Magazine*, XXV, p. 563-564, December 1755

Gent's Mag., 1756b (Anonymous). *Gentleman's Magazine*, XXV, Supplement, p. 589-591

Godin L., 1755. Sur le Tremblement de Terre qui s'est fait sentir à Cadix le 1er novembre 1755 (lu par Bouguer). *P. -V. Acad. Roy. Sci. Paris,* 74, 772-781, 3 & 6 décembre 1755.

Heberden Th., 1756. LIX. LETTER XVII. An account of the Earthquake in the Island of Madeira, Nov. 1, 1755. *Philos. Trans. Roy. Soc.* XLIX (I) 432-434.

Hen. Tom., 1756. LVIII. LETTER XXIV. At Cork. *Philos. Trans. Roy. Soc.* XLIX (I) 393-394.

Holdsworth Revd, 1756. XCVII. At Dartmouth, relating to the Agitation of the Waters observed there on the 1st of November, 1755. *Philos. Trans. Roy. Soc.* XLIX (II) 643-644.

Hillary W., 1756. Observations on the Changes of the Air and the Concomitant epidemical Diseases in the Island of Barbadoes. *London,* 1756.

Huxham J., 1756. LVIII. LETTER XIV. In Devonshire and Cornwall; At Plymouth, Mount-Bay, Penzance & c. *Philos. Trans. Roy. Soc.* XLIX (I) 371-373.

JHMT, 1755 (Anonyme). *Jour. Hist. Mat. tems.* LXXVIII, p. 473-474. , Déc. 1755

JHMT, 1756a (Anonyme). D'Espagne. *Jour. Hist. Mat. tems.* LXXIX, p. 42-46. janv. 1756.

JHMT, 1756b (Anonyme). D'Espagne… Afrique… Amérique. *Jour. Hist. Mat. tems.* LXXIX, p. 131-139, fév. 1756.

JHMT, 1756c (Anonyme), 1756. Iles Barbades. *Jour. Hist. Mat. tems,* LXXIX, p. 463-464, Juin 1756.

Kaabouben F., M. A. Baptista, A. Iben Brahim, A. Toto &, A. El Mouraouah, 2009. On the Moroccan Tsunami Catalogue. *Nat. Hazards Earth Syst. Sci.*, 9, 1227-1236.

Leidse Courant, 1756a (Anonymous). *Leidse Vrydagse Courant,* n° 4, 9 January 1756.

Leidse Courant, 1756b (Anonymous). *Leidse Vrydagse Courant,* n° 9, January 1756, *fide* J. Vogt.

Lettée Mr, 1756. Extrait d'une lettre écrite par M Lettée dattée de la Martinique du Fort St.Pierre le 5 novembre 1755. *P. V. Séances Acad. Sci. Paris, 75, 48-49, 28 janvier 1756.*

Lima V.V., J. M. Miranda, M. A. Baptista, J. Catalão, M. Gonzalez, L. Otero, M. Olabarrieta, J. A. Álvarez-Gómez & E. Carreño, 2010. Impact of a 1755-like tsunami in Huelva, Spain. *Nat. Hazards Earth Syst. Sci.*, 10, 139–148.

Lond. Chron., 1761 A (Anonymous). Extract of a letter from Cork, March 31. *Lond. Chron. IX* (670), p. 350.

Lond. Chron., 1761 B (Anonymous). A letter from Kinsale… *Lond. Chron.* IX (671), p. 358.

Lond. Gaz., 1755 (Anonymous). Madrid, Nov. 10. *London Gazette* n° 9532. Nov. 25-29, 1755.

Lyell Ch., 1830. Principles of Geology. *John Murray, Albemarle-Street,* vol 1, 511 p.

Martínez Solares J.M. & A. López Arroyo., 2004. The great historical 1755 earthquake. Effects and damage in Spain. *Journal of Seismology*. 8 (2), 275-294.

Mendes-Victor L., A. Ribeiro, L. Matias, M.A. Baptista, J.M. Miranda, P. Miranda, N. Zitellini, E. Garcia, C. Corela, P. Terrinha, M. Rovere & F. Teixeira, 2005. Progresses in the Assessment of Tsunami Genesis and Impacts around the Portuguese Coasts, 217-230 in Tsunamis, Case Studies and Recent Developments, *Adv. Nat. and Technol. Hazards Res.*, Vol. 23, Satake, K. (Ed.), 2005, VIII, 356 p.

Mercure de France, 1755 (Anonymous). Nouvelles étrangères. De Madrid, le 10 Novembre. *Merc. France*, Vol. I, p. 241-244

Muysson M[ons]., 1756. LIX. LETTER XII. Concerning the Earthquake at Oporto, Nov. 1, 1755. *Philos. Trans. Roy. Soc.* XLIX (I) 422-423.

Nicola L., 1756. LVIII. Letter XXIII. At Kingsale. *Philos. Trans. Roy. Soc.* XLIX (I) 391-393.

Partyet M[r], 1755a. Lettre consulaire de Madrid du 10 Novembre 1755. *Archives Nationales de France/ Marine B7/ 396.*

Partyet M[r], 1755b. Extrait de plusieurs lettres que le Général de St François a reçues des Missionnaires de son ordre qui se trouvent Dans le Royaume de Maroc. Annexe à la lettre consulaire de Madrid du 22 décembre 1755. *Archives Nationales de France/ Marine B7/396.*

Pérez-Gómez B., 2009. Tsunamis. *Puertos informa* 154, 5-19. (excerpts from L. Godin and Ch. Lemaur)

Philos. Trans., 1756a (Anonymous). *Philos. Trans. Roy. Soc.*, LVIII. Extract of LETTER XVI from Swansey. Vol. XLIX (I) 379-379.

Philos. Trans., 1756b (Anonymous). *Philos. Trans. Roy. Soc.*, LIX. Extract of LETTER VIII concerning the earthquake at Oporto in Portugal, Nov. 1, 1756. Vol. XLIX (I) 418-419

Philos. Trans., 1756c (Anonymous). LIX. Extract of LETTER X from Oporto, dated Novem. 1, 1756. *Philos. Trans. Roy. Soc.*, Vol. XLIX (I) 421

Philos. Trans., 1756d (Anonymous). Extract of LETTER XI From Oporto, dated Novem. 1, 1756. *Philos. Trans. Roy. Soc.*, LIX. Vol. XLIX (I) 421-422

Philos. Trans., 1756e (Anonymous). LIX. LETTER XXI*[A]* An account of the Earthquake felt at Boston in New-England, Novem.18, 1755. *Philos. Trans. Roy. Soc.*, Vol. XLIX (I) 439-442.

Philos. Trans., 1756f (Anonymous). LIX. LETTER XXII An account of the Earthquake felt at Boston in New-England, Novem. 18, 1755. *Philos. Trans. Roy. Soc.*, Vol. XLIX (I) 444.

Reid H.F., 1914. The Lisbon Earthquake of November 1, 1755. *Bull. Seismological Soc. America*, IV (2), 54-80.

Roger. J., S. Allgeyer, H. Hébert, M.A. Baptista, A. Loevenbruck & F. Schindelé, 2010. The 1755 Lisbon Tsunami in Guadeloupe Archipelago: Source Sensitivity and Investigation of Resonance Effects. *Open Oceanogr. Journ.*, 4, *Special Issue 1*. 58-70.

Rothman, R. L. 1968. A Note on the New England Earthquake of November 18, 1755. *Bull. Seismological Soc. America,* 50 (5), 1501-1502.

Rufz de Lavison É., 1850. Études historiques et statistiques sur la population de la Martinique. *à Saint-Pierre*. T. 1, p. 394.

Tedim Pedrosa F. & J. Gonçalves 2008. The 1755 earthquake in the Algarve (South of Portugal) : what would happen nowadays? *Adv. in Geosci.*, 14, 59–63.

Ulloa A. de, 1756a. LIX. LETTER XV. An account of the Earthquake at Cadiz. *Philos. Trans. Roy. Soc.* XLIX (I) 427-428.

Ulloa A. de, 1756b. Utdrag as en berättelse, om Jordbäsningen, som kändes i Cadix, den 1 :sta November År 1755. *Kongl. Svensk. Vetensk. Acad. Handlingar*, 17, 139-145.

Winthrop Prof., 1757. An Account of the Earthquake felt in New England, and the neighbouring Parts of America, on the 18th of November, 1755. *Philos. Trans. Roy. Soc.*, L (I) 1-18.

Wolsall M[r]. 1756. LIX. LETTER II. An Account of the Earthquake at Lisbon, Nov. 1. 1755. *Philos. Trans. Roy. Soc.* XLIX (I) 402-407.

21

The 1889 Johnstown, Pennsylvania Flood - A physics-based Simulation

Steven N. Ward
Institute of Geophysics and Planetary Physics
University of California,
Santa Cruz,
USA

1. Introduction

1.1 Thumbnail review of the 1889 flood.

In 1889 Johnstown Pennsylvania was an industrial workingman's town, sited on a flat triangle of ground at the confluence of the Little Conemaugh and Stoney Creek Rivers (*Figure 1*). Johnstown and its neighbors in the valley—Cambria, Woodvale and E. Conemaugh—housed 30,000 residents. Years of industrialization narrowed and channelized the valley's three rivers to a fraction of their natural widths (*Figure 2*). As a consequence, nearly every spring Johnstown flooded. Even so, the water was little more than aggravation to town folk who simply relocated carpets and furniture to upper floors to wait out the water's retreat.

On May 30 and 31, 1889 the upper reaches of the Little Conemaugh and Stoney Creek basins experienced the strongest rains in anyone's memory. The Conemaugh River at Johnstown rose from 1 foot to 23 feet in just over 24 hours. By mid-day on the 31st, water crept over most of the lower town reaching to Central Park. Although everyone agreed that this flood was worse than usual, residents nonchalant about the affair, yet again moved carpets and furniture upstairs.

Like floods of the past, the 1889 one would have faded from history if not for South Fork Lake, 24 km up the Little Conemaugh River and 140 meters higher than Johnstown (*Figure 1*). Also swollen by the strong rains, South Fork Lake rose steadily throughout May 31 and water began running out its spillways at mid-day. Had it been rebuilt to its original 1853 specifications, South Fork Dam may have withstood even this torrent. However, blunders in the dam's 1886 reconstruction and later management led to its overtopping and failure around 2:45 PM. Collapse of the 22 m high dam released 16 million cubic meters of water into the Little Conemaugh River targeting Johnstown and neighboring boroughs. When the surge arrived eighty minutes later, it found no match for the wood frame construction in the valley. Hundreds of families waiting out high water in upper floors were carried away -- house and all. Until the terrorist attacks on September 11, 2001, the 2200 casualties in the 1889 Johnstown Flood amounted to the largest single-day disaster in United States history.

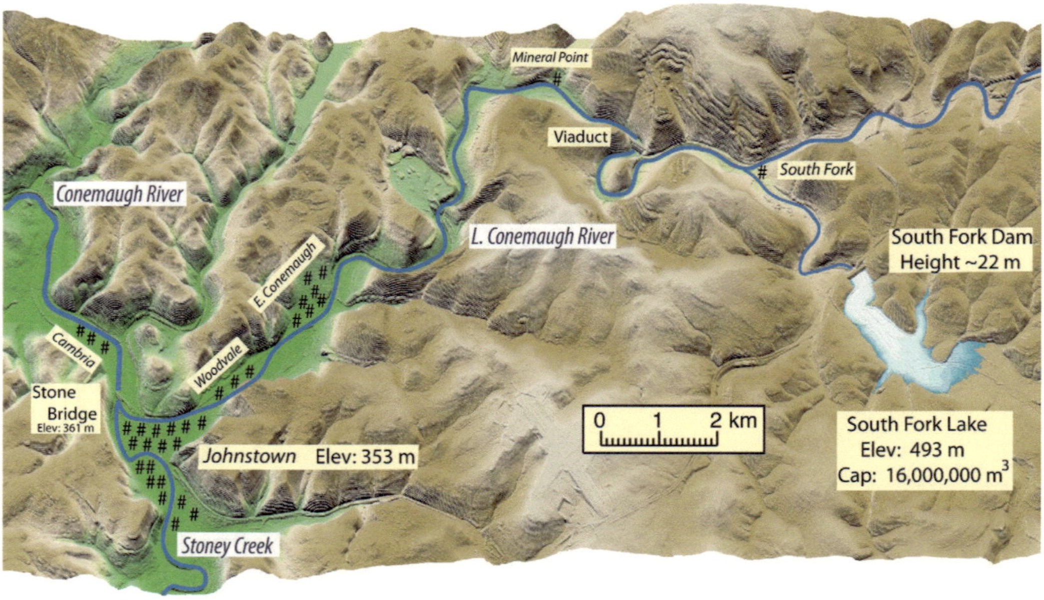

Fig. 1. Map of the Johnstown region showing the principal locations and players in the events of May 31, 1889.

1.2 Physics-based simulation versus hollywood animation:

Dozens of published accounts, memoirs, summaries *(e.g. Johnson, 1889; Beale, 1890; McCullough, 1968.)* and Hollywood films chronicle the Johnstown Flood. This article breaks from storytelling; instead, it constructs a physics-based simulation of the dam break and flood. In contrast to artistic depictions, physics-based simulations take care to discern anecdotal reports from quantitative information. Times, locations, elevations, volumes, velocities, flow rates, depth and flood extent comprise particularly valuable constraints. Unfortunately, the majority of information that one finds in researching the Flood is sketchy. Pinning even the basics, like the time sequence of events can be challenging. In that era, few people owned, much less carried, watches. "Somewheres around 4 PM" is a typical reckoning. Accounts of flood size too make for good reading—"It looked like a 40 foot high wall of water"—but scantly find application in physical modeling. Even tantalizing specifics —"Afterwards I measured water marks eight feet high on my rail car"— are not much help if they lack critical details; in this case, the location of the car and the level of grade where it stood.

In the face of hit-or-miss data, physics-based simulations hold special worth in that they can accept input from a wide range of observations. Too physics-based simulations can output quantitative predictions on aspects that were not, or could not, be measured. Truth to say, physics-based simulations also have limitations. In the 121 years since the Flood, things have changed —rivers have been channelized, flow controls constructed, various fills and excavations have forever changed the lay of the land. Spot elevations today may not reflect 1889 elevations, indeed "the spot" may no longer exist. It may not be possible to re-create events with the desired level of detail even with a numerical approach. Still, some quantitative information exists and not every version of the flood story is physically likely within those constraints. Sorting historical inconsistencies through the screen of physical laws, partly justifies computer simulation. Too, by wearing physics-based eyeglasses, new insights and connections may appear in the Johnstown Flood story that had been invisible before.

Fig. 2. Johnstown before the flood. View looks down Stoney Creek River toward the Stone Bridge. Note the infilling at the river banks and the construction virtually in the river bed. You can imagine that it would not take much of a flood to float these wood frame structures from their foundation can carry them off.

2. Tsunami Ball approach.

This article closely follows *Ward* and *Day* (2010a, 2010b) who simulated wave runup and inundation from the 1958 Lituya Bay Alaska and the 1963 Viaont, Italy landslide disasters. Simply put, tsunami balls are pencil-like columns of water gravitationally accelerated over a 3-D surface. The volume density of tsunami balls at any point equals the thickness of the water column there. Tsunami balls care little if they find themselves on parts of the surface originally under water or on parts originally dry. Such indifference is made-to-order for flood-like inundations.

2.1 Tsunami Ball flow chart.

The tsunami ball method distills to six flow chart steps.

A. Distribute N tsunami balls over initially wet areas at initial positions $\mathbf{r}_j$ with regular spacing $\Delta x \Delta y$.

B. Assign each ball a zero initial velocity and a volume $V_j = \Delta x \Delta y H(\mathbf{r}_j)$ based on the initial water thickness at the site $H(\mathbf{r}_j)$.

C. Evaluate the water surface $\zeta(\mathbf{r}_g, t)$ at time t on a fixed set of grid points $\mathbf{r}_g$ (*Figure 3*)

$$\begin{aligned}\zeta(\mathbf{r}_g,t) &= \sum_{j=1}^{N} V_j\, A(\mathbf{r}_g,\, \mathbf{r}_j(t)) + \mathrm{h}(\mathbf{r}_g) \\ &= H(\mathbf{r}_g,t) + \mathrm{h}(\mathbf{r}_g)\end{aligned} \tag{1}$$

In (1), V_j is the fixed water volume of the j-th ball, $\mathbf{r}_j(t)$ is the ball's location at time t, $h(\mathbf{r}_g)$ is topographic elevation and $A(\mathbf{r}_g,\mathbf{r}_j(t))$ is an averaging function with m^{-2} units such that when integrated over all (x,y) space

$$\int A(\mathbf{r},\mathbf{r}')d\mathbf{r}' = 1 \tag{2}$$

Operation (1) smoothes tsunami ball volume at arbitrary positions $\mathbf{r}_j(t)$ to grid positions $\mathbf{r}_g$.

D. Accelerate each tsunami ball for a short duration Δt using

$$\frac{\partial \mathbf{v}_j(t)}{\partial t} = -g\nabla_h \zeta_{smooth}(\mathbf{r}_g,t) - C_d \mathbf{v}_j(t)\left|\mathbf{v}_j(t)\right| \tag{3}$$

The $\mathbf{r}_g$ here is the grid point closest to the ball's position, ∇_h is the horizontal gradient and $\zeta_{smooth}(\mathbf{r}_g,t)$ is a version of (1) smoothed over time and space.

E. Update the ball position,

$$\mathbf{r}_j(t+\Delta t)=\mathbf{r}_j(t)+\mathbf{v}_j(t)\Delta t+\frac{\Delta t^2}{2}\frac{\partial \mathbf{v}_j(t)}{\partial t} \tag{4}$$

ball velocity,

$$\mathbf{v}_j(t+\Delta t)=\mathbf{v}_j(t)+\Delta t\frac{\partial \mathbf{v}_j(t)}{\partial t} \tag{5}$$

and time $t=t+\Delta t$.

F. Here you might return to Step (**D**) several times keeping the surface fixed at its most recent evaluation (inner loop); or, return to Step (**C**) to compute a fresh surface (outer loop). Regardless if the tsunami ball is in water or has been tossed onto land, steps (**C**) to (**F**) hold. Note that acceleration (3) includes a dynamic drag friction to control balls that might encounter unreasonably steep surface gradients and run away. I specify C_d in Section 4.

What in the flow chart makes the flood? Floods and waves get stirred if the underlying topography $h(\mathbf{r})$ in (1) becomes a function of time

$$h(\mathbf{r}) \Rightarrow h_0(\mathbf{r})+\Delta h(\mathbf{r},t) \tag{6}$$

where $h_o(\mathbf{r})$ is the original value in flow chart step (**B**) and $\Delta h(\mathbf{r},t)$ is the bottom uplift or subsidence due to a earthquake, landslide or in this case, failure of the South Fork dam.

2.2 Stabilization.

Numerical stability is the primary concern in any approach that employs granular material, particularly with respect to evaluating the gradient operator in (3) that drives the balls. Fluid simulations especially, because they have both flow-like and wave-like behaviors, can run amok rapidly. No single recipe guarantees stability, but I find that combinations of the three methods below usually can generate simulations that "look right".

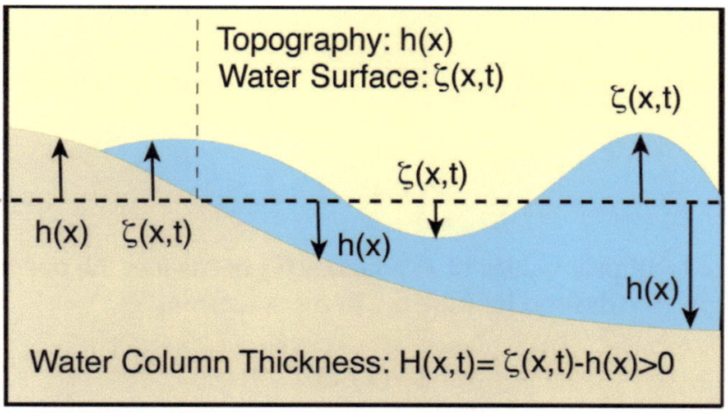

Fig. 3. Geometry for tsunami ball calculations. Topography h($\mathbf{r}$) and water surface $\zeta(\mathbf{r},t)$ are measured positive upward from some arbitrary level.

(a) Spatial Smoothing. Water surface $\zeta(\mathbf{r}_g,t) = H(\mathbf{r}_g,t) + h(\mathbf{r}_g)$ at grid points $\mathbf{r}_g$ is the sum of water thickness and topographic elevation. Both functions need to be spatially smoothed. The fastest means for this is rectangular smoothing, first applied to all grid rows and then to all grid columns. The q-th smoothed value in a grid row or column might be

$$\bar{G}_q = \frac{\Sigma_q}{Norm}; \quad \Sigma_q = \sum_{r=q-w}^{q+w} G_r \tag{7}$$

where G_r are raw values in the row or column, W is the half width of smoothing and Norm is some normalization. The q+1-th smoothed value is

$$\bar{G}_{q+1} = \frac{\Sigma_{q+1}}{Norm}; \quad \Sigma_{q+1} = \Sigma_q + G_{q+w+1} - G_{q-w} \tag{8}$$

You can see that each value (8) derives from the previous value (7) by only one addition, one subtraction and one division. The number of numerical operations in rectangular smoothing is proportional to the number of grid points only, and independent of both the number of balls and the smoothing half width. It takes experimentation to select an appropriate smoothing width. Smoothing width for the topography may not be the same as that for the water thickness. The goal is to stabilize the calculation, not to lose interesting aspects nor add many artifacts. I run the simulation on the smoothed topography, but display it on unsmoothed topography.

(b) Temporal Smoothing. Water thickness H($\mathbf{r}$,t) is also function of time, so temporal smoothing helps to damp waves versus flows. Smoothing any time series F(t) usually means storing many past values of the function and then averaging. If F(t) is also a function of space, storage may be overwhelming. One way to smooth in time without storing long-past values evaluates

$$F_{smooth}(t) = qF(t) + (1-q)F_{smooth}(t-\Delta t) \tag{9}$$

where q is less than one. For $t=N\Delta t$, it is easy to see that

$$F_{smooth}(N\Delta t) = \sum_{n=0}^{N} q(1-q)^n F([N-n]\Delta t) \tag{10}$$

and that the sum of the weights

$$\sum_{n=0}^{n\to\infty} q(1-q)^n = 1 \tag{11}$$

Evaluating (10) needs all past values of F, whereas (9) needs just the one current value. You choose q in (9) to give the desired half-life $t_{1/2}$ to the smoothing

$$q = 1-(1/2)^{\Delta t/t_{1/2}} \tag{12}$$

I employ both spatial and temporal smoothing to generate the water surface $\zeta_{smooth}(\mathbf{r}_g, t)$ for use in (3) and for display in the figures to follow.

(c) One Way Gravity. "One-way" gravity is yet another means to keeps accelerations (3) under control. One way gravity picks the value of g in equation (3) at each location and time based on the current velocity $\mathbf{v}_j(t)$ of the tsunami ball being accelerated and the gradient of the surface slope $\zeta_{smooth}(\mathbf{r}_g, t)$ at the nearest grid location

$$\begin{aligned} &-\nabla_h \zeta_{smooth}(\mathbf{r}_g, t) \bullet \mathbf{v}_j(t) < 0; \text{ then } g = 9.8 \text{ m/s} \\ &-\nabla_h \zeta_{smooth}(\mathbf{r}_g, t) \bullet \mathbf{v}_j(t) > 0; \text{ then } g = \varepsilon \times 9.8 \text{ m/s} \end{aligned} \tag{13a,b}$$

In words, if acceleration $-g\nabla_h \zeta_{smooth}(\mathbf{r}_g, t)$ opposes ball velocity $\mathbf{v}_j(t)$, then gravity acts in full force. If acceleration $-g\nabla_h \zeta_{smooth}(\mathbf{r}_g, t)$ reinforces velocity $\mathbf{v}_j(t)$, then fractional gravity ε x g, acts. If (13b) is persistently met over several time steps, the tsunami ball accelerates to terminal velocity soon enough, whereas reaction to short lived stimulus is less. Herein lies the temporal smoothing.

3. 1889 flood simulation: input information

As mentioned earlier, delving into the Johnstown Flood reveals lots of information -- much of it mistaken, dramatized, conflicting, or just not quantitative. I found certain post-event investigations and records of the Pennsylvania Railroad inquiry to be most helpful. Railroad engineers, conductors, station agents, yard masters, signal tower and telegraph operators were competent people accustomed to detailed observation and living according to schedules. In the course of their jobs, many had access to accurately set clocks.

South Fork Dam. While the center half of the South Fork Dam vanished in 1889, its two ends exist to this day largely unaltered from their original configuration. Engineering diagrams from a 1891 investigation fix the elevation of the eastern end of the original 1853 dam at 1613.34 (491.9 m). It is generally agreed that the center of the 1887 dam was rebuilt lower than the original one by several feet to accommodate two-way carriage traffic. (This decision reduced spillway capacity and largely contributed to the 1889 overtopping.) The 1891 engineering diagram places the center of the rebuilt dam at 1610.76 feet (491.1 m). Hours before failure, workers hastily ran a two foot high dirt bank across the top of the dam attempting to stem the overflow. At failure, water ran one or two feet over that bank. I put

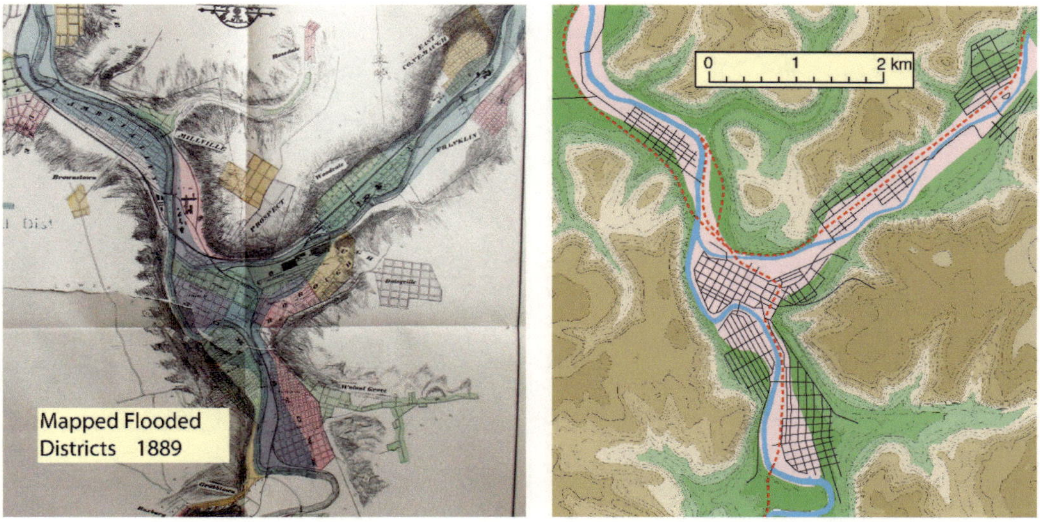

Fig. 4. (*Left*) Plate from Cadwell's 1890 Atlas of Cambria County. Bluish-green tint covers the "flooded districts" of the Conemaugh Valley. (*Right*) My translation of Cadwell's flood zone over today's topography and a 1900 street map. Ideally, the simulated flood will cover the pink area. Dashed red lines are railroads.

the final lake level at 1614 feet (492 m). Reportedly, the lake was rising about 1 foot per hour at the time of failure. Supposing that it takes the lake one and one-half or two hours to drain, inflows would add an equivalent 1-2 feet to lake level. For purposes here, I fix "Effective Lake Level" at 493 m. Adding a virtual dam to a modern Digital Elevation Map and refilling the basin to 493 m, reconstructs South Fork Lake containing 16 million cubic meters of water (*Figure 5*).

Timing- South Fork Arrival. Testimony by Emma Erhefield, telegraph operator at South Fork, specifically times the flood. It first reached the junction of South Fork Creek and the Little Conemaugh River (*Figure 5*) at "At 3 O'clock, probably a minute or two after". C. P. Dougherty, Railroad Agent at South Fork confirms this timing. He testified that he found the South Fork station clock reading 3:08, stopped when the flood knocked the place off its foundation a few minutes after first arrival.

Timing- Dam Failure. Timings of the dam failure vary, but both E. J. Unger and W. T. Sherman Showers testified that major water began to flow over the breast at 2:45 PM and that it was only minutes until the gap reached full size. My simulations require about 10 minutes for water to run the 3.6 km from the Dam to South Fork Junction. Given the firm 3 PM arrival time at South Fork, I place dam failure at 2:50 PM.—roughly 20 minutes earlier than often reported.

Timing- East Conemaugh Arrival. Telegraph operator Charles V. Haak was specific about the flood's arrival at East Conemaugh Railway Yard (*Figure 7a*). Watching the clock for a 4 PM shift change he says, "It came at exactly ten minutes of four." Going outside, he states that water jumped bank onto the railroad tracks a minute or two later. This timing is supported by Engineer V. Wierman who was walking the tracks about 2 km upstream of the Yard toward "AO" signal tower. He met the tower operator walking downstream on the track. Mr. Wierman recalls that the operator carried the AO station clock—stopped at 3:40 PM. The operator said that he had removed the clock just as flood took the tower into the river.

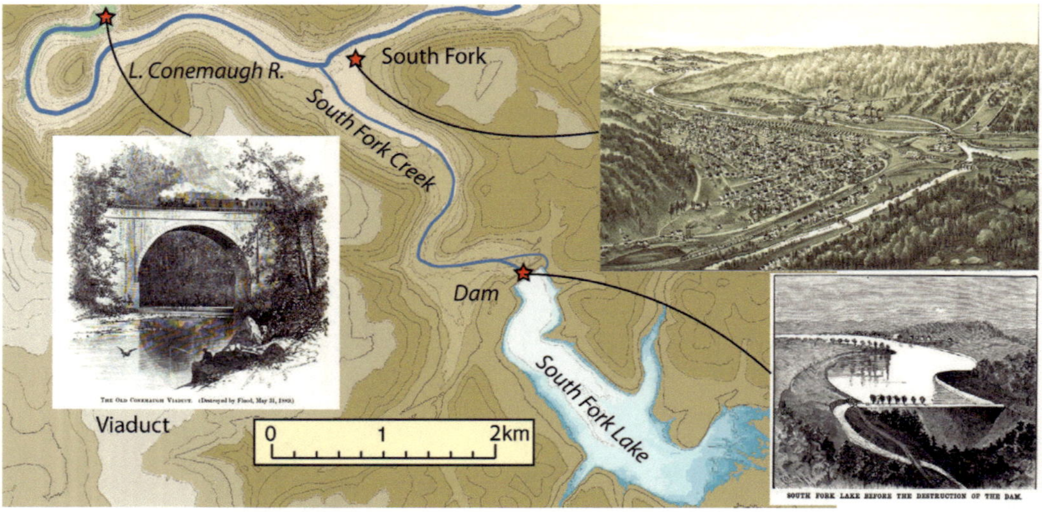

Fig. 5. Overview of the Little Conemaugh River near South Fork. The three historical prints show the Viaduct Arch, The South Fork Dam and Lake, and A panorama of South Fork Town Site looking up South Fork Creek toward the Lake. The star at South Fork marks the railroad station.

Timing- Johnstown and Stone Bridge Arrival. T. H. Watt, Johnstown ticket agent described conditions at his house which stood "a great deal lower than the main Johnstown station and directly opposite. The flood struck us at 4:07." Corroboration comes from District Supervisor W. M. Hays who was on a railroad work party at the Stone Bridge attempting to clear the arches—already beginning to clog with debris. (Booms at the lumber storage yard in Stoney Creek River above Kernville had broken earlier that day piling hundreds of logs under the Bridge.) "About 4:10 our attention was attracted by people shouting, and I saw this bank of water and drift coming down the Conemaugh, almost like a wall. About 4:10, it crossed the town before it reached our bridge."

Timing and Size of Causeway Breach. Period photographs show the causeway breach at the east end of the Stone Bridge repaired with temporary trestle. I measure the width of the breach to be well less than one-half the 170 m length of the bridge. Time of the causeway breach is indefinite, but "Somewhere past 5 PM". For this simulation, the Stone Bridge and Causeway act like dam until I introduce a 70 m wide causeway gap at 5:10 PM.

Timing- Lake Drainage and Flood Durations. Durations are elusive because while the flood initiated rapidly, it decayed away slowly. Rarely do accounts specify what defines the beginning and end of the interval. W. Y. Boyer, present at the dam, testified that it took "about an hour and ten minutes- for the lake to drain to creek bottom, but that others have different opinions." W. T. Sherman Showers states "about 2:45 the wearing away of the dam became more rapid, so that it was cut out quite fast and the water began to go through in great and still increasing quantities, so that by about four o'clock the main body of water had gone out." Placing full breakage at 2:50 PM, 4 o'clock equates to one hour and ten minutes for the Lake to drain, the same as Boyer's estimate. Because the tsunami balls follow longer and shorter paths in transit, certainly the duration of flood downstream must exceed the one hour ten minutes it took to pass the dam. Accounts that give a shorter flood duration must be suspect.

Fig. 6. Snapshots of the early phase of the flood simulation. The yellow backed clock gives elapsed time T= 00:00, 02:40, 06:00, 10:00 and 00:19:20. The red backed clock gives clock time T= 2:50, 2:52, 2:56, 3:00 and 3:19 PM. Note up-river counter flow at the junction of the South Fork and Little Conemaugh Rivers (Panel D) and the floodwater cutting across the oxbow in front of the Viaduct (Panel E). (Movie at http://es.ucsc.edu/~ward/jtown-lake-close-s.mov).

Existing conditions water in channel. It is agreed that the lower sections of Johnstown stood awash waist high and more before the flood's arrival. Elsewhere, witnesses described the rivers as "bank full". In the simulation, the Stone Bridge and Causeway act like a dam early on, so trapping additional tsunami balls (1.8×10^6 m^3) in river channels nearby submerges lower Johnstown under a meter or so of water. It is tougher to simulate bank full conditions in the upper reaches of the Little Conemaugh. Water placed there will simply run to Johnstown ahead and with the flood. While it is possible to start the simulation with tsunami balls in the upper channel, this simulation assumes it to be dry.

Inundation Zone. A. A. Cadwell (1890) was in the final stages of assembling his "Atlas of Cambria Country, Pennsylvania" when the Flood struck. He quickly commissioned several existing Atlas' plates to be revised with hand coloring covering flooded districts of the Conemaugh Valley (*Figure 4, left*). Figure 4 (*right*) shows my interpretation of the flooded zone overlain on today's topography and a street map of 1900 vintage. *Cadwell* likely mapped the flood damage zone rather than the inundation zone. Too, I find that some edges of his zone cut oddly across topographic contours. Still, *Cadwell's* plates represent the best information available to bound flood extent.

4. 1889 flood simulation:

4.1 DEM and dam failure conditions.

I started this exercise by extracting a 1/9 arcsec (~3 m spacing) DEM from the USGS National Elevation Dataset and then down sampling to every third point (~10 m spacing). For the 1889 flood simulation, 1 million tsunami balls filled a reconstructed South Fork Lake. Behind a virtual dam of 493 m elevation sat 16 x10^6 m^3 of water. By all accounts, the dam failed within minutes after first breach. A few test runs suggest that whether dam failure happened instantly or over a few minutes does not alter significantly the flood history down stream. For my purposes, there is no need for additional model assumptions regarding timing of failure. There is also no need for additional model assumptions regarding the geometry of the breach because the immediate landscape has not been altered significantly since 1889. The simulations assume that the virtual dam instantly reverts to remnant dam's current shape. After releasing the tsunami balls, I follow the flow chart in Section 2, and update their acceleration, position and velocity every Δt=0.1s (inner loop), reform a new surface every 4s (outer loop) and generate a movie frame at 40s intervals.

4.2 Dynamic drag

The only remaining physical element to be fixed is the dynamic drag C_d coefficient. C_d imparts a terminal velocity to overland flow of

$$v_{term}(\mathbf{r}) = \sqrt{\frac{g\left|\nabla_h \zeta_{smooth}(\mathbf{r})\right|}{C_d}} \tag{14}$$

so the flood timings established in Section 3 heavily influence its selection. With C_d too small, the flood arrives early. With C_d too big, the flood arrives late.

Another flow related aspect, is that by all accounts the flood onset rapidly in step-like pulses or waves. With water's natural tendency to spread out, how could it maintain a steep onset after an hour or more of travel? People speculated almost immediately that the flow dammed and re-dammed itself several times with transported debris and fallen obstacles (like Bridge #6, ~500 m east of AO tower). Each short-lived blockage, presumably trapped water behind and sent forward a pulse when it broke. While such things are possible, it is hard to say if multiple dams could survive long enough to have much effect or how much volume the blockages could retain to release as pulses.

Another means to maintain steep onsets without multiple blockages lets C_d depend upon the flow thickness $H(\mathbf{r},t)$. If C_d increases with $H(\mathbf{r},t)$, thin flows that outrace the main body suffer more drag and lower terminal velocity (14). Thin vanguards will be forced to slow and wait for the main body. Thickness-dependent friction keeps the flow together and its onset steep. After several trial and error runs, the following form of $C_d(\mathbf{r},t)$ seemed to match all the information:

$C_d(\mathbf{r},t)=0.0045/\text{m}$ if $H(\mathbf{r},t)<2\text{m}$,
$C_d(\mathbf{r},t)=0.0022/\text{m}$ if $2\text{m}<H(\mathbf{r},t)<3\text{m}$, (15)
$C_d(\mathbf{r},t)=0.0011/\text{m}$ if $3\text{m}<H(\mathbf{r},t)<4\text{m}$ and
$C_d(\mathbf{r},t)=0.0001/\text{m}$ if $H(\mathbf{r},t)>4$ m.

To get a feeling for these values, let's compute representative terminal velocities. From South Fork Lake to Johnstown, floodwater dropped 140 m over a 24 km run, so $\left|\nabla_h \zeta_{smooth}(\mathbf{r})\right| \sim 0.00583$ and average terminal velocities (14) would be

v_{term}= 3.6 m/s if H(**r**,t)<2m,
v_{term} =5.1 m/s if 2m<H(**r**,t)<3m, (16)
v_{term} =7.2 m/s if 3m<H(**r**,t)<4m and
v_{term} =23.9 m/s if H(**r**,t)>4 m.

Naturally v_{term} would be higher than (16) on the steeper parts of the water course near the Dam, and less than (16) on the flatter sections near Johnstown. Be aware that the velocity of flood advance v_{flood} generally lags v_{term}. Rarely does $\nabla_h \zeta_{smooth}(\mathbf{r})$ point directly down river. Too, in a contorted channel, the tsunami balls bounce back and forth between the walls slowing the advance of the flood. Lower v_{term} can actually increase v_{flood} because slower speeds reduce the bouncing and make for a shorter flow path.

4.3 Output products.

Figure 6 shows the initial phases of the 1889 flood in 2D map view at T= 00:00:00, 00:04:00 00:12:00, and 00:19:20. (Quicktime movie animations accompany all of the simulations presented in this article. See Appendix A. View the movie of Figure 6 at http://es.ucsc.edu/~ward/jtown-lake-close-s.mov). In Figure 6 and those that follow, the circled numbers measure peak flow depth to that time in meters anywhere along a vertical line passing through the circle. The small numbers in the thin white boxes list peak flow depth to that time at that position. The numbers in the thin yellow boxes list two-minute average flow rates past that location. Thin lines in the flow give time averaged ($t_{1/2}$=80s) speed and direction.

⇒*T: 0:00-3:00 (2:50–2:53PM)* (*Figure 6: Frames A & B*): Two minute average flow through the dam gap quickly increases to a peak of about 12,000 m^3/s. Multiple bends in the channel hold the flood advance to 5-6 m/s.

⇒*T: 3:00-6:00 (2:53-2:56PM)* (*Figure 6: Frame C*): Flow rates through the breach decrease to 7000 m^3/s. $3x10^6$ m^3 has already passed the dam. Peak flow depth at the second bend below the dam hits 9.2 m. Water has now traversed two-thirds the distance to South Fork. Flow speeds increase to about 7 m/s.

⇒*T: 6:00–12:00 (2:56-3:02PM)* (*Figure 6: Frame D*): After eight minutes, water reaches South Fork Junction 3.6 km below the dam, hits the hill on the north bank and splits into east and west streams. The westward one continues down river toward the Viaduct. The counter flow pushes eastward up the Little Conemaugh River toward South Fork Station. Mean velocity of flood advance to now is 3590m/(8*60s)= 7.5 m/s. By T=12:00, two minute average flow rates through the breach slacken to 5000 m^3/s. More than ¼ of the lake has drained.

⇒T: *12:00-30:00 (3:02-3:20PM)* (*Figure 6: Frame E*): The counter flow reaches its maximum up stream extent (900 m east of the river junction) putting 6.4 m of water in the channel behind South Fork Station, 400 m east of the junction. Nineteen minutes after the breach, part of the flow cuts the oxbow narrows and fills the pool in front of the Viaduct. (It is said that such a division of flow did occur.) The majority of the flood runs around oxbow and reaches the Viaduct in 20-22 minutes. Mean velocity of flood advance from the Dam to the Viaduct is 8390m/(22*60s)= 6.4 m/s. In the simulation, the Viaduct acts like a barrier until T=30:00 when it fails completely (see next section). Of the $16.0x10^6$ m^3 of water originally in South Fork Lake, $7.0x10^6$ m^3 and $9.2x10^6$ m^3 have escaped by 20 and 30 minutes respectively.

Figure 7 maps the flood as it passes "AO" Tower, E. Conemaugh and Woodale.

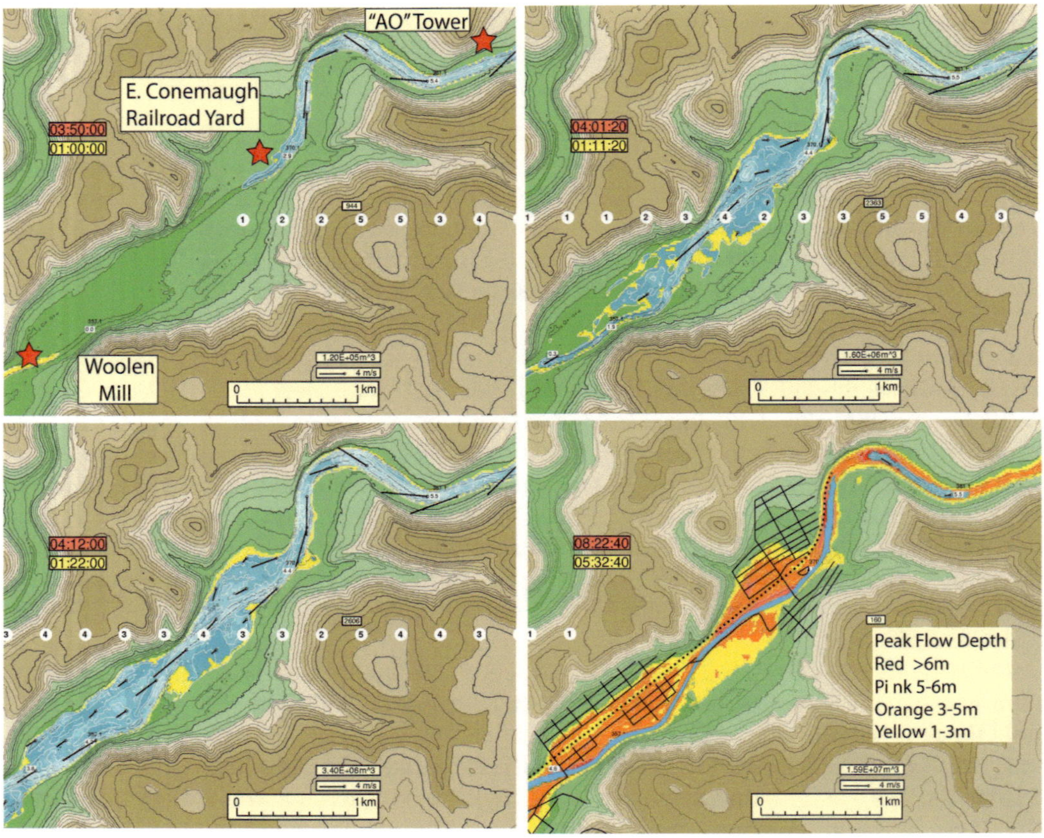

Fig. 7. Snapshots of the flood simulation through E. Conemaugh and Woodvale at T= 1:00:00, 1:11:20, 1:22:00 and 5:32:40 (3:50, 4:01, 4:12 and 8:22PM). Panel A locates "AO" signal tower, the E. Conemaugh Rail Yard and the Woolen Mill at the south end of Woodvale. Panel D shows the final extent and depth of flooding and a 1900 vintage street map. Note that the flood expands to cover the full width of the valley. Compare Panel D with Figure 4, right. (Movie at http://es.ucsc.edu/~ward/conemaugh-close-s.mov).

⇒*T: 50:00-60:00 (3:40–3:50PM)* (*Figure 7: Frame A*): The flood passes AO tower and arrives at E. Conemaugh rail yard. The simulation gives 5.5 m and 4.7 m of peak flow in the channel at the two locations.

⇒*T: 1:00:00-1:10:00 (3:50–4:00PM)* (*Figure 7: Frame B*): Upon rounding the bend above the rail yard, the flow jumps the north and south banks into the streets of E. Conemaugh and Franklin respectively. From there southward, water fills the whole 500 m width of the valley. Two minute average flow rates passing the rail yard exceed 3000 m^3/s. To put this flow into perspective, consider information from the U.S. National Weather Service who has operated a river gauge close to the E. Conemaugh rail yard since the 1930's. Their web site (http://www.erh. noaa.gov/ctp/hydro) calculates that the Little Conemaugh River suffers major double bank flooding here when flow hits 950 m^3/s. Only once during the gauge's existence did the river run this high, in 1936, during the Valley's worst flood of the 20-th century. According to the simulation, flow rate passing the rail yard in 1889 exceeded by three times that experienced in 1936. By 4:00 PM, one hour and ten minutes after the breach, 85% of the reservoir has drained. I take this stage to be "down to creek bottom" as described by W. Y. Boyer.

⇒*T: 1:10:00-1:20:00 (4:00 – 4:10PM)* (*Figure 7: Frame C*): The valley-wide flow tears through Woodvale, wiping most of the town clean. Only the brick Woolen Mill at the south end of town survived. The simulation puts 5.5 m of water in the channel there and 4.4 m at the building site. Because the lower Valley slopes just slightly compared to its upper reaches, velocity of flood advance slows to 3.6 m/s.

⇒*T: 1:30:00-past 5:00:00 (4:20-7:50PM)* (*Figure 7: Frame D*): About 2 hours after it arrived, the flood fully passed Woodvale and moved down stream. Sometime during that interval, the whole valley had been covered with 3-5 meters of water (orange color, Panel D, Figure 7). In the simulation, water reached to the third street of E. Conemaugh and it took out all but the two highest streets in Woodvale. In Panel D you can clearly the see the raised railroad bed (yellow stripe) running up from Johnstown. Flooding on rail beds was 2-3 m less than on spots adjacent. Raised rail beds helped save a few parked trains. Still, virtually all of the track and rolling stock from the Railway Yard to Johnstown Station washed away—many passengers included.

Figure 8 pictures the flood as it rolls into Johnstown and then Cambia.

⇒T: *1:00:00-1:15:00 (3:50-4:05PM)* (*Figure 8: Frame A*): The flood passes below the Main Station and into Johnstown, already under 1- 2 m of water. Although the mean velocity of flood advance from the Dam to the Woolen Mill is 5.3 m/s (22.2km/70*60s), water entered town more slowly, 2 to 3 m/s. I imagine that structural damage and causalities would have been worse still had the water maintained its previous 5 or 7 m/s speeds.

⇒T: *1:15:00-1:30:00 (4:05-4:20PM)* (*Figure 8: Frame B*): The flow splits again, one stays in the channel running below the Main Station toward the Stone Bridge. The other cuts directly across town and down Clinton Street toward Stoney Creek. Upper Clinton Street tracks a reclaimed section of the old portage railroad canal basin. The street occupies a slight linear depression and offered an attractive path for the flood.

⇒T: *1:30:00–2:00:00 (4:20-4:50PM)* (*Figure 8: Frame C*): Trapped behind the Stone Bridge now acting like a dam, floodwater forms "Lake Johnstown". Eventually the lake submerges everything at elevation less than 361 meters - a bit higher than the level of the bridge rails. Velocity arrows in Frame C show the "counter flow" as water backs up Stoney Creek River. This counter flow heavily damaged Kernville on the west bank; however, it saved many floating houses (and fortunate occupants) by directing them upstream away from sure destruction at the Stone Bridge. Many of the buildings swept down the valley came to rest in Kernville and were later relocated back to their original locations. Some eloped houses are still in use today.

⇒T: *2:00:00–4:00:00 (4:50-6:50PM)* (*Figure 8: Frame D*): At T=2:20:00 (5:10 PM) the simulation extracts a 70 m wide piece from the causeway at the east end of the Stone Bridge. Lake Johnstown drains, but progress is slow because considerable volume (>1000 m^3/s) continues to arrive down the Little Conemaugh. The town remained several meters submerged when the simulation stopped at 8:20 PM. Many hours more pass before water vacates the streets. The 70 m wide causeway break dumped far more water into the downstream channel than it could hold. The excess discharged through the large buildings of Cambria Iron Works on the east bank and the homes of Cambria town on the west bank.

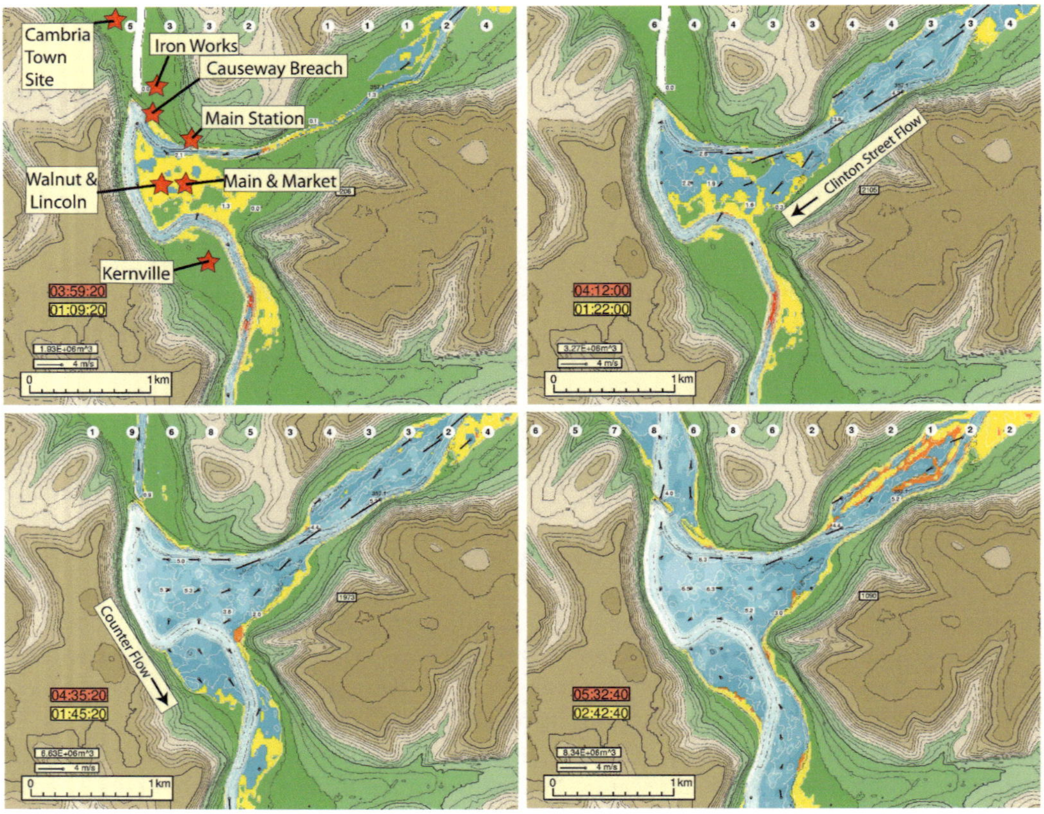

Fig. 8. 2D view of the flood at Johnstown at T= 1:09:20 (3:50 PM), 1:22:00 (4:12 PM), 1:48:00 (4:38 PM) and 2:36:40 (5:26 PM). Note the back current in Panel C pushing water up Stoney Creek and washing Kernville on its west bank. After the causeway failure at 5:10 PM, water rushing through the breach floods the Iron Works and Cambria town site (Panel D). (Movie at http://es.ucsc.edu/~ward/jtown-close-s.mov).

5. Assessment of the 1889 flood simulation:

Having run the 1889 flood simulation, one asks -- How well did it do? Did it match observations?

<u>Duration of Lake Drainage.</u> The red line in the bottom panel of Figure 9 graphs volume past the dam versus time. About 50% of the lake volume evacuated in 25 minutes and 85% in 70 minutes. Earlier I recounted firm testimony that the lake drained "to creek bottom" in 1 hour-ten minutes. Simulation and observation coincide here.

<u>Timing and Flood Heights at South Fork.</u> The simulated flood arrives at South Fork at 2:58PM (*Figure 6, Frame C*) and fully pushes upstream behind the South Fork Station by 3:09 PM (*Figure 6, Frame D*). These times coincide with the telegraph operator's statement and the 3:08 time of the flood-stopped station clock.

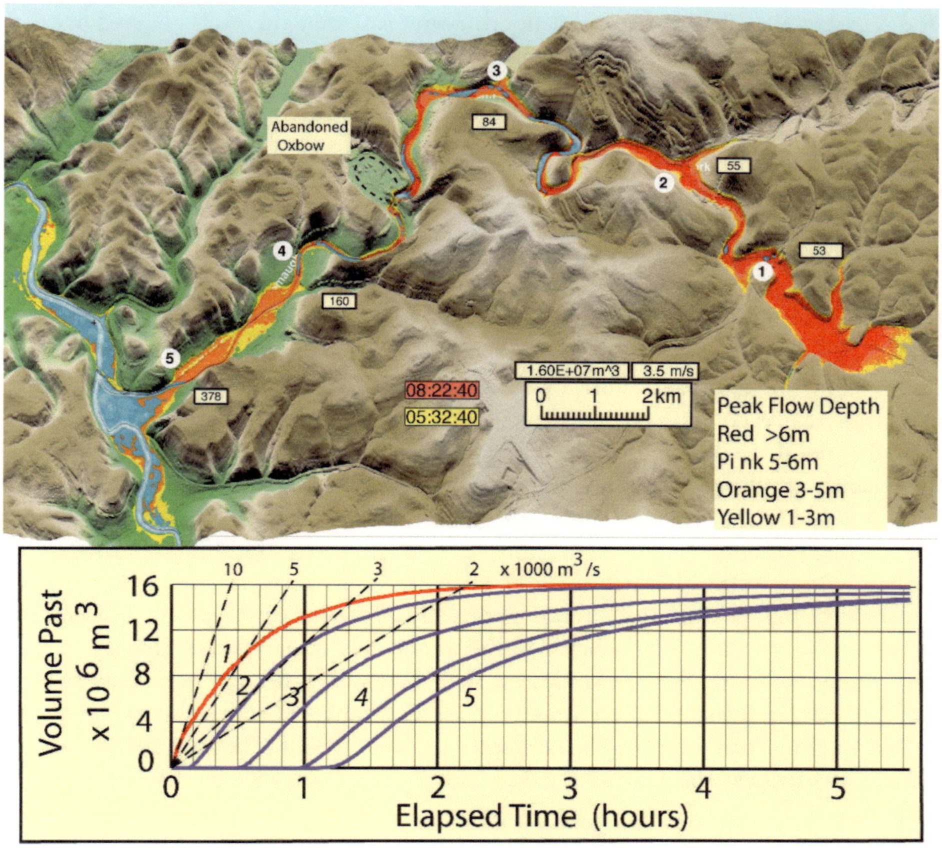

Fig. 9. (*top*) 3D-like view of the flood at 8:22 PM. The Lake had long drained by now and the Little Conemaugh River returned to pre-flood conditions. Water still covers much of Johnstown and will do so for several more hours. (*bottom*) Total water volume versus time past five locations in the valley marked with white circles: 1. The dam, 2. South Fork, 3. Mineral Point, 4. E. Conemaugh, 5. E- Johnstown. Slopes of these curves reflect average flow rate. Dashed lines mark 10,000, 5,000, 3,000 and 2,000 m^3/s. Nominal flow over Niagara Falls is 2,800 m^3/s. The 1936 flood rate was 950 m^3/s at E. Conemaugh. Note the now-abandoned oxbow between Mineral Point and E. Conemaugh. The river currently runs through a large cut, parallel to the railroad cut extant in 1889.

C. P. Dougherty further testifies that the counter flow at South Fork went ½ mile (800 m) up stream from the junction and that engineers later measured a 23 foot (7 m) increase in water level due to the dam break alone – high water already in the river being accounted for. No conflict here. In the model, peak flow depth reached 6.4 m in the riverbed behind the station and the counter flow stretched 900 m up stream from the junction.

Timing and Flood Heights at the Viaduct. One of the stories dealing with the 1889 flood concerns the Stone Viaduct located at the oxbow below South Fork. The Viaduct was a massive stone structure built in 1853 as part of the portage railroad (*Figure 5*). It stood 22 meters above river bed atop a single 80 foot diameter arch. All that we know for certain is that the Viaduct vanished during the flood. Some say that the arch clogged with debris when the flood cut across the oxbow (*Figure 6, Panel E*). Water then dammed behind, flowed over top by several feet, and eventually knocked down or undermined the structure. I have not discovered any definitive accounts to support or quantify the Viaduct story. Others suggest that, like the Stone Bridge, the Viaduct was far too massive to succumb to flood alone. Dynamite, they say, packed in a railroad boxcar washed down from South Fork struck and blew up the structure.

To interpret the Viaduct story from a physics-based perspective, first consider water volume and timing. From the modern Digital Elevation Map, I find that damming the river at the Viaduct to the 444 m elevation of the railroad bed would impound 4.4 million cubic meters of water. Raising a temporary lake several feet over the tracks (to 446 m elevation) requires 5.6 million cubic meters. South Fork Lake held considerably more water than this, so if the Viaduct arch really could have dammed tight, and if the structure held in place long enough, the flood could have overtopped it. Timings remain an issue however, "Can sufficient water reach the Viaduct rapidly enough so that the well documented arrivals of the flood downstream can still be met?" Figure 9 (*bottom*) says that $4x10^6$ m^3 of water passed South Fork in 25 minutes and $6x10^6$ m^3 million in 34 minutes. Flood travel time from there, around the oxbow to the Viaduct is about 12 minutes. So, four million cubic meters could back up behind the Viaduct in 37 minutes or (3:27 PM) and 6 million by 46 minutes (3:36 PM). These volumes would be achieved a bit sooner if I added water already in the river channel to that of South Fork Lake. Recall that the flood must reach E. Conemaugh Rail Yard by 3:50 PM. In the simulation it takes about 25 minutes for the flood to travel from the Viaduct to the Rail Yard. Although the flood has a very tight schedule, conceivably water could have overtopped the Viaduct by 3:25-3:30 PM and still met the timing requirements down stream.

Whether water really overtopped the Viaduct and whether flood or dynamite caused its demise, will take further forensic investigation. In any case, the simulation breaks the Viaduct at T=30:00 (3:20PM) after $5.06x10^6$ m^3 have passed South Fork on the way to the short-lived lake.

Timing and Flood Extent at E. Conemaugh and Woodvale. The simulated flood reached "AO" tower and the Rail Yard at 3:40 and 3:50 PM as the accounts state. Granted, I adjusted the dynamic drag coefficient C_d and Viaduct failure time to generally meet these constraints.

It might be noted that in 1889, the Little Conemaugh River circled a large oxbow between Mineral Point and the Rail Yard (*Figure 9*). The oxbow is abandoned now and the river course lost to mining operations. The modern river runs through a man made cut just west of the railroad cut extant in 1889. Eyewitnesses state that 1889 flood flowed many feet deep through the railroad cut and helped knock out Bridge #6 on the west side. It seems that much of the 1889 water bypassed the oxbow and followed a path not too different from the current one. Although the oxbow path is not included in the simulation, I do not believe that this substantially alters flood circumstances down stream. Still, a reconstruction of the old oxbow might be worth considering.

The simulation puts 3 to 5 m of water over the Rail Yard (*Figure 7, Panel D*). I have not discovered any definitive measurements of flood heights there for comparison. References to water marks on surviving locomotives do exist but, railroad beds stand a few to several meters higher than surroundings. The Yard's multiple parallel sidings of various elevations further complicate the interpretation of locomotive watermarks. Yard managers in fact did everything possible to park delayed trains as far up and away from the river as possible. Adjacent trains, even adjacent cars, fared better or worse depending on location.

Although the depth of flooding here was uncertain, the extent was well indicated. Both *Cadwell's* Atlas and Railroad Testimony agree that flooding reached the third street in Conemaugh. The simulation (*Figure 7, Panel D*) finds no disparity.

Timing and Flood Extent at Johnstown. The simulated flood passes the Main Station at 4:08 PM and reaches the Stone Bridge at 4:11 PM, in accord with the accounts.

The models puts 3-7 m water over the town, nearly replicating the pattern of depths expected in a static "Lake Johnstown" filled to 361 m (*Figure 10*). Few quantitative observations of water depth in the city exist. Most of the structures upon which one might measure it were destroyed. City Hall at Main and Market provides one definitive datum - a plaque marking high water, 6.4 m above grade. Nearby at Walnut and Lincoln, photographs show a clear high water mark on a surviving house. Judging from the size of a person standing in the photo, I fix flood height there at 6.7 m. The simulation predicts 6.1 m and 6.5 m at these two places - close enough. Further investigation should uncover other quantitative flood heights in the city and better test the "Lake Johnstown" hypotheses.

I see no glaring mismatch between the predicted inundation zone in the city (*Figure 10*) and the flooded areas mapped in *Cadwell's* Atlas (*Figure 4*). In Kernville and South Johnstown the simulated flood does extend a few streets beyond the mapped zone. I note that *Cadwell's* inundation limit at these places runs oblique to, rather than parallel with topographic contours. Inundation need not parallel topography in high-speed flows, but flooding oblique to topography is difficult to reconcile in this situation where water moves very slowly. Possibly, the artists painting the Atlas' plates were a little interpretative. Also as I mentioned, *Cadwell's* mapping likely speaks to the damage zone rather than inundation zone, the latter being more extensive. At the edges of a flood, water stands in the streets but inflicts no damage.

Timing and Flood Extent at Cambria. The timing of causeway failure dictates the onset of the simulated flood at Cambria Iron Works and Cambria Town. Time of causeway failure (5:10 PM) is a kinematic input parameter. I do not have any independent timings of the flood in Cambria, but there may be quantitative information from witnesses further down river that could be brought to bear.

Comparing inundation zones in Figure 10 with Figure 4, I see that the simulation puts a considerably wider flood track over the Iron Works than did *Cadwell*. Compared with other locations in the Valley, the Iron Works was rather featureless (few roads) and its steel construction distinct from and perhaps stronger than the wood frame construction elsewhere. I do not know how well the Atlas mapped inundation in the industrial zone. In Cambria Town, the residential area across the river, the simulated flood extent agrees with the Atlas within a street or two.

Overall Assessment. In my opinion, the Johnstown Flood simulation successfully reproduces the bulk of credible historical information to the extent that that information is quantified. Additional facts affirming or contradicting its predictions will certainly come to light, and the model may need modification. Regardless, the exercise demonstrates that realistic physics-based simulations of long-run river floods are achievable with modest laptop-computer resources. When applied in an historical context, physics-based models add new dimensions and flavors to "storytelling" accounts and they help contemporary folk better understand and appreciate the drama of that day.

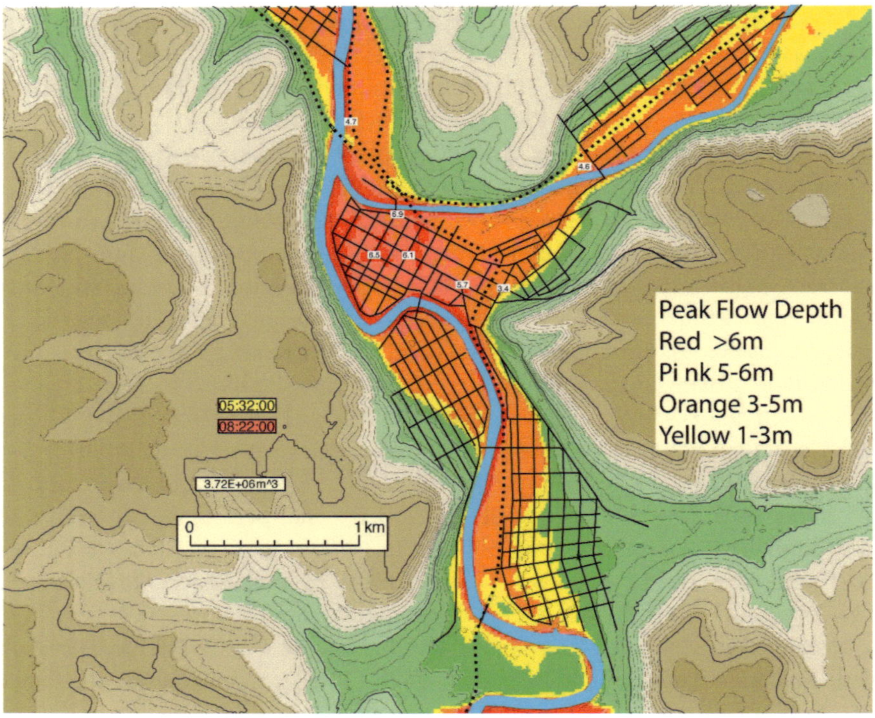

Fig. 10. Map of inundation zone and peak flow depth flood extent at T=5:32:00 (8:22PM). Compare with the observed inundation extent in Figure 4 (*left*).

6. Conclusions

The May 31, 1889 Johnstown Flood was a tragic engineering and societal failure. I approach the story for the first time from a science-based perspective. By combining text and photographic information to constrain dynamic and kinematic parameters, the simulations successfully capture in time and space the full sequence of events including: the breaking of and initial high flow rate through the dam, the down draw of the lake, the counter flow flooding at South Fork, the overtopping of the Viaduct, the valley-filling expansion of the flood at E. Conemaugh, the creation of "Lake Johnstown" behind the Stone Bridge, the back flow up Stoney Creek River, the failure of the causeway, the flooding of Cambria and the eventual passage of the flood.

Being more than artist's conceptions, physics-based simulations can accept input from a wide range of observational information and they can output quantitative predictions on many aspects that could not be measured. For instance, the simulation estimates that the

peak flow past East Conemaugh in 1889 surpassed by three times, that of the 1936 flood, the greatest of the 20-century. As by-products, physics-based simulations build support for, or cast shadows over, anecdotal accounts such as the overtopping of the Viaduct.
Given the speed, flexibility and modest computational requirements of the tsunami ball approach, the method should continue to find application in dam break and flood situations. Perhaps the technique might help to mitigate losses in future Johnstown style disasters.

7. Acknowledgement

Much of the historical information quoted in this article came from websites hosted by *The National Park Service Johnstown Flood National Memorial*:
http://www.nps.gov/archive/jofl/home.htm
and the *Johnstown Flood Museum*
http://www.jaha.org/FloodMuseum/oklahoma.html

8. References

Beale, Reverend D., 1890. *Through the Johnstown Flood.* Hubbard Brothers Publishing Co., Philadelphia.

Caldwell, A. A., 1890. Atlas of Cambria County Pennsylvania. Atlas Publishing Co.

Johnson, W. F., 1889. *History of the Johnstown Flood.* Edgewood Publishing Co., Philadelphia.

McCullough, D. G., 1968. *The Johnstown Flood.* Simon and Schuster Publishing Co., New York.

Ward, S. N. and S. Day, 2010a. The 1958 Lituya Bay Landslide and Tsunami - A tsunami ball Approach. *J. Earthquake and Tsunami, 4*, p1-35, DOI:10.1142/S1793431110000893, *http://es.ucsc.edu/~ward/papers/lituya-R.pdf*

Ward, S. N. and S. Day, 2010b. The 1963 Viaont Landslide and Tsunami - A tsunami ball Approach. Italian Journal of Geosciences, Vol 129, No. 3, in press. http://es.ucsc.edu/~ward/papers/Vaiont.pdf

Appendix A.

Quicktime movie links to the simulations presented in this article.

Composite with Text and Photos (Figure 1)
http://www.youtube.com/watch?v=tMc9kP9q-d8

Lake View Close (Figure 6):
http://es.ucsc.edu/~ward/jtown-lake-close-s.mov

East Conemaugh View Close (Figure 7):
http://es.ucsc.edu/~ward/conemaugh-close-s.mov

Johnstown View Close (Figure 8):
http://es.ucsc.edu/~ward/jtown-close-s.mov

Flood in Map View (Figure 9-10):
http://es.ucsc.edu/~ward/jtown-map-s.mov

Part 3

Technology

22

Land Cover Mapping and Change Detection Analysis over Nanggroe Aceh (NAD) Province after Tsunami

K. C. Tan, H. S. Lim, M. Z. MatJafri, and K. Abdullah
School of Physics, Universiti Sains Malaysia
11800 USM, Penang,
Malaysia

1. Introduction

Land cover as the basic parameter to evaluate the content of Earth surface, are important factors that affect ecosystem condition and function (Lunetta et al., 2006). In addition, it is also as a biophysical state of the Earth, which can be used to estimate the interaction in form of biodiversity, biosphere-atmospheric and geosphere-atmospheric interactions. Hence, land cover analysis plays an important role in many environmental applications nowadays. There has been growing interest in the use of remote sensing systems for a regular monitoring of the Earth's surface. The availability of remote sensing data applicable for global, regional and local environment monitoring has greatly increased over recent years (Ehlers et al., 2003). Availability of accurate and up-to-date land cover information is central to much resource management, planning and monitoring program (Kasetkasem at al., 2005). Remote sensing is an attractive source to produce thematic maps such as those depicting land cover as it provides a map-like representation of the Earth's surface that is spatially continuous, as well as available at a range of spatial and temporal scales. Thematic mapping from remotely sensed data is typically based on an image classification (Foody, 2002). Satellite remote sensing methods, which can in principle, be used to monitor very large areas in a reasonably short period of time, have notable potential for monitoring this high-latitude vegetation, although the techniques are in general less well established than at lower latitudes (Rees et al., 2003). Information on land-cover status at the regional scale is needed for natural resource management, carbon cycle studies, and modeling of biogeochemistry, hydrology, and climate. Satellite-based remote sensing products can meet these data needs in a timely and consistent manner (Boles, et al., 2004). Besides that, change detection analysis becomes extremely important in order to understand the interactions and relationships between natural phenomena and human, through the accurately change detection of Earth's surface features (Lu et al, 2003). Usually, change detection analysis involved utilization of multi-temporal satellite images to determine the temporal effects for every phenomenon in quantitatively. One of the aspect of change detection applications using remote sensing technique, such as land-use and land-cover changes. Many researchers used remotely sensed images in their land use and land-cover studies (Friedl et al., 2002). This is due to multi-temporal maps of land-use and land-cover changes are useful in identifying temporal changes, which have implications for local, regional or global environmental changes. Furthermore, abrupt changes in surface content may be due to the natural environment or anthropogenic activities (Fernandes et al, 2004). All of these may be interrelated. From the study done on Nanggroe Aceh Darussalam (NAD), the study focuses on the land-cover mapping and change detection analysis to identify the coastal change before and after the Tsunami effect. Using remote sensing data to detect coastal line change requires high spatial resolution data. In this study, two high spatial data with 30 meter resolution of Landsat TM images captured before and after the Tsunami event were acquired for this purpose. The two satellite images was overlain and compared with pre-Tsunami imagery and with after Tsunami. The two Landsat TM images also were used to generate land cover classification maps for the 24 December 2004 and 27 March 2005, before and after the Tsunami event respectively. The standard supervised classifier was performed to the satellite images such as the Maximum Likelihood, Minimum Distance-to-Mean and Parallelepiped. High overall accuracy (>80%) and Kappa coefficient (>0.80) was achieved by the Maximum Likelihood classifier in this study. Estimation of the damage areas between the two dated was estimated from the different between the two classified land-cover maps. Visible damage could be seen in either before and after image pair. The visible damage land areas were determined and draw out using the polygon tool included in the PCI Geomatica image processing software. The final set of polygons containing the major changes in the coastal line. An overview of the coastal line changes using Landsat TM images is also presented in this study. This study provided useful information that helps local decision makers make better plan and land management choices.

2. Change detection analysis

Change detection is the process of identifying differences in the state of an object or phenomenon by observing it at different time (Singh, 1989). Processing of multi-temporal images and change detection techniques has been developed in the past three decades. Change information of the Earth's surface is becoming more and more important in monitoring the local, regional and global environment. There are a lot of successful studies found from many research in application cases, especially on the detecting and monitoring environmental change.

The remote sensing imagery from previously till presently can provide valuable data through timely and accurate change detection of Earth's surface features, by providing the basic for better understanding the relationships between human and natural phenomena to better manage and use resources (Lu et al., 2003). Basically, the different algorithms can be grouped into the following categories: algebra (differencing, rationing, and regression), change vector analysis, transformation (e.g. principal component analysis, multivariate alteration detection, Chi-square transformation), classification (post-classification comparison, unsupervised change detection, expectation-maximization algorithm) and hybrid methods. Since monitoring change of Earth's surface features is very important, research of change detection techniques is an active topic, and new techniques are steadily developed.

The purpose of conducting change detection is to compare spatial representation of two points in time by measuring changes caused by differences in the variables of interest and controlling all variables caused by differences in variables that are not of interest (Green et al., 1994). Normally, change detection involves the utilization of multi-temporal satellite imagery to quantitatively analyze the temporal effects of the phenomenon. Therefore, the available remotely sensed data, such as radar, Advanced Very High Resolution Radiometer (AVHRR), Thematic Mapper (TM), and Satellite Probatoire d'Observation de la Terre (SPOT), play an important role for different change detection applications. Ten aspects of change detection applications using remote sensing technologies are summarized:

1. Land use and land cover (LULC) change (Lunetta et al., 2002; Read and Lam, 2002; Weng, 2002)
2. forest or vegetation change (Woodcock et al., 2001; Lu et al., 2002)
3. forest mortality, defoliation and damage assessment (Rigina et al., 1999)
4. deforestation, regeneration and selective logging (Asner et al., 2002; Wilson and Sader, 2002)
5. wetland change (Munyati, 2000)
6. forest fire (Bourgeau-Chavez et al., 2002)
7. landscape change (Taylor et al., 2000)
8. urban change (Zhang et al., 2002)
9. environmental change (Kimura and Yamaguchi, 2000)
10. other application such as crop monitoring (Engeset et al., 2002)

A good and successful change detection research should provide the following information: (1) accuracy assessment of change detection results; (2) area change and change rate; (3) spatial distribution of changed types; and (4) change trajectories of land cover types (Lu et al., 2003). In addition, there are four major steps are involved when implementing a change detection analysis: (1) precise radiometric and atmospheric calibration or normalization between multi-temporal images; (2) precise registration of multi-temporal images; (3) selection of the same spatial and spectral resolution images if possible; and (4) similar phonological states between multi-temporal images.

One important area where remote sensing plays a key role in change detection analysis is the study of land cover change. The study on land cover change is an important finding in the Earth Science domain because of its impacts on local climate, biogeochemistry, hydrology, radiation balance, and the diversity and abundance of terrestrial species (Mas, 1999). Definitely, the conversion of natural land cover types to human-induced cover types continues to be a change and bring an environmental problem for local, regional or global scales (Weng, 2002). The land cover change detection problem is essentially one of the detecting when the land cover at a given location has been converted from one type to another.

3. Land cover mapping

The main purpose of land cover mapping is to serve as a main reserve of land resources in local, regional or global levels in the world for all levels of environmental agencies, government, and private industry. Remote sensing technique provides valuable land cover data for local or regional level, from the past history till recently. Land cover composition and change are important manners for many scientific research and socioeconomic assessments (Xian et al., 2009). Information regarding the land cover types and distributions can be utilized to monitor the status of ecosystems change over desired period and assess landscape condition (Coppin et al., 2004). Continuous monitoring of the types and locations of LULC change gives us an obvious view on change mechanisms and to model impacts on the environment and associated ecosystems (Homer et al., 2004; Xian and Crane, 2005). Many studies have been

done on the regional land cover mapping and it can be used as a guideline for future monitoring and land management by policy-makers or local authorities in order to avoid any environmental change caused by improper planning in LULC (Kumar et al., 2009).
Usually, the process of land covers mapping involved image classification. The aim of classification analysis is to categorize all the pixels in the digital satellite imageries into land-cover classes. Basically, the process can be divided into three simple steps, the pre-processing step, data classification and output. For the first step of pre-processing, the digital satellite images were chosen for land-cover mapping. For the second step of data classification, the digital satellite images were processed using PCI Geomatica image processing software. Supervised classifications operate in three basic steps: training, classification, and accuracy assessment. Training sites are needed for supervised classification. The areas were established using polygons. They are delineated by spectrally homogenous sub areas, which have, class name given. The satellite images were the classified using the three supervised classification methods (Maximum Likelihood, Minimum Distance-to-Mean and Parallelepiped).

4. Study area

The Landsat TM imageries used in this study were acquired on 24 December 2004 and 27 March 2005. The selected study area was Nanggroe Aceh Darussalam (NAD) province, Indonesia. Figure 1 shows the study area. The study area is in the Sumatra Peninsula, located between latitudes 2ºN and 6ºN and longitudes 95ºE and 99ºE (Figure 1), at the nortern tip of the island of Sumatra. NAD province is located in equatorial region and enjoys a warm equatorial weather the whole year. Therefore, it is impossible to get the 100 % cloud free satellite image over NAD province. But, the satellite image chosen is less than 10 % cloud coverage over the study area.

Fig. 1. The study area of NAD province, Indonesia.

There are five main islands (Sumatra, Java, Kalimantan, Sulawesi, and Irian Jaya) in Indonesia. Indonesia covers total land area for approximately 1919317 km^2 (Banda Aceh-Wikipedia, 2010). While NAD province covers an area of 57365 km2. Total population in NAD province is about 3930000, with the major ethic of acehnese. It is western most province of the Indonesia with Indian Ocean to the west, the strait of Malacca to the east. Bukit Barisan mountain ranges with tangse, gayo and Alas upland is located in the central part of this province. The highest peaks are Leuser (3466 m), Ucop Molu (3187 m), Abong - abong (3015 m), Peut Sago (2786 m), and Geureudong (2295 m) and Burni Telong (2566 m).
The main variables of NAD province's climate are rainfall, rather than air pressure or temperature, with the weather of tropical, hot and humid throughout the year. NAD province has a dry and wet season. The dry season is from March through to August and the wet season from September to February. The average annual rainfall ranges from 2000 mm to 3000 mm, with tempearature ranging from 25 to 30° Celsius. The mean daily humidity varies between 65 % to 75 %.

5. Satellite remote sensing data

The research has been carried out in USM with the utilization of satellite images from passive sensors in our analysed. Images acquired by Landsat TM 5. Landsat 5 was launched by NASA on March 1 1985 in order to replace Landsat 4. Landsat 5 was designed and built at the same time as Landsat 4 and carried out the same payload: the Multispectral Scanner System (MSS) and the Thematic Mapper (TM) instruments.

Fig. 5. Landsat Satellite TM.

6. Data analysis and results

On 26 December 2004 an underwater earthquake, measured at 9.0 on the Moment Magnitude scale, occurred off the coast of NAD on the island of Sumatra, in the Indonesian Archipelago. The resulting tsunamis have killed tens of thousands, with many more people missing, feared to be washed out to sea and drowned. Areas near to the epicentre in Indonesia, especially Banda Aceh, were devastated by the earthquake and tsunami.
The final death toll, taking into account missing persons, is likely to exceed 100,000 with up to two million people displaced and made homeless. The effects of the tsunami were also strongly felt in the coastal regions of Thailand, Malaysia, Myanmar (Burma) (UNICEF report), Sri Lanka (UNICEF report), India (UNICEF report), Maldives, and as far away as the coasts of Africa.
Many town areas were destroyed or devastated. The Indian Ocean tsunami was one of the greatest tragedies of recent history. The toll in deaths, injuries and human suffering was enormous. Whole communities were destroyed and displaced. Damage to property, infrastructure and the environment was also immense. In parts of South East Asia and Africa the tsunami caused many deaths yet left no lasting scar on the landscape. Elsewhere towns were devastated, entire villages destroyed and wide swaths of agriculture, forest and grazing- land washed away.
A full TM scene (path 131, row 56) recorded on 24 December 2004 was used as a reference image before the Tsunami event. The 27 March 2005 was the only cloud free image recorded after the Tsunami event. Figure 2 shows the raw satellite images for 24 December 2004 and 27 March 2005. All image-processing analysis was carried out using PCI Geomatica version 9.1.8 software at the School of Physics, University of Science Malaysia (USM). Figure 3 shows the coastal line change map over Indonesia. The white line on the figure depicts the original coastline, the red represents the post-tsunami situation.
From the two images displayed, we can clearly see that the area near the coastal line on 27 March 2005, viz after the Tsunami, was previously vegetated on 24 December 2004, viz before the Tsunami. The distributions of seriously damaged land cover along coastal line of the North West Sumatra, Banda Aceh is shown in Figure 3.
Land cover classification was accomplished in three steps. The first step in pre-processing was to apply a brightness correction technique to the digital images. The second step was to classify all the pixels in the digital images into land cover classes. Training sites were needed for supervised classification. A total of 45 training sample areas were established in this study. The digital image was classified into five classes using all three visible bands and three infrared bands (water, urban, vegetation, cloud, unprocessed area). The final step is the accuracy assessment. The available ground truth data were used to derive an accuracy assessment analysis for the classified map. The overall accuracy is expressed as a percentage of the test-pixels successfully assigned to the correct classes. The high Kappa coefficient suggests a good relationship between the classified image and the reference data. A total of 200 samples were chosen randomly for the accuracy assessment in this study. The assessment results showed a reasonably good agreement between the land cover data set and the reference data. The produced results in this study are shown in Table 1. The accuracy assessment results are shown in Table 2 and 3. Finally, the land cover damaged areas were calculated from two classified maps. Table 4 shows the area of LULC changes in

NAD province, which was generated based on the data acquired from image classification. From this table, all the land cover types have been compared within the area covered by them, in 24 December 2004 and 27 March 2005, before and after Tsunami event.

(a)

(b)

Fig. 2. Raw satellite images used in this study, (a) 24 December 2004 (b) 27 March 2005.

(a)

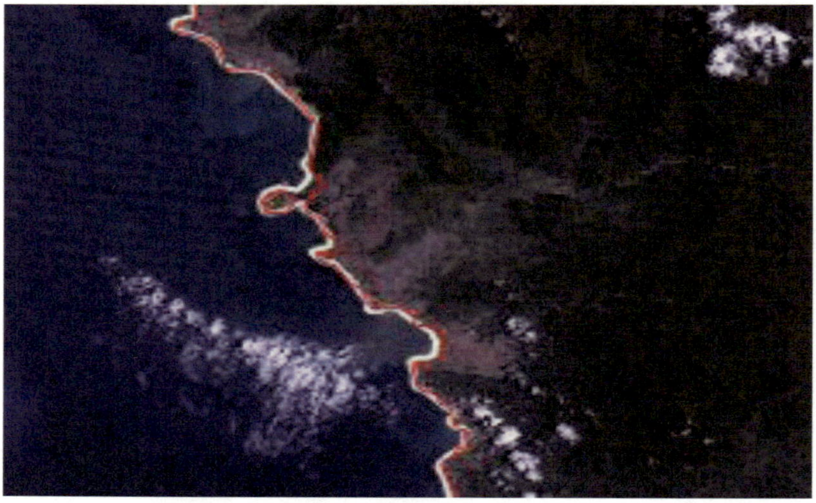

(b)

Fig. 3. Illustration of the original coastline and the post-tsunami situation, a) 24th December 2004 b) 27th March 2005

Classes	24 December 2004	27 March 2005
Water	20768.1921	17302.0662
Urban	4035.3732	3525.2577
Vegetation	8798.6754	8692.7544
Cloud	1996.1550	2099.3643
Unprocessed Area	6449.6925	10428.6456
Total	42048.0882	42048.0882

Table 1. Statistic analysis for the areas (in km).

Classification method	Kappa coefficient	
	24 December 2004	27 March 2005
Maximum Likelihood	0.8125	0.8236
Minimum Distance-to-Mean	0.6525	0.6895
Parallelepiped	0.4258	0.4582

Table 2. The Kappa coefficient for the two images.

Classification method	Overall classification accuracy (%)	
	24 December 2004	27 March 2005
Maximum Likelihood	89.25	91.25
Minimum Distance-to-Mean	72.21	78.06
Parallelepiped	55.36	59.36

Table 3. The overall classification accuracy for the two images.

Classes	Area changes from 24 December 2004 to 27 March 2005 (km^2)	Percentage (%)
Water	-3466.1259	-16.6895
Urban	-510.1155	-12.6411
Vegetation	-105.921	-1.2038

Table 4. Results of land use classification for 24 December 2004 and 27 March 2005 images showing the area changes of each class.

Accuracy assessment of coarse-resolution land cover products is a critical and challenging task, as these maps can overestimate or underestimate cover types due to the fragmentation and sub-pixel proportion of each cover type. As an alternative approach to field surveys, fine-resolution images and derived land cover maps have been used for validation of coarse-resolution thematic maps. In this study, validation was performed using the in situ data collected during the fieldwork in the study area. The classified land cover maps using Maximum Likelihood Classifier are shown in Figures 4 and 5 for 24 December 2004 and 27 March 2005 respectively.

7. Conclusion

This study indicated that the digital satellite imagery could provide useful information for remotely sensed data. The Maximum Likelihood classifier classification method proved to be very efficient to classify land cover over Nanggroe Aceh Darussalam (NAD) province, Indonesia. An overall accuracy and overall Kappa higher than 80 % were achieved using Landsat TM image. This land cover map was very useful for urban planner especially after the Tsunami event that occurs on the 26 December 2004. Recent satellite based instrument and digital image-processing techniques now make the coastal line changes over a big coverage of study area possible. Further studies have to be carried out to investigate the coastal line features and their spatial relationship relative to the physical land-water boundary.

Fig. 4. The land cover map using Landsat TM image (24 December 2004) [Color Code: Green = vegetation, Yellow = land/urban, blue = water, White =cloud and Black = Unprocessed area]

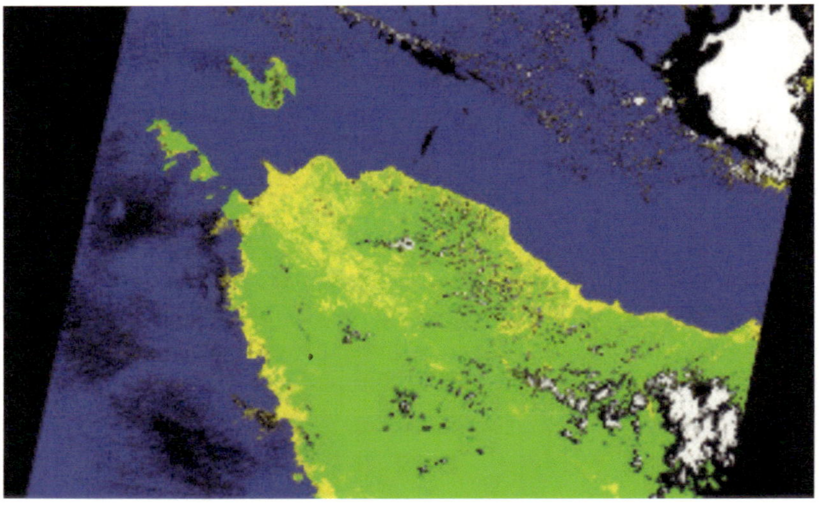

Fig. 5. The land cover map using Landsat TM image (27 March 2005) [Color Code: Green = vegetation, Yellow = land/urban, blue = water, White =cloud and Black = Unprocessed area].

8. Acknowledgements

The authors gratefully acknowledge the financial support from the Simulation and Modeling of the Atmospheric Radiative Transfer of Aerosols in Penang, account number: 203/PFIZIK/6711111 and Universiti Sains Malaysia - Short term grant. We would like to thank the technical staff who participated in this project. Thanks are also extended to USM for support and encouragement.

9. References

Asner, G. P.; Keller, M.; Pereira, R. & Zweede, J. C. (2002). Remote sensing of selective logging in Amazonia—assessing limitations based on detailed field observations, Landsat ETM+, and textural analysis. Remote Sensing of Environment, 80: 483–496.

Boles, S. H.; Xiao, X. M.; Liu, J. Y.; Zhang, Q. Y.; Munkhtuya, S.; Chen, S. Q. & Ojima, D. (2004). Land cover characterization of Temperate East Asia using multi-temporal vegetation sensor data. Remote Sensing of Environment, 90: 477-489.

Bourgeau-Chavez, L. L.; Kasischke, E. S.; Brunzell, S.; Mudd, J. P. & Tukman, M. (2002). Mapping fire scars in global boreal forests using imaging radar data. International Journal of Remote Sensing, 23: 4211–4234.

Coppin, P.; Jonckheere, I.; Nackerts, K. & Muys, B. (2004). Digital change detection methods in ecosystem monitoring: An review. International Journal of Remote Sensing, 25(9): 1565-1596.

Ehlers, M.; Gahler, M & Janowsky, R. (2003). Automated analysis of ultra high resolution remote sensing data for biotope type mapping: new possibilities and challenges. ISPRS Journal of Photogrammetry & Remote Sensing, 57: 315– 326.

Engeset, R. V.; Kohler, J.; Melvold, K. & Lunden, B. (2002). Change detection and monitoring of glacier mass balance and facies using ERS SAR winter images over Svalbard. International Journal of Remote Sensing, 23: 2023–2050.

Fernandes, R.; Fraser, R.; Latifovic, R.; Cihlar, J.; Beaubien, J. & Du, Y. (2004). Approaches to fractional land cover and continuous field mapping: A comparative assessment over the BOREAS study region. Remote Sensing of Environment, 89: 234-251.

Foody, G. M. (2002). Status of land covers classification accuracy assessment. Remote Sensing and Environment, 80: 185-201.

Friedl, M. A.; Mclver, D. K.; Hodges, J. C. F.; Zhang, X. Y.; Muchoney, D.; Strahler, A. H.; Woodcock, C. E.; Gopal, S.; Schneider, A.; Cooper, A.; Baccini, A.; Gao, F. & Schaaf, C. (2002). Global land cover mapping from MODIS: algorithms and early results. Remote Sensing of Environment, 83: 287–302.

Green, K.; Kempka, D. & Lackey, L. (1994). Using remote sensing to detect and monitor land-cover and land-use change. Photogrammetric Engineering & Remote Sensing, 60: 331–337.

Homer, C.; Huang, C.; Yang, L.; Wylie, B. & Coan, M. (2004). Development of a 2001 national land cover database for the United States. Photogrammetric Engineering & Remote Sensing, 70: 829-840.

Kasetkasem, T.; Arora, M. K. & Varshney, P. K. (2005). Super-resolution land cover mapping using a Markov random field based approach. Remote Sensing of Environment, 96: 302-314.

Kimura, H. & Yamaguchi, Y. (2000). Detection of landslide areas using satellite radar interferometry. Photogrammetric Engineering & Remote Sensing, 66: 337–343.

Kumar, M.; Mukherjee, N.; Sharma, G. P. & Raghubanshi, A. S. (2009). Land use patterns and urbanization in the holy city of Varanasi, India: a scenario. Environmental Monitoring and Assessment. [Online], [Accesses 19th July 2009], p.1-6. Available from World Wide Web:
http://www.springerlink.com/content/m223g540364213h2/fulltext.pdf

Lu, D. S.; Mausel, P.; Brondízio, E. S. & Moran, E. (2002). Change detection of successional and mature forests based on forest stand characteristics using multitemporal TM data in the Altamira, Brazil. XXII FIG International Congress, ACSM–ASPRS Annual Conference Proceedings, Washington, DC, USA, 19–26 April (Bethesda, MD: American Society of Photogrammetry and Remote Sensing) (CD-ROM).

Lu, D.; Mausel, P.; Brondízio, E. & Moran, E. (2003). Change detection techniques. International Journal of Remote Sensing, 25(12): 2365-2407.

Lunetta, R. S.; Ediriwickrema, J.; Johnson, D. M.; Lyon, J. G. & Mickerrow, A. (2002). Impacts of vegetation dynamics on the identification of land-cover change in a biologically complex community in North Carolina, USA. Remote Sensing of Environment, 82: 258–270.

Lunetta, R. S.; Knight, J. F.; Ediriwickrema, J.; Lyon, J. G. & Worthy, L. D. (2006). Land-cover change detection using multi-temporal MODIS NDVI data. Remote Sensing of Environment, 105: 142-154.

Mas, J. F. (1999). Monitoring land-cover changes: a comparison of change detection techniques. International Journal of Remote Sensing, 20: 139–152.

Munyati, C. (2000). Wetland change detection on the Kafue Flats, Zambia, by classification of a multitemporal remote sensing image dataset. International Journal of Remote Sensing, 21: 1787–1806.

Read, J. M. & Lam, N. S.-N. (2002). Spatial methods for characterizing land cover and detecting land cover changes for the tropics. International Journal of Remote Sensing, 23: 2457–2474.

Rees, W. G.; Williams, M. & Vitebsky, P. (2003). Mapping land cover change in a reindeer herding area of the Russian Arctic using Landsat TM and ETM+ imagery and indigenous knowledge. Remote Sensing of Environment, 85: 441-452.

Rigina, O.; Baklanov, A.; Hagner, O. & Olsson, H. (1999). Monitoring of forest damage in the Kola Peninsula, Northern Russia due to smelting industry. Science of the Total Environment, 229: 147-163.

Singh, A. (1989). Digital change detection techniques using remotely sensed data. International Journal of Remote Sensing, 10: 989–1003.

Taylor, J. C.; Brewer, T. R. & Bird, A. C. (2000). Monitoring landscape change in the national parks of England and Wales using aerial photo interpretation and GIS. International Journal of Remote Sensing, 21: 2737–2752.

Weng, Q. (2002). Land use change analysis in the Zhujiang Delta of China using satellite remote sensing, GIS and stochastic modeling. Journal of Environmental Management, 64: 273–284.

Wilson, E. H. & Sader, S. A. (2002). Detection of forest harvest type using multiple dates of Landsat TM imagery. Remote Sensing of Environment, 80: 385–396.

Woodcock, C. E.; Macomber, S. A.; Pax-Lenney, M. & Cohen, W. B. (2001). Monitoring large areas for forest change using Landsat: generalization across space, time and Landsat sensors. Remote Sensing of Environment, 78: 194–203.

Xian, G.; Homer, C. & Fry, J. (2009). Updating the 2001 National Land Cover Database land cover classification to 2006 by using Landsat imagery change detection methods. Remote Sensing of Environment, 113: 1133-1147.

Xian, G. & Crane, M. (2005). Assessment of urban growth in the Tampa Bay watershed using remote sensing data. Remote Sensing of Environment, 97: 203-215.

Zhang, Q.; Wang, J.; Peng, X.; Gong, P. & Shi, P. (2002). Urban build-up land change detection with road density and spectral information from multitemporal Landsat TM data. International Journal of Remote Sensing, 23: 3057–3078.

23

Numerical Model for Multi-layered Tsunami Waves

Monzur Alam Imteaz[1], Fumihiko Imamura[2] and Jamal Naser[1]

[1]*Swinburne University of Technology,*
[2]*Tohoku University,*
[1]*Australia*
[2]*Japan*

1. Introduction

In order to develop a generalized numerical model for multi-layered tsunami wave system, a three-layer system was considered. Six governing equations, two for each layer were derived from Euler equations of motion and continuity for three layers, assuming long wave approximation, negligible friction and interfacial mixing. From derived equations, it is found that only top layer equations are independent of number of intermediate layers; equations for all other layers are dependent on number, extent and density of intermediate layer(s). Momentum and continuity equations for the top layer are exactly same as in the case of earlier developed governing equations for two-layered system. Continuity equation for the bottom layer is also exactly same as in the case of two-layered system. Momentum equation for the bottom layer is dependent on extent and density of top layer as well as all intermediate layers. Continuity equation for intermediate layer is affected by levels of immediate bottom layer. Momentum equation for the intermediate layer is affected by extent and density of upper layer(s). Developed governing equations were converted to a numerical model using staggered Leap-Frog scheme for the computations of water level and discharge in each layer in one-dimensional propagation. Developed numerical model results were compared with an earlier developed model for two layers, which was rigorously verified by analytical solution. It was found that this three-layer model produces same results when it is converted to two-layer through mathematical manipulation (i.e. by assuming a negligible/zero depth or similar density of adjacent layer for any layer). The details properties of three-layer model were discussed through numerical simulations for different scenarios. The developed model can be easily converted to a multi-layer (any number) model and can be applied confidently to simulate the basic features of different practical tsunami problems similar to that investigated in this study.

2. Background

Multi-layered flow is related with many environmental phenomena. Thermally driven exchange flows through doorways to oceanic currents, salt water intrusion in estuaries, spillage of the oil on the sea surface, spreading of dense contaminated water, sediment laden discharges into lakes, generation of lee waves behind a mountain range and tidal flows over sills of the ocean are examples of multi-layered flow. In hydraulics, this type of flow is often termed as gravity current. An extensive review on hydrodynamics of various gravity currents was provided by Simpson, J.E. (1982).

Tsunamis are generated due to disturbances of free surfaces caused by not only seismic fault motion, but also landslide and volcanic eruptions (Imamura and Imteaz, 1995). Tsunami waves are also affected by density differences along the depth of ocean. There are some studies on two-layered long waves or flows in the case of underwater landslide generated tsunamis (Hampton, 1972; Parker, 1982; Harbitz, 1991; Jiang & Leblond, 1992). Imamura & Imteaz (1995) developed a linear numerical model on two-layered long wave flow, which was successfully validated by a rigorous analytical solution. Later the linear model was extended to a non-linear model and effects of non-linearity were investigated (Imteaz & Imamura, 2001b). Madsen et al. (2002) developed a model of multi-layered flow based on Boussinesq-type equations, which are suitable for shallow depth flow. Lynett and Liu (2004) developed another model of multi-layered flow using piecewise integration of Laplace equation for each individual layer and expanded the model for deep water.

Choi and Camassa (1996) derived two-dimensional non-linear equations for two-layered fluid system and presented some numerical simulations of their model for one-dimensional unidirectional wave propagation. Later Choi and Camassa (1999) further developed governing equations for the unidirectional propagation of internal gravity waves at the interface of two immiscible inviscid fluids. They have compared their numerical results with available experimental data for solitary waves of large amplitude in two-fluid system. Liska and Wendroff (1997) derived one-layer and two-layer classic shallow water equations for flow over topography, as well as one-layer and two-layer non-hydrostatic equations. They have compared their numerical results with the numerical computations obtained by others. Percival et al. (2008) presented a multi-layer extension of Green-Naghdi equations (Green and Naghdi, 1976) using a special framework based on the Euler-Poincare theory. Through numerical simulations they have shown that free surface of a multi-layer model can exhibit intriguing differences compared to the results of single layer model. Cotter et al. (2010) modified the multi-layer Green-Naghdi equations to incorporate effects of shear stress. They have presented numerical simulations for the wave propagation and interactions between two layers, with and without shear considerations.

All the above mentioned models, presented their results in two-layer case only. However, interactions within the layers for a two-layer fluid are significantly different than a three (or more) layer fluid. Top surface is having effect from immediate lower layer only, which is same for both the models; however, intermediate layer is having effects from both the lower layer and upper layers and lower layer is having effects from all the upper layers (Imteaz et al., 2009). Imteaz et al. (2009) provided detailed derivation of multi-layered tsunami wave and flow equations based on Navier-Stokes equation. Also, properties of multi-layered equations were discussed in details. Present paper describes conversion of developed multi-layered tsunami equations in to numerical form, comparison of developed model with earlier validated model and several simulations scenario.

3. Governing equations

Figure 1 shows the schematic diagram of three layer propagation having different densities and depths. For a three layered one-dimensional propagation, considering Euler's equations of motion and continuity for each layer in a wide channel with non-horizontal bottom, through integration and rigorous formulation assuming a hydrostatic pressure distribution, negligible friction, negligible interfacial mixing and proper boundary conditions, mathematical continuity and momentum equations were derived for each layer. Also uniform density and velocity distributions in each layer were assumed. Original derived equations are further simplified considering horizontal bottom (i.e. no variations of 'h' along x direction, ∂h/∂x=0).

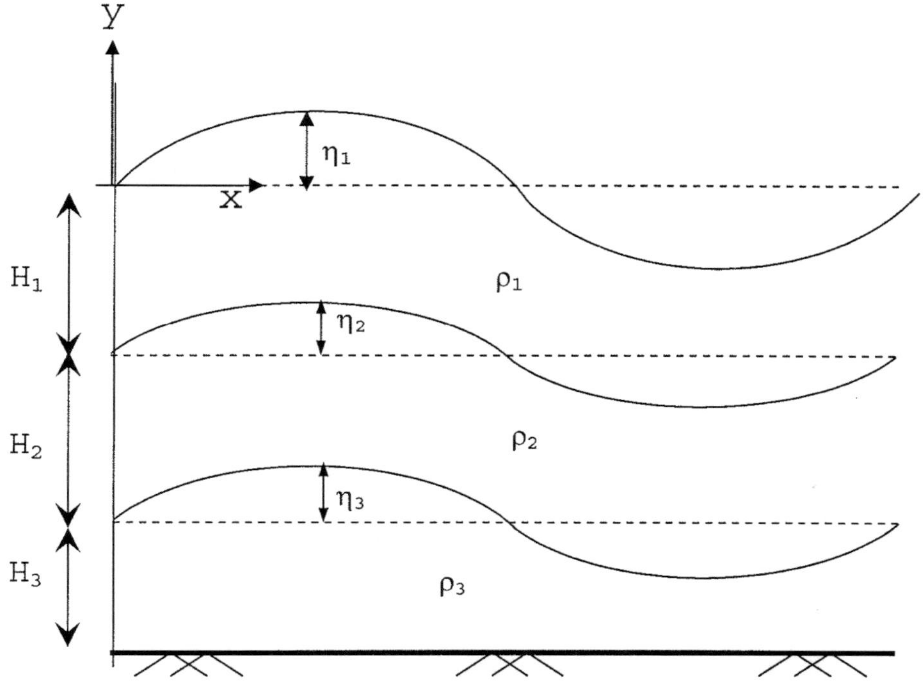

Fig. 1. Schematic diagram of three layer profile

Detailed derivations of mathematical equations are described by Imteaz et al. (2009). The equations are as follows:

For the upper layer-

Continuity equation,

$$\frac{\partial M_1}{\partial x} + \frac{\partial(\eta_1 - \eta_2)}{\partial t} = 0 \tag{1}$$

Momentum equation,

$$\frac{\partial M_1}{\partial t} + \frac{\partial(M_1^2 / D_1)}{\partial x} + gD_1 \frac{\partial \eta_1}{\partial x} = 0 \tag{2}$$

For the intermediate layer-

Continuity equation,

$$\frac{\partial M_2}{\partial x} + \frac{\partial(\eta_2 - \eta_3)}{\partial t} = 0 \tag{3}$$

Momentum equation,

$$\frac{\partial M2}{\partial t}+\frac{\partial (M_2^2/D2)}{\partial x}+gD2\left\{\frac{\alpha_1}{\alpha_2}\left(\frac{\partial \eta_1}{\partial x}-\frac{\partial \eta_2}{\partial x}\right)+\frac{\partial \eta_2}{\partial x}\right\}=0 \quad (4)$$

For the lower layer-
Continuity equation,

$$\frac{\partial M3}{\partial x}+\frac{\partial \eta 3}{\partial t}=0 \quad (5)$$

Momentum equation,

$$\frac{\partial M_3}{\partial t}+\frac{\partial (M_3^2/D_3)}{\partial x}+gD_3\left\{\alpha_1\left(\frac{\partial \eta_1}{\partial x}-\frac{\partial \eta_2}{\partial x}\right)+\frac{\partial \eta_3}{\partial x}+\alpha_2\left(\frac{\partial \eta_2}{\partial x}-\frac{\partial \eta_3}{\partial x}\right)\right\}=0 \quad (6)$$

Where,
η_1 = Water surface elevation above still water level of layer '1'
η_2 = Water surface elevation above still water level of layer '2'
η_3 = Water surface elevation above still water level of layer '3'

$$D_1 = \eta_1 + h_1 - \eta_2,\ D_2 = h_2 + \eta_2 - \eta_3,\ D_3 = h_3 + \eta_3,\ \alpha_1 = \rho_1/\rho_3,\ \alpha_2 = \rho_2/\rho_3$$

h_1 = Still water depth of layer '1'
h_2 = Still water depth of layer '2'
h_3 = Still water depth of layer '3'

$$M_1=\int_{-h1+\eta_2}^{\eta_1} u_1\,dy,\quad M_2=\int_{-h1-h2+\eta_3}^{-h1+\eta_2} u_2\,dy,\quad M_3=\int_{-h1-h2-h_3}^{-h1-h_2+\eta_3} u_3\,dy$$

From the derived equations, it is found that momentum equation for upper layer is not affected by the properties of adjacent layer (layer underneath). However, continuity equation of upper layer is affected by surface elevation of intermediate layer. Continuity equation for intermediate layer is affected by the surface elevation of bottom layer. Momentum equation for intermediate layer is affected by density and spatial change of surface elevation of upper layer. Continuity equation for bottom layer is not affected by either uppermost layer or intermediate layer. However, momentum equation of bottom layer is affected by densities and spatial changes in surface elevations of all the layers above it. Properties of all these equations were described in detailed by Imteaz et al. (2009). Also, wave celerity of each layer was deduced as follows:

$$C_1=\sqrt{gh_1\left(1+\alpha_3\beta_1\right)}$$

$$C_2=\sqrt{gh_2\left(1+\beta_2(\alpha_2-\alpha_1)\right)}$$

$$C_3=\sqrt{gh_3\left(1-\alpha_2\right)}$$

4. Numerical model

Developed governing equations are non-linear. It is very difficult to solve the non-linear governing equations analytically, but it can be solved numerically using proper finite difference scheme. Staggered Leap-Frog scheme has been used to solve the governing equations numerically, as it was found producing very good results in earlier developed models (Imamura & Imteaz, 1995; Imteaz & Imamura, 2001a and Imteaz & Imamura, 2001b). Figure 2 shows the schematic diagram of the staggered Leap-Frog scheme. This scheme is one of the explicit central difference schemes with the truncation error of second order. The staggered scheme considers that the computation point for one variable (η) does not coincide with the computation point for other variable (M). There are half step differences (½Δt and ½Δx) between computation points of two variables (as shown in Figure 2). Using this scheme, finite difference equations for the derived governing equations are as follows:

For the upper layer-
Continuity equation,

$$\frac{\eta_{1,i}^{n+1/2}-\eta_{1,i}^{n-1/2}-\eta_{2,i}^{n+1/2}+\eta_{2,i}^{n-1/2}}{\Delta t}+\frac{M_{1,i+1/2}^{n}-M_{1,i-1/2}^{n}}{\Delta x}=0 \tag{7}$$

Momentum equation,

$$\frac{M_{1,i+1/2}^{n}-M_{1,i+1/2}^{n-1}}{\Delta t}+g\frac{D_{1,i+1}^{n-1/2}+D_{1,i}^{n-1/2}}{2}\frac{\eta_{1,i+1}^{n-1/2}-\eta_{1,i}^{n-1/2}}{\Delta x}+$$

$$\frac{\dfrac{(M_{1,i+1/2}^{n-1})^2}{(D_{1,i+1}^{n-1/2}+D_{1,i}^{n-1/2}+D_{1,i+1}^{n-3/2}+D_{1,i}^{n-3/2})/4}-\dfrac{(M_{1,i-1/2}^{n-1})^2}{(D_{1,i}^{n-1/2}+D_{1,i-1}^{n-1/2}+D_{1,i}^{n-3/2}+D_{1,i-1}^{n-3/2})/4}}{\Delta x}=0 \tag{8}$$

For the intermediate layer-
Continuity equation,

$$\frac{\eta_{2,i}^{n+1/2}-\eta_{2,i}^{n-1/2}-\eta_{3,i}^{n+1/2}+\eta_{3,i}^{n-1/2}}{\Delta t}+\frac{M_{2,i+1/2}^{n}-M_{2,i-1/2}^{n}}{\Delta x}=0 \tag{9}$$

Momentum equation,

$$\frac{M_{2,i+1/2}^{n}-M_{2,i+1/2}^{n-1}}{\Delta t}+g\frac{D_{2,i+1}^{n-1/2}+D_{2,i}^{n-1/2}}{2}\left[\frac{\alpha_1}{\alpha_2}\left(\frac{\eta_{1,i+1}^{n-1/2}-\eta_{1,i}^{n-1/2}}{\Delta x}-\frac{\eta_{2,i+1}^{n-1/2}-\eta_{2,i}^{n-1/2}}{\Delta x}\right)+\frac{\eta_{2,i+1}^{n-1/2}-\eta_{2,i}^{n-1/2}}{\Delta x}\right]$$

$$+\frac{\dfrac{(M_{2,i+1/2}^{n-1})^2}{(D_{2,i+1}^{n-1/2}+D_{2,i}^{n-1/2}+D_{2,i+1}^{n-3/2}+D_{2,i}^{n-3/2})/4}-\dfrac{(M_{2,i-1/2}^{n-1})^2}{(D_{2,i}^{n-1/2}+D_{2,i-1}^{n-1/2}+D_{2,i}^{n-3/2}+D_{2,i-1}^{n-3/2})/4}}{\Delta x}=0 \tag{10}$$

For the lower layer-
Continuity equation,

$$\frac{\eta_{3,i}^{n+1/2}-\eta_{3,i}^{n-1/2}}{\Delta t}+\frac{M_{3,i+1/2}^{n}-M_{3,i-1/2}^{n}}{\Delta x}=0 \tag{11}$$

Momentum equation,

$$g\frac{D_{3,i+1}^{n-1/2}+D_{3,i}^{n-1/2}}{2}\left[\alpha_1\left(\frac{\eta_{1,i+1}^{n-1/2}-\eta_{1,i}^{n-1/2}}{\Delta x}-\frac{\eta_{2,i+1}^{n-1/2}-\eta_{2,i}^{n-1/2}}{\Delta x}\right)\right]+$$

$$g\frac{D_{3,i+1}^{n-1/2}+D_{3,i}^{n-1/2}}{2}\left[\frac{\eta_{3,i+1}^{n-1/2}-\eta_{3,i}^{n-1/2}}{\Delta x}+\alpha_2\left(\frac{\eta_{2,i+1}^{n-1/2}-\eta_{2,i}^{n-1/2}}{\Delta x}-\frac{\eta_{3,i+1}^{n-1/2}-\eta_{3,i}^{n-1/2}}{\Delta x}\right)\right]+$$

$$\frac{\dfrac{(M_{3,i+1/2}^{n-1})^2}{(D_{3,i+1}^{n-1/2}+D_{3,i}^{n-1/2}+D_{3,i+1}^{n-3/2}+D_{3,i}^{n-3/2})/4}-\dfrac{(M_{3,i-1/2}^{n-1})^2}{(D_{3,i}^{n-1/2}+D_{3,i-1}^{n-1/2}+D_{3,i}^{n-3/2}+D_{3,i-1}^{n-3/2})/4}}{\Delta x}$$

$$+\frac{M_{2,i+1/2}^{n}-M_{2,i+1/2}^{n-1}}{\Delta t}=0 \tag{12}$$

where, 'n' denotes the temporal grid points and 'i' denotes the spatial grid points as shown in Figure 2. To calculate 'D' values at the computation point of 'M', average of four surrounding 'D' values were taken.

In spatial direction all η_1, η_2 and η_3 at step 'n-1/2' and all M_1, M_2 and M_3 at step '(n-1)' are given as initial conditions. For all later time steps at right boundary all values of M_1, M_2 and M_3 are calculated by characteristic method, using the values of previous time step and wave celerity. By using deduced finite difference Momentum equations for upper, intermediate and lower layer all M_1, M_2 and M_3 values at step 'n' are calculated. Then using the latest values of M_3 and deduced finite difference continuity equation for the lower layer all the values of η_3 at step '(n+1/2)' are calculated. Then using the latest values of η_3, M_2 and deduced finite difference continuity equation for the intermediate layer, all the values of η_2 at step '(n+1/2)' are calculated. Again, using the latest values of η_2, M_1 and deduced finite difference continuity equation for the upper layer, all the values of η_1 at step '(n+1/2)' are calculated. Similarly, using new values of η_1, η_2, η_3, M_1, M_2 and M_3 as initial conditions calculations proceeded in time direction up to the desired time step.

As initial condition (i.e. at t=0) all η_1, η_2, M_1 and M_2 values are taken as zero. For interface (between bottom layer and intermediate layer), assumed initial conditions are shown in Equations 13 & 14, which are based on the initial formation of tsunami wave. Expression of η_3,

$$\eta_3=a_3\mathrm{Sin}(kx)=a_3\mathrm{Sin}\left(\frac{2\pi}{L}x\right) \tag{13}$$

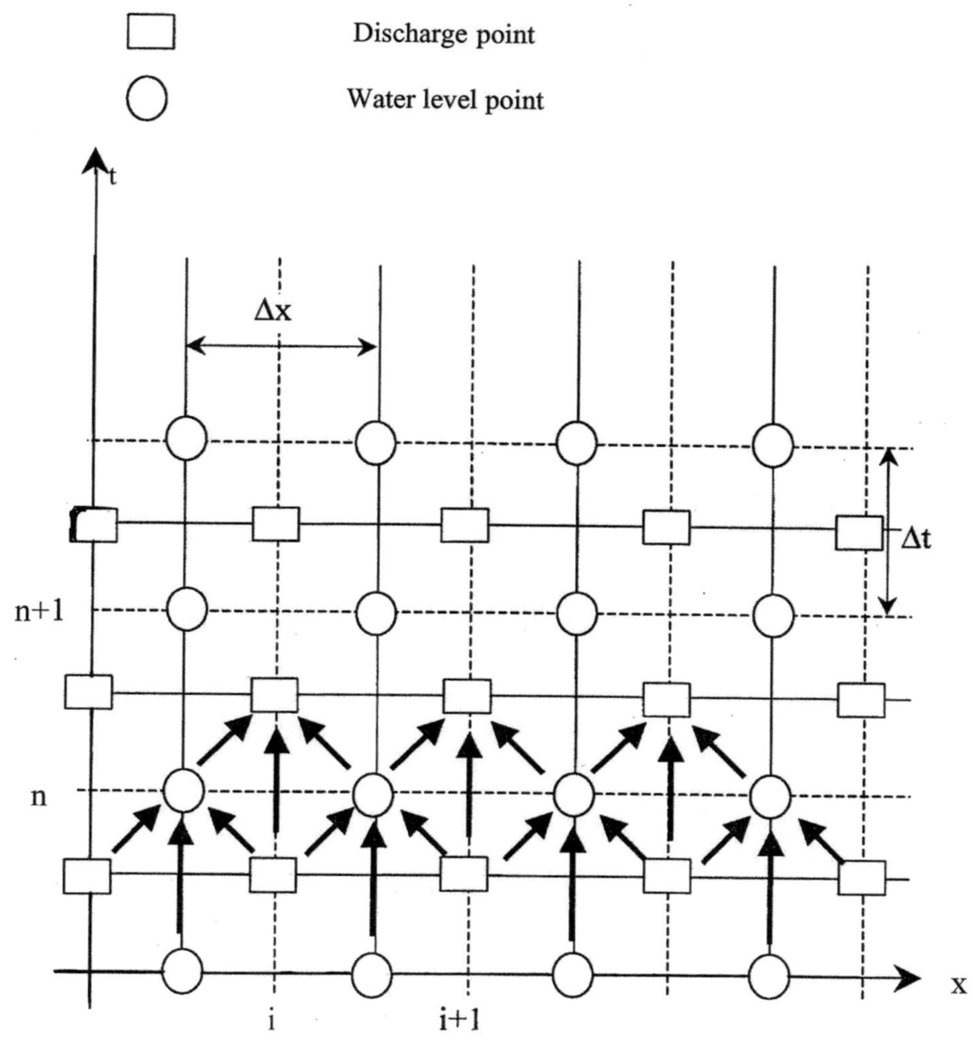

Fig. 2. Schematic diagram of the staggered Leap-Frog scheme

and corresponding discharge,

$$M_3 = \sqrt{(g / h_3)}\eta_3(h_3 + \eta_3) \tag{14}$$

where, a_3, k and L are the wave amplitude, wave number and wave length for the interfacial surface of intermediate layer and the bottom layer.

While computing water levels and discharges using Staggered Leap-Frog scheme, it is required to calculate boundary values (right and left boundaries) of water levels and discharges, using appropriate boundary conditions. Imamura and Imteaz (1995) discussed several possible boundary conditions. Among all the possible boundary conditions, it was found that the following provides good results for non-linear model simulations (Imteaz & Imamura, 2001b): M_1, M_2 and M_3 at the right boundary are calculated using characteristic method; constant wave celerities (which were estimated using analytical expressions) were used throughout the computational domain. Finally, using periodic condition, water levels at left boundary was used same as the right boundary.

5. Model comparison

It is found that the developed governing equations are complicated having influence from upper and/or lower layer(s) flow. Analytical solutions for such complicated differential equations are yet to be achieved. However, numerical models for two layer tsunami wave were validated with analytical solutions for known interface propagation with unknown top surface propagation by Imamura & Imteaz (1995) and for known top surface propagation with unknown interface propagation by Imteaz & Imamura (2001a). As the present developed model could not be verified with an analytical solution, it is indirectly compared with an earlier validated numerical model. Developed model was converted to a pseudo two-layer by assuming very close densities (1.0 and 0.99) for the intermediate and bottom layers respectively, with an upper layer density of 0.90. This numerical manipulation is supposed to produce similar results with a two-layer model having densities of 0.90 and 1.0 for the upper layer and lower layer respectively, provided depth of upper layer in both the two-layer and three-layer models are same and total depth of lower layers (intermediate layer and bottom layer) in three-layer model is same as the bottom layer of the two-layer model. With the above-mentioned conditions two separate models (two-layer model and three-layer model) were simulated with the same boundary conditions and having other properties as:

Wave length: 395.0 m
Wave amplitude: 2.0 m
DX = 10.0 m
DT = 0.20 sec

Models were simulated for a period of 4 seconds. Figures 3 & 4 show the pseudo three-layer model results comparison with the two-layer model results for similar input data. Figure 3 shows the comparison of model results for the top surface. It is found that the pseudo three-layer model simulates exactly same top surface level as the original two-layer model. Figure 4 shows the comparison of model results for the interface (interface between the intermediate layer and the upper layer for three-layer model). Again it is found that the pseudo three-layer model simulates almost same interface level as the original two-layer model. However, there are slight variations in the simulation of interface level, compared to the original two-layer model simulation. The reason for this variation is the fact that the lower layer density for the two-layer model is 1.0; however for a three-layer model this is not a single layer having a same density of 1.0, rather it is a combination of two layers having densities of 1.0 and 0.99. However, the variations are very insignificant and agreements of models results can be termed as very good.

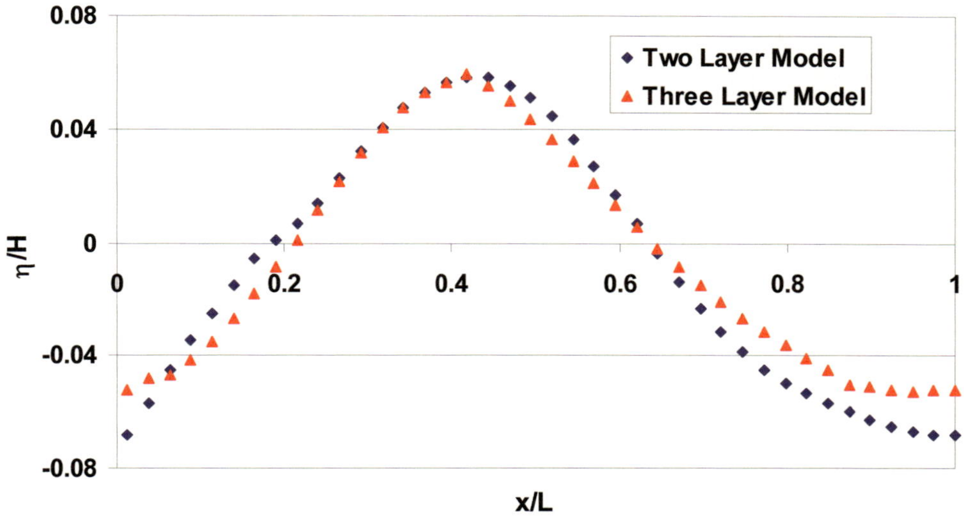

Fig. 3. Comparison of model results for the top surface

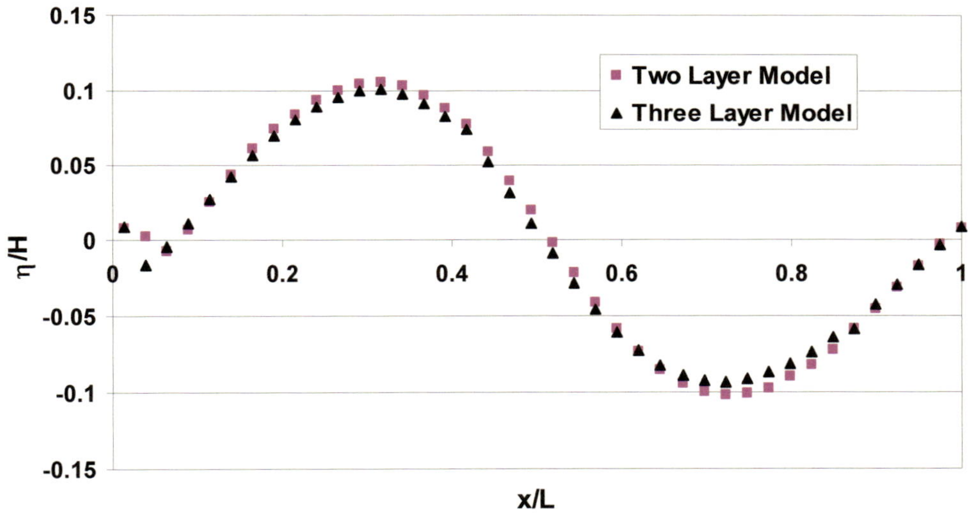

Fig. 4. Comparison of model results for the interface

6. Scenario simulations

The developed model was used for several scenario simulations. Model was simulated for the following conditions:

Wave length: 395.0 m
Lower layer wave amplitude: 2.0 m
DX = 10.0 m, DT = 0.20 seconds

$h_1 = h_2 = h_3 = 25.0$ m
$\rho_1 = 0.8, \rho_2 = 0.9, \rho_3 = 1.0$

For simplicity each layer depth was assumed as 25.0m, however the model can simulate for any layer depth. For the previous two-layered models, it was found that with the assumed DX/DT ratio, model's stability condition is good. Imteaz (1994) has presented detailed stability criteria analysis for the two-layered tsunami waves with Staggered Leap-Frog scheme. Eventually, Imteaz (1994) has proposed several regimes of DX and DT for the numerical stability of the model. A separate stability criteria analysis for the current three-layered model needs to be performed to achieve best numerical stability.
Figure 5 & 6 show the model simulations for all the three layer surfaces after 4 seconds and 12 seconds respectively. From the figures it is found that the surface of the lower layer gets amplified with the course of time, whereas the surface of the intermediate layer gets dampen down with the course of time. These are because of the complicated interactions from the adjacent layers.
Figures 7, 8 & 9 show the wave propagation patterns for lower surface, intermediate surface and top surface respectively. From the figures it is clear that propagations of lower layer and intermediate layers are relatively smooth. However, propagation of top surface is faster and having dramatic changes. For the top surface, wave phase changes from 4 seconds to 12 seconds, i.e. wave moves a distance of half of the wave length within 8 seconds. Also, within 4 seconds to 12 seconds, two wave crests of smaller amplitudes were formed.
Till now analytical solution of such complicated flow was not successfully achieved. However, in the future if such analytical solution is achieved, numerical results of the current model can be verified.

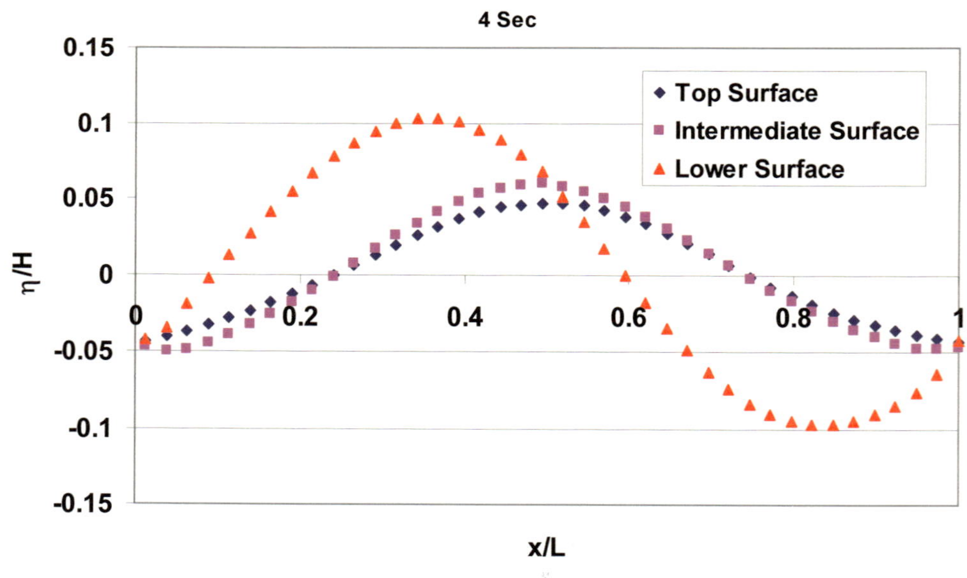

Fig. 5. Three layer model simulation results after 4 seconds

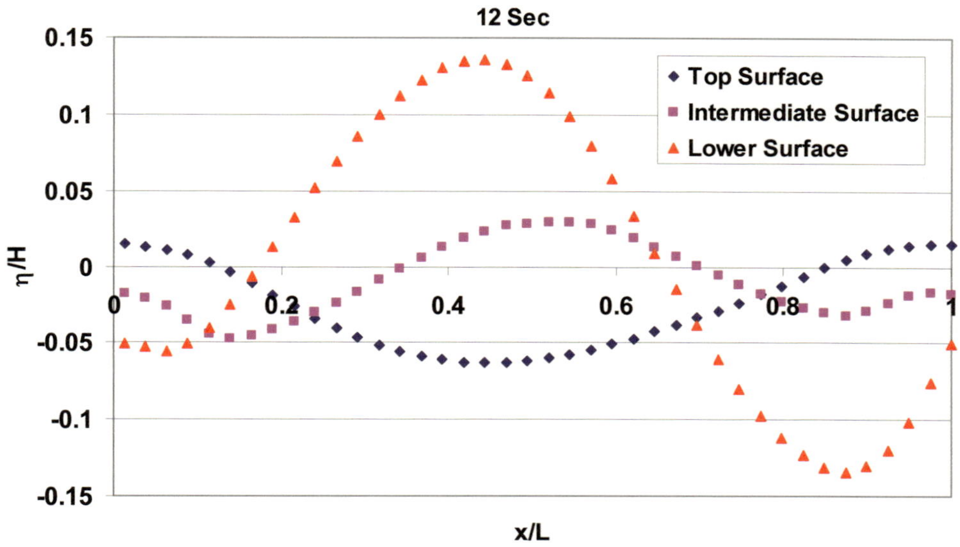

Fig. 6. Three layer model simulation results after 12 seconds

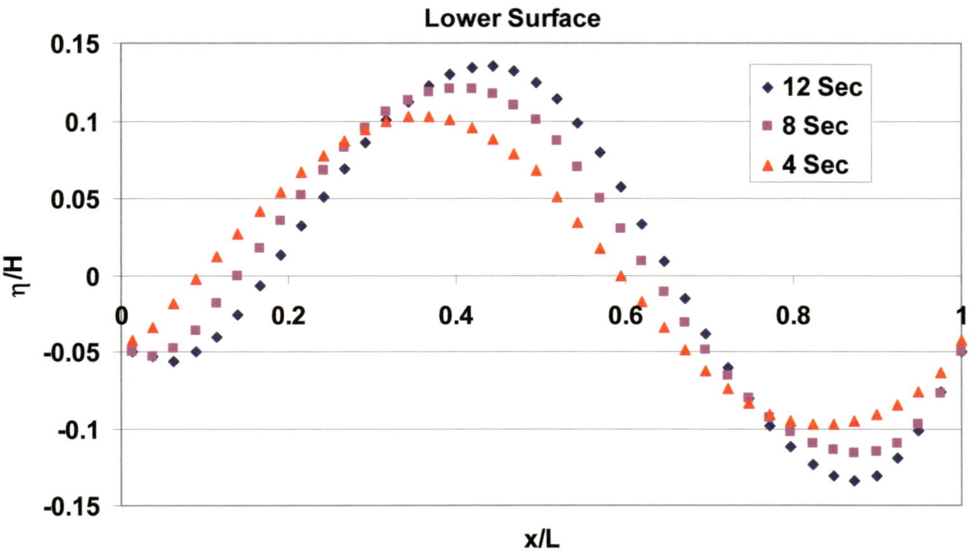

Fig. 7. Simulation results for the propagation of lower surface

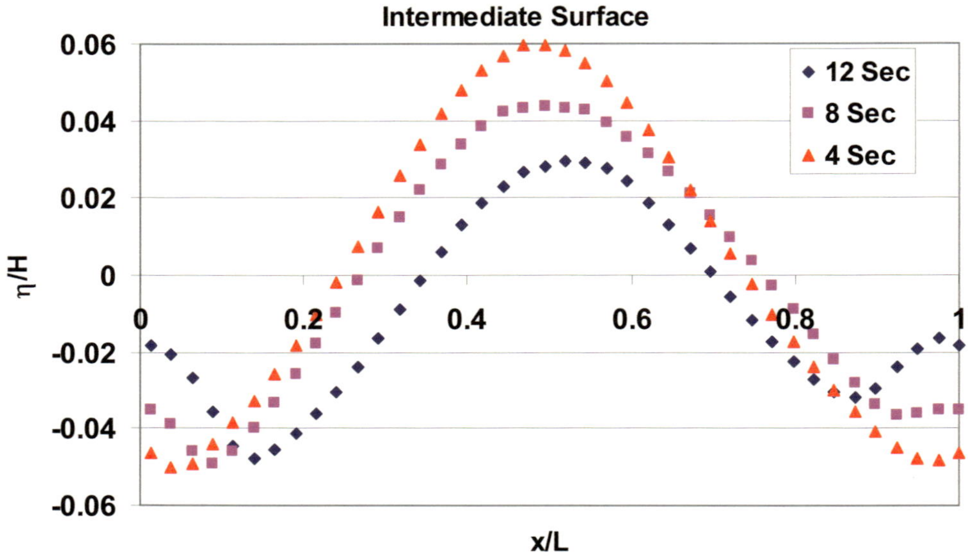

Fig. 8. Simulation results for the propagation of intermediate surface

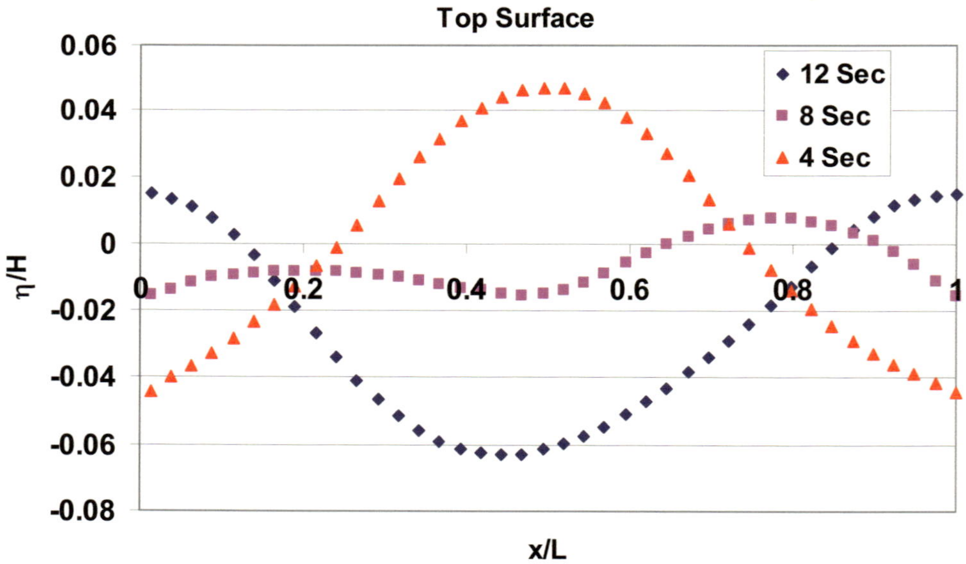

Fig. 9. Simulation results for the propagation of top surface

7. Conclusion

Earlier developed governing equations for three layered long waves are complex having several interactions from adjacent/all layers, which make those equations very difficult to solve analytically. Analytical solutions for these sorts of equations are yet to be succeeded by anyone. To achieve a numerical solution, governing equations were transformed into numerical formulations. Original derived equations were simplified considering horizontal bottom (i.e. no variations of 'h' along x direction, $\partial h/\partial x=0$). Numerical model was developed using staggered Leap-Frog scheme, as the same scheme produced good results for two layer numerical models.

At the beginning of the computations, all the initial values of η_1, η_2, η_3, M_1, M_2 and M_3 were given as initial conditions. All the variables for the later time steps were computed as follows:

- using deduced finite difference Momentum equations for the upper, intermediate and lower layer all the M_1, M_2 and M_3 values were calculated
- then using the latest values of M_3 and deduced finite difference continuity equation for the lower layer all the values of η_3 were calculated
- then using the latest values of η_3, M_2 and deduced finite difference continuity equation for the intermediate layer, all the values of η_2 were calculated
- using the latest values of η_2, M_1 and deduced finite difference continuity equation for the upper layer, all the values of η_1 were calculated
- at right boundary all the values of M_1, M_2 and M_3 were calculated by characteristic method, using the values of previous time step and wave celerity
- using periodic condition, water levels at left boundary was used same as the right boundary

Model results for a pseudo two-layer case were compared with an earlier validated model for real two-layer mode. Agreements are very good and it can be concluded that developed three-layer model is capable to produce realistic results. Using the developed model, some scenarios were presented. From scenario simulations it is found that the lower surface gets amplified with the course of time, whereas the intermediate surface and top surface get dampen down with the course of time. Also, it is found that the propagations of lower and intermediate surfaces are relatively smooth. However, propagation of top surface is faster and having dramatic changes. These are because of the interactions from the adjacent layers.

8. References

Choi, W. & Camassa, R. (1996) Weakly nonlinear internal waves in a two-fluid system. J. Fluid Mech., Vol. 313, pp. 83–103.

Choi, W. & Camassa, R. 1999 Fully nonlinear internal waves in a two-fluid system. J. Fluid Mech., Vol. 396, pp. 1–36.

Cotter, C. J., Holm, D. D. & Percival, J. R., (2010) The square root depth wave equations. Proc. of the Royal Society A. doi: 10.1098/rspa.2010.0124.

Hampton, M. H. (1972) The Role of Subaqueous Debris Flow in Generating Turbidity Currents, J. *Sedimentary Petrology*, Vol. 42, No. 4, pp.775-793.

Green, A. E. & Naghdi, P. M. (1976) A derivation of equations for wave propagation in water of variable depth. J. Fluid Mech. Vol. 78, pp. 237–246.

Harbitz, C. (1991) Numerical Simulations of Slide Generated Water Waves, *Sci. of Tsunami Hazards,* Vol. 9, No. 1, pp. 15-23.

Imamura, F. & Imteaz, M. A. (1995) Long Waves in Two-Layers: Governing Equations and Numerical Model, *Sci. of Tsunami Hazards,* Vol. 13, No. 1, pp. 3-24.

Imteaz, M. A. (1994) Numerical Model for the Long Waves in the Two Layers, Master's Thesis, *Asian Institute of Technology*, Bangkok, 86p.

Imteaz, M. A. & Imamura, F. (2001a) Analytical Solution and Numerical Model for the Interface in a Stratified Long Wave System, *Sci. of Tsunami Hazards,* Vol. 19, No. 1, pp. 39-54.

Imteaz, M. A. & Imamura, F. (2001b) A Non-linear Numerical Model for Stratified Tsunami Waves and its Application, *Sci. of Tsunami Hazards,* Vol. 19, pp. 150-159.

Imteaz, M. A., Imamura, F. & Naser, J. (2009) Governing Equations for Multi-layered Tsunami Waves, *Sci. of Tsunami Hazards,* Vol. 28, No. 3, pp. 171-185.

Jiang, L. & Leblond, P. H. (1992) The Coupling of Submarine Slide and the Surface Waves which it Generates, J. Geophys. Res., Vol. 97, No. C8, pp.12713-12744.

Liska, R. & Wendroff, B. (1997) Analysis and computation with stratified fluid models. J. Comput. Phys. Vol. 137, No. 1, pp, 212–244.

Lynett, P. & Liu, P.L.-F. (2004) Linear Analysis of the Multi-layer Model, Journal of Coastal Engineering, Vol. 51, pp. 439-454.

Madsen, P.A., Bingham, H.B. & Liu, H. (2002) A New Boussinesq Method for Fully Nonlinear Waves from Shallow to Deep Water, J. Fluid Mech., Vol. 462, pp. 1-30.

Parker, G. (1982) Conditions for Ignition of Catastrophically Erosive Turbidity Currents, *Marine Geology,* Vol. 46, pp. 307-327.

Percival, J. R., Cotter, C. J. & Holm, D. D. (2008) A Euler–Poincaré framework for the multilayer Green–Naghdi equations. *J. Phys. A: Math. Theoret.* Vol. 41, No. 34.

Simpson, J.E. (1982) Gravity Currents in the Laboratory, Atmosphere and Ocean, Ann. Rev. Fluid Mech., Vol. 14, pp. 213-234.

NOTATIONS

ρ = Density of fluid
M = Discharge per unit width of flow
η = Water surface elevation above still water level
h = Still water depth for a particular layer
D = Total depth of layer
α = Ratio of density of upper layer fluid to lower layer fluid
x = Distance along downstream direction
y = Distance perpendicular to x-direction
u = Uniform velocity over the depth along x-direction
v = Uniform velocity along y-direction
P = Hydrostatic pressure of fluid
β = Ratio of depths of lower layer to upper layer
C = Wave celerity
't' represents for time and subscripts '1', '2' and '3' denotes for upper layer, intermediate layer and bottom layer respectively.

24

Estimation of Tsunami Initial Displacement of Water Surface Using Inversion Method with a priori Information

Nobuaki Koike
Wakayama National College of Technology
Japan

1. Introduction

Up to now, because the tsunami initial displacement of water surface cannot be observed directly, the approach of presuming it from the fault model of the earthquake has been taken. Thanks to Yamashita and Sato (1974), it is established that an amount of the water level change in the surface of the sea is equal to the amount of the perpendicular ground change in the bottom of the sea caused by the earthquake. And, thanks to Aida (1969), the style of a present tsunami numerical calculation is established. The analytical solution proposed by Mansinha and Smylie (1971) is widely used, in order to presume the amount of the perpendicular ground change in bottom of the sea from the fault model. Still, for the risk evaluation of the tsunami, magnitude scales of earthquakes and those locations are assumed, the earthquake fault model is presumed, and a tsunami numerical calculation is done by the method as the above-mentioned.

On the other hand, the research on the estimation process of the tsunami initial displacement of water surface, that is the initial-value problem of the tsunami, includes pioneering research of Aida (1969). Here, the coefficient that becomes each amount of the change is requested by the least squares method, when setting up the unit source of wave in the sea area where the initial water level distribution is expected. However, it is necessary to note that the location of the unit source of wave is assumed from earthquake information beforehand. In the research on the tsunami inversion, the research of Satake (1986) is typical. Here the tsunami record of the tidal station is used by the research. However, Satake only used the tsunami record of the tidal station, not to have done the tsunami inversion in order to presume the tsunami initial displacement of water surface, but in order to examine no homogeneity of the fault movement. Actually, the amount of slip of each small fault that divides the fault model into some models is chosen as an unknown parameter of the inverse problem. Therefore, it can be said that the style that essentially assumes the fault model and calculates the numerical value of the tsunami will not have changed.

However, Koike et al. (2003) proposed a new method. Here, the tsunami initial displacement of water surface is presumed directly by the tsunami inversion. It is devised to reduce the number of the unknown parameters by using the base of wavelet in order to evade non-appropriateness of the inverse problem, and to solve the inverse problem by the least squares method uniquely. But, at the present stage theoretically setting the standard of the

selection of the base of wavelet is extremely difficult, and the theoretical consideration concerning the presumption accuracy becomes impossible.
In this study, we propose the method of solving the inverse problem by giving new a priori information. This inverse problem is that the unknown parameter is the tsunami initial displacement of water surface. This has seemed to be impossible in the tsunami inversion that has increased too much up to now the unknown parameter and uses the tsunami waveform at the tidal station. That is, it is clarified to be able to solve the inverse problem uniquely by newly giving a priori information, though the inverse problem cannot be uniquely solved by the least squares method without a priori information because there are a lot of unknown parameters.

2. Method

2.1 Superposition principle

When the linear long wave theory consists in the numerical calculation of a tsunami, the superposition principle of the output from the numerical calculation consists also. When the tsunami initial displacement of water surface assumes $\mathbf{e}_0$, this can be developed as follows, as the superposition of basis vector $\mathbf{e}_k$(k=1, 2,..., n).

$$\mathbf{e}_0 = c_1\mathbf{e}_1 + c_2\mathbf{e}_2 + \ldots + c_n\mathbf{e}_n \tag{1}$$

Here, c_k (k=1, 2,..., n) is a development coefficient.
It is $a_k(t)$ (k=1, 2,..., n: t is time) respectively as for the output of assuming basis vector $\mathbf{e}_k$ (k=1,2,...,n) to be an initial condition and calculating the numerical value. The numerical calculation output a(t) in an observation station to the former tsunami initial displacement of water surface $\mathbf{e}_0$, can be shown by the superposition principle as follows.

$$a(t) = c_1\, a_1(t) + c_2\, a_2(t) + \ldots + c_n\, a_n(t) \tag{2}$$

Thus, the superposition principle consists strictly theoretically. However, it is realistic in an actual numerical calculation to select the some among bases developed as Eq. (1) and to apply the superposition principle of Eq. (2). The problem of the selection of the base is important for the tsunami inversion. Actually, the number of unknown parameters that should be presumed, changes by the way how to select the base.

2.2 Formulation of tsunami inversion

Here, a tsunami inversion is formulated based on Eq. (2). A tsunami inversion presumes the tsunami initial displacement of water surface from the observation waveform of the tidal stations. In Eq. (2), the numerical calculation output a(t) thinks of the observation waveform d(t) in the tidal stations and development coefficient c_k(k=1, 2,..., n) thinks of unknown development coefficient m_k(k=1 ,2 ,..., n). A tsunami inversion can be formulated as follows.

$$\mathbf{Gm} = \mathbf{d} \tag{3}$$

$$G = \begin{bmatrix} a_1(1) & a_2(1) & \ldots & a_n(1) \\ a_1(2) & a_2(2) & \ldots & a_n(2) \\ \ldots & \ldots & \ldots & \ldots \\ a_1(t) & a_2(t) & \ldots & a_n(t) \end{bmatrix}$$

$$\mathbf{m} = [\, m_1, m_2, ..., m_n \,]^T$$

$$\mathbf{d} = [\, d(1), d(2), ..., d(t) \,]^T$$

When there are two or more numbers of tidal stations, it only has to increase the number of t of the observation waveform d(t). And, this inverse problem is able to be solved by some methods. We assume that unknown development coefficient m_k (k= 1, 2,..., n) are able to be presumed as follows.

$$\mathbf{m}^{est}=\mathbf{G}^{-g}\mathbf{d} \tag{4}$$

Here, $\mathbf{G}^{-g}$ is called as general inverse matrix, but isn't ordinary inverse matrix.
When the unknown development coefficient m_k^{est}(k= 1, 2,..., n) presumed thus are substituted for the development coefficient of the base that Eq. 1 corresponds, the estimation of the tsunami initial displacement of water surface $\mathbf{e}_0^{est}$ is finally requested as follows.

$$\mathbf{e}_0{}^{est}=m_1^{est}\,\mathbf{e}_1+m_2^{est}\,\mathbf{e}_2+...+\,m_n^{est}\,\mathbf{e}_n \tag{5}$$

2.3 Least squares method

Eq. (3) is the simplest equation in the inverse problems, and it is called the linear inverse problem. The linear inverse problem is often solved by the least squares method that generally minimizes the prediction error (ex., Menke, 1989). Here, we think how the linear inverse problem of Eq. (3) is solved by the least squares method. First of all, error E can be shown in the linear inverse problem as follows.

$$E=\mathbf{e}^T\mathbf{e}=(\mathbf{d}-\mathbf{Gm})^T(\mathbf{d}-\mathbf{Gm}) \tag{6}$$

And, each of partial differential of error E by m_k (k= 1, 2,..., n) are assumed to be 0. The following equation obtains.

$$\mathbf{G}^T\mathbf{Gm}-\mathbf{G}^T\mathbf{d}=0 \tag{7}$$

Here, if $[\mathbf{G}^T\mathbf{G}]^{-1}$ assumes to exist, the solution of the linear inverse problem by the least squares method can be requested as follows.

$$\mathbf{m}^{est}=[\mathbf{G}^T\mathbf{G}]^{-1}\mathbf{G}^T\mathbf{d} \tag{8}$$

Here, the inverse matrix $[\mathbf{G}^T\mathbf{G}]^{-1}$ assumes to exist, the solution by the least squares method has been derived. Because this assumption is an important problem, we examine it later.

2.4 Uniqueness of solution and selection of base

In Eq. (8), when the inverse matrix $[\mathbf{G}^T\mathbf{G}]^{-1}$ exists, the solution of the inverse problem has uniqueness. Therefore, the existence of the inverse matrix becomes important. This inverse matrix depends on only the data nucleus **G**, but not on the observed value **d**. And, as being able to understand from Eq. (3), the data nucleus **G** are the sets of Green's functions $a_k(t)$ (k= 1, 2,..., n). Therefore, the existence of the inverse matrix $[\mathbf{G}^T\mathbf{G}]^{-1}$ depends on Green's functions. In addition, it also depends on the selection of base which becomes the origin.
The study of Koike et al. (2003) had selected the base of wavelet as a base before. The method used in that study was the method derived for the inverse matrix $[\mathbf{G}^T\mathbf{G}]^{-1}$ to exist, in order to solve the tsunami inversion by least squares method with uniqueness. However, when the base by wavelet is used, because calculating all Green's functions in Eq. (2) is

impossible, they should actually discontinue the number n of bases in a suitable point. But, it is a current state that it is difficult to derive the standard theoretically.
In this study therefore, we will adopt the base in which only the 10 times 10 meshes have 1, but all the remainder have 0, that understands easily and knows by intuition. Though this is not a basis vector precisely, all the discussions of the above-mentioned consist, because the superposition principle consists even in this case. This also will be called a base in this study. Actually, if we will select such base, the character of data nucleus **G** becomes extremely worse and the inverse matrix $[\mathbf{G}^T\mathbf{G}]^{-1}$ doesn't exist. Therefore, the inverse problem cannot be uniquely solved by the least squares method. Then, it is necessary to give the condition in addition.

2.5 A priori information

If in the inverse problem the character of data nucleus **G** is bad and the inverse matrix $[\mathbf{G}^T\mathbf{G}]^{-1}$ doesn't exist, this problem is called as inferior decision. The inferior decision happens when many solutions whose prediction error E becomes 0 exist. Usually, when the number of unknown parameters is more than that of data, inferior decision problem is caused. To obtain the solution $\mathbf{m}^{est}$ to the inverse problem, the method of choosing only one appropriate solution among the infinite solutions whose prediction error becomes 0 is needed. For that, it is necessary to add other information not included to the inverse problem in Eq. (3). This additional information is called as a priori information (Jackson, 1979). That is quantitatively the expressions of the feature expectation of the solution beforehand, but no one is based on actual data.
In this study, it decides to be taken that the solution is smooth as a priori information. That is, tsunami initial displacement of water surface is expected to change smoothly. This a priori information can be formulated as follows. The smoothness of the continuous function of the space can be quantified with norm of the first derived function. It only has to use the difference between unknown development coefficients that are physically adjacent as the approximation of the differentiation for a discrete, unknown development coefficient. That is, the smoothness **I** of unknown development coefficient **m** is shown as follows (For one dimension).

$$\mathbf{I}=\begin{bmatrix} -1 & 1 & & & \\ & -1 & 1 & & \\ & & \cdots & & \\ & & & -1 & 1 \end{bmatrix}\begin{bmatrix} m_1 \\ m_2 \\ \cdots \\ m_n \end{bmatrix}$$

$$= \mathbf{Dm} \tag{9}$$

Here, **D** is the matrix who gives smoothness. The smoothness of the solution as the whole is shown as follows in the form of norm L.

$$L = \mathbf{I}^T\mathbf{I} = [\mathbf{Dm}]^T[\mathbf{Dm}]$$

$$= \mathbf{m}^T\mathbf{D}^T\mathbf{Dm} = \mathbf{m}^T\mathbf{W}_m\mathbf{m} \tag{10}$$

Here, the matrix $\mathbf{W}_m = \mathbf{D}^T\mathbf{D}$ can be interpreted as the weight coefficient, when norm of vector **m** is calculated.
And, the solution of the inverse problem when a priori information is given to the least squares method can be solved as follows by the minimization of $E+p^2L$ that is the combination of the prediction error and norm of the solution.

$$\mathbf{m}^{est}=[\mathbf{G}^T\mathbf{G}+p^2\mathbf{W}_m]^{-1}\mathbf{G}^T\mathbf{d} \tag{11}$$

Here, p is the weight coefficient. That is a coefficient by which a relative ratio of the importance is decided which the prediction error E and norm L of the solution are to put the emphasis. When p is large enough, the inferior decision part of the solution is minimized. However, the superior decision part of the solution tends also to be minimized at the same time. Consequentially, the solution doesn't become to minimize the prediction error E, and not become good presumption too much about the parameter of a true development coefficient. Oppositely, though the prediction error is minimized when putting p to 0, a priori information will not be considered at all, to decide only one parameter of the development coefficient of the inferior decision. Thus, when this equation regards as dumping least squares method, the theoretical method to decide the value of p isn't at present (Menke). However, from the viewpoint of the theory of probability idea, this weight coefficient p is not only to stabilize the solution, but also to decide that value by the ratio of the dispersion of the error in observational data and a priori data (Franklin, 1970). Here, Eq. (11) is regarded as dumping least squares method.

Thus, if the 10 times 10 mesh base is taken as a base even, the inverse problem can be solved by adding a priori information. However, when to doing the discussion related to uniqueness of the estimation $\mathbf{m}^{est}$ in Eq. (11), it becomes extremely important whether a priori assumption that the solution is smooth is appropriate

3. Results

Here, we will plainly explain the results of the theory in the preceding chapter through a concrete numerical calculation example of the tsunami inversion.

3.1 Location of computational domain and tidal stations

As the numerical calculation area where the tsunami inversion is done, the sea area off the coast of Japan shown in Fig.1 is taken. In addition, the tidal staions of nine points that actually exist as shown in Fig.1 are chosen.

3.2 10 times 10 mesh base and Green's function

When the 10 times 10 mesh bases (10km times 10km) shown in Fig.2 in which 10 times 10 meshes only take the value of 1 but remainder is all 0 are the initial condition, the waveforms (Green's function) in the tidal stations of nine points are calculated by the usual linear tsunami numerical calculation. In the computational domain of Fig.1, because the number of total meshes is 400 times 400, 10 times 10 mesh bases will be 40 times 40 = 1,600 pieces, and we have to calculate a tsunami numerical simulation 1,600 times to request all Green's functions. Actually, the number of calculating is fewer than 1,600, because we may not calculate the part in the land shown in Fig.1. However, here, the development coefficient (actually, tsunami initial displacement of water surface in 10 times 10 meshes) to 40 times 40 = 1,600 bases including the part in the land will be taken as unknown parameters.

3.3 A priori information

In the preceding chapter, when an unknown development coefficient is one dimension, the matrix which gives a priori information on the smoothness of the solution is shown. However, here, when an actual, unknown development coefficient is two dimensions, the matrix which gives the smoothness of the solution will be shown.

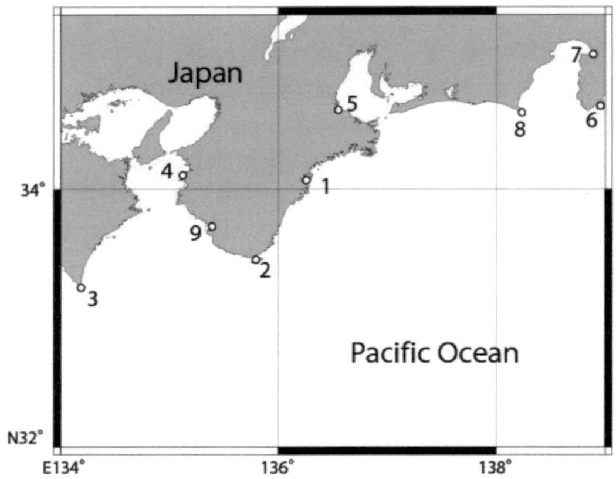

Fig. 1. Location of computational area and tidal stations

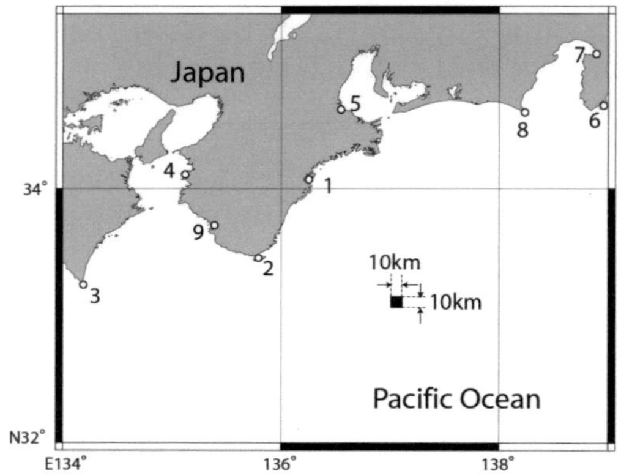

Fig. 2. The 10 times 10 mesh base

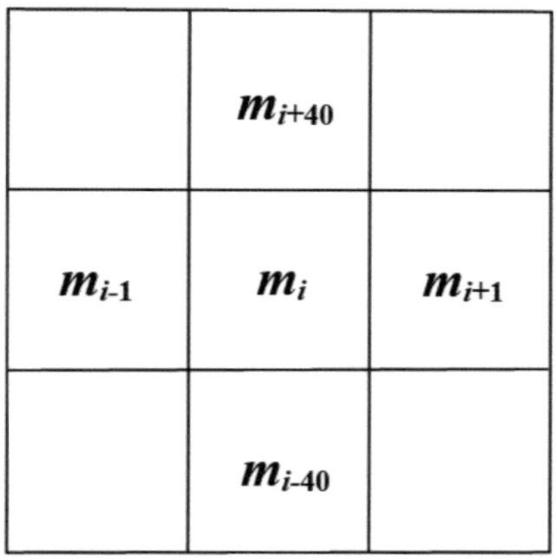

Fig. 3. Two dimension's arrangement of unknown development coefficient

Though Fig.3 is two dimension's arrangement of unknown development coefficient of 10 times 10 mesh bases, we think about the condition concerning development coefficient m_i. Though the development coefficient m_i is adjacent to the development coefficients m_{i+1}, m_{i-1}, m_{i+40}, and m_{i-40}, we has to think only m_{i+1} and m_{i+40}, because we think about m_{i-1} and m_{i-40} when each of m_{i-1} and m_{i-40} is the centre of Fig.3. The smoothness of the solution can be quantified with norm of the first derived function as described in the preceding chapter, and because we only has to use the difference instead of the derivation, the row corresponding to the development coefficient m_i of the matrix **D** in Eq. (9) is as follows.

$$[\ -2 \quad 1 \qquad 1\ ] \tag{12}$$

Here, -2 is in i column, and just right 1 is in i+1 column, and in addition right 1 is in i+40 column.
In addition, a priori information that the unknown development coefficient located in the land is 0 is given. Though the development coefficient located in the land do not have to be put in the unknown parameter of the inversion originally, giving the matrix becomes difficult if only the part of the land will be excluded when the matrix which gives the above-mentioned smoothness of the solution is considered. So, a priori information that the value is sure to become 0 is decided to be programmed by putting the development coefficient located in the land into the unknown number here. The condition of unknown development coefficient m_i located in the land being equal to 0 can be shown as follows.

$$[0 \ldots 0\,1\,0 \ldots 0]\begin{bmatrix} m_1 \\ m_2 \\ \ldots \\ m_i \\ \ldots \\ m_n \end{bmatrix}=0 \tag{13}$$

In order to put in such a constraint condition, we should include this equation of constraint condition in the row of the inverse problem **Gm=d**. And, as putting of infinitely large weight compared with other equations, we only has to adjust the weight matrix $\mathbf{W}_e$ of the prediction error $E=\mathbf{e}^T\mathbf{W}_e\mathbf{e}$ that considers weight (Lawson and Hanson, 1974).
Finally, when these a priori information is given to the least squares method, the solution of the inverse problem **Gm=d** can be solved as follows by the minimization of $E+p^2L=\mathbf{e}^T\mathbf{W}_e\mathbf{e}+p^2\mathbf{m}^T\mathbf{W}_m\mathbf{m}$ that is the combination of the prediction error and norm of the solution. That is, the combination of the prediction error and norm of the solution is shown as follows.

$$E+p^2L=(\mathbf{d}-\mathbf{Gm})^T\mathbf{W}_e(\mathbf{d}-\mathbf{Gm}) +p^2\mathbf{m}^T\mathbf{W}_m\mathbf{m} \tag{14}$$

And, each of partial differential of this equation by m_k (k= 1, 2,..., n) are assumed to be 0. The following equation obtains.

$$[\mathbf{G}^T\mathbf{W}_e\mathbf{G}+p^2\mathbf{W}_m\ ]\mathbf{m}-\mathbf{G}^T\mathbf{W}_e\mathbf{d}=0 \tag{15}$$

And, if the inverse matrix $[\mathbf{G}^T\mathbf{W}_e\mathbf{G}+p^2\mathbf{W}_m]^{-1}$ exists, the solution is required as follows.

$$\mathbf{m}^{est}=[\mathbf{G}^T\mathbf{W}_e\mathbf{G}+p^2\mathbf{W}_m\ ]^{-1}\mathbf{G}^T\mathbf{W}_e\mathbf{d} \tag{16}$$

3.4 Artificial observation data

Actually, here, we try to test numerically whether tsunami initial displacement of water surface can be presumed by the tsunami inversion using this method. That is, we apply this method to the artificial observation data, but not to the tsunami record in actual tidal stations. That is, we apply to the artificial observation data calculated from artificial tsunami initial displacement of water surface, and how much former tsunami initial displacement of water surface can be reproduced is examined.

First of all, the tsunami initial displacement of water surface (Fig.4) calculated from the fault model of the Tonankai earthquake in 1944 by the method of Mansinha and Smylie (1971) is used. And, this is assumed to be an initial condition, the usual linear tsunami numerical calculation is calculated, and the calculation waveform in the nine tidal stations of Fig.1 is assumed to be artificial observation data.

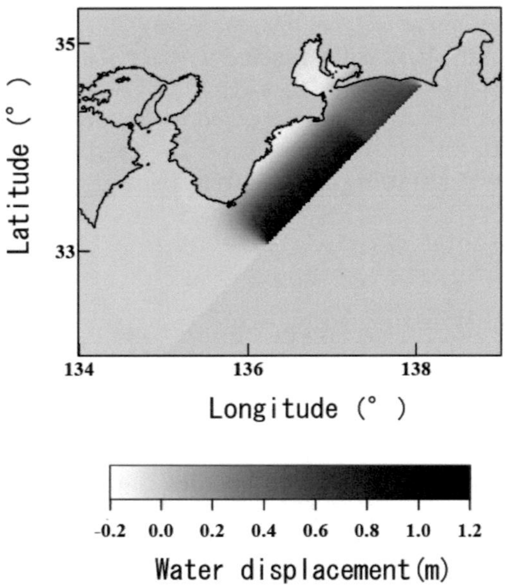

Fig. 4. The artificial tsunami initial displacement of water surface

3.5 Tsunami inversion

Here, we executed the tsunami inversion based on the artificial observation data of every 20 seconds until after two hours from the earthquake generation. Moreover, the number of tidal stations used all of the nine points. In addition, the value of p in Eq. (16) was set to 1.0. The tsunami initial displacement of water surface as shown in Fig.5 can be presumed by doing the tsunami inversion on the above-mentioned condition. When this is compared with former artificial tsunami initial displacement of water surface (Fig.4) that is the correct answer, it is understood to be corresponding by considerable accuracy if the notched part of 10 times 10 meshes that cannot be avoided because of the setting of an unknown development coefficient is excluded. If Root Mean Square Error (RMSE) is actually calculated, it becomes 0.070m.

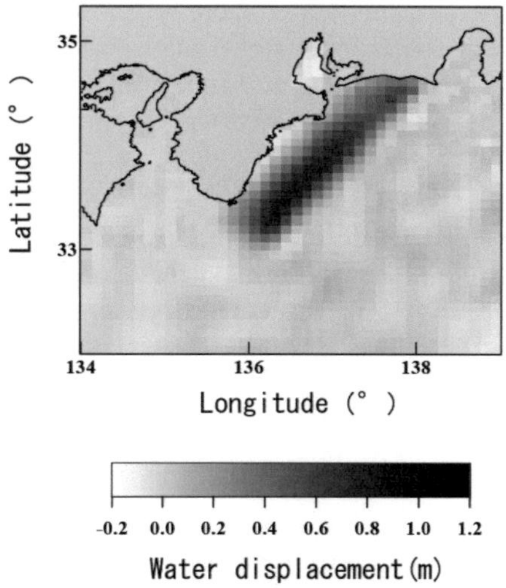

Fig. 5. Estimation result of tsunami initial displacement of water surface by tsunami inversion

4. Consideration about estimation error

When the tsunami inversion is done, it often becomes a problem which tide stations' waveform records should be used and how much observation time should be used. Because the influence of the observation error and the nonlinearity of a tsunami are included in the tide record of a tsunami in the real problem, consideration of the record of tide stations is necessary. However, here, uniqueness of the solution and the estimation error clarified when thinking about the tsunami inversion as an ideal linear inverse problem that the observation error is not included will be considered.

4.1 Model resolution matrix and uniqueness of solution

First of all, it is assumed that true development coefficient $\mathbf{m}^{true}$ of $\mathbf{Gm}^{true}=\mathbf{d}^{obs}$ exists. And, we examines whether the estimation of an unknown development coefficient is how much close to this true solution. When it substitutes the equation of $\mathbf{Gm}^{true}=\mathbf{d}^{obs}$ to the observation value for the equation of $\mathbf{m}^{est}=\mathbf{G}^{-g}\mathbf{d}^{obs}$ that shows the estimation of an unknown development coefficient, it becomes the following.

$$\mathbf{m}^{est}= \mathbf{G}^{-g}\mathbf{d}^{obs}=\mathbf{G}^{-g}[\mathbf{Gm}^{true}]$$

$$= [\mathbf{G}^{-g}\mathbf{G}]\mathbf{m}^{true}=\mathbf{Rm}^{true} \tag{17}$$

Here, the matrix of $\mathbf{R}=\mathbf{G}^{-g}\mathbf{G}$ is called as the model resolution matrix of n times n (n is a number of unknown development coefficients) (Wiggins, 1972). In this method, because the generalized inverse matrix $\mathbf{G}^{-g}$ can be regarded as $\mathbf{G}^{-g} = [\mathbf{G}^T\mathbf{W}_e\mathbf{G}+p^2\mathbf{W}_m]^{-1}\mathbf{G}^T\mathbf{W}_e$ from Eq. (16), the model resolution matrix is requested as follows.

$$\mathbf{R}=[\mathbf{G}^T\mathbf{W}_e\mathbf{G}+p^2\mathbf{W}_m]^{-1}\mathbf{G}^T\mathbf{W}_e\mathbf{G} \tag{18}$$

If $\mathbf{R}=\mathbf{I}$ ($\mathbf{I}$ is the identity matrix) is true, the value of each unknown development coefficient can be uniquely decided (uniqueness of the solution). If $\mathbf{R}$ is not the identity matrix, the estimation of an unknown development coefficient will actually show the average with the weight of a true development coefficient. The model resolution matrix is a function of a priori information added to data nucleus **G** and the problem. The model resolution matrix is irrelevant to an actual value of the observation data.

4.2 Introduction of parameter resolution

By the foregoing paragraph, it has been understood to be able to discuss uniqueness of the solution by examining the model resolution matrix. That is, when a row of the model resolution matrix **R** is taken out, if only the value of a certain column is 1 and the value of the column of the remainder is all 0, a corresponding unknown development coefficient is uniquely decided. However, when it is not so, an unknown development coefficient cannot be uniquely decided. But, if the value of the column corresponding to an unknown development coefficient is close to 1 and the difference of the value of the row is small, it can be interpreted that it is case when the unknown development coefficient is decided almost uniquely and the character as the inverse problem is good. Therefore, if the difference condition of the value of the row corresponding to an unknown development coefficient in the model resolution matrix **R** can be quantified, it is convenient to do a discussion about uniqueness of the solution. So, we think introduction of the index of the parameter resolution **C**. First of all, from the value (r_1, r_2, ..., r_i, ..., r_n) of the row corresponding to an unknown development coefficient m_i of the model resolution matrix **R** of Eq. (17), the estimation m_i^{est} of an unknown development coefficient can be shown as follows.

$$m_i^{est}=r_1 m_1^{true}+r_2 m_2^{true}+\ldots$$

$$+r_i m_i^{true}+\ldots+r_n m_n^{true} \tag{19}$$

Here, if r_i=1 is and r of the remainder is all 0, unknown development coefficient m_i is uniquely decided. Moreover, when r_i is close to 1, only the one that the value of other r is close to m_i is large and the difference condition is small, it is thought that the estimation is also very close to the true value. In order to express this quantitatively, it only has to think about the weighed mean average, by giving big weight to the value of the development coefficient to which the distance is close, and by giving only small weight to a distant development coefficient. Then, parameter resolution C_i to estimation m_i^{est} of an unknown development coefficient will be shown as follows.

$$C_i = \sum_{l=1}^{n}\left(r_l \times \frac{1}{1+\sqrt{\Delta x_{l-i}^2 + \Delta y_{l-i}^2}} \right) \tag{20}$$

Here, Δx_{l-i} is the distance in the direction of x of development coefficient m_i and m_l, and Δy_{l-i} is the distance in the direction of y. For instance, when an unknown development coefficient is arranged shown as Fig.3, the distance of m_i and m_{i+1} is 1, and the distance of m_i and m_{i+41} is $2^{1/2}$. The parameter resolution C_i becomes 1 when the value of $0 < C_i < 1$ is taken and an unknown development coefficient is decided uniquely, and the more approaches 0, the more the difference condition grows and the character worsens. Therefore, it is thought that the estimation error is small when resolution C_i is close to 1, but the estimation error is big when it is close to 0.

Though there is Dirichlet spread function based on L_2 norm of the difference of the resolution matrix and the identity matrix as an amount that shows the goodness of the character of the entire of the parameter resolution matrix. This is different from the parameter resolution introduced in this chapter. The parameter resolution is an index that pays attention to one parameter (development coefficient).

4.3 Relationship between parameter resolution and estimation error

Here, the influence that the parameter resolution C_i affects the estimation error will be seen through the concrete example. The estimation error is defined as follows.

$$\text{estimation error (m)} = |\text{true value (m)} - \text{estimation value(m)}|$$

Fig.6 shows the estimation result that executes the tsunami inversion used of tide stations No.1, 2 in Fig.1 as well as Chapter 3, and Fig.7 is used of No.2, 3. The value of root mean square error RMSE is 0.092m, and 0.119m respectively. Compared with Fig.4 that is the correct answer, it is understood that estimation error of Fig.6 is small than that of Fig.7. In these two cases, the condition of inversion is the same, which is 20 seconds at observation intervals, is two hours at observed time and is p=1.0, but the location of the observation stations is only different. Where is the difference of this estimation error caused from? By examining the parameter resolution, the difference is possible to explain to some degree.

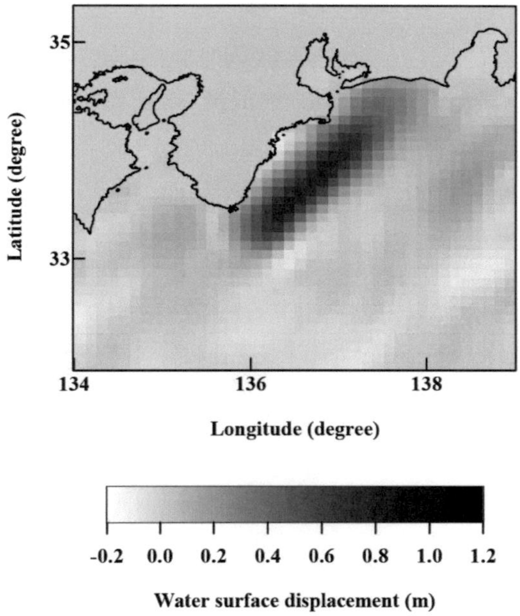

Fig. 6. Estimation of tsunami initial displacement of water surface (in the case used of No.1, 2 stations)

Fig.8 and Fig.9 are figures drawn by arranging the parameter resolution, when the tsunami inversion is done by using the combination of observation stations No.1, 2 and No.2, 3, at the location of an unknown development coefficient. From these figures, while in Fig.8 the area where the parameter resolution is 0.5 or more covers the area of tsunami initial displacement of water surface of the correct answer of Fig.4, in Fig.9 the area where the parameter resolution is 0.5 or more moves to the west side and the parameter resolution has

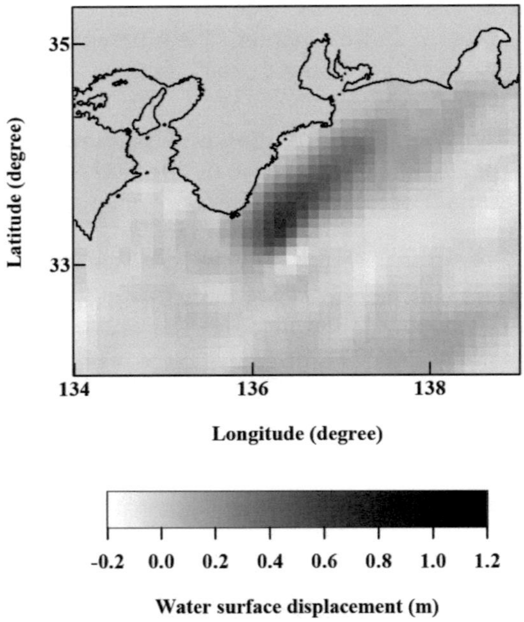

Fig. 7. Estimation of tsunami initial displacement of water surface (in the case used No.2, 3 stations)

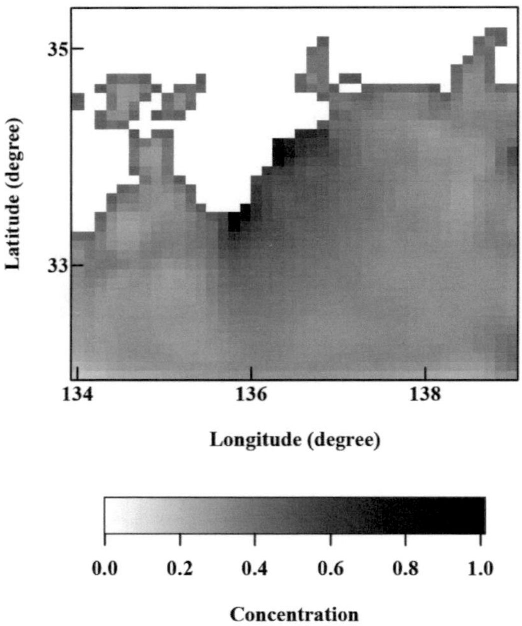

Fig. 8. Parameter resolution when the tsunami inversion is done (in the case used of No.1, 2 stations)

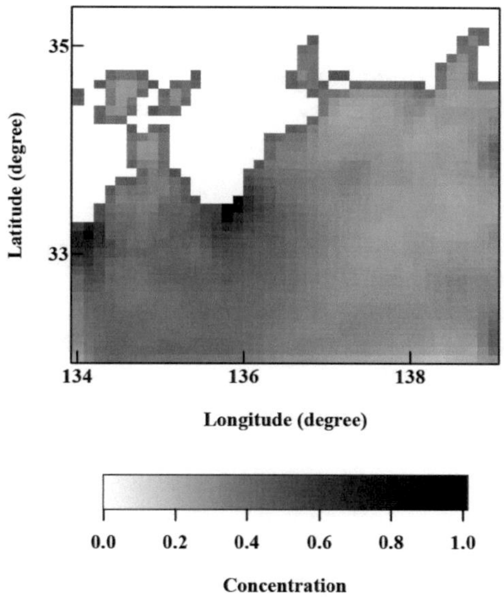

Fig. 9. Parameter resolution when the tsunami inversion is done (in the case used of No.2, 3 stations)

lowered in half an east side of the area of tsunami initial displacement of water surface of the correct answer of Fig.4. When Fig.8 and Fig.9 of distribution of these parameter resolutions are made to correspond to Fig.6 and Fig.7 respectively, it is understood that the estimation accuracy has lowered in the area where the parameter resolution is low, though the estimation accuracy is high in the area where the parameter resolution is high. Thus, the estimation error of an unknown development coefficient understands that the expectation is roughly applied by examining the parameter resolution.

5. Conclusion

Here, the condition that the solution is smooth as a priori information is taken to the tsunami inversion, and we proposed a new method of presuming the tsunami initial displacement of water surface directly from the observed waveforms of a tsunami in the tidal stations. And, when the artificial observation data is set and the condition of the tsunami inversion is changed, it was considered how the estimation error became. The conclusion obtained in this study is as follows.

1. When the tsunami initial displacement of water surface is taken as an unknown parameter, the number of the unknown parameters increases more than the number of observation data and the unknown parameters cannot be specified by the tsunami inversion method that uses only a usual least squares method. Then, we proposed a new method of specifying an unknown development coefficient by taking to the inversion a priori information which the development coefficients of the adjoined unknown parameters are smooth. And, it was shown that the tsunami initial displacement of water surface can be estimated by using this method due to the error with a small.

2. In order to discuss uniqueness of the unknown development coefficient requested by the tsunami inversion, the index of parameter resolution based on the model resolution matrix was introduced. Though uniqueness of an unknown development coefficient consists only when the parameter resolution is 1, when the parameter resolution is 0.5 or more, it showed that the estimation error lowered to 0.1m or less from consideration when the artificial observation data was set.
3. The parameter resolution is an index that doesn't depend on the observation data, only depending on the condition of Green's function and the inversion. Therefore, when it tries to presume the tsunami initial displacement of water surface that actually happens, it can be discussed how accuracy the tsunami initial displacement of water surface can be presumed by firstly examining the distribution of the parameter resolution. Thus, the index of parameter resolution is an index very useful for the discussion about the estimation error of the tsunami inversion.
4. As an application of this study, the tsunami initial displacement of water surface is presumed by using the data of the tsunami recorder set up offshore, and the tsunami numerical value forecast is that the height of tsunamis of various places in the coast part is forecast by the numerical calculation. The installation plan of the tsunami recorder is able to discuss by examining the introduced parameter resolution in this study, for example, where it is necessary to set up the tsunami recorders of how much number.

6. References

Aida, I. (1969). Numerical experiments for tsunamis caused by moving deformation of the sea bottom, *Bull. Earthq. Res. Inst.*, Vol.47, pp. 849-862.

Franklin, J. N. (1970). Well-posed stochastic extensions of ill-posed linear problems, *J. Math. Anal. Appl.*, Vol.31, pp. 682-716.

Jackson, D. D. (1979). The use of a priori data to resolve non-uniqueness in linear inversion, *Geophys. J. Roy. Astron. Soc.*, Vol.57, pp. 137-157.

Koike, N., Kawata, Y. & Imamura, F. (2003). Far-Field Tsunami Potential and A Real-Time Forecast System for the Pacific Using the Inversion Method, *Natural Hazards*, Kluwer Academic Publishers, pp.423-436.

Lawson, C. & Hanson, D. J. (1974). *Solving Least Squares Problems*, Prentice-Hall, Englewood Cliffs, New Jersey., Vol.61, No.5, pp.1433-1440.

Mansinha, L. & Smylie, D. E. (1971). The displacement fields of inclined faults, *Bull. Seism. Soc. Am.*, Vol.61, No.5, pp.1433-1440.

Satake, K. (1989). Inversion of tsunami waveforms for the estimation of heterogeneous fault motion of large submarine earthquake, *J. Geophys. Res.*, pp. 5627-5636.

Menke, W. (1989). Geophysical Data Analysis: *Discrete Inverse Theory*, Academic Press.

Wiggins, R. A. (1972). The general linear inverse problem : Implication of surface waves and free oscillations for earth structure, *Rev. Geophys. Space Phys.*, Vol.10, pp.251-285.

Yamashita, T. & Sato, R. (1974). Generartion of tsunami by a fault mdel, *J.Phys.Earth,* Vol.22, pp.415-440.

25

The All-Source Green's Function of Linear Shallow Water Dynamic System: Its Numerical Constructions and Applications to Tsunami Problems

Zhigang Xu
Modelling and Operational Oceanography, Canadian Hydrographic Service, Maurice Lamontagne Institute, Fisheries and Oceans
Canada

1. Introduction

Propagation of tsunamis in deep water (>50 m, Shuto, 1991) obeys linear dynamics. For a linear dynamics system, any response of the system to arbitrary external forcing can be expressed as a linear combination of the Green's functions. The Green's functions can be pre-calculated for fixed physical parameters and geometry of the system, whereas the pre-calculation implies a great time saving, which is essential when a catastrophe occurs and an immediate solution is required.

A Green's function has traditionally been calculated as a domain wise solution to a unit-impulse acting at a grid point. Such a Green's function may be referred as a single-source Green's function (SSGF). However a tsunami is rarely triggered at a single source point. A reasonable extension is therefore to pre-calculate a set of such SSGFs corresponding to a group of source points in a pre-assumed tsunami source region. Nevertheless when a future tsunami happens outside of the pre-assumed source region, the pre-calculated SSGFs will not be helpful.

Xu (2007) proposed a new type of Green's function, an all-source Green's function (ASGF). An ASGF focuses on a receiver point, regarding all the model grid points as potential sources. The computational cost for an ASGF is the same as that for an SSGF. However an ASGF provides thorough preparedness for a point of interest (POI) against all possible sources. Tsunamis from different sources may arrive at a POI at different times and may have different maximum wave amplitudes within certain period. Xu showed that with an ASGF one could make an arrival time map and a map of relative maximum amplitudes for a POI. Xu further demonstrated that with the ASGFs, one could also establish a real-time simulation system, which would give tsunami arrival time series at one or more POIs immediately.

This chapter will describe in details the ASGF in terms of its numerical constructions and its applications to tsunami problems. It will first present the governing equations and boundary conditions. It will move on to grid auto-generations from a realistic topographic dataset. It will then show how to build the global operators, with which the shallow water equations can be recast as a single matrix equation, with all the dynamics and boundary conditions encapsulated. Considerations will be given on how to stabilize the matrix before it can be used iteratively. With all these preparations, the chapter will then show how the ASGF can easily be calculated from the dynamics matrix, as easy as to calculate the traditional SSGF.

Section 7 will discuss applications of the ASGF to tsunami problems in terms of both tsunami preparedness and real-time simulations. The section will show how the ASGFs can be used for making tsunami arrival time maps and tsunami amplitude maps for a POI. A web-based real-time simulation system for the India and Pacific Oceans will also be demonstrated. The simulation system, which can be found at odylab.uqr.ca[1], allows its users to see the tsunami arrival time series at one or several POIs, upon receiving the source region specified by users on its interactive map.

2. Linear Shallow Water Equations (LSWE) and boundary conditions

2.1 Continuous form of LSWE

The governing equations consist of

$$\frac{\partial \eta}{\partial t} = -\frac{1}{R\cos\varphi}\left(\frac{\partial U}{\partial \lambda} + \frac{\partial V\cos\phi}{\partial \varphi}\right) \tag{1}$$

$$\frac{\partial U}{\partial t} = -\frac{gh}{R\cos\varphi}\frac{\partial \eta}{\partial \lambda} + fV - \kappa\frac{U}{h} \tag{2}$$

$$\frac{\partial V}{\partial t} = -\frac{gh}{R}\frac{\partial \eta}{\partial \varphi} - fU - \kappa\frac{V}{h} \tag{3}$$

The meaning of the notation is listed below:

λ , φ, t	Longitude, latitude, and time variables respectively.
η, U, V,	Sea surface elevation, and components of total stream in longitudinal and latitudinal directions.
f, g, R,	Coriolis parameter, gravity acceleration, and the Earth's mean radius.
h, κ	Water depth and bottom frictional coefficients.

The components of total stream are defined as the following depth integrals

$$U = \int_{-h}^{0} u dz, \qquad V = \int_{-h}^{0} v dz, \tag{4}$$

where u and v are eastward and northward velocity components respectively. The total stream may also be referred as the mass transport.

2.2 Boundary conditions

The boundary conditions consist of zero flows normal to the coasts,

[1] If you find this becomes inaccessible, please contact me at zhigang.xu@dfo-mpo.gc.ca or at zhigang_xu_98@yahoo.com.

$$U = 0 \qquad \text{at the west and east coasts} \tag{5}$$
$$V = 0 \qquad \text{at the south and north coasts} \tag{6}$$

and the Sommerfeld radiation conditions at open water boundaries,

$$\frac{\partial p}{\partial t} + c\frac{\partial p}{\partial n} = 0 \qquad \text{at open water boundaries} \tag{7}$$

where p may stand for U, V or η, c is a phase speed, and ∂n means an infinitesimal line segment along the directions outward and normal to the open boundaries. Chapman (1985) discussed various forms of the phase speed and their numerical effects on the quality of the model solutions inside of the model domain. For this study, the long gravity wave $c = \sqrt{gh}$ is taken.

2.3 Frictional coefficient as a function of water depth

The last terms on RHS of Eqs. (2) and (3) represent the frictional effects of the seabed on the water column above. Their forms imply that the bottom frictional force is linearly proportional to the depth averaged velocities. The coefficient κ reflects the proportionality. Heaps (1969) simply took a constant κ, 0.0024 m/s, for his two-dimensional numerical model in the North Sea. It is desirable to have a spatially varying κ since there is less friction in deep water than in shallow water. To derive κ as a function of h, we may go back to a more general frictional parameterization as follows (e.g., Brummelhuis *et al*, 1993)

$$\left(\tau_x^b, \tau_y^b\right) = \frac{g\sqrt{\overline{u}^2 + \overline{u}^2}}{\mathcal{C}^2}(\overline{u}, \overline{v}) \tag{8}$$

where $\mathcal{C}$ is the Chezy's coefficient, whose unit is $m^{1/2}s^{-1}$. This is a non-linear form in terms of the velocities, the stresses being a quadric function of the velocity components. $\sqrt{\overline{u}^2 + \overline{u}^2}$ is usually replaced by an estimate of the magnitude of background currents, in order to preserve the linearity of the whole system. Let $|w|$ represents such an estimate. Further the Chezy's coefficient can be linked with water depth as $\mathcal{C} = \frac{1}{\mathcal{N}}h^{1/6}$ where $\mathcal{N}$ represents Manning's roughness. Now Eq.(8) can be transformed as

$$\left(\tau_x^b, \tau_y^b\right) = \kappa(\overline{u}, \overline{v}) \tag{9}$$

with

$$\kappa = \frac{g\mathcal{N}^2\,|w|}{h^{1/3}} \tag{10}$$

as desired.

The tidal currents are usually taken as the background currents in oceans, whose typical value is of order of 0.1m/s. The Manning's roughness in ocean appears to be on a range of 0.01 to 0.4 (Ding et al 2004; Fujima 2001). With $|w| = 0.1$m/s, if we let $\mathcal{N}$ be 0.1 and h be 65m to represent the mean water depth of the North Sea (Heaps, 1971), we can recover $\kappa = 0.0024$m/s as used by Heaps (1969). If we still use $\mathcal{N} = 0.1$ but let $h = 4000$m as typical water depth in deep oceans, we can have $\kappa = 6 \times 10^{-4} \mathrm{m/s}$. For the India-Pacific model domain, the minimum, mean and max values of κ are $4.5 \times 10^{-4} \mathrm{m/s}$, $7.6 \times 10^{-4} \mathrm{m/s}$ and $4.6 \times 10^{-3} \mathrm{m/s}$.

2.4 A matrix equation to be derived for hosting discrete LSWE

It is necessary to discretize the continuous form of LSWE and boundary conditions in order to be able to solve them numerically. Discretization for such a purpose is a common practice, but what is pursued here is beyond. The goal here is to recast the shallow water equations in the following matrix form:

$$\begin{bmatrix} \eta \\ U \\ V \end{bmatrix}^{(\tau+1)} = A \begin{bmatrix} \eta \\ U \\ V \end{bmatrix}^{(\tau)} \qquad (11)$$

where τ is an index of the time step. The coefficient matrix A, encapsulates the model physics and geometry, hence it may also be called the dynamics matrix. It is also called propagation matrix because it propagates the solutions from the current time step to the next. The column vector $[\eta\ U\ V]^T$ consists of all the model variables defined on the grid points.

There are advantages in having such a matrix form, especially for tsunami problems, as will be shown later on. In order to arrive at Eq.(11), we have to first construct some global operators, which in turn require an organized grid beforehand.

3. An organized rectangular grid

A rectangular grid has been the simplest to construct comparing with the other types of grids (e.g. a triangular grid): two sets of straight lines plus a land masking matrix equals a grid. However for assembling the dynamics matrix A, such a simple approach is not good enough, because all the entries into the matrices and vectors must be valid. We have to go above this basic level.

3.1 Issues encountered in generating grid from a topographic dataset

The starting point for generating a grid is a topographic dataset. From a topographic dataset to a valid grid, we will encounter some issues and have to address them well before we can have a valid grid.

Shown in Panel (a) of Figure 1 is a mini topographic dataset. The dataset is cell oriented, meaning that the water depths or land heights are defined on the centers of the cells. The cells are outlined by the two sets of thick straight lines. The centers of the cells are intersected by the two sets of thin straight lines and also marked by the blue dots. The gray cells are land cells and white ones are water cells. This mini topographic dataset contains all the issues that have to be addressed during the grid generation:

1. *Singletons.* A singleton is a single water cell not connected with others. Examples of singletons in Panel (a) are the water cells centered at (i=14, j=34), and the one at (i=38, j=14). There is no need to retain singletons. Retaining them not only wastes computational resources, but may also be harmful because they may mislead the judgment of minimum of the widths of the water cells. When a singleton lies on higher latitude than all other water cells, the width of that singleton would dictate the minimum value of the horizontal grid spacing, which would in turn cause the time step unnecessarily smaller than it could be without that singleton.

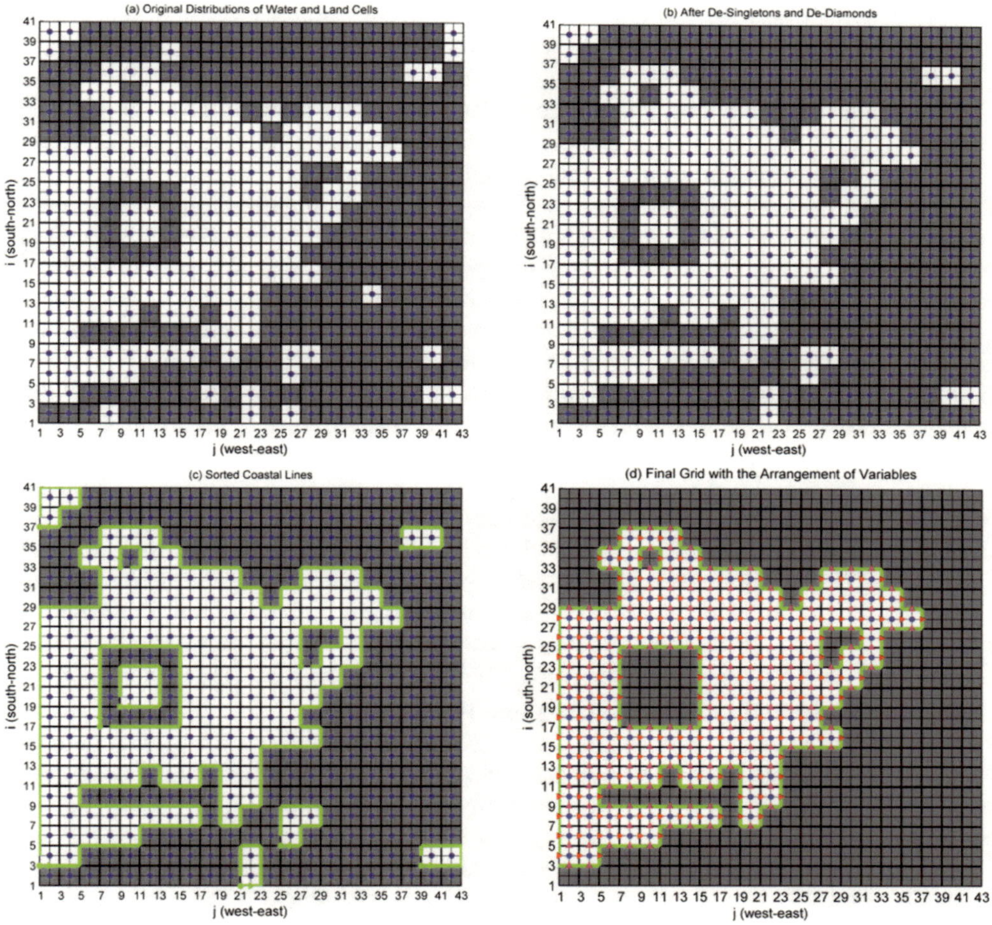

Fig. 1. Procedure from a topographic dataset to a model grid. Panel (a) shows the original distributions of water cells and land cells (in gray) in a topographic dataset. Panel (b) shows the results after removal of the singletons and the diamonds. Panel (c) shows the results after sorting out the coastal lines. Panel (d) defines the model domain with model variables arranged on the grid points.

2. *Diamonds*. A diamond is a pair of water cells connected by a common corner point. The water cells centered at (i=8, j=16) and (i=10, j=18) in Panel (a) of the figure forms a diamond, sharing a corner point at (i=9, j=17). Diamonds are unwanted because they can cause ambiguity in tracing coastal lines. For example, starting from the point of (i=3, j=1), one can easily follow the coastal line until he reaches the point of (i=9, j=17), where he will have an ambiguity in deciding which direction for his next move. It is necessary to trace coastal lines so that we can well define a model domain.
3. *Coastal lines*. A coastal line consists of a series of common edges shared by water cells and land cells. In order to well define a model domain, we need an algorithm to automatically find out all of the coastal points and sort them into loops. The longest coastline and any others nested in it (excluding those nested doubly or more) form a multiply connected

region, which defines the model domain. Water cells that are outside of the multiply connected region can be disregarded (by turning them into land cells).

4. *Lagoons*. A lagoon consists of a group of water cells which are connected with themselves only, but not with the rest. The water cells centered at (i=20, j=10), (i=22, j=10), (i=21, j=12) and (i=22, j=12) is one of the lagoons presented in Panel (a). Retaining lagoons are as undesirable as retaining singletons.

These issues can be well addressed and the auto-generation of organized grid can be achieved. Limited by the volume of the chapter, details of the implementations are omitted. Only the basic procedure is outlined in Figure 1: Panel (a) is what the original topographic data look like. Panel (b) is after removal of singletons and diamonds. Panel (c) is after sorting out all the coastal points. The coastal lines are shown in green. The longest one and those nested in only once form a multiply connected region as the model domain. The coastal line whose starting point is at (i=19, j=9) is excluded. It is nested within a nested coastal line, it is nested doubly. Panel (d) is the final stage where all the grid points that are outside of the multiply connected region are removed, and the model variables are arranged on the grid points, as indicated by the blue dots for η and red arrows for U and V. Here the variables are arranged according to Arakawa C-grid arrangement (Arakawa and Lamb, 1977).

All the η's, U's and V's defined on the grid can be viewed as elements of three long vectors, [η], [U] and [V]. The square bracket is used to explicitly indicate that the single letter inside is a vector, but may be dropped off in a context where it is clear that the single letter indicates a vector. From the three vectors, we can also form a super long vector [η U V].

3.2 An auto-generated grid for the India and Pacific Oceans

The General Bathymetric Chart of the Oceans (GEBCO[2]) provides a fine global bathymetry dataset of half minute resolution (GEBCO_08). With this dataset and the grid generation method mentioned above, we can automatically generate a model grid for any part of the world oceans. Shown in Figure 2 is a model domain for the India and Pacific Oceans with the grid generated from GEBCO_08 decimated to a 5-minute resolution. For clarity, only the coastal lines are shown. The lines in red are the coastal lines, which define the model domain. The blue lines would be the coastal lines if the lagoons were not removed. Note, according to the lagoon definition given above, parts of other big oceans, such as of the Atlantic Ocean, are also viewed as lagoons, because they are outside of the multiply connected region in question.

There are 3,771,781 η-points, 3,782,233 U-points, 3,784,476 V-points, and 970 coastal lines. The maximum water depth is 10545m, and the minimum water depth is set to be 10m. The mean water depth of the domain is 3806m. In the figure, the green dots show the DART stations (Deep-ocean Assessment and Reporting of Tsunamis[3]). The three gray zones are the sponge zones next to the three open water boundaries. Both the DART stations and the gray zones will be used later on. A sponge zone is designed to absorb exiting waves so that the unwanted wave reflections can be reduced to minimum. Such a sponge effect has been customarily achieved by setting up much larger than normal values of frictional parameter values. In this study, however, the sponge effect will be achieved through a two-dimensional Lax-Friedriches scheme, more details of which will be given in Section 5.2.

---

[2] http://www.gebco.net

[3] http://www.ndbc.noaa.gov/dart/dart.shtml

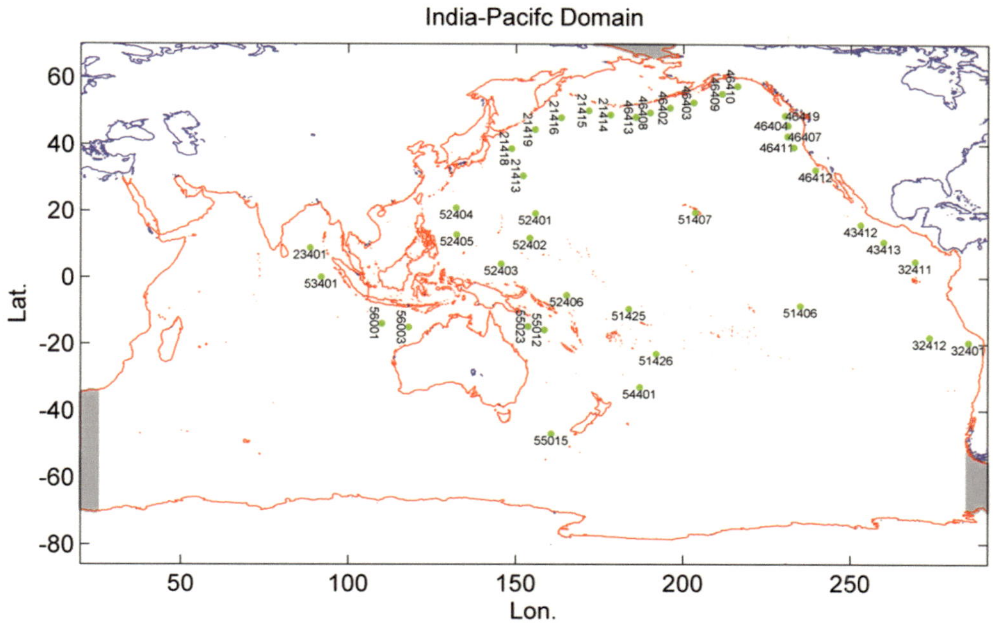

Fig. 2. An India-Pacific grid auto-generated from GEBCO_08 topographic dataset. The coastal lines in red define the model domain. The DART stations are indentified by the green dots and their station numbers. The three gray zones are sponge zones next to the three open water boundaries.

3.3 K-indexes of model variables

The model variables defined on the grid points can have their own indexes. For example, there are 157 η-points in Panel (d) of Figure 1, and they can be numbered from 1 to 157, counted from the lower left corner upwards and rightward. Therefore, the first η is at *(i=4, j=2)* and the second one is at *(i=6, j=2)*, and the last one is at *(i=28, j=36)*. We may use symbols k_e, k_u and k_v to denote the linear indexes of the variables η, U and V. respectively. In a context where there is no ambiguity, the subscripts with the k's may be dropped off.

Therefore a variable defined on a grid point can be addressed either through an (i, j)-index or a k-index. Obviously there is one-to-one relationship between the two indexing systems. Their conversions will be needed in the next section. Thus let us assume here that the following three index conversion functions have been made:

$$k_e = \text{KeIndex}(i,j) \tag{12}$$

$$k_u = \text{KuIndex}(i,j) \tag{13}$$

$$k_v = \text{KvIndex}(i,j) \tag{14}$$

which convert the *(i, j)*-indexes to the *k*-indexes. These functions are easy to make, so the details on their implementation are omitted here.

4. Global operators

This section will discuss how to assemble the global operators. The global operators, which come as sparse matrices, will be integrated into the sub-matrices of the dynamic matrix A as in Eq.(11).

4.1 Sea surface gradient operators, G_x and G_y

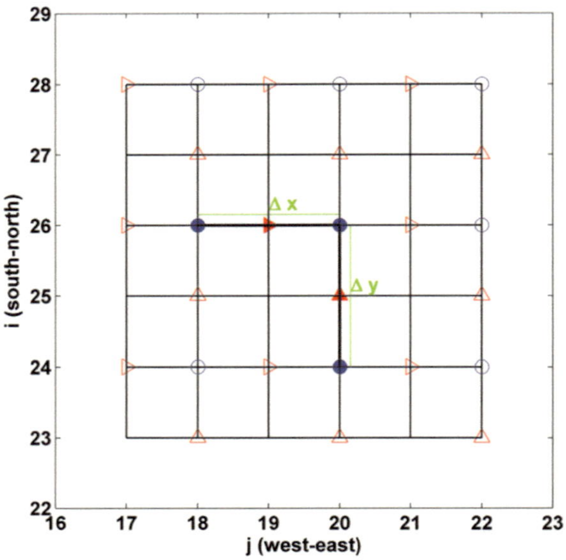

Fig. 3. Grid stencil for consideration of the sea surface gradients. The grid points indicated by the horizontal arrows are the U-points, those indicated by the vertical arrows are the V-points, and those by the blue circles are the η -points.

Extracted from Panel (d) of Figure 1, Figure 3 serves a grid stencil to construct the operators for the sea surface gradients. The x-directional sea surface gradient at the point (i, j) for an interior and non coastal U-point, indicated by the solid horizontal red arrow in the figure, can be approximated as

$$\frac{gh}{R\cos\phi}\frac{\partial\eta}{\partial\lambda}\bigg|_{i,j} \approx \frac{gh_{i,j}}{\Delta x_{i,j}}\begin{bmatrix}-1 & 1\end{bmatrix}\begin{bmatrix}\eta_{i,j-1}\\ \eta_{i,j+1}\end{bmatrix} \tag{15}$$

where $\Delta x_{i,j} = x(i, j + 1) - x(i, j - 1)$ with $x = R\lambda\cos\phi$. Note that coastal U-points have been excluded from consideration. We need not to assign a force for a coastal U-point, because the value there should be zero all the time.
The coefficient row vector on the RHS of the equation is a local operator, since it operates only on the two neighbouring η-points. We need to assemble it into a global operator. With aid of the index conversion functions given by Eqs. (12) to (14), we can have

$$G_x(k_0, [k_1\ k_2]) = \frac{gh_{i,j}}{\Delta x_{i,j}}\begin{bmatrix}-1 & 1\end{bmatrix} \tag{16}$$

where $k_0 = \text{KuIndex}(i,j)$ and $k_{1,2} = \text{KeIndex}(i, j \mp 1)$. Thus G_x is a global gradient operator, which upon operating on the elevation column vector $[\eta]$, produces a column vector of sea level gradient in the x-direction. Here the word *x-direction* is synonymous to the *longitudinal direction*.

Matlab users should avoid looping through individual (i, j)-points for the assembly. Vectorizing the (i, j)-indexes will greatly enhance the speed of the assembly. Also G_x should be treated as a sparse matrix, and use the following syntax in calling the Matlab sparse function for the assembly:

$$G_x = G_x + \text{sparse}(k_0, k_1, \frac{-gh_{i,j}}{\Delta x_{i,j}}, \text{kumax}, \text{kemax}) \tag{17}$$

$$G_x = G_x + \text{sparse}(k_0, k_2, \frac{gh_{i,j}}{\Delta x_{i,j}}, \text{kumax}, \text{kemax}) \tag{18}$$

where kumax and kemax means the maximums of k_u- and k_v-indexes. This syntax avoids indexing into G_x on the LHS of the equation for the sake of a big difference in the computational speed.[4] If G_x starts to be a zero sparse matrix, all the entries will be accumulated correctly. The same advices apply to the assembly of the other global operators to be discussed below.

The y-directional sea surface gradient at the point (i, j) for an interior and non-coastal V-point, indicated this time by the solid vertical red arrow in Figure 3, can be approximated by

$$\frac{gh}{R}\frac{\partial\eta}{\partial\phi}\bigg|_{i,j} \approx \frac{gh_{i,j}}{\Delta y_{i,j}}\begin{bmatrix}-1 & 1\end{bmatrix}\begin{bmatrix}\eta_{i-1,j}\\ \eta_{i+1,j}\end{bmatrix} \tag{19}$$

where $y = R\phi$, and the local operator the RHS can be the assembled in to a global one as

$$G_y(k_0, [k_1\ k_2]) = \frac{gh_{i,j}}{\Delta y_{i,j}}\begin{bmatrix}-1 & 1\end{bmatrix} \tag{20}$$

where $k_0 = \text{KvIndex}(i,j)$ and $k_{1,2} = \text{KeIndex}(i \mp 1, j)$. G_y is global operator, when it operates on $[\eta]$, it produces the sea level gradients in the y-direction (latitudinal direction).

4.2 Mass divergence operators, D_x and D_y

Figure 4 gives a cell for consideration of the mass divergence through the cell. An approximation for the mass divergence at (i, j) due to the fluxes associated with U is

$$\frac{\partial U}{\partial x}\bigg|_{i,j} \approx \begin{bmatrix}-\frac{\Delta y_{i,j}}{e} & \frac{\Delta y_{i,j}}{e}\end{bmatrix}\begin{bmatrix}U_{i,j-1}\\ U_{i,j+1}\end{bmatrix} \tag{21}$$

where $x = R\lambda\cos\phi$, and $\Delta y_{i,j} = y_{i+1,j} - y_{i-1,j}$ with $y = R\phi$, and

$$e = 2R^2(\lambda_e - \lambda_w)\cos\frac{\phi_n + \phi_s}{2}\sin\frac{\phi_n - \phi_s}{2} \tag{22}$$

[4] http://blogs.mathworks.com/loren/2007/03/01/creating-sparse-finite-element-matrices-in-matlab/

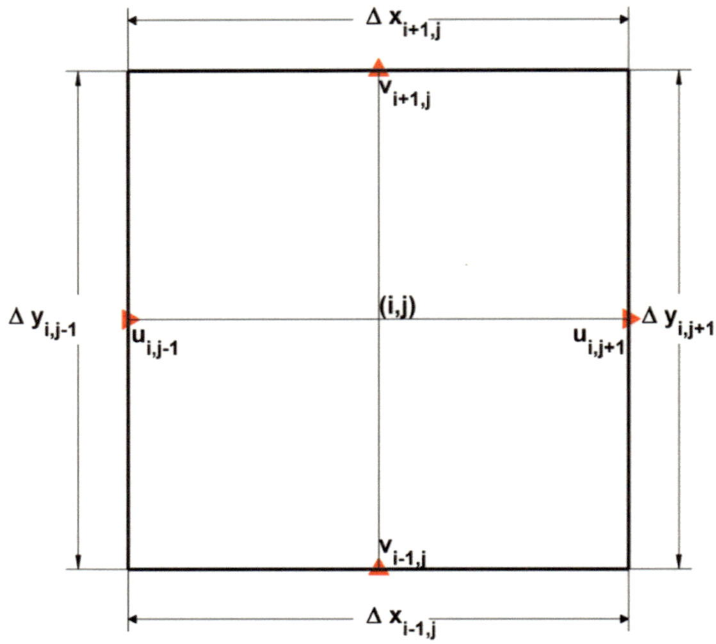

Fig. 4. A cell for the consideration of water mass divergence. The west and east side lengths of the cell are denoted by $\Delta y_{i,j-1}$ and $\Delta y_{i,j+1}$, which are equal. The south and north side lengths are denoted by $\Delta x_{i-1,j}$ and $\Delta x_{i+1,j}$ which are not equal on a spherical surface. The streams normal to the four sides are denoted by $U_{i,j-1}$, $U_{i,j+1}$, $V_{i-1,j}$, and $V_{i+1,j}$.

is the area of the cell. The 2-element row vector on the RHS of Eq.(21) is a local mass divergence operator, which operates on the two neighbouring U's and produces a cell-mean mass divergence due to the fluxes through the west and east sides of the cell.
The mass divergence at (i, j) due to fluxes caused by V can be approximated by

$$\frac{1}{R\cos\phi}\frac{\partial V\cos\phi}{\partial\phi}\bigg|_{i,j} \approx \begin{bmatrix}-\frac{\Delta x_{i-1,j}}{e} & \frac{\Delta x_{i+1,j}}{e}\end{bmatrix}\begin{bmatrix}V_{i-1,j}\\ V_{i+1,j}\end{bmatrix} \tag{23}$$

where $\Delta x_{i\pm1,j} = x(i \pm 1, j+1) - x(i \pm 1, j-1)$ are the lengths of the north and south sides of the cell. The 2-element row vector on the RHS of Eq.(23) is another local mass divergence operator, which operates on the two neighbouring V's and produces a cell-mean mass divergence due to the fluxes through the south and north sides of the cell.
The two local divergence operators need to be assembled into their global counterparts

$$D_x(k_0, [k_1\ k_2]) = \begin{bmatrix}-\frac{\Delta y_{i,j}}{e} & \frac{\Delta y_{i,j}}{e}\end{bmatrix} \tag{24}$$

$$D_y(k_0, [k_3\ k_4]) = \begin{bmatrix}-\frac{\Delta x_{i-1,j}}{e} & \frac{\Delta x_{i+1,j}}{e}\end{bmatrix} \tag{25}$$

where $k_0 = \mathrm{KeIndex}(i,j)$, $k_{1,2} = \mathrm{KuIndex}(i, j \mp 1)$, $k_{3,4} = \mathrm{KvIndex}(i \mp 1, j)$

4.3 Coriolis Operator, C_u and C_v

Since the staggered Arakawa C-grid does not co-locate the U- and V-points, for evaluations of the Coriolis force we have to resort to a kind of mean values of U's and V's. For an interior V-point, which is neither on open water boundary nor on coast, there are always four next neighbouring U-points. The four neighbouring U's can be somehow averaged and the result, denoted as $\overline{\mathrm{U}}$, can then be used for the evaluation of the Coriolis force, $f\overline{\mathrm{U}}$, at the V-point.

As often seen, a straight forward way to calculate an $\overline{\mathrm{U}}$ is to sum over the four surrounding U's and then divide the result by 4. However, Espelid *et al* (2000) pointed out that such an equally weighted averaging could result in an artificial energy growth or decay in an environment where the Coriolis parameter and/or the water depth varied spatially. To see this point, we can first multiply Eqs.(2) and (3) by U/h and V/h respectively and then add them to have an equation for the kinetic energy over the water column. In doing so, we can see that the contributions to the kinetic energy by the Coriolis force in the two directions cancel each other: $\frac{fVU}{h}-\frac{fUV}{h}=0$, a well known property in geo-fluid dynamics that Coriolis force does not do any work. However the spatial discretization can destroy this property. On a staggered grid such as the C-grid where U and V-points are not collocated, the two terms may not cancel each other because they are not evaluated at the same point, giving rise to increasing or decaying in energy artificially.

As a remedy to this problem, Espelid *et al* (2000) proposed the following weighted average scheme,

$$\overline{\mathrm{U}}\Big|_{i,j} = \frac{w_{i,j}}{4}\sum_{m=j-1}^{m=j+1}\sum_{l=i-1}^{l=i+1}\frac{U_{l,m}}{w_{l,m}} \tag{26}$$

where $w_{i,j}=\sqrt{h_{i,j}/|f_{i,j}|}$ is the weight for the averaging. The weighted average $\overline{V}$ can be calculated in the same way. With this averaging scheme, one can verify that the work done by the two Coriolis force components cancel each other point wise in the C-grid, if the Coriolis parameter does not change its sign. It is interesting to note that the $1/w_{i,j}$ is the square-root of the planetary vorticity, f/h, (e.g. Pedloski, 1979), which is also known as the geostrophic contours. In the absence of the external force and internal friction, the circulations in the ocean tend to flow the geostrophic contours.

However it seems that the above scheme was designed only for a regional model where there is no need to be concerned with sign changes of the Coriolis parameter. For a global mode, or a regional model whose domain is across the equator, or a model whose spherical north pole is shifted on a land to avoid a mathematical singularity which otherwise would arise at the geographic north pole, we do need to be concerned with the case where the discretized Coriolis parameters, $f_{i,j}$, at the four neighbouring points do not shear the same signs. To address this concern, the averaging weight should be modified as $w_{i,j}=\mathrm{sgn}(f_{i,j})\sqrt{h_{i,j}/|f_{i,j}|}$.

Therefore, with the modified averaging weight, we can have the following formula to evaluate the Coriolis force at a V-point,

$$\left(f\overline{U}\right)_{i,j} = \frac{\sqrt{|\mathrm{f}_{i,j}|\mathrm{h}_{i,j}}}{4}\begin{bmatrix}\frac{\mathrm{s}_1\sqrt{|\mathrm{f}_{i-1,j-1}|}}{\sqrt{\mathrm{h}_{i-1,j-1}}} & \frac{\mathrm{s}_2\sqrt{|\mathrm{f}_{i+1,j-1}|}}{\sqrt{\mathrm{h}_{i+1,j-1}}} & \frac{\mathrm{s}_3\sqrt{|\mathrm{f}_{i-1,j+1}|}}{\sqrt{\mathrm{h}_{i-1,j+1}}} & \frac{\mathrm{s}_4\sqrt{|\mathrm{f}_{i+1,j+1}|}}{\sqrt{\mathrm{h}_{i+1,j+1}}}\end{bmatrix}\begin{bmatrix}\mathrm{U}_{i-1,j-1}\\ \mathrm{U}_{i+1,j-1}\\ \mathrm{U}_{i-1,j+1}\\ \mathrm{U}_{i+1,j-1}\end{bmatrix} \tag{27}$$

where $s_1 = \mathrm{sgn}(f_{i-1,j-1})$, $s_2 = \mathrm{sgn}(f_{i+1,j-1})$, $s_3 = \mathrm{sgn}(f_{i-1,j+1})$ and $s_4 = \mathrm{sgn}(f_{i+1,j+1})$. The 4-element row vector on the RHS is a local Coriolis operator, which operates on the neighbouring U's and produces Coriolis force at the center V-point. We need to assemble this local Coriolis operator into its global counterpart C_u, as

$$C_u(k_0, [k_1\ k_2\ k_3\ k_4]) = \frac{\sqrt{|f_{i,j}|h_{i,j}}}{4}\left[\frac{s_1\sqrt{|f_{i-1,j-1}|}}{\sqrt{h_{i-1,j-1}}} \quad \frac{s_2\sqrt{|f_{i+1,j-1}|}}{\sqrt{h_{i+1,j-1}}} \quad \frac{s_3\sqrt{|f_{i-1,j+1}|}}{\sqrt{h_{i-1,j+1}}} \quad \frac{s_4\sqrt{|f_{i+1,j+1}|}}{\sqrt{h_{i+1,j+1}}}\right] \tag{28}$$

with $k_0 = \mathrm{KvIndex}(i,j)$, $k_{1,2} = \mathrm{KuIndex}(i \mp 1, j-1)$, $k_{3,4} = \mathrm{KuIndex}(i \mp 1, j+1)$.
When the positions of the two indexing functions, KvIndex and KuIndex, in the above are switched, Eq.(28) can be equally well used to calculate C_v, another global Coriolis operator which operates on the V-variables and produced the Coriolis force on all the U-points.

4.4 Frictional operators, F_u and F_v

The frictional operators are easiest to assemble. They are diagonal, with elements being κ/h. Let F_u denote a global frictional operator, which operates on U and produces frictional force in x-direction. It can be given by

$$F_u(k,k) = \kappa_{i,j}/h_{i,j} \tag{29}$$

with $k = \mathrm{KuIndex}(i,j)$ for a (i, j)-point where U is defined. Similarly we can have a frictional operator F_u which operates on V and produces frictional force in the y-direction,

$$F_v(k,k) = \kappa_{i,j}/h_{i,j} \tag{30}$$

with $k = \mathrm{KvIndex}(i,j)$ for a (i, j)-point where V is defined.

4.5 Open boundary operators $\mathbb{O}_u$and $\mathbb{O}_v$

Since the most outside variable points are velocity points, the Sommerfeld radiation boundary condition are thus applied at the U- and V-points on open water boundaries. Substitute U and V for p in Eq.(7), we can have

$$\frac{\partial U}{\partial t} + c\frac{\partial U}{\partial n} = 0 \qquad \text{at east and west open water boundaries} \tag{31}$$

$$\frac{\partial V}{\partial t} + c\frac{\partial V}{\partial n} = 0 \qquad \text{at south and north water boundaries} \tag{32}$$

where $c = \sqrt{g}h$.
For a $U_{i,j}$ on an open water boundary, its outward and normal derivative can be approximated as,

$$c\frac{\partial U}{\partial n}\bigg|_{i,j} \approx \begin{cases} \dfrac{c}{\Delta x_{i,j}}[1 \quad -1]\begin{bmatrix} U_{i,j} \\ U_{i,j+2}\end{bmatrix} & \text{for (i, j) on a west open boundary, and } \Delta x_{i,j} = x(i,j+2) - x(i,j), \\ \dfrac{c}{\Delta x_{i,j}}[-1 \quad 1]\begin{bmatrix} U_{i,j-2} \\ U_{i,j}\end{bmatrix} & \text{for (i, j) on a east open boundary, and } \Delta x_{i,j} = x(i,j) - x(i,j-2). \end{cases} \tag{33}$$

where $c = \sqrt{gh_{i,j}}$. Similarly, for a $V_{i,j}$ on an open water boundary,

$$c\frac{\partial V}{\partial n}\bigg|_{i,j} \approx \begin{cases} \dfrac{c}{\Delta y_{i,j}}[1 \quad -1]\begin{bmatrix} V_{i,j} \\ V_{i+2,j} \end{bmatrix} & \text{for (i, j) on a south open boundary, and } \Delta y_{i,j} = y(i+2,j) - y(i,j), \\ \dfrac{c}{\Delta y_{i,j}}[-1 \quad 1]\begin{bmatrix} V_{i-2,j} \\ V_{i,j} \end{bmatrix} & \text{for (i, j) on a north open boundary, and } \Delta y_{i,j} = y(i,j) - y(i-2,j). \end{cases} \tag{34}$$

The row vectors on the RHS of the above expressions can be assembled into a global operator $\mathbb{O}_u$ as

$$\mathbb{O}_u(k_0, k_1) = \frac{c}{\Delta x_{i,j}}[1 \quad -1] \tag{35}$$

with $k_0 = \text{KuIndex}(i,j)$ and $k_1 = \text{KuIndex}(i,j+2)$, and

$$\mathbb{O}_u(k_0, k_1) = \frac{c}{\Delta x_{i,j}}[-1 \quad 1] \tag{36}$$

with $k_0 = \text{KuIndex}(i,j-2)$ and $k_1 = \text{KuIndex}(i,j)$ when point (i, j) is on a east open boundary. Upon operating on the U-vector, $\mathbb{O}_u$ produces a vector of outward gradients in U along the east and west open boundaries.
$\mathbb{O}_v$ can be assembled in the save way but with the KvIndex function instead.

5. Linear shallow water equations in matrix form

All the operators constructed in the proceeding section can now be put together to discretize in space the governing equations and boundary conditions for the linear shallow water dynamics. We also have to be concerned with how to discretize the equation in time, and how to solve the total discretized equations in a manner that is computationally stable. These issues are the content of this section.

5.1 Governing equations and boundary conditions in one matrix equation

With the global operators constructed in the proceeding section, we can now discretize the RHS of Eqs.(1) to (3) and Eqs (5) to (7), and present the result in a single matrix equation as follows:

$$\frac{\partial}{\partial t}\begin{bmatrix} \eta \\ U \\ V \end{bmatrix} = \begin{bmatrix} 0 & -D_x & -D_y \\ -G_x & -\mathbb{O}_u - F_u & C_v \\ -G_y & -C_u & -\mathbb{O}_v - F_v \end{bmatrix}\begin{bmatrix} \eta \\ U \\ V \end{bmatrix} \tag{37}$$

where 0 represents a zero matrix.
The zero normal flow conditions at coasts have accommodated in the U- and V- vectors. All the stream points on the coasts are included as the elements of the two stream vectors. The coastal elements start to be zero and will remain to be so. They are guaranteed to remain to be zero because F_u and F_v are diagonal, whereas the corresponding rows of the other sub-matrices are all zero. In fact, the coastal elements can be deleted from the U- and V-vectors, and the related sub-matrices can shrink just to include the interior and open boundary points in order to reduce the computations. This can be automatically cared within a program running the matrix-vector multiplications. Conceptually however, we had better regard that the U- and V-vectors include their coastal elements. The radiation conditions at open water boundaries are implemented through $\mathbb{O}_u$ and $\mathbb{O}_v$. In case there are no open water boundaries, they become zero matrices.

5.2 Time discretization schemes and sponge zones

The time derivatives remain to be discretized, for which we have two schemes. Take the discretization for $\partial p/\partial t$ as example, where p may represent η, U, or V, we can have

$$\left.\frac{\partial p}{\partial t}\right|_{i,j} \approx \frac{p_{i,j}^{(\tau+1)} - p_{i,j}^{(\tau)}}{\Delta t}, \qquad (p=\eta,\ U \text{ or } V) \tag{38}$$

or

$$\left.\frac{\partial p}{\partial t}\right|_{i,j} \approx \frac{p_{i,j}^{(\tau+1)} - \overline{p}_{i,j}^{(\tau)}}{\Delta t}, \qquad (p=\eta,\ U \text{ or } V) \tag{39}$$

where Δt is the time step, the superscript τ is a time step index such that $t = \tau\Delta t$ for $\tau = 0,1,2,\cdots$. The second scheme differs from the first one in its second term, where an averaged elevation, $\overline{p}_{i,j}$, instead of $p_{i,j}$, is used. For variable on an interior grid point (i, j), the average is defined as follows:

$$\overline{p}_{i,j} = \begin{cases} \dfrac{p_{i,j-2} + p_{i,j+2} + p_{i-2,j} + p_{i+2,j}}{4} & \text{if 2 pairs of neighbours which are non-boundary nodes.} \\ p_{i,j} & \text{otherwise, no averaging.} \end{cases} \tag{40}$$

In other words, $\overline{p}_{i,j}$ is obtained by averaging over possible neighbours. The neighbours here are all in the sense of two grid-point apart, i.e., in the sense of $i \pm 2, \text{and/or}\ j \pm 2$. The neighbours must also be of the same types, e.g., the neighbours of an η-point must be η-points too. The neighbours have to be aligned either in the i-direction or in the j-direction. Most interior points have two lined pairs, such as those away from coasts. For an interior point near a coast, it may have one or zero lined pairs. In Figure 1, the η-point at (i=28, j=34) has three neighbouring η-points: (i=28, j=32), (i=28, j=36), and (i=30, j=34), and only the first two points forms a lined pair. The η-point at (i=24, j=32) has zero lined pair, because its two neighbours at (i=24, j=30) and (i=26, j=32) are not aligned.

The above averaging scheme has excluded certain types of points: Firstly only the interior points are assigned with the averages, the boundary nodes do not need the averages. This exclusion assures that the condition of zero normal flow at coasts will not be destroyed. Secondly, in assigning an average for an interior point, its boundary neighbours are always excluded even when they are part of lined neighbours. This exclusion is to avoid a zero averaged value at a place where the actual value should not. The averaged U at (i=24, j=31) would be zero if its four coastal neighbours were not excluded. Thirdly, open water boundary nodes are also excluded from averaging. Updating on the open boundary nodes is cared by the radiation boundary conditions.

Scheme (39) is an extension of the well known Lax-Friedrichs scheme in one dimension to the two dimensional case here. It helps to stabilize the computation, because it equivalently introduces a horizontal frictional term. To see this point, let us first note the following identities and approximations

$$\frac{p_{i,j}^{(\tau+1)} - \overline{p}_{i,j}^{(\tau)}}{\Delta t} \equiv \frac{p_{i,j}^{(\tau+1)} - p_{i,j}^{(\tau)} - (\overline{p}_{i,j}^{(\tau)} - p_{i,j}^{(\tau)})}{\Delta t} \tag{41}$$

$$\equiv \frac{p_{i,j}^{(\tau+1)} - p_{i,j}^{(\tau)}}{\Delta t} - \left(\frac{\Delta x^2}{4\Delta t}\right)\frac{p_{i,j+1}^{(\tau)} - 2p_{i,j}^{(\tau)} + p_{i,j-1}^{(\tau)}}{\Delta x^2} - \left(\frac{\Delta y^2}{4\Delta t}\right)\frac{p_{i+1,j}^{(\tau)} - 2p_{i,j}^{(\tau)} + p_{i-1,j}^{(\tau)}}{\Delta y^2} \tag{42}$$

$$\approx \frac{\partial p}{\partial t} - \left(\frac{\Delta x^2}{4\Delta t}\right)\frac{\partial^2 p}{\partial x^2} - \left(\frac{\Delta y^2}{4\Delta t}\right)\frac{\partial^2 p}{\partial y^2} \tag{43}$$

Now the time difference has been the split into three terms. The first term approximates the time derivative, the second and the third terms resemble the horizontal momentum diffusion terms, with the $\left(\frac{\Delta x^2}{4\Delta t}\right)$ and $\left(\frac{\Delta y^2}{4\Delta t}\right)$ being the diffusivities in the x- and y-directions when p stands for U and V respectively. We started with a set of governing equations, i.e., Eqs (1), (2) and (3), which left out the horizontal friction, now consideration for the computational stabilization has led us to re-pick up what was left out.

However numerical experiments with the India-Pacific grid show that two dimensional Lax-Friedrich scheme brings in too strong numerical dissipations. It not only desirably suppresses the small scales noises, but also undesirably smears out or even kills wave forms. Thus scheme (39) will only be used as a complement to scheme (38). We need two parameters to reflect the proportionality of the two schemes. Let $\alpha, \beta \in [0,1]$, and $\alpha + \beta = 1$, the two schemes can be joined together as

$$\left.\frac{\partial p}{\partial t}\right|_{i,j} = \frac{p_{i,j}^{(\tau+1)} - \alpha p_{i,j}^{(\tau)} - \beta \overline{p}_{i,j}^{(\tau)}}{\Delta t} \tag{44}$$

When α is 1 , β is 0 and the joined scheme recovers scheme (38); when α is 0, β is 1 and the joined scheme recovers scheme (39). Note that α and β are not independent. They are both introduced just for brevity of the mathematical expressions that involve them.

The averaging scheme as in Eq.(40) has to be assembled into three global averaging operators, Q_e, Q_u and Q_v for p=η, U or V. This can be done in the same way as for assembling the other global operators. Replacing p respectively by vectors η, U and V, we can now express Eq. (44) in matrix form,

$$\frac{\partial}{\partial t}\eta \approx \frac{\eta^{(\tau+1)} - \mathbb{I}_e \eta^{(\tau)}}{\Delta t} \tag{45}$$

$$\frac{\partial}{\partial t}U \approx \frac{U^{(\tau+1)} - \mathbb{I}_u U^{(\tau)}}{\Delta t} \tag{46}$$

$$\frac{\partial}{\partial t}V \approx \frac{V^{(\tau+1)} - \mathbb{I}_v V^{(\tau)}}{\Delta t} \tag{47}$$

where $\mathbb{I}$'s are

$$\mathbb{I}_e = (\alpha_e I_e + \beta_e Q_e) \tag{48}$$

$$\mathbb{I}_u = (\alpha_u I_u + \beta_u Q_u) \tag{49}$$

$$\mathbb{I}_v = (\alpha_v I_v + \beta_v Q_v) \tag{50}$$

and the I's are identity matrices, with their subscripts indicating with which vectors, η, U or V, their sizes are conformable. The α and β parameters are now also presented with appropriate subscripts too. As we can see, the $\mathbb{I}$'s are not identity matrices, but are linked to the latter. When α's are 1 and β's are 0, $\mathbb{I}$'s become I's.

Adjusting the value of β, and hence of α, between 0 and 1, gives a control to balance between the sub-grid noise suppressing and wave form preserving. If let α and β be diagonal matrices, we even can have a control point wisely. The balance experiments have been performed in in the India and Pacific domain, and the following are the experiences:

1. Model solutions are very sensitive to the values of β's, especially of β_e. When β's are set to be ≥0.1, strong numerical dispassion results; waves die quickly.
2. In the interior region away from the open water boundaries, assigning 0 to β_e and 0.005 to both β_u and β_v can satisfactorily filter out noises at sub-grid scales, but keep information on larger scales almost un-touched.
3. A radiation boundary condition will more or less reflect an outgoing signal back into the model domain (Chapman, 1987), which is undesirable. To suppress this kind of reflections, a *sponge* zone can set up where large values of β_e, β_u, β_v, can be assigned. Values of 0.1 for all β_e, β_u, and β_v are large enough. Traditionally a sponge zone means a zone where the frictional parameters (such as κ in the dynamic system considered here) are set up with values much larger than the realistic values. Here the Lax-Freidrich's scheme is used as an effective frictional source. The sponge zones shown in Figure 2 are made for this purpose.

Setting β_u and β_v to be 0.005 for the interior region is equivalent to setting the horizontal momentum diffusivities to be $0.005\left(\frac{\Delta x^2}{4\Delta t}\right)$ and $0.005\left(\frac{\Delta y^2}{4\Delta t}\right)$. For the India-Pacific grid used for this chapter, which has a 5-min spatial resolution longitude and latitude, Δy=9266 m and Δx varies from $1815m$ to $9266m$. The Δt is taken as 5 seconds. With these numbers, the numerical horizontal momentum diffusivities vary from $8 \times 10^2\ m^2/s$ to $2 \times 10^4 m^2/s$, which are well within the ranges intended to represent the physical horizontal momentum diffusivity in the oceans (e.g, Brian *et al* 1975, Bryan 1986).

If the horizontal momentum diffusivities can be well estimated, one may use the following formula to calculate β_u and β_v

$$(\beta_u, \beta_v) = 4\Delta t\left(\frac{Ah_u}{\Delta x^2}, \frac{Ah_v}{\Delta y^2}\right) \tag{51}$$

where Ah_u and Ah_v represent the horizontal momentum diffusivities, which are usually taken to be equal in the oceans.

5.3 Combination of explicit and implicit schemes for stability

With the time derivatives approximated as in the Eqs. (45) to (47), Eq. (37) can now be discretized in time

$$\begin{bmatrix}\eta\\U\\V\end{bmatrix}^{(\tau+1)} - \begin{bmatrix}\mathbb{I}_e\eta\\\mathbb{I}_u U\\\mathbb{I}_v V\end{bmatrix}^{(\tau)} = \Delta t\begin{bmatrix}0 & -D_x & -D_y\\ -G_x & -\mathbb{O}_u - F_u & C_v\\ -G_y & -C_u & -\mathbb{O}_v - F_v\end{bmatrix}\begin{bmatrix}\eta\\U\\V\end{bmatrix}^{(\tilde{\tau})} \tag{52}$$

where $\tilde{\tau}$ indicate a time step yet to be determined. If $\tilde{\tau} = \tau$, the above scheme is explicit. If let $\tilde{\tau} = \tau+1$, the scheme is implicit. An implicit scheme is usually good for the computational stability, but is expensive because the equations have to be solved simultaneously. An explicit scheme is economic in computation but it risks the computational stability. To balance between the computational cost and stability, the following combination of $\tilde{\tau} = \tau$ and $\tilde{\tau} = \tau + 1$ becomes a reasonable choice,

$$\begin{bmatrix}\eta\\U\\V\end{bmatrix}^{(\tau+1)} - \begin{bmatrix}\mathbb{I}_e\eta\\\mathbb{I}_u U\\\mathbb{I}_v V\end{bmatrix}^{(\tau)} = \Delta t\begin{bmatrix}0 & -D_x & -D_y\\0 & -\mathbb{O}_u & C_v\\0 & 0 & -\mathbb{O}_v\end{bmatrix}\begin{bmatrix}\eta\\U\\V\end{bmatrix}^{(\tau)} + \Delta t\begin{bmatrix}0 & 0 & 0\\-G_x & -F_u & 0\\-G_y & -C_u & -F_v\end{bmatrix}\begin{bmatrix}\eta\\U\\V\end{bmatrix}^{(\tau+1)} \tag{53}$$

which split the coefficient matrix into two matrices. The splitting is based on the consideration that the matrix associated with $\tau+1$ time step can be economically inverted. Eq. (53) can be rearranged as

$$\begin{bmatrix}I_e & 0 & 0\\\Delta tG_x & E_u & 0\\\Delta tG_y & \Delta tC_u & E_v\end{bmatrix}\begin{bmatrix}\eta\\U\\V\end{bmatrix}^{(\tau+1)} = \begin{bmatrix}\mathbb{I}_e & -\Delta tD_x & -\Delta tD_y\\0 & \mathbb{I}_u-\Delta t\mathbb{O}_u & \Delta tC_v\\0 & 0 & \mathbb{I}_v-\Delta t\mathbb{O}_v\end{bmatrix}\begin{bmatrix}\eta\\U\\V\end{bmatrix}^{(\tau)} \tag{54}$$

where $E_u = I_u + \Delta tF_u$ and $E_v = I_v + \Delta tF_v$, which are diagonal. The inverse of the coefficient matrix on the LHS of the above equation can be easily found as

$$\begin{bmatrix}I_e & 0 & 0\\\Delta tG_x & E_u & 0\\\Delta tG_y & \Delta tC_u & E_v\end{bmatrix}^{-1} = \begin{bmatrix}I_e & 0 & 0\\-\Delta tE_u^{-1}G_x & E_u^{-1} & 0\\-\Delta tE_v^{-1}G_y+(\Delta t)^2E_v^{-1}C_uE_u^{-1}G_x & -\Delta tE_v^{-1}C_uE_u^{-1} & E_v^{-1}\end{bmatrix}. \tag{55}$$

What have been actually inverted are the diagonal matrices E_u and E_v. Their inversions are not expensive at all.

Eq. (54) can now be solved as

$$\begin{bmatrix}\eta\\U\\V\end{bmatrix}^{(\tau+1)} = \begin{bmatrix}\mathbb{L}_{11} & 0 & 0\\\mathbb{L}_{21} & \mathbb{L}_{22} & 0\\\mathbb{L}_{31} & \mathbb{L}_{32} & \mathbb{L}_{33}\end{bmatrix}\begin{bmatrix}\mathbb{W}_{11} & \mathbb{W}_{12} & \mathbb{W}_{13}\\0 & \mathbb{W}_{22} & \mathbb{W}_{23}\\0 & 0 & \mathbb{W}_{33}\end{bmatrix}\begin{bmatrix}\eta\\U\\V\end{bmatrix}^{(\tau)} \tag{56}$$

$$= \begin{bmatrix}\mathbb{L}_{11}\mathbb{W}_{11} & \mathbb{L}_{11}\mathbb{W}_{12} & \mathbb{L}_{11}\mathbb{W}_{13}\\\mathbb{L}_{21}\mathbb{W}_{11} & \mathbb{L}_{21}\mathbb{W}_{12}+\mathbb{L}_{22}\mathbb{W}_{22} & \mathbb{L}_{21}\mathbb{W}_{13}+\mathbb{L}_{22}\mathbb{W}_{23}\\\mathbb{L}_{31}\mathbb{W}_{11} & \mathbb{L}_{31}\mathbb{W}_{12}+\mathbb{L}_{32}\mathbb{W}_{22} & \mathbb{L}_{31}\mathbb{W}_{13}+\mathbb{L}_{32}\mathbb{W}_{23}+\mathbb{L}_{33}\mathbb{W}_{33}\end{bmatrix}\begin{bmatrix}\eta\\U\\V\end{bmatrix}^{(\tau)} \tag{57}$$

where $\mathbb{L}_{i,j}$ represents the sub-matrices of the lower triangular matrix given by the RHS of Eq. (55) and $\mathbb{W}_{i,j}$ represents the sub-matrices of the upper triangular coefficient matrix given by the RHS of Eq. (54).

If let A be the RHS of Eq. (57), we can have an updating equation in a compact matrix form as

$$\begin{bmatrix}\eta\\U\\V\end{bmatrix}^{(\tau+1)} = A\begin{bmatrix}\eta\\U\\V\end{bmatrix}^{(\tau)}, \qquad \tau = 0,1,2,\cdots\tau_{max} \tag{58}$$

which is the target form as was aimed in Section 2.4.

Heaps (1969) suggested the time step as $\Delta t = r[2\min(R\Delta\phi, R\cos\phi\ \Delta\lambda)/\sqrt{gh}]$ where r is a lower bound to bring the time step Δt within the maximum value allowed by the CFL condition. For the India and Pacific domain, r is set as 0.95.

In the above, the explicit and implicit schemes have been combined into one, which was originally proposed by Sielecki (1968). In the absence of frictions, and without the Lax-Friedrichs averaging, according to Sielecki, the scheme preserves the energy. When frictions or Lax-Friedrichs averaging is present, the energy decays with the time.

Notice that in the above updating scheme, updating V can always "enjoy" the most recently updated U, but not the other way around. If this kind of asymmetry in updating U and V is undesirable, it can be remedied by developing a counterpart to Eq. (58) by following the same route from Eq. (53) to Eq. (58) but with the positions of U and V exchanged. The two updating schemes can then be used alternatively in time. That is to say, if Eq. (58) is used from τ to $\tau + 1$, then its counterpart should used from $\tau + 1$ to $\tau + 2$. In order to have one matrix equation, the two updating schemes should be further combined into one by multiplying the two updating matrices. Before the multiplication, the counterpart matrix has to be permutated to restore the same order as of $[\eta\ U\ V]^T$. The matrix multiplication will result in a two-step updating scheme, covering from τ to $\tau + 2$.
The author of this paper tried both asymmetric and symmetric schemes, but did not find noticeable difference in the solutions with the India-Pacific grid. Therefore, only the asymmetric scheme, i.e. Eq. (57), is used in the simulations, of which the results are going to be presented in the rest of this chapter. However it is worthwhile to note here on how we can derive a symmetric updating in matrix form, in case we will need it in some other modelling settings.

6. All-source Green's functions

Having derived the LSWE in matrix form, we are now ready to define the ASGF and to give its algorithm. We can then compare the ASGF and the SSGF side by side in terms of their computational costs and their scopes of applications. An important part of the ASGF is the part that operates on the initial sea level distribution. This part will be called the amplitude weights. Visualization of the amplitude weights brings up a classical concept in wave dynamics to the surface: the *domain of dependence*. The boundary of the domain of dependence serves a co-time line. The trace left by the ever growing boundary is a map of tsunami arrival times at a POI. Recordings of the history of the amplitudes weights at a particular place link to the responses in sea level at a POI to the initial forcing exerted at that particular place.

6.1 Definition and algorithm of the ASGF

Eq. (58) can be equivalently expressed as

$$\begin{bmatrix}\eta\\U\\V\end{bmatrix}^{(\tau)} = A^{\tau}\begin{bmatrix}\eta\\U\\V\end{bmatrix}^{(0)}, \quad \tau = 1,2,3,\cdots\tau_{max} \tag{59}$$

where A^{τ} represent the τth power of the matrix A. However the two equations differ dramatically in their computational costs. If the solutions at all the model grid points are required, Eq. (58) should be the choice for computational efficiency, since each time step only involves a matrix-vector multiplication. In this case, to choose (59) would be unwise, since, when $\tau \geq 2$, it would require the calculations of all the powers of the matrix A, which would be very expensive or infeasible when the size of the matrix is large. However, if only solutions at a few model grid points are interested, Eq. (59) provides a distinct advantage as shown below.
Assume that there is only one POI, say the *kth* η-point, where the sea surface elevation is interested. According to Eq. (59), the time series of the elevation at the *kth* η-point can be written as

$$\eta_k^{(\tau)} = A^\tau(k,:)\begin{bmatrix}\eta\\U\\V\end{bmatrix}^{(0)}, \quad \tau = 1,2,3,\cdots\tau_{max} \tag{60}$$

The crux here is how to compute the matrix power $A^\tau(k,:)$ economically for a series of τ's. If we first calculate the power of the whole matrix, and then extracts the *kth* row from the result, the calculation would be too expensive. Can we obtain the value of just the *kth* row of the matrix power without calculations for the other rows? Once the question is asked like this, the answer follows naturally. The following simple algorithm allows us to achieve what we want:

$$G^{(0)} = I(k,:) \tag{61}$$

$$G^{(1)} = G^{(0)}A \tag{62}$$

$$G^{(2)} = G^{(1)}A \tag{63}$$

$$\vdots$$

$$G^{(\tau)} = G^{(\tau-1)}A \tag{64}$$

$$\vdots$$

$$G^{(\tau_{max})} = G^{(\tau_{max}-1)}A \tag{65}$$

Each of the series $G^{(\tau)}, (\tau = 0,1,2,\cdots)$, is a row vector. The initial one, $G^{(0)}$, is a row vector of the identity matrix. $G^{(1)}$ is simply a copy of the *kth* row of the matrix A. $G^{(2)}$ is the *kth* row of the second power of the matrix A. Eq. (63) avoids the calculation of $A^2 = A \times A$, it only calculates the second power for the needed row. The multiplication of the vector-matrix iterates to produces higher orders of the matrix powers for the needed row only. Since the identity matrix can be viewed as the zeroth power of the matrix A, and the matrix A can be viewed as the first power of itself, the series $G^{(\tau)}, (\tau = 0,1,2,\cdots)$, can be uniformly referred as the *kth* row of the *τ'th* power of the matrix A.
The series $G^{(\tau)}, (\tau = 0,1,2,\cdots)$ define the ASGF, evaluated at $t = \tau\,\Delta t$ (τ =0,1,2, ⋯). Each $G^{(\tau)}$ has N columns, with N being the number of total grid points where η-, U- or V-variables are defined. A column can be viewed as an information channel. Having N channels means that signals from N different sources can be simultaneously received and stored separately by a receiver. The superposition principle will allow the N-channel signals to be linearly combined to quickly yield a full solution when a real-time event happens.
We can now form a new matrix G as a collector of the individual row vectors, i.e.

$$G = [G^{(0)}; G^{(1)}; \cdots; G^{(\tau_{max})}] \tag{66}$$

where the semicolon inside the square bracket indicate the end of a previous row and the beginning of the next. A simple matrix-vector multiplication as

$$\eta_k(\tau) = G \times \begin{bmatrix}\eta\\U\\V\end{bmatrix}^{(0)} \tag{67}$$

can then provide the solution times series at the POI in response to an initial setup anywhere in the model domain. The collector G is also called the ASGF.
In the above, $\eta_k(\tau)$ represents the time series of the elevations at the *kth* η-node. Note that the parenthesized τ here is not in a superscript position. A parenthesized τ in a normal

position means a vector of time indexes, $\tau = [1\ 2\ \cdots \tau_{max}]$. A parenthesized τ is in a superscript position means one of the possible time steps, i.e., $\tau = 1, 2, \cdots, \tau_{max}$. A non-parenthesized τ in a superscript position means the exponent of the power.

For a different initial setup, corresponding to a new tsunami, one can simply substitute a new column vector for $[\eta^{(0)}\ U^{(0)}\ V^{(0)}]^{T}$, without modifying the matrix G . That is, once G has been calculated, it can be repeatedly used. The evolution from any specified initial state is easily calculated from a simple matrix-vector multiplication. The determination of the matrix G may require significant computer resources, but the matrix-vector multiplication can be quickly performed even on an ordinary personal computer or on a web server.

In the above discussions, the k-index was taken to indicate a single POI. This was just for simplicity of presentation; nothing prevents use of the k-index to represent several distinct POIs. One can thus use the same algorithm to calculate ASGFs for several POIs simultaneously. Also, the algorithm is completely ready for parallel computations. When there are many POIs, we can assign different values of the k-indexes to different processors or even different computational platforms without any need for exchange of information between processors or computational platforms. It should be feasible to calculate and store ASGFs for worldwide important cities as POIs and the entire global ocean as the potential sources.

6.2 Comparison with the single-source Green's functions

It is convenient at this stage to make a comparison between the ASGF and the SSGF. The procedure to calculate an SSGF can be recaptured essentially as follows:

$$F^{(0)} = I(:,k) \tag{68}$$

$$F^{(1)} = AF^{(0)}$$

$$F^{(2)} = AF^{(1)} \tag{69}$$

$$\vdots$$

$$F^{(\tau)} = AF^{(\tau-1)} \tag{70}$$

$$\vdots$$

$$F^{(\tau_{max})} = AF^{(\tau_{max}-1)} \tag{71}$$

where $F^{(\tau)}(\tau = 1,2,3,\cdots)$ defines the SSGF.

Comparing the algorithm listed above and the one listed in Eqs. (61) to (65), we can contrast the differences the ASGF and the SSGF side by side as below:

• The ASGF starts with a row of the dynamic matrix A. The row corresponds to a receiver point.	• The SSGF starts with a column of the dynamic matrix A. The column corresponds to a source point.
• The ASGF grows in time when a vector-matrix multiplication iterates.	• The SSGF grows in time when a matrix-vector multiplication iterates.
• The ASGF concentrates the information from N-sources on one receiver	• The SSGF spreads information from one source over N receivers.

A full name for the ASGF may read as all-source-one-receiver Green's function; a full name of the SSGF may read as single-source-all-receiver Green's function. It should be clear now that the ASGF and the SSGF cost the same computationally, but differ dramatically in terms of the hazard preparedness: the ASGF prepares a single POI against the entire ocean as potential hazard sources; the SSGF prepares the entire ocean against a single potential source. Obviously we only need to care a place where we live, not the entire ocean, against potential hazards. Since we cannot know where a future hazard source point will be, we had better assume that any parts of the ocean can be the potential source points.

6.3 Amplitude weights and domain of dependence

The matrix G, as defined by Eq. (66), can be split into three groups of columns, according to their dynamic roles, as follows:

$$G = [W \quad M_u \quad M_v]. \tag{72}$$

Accordingly, Eq. (67) can be re-expressed as

$$\eta_k(\tau) = [W \quad M_u \quad M_v]\begin{bmatrix}\eta\\U\\V\end{bmatrix}^{(0)} \tag{73}$$

The above expression says that the elevations at a POI is weighted averages of the initial sea level distribution, $\eta^{(0)}$, plus mass fluxes due to the initial stream component vectors, $U^{(0)}$ and $V^{(0)}$. The matrices of M_u and M_v capture the compound effects of many cycles of *mass divergence → sea surface gradients→ mass divergence* during the propagations of waves from source points all the way to a destination point.
If the initial stream components are zero, Eq. (70) can be reduced to

$$\eta_k(\tau) = W\eta^{(0)}. \tag{74}$$

This equation says that the matrix W transmits a tsunami wave from its source region to the sea levels at a destination. The elements in columns of W specify the weights exerted by the individual source points, and hence may be referred as *amplitude weights*. For a particular time $t = \tau\Delta t$ $(\tau = 0,1,2,\cdots)$, $W(\tau, :)$ is a row vector. The initial one, $W(0, :)$, has only one non-zero element in the *kth* column, whose value is 1. At the initial time, the only disturbance that can affect the *kth* point is the disturbance at the *kth* point itself. As τ progresses, the number of the non-zero elements of $W(\tau, :)$ increases, which means that any disturbance within the vicinity can affect the *kth* point, and such vicinity grows with the time.
It is worth of pointing out that the amplitude weights are always summed up to 1. This means that the initial unit weight is merely fractioned and redistributed by the subsequent operations, reflecting the law of mass conservation. It is also worth noting that while ordinary averaging weights are always positively valued between 0 and 1, the amplitude weights are valued between -1 and 1. A negative amplitude weight means that a positive/negative initial sea level at a source point results in a negative/positive sea level response at a receiver point.
Coloring the η-points according to the values of the amplitude weights depicts a *domain of dependence* , a key concept in the classical wave dynamics (e.g., Knobe, 2000): the elevation at the *kth* η-point at a particular time only depends on the distribution of the initial sea level within the domain. Any $\eta^{(0)}$ distributed outside of the domain has no effect on the elevation at that time. The three panels in Figure 5 show the amplitude weights at t=2, 4 and 6 hours, where a DART buoy (51407) offshore Hawaii is taken for the POI, as marked by the crosses in purple in the panels.
The domain of dependence grows with time, as can be seen clearly in the figure. A tsunami triggered offshore California will not hit Hawaii in two hours and will for sure in 6 hours. Whether this tsunami will hit Hawaii in 4 hours will depend on how far offshore this tsunami is initially distributed.

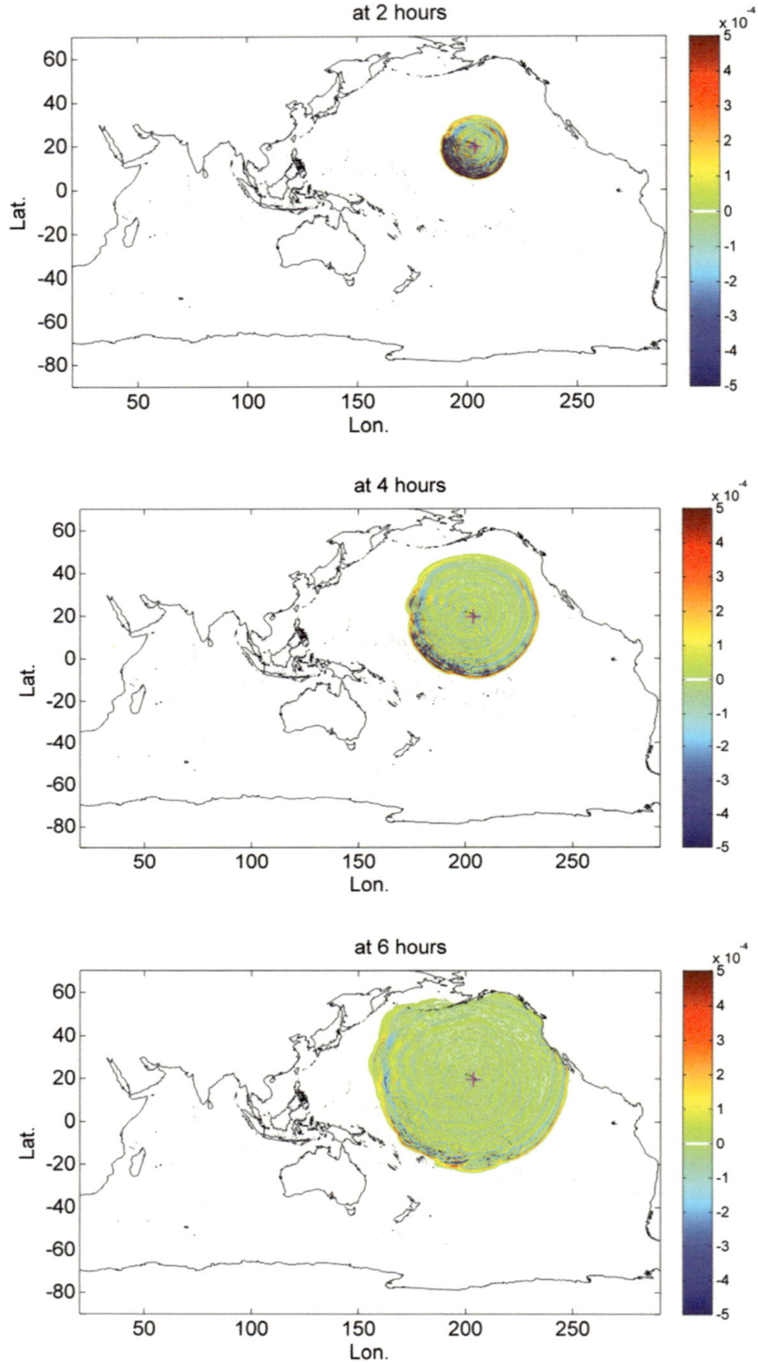

Fig. 5. Amplitude weights at 3 different times. The receiver point is marked by the crosses in purple, where a DART buoy (51407) is deployed.

The boundary of the domain of dependence at a specific time can serve a co-time line. A disturbance on the boundary or within will arrive at the POI in time equal to or less than that specific time. Any disturbances outside the boundary have to take longer time to reach the target. A disturbance within a zone between the boundaries of *domain of dependence* corresponding to two specific times will arrive at the POI only between the two specific times.

6.4 Response curves, Arrival Times (ATs), and Maximum Relative Amplitudes (MRAs)

While a row of the matrix W is a snapshot of the effects of all the cells at a particular time, a column of the matrix W contains the history of the effects by a particular cell. Shown in Figure 6 are three such histories of the effects on the same receiver by three distinct sources. In this example, the receiver is the DART station (51407) offshore Hawaii; the sources are three other DART stations (21414, 55015, 56001). The duration of the time series is 36 hours. The vertical axes of the curves indicate the amplitudes relative to the source amplitudes. The green circles indicate the arrival times (ATs) of the first signal whose absolute value is above a threshold (10^{-10}). The red circles indicate the values of the *maximum relative amplitudes* (MRAs) and their occurrence times as well. An MRA is the maximum absolute value seen in the curve for the entire duration. If the duration can be viewed long enough, an RMA then represents the maximum relative amplitude that can never be exceeded. From the curves, we can see that that the first waves are not always the largest.
The time step used in calculating the ASGFs is 5 seconds. Therefore one could store the results of the calculations as fine as 5 seconds, if this is needed. Or for the sake of disk space and the time needed for retrieving the data files, one may choose to save the results at a coarser interval such as 5 minutes, which is the case for the curves seen in Figure 6.

7. Applications for tsunami preparedness and real-time simulations

The time series as those in Figure 6 contain the details on the long wave dynamics. For a real-time simulation of the arrivals of an eventual tsunami, every bit of the information will be useful. In terms of tsunami preparedness, we only need to extract some pieces of information from the curves. The most important pieces of information we may read off are the ATs, the MRAs and MTs (maximum amplitude times). Each cell can be assigned with these values. Contouring and coloring the cells according to the assigned values give useful maps for tsunami preparedness.

7.1 Arrival time map

Figure 7 is an AT map for Hawaii. The destination point is marked by the white cross inside the 1-hour contour. This AT map tells when a tsunami will arrive at Hawaii should the tsunamis triggered somewhere in the India and Pacific Oceans. A traditional tsunami time map is a travel time map. It assumes a source point and contours the travel times to anywhere from the source point. However no one knows if a future tsunami will be indeed triggered at a pre-assumed source point. The arrival time map presented here focus on a point of destination instead, allowing source points to be everywhere.

7.2 Maximum Relative Amplitude Map

An MRA gives an upper bound on the relative wave amplitudes during a sufficient long time of a tsunami sourced at a particular point. Shown in Figure 8 is an MRA map, made for Hawaii. The map can be read as follows:

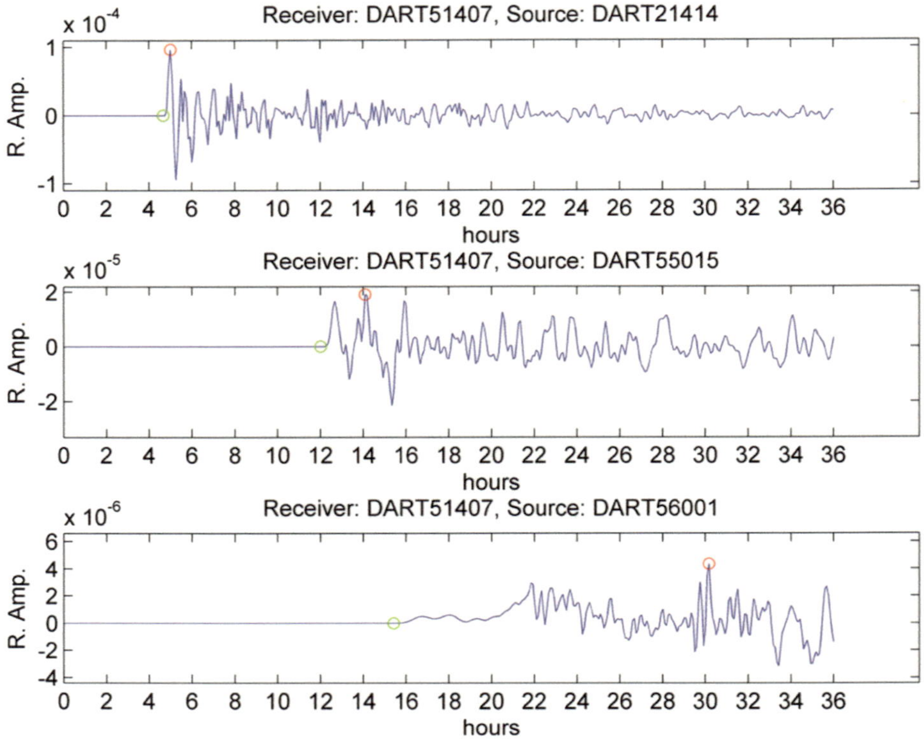

Fig. 6. Time series of relative amplitudes (R. Amp) at a receiver, DART 51407, as responses to the three impulse signals placed at three other DART stations (21414, 55015, 56001).

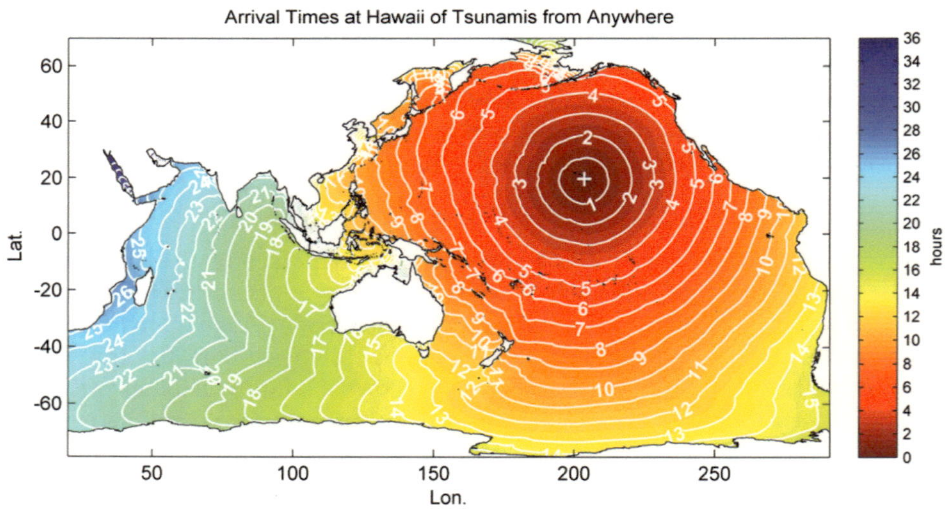

Fig. 7. A map of arrival times at Hawaii of tsunamis from anywhere in the India and Pacific Oceans. The destination point is marked by the white cross inside the 1-hour contour where the DART buoy 51407 is deployed.

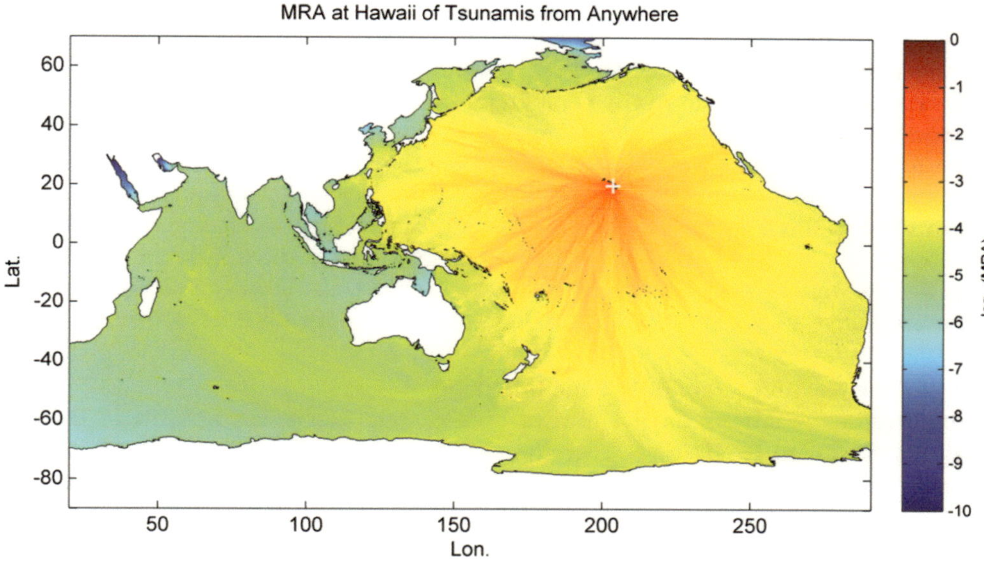

Fig. 8. Maximum relative amplitudes at Hawaii of tsunamis sourced anywhere in the India and Pacific Oceans. The destination point is marked by the white cross where the DART buoy 51407 is deployed.

Should someone be able to raise the mean sea level somewhere by one meter over the area occupied by a grid cell there and then suddenly release it, the dropping water column would generate waves propagating out in all directions. The waves would reach Hawaii eventually, and could be recorded by the DART buoy deployed there (please forget the sensitivity of the instrument in this thought experiment). As time goes by, the DART would yield a time series of the wave amplitudes. The largest amplitude one would see from the time series has already been predicted by the color on the map over the place where the water column was raised and dropped off.

The values of MRAs are all very small, ranging from 0 to 1. This is because each of them represents the impact of a very small water column, whose height is 1m and whose base area is the area of a cell, which is less than $10 \times 10\ km^2$ on the grid with a 5-minute spacing. The potential energy stored in the water column has to spread out over the entire ocean. In order to see the small MRA values well, their log10 values are mapped to colors instead.

The values below 10^{-10} are not shown on the map. Any values below this threshold can be safely discarded for the grid in question. This is because that the total number of the grid cells is less than 4×10^6. If the maximum source amplitude was $10^2 m$ and evenly distributed over a source region which occupies $1/100$ of the entire India-Pacific Oceans (which is as large as $1/3$ of the area of Australia), the contribution by the discarded MRA part would only account for not more than 4×10^{-4} m at the POI.

A tsunami triggered nearby a target has more impact than those triggered in a far field. The colors on the map visualize such differences in details. In contrast to the more or less co-circular contours seen in the arrival time map (Figure 7), the MRA map shows a strip pattern. This indicates that not all the points with the equal distances to a target have the same impact on the target. The variations in the water depths and the other factors such as wave reflections/refractions/interference all contribute to the differences. Knowledge on tsunami arrival times and knowledge on tsunami amplitudes are equally important. A time map and an amplitude map together offer us complementary information.

The concept of the maximum relative amplitude (MRA) was referred as a *gain* in Xu (2007). The term was coined because it had a sense of percentage of wave amplitude left when an initially unit amplitude wave arrived at a destination. Now the matrix form of the shallow water equations, as shown by Eq. (55), has rendered itself as a natural connecting point to the Kalman filter method. This is good for further studies on the assimilations of data coming from observatory stations such as DART buoys. In Kalman filter, there is an important concept called *gain* as well. In view of the potential name conflict, the term *gain* is replaced by the maximum relative amplitude, which is also more self-descriptive albeit it is longer.

7.3 Maximum Relative Amplitude Time (MT) Map

The occurrences of MRAs do not always immediately follow the arrivals of the initial signals, as illustrated by Figure 6. Therefore it is desirable to have also a map of the times when the MRAs occur. For brevity, let a map of this type called an MT map. Shown in Figure 9 is an MT map for Hawaii. In comparison with the AT map shown in Figure 7, the color bands in the MT time map are much less smooth even very rough in far-fields. This is because the MRAs are subjected to more complicated wave interactions. The initial arrival time is basically a kinematic issue. One could use the $\sqrt{gh}$ as the wave speed to make time contours and the results would be very close to what is shown in Figure 7. The time of the occurrence of an MRA is a dynamic issue. There is no simple rule as $\sqrt{gh}$ one could use to approximate the results as seen in Figure 9. In terms of time recording, the initial arrival times will never change once they are recorded. By contrast, maximum amplitude times are subjected to update if any subsequent waves exceed the previous ones.

It should be useful to know when a tsunami will initially arrive and when it will reach the maximum. Comparison of an AT map with an MT map gives us a bird's view of the difference between the two time points at a POI for tsunamis coming from anywhere in the oceans.

7.4 Density of Maximum Relative Amplitudes

The MRAs discussed above bear the influence of the grid cell sizes. Two water columns raised to the same height but over two cells of different sizes would have different impacts at the same destination point. If all the other affecting factors are the same on the their trips to the destination, the waves generated by the water column over the larger cell would impact more than the waves generated by the columns over the smaller cell. It is not necessary to pass on such cell size dependence to users of an MRA map, albeit the dependence may not be significant when the cells are fine.

To make an amplitude map free of the influences of cell sizes, we may normalize the MRAs by the cell areas. Let E be a diagonal matrix whose diagonal elements consist of the areas of all the water cells, we can transform Eq. (74) as

$$\eta_k(\tau) = (WE^{-1})E\eta^{(0)} \tag{75}$$

The largest values along the columns of WE^{-1} are the maximum relative amplitude per unit area, and may thus be referred as the densities of maximum relative amplitudes (DMRAs).

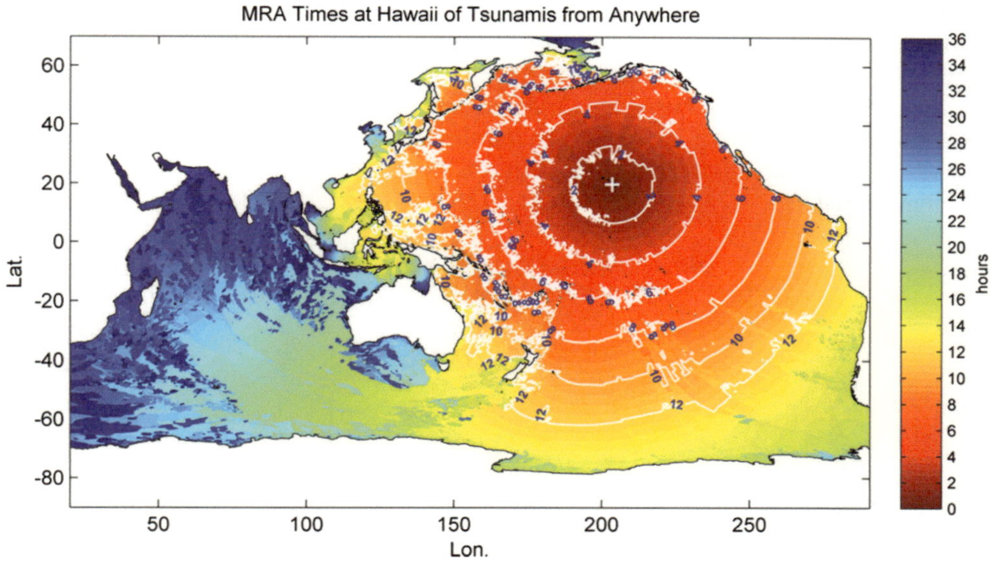

Fig. 9. A map of times when the maximum relative amplitude occur. For clarity, the contour lines are not superimposed as many as those in Figure 7.

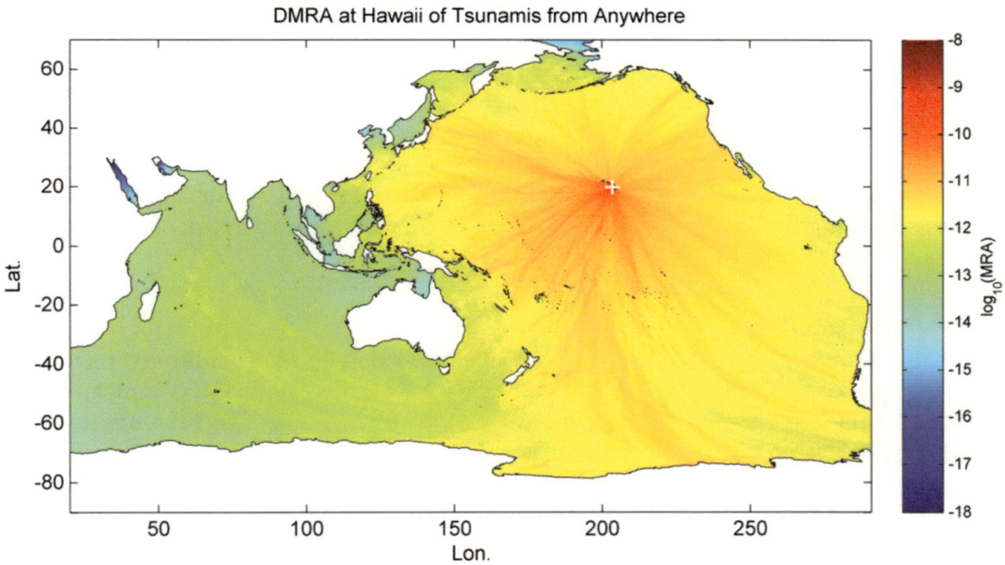

Fig. 10. A DMRA map showing the densities of maximum relative amplitudes.

A DMRA map can be read as follows: should someone be able to raise a 1m water column over a 1-m^2 area somewhere on the map, the color on that spot indicates the largest amplitude that Hawaii would receive. The area-normalization makes the map *area integrable,* which means a user can integrate the DMRA over the area of potential hazard regions, and come up with an upper bound estimate of the *largest relative amplitude* or even better, the *largest absolute amplitude* if he can also estimate on the distribution of the wave amplitudes in the source region.

Note that maps as presented above are not area preserving. The spherical surface areas at high latitudes (both south and north) are distorted on the paper. Therefore when a user exercises an area integration, he has to correct for the area distortion. To relieve users from this burden, the DMRA had better be presented on an area-preserving map. Figure 11 shows the same DMRAs but on a map made with the sinusoidal projection. Note that the units in x and y are all in meters on the map, and that the line of x=0 on the map corresponds to the 155° longitudinal line.

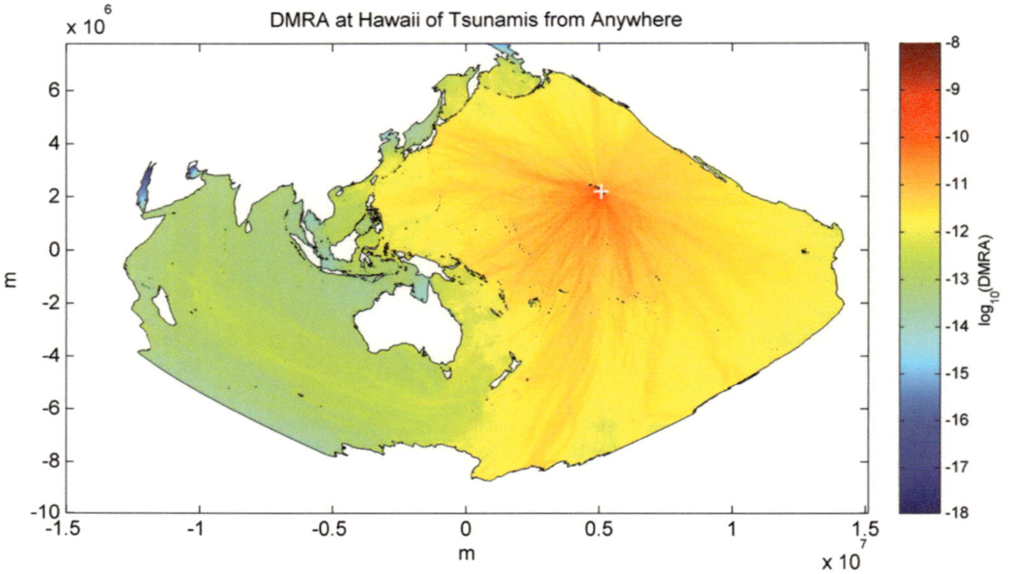

Fig. 11. Densities of maximum relative amplitudes shown an area-preserving map. The map results from using the sinusoidal projection. The line of x=0 corresponds to 155° longitude line.

7.5 Real-time simulations

So far the amplitude weight matrix W has been used only in a summary manner: the important pieces of information are summarized into maps. With the complete dynamic information the weight matrix contains, we can go further: the real time simulations of tsunami arrivals from anywhere in the oceans.

To simulate tsunami propagations in real-time has been a great challenge. When a tsunami is triggered, it is very difficult to run a model on the fly in a vast transoceanic domain and yet still win the race for time against a fast moving tsunami (~700km/hour) even with a super computer and massively parallel computations. Now with Eq. (74), we can instantaneously transfer any disturbance at a source region to the sea level responses at a destination, no matter how far apart the source and the destination points are. This will give a valuable time window for tsunami warnings.

The weight matrix W in Eq. (74) has been pre-calculated and stored in a hard disk. If the matrix W has already been in the RAM of a computer, which may well be the case in an institute designated for tsunami warnings, it will take no time to compute just one matrix-vector multiplication. If the weight matrix W has not been in the RAM, it may take a few seconds to retriever it from the disk. Thus, with a tolerance of a few seconds, we can say that simulations of tsunami arrivals at a POI can always be performed instantaneously in real-time.

Based on Eq. (74), one can develop a desktop or web-based application for simulations of the arrivals at one or more POIs of tsunamis from anywhere in the oceans. Shown in Figure 12 is such a web-based application. The graphical user interface (GUI) shows a map with all the 40 DART stations as POIs. Forty weight matrices have been pre-calculated for a duration of 36 hours, and can be called for to deliver the responses upon receiving users' instructions through the interactive map.

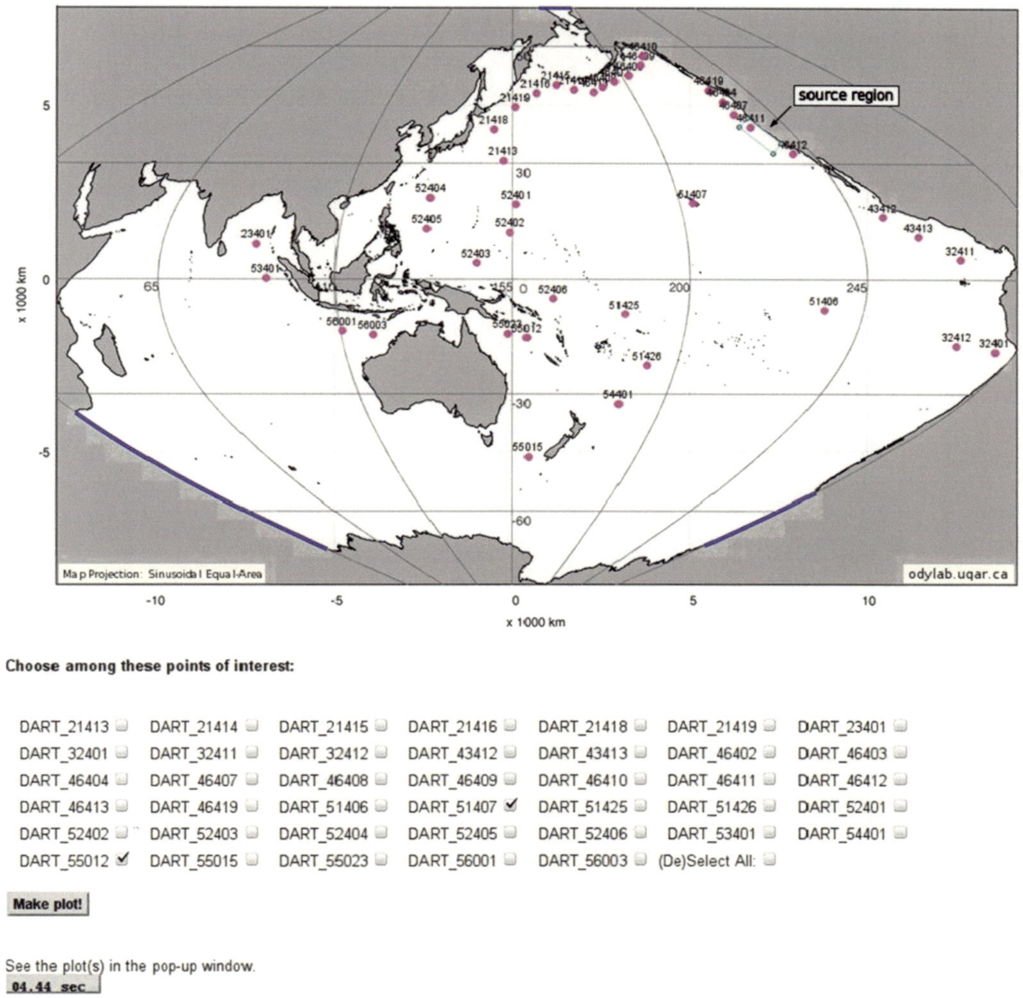

Fig. 12 A Real-time simulation system for tsunami arrivals at DART stations from anywhere in the India and Pacific Ocean. On the map, a source region offshore California is specified, and two POIs are selected (DART 55012 and 51407). It takes less than 5 seconds for the simulation system to deliver the response curves (see Figure 13).

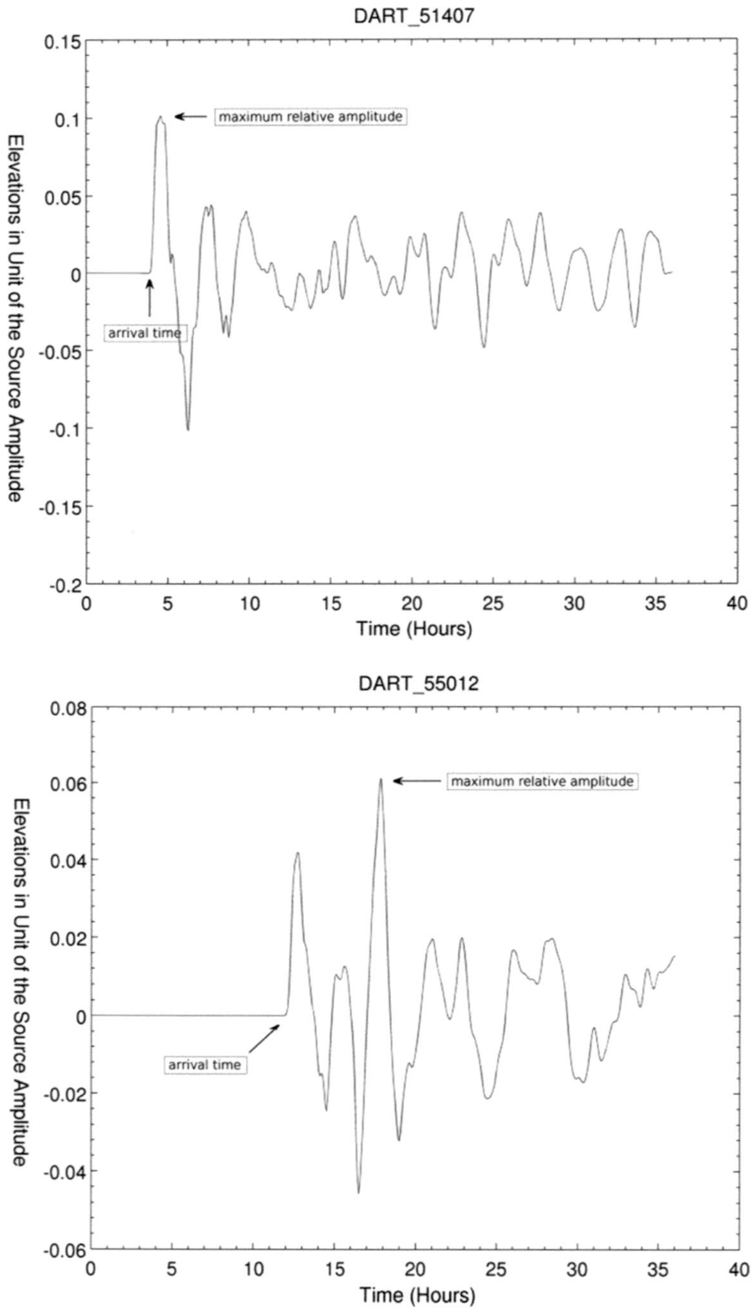

Fig. 13. Response curves by DART 51407 and DART 55012 to a tsunami source region offshore California (cf. Figure 12).

A user can arbitrarily click some points on the map to create a polygon defining a tsunami source region. Then he can choose one or more points of interest by checking the boxes in the list beneath the map. In a few seconds after pressing the "Make plot!" button on the lower left part of the GUI, he will see the tsunami arrival curves in a pop-up window. For the example shown in the figure, a rectangular source region was specified offshore California, and two POIs were chosen, DART 51407 and DART 55012. As we can see from the little timing box below the "Make plot" button, it took less than 5 seconds to deliver two response curves, which are shown in Figure 13.

For the above example, it was assumed that sea surface in the source region was uniformly raised by 1m above the mean sea level. In other words, the sea level heights shown in Figure 13 are only relative the sea levels in the source region. If a realistic source distribution could be estimated or supplied by a tsunami genesis model, the units of response curves would then become absolute.

To reduce the time needed to retrieve them back from a hard disk, the amplitude weights should not be saved in a single large data file. Instead, they should be saved in many small data files. One may achieve this is by first tiling the model domain, then regrouping the columns of the weight matrix W such that each group of the columns corresponds to a tile. A group of the columns can then be saved as a small file. When a tsunami source region is specified, the tiles that the source region falls into can be detected, and hence only the data files belonging to those tiles need to be read off the disk. For a 5 minute grid spacing, a tile size of $10° \times 10°$ seems a good balance between the number and the sizes of the data files.

8. Conclusions and future work

A tsunami has three life phases: the genesis, the open-ocean propagation and the onshore run-up. Towards a complete real time simulation system which covers the whole life of a tsunami, there are distinct challenges in each of the three phases to overcome. The challenges in the first and last phases lie in the dynamics complexity. Not all the earthquakes underneath the oceans can generate tsunamis. For simulations of onshore run-ups on the other end, the highly non-linear tsunami bores over a complex onshore land and lack of timely and reliable open water boundary conditions are the conceivable challenges.

The challenges in the second phase are not due to dynamics complexity. Rather the dynamics in this phase is fortunately simple. The biggest challenge in this stage come from the fact that tsunamis run too fast in oceans that are too big for a propagation model to run on the fly and yet still be able to win a sufficient time window for tsunami warnings. The all-source Green's function (ASGF) offers a new approach to cope with this challenge. With a pre-calculated ASGF, simulations of tsunami arrivals become instantaneously realizable, no matter where and how far a tsunami comes from. As long as supplied with a reliable source function (the $\eta^{(0)}$ in Eq. (74)), a set of ASGFs can predict sea level anomalies for a set of points along a line offshore a coastal area, furnishing a local run-up model with necessary open water boundary conditions to simulate the onshore run-ups for that coastal area. The ASGF approach offers a good and fast link between the first and the third phases.

The web-based real time simulation system demonstrated in the chapter is just a starting point towards a complete system. Much more work can be done to improve it: linking with a genesis model and a run-up model on the two ends; linking with and assimilation of the live data from the DART buoys and other observational stations; important coastal cities from the Pacific Rim and countries along the India Ocean can be added as POIs in the system. The domain of the system can be increased too to include all the oceans on the planet.

For tsunami preparedness, the ASGF also brings in new types of tsunami maps: maps of tsunami arrival times, maps of maximum relative amplitudes and their occurrence times.

These maps can also be helpful in a real-time when a tsunami is triggered. For example, whether a number read off an MRA map is above 70% or below 10% will make a big different to an emergence management officer in making his decision. An atlas of such maps can be made in the future for importance coastal cities around the world.

9. Acknowledgement

The author wishes to sincerely thank Dr. Dave Brickman for his careful review of the manuscript and valuable comments, Prof. Ted Murty for his very encouraging comments, and Mr. Michel Beaulieu for his help on the ODYLAB web page development. Thanks also go to Prof. Jarle Berntsen for the personal communications on the Coriolis weighting scheme.

10. References

Arakawa, A. & Lamb, V. R. (1977). Computational design of the basic dynamical process of the UCLA general circulation model. *Methods in Computational Phsics.* 17, Academic Press, 173-265.

Bryan, K.; Manabe S. & Pacanowski R. C. (1975). A global ocean-atmosphere climate model. Part II. The oceanic circulation, *Journal of Physical Oceanography, 5,* 30-46.

Bryan, F. 1987. Parameter sensitivity of primitive equation ocean general circulation models. *J. Phys. Oceanogr.,* 17, 970 985

Brummelhuis, P. G. J.; Heemink, A. W. & Van Den Boogaard, H. F. P. (1993) Identification of shallow sea models. *International journal for numerical methods in fluids.* vol. 17, 637-665.

Chapman, D. C. (1985). Numerical Treatment of Cross-Shelf Open Boundaries in a Barotropic Coastal Ocean Model. *Journal of Physical Oceanography,* 15: 1060-1075

Ding, Y.; Jia Y.; ASCE, M.; Wang, S. S. Y. & ASCE, F. (2004). Identification of Manning's Roughness Coefficients in Shallow Water Flows. *Journal of Hydraulic Engineering, ASCE,* June 2004.

Espelid, T. O; Berntsen J. & Barthel K. (2000): Conservation of Energy for Schemes Applied to the Propagation of Shallow-Water Inertia-Gravity Waves in Regions with varying Depth. *Int. J. Numer. Engng 2000;* 49:1521-1545

Fujima, K. (2001). Long wave propagation on large roughness, ITS Proceedings. Session 7, No. 7-22. pp. 891-895.

Heaps, N. S. (1969). A Two-Dimensional Numerical Sea Model. *Phil. Trans. R. Soc. Lond. A* October 30, 1969265:93-137.

Heaps, N.S. (1971). On the numerical solution of the three-dimensional hydrodynamical equations for tides and storm surges, *in*: *Mémoires de la Société Royale des Sciences de Liège. 6e Série,* Tome II: pp. 143-180.

Knobe, R. (2000) *An introduction to the mathematical theory of waves.* American Mathematical Society; 1 edition (2000). ISBN-10: 0821820397.

Pedloskey, J. (1979) *Geophysical Fluid Dynamics.* Springer Verlag, 624 pp. ISBN: 3-540-90368-2.

Shuto, N. (1991). Numerical Simulation of Tsunamis --- Its Present and Near Future. Natural Hazards 4: 171-191.

Sielecki, A. (1968): An Energy-Conserving Difference Scheme for the storm surge equations Mon. Wea. Rev., 96, 150-156.

Sommerfeld, A. (1949). *Partial Differential Equations in Physics.* Academic Press, New York, New York, 1949.

Xu, Z. (2007). The all-source Green's function and its applications to tsunami problems. *Science of Tsunami Hazards,* 26(1), 59-69.

26

A Multi-Sensor Research Program to Improve Tsunami Forecasting

George D. Curtis,
University of Hawaii, Honolulu
Hawaii

1. Introduction

While the warning systems do an increasingly good job of warning of all probable events, there are still apparent false alarms and a lack of magnitude prediction. History shows that major advances in technology occur when several components mature to be combined into a final system. An example is the communications satellite: around 1960, the silicon solar cell, the RF transistor, the integrated circuit, the large rocket, knowledge of the synchronous orbit, reliable components, and supporting systems came together to produce systems so workable that the early ones are still usable and the current ones just have more channels.

We are now in a relative 1960 position in tsunami work. Deep ocean sensors are in place, numerous tide gauges and sea level gauges are telemetered (via satellite), real-time numerical analysis programs are under trial. Mw can now be calculated in an hour and Mwp in a minute, a global infrasonic network has been deployed, T-phase data are readily available, and so are ionospheric sounders. We have the capability to cheaply and quickly record, correlate, analyze, transmit, and discard data from all these sensor systems. Thus, we should have the means to develop the ability to evaluate a tsunamigenic earthquake and early tsunami waves, reliably and even quantitatively.

I suggest the ideal way to accomplish this is to establish a small, dedicated laboratory which will bring together a selection of data from these sensor systems to evaluate every possible tsunamigenic event. It is now easy to temporarily store these data in volatile memory for analyses, retention, and discard. An earthquake that would be checked and discarded by the warning center can provide the trigger for storage and evaluation. Most of the data are low frequency, easy to transmit (part of it is on the Internet now), and to store.

Information on prior experiments with some of the components, current status, and processing estimates is provided, along with ample references.

2. The warning problem

While the warning systems do an increasingly good job of warning of all probable events, there are still false alarms and a lack of magnitude prediction. Warnings are binary, with a very fine line between "warn" and "no warn"; the decision is usually conservative and produces false alarms from the public viewpoint if not in fact. This occurs because we still have not achieved the ability to forecast a damaging event accurately and rapidly. Yet, we have more seismic stations than ever, more sea level stations, more computers with faster,

more capable programs, and better communication systems. Our use of modern technology is good, but more can be incorporated into a systems approach. Curtis (1993) reviewed the progress at that time and pointed out some research which might be cost effective. How can we accomplish more improvements?
History shows that major advances in technology occur when several components mature to be combined into a final system (Curtis et al., 1986).

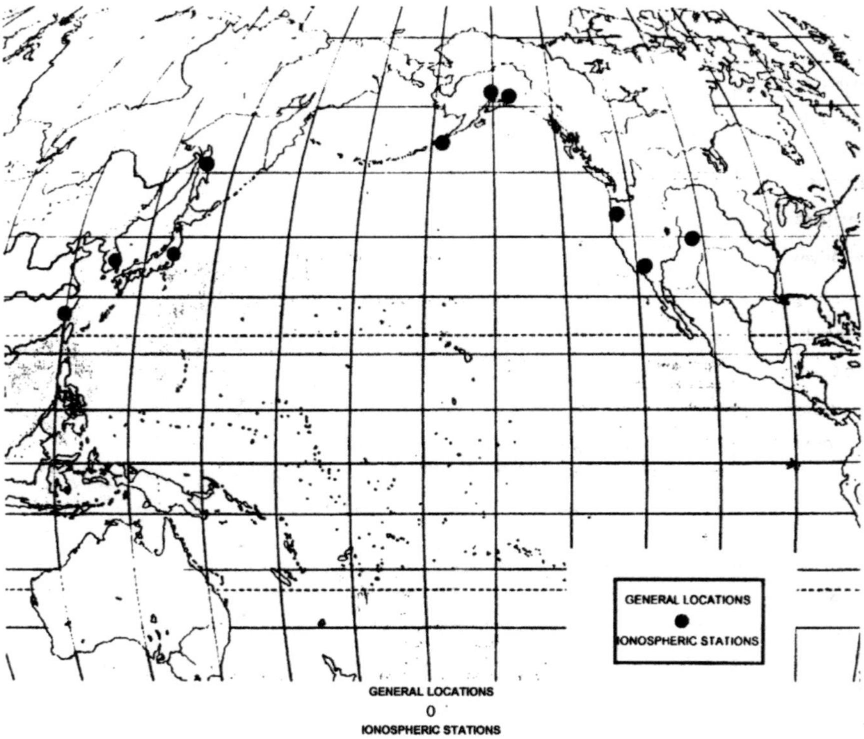

Fig. 1. General locations, ionospheric stations.

An example is the communications satellite: around 1960 the silicon solar cell, the RF transistor, the integrated circuit, the large rocket, knowledge of the synchronous orbit, and highly reliable components, came together 2 to produce systems so workable that the early ones are still usable and the current ones just have more channels.
We are now in a "1960 position" in tsunami work. Deep ocean sensors are in place, numerous tide gauges and sea level gauges are telemetered (via satellite), real-time numerical analysis programs are under development, Mw can now be calculated in an hour and M_{wp} in a minute, a global infrasonic network has been deployed, T-phase data are readily available, and so are ionospheric sounders. We have the capability to cheaply and quickly record, correlate, analyze, transmit, and discard data from all of these sensor systems. Thus, we should have the means to develop the ability to evaluate a tsunamigenic earthquake and early tsunami waves, reliably, and even provide a quantitative warning. If we bring all appropriate technology to bear on the problem, we can surely advance that capability nearer to the goal.

3. The concept

I suggest that the ideal way to accomplish this is to establish a small, dedicated laboratory which will bring together a selection of data from these sensor systems to evaluate every possible tsunamigenic event. It is now easy to temporarily store these data in volatile memory for analyses, retention, and discard. An earthquake that would be checked and discarded by the warning center can provide the trigger for storage and evaluation. Most of the data are low frequency, and easy to transmit (part of it is on the Internet now) and store. A precept was provided by Najita and Yuen (1978). The reference includes a full description of tsunami generation by earthquakes and of detection of tsunamigenic events by HF ionospheric sounding. The laboratory they set up at the University of Hawaii (UH) is described; the sounding was continuously plotted on a chart recorder at slow speed activated to a high speed when seismic waves from the UH seismometer exceeded a threshold similar to the alarm at the Pacific Tsunami Warning Center (PTWC). A significant problem was in disregarding the natural perturbations occurring at dawn and dusk. These could only be partially removed by the band pass filter used, and the project was eventually discontinued.

Was an important opportunity missed when this work was not carried further? Most of the technology involved has now grown and advanced so it will be much easier to carry out the work in the future. Plus, with more geophysical sites in use, far more good data are readily available. In fact, it is likely that this and other ideas can be done as a hindcast experiment with the large amount of sensors and recordings from them currently stored in accessible data banks.

4. Sensors and sources

4.1 Ionospheric

Ionospheric sounders indicate the height of the ionosphere, the upper portion of the atmosphere that is heavily ionized and which enables "short wave" radio transmission by reflection of the signals. Earthquakes with a significant component of vertical deformation perturb this layer in a manner that is readily measured using precise radio signals in the high frequency region broadcast by standard stations around the world. The vertical displacement travels toward the ionosphere (100 m altitude or more) as a Rayleigh wave, faster than the speed of sound; the radio measurements are of course instantaneous.

Weaver et al. (1969) report such measurements for an earthquake in the Kuriles and Furumoto (1970) discusses using them to evaluate the source mechanism of tsunamis. These investigators used 5 and 10 MHz transmissions from WWVH on Kauai, received in Honolulu and utilized as described by Najita. Fitzgerald and Walcott (1985) reported the ionospheric disturbances from the Coalinga earthquake; there are numerous other similar recordings. The actual data of interest is between 2 seconds and 300 seconds. It appears that there are at least 12 HF sounding stations around the Pacific in the U.S., Canada, and Japan which can provide real-time or recorded data for this purpose. Figure 1 shows the location of some.

A newer but widely used method of measuring ionospheric changes is by their subtle changes in SHF radio transmissions from satellites; Navstar GPS is usually used. Calais et al. (1997) obtained signatures from a mine blast by this method while others have described the effects of earthquakes on the signals. Romans and Hajj (1996) give a general view of the methodology. Since the satellites transmit on two frequencies, it is feasible to obtain an excellent differential measurement. These receivers can be set up for real-time use wherever telemetry is available.

4.2 Infrasonic sensors

Vela, a set of ARPA projects to detect nuclear explosions produced many standard seismic systems around the world, but also a number of infrasonic receiving systems (microbarographs) were established. (Of course many such geophysical labs had existed for years.) These sensors added to the ability of research in atmospheric detection of events such as hurricanes, tornados, earthquakes, and other disturbances. It was known that earthquakes sometimes were reported to produce a noise, and Nakamura (1988) described the sound of an approaching tsunami and analyzed how it might be produced. Miller (1968) had earlier brought up the possibility of predicting tsunami height from atmospheric wave data, mentioning the effect of the vertical component. In the same symposium, Donn (1968) discussed sources of infrasonic waves and their detection. (This writer developed the sensors then in use by Donn at Lamont Geophysical Observatory and recalls their detection of hurricanes as well as nuclear explosions.)

Bedard and Georges (2000) and Bedard (1998) discuss the varied applications of infrasonic detection (the 1998 paper alone has 31 references), including earthquake detection and analysis. An example of recording both infrasonic and acoustic waves for geophysics is provided by Tihara et al. (1997). There area number of microbarographs around the Pacific of varying capability whose records might be used for analysis, if not in real time. Ironically, the Honolulu Observatory (which became the Pacific Tsunami Warning Center) had a microbarograph, which apparently was discarded long ago.

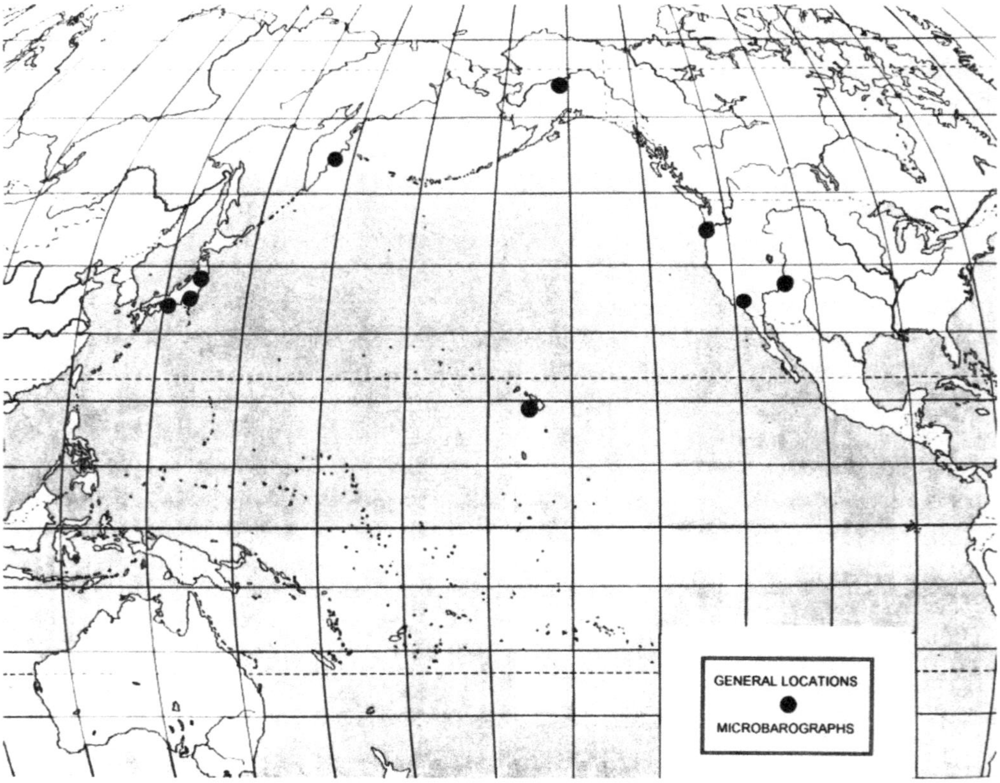

Fig. 2. B Microbarographic stations.

The Comprehensive Test Ban Treaty (CTBT) has resulted in the establishment of an enhanced Vela-type global network by the U.S. Department of Defense. These microbarographic stations are extremely sophisticated, have a large wind noise reduction system (detection is limited by wind noise, not the sensor elements), and are well equipped and manned. The pass band of the microphone is 8 to 0.02 Hz (period of 0.125 to 50 s), less some noise reduction effect, and the data are telemetered to a central location in Virginia. There are systems now reported to be in operation in the Pacific region, as part of the International Monitoring System (IMS). Because they are wide band, low noise, and on line, they are probably the best sources of earthquake-effect data. Figure 2 maps some of the various microbarographs in the Pacific. Since the signals travel at roughly the speed of sound, it is necessary to utilize sensors as near the source as possible.

4.3 T-phase sensors

The T-phase signal from an earthquake is propagated via the SOFAR (SOund Frequency And Ranging Fixing And Ranging) channel in the ocean and couples into the earth near the coast to be detected by a seismograph. Since most of the path is acoustic (1500 m/s) it arrives after the P and S waves and so is referred to as the tertiary wave. It can, of course, be readily detected by hydrophones in the sound channel (U.S. Navy) and also by other hydrophones. The use of the T-phase signals in the warning system was proposed by Ewing et al. (1950). Considerable work was done on seismic T-phase evaluation for that purpose (Johnson, 1970) but with limited success. Johnson pointed out that if an array of hydrophones were available, he could have combined directivity with duration and accurately determined rupture length. The SOSUS (SOund SUrveillance System) can now provide just what Johnson needed, in the desired band of 2 to 30 Hz.
They cover the North Pacific. The several SOSUS stations, each with an array of many hydrophones, offers a means to deal with the path discrepancies encountered with a single location. It should be possible to evaluate predetermined paths to likely seismic locations and make a more valid analysis of the tsunamigenesis of an event in near-real time. Figure 3 shows the approximate locations of some SOSUS stations; some are part of the IMS and some are connected to NOAA via a Navy processing station at Whidbey Island, Washington.

4.4 Other sensors

An open scientific mind may suggest other sensors not yet considered or which, though not feasible in the past, may now be possible. Satellite altimetry has improved since Seasat and the fortuitous detection of waves may occur if we look and analyze carefully. The use of differential GPS on a buoy with satellite transmission is a possible method of observing a wave. Improvements in evaluating the source in real time: the seismic moment, the rupture length from wave arrival time at area gauges, the deployment of more deep-ocean sensors will certainly continue and will offer more means to promptly evaluate a possible tsunami, while at the same time make the job more complex.

5. The Tsunamigenesis experiment

The above descriptions of past and current programs related to tsunamigenesis indicate that (a) there are technologies available or known that can help define the onset of a significant tsunami, but (b) we have not combined them into a working system.

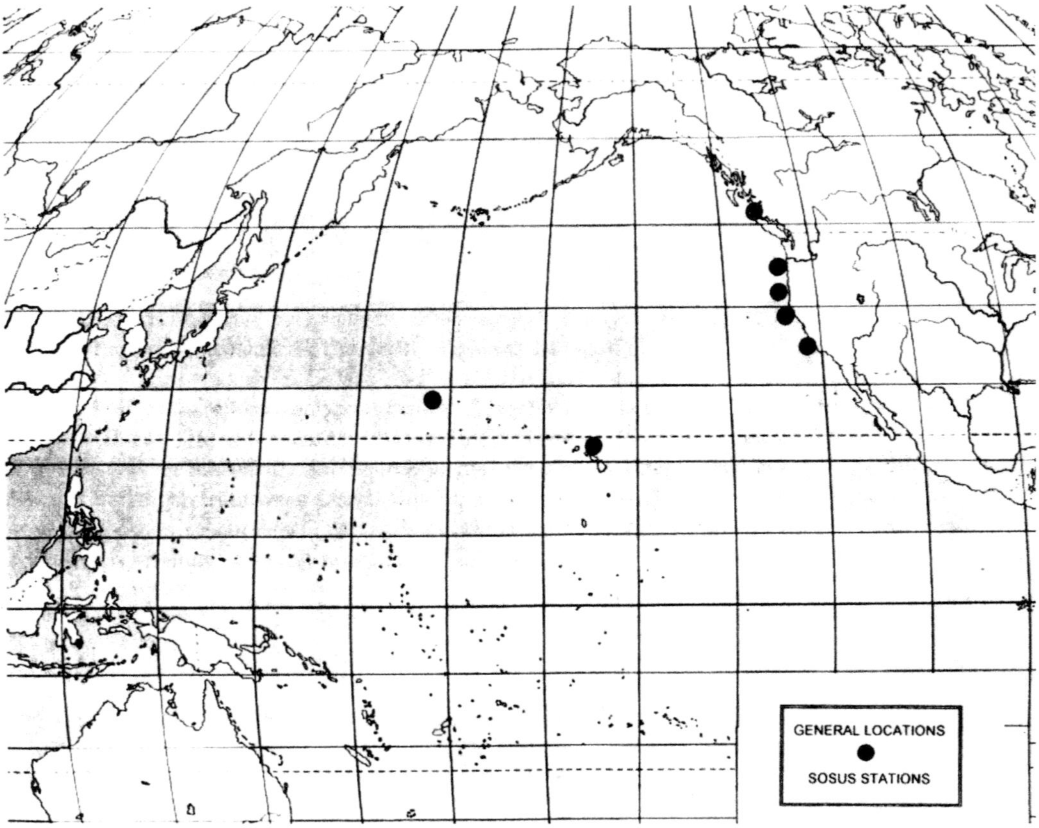

Fig. 3. B Navy Hydroacoustic Stations

The other vital technology available but not mentioned as it is known to all is rapid data handling systems. We now have relatively cheap, fast, high volume, digital communication, storage, and analytical equipment not available to many of the investigators cited above. We should utilize these capabilities along with careful consideration of the science that has been done and that is now in use, to develop a more reliable tsunami forecasting system. This is the sort of concurrent approach outlined in the introduction and can move the 1960's point into this century.

I suggest that the ideal way to carry this out is to establish a small, dedicated laboratory that can gather, correlate, and evaluate a selection of data from every possible tsunamigenic event. The extensive and very capable seismic network now deployed provides the basis for experiments by triggering the data collection scheme. Inputs from several sensors temporarily stored in volatile memory of ample duration may then be stored for analysis, archiving, or discard. It is easy to save before as well as after. This writer designed a system 20 years ago that keeps four channels of data plus WWVH time for 4 minutes before and 6 minutes after an event. It is still in use but could be done much better with present technology. This is a good approach to our data problem. As in any good experiment, the negative data (no significant tsunami) can be as important as the positive results.

In addition, very productive work can be accomplished by hindcasting with existing data. A good example already accomplished is the evaluation of moment analysis as a tsunamigenic indicator (Walker et al., 1991; Tsuboi et al., 1999). This has not provided a satisfactory answer yet, but if it is correlated with other data it may eventually yield a better solution. Other examples might be microbarographic data after an earthquake, from Japanese and U.S. archival sources.

The optimum approach is to establish a small, dedicated laboratory at a location where essential real-time data are available and staff and analytical capability exist--even part time. Because most pertinent data can be saved automatically by this system for later analysis, attendance would not be mandatory. Certainly, a combination of on-line data, recorded data, and retrievable data can be used to present the coordinated information to knowledgeable scientists, present or not.

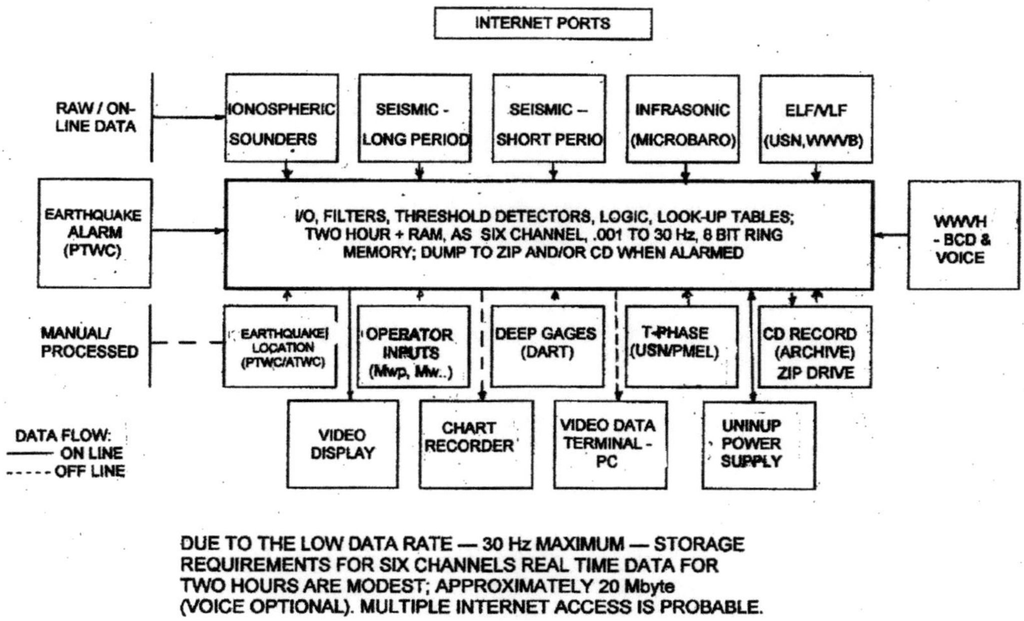

Fig. 4. Data flow, Tsunamigenesis experiment laboratory.

Figure 4 outlines a system intended to be comprehensive; what should be planned for, not necessarily what can be done this year. But none of it is technically difficult. Funding for new projects is often difficult but we have far more available to solve this problem than we had 10 or 20 years ago (Curtis, 1993). Much of the data are "free" for the connection or retrieval, now even from the Internet.

6. References

Bedard, A.J., Jr. (1998): Infrasonic and near-infrasonic atmospheric sounding and imaging. Proceedings of the Progress in Electromagnetic Research Symposium, Nantes, France.

Bedard, A.J., and T.M. Georges (2000): Atmospheric infrasound. Physics Today, March 2000.

Curtis, G.D. (1993): The Tsunami Research Problem. IAPSO XXI General Assembly, Honolulu, Science of Tsunami Hazards, 11(2).

Curtis, G., and C. Mader (1986): Real-time monitoring and modeling for tsunami threat evaluation. The Science of Tsunami Hazards, 5(1).

Curtis, G.D., and P. Takahashi (1986): Engineering and humane values. In Essays on Creativity and Science, Hawaii Council of Teachers of English, Honolulu, Hawaii.

Donn, W.L., and E. Posmentier (1968): Infrasonic air waves from natural and artificial sources. In Acoustic- Gravity Waves in the Atmosphere Symposium, ESSA/ARPA, Boulder, Colorado, July 1968.

Ewing, M., I. Tolstoy, and F. Press (1950): Proposed use of the T-phase in tsunami warning systems. Bull. Seismol. Soc. Amer., 40, 53-58.

Fitzgerald, T.J., and J.H. Walcott (1985): E-layer ionospheric disturbances following the Coalinga earthquake. Los Alamos National Labs report.

Fox, C.G., and S. Hammond (1994): The VENTS program T-phase project. Mar. Tech. Soc. J., 27(4).

Furumoto, A.S. (1970): Ionospheric recordings of Rayleigh waves, for the source mechanism of tsunamis. In Tsunamis of the Pacific Ocean, edited by W.M. Adams, East-West Center, Honolulu, Hawaii.

Johnson, R.H. (1970): Estimating earthquake rupture length from T-waves. In Tsunamis of the Pacific Ocean, edited by W.M. Adams, East-West Center, Honolulu, Hawaii.

Miller, G.R. (1968): On the question of tsunami height predictions based on atmospheric wave data. In Acoustic- Gravity Waves in the Atmosphere Symposium, JTRE/UH, ESSA/ARPA, Boulder, Colorado, July 1968.

Murty, T. (1977): Seismic Sea WavesCTsunamis. Canada Department of Fisheries and Environment, Ottawa.

Najita, K., and P.C. Yuen (1978): Ionosphere method of detecting tsunami-generating earthquakes. The Physics Teacher, December 1978.

Nakamura, S. (1988): On audible tsunami on the coast. Science of Tsunami Hazards, 6(1).

Romans, L.J., and G.A. Hajj (1996): Ionospheric imaging with low-earth orbiters tracking GPS. AGU Spring Meeting, Baltimore, G1113-03Tihara, M., et al. (1997): Infrasonic and acoustic waves from Mt. Pinatubo. In Fire and Wind, edited by C.G. Newhall, University of Washington Press, Seattle.

Tsuboi, S.Whitmore, and T. Sokolowski (1999): Application of MWP to deep and teleseismic earthquakes. Bull. Seismol. Soc. Amer., 89(5).

Walker, D.A., and E.N. Bernard (1993): Comparison of T-phase spectra and tsunami amplitudes for tsunamigenic and other earthquakes. J. Geophys. Res., 98(C7).

Walker, D.A., C.S. McCreery, and Y. Hiyoshi (1992): T-phase spectra, seismic moments, tsunamigenesis. Bull. Seismol. Soc. Amer., 82, 1275-1305.8Weaver, P., C. Yuen, G.W. Proliss, and A.S. Fururnoto (1969): Acoustic coupling into the ionosphere from seismic waves of the earthquake of the Kuriles, August 11, 1969. Nature, 226, 1239-1241.

Note: copies of most of the above references are in this writer's files, and can be made available to other users.

27

Algorithms for Automatic, Real-Time Tsunami Detection in Sea Level Measurements

Gian Mario Beltrami, Marcello Di Risio and Paolo De Girolamo
DISAT-LIAM - Università di L'Aquila
Italy

1. Introduction

Automatic, real-time tsunami detection in sea-level measurements is a main component of a tsunami early warning system (TEWS). Although a great effort has been recently undertaken by the scientific and engineering community in developing new technologies (e.g. satellite altimetry, detectors of low-frequency elastic oscillations associated to a tsunami) capable of increasing the awareness of potential tsunamis in the minimum amount of time, at present direct detection in sea level measurements is still the main mean to confirm their actual generation and propagation.
Clearly, the device used to collect these measurements has to be chosen between those equipped with sensors capable of detecting sea-level oscillations within the tsunami frequency band. At present, the main devices that can be actually used should belong to the following classes:

i) Bottom pressure recorders (BPRs), or tsunamometers;
ii) Tidal gauges (TGs); and
iii) Wind-wave gauges (WWGs) equipped with either pressure, or acoustic, or optical sensors.

As far as the third class is concerned, it is to be noticed that tsunami detection cannot be performed by an accelerometric 'wave buoy', i.e. the main device used to collect wind wave measurements. Actually, a buoy is not capable of detecting the propagation of very long waves such tsunamis.
As a general rule, the choice of the device to be effectively used mainly depends on its operational location. This location should be determined on the basis of a tsunami hazard assessment (i.e. a previous knowledge both of probable tsunami sources and of places at risk along the coast). To be effective, measurements should be collected either in close proximity of probable tsunami sources, or between these sources and the coast at risk (at a distance from the coast capable of guaranteeing timely warning of actually approaching tsunamis). For the purpose of the present chapter, the probable tsunami sources can be listed on the basis of their location with respect to the coast of interest, i.e.:

i) sources located far away from the coast (at a distance on the order of thousands of kilometers);
ii) sources located at the coast; and

iii) sources located 'near' the coast (at a distance ranging from tens to few hundreds of kilometers).

When the tsunami sources are far away from the coast of interest, it will be convenient to collect sea-level measurements far out at sea, between these sources and the coast at risk. Such a location implies great water depths, and makes bottom pressure recorders (BPRs) with autonomous power supply the more suitable devices to be used. A BPR (Eble & Gonzlez, 1991) actually measures pressure fluctuations indirectly by evaluating the vibrational variations that these fluctuations cause in a quartz-bean piezometrically induced to oscillate in its lowest resonant flexural mode. Its sampling interval is generally set equal to 15 s. Such a technology makes possible to locate the device on the sea floor, at water depths ranging from hundreds to some thousands of meters (i.e. at water depths at which only pressure fluctuations induced by propagating waves within the tsunami and tidal frequency band can be detected). These characteristics gain BPR the name of *tsunamometer*, and make it undoubtedly the more suitable device to confirm the propagation towards a coast of a tsunami generated far out at sea by either an earthquake, or a submarine landslide, or the eruption of a submerged volcano, or a meteorite impacting the ocean. Indeed, wind-wave gauges (WWGs) mounted above or below mean sea level (at a distance from it around 10 m) on preexisting offshore platforms may also be used in this case, depending on the platform distance from the coast of interest. These devices can be equipped with either pressure, or acoustic, or optical sensors, and usually collect measurements at a sampling frequency equal to 2 Hz. If the sea surface is continuously monitored, these devices will be perfectly able to detect a tsunami.
When tsunamis may be generated at the coast of interest (either by landslides, or coastal earthquakes, or coastal volcano eruptions), it will be essential to collect sea-level measurements in close proximity of the sources. Such a location implies water depths in the order of tens of meters, and makes WWGs such as the standard pressure transducers (SPTs) a suitable and economically convenient choice. In this case, a further important issue will be whether or not the source distance from the closest inhabited location makes the elapsed time between the tsunami generation and arrival sufficient to provide a warning. An example is given by the island of Stromboli, in the southern part of the Tyrrhenian Sea (Italy). On December 30, 2002 two tsunamis were generated by landslides detached from the 'Sciara del Fuoco' on the north-west flank of the volcano of Stromboli (Tinti et al., 2003). The tsunamis propagated both towards the coasts of Sicily, and around the island, hitting the village of Stromboli few minutes after their generation. The landslides and the tsunamis represented the most threatening episodes of a strong eruption started on December 28. In spite of the intensified seismic monitoring carried out in response to the eruption, nor the landslides nor the tsunamis were foreseen. Following these events, the 'Dipartimento della Protezione Civile', DPC (Italian Civil Protection Department), set up a specific TEWS which relies on real-time monitoring not only of the volcano seismic activity (e.g. Langer & Falsaperla, 2003), but also of the 'Sciara del Fuoco' ground movements (e.g. Corsini et al., 2003), and of sea-level oscillations. In particular, two land-cabled SPTs with a sampling frequency equal to 2 Hz were initially deployed in the sea at the two sides of the 'Sciara del Fuoco'. Both SPTs lay on the sea bottom at a water depth around ten meters. At present, sea-level oscillations are monitored by a SPT installed at a depth equal to 14 m on an elastic beacon deployed in the sea opposite the 'Sciara del Fuoco' at a water depth equal to 43 m.
When tsunamis may be generated 'near' the coast of interest, the main problem won't be the type and the location of the device used for detection. The main problem will be whether or not the source distance from the coast of interest makes the elapsed time between the

tsunami generation and arrival sufficient to provide a warning. The celerity of tsunamis is on the order of hundreds of kilometers per hour, so a time-window around some tens of minutes will be available for a warning in the case of a tsunami generated few hundreds of kilometers away, whilst a tsunami generated some tens of kilometers offshore will impact the adjacent coast in minutes. In the first case, either specifically deployed BPRs or WWGs mounted on opportunely located preexisting offshore platforms can be effectively included in a TEWS. As far as the second case is concerned, the only chance is that, due to the frequency dispersion of the tsunami, the time elapsed between the arrival of its leading wave and its peak amplitude (McGehee & McKinney, 1996) would be sufficient to provide some degree of warning. Either SPTs or other coastal WWGs (mounted on either elastic beacons or preexisting offshore platforms) should be use in this case.

Whatever the sensor type and location, the optimal use of sea-level measurements depends on the effectiveness of the detection algorithm implemented in the software of the sensor. It has been recognized (Beltrami, 2008) that the requisites for an effective algorithm can be identified as:

a) an ability to discriminate a tsunami from other sea-level oscillations that, falling within the frequency band detected by the sensor, are 'disturbances' in the context of tsunami detection;

b) an ability to identify the waveform of a tsunami and to characterize it in terms of both amplitude and period; and

c) a fast computational time.

In the case of BPRs, a fourth requisite should also be considered and met (Beltrami, 2008). Owing to their location, these sensors must have an autonomous power supply. Therefore, the algorithm should run with the lowest possible power consumption. Such a requisite has not necessarily to be met by an algorithm implemented in the software of either a WWG or a tidal gauge. Near-shore SPTs and WWGs located either on offshore platforms or elastic beacons, as well as tidal gauges, can actually be connected to a source of power by a cable. It is furthermore to be noticed that the characteristics of the algorithm may also depend on the apparatus for device to land data transmission.

As far as requisite (a) is concerned, it should be noted that only filtering out non tsunami waves makes it possible to monitor the actual propagation of a tsunami by checking either the amplitude (*amplitude-discriminating algorithm*) or the slope (*slope-discriminating algorithm*) of the recorded signal against a prescribed threshold. Indeed, the presence of several frequencies in a sensor recorded signal guarantees the instrument's polifunctionality. It is in-fact economically desirable to have a device that can measure the wider range of oscillation phenomena.

While the 'disturbance' recorded by a BPR is caused by the superposition of actual sea-surface fluctuations (e.g. planetary waves, astronomical and meteorological tides, gravitational normal modes) and background sea noise (Mofjeld, 1997; Beltrami, 2008), several other 'disturbances' may affect WWG measurements (Beltrami & De Girolamo, 2006). In addition to the above listed wave patterns and to wind waves (for detecting which these sensors are designed), a near-shore WWG can actually measure infragravity waves (e.g. bound/free-long waves, edge waves, surf beats).

It has been shown (Mofjeld, 1997; Beltrami, 2008) that the 'disturbance' recorded by a BPR can be closely predicted, making possible to filter it out simply by subtracting the values observed from those predicted. On the contrary, a close prediction of a wind-wave signal is not possible, therefore some form of low-pass digital filter has to be implemented in the tsunami detection algorithm of a WWG in order to filter out these waves. It is well known (e.g. Emery & Thomson, 2001), that an effective low-pass digital filter should posses five, often mutually exclusive, essential qualities:

1) a sharp cut-off, so that unwanted high-frequency components are effectively removed;
2) a comparatively flat pass-band that leaves the low frequencies unchanged;
3) a clean transient response so that rapid changes in the signal do not result in spurious oscillation or 'ringing' within the filtered record;
4) a zero phase shift; and
5) an acceptable computational time.

A sixth requisite should also be met in the context of automatic, real-time tsunami detection. The filter to be used should belong to the class of causal or physically realizable filters, i.e. to the class of filters that use as input only actually available or 'past' signal samples. It is to be noticed that the use of only 'past' samples makes a zero-phase response not possible for causal filters. A way to get round this problem is to design a filter that has, at least, a linear phase response, i.e. a symmetrical impulse response with location of symmetry shifted from the sample to be filtered. In-fact, since this shift does nothing but produce an identical shift in the output signal, a linear phase filter is equivalent to a zero phase one for most purposes (e.g. Smith, 1997; Shenoi, 2006). It is finally to be stressed that, in the context of real-time tsunami detection, the time-domain approach to filter design undoubtedly posses the appealing pro of making the filtered signal immediately available.

At present, two algorithms expressly designed to detect a tsunami in real-time within a BPR recorded signal have been already published: the one developed by Mofjeld (1997) under the NOAA's DART program; and that proposed by Beltrami (2008). These algorithms use a cubic polynomial and an artificial neural network respectively in order to closely predict and filter out the 'disturbance' recorded by a BPR. Currently, two further algorithms are being developed (Pignagnoli et al., 2010; Tinti et al., 2009; Bressan & Tinti, 2010). These last algorithms have been presented at the EGU General Assembly 2010, but have not been published yet.

As far as algorithms designed to detect a tsunami within a WWG's signal are concerned, those already published are that developed by McGehee & McKinney (1996), that used by the Italian Department for Civil Protection (DPC) within the Stromboli's TEWS (Leva, 2004), and that developed by the Port and Airport Research Institute of Japan - PARI (Shimizu et al., 2006; Nagai & Shimizu, 2009). While the first two algorithms are based on a cascade of time domain moving average filters, the third one uses a finite impulse response (FIR) time domain filter (TDF). The algorithm by McGehee & McKinney (1996) and by PARI (Shimizu et al., 2006) are suitable for automatic, real-time detection of tsunamis with periods greater than 10 minutes. On the other hand, that by DPC is limited by its poor performance in removing wind-waves. In order to overcome the constraints of the preceding listed algorithms Beltrami & Di Risio (2010) have developed a new algorithm based on a cascade of causal filters, the main one being an infinite impulse response (IIR) time domain digital filter (TDF).

The aim of this chapter is to present a review of all the available (either already published or in phase of development) algorithms. Form and characteristics of the algorithms are either recalled or presented, discussing their field of application. Performance and efficiency of the illustrated algorithms are analyzed and compared using both synthetic and actually measured time series.

The chapter is structured as follows. The algorithms for automatic, real-time detection in deep sea level measurements (i.e. in BPR observations) are recalled in section 2, comparing their ability to meet the requisites set out above. Detection in tidal gauge measurements is discussed in section 3. As far as the detection in wind wave measurements is concerned (section 4), the algorithms based on moving average filters are recalled in section 4.1, whilst those based on time domain digital filters such as FIR or IIR filters in section 4.2. Performance and efficiency of these algorithms are analyzed and compared in section 4.3 using synthetic time series, while observations are made and conclusions drawn in Section 5.

2. Detection in BPR measurements

2.1 The DART algorithm

The tsunami-detection algorithm developed by Mofjeld (1997) under the Deep-ocean Assessment and Reporting of Tsunami (DART) program uses a cubic polynomial to predict disturbing wave patterns. The filtered signal is obtained by subtracting at each new time step the prediction from the observation. The algorithm monitors the actual propagation of a tsunami by checking the amplitude of the filtered signal against a prescribed threshold (TS_{amp}), and therefore belongs to the class of the amplitude-discriminating ones.
The predictions ζ_p are updated every sampling interval (i.e. every 15 s), and the prediction time is set equal to 15 s in the future with respect to the actual time t_i. The polynomial is fitted to p-minute averages $\bar{\zeta}$ (centered at the $p/2$ minute) of observations ζ collected over the preceding three hours and p minutes and can be expressed as

$$\zeta_p(t_{i+1}) = \sum_{j=0}^{3} w_j \bar{\zeta}(t_i - 0.5p - 60j) \tag{1}$$

where t_i is the actual time expressed in minutes. The coefficients w_i are calculated by applying the Newton's forward divided difference formula, using the preceding temporal parameters. Mofjeld (1997) suggests setting the time interval p at 10 min so as to:

a) regularize the polynomial fitting point pattern by filtering out background noise typically recorded by BPRs; and
b) minimize the extent to which an actually detected tsunami indirectly influences the filtered signal by affecting the observation averages (Mofjeld, 1997).

As already stated, in this case (p=10 min) the prediction time is set equal to 15 s in the future with respect to the actual time t_i, i.e. 5 minutes and 15 seconds in the future with respect to the time at which is centered the first average.
As shown by Beltrami (2008), the algorithm's prediction error depends both on the time interval p and on the magnitude of the disturbance to be filtered out. In the absence of background sea-noise, this disturbance is mainly caused by the tide. Departure from a perfectly filtered signal (a zero signal) therefore depends on the measurement location. For example, the filtered signals obtained by testing the algorithm on M_2 tides of equal phase and different amplitudes show a residual oscillation of sinusoidal shape. When p is equal to 10 min, the range of this oscillation (i.e. the prediction error) will be approximately 0.26 % of the tidal one. At a location that experiences a tidal range of 2.0 m, an oscillation with a range of 5.0 mm is therefore expected to persist in the filtered signal.
Whatever the time interval p, fixing all the preceding temporal parameters makes it possible to calculate the w_i coefficients *once and for all* and *a priori*. The implementation of equation

(1) is therefore particularly simple. Furthermore, the coefficients w_i are calculated on the sole basis of temporal parameters. This makes the same set applicable to all BPRs, whatever their location. Such characteristics, together with a filtering performance that is effective from a practical point of view, constitute the algorithm's main strengths.

2.2 The ANN algorithm

Similarly to the algorithm by Mofjeld (1997) that by Beltrami (2008) filters out the astronomical and meteorological tides recorded by a BPR by subtracting at each new time step the prediction from the observation, and monitors the actual propagation of a tsunami by checking the amplitude of the filtered signal against a prescribed threshold (TS_{amp}). Therefore, also this algorithm belongs to the class of the amplitude-discriminating ones.
In order to update the prediction of disturbing fluctuations every 15 s, the ANN algorithm proposed by Beltrami (2008) uses a two adaptive-weight (feed-forward) network characterized by 4 input units plus bias, 6 hidden units plus bias and one output unit ($I_4H_6O_1$). The network's inputs consist of the same p-minute averages $\bar{\zeta}$ of bottom pressure observations ζ as those used by the DART algorithm. These values are preprocessed so as to re-scale them linearly in the range $[0;1]$. The network function can be expressed as

$$\zeta_p(t_{i+1}) = \tilde{g}\left\{ w_b^{(2)} + \sum_{h=1}^{6} w_h^{(2)} g\left[w_{bh}^{(1)} + \sum_{j=0}^{3} w_{jh}^{(1)} \bar{\zeta}(t_i - 0.5p - 60j) \right] \right\} \tag{2}$$

where t_i is the actual time expressed in minutes, $w_{jh}^{(1)}$ and $w_{bh}^{(1)}$ are the adaptive weights connecting input units and bias to the hidden units, and $w_h^{(2)}$ and $w_b^{(2)}$ those connecting hidden units and bias to the output unit. Moreover, $g(\cdot)$ and $\tilde{g}(\cdot)$ represent the hidden and output unit activation functions. In particular, the logistic-sigmoid and the linear activation functions characterize the hidden units and the output unit respectively.
Unlike the coefficients of the cubic polynomial (1), the adaptive weights of equation (2) result from the network's *supervised learning*. If a time series of actual observations ζ is available, it will be possible to present the network with an input array $[\bar{\zeta}]$ and a corresponding output vector $\{\zeta\}$, i.e. with a *training set*. The adaptive weights $[w]$ result from minimizing the error function chosen to express the difference between the calculated $\{\hat{\zeta}\}$ and the actually observed outputs $\{\zeta\}$. In particular, the network proposed by Beltrami (2008) uses:

a) the mean square error (MSE) as the error function (Bishop, 1995);

b) the standard back-propagation technique to calculate the MSE derivatives with respect to weights (Rumelhart et al., 1986); and

c) the Levenberg-Marquardt optimization scheme for iteratively adjusting the network weights at the end of each *epoch* by means of the calculated derivatives (Levenberg, 1944; Marquardt, 1963; Bishop, 1995).

The way in which the adaptive weights are calculated makes the network performance dependent neither on the time interval p, nor on the observed signal range. Actually, the efficiency of the supervised learning relies totally upon how accurately the training set represents all the possible disturbing fluctuations and their composition. If it succeeds in this, the ANN algorithm error is expected to be nearly zero. This is especially so in the case of disturbing fluctuations consisting of regular wavy patterns such as the tidal one. On the other hand, the network training relies on actual data. This makes the resulting adaptive weights

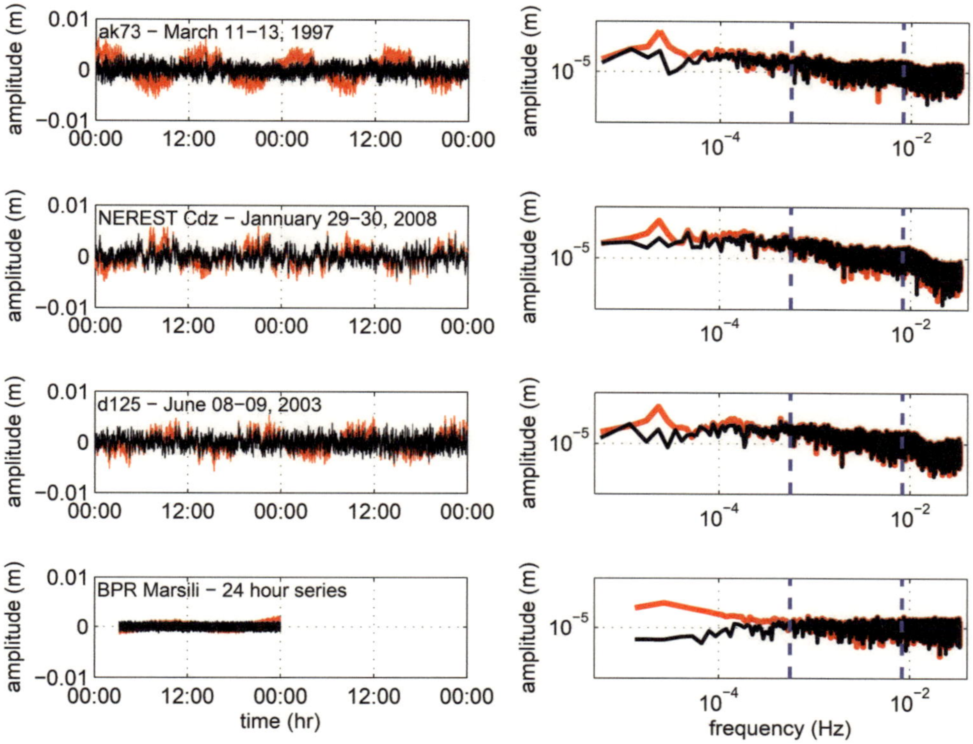

Fig. 1. A comparison of the filtering performance of the DART algorithm (red line) with that of the ANN one (black line). Filtered time series (left graphs) and corresponding amplitude spectra (right graphs). Blue dashed lines in the amplitude spectrum graphs show the limit of the period band 2-30 minutes.

mainly tailored to a specific BPR location. In general, therefore, each BPR is characterized by a specific set of weights, although the same set might be used effectively in different BPRs if their locations are close enough to ensure that the devices are exposed to almost equal tidal and meteorological conditions.

2.3 Filtering and detection performances of DART and ANN algorithms

As far as the filtering performance is concerned, it has been shown (Mofjeld, 1997; Beltrami, 2008) that both the DART and the ANN algorithm are quite effective in predicting, and therefore filtering out, tidal wave patterns. In particular, Beltrami (2008) has shown that the ANN algorithm's closer prediction of both tide and other regular patterns makes it capable of a slightly better filtering performance with respect to the DART algorithm. Such an improvement may be more or less significant, depending on the range of the tide and the characteristics of the background sea noise. In particular, the higher the tidal range at the location of interest (and the lower the background sea noise), the greater the improvement in filtering performance will be. Nevertheless, Beltrami (2008) has also shown that neither the DART nor the ANN algorithm is capable of capturing the feature of the background sea noise. The improved filtering performance that can be achieved using the ANN algorithm therefore relates solely to the astronomical and meteorological tide.

Four different sets of observations (Tab. 1) were analyzed in order to compare the filtering performance of the DART algorithm with that of the ANN one. Two sets were collected during the testing phase of the DART program (retrospective data) by two BPRs (ak73 and d125) respectively deployed in the Northern (Alaska) and Equatorial Pacific (http://nctr.pmel.noaa.gov/Dart/). Both the data-sets cover a period of observations of around one year. The third set refers to the observations collected by the BPR deployed in the Gulf of Cadiz (Atlantic Ocean) in the context of the NEAREST program funded by the European Union (http://nearest.bo.ismar.cnr.it/). The available set was collected during January, 2008. Finally, a 24-h record of pressure observations collected at a BPR located in the Mediterranean Sea was also analyzed. The considered BPR is located 2040 m below the mean sea level on the slopes of Marsili (a submerged volcano north of Stromboli in the Tyrrhenian Sea).

All pressure data - recorded in either PSIA, mmBar or mH_2O - were firstly converted in Pascals (Pa), and then in meters (i.e. in sea levels) by dividing them for the mean specific (unit) sea-water weight $\overline{\gamma} = \overline{\rho} g$ at the BPR location, being $\overline{\rho}$ the mean sea water density that should be applied in order to make the mean sea level resulting from the recorded data equal to the declared BPR water depth.

Fig. 1 shows the filtered signal resulting from the application of the DART and ANN algorithms (left graphs), and the corresponding amplitude spectra (right graphs). Blue dashed lines in the amplitude spectra show the limits of the period band 2-30 minutes. The ANN algorithm clearly performs better than the DART one in removing regular wave patterns such as the astronomical and the meteorological tide. The amplitude of the components of the ANN filtered signal within the period band 1-26 h ($1 \cdot 10^{-5}$ Hz - $3 \cdot 10^{-4}$ Hz) is invariably lower than that of the components of the DART one. This is particularly evident for the astronomical-tide components with period around 12 h ($2.315 \cdot 10^{-5}$ Hz). These are the components that are responsible for the residual oscillation of sinusoidal shape still present in the signal filtered by the DART algorithm. The amplitude spectra show that neither the DART nor the ANN algorithm is capable of filtering out the components with frequency greater than $3 \cdot 10^{-5}$ Hz. A close prediction of these components is in-fact not possible. Indeed, the components within the period band 2-40 minutes ($4.17 \cdot 10^{-4}$ Hz - $8.33 \cdot 10^{-3}$ Hz) cannot be filtered out using some form of digital filter without removing a possible tsunami.

As far as the detection performance is concerned, it should be noticed that the DART and the ANN algorithm were designed in order to detect an anomaly in the signal resulting from the difference between the observations and the predictions. A tsunami will be detected if the amplitude of such an anomaly is greater than a preselected thresholds. In this regard, both the algorithm guarantee quite an effective detection performance. Nevertheless, although these algorithms can detect a tsunami, they can neither properly identify its waveform nor characterize it in terms of amplitude and period. In other words, neither the DART nor the ANN algorithm can meet the second of the three requisites set out in the introduction.

Station ID	Program	Depth (m)	sample rate (s)	Location	Unit
ak73	DART	4575	15	Alaska	Psia
d125	DART	4500	15	Equatorial Pacific	Psia
cdz	NEAREST	3207	15	Gulf of Cadiz	mmBar
mrs		2040	15	Mediterranean Sea	mH_2O

Table 1. Analyzed sets of pressure observations.

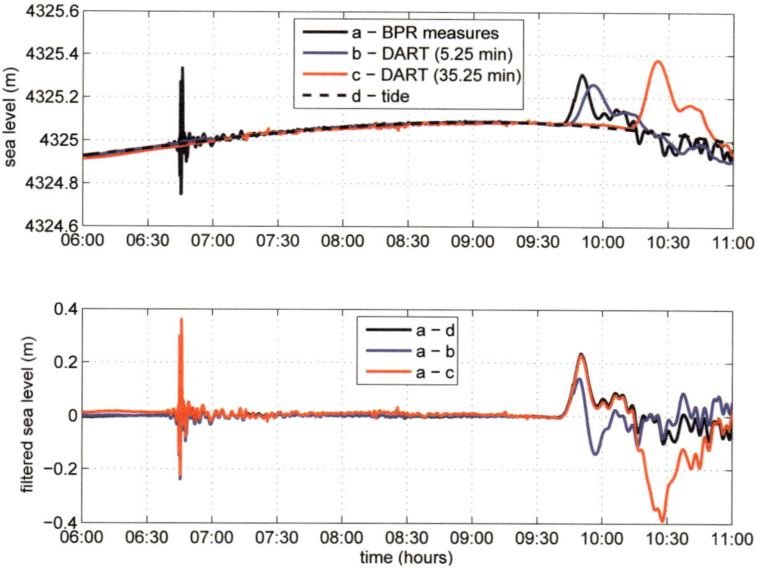

Fig. 2. DART station 32412, February 27, 2010. In the upper graph the actual BPR measurements (black continuous line) are plotted together with DART algorithm predictions (blue and red continuous lines - prediction time respectively set at 5.25 and 35.25 minutes) and the tidal wave (black dashed line) obtained by filtering out from the BPR measurements the components with period lower than 8 hours using a low-pass digital filter. In the lower graph, the de-tided series (black line) is plotted with the series obtained by subtracting the predictions from measurements (blue and red lines).

In order to show this, the data recorded by the BPR of the DART station 32412 (located 630 nautical miles Southwest of Lima, Peru, at a water depth equal to 4325 m) on February 27, 2010 were analyzed. A 8.8 Mw-magnitude earthquake occurred off the coast of the Maule Region of Chile on that date. Following the earthquake, a tsunami was generated, propagating towards the Chilean coasts and spreading in the Pacific.

Pressure data - recorded in PSIA and displayed in meters of water mH_2O - were downloaded from the DART program official site (http://nctr.pmel.noaa.gov/Dart/). The data are available with a sampling interval equal to 1 min. A spline was therefore fitted to the original time series in order to synthesize a series with a sampling rate of 15 s.

The data recorded between 06:00 and 11:00 UTC on February 27, 2010 are shown in the upper graph of Fig. 2 by a black continuous line. As before, pressure data were converted in sea level data (i.e. in meters) by dividing them by the mean local specific sea-water weight $\overline{\gamma} = \overline{\rho}g$. At 06:40 UTC, Rayleigh waves are evident in the recorded signal. The propagating tsunami arrived at the BPR location around 09:40 UTC, reaching its maximum amplitude (around 0.24 m over mean sea level) around 09:50 UTC. In the graph, the black dashed line refers to the tidal wave. These data were obtained by filtering out the components with period shorter than 8 hours from the original record. The blue continuous line represents the DART algorithm prediction, being the prediction time set equal to 5.25 minutes (i.e. 15 s in the future with respect to the actual time). As it can be noticed, after few minutes, the DART predictions tend to follow the observations. This is due to the fact that the tsunami has started to influence the first of the observation averages to which is fitted the cubic polynomial.

The black continuous line in the lower graph of Fig. 2 shows the signal obtained by subtracting the tidal pattern from the recorded data. This is the signal that shows the actual tsunami. The blue continuous line shows the tsunami detected by the DART algorithm, and is obtained by subtracting the DART predictions from the observations. The tsunami reconstructed by the DART algorithm has the form of a N-wave (Tadepalli & Synolakis, 1994). Although the departure from a zero signal represented by this wave is sufficient for detection, i.e. is sufficient to exceed a preselected threshold, the form of the tsunami is not properly reconstructed. This is a consequence of the cubic polynomial prediction, i.e. of the fact that the tsunami has started to influence the first of the observation averages to which is fitted the cubic polynomial. It is to be noticed that, given its characteristics, an equal behavior would be observed in the case of the application of the ANN algorithm.
A possible and simple way to overcome this limit is that of lengthening the DART and ANN algorithm prediction time (Beltrami, 2010). By way of example, setting the prediction time to 30.25 minutes (i.e. 1815 s in the future with respect to the actual time, or 35.25 minutes in the future with respect to the time at which is centered the first average), the resulting DART algorithm predictions take the form of the red continuous line of the upper graph of Fig. 2, while the filtered signal is that represented by the red continuous line of the lower graph. As it is shown by the figure, in this case the algorithm is capable of a close reconstruction of the tsunami, at least within the interval of the prediction time (i.e. 30 minutes). It is to be stressed that, in the case of the DART algorithm, such a lengthening has the cost of augmenting the already seen residual oscillation of sinusoidal shape. When the prediction time is set equal to 30.25 minutes the range of this oscillation (i.e. the prediction error) will be approximately 2.54 % of the tidal one. The ANN algorithm with a lengthened prediction time should be therefore preferred at a location that experiences a considerable tidal range.
Clearly, if the prediction time is set equal to 30.25 minutes, equations (1) and (2) will be expressed accordingly, and either the coefficients w_i of the DART algorithm (Tab. 2), or the adaptive weights of the ANN one will be calculated considering this temporal parameter.

3. Detection in TG measurements

Tidal gauges, although important in the context of tsunami detection, cannot be the main devices on which rely a Tsunami Early Warning System (TEWS). In-fact, these gauges are located at the coast, usually within harbor basin, and therefore can be damaged or swept away by the outgoing wave before the onslaught of the first tsunami wave. Furthermore, their readings can be complicated by shape and size of the basins where they are located.
Nevertheless, although affected by these limitations, tidal gauges can be undoubtedly useful for confirming that a coast has been actually hit by a tsunami (and therefore provide a warning to places further away), and for a post-processing analysis of the tsunami's characteristics.

Coefficient	Prediction time 5.25 min	Prediction time 35.25 min
w_0	+ 1.16818457031250	+2.45603613281250
w_1	- 0.28197558593750	-2.72678027343750
w_2	+ 0.14689746093750	+1.67295214843750
w_3	- 0.03310644531250	-0.40220800781250

Table 2. The DART algorithm's coefficients for different settings of prediction time (sampling interval equal to 15s).

Therefore, provided that the sea level is continuously monitored with a suitable sampling interval ($\leq$1 minute), they should be equipped in order to perform the automatic, real-time detection of possible tsunamis.
Algorithms for real-time detection of tsunamis in coastal TG measurements were developed for the US network (Mero, 1998), and for the gauges located in British Columbia, Canada (Rabinovich and Stephenson, 2004). As far as the DART and ANN algorithm are concerned, the second one may be effectively used for real-time detection in TGs measurements. As shown by Beltrami (2008), a convenient choice of the training set can actually make the ANN algorithm capable of predicting not only the 'disturbance' due to the astronomical tide, but also other regular disturbing wave patterns such as the meteorological tide and the gravitational normal modes of the harbor basins (i.e. the harbor seiches).
Tinti et al. (2009) and Bressan & Tinti (2010) have proposed a further algorithm that could be useful in order to perform the automatic, real-time detection of a tsunami in tidal gauge measurements. The Tsunami Early Detection Algorithm (TEDA) has been presented at the EGU General Assembly 2010, but has not been published yet. The reader is referred to the available abstract (Bressan & Tinti, 2010) and presentation (Tinti et al., 2009) for more information.

4. Detection in WWG measurements

4.1 Algorithms based on moving-average filters

Two algorithms based on moving-average filters have been already presented. That by McGehee & McKinney (1996) and that used by the DPC (Dipartimento della Protezione Civile - Italian department for civil protection) within the Stromboli's TEWS (Leva, 2004).
The algorithm by McGehee & McKinney (1996) relies on the assumption that a tsunami is defined as an unusually large change in the mean sea surface slope. In this regard, the algorithm belongs to the class of the slope-discriminating ones. The algorithm works as follows. The filtered sea-level $\widehat{\zeta}(t_i)$ is obtained by taking the arithmetic mean of the vector $\{\zeta\} = \{\zeta(t_i), ..., \zeta(t_{i-n+1})\}$ consisting of n samples collected in the time interval defined by the duration Δt of the averaging window. It is clear that n is equal to $\Delta t \cdot f_s$, being f_s the sampling frequency. The averaging window is slid forward by an increment $s\Delta t_s$=s/f_s (with s an integer value), and the slope of the filtered sea-level between successive intervals is calculated. A tsunami will be triggered if the absolute value of the calculated slope exceeds a preselected threshold TS_{slp} for some m number of successive intervals $s\Delta t_s$. The filter on which is based the algorithm therefore belongs to the class of causal moving-average digital filters which produce a linear phase alteration.
The choice of the values of the algorithm parameters is clearly site dependent, since it depends on both the wind-wave climate and the tidal range at the WWG's location. The issue is clearly a trade-off between opposite demands. The duration Δt of the averaging window should be long enough to filter out the effects of wind waves on the mean. The longer this window duration - and therefore the better the filtering performance - the higher the period of the shorter detectable tsunami. The lower bound of the band of the detectable-tsunami periods is, in-fact, determined by the value chosen for Δt. The upper bound depends on the amplitude of the minimum tsunami to be detected. Given this amplitude, the upper bound period should make the slope of the tsunami steeper than that of the steepest tidal wave measured at the location of interest. As far as the increment $s\Delta t_s$ is concerned, too short a value may lead to a not measurable change in the slope between successive windows for an event of interest, while too long a value may lead to a not usable warning (i.e. a late warning). Finally, the

parameter	*value*	*units*
Δt	180-600	s
$s\Delta t_s$	10-50	s
m	1- 4	
TS_{slp}	0.25-1	mm/s

Table 3. Values for the algorithm parameters suggested by McGehee & McKinney (1996) in the case of a sampling frequency equal to 2Hz.

threshold TS_{slp} should be small enough to detect the minimum tsunami considered a threat, and large enough to reject usual changes in slope, as well as unusual but not-threatening ones. Table 3 shows the range of values suggested by McGehee & McKinney (1996) in the case of f_s=2 Hz. These values steams out from the assumptions that: (i) the range of the maximum peak period of the wind-waves observed at the generic location of interest is equal to 18-60 s (the duration Δt is therefore on the order of 10 times this period); (ii) the range of the semi-diurnal tide at the same location is on the order of 10 m; and (iii) the amplitude of the minimum tsunami to be detected is equal to 0.5 m. These assumptions constrain the period of the tsunami to be detected to the band 10-40 minutes.

The algorithm used by the DPC (Dipartimento della Protezione Civile - Italian department for civil protection) within the Stromboli's TEWS is characterized by a *cascade* of filters belonging to the class of causal moving-average digital filters and producing a linear phase alteration (Leva, 2004). In particular, given the SPT's sampling interval (0.5 s), it works as follows. A vector $\{\zeta\}$ consisting of the 120 samples collected during the preceding 60 s (i.e. $\{\zeta(t_{i-1}),...,\zeta(t_{i-120})\}$) is updated at each time step t_i. The arithmetic mean $\overline{\zeta}$ of this vector is computed and updated every 60 s. This arithmetic mean is subtracted from each new sample $\zeta(t_i)$ in order to filter out tidal oscillations from the signal. The result $\zeta'(t_i) = \zeta(t_i) - \overline{\zeta}$ of this subtraction is stored in the vector $\{\zeta'\} = \{\zeta'(t_i),...,\zeta'(t_{i-29})\}$, consisting of the 30 values stored during the last 15 s. The filtered sea-level $\widehat{\zeta}(t_i)$ at each new time step t_i is then obtained by taking the mean $\overline{\zeta'_i}$ of $\{\zeta'\}$. The duration of the wind-wave filtering window is therefore equal to 15 s. A tsunami will be triggered if $|\widehat{\zeta}(t_i)|$ exceeds a preselected threshold TS_{amp}. The algorithm therefore belongs to the class of the amplitude-discriminating ones.

Given the duration of the filtering window, such an algorithm has a poor performance in removing wind-waves. Nevertheless, it is important to stress that this algorithm was specifically designed for the TEWS operating at the island of Stromboli. In other words, this algorithm was designed for detecting (within measures taken, close to the source, at water depths around 10-20 m) a tsunami generated by a fast-falling landslide detached from the 'Sciara del Fuoco'. To be actually threatening such a tsunami should have - close to the source - an amplitude of several meters and a period from some tens of seconds to few minutes Di Risio et al. (2009). The detection of a tsunami characterized by such an amplitude is not influenced by the residual noise present in the filtered signal resulting from the application of the DPC's algorithm.

4.2 Algorithms based on time-domain, finite or infinite impulse response, digital filters

4.2.1 The algorithm by PARI

The algorithm by PARI (Port and Airport Research Institute of Japan) has been developed in order to perform the automatic, real-time detection of a tsunami in the sea-level measurements collected by both the Doppler-typed Wave Directional Meters (DWDM), and the newly developed GPS buoy system (Nagai & Shimizu, 2009) of the Japanese Nationwide Ocean Wave information network for Ports and HArbourS (NOWPHAS).

In order to filter out wind-wave components, the algorithm (Shimizu et al., 2006) uses a finite impulse response (FIR) time domain digital filter (TDF) which relies on the direct use of a Hamming window as impulse response. The filter length is fixed to 120 s. Therefore, given the sampling frequency f_s, it spans $2M+1$=120· f_s+1 samples. Such a length implies that an output delay of 60 s with respect to the actual time should be accepted in order to make the filter physically realizable (causal). Indeed, this is usual for a FIR filter for which half of the impulse response should be applied to samples collected after the filtered one. At the actual time step t_i, the output of the filter therefore refers to time step t_j=t_{i-M} (M=60· f_s), and can be expressed as

$$\zeta'(t_j) = \zeta'(t_{i-M}) = \sum_{k=-M}^{+M} h_k \zeta(t_{j-k}) \tag{3}$$

where $\zeta(t)$ is the measured signal, and $\{h\}$ the set of invariant weights known as impulse response. As already stated, Shimizu et al. (2006) suggest the use of a normalized Hamming window as impulse response.

$$h_k = A\left[0.54 + 0.46\cos\left(\frac{2\pi k}{2M+1}\right)\right] \tag{4}$$

where $-M \leq k \leq +M$, and

$$A = \frac{1}{\sum_{k=-M}^{+M} 0.54 + 0.46\cos\left(\frac{2\pi k}{2M+1}\right)} \tag{5}$$

is a constant value used to normalize the output signal level.

Fig. 3 shows the gain or magnitude response function $G(f)$ of the FIR-TDF. The figure shows that $G(f) \simeq 1$ for $f < 1/1000$ Hz, i.e. the filter leaves unchanged the components with periods longer than 1000 s. On the other hand, the components with $1/1000$ Hz $\leq f \leq 1/60$ Hz are increasingly attenuated, while those with $f > 1/60$ Hz completely removed from the signal. The FIR-TDF therefore makes the algorithm by PARI capable of detecting and completely characterizing tsunamis with periods greater than 16 minutes (i.e. about 1000 s). Tsunamis with periods ranging from about 16 minutes down to 1 minute, although still detectable, are progressively less characterizable in terms of both amplitude and period.

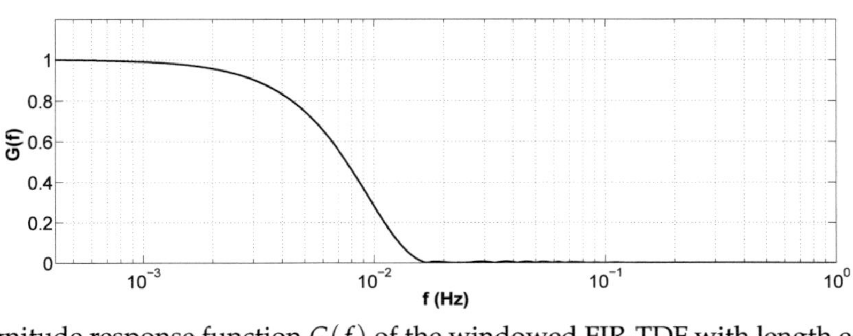

Fig. 3. Magnitude response function $G(f)$ of the windowed FIR-TDF with length equal to 120 s (sample frequency equal to 2Hz).

In order to filter out the astronomical tide, Shimizu et al. (2006) propose to subtract its harmonic prediction $\zeta_p(t_j)$ from the output $\zeta'(t_j)$ of the FIR-TDF. In particular, it is suggested to carry out such a prediction by using the main four tidal constituents resulting from the harmonic analysis of a 6-day record of previous measurements. Relying on harmonic analysis clearly makes the algorithm site dependent.

The filtered sea-level $\widehat{\zeta}(t_j)$ at each new time step t_i is finally obtained by carrying out a moving-average analysis (with one- or half-day time scale) of the resulting signal $\zeta''(t) = \zeta'(t) - \zeta_p(t)$. Such an analysis is suggested in order to correct the mean sea level filtering out the meteorological tide (see Shimizu et al., 2006). As in the case of the DPC algorithm, the algorithm by PARI belongs to the class of the amplitude-discriminating ones. Therefore, it monitors the actual propagation of a tsunami by checking the amplitude of the filtered sea-level against a prescribed threshold (TS_{amp}). In other words, a tsunami will be triggered if $|\widehat{\zeta}(t_j)|$ exceeds such a threshold.

The FIR-TDF guarantees the actual effectiveness of the algorithm in filtering out wind waves. On the other hand, the shorter the period of the tsunami, the lower the capability of the algorithm of detecting and fully characterizing it. The algorithm, in-fact, has been mainly designed for detecting long-period tsunamis (>10 minutes), and providing timely warning of the actually approaching ones not influenced by a delay $\geq$60 s between tsunami measurement and detection.

4.2.2 An algorithm based on a IIR-TDF

In order to overcome the constraints shown by the previously recalled algorithms, Beltrami & Di Risio (2010) have proposed a new algorithm based on a *cascade* of causal filters, the main one being an infinite impulse response (IIR) time domain digital filter (TDF). In particular, the causal IIR-TDF on which relies the algorithm uses the Butterworth (1930) approximation of the gain or magnitude response function (e.g. Smith, 1997; Emery & Thomson, 2001; Shenoi, 2006). This type of filter has a nonlinear phase response. Therefore, a *bidirectional* filtering (e.g. Smith, 1997) of the considered signal segment has to be applied in order to avoid the output phase shift. The combination of forward and reverse filtering actually produces a zero phase response at the sole cost of doubling the filter execution time.

Given a sampling interval equal to 0.5 s (2 Hz), the algorithm works as follows (Fig. 4). A vector $\{\zeta\}_i$=$\{\zeta(t_i),...,\zeta(t_{i-n})\}_i$ consisting of the n samples collected during the time interval Δt_a is updated at each time step t_i. In order to filter out wind waves, the signal stored in $\{\zeta\}_i$ should be previously lengthened using a fictitious series made of a transition signal $\{\zeta_t\}_i$ and by the mirror of $\{\zeta\}_i$ (see Beltrami & Di Risio, 2010). The procedure of lengthening is necessary in order to reduce the effect of Gibbs' phenomenon (Gibbs, 1899), i.e the presence of a succession of overshoot ripples at both ends of the filtered signal. In the context of real-time tsunami detection this is a major problem since ringing affects the most important part of the filtered signal, i.e. the output at current time. The lengthened signal is stored in vector $\{\zeta'\}_i$ of duration Δt=Δt_a+Δt_f, being the duration Δt_f of the fictitious signal given by the sum of the duration of transition and mirror signal Δt_f=Δt_t+Δt_a).

A seventh order IIR-TDF, i.e.

$$\zeta''(t_i) = \sum_{k=0}^{7} a_k \zeta'(t_{i-k}) - \sum_{h=1}^{7} b_h \zeta''(t_{i-h}), \tag{6}$$

is then applied in order to perform the bidirectional filtering of all the samples contained in vector $\{\zeta'\}_i$. In the equation (6), a_k and b_h are the weights or *taps* of the filter. Given the

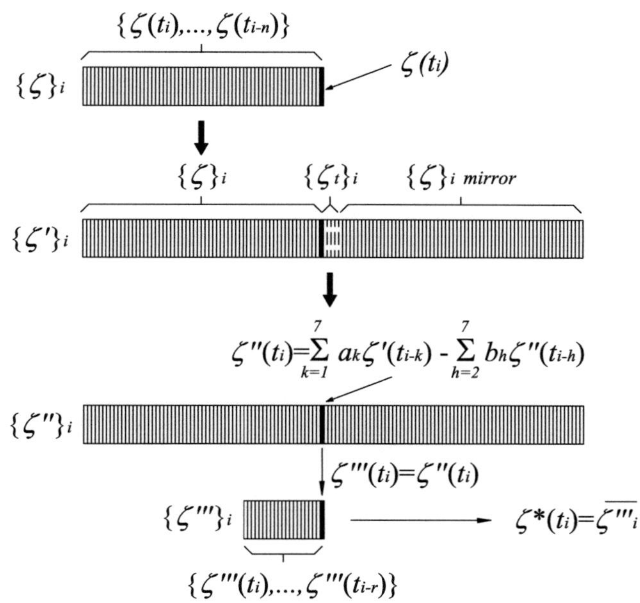

Fig. 4. Proposed algorithm scheme: wind-wave filtering part.

sampling rate and the pass and transition band of the filter, these taps can be calculated *once and for all* and *a priori*. Fig. 5 shows the frequency magnitude response of the chosen IIR-TDF, i.e. a seventh order filter characterized by a pass-band up to 1/60 Hz and a transition band from 1/60 Hz up to 1/30 Hz.

The results of the combination of forward and reverse filtering are stored in vector $\{\zeta''\}_i$. The filtered sample at time t_i, is then extracted from $\{\zeta''\}_i$ and stored in a further vector $\{\zeta'''\}_i = \{\zeta''(t_i),...,\zeta''(t_{i-r})\}_i$, consisting of the $(r-1)$ values obtained in the last r/f_s seconds. At each new time step t_i, the sea-level sample $\zeta^*(t_i)$ filtered for the 'disturbance' due to wind-waves is finally obtained by taking the mean $\overline{\zeta'''}_i$ of $\{\zeta'''\}_i$.

The application of the described part of the proposed algorithm implies a choice for the values of parameters such as the duration of time intervals Δt_a, Δt_t and the number r of samples stored in vector $\{\zeta'''\}_i$. This choice clearly depends on the range of periods of the tsunamis to be detected. By way of example, Tab. 4 shows the values chosen in order to give the algorithm

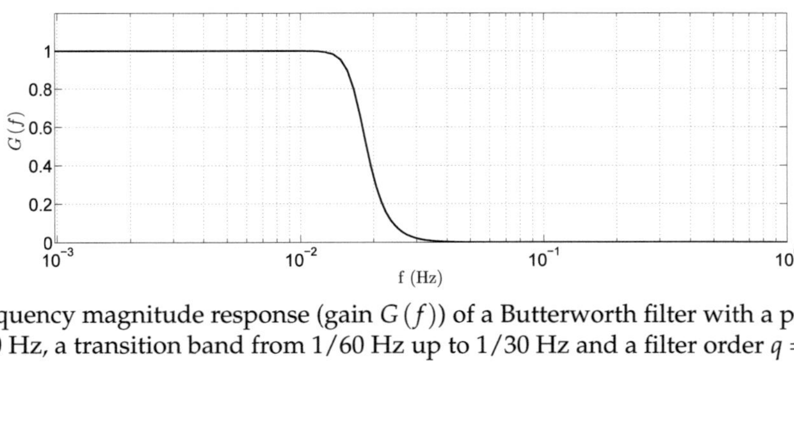

Fig. 5. Frequency magnitude response (gain $G(f)$) of a Butterworth filter with a pass-band up to 1/60 Hz, a transition band from 1/60 Hz up to 1/30 Hz and a filter order q =7.

parameter	*value*	*units*
Δt_a	30	minutes
Δt_t	6	s
r	45	samples

Table 4. Values for the algorithm parameters suggested in the case of a sampling frequency equal to 2 Hz.

the wider range of application, i.e. in order to make the algorithm capable of detecting and fully characterizing tsunamis with periods $\geq$ 60 s.
Either the cubic polynomial on which rely the tsunami-detection algorithm developed by Mofjeld (1997) or the artificial neural network (ANN) proposed by Beltrami (2008) is then used in order to filter out the tidal wave patterns from the signal $\{\zeta^*\}_i$. Given the sampling interval (0.5 s) and an averaging interval p=10 minutes, the prediction time is set equal to 35.00833 minutes (i.e. 1800.5 s in the future with respect to the actual time). As shown in section 2.3, this choice makes the algorithm capable not only of detecting a tsunami, but also of characterizing it in terms of both amplitude and period. Clearly, the polynomial coefficients (Tab. 5) or the ANN's weights should be calculated on the basis of the preceding temporal parameters. As already stated, in the case of the cubic polynomial, such a lengthened prediction time has the cost of augmenting the range of the residual oscillation still present in the filtered signal. In particular, the greater the tidal range at the location of interest, the greater the range of the residual oscillation. The ANN algorithm should be therefore preferred at a location that experiences a considerable tidal range. Nevertheless, while the ANN algorithm can guarantee a better filtering for tidal 'disturbance', its implementation makes the overall algorithm site dependent (Beltrami, 2008). Furthermore, at a location that experiences a moderate tidal range, the gain in terms of tidal wave removal from the signal can be masked by the magnitude of the residual noise still present in the signal filtered for wind waves (in particular, in the case of rough and high sea states). In this case, the varied version of the algorithm by Mofjeld (1997) should be preferred.
The tidal wave pattern is filtered out by subtracting at each new time step t_i the polynomial or the ANN output $\zeta_p(t_i)$ from the corresponding sample $\zeta^*(t_i)$. The filtered sea-level at each new time step t_i is therefore expressed as $\widehat{\zeta}(t_i)$=$\zeta^*(t_i)$-$\zeta_p(t_i)$. As in the case of the DPC and PARI algorithms, a tsunami will be triggered if $|\widehat{\zeta}(t_i)|$ exceeds a preselected threshold TS_{amp}.
In conclusion, given a sampling frequency equal to 2 Hz and the suggested parameters (Tab. 4), the following data should be stored and progressively updated in the RAM of the measurement device:

1. the 3600 progressively updated measure samples stored in vector $\{\zeta\}_i$;
2. the 7200 progressively updated samples resulting from the lengthening of vector $\{\zeta\}_i$ and stored in vector $\{\zeta'\}_i$;

p (minutes)	10
w_0	+ 2.4432451059353566
w_1	- 2.7008348451980630
w_2	+ 1.6554065948122720
w_3	- 0.3978168555495660

Table 5. Polynomial coefficients for 10-minute averages of observations, f_s=2 Hz.

3. the 7200 progressively updated samples stored in vector $\{\zeta''\}_i$ resulting from the bidirectional filtering of the samples stored in vector $\{\zeta'\}_i$;
4. the 44 progressively updated samples stored in vector $\{\zeta'''\}_i$.
5. the 26400 samples filtered for wind-waves (stored and progressively updated in vector $\{\zeta^*\}_i$) during the preceding 3 h and 40 minutes that are necessary to apply the cubic polynomial (or the ANN algorithm), and therefore to filter out tidal wave patterns.

The 4 invariant coefficients of the cubic polynomial (or the 37 ANN weights) and the 15 invariant weights of the IIR-TDF should be clearly stored in the RAM as well. Of course, also the filtered vector $\widehat{\zeta}(t_i)$ will be stored, when the filtered signal exceeds the detection threshold TS_{amp}.

4.3 Filtering and detection performances of moving average and time domain FIR or IIR digital filters

A first set of tests was carried out in order to analyze how well the illustrated algorithms remove wind-waves (filtering performance). A total of 6861 different Jonswap wave-spectra, each characterized by a standard peak enhancement factor γ=3.3 and by a randomly-chosen pair of significant wave height H_{m0} and peak period T_p (Fig. 6), were used to synthesize an equal number of sea-level time-series using the random-phase method (Tuah & Hudspeth, 1982). Each time-series is made of 1024 points, corresponding to a duration of 512 s. No off-set representing the long tidal wave is present in the time-series.

Fig. 7 shows the scatter plots of the maximum amplitude of the filtered signals as a function of the significant wave height H_{m0} of the original synthetic time-series. An upper bound of each scatter corresponding to a specific algorithm has been also assumed and drawn. Given a measurement location characterized by a specific wave climate, such an upper bound gives an indication of the lower threshold that can be selected in the case of amplitude-discriminating algorithms such as that by DPC and PARI, and that based on the IIR-TDF.

Tab. 6 shows the percentages of the maximum amplitude of the filtered signals relative to the significant wave height H_{m0} of the original ones. The table shows that the maximum amplitude of the 'disturbance' still present in the filtered signal is actually only a small fraction of the significant wave height H_{m0} of the original signal, particularly in the case of both the algorithm by PARI, and that based on the IIR-TDF.

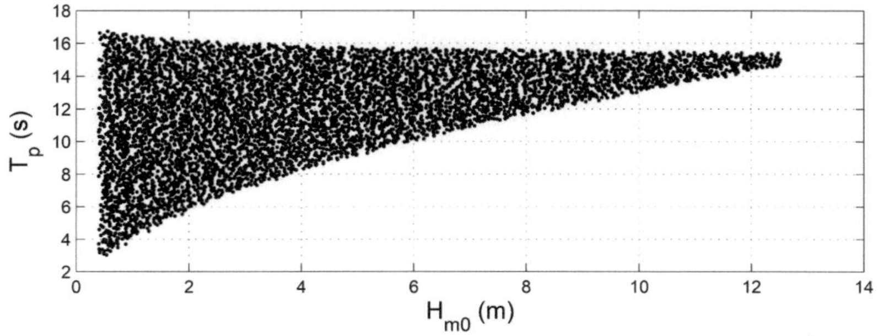

Fig. 6. Scatter plot of the randomly-chosen pair (H_{m0}, T_p) used to characterize 6861 different Jonswap wave-spectra and to synthesize an equal number of wind-wave time series by means of the random-phase method.

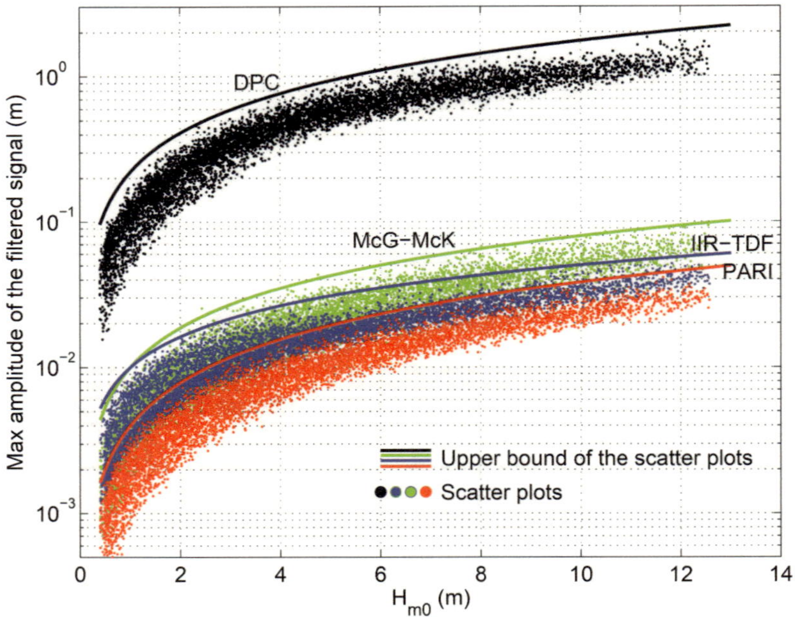

Fig. 7. Scatter plots of the maximum amplitude of the filtered signals as a function of the significant wave height H_{m0} of the original synthetic time-series. Solid lines identify the assumed upper bound of each scatter.

Inspection of both Fig. 7 and Tab. 6 reveals that the best filtering performance is achieved by both PARI and IIR-TDF algorithm. Using these algorithms makes it possible to select a lower threshold TS_{amp} with respect to that needed by the DPC one. In-fact, the lowest detectable tsunami amplitude for PARI and IIR-TDF algorithm will be about 0.05 m, even if the time series characterizes a severe wave climate (i.e. wind-induced significant wave height greater than 10 m). In other words, the algorithm by PARI and that based on the IIR-TDF are equivalent in terms of filtering performance, whilst the one by DPC shows a poor filtering capability. However, as already stated, the DPC algorithm was specifically designed for detecting tsunamis generated by fast-falling landslides detached from the 'Sciara del Fuoco' at Stromboli. To be actually threatening such tsunamis should have - close to the source - an amplitude of several meters, making the wind-wave filtering performance of the DPC's algorithm sufficient. In conclusion, the DPC algorithm can be applied to cases similar to the one of Stromboli, but it can be hardly applied to a more general situation.

$H_{m0} \backslash Alg$ (m)	DPC (%)	$McG - McK$ (%)	$PARI$ (%)	$IIR - TDF$ (%)
1	22.00	1.00	0.40	1.00
3	19.71	0.90	0.39	0.72
6	18.39	0.84	0.39	0.58
9	17.66	0.80	0.38	0.52
12	17.16	0.78	0.38	0.48

Table 6. Percentages of the maximum amplitude of the filtered signals relative to the significant wave height H_{m0} of the original synthetic time-series.

A second set of tests was carried out in order to analyze how well the considered algorithms detect a propagating tsunami within a sea-level signal given by the superposition of a wind-wave and an astronomical tide record. Only the amplitude-discriminating algorithms were considered, i.e. the algorithm by DPC, by PARI, and that based on the IIR-TDF. In-fact, these algorithms will trigger a tsunami if an amplitude threshold TS_{amp} is exceeded. On the contrary, the algorithm by McGehee & McKinney (1996) checks against a prescribed threshold the *slope* of the filtered signal. Therefore, its results are not directly comparable with that of an amplitude-discriminating algorithm. Furthermore, due to its characteristics, it can neither guarantee the characterization of the waveform of the detected tsunami, nor detect tsunamis with period lower than 10 minutes.

Different sinusoidal tsunamis - each characterized by a different amplitude and period - were superposed to different 9 hour-long sea-level time-series, each resulting from the superposition of the same track of the Venice astronomical tide and different wind-wave signals synthesized by means of the random-phase method (Tuah & Hudspeth, 1982). In particular, the method was applied using different Jonswap wave-spectra, each characterized by a standard γ=3.3, and by a selected significant wave height H_{m0} and a peak period T_p=4$\sqrt{H_{m0}}$.

Figures from 8 to 10 show a selection of the results obtained by performing the real-time analysis of different synthesized sea-level records. Two parameters are particularly useful to assess the detection performance of the algorithms, i.e.

i) the dimensionless detected-tsunami amplitude a^* expressed as the ratio between the detected $\tilde{a}$ and the actual tsunami amplitude a; and

ii) the dimensionless detection delay t^* expressed as the ratio between the detection delay $\tilde{t}$ and the tsunami period T.

It is to be noticed that the detection delay $\tilde{t}$ is defined as the elapsed time between the instants at which the actual and the detected (by the algorithm) tsunami exceed the preselected threshold.

Fig. 8 shows the result of the analysis carried out on a sea-level record resulting from the superposition of the Venice astronomical tide, a wave signal characterizing a slight sea state (H_{m0}=1.0 m, T_p=4.0 s) and a sine-like tsunami of amplitude a= 0.4 m and period T=3.0 minutes. In this case, the DPC algorithm is still capable of detecting such a small sinusoidal tsunami, and to identify its waveform, at least roughly. Furthermore, the DPC algorithm can spread the alarm 21.5 s after the tsunami starting time (detection delay $\tilde{t} = 6.0$ s, dimensionless detection delay t^*=0.03). Although still capable of detecting such a small and short tsunami, the PARI algorithm shows a reduced detection performance in terms of both detected tsunami amplitude (a^*=0.72) and detection delay ($\tilde{t} = 60.0$ s, t^*=0.33). In particular, the PARI algorithm can spread an alarm only 75.5 s after the tsunami starting time. As far as the IIR-TDF algorithm is concerned, it is shown to be capable of both a good characterization of the tsunami waveform (a*=0.87) and a timely warning. In-fact, the IIR-TDF algorithm can spread an alarm 37.0 s after the tsunami starting time ($\tilde{t} = 21.5$ s, t^*=0.12).

Fig. 9 shows the result of the analysis carried out on a sea-level record resulting from the superposition of the same track of the Venice astronomical tide previously used, a wave signal characterizing a very high sea state (H_{m0}=10.0 m, T_p=12.65 s), and a sine-like tsunami of equal amplitude (a= 0.4 m) and longer period (T=8.0 minutes). Due to its poor filtering performance, the DPC algorithm cannot detect such a small sinusoidal tsunami in this case. On the contrary, both PARI and IIR-TDF algorithm are capable of detecting and characterizing it in terms of

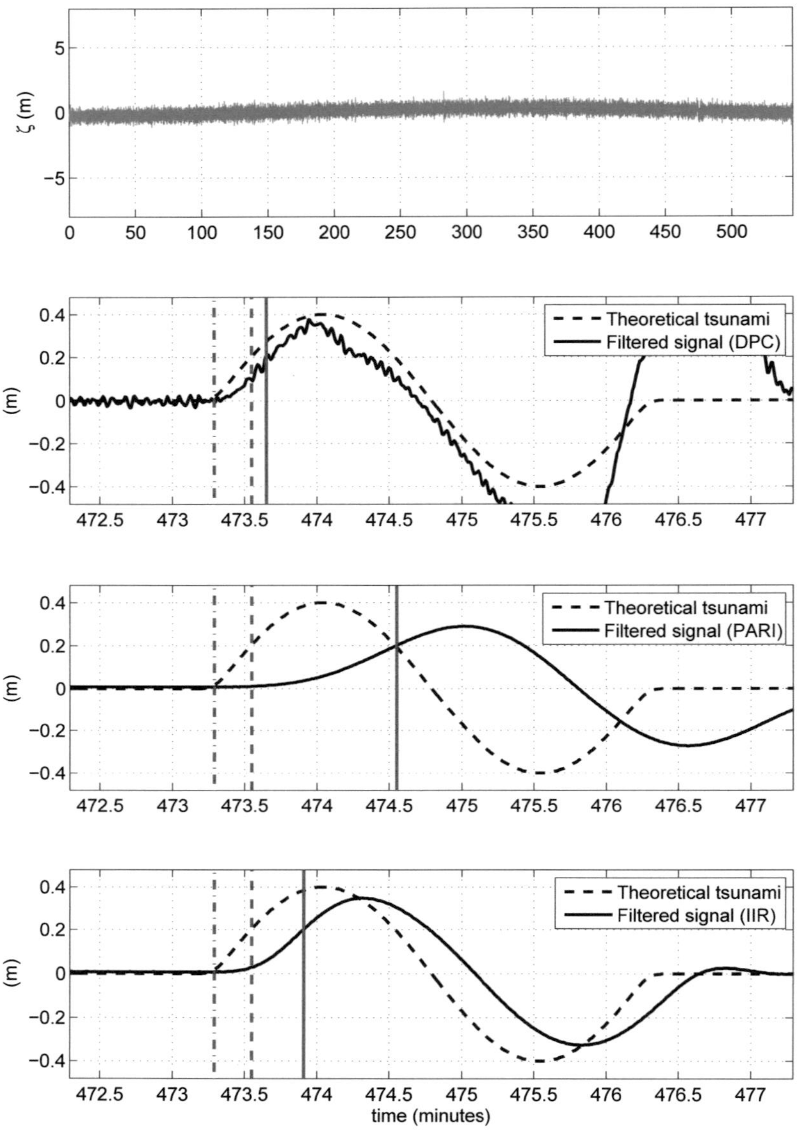

Fig. 8. Upper panel: analyzed sea-level record (resulting from the superposition of Venice tide, a wind-wave signal characterized by a significant height H_{m0}=1.0 m and a peak period T_p=4.0 s and a sinusoidal tsunami of amplitude a= 0.4 m and period T=3.0 minutes). Lower panels: comparison between the filtered signals resulting from the application of DPC, PARI and IIR-TDF algorithm, and the theoretical sine-like tsunami. Vertical dot-dashed and dashed lines represent the instants at which the sine-like tsunami starts and exceeds the selected threshold respectively (TS_{amp}=0.2 m). Vertical continuous line represents the instants at which the filtered signal exceeds the same threshold.

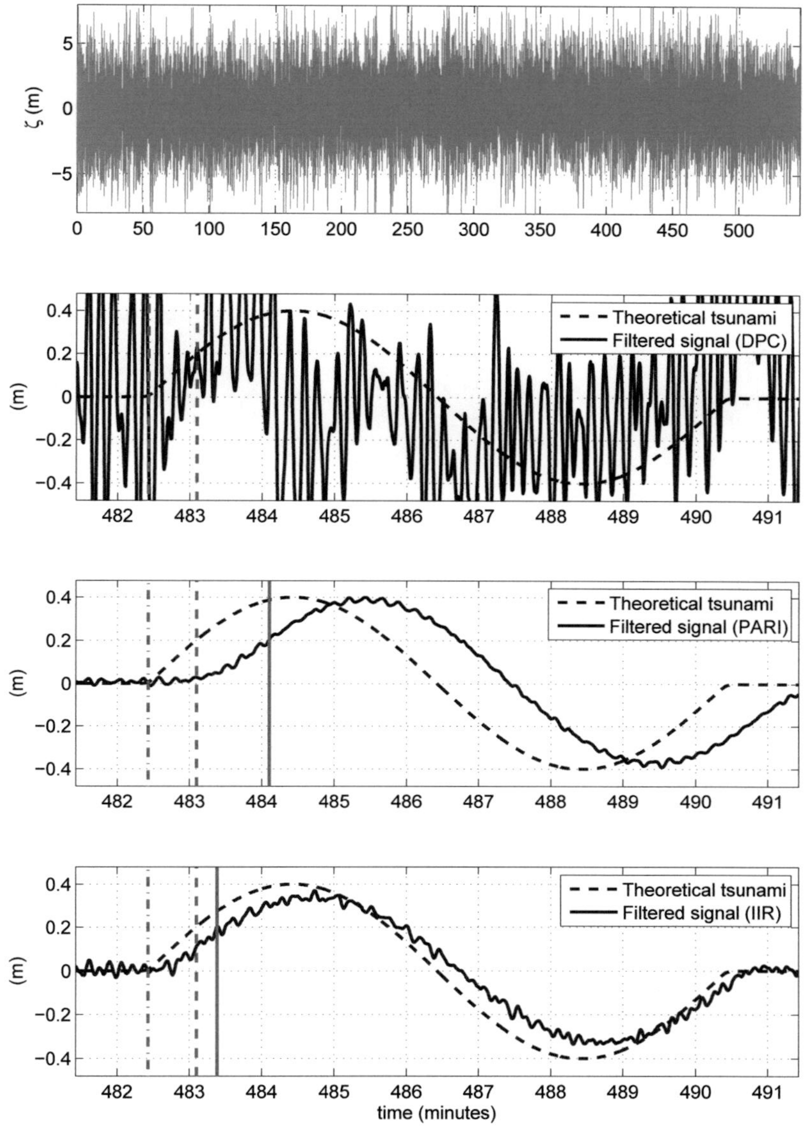

Fig. 9. Upper panel: analyzed sea-level record (resulting from the superposition of Venice tide, a wind-wave signal characterized by a significant height H_{m0}=10.0 m and a peak period T_p=12.65 s and a sinusoidal tsunami of amplitude a= 0.4 m and period T=8.0 minutes). Lower panels: comparison between the filtered signals resulting from the application of DPC, PARI and IIR-TDF algorithm, and the theoretical sine-like tsunami. Vertical dot-dashed and dashed lines represent the instants at which the sine-like tsunami starts and exceeds the selected threshold respectively (TS_{amp}=0.2 m). Vertical continuous line represents the instants at which the filtered signal exceeds the same threshold.

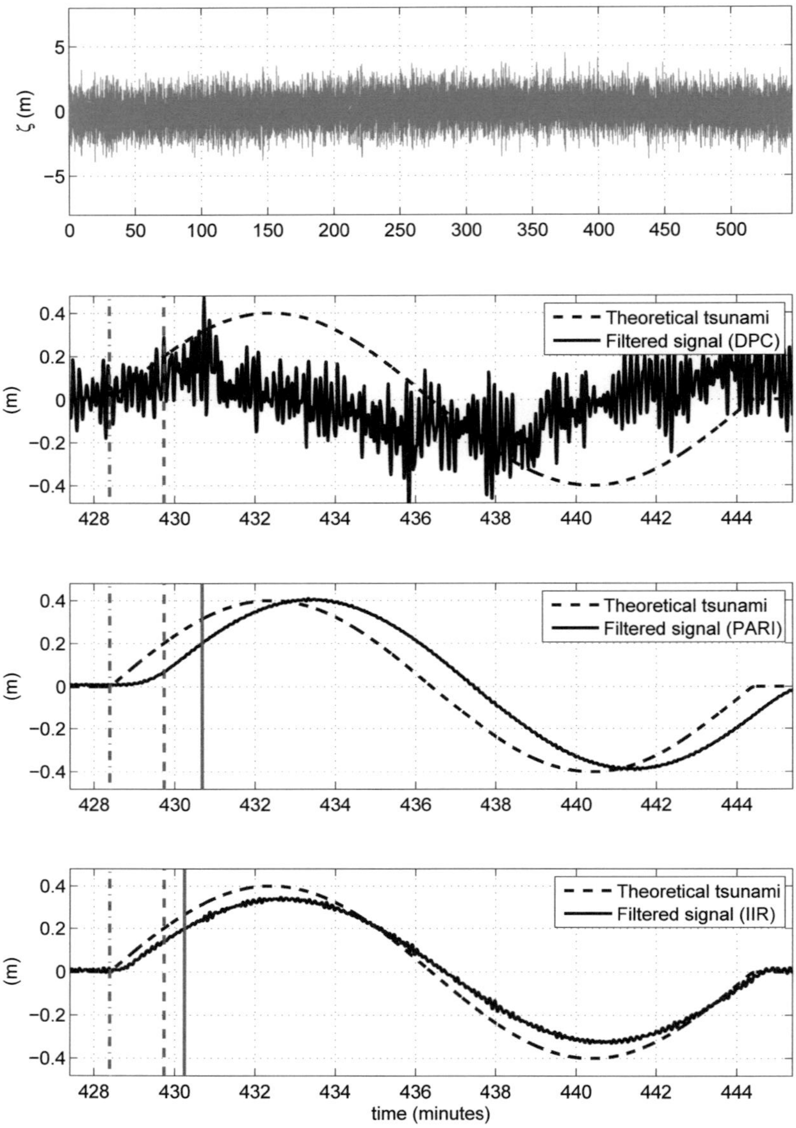

Fig. 10. Upper panel: analyzed sea-level record (resulting from the superposition of Venice tide, a wind-wave signal characterized by a significant height H_{m0}=4.0 m, peak period T_p=8.0 s and a sinusoidal tsunami of amplitude a= 0.4 m and period T=16.0 minutes). Lower panels: comparison between the filtered signals resulting from the application of DPC, PARI and IIR-TDF algorithm, and the theoretical sine-like tsunami. Vertical dot-dashed and dashed lines represent the instants at which the sine-like tsunami starts and exceeds the selected threshold respectively (TS_{amp}=0.2 m). Vertical continuous line represents the instants at which the filtered signal exceeds the same threshold.

amplitude and period. In this regard, a close analysis of the figure shows that the IIR-TDF algorithm has a slightly worst performance with respect to the PARI one in characterizing the tsunami in terms of amplitude. This is due to the mean $\overline{\zeta^*}_i$ that is taken of $\{\zeta^*\}_i$ in order to improve the filtering performance (slightly lower with respect to the PARI algorithm as shown in Fig. 7) of the IIR-TDF algorithm (see par. 4.2.2). On the other hand, the IIR-TDF algorithm shows a better performance with respect to the PARI one in terms of detection delay. In-fact, the detection delay with respect to tsunami starting time is equal to 57.5 s in the case of the IIR-TDF algorithm ($\tilde{t} = 17.0$ s, t^*=0.04), and equal to 100.5 s ($\tilde{t} = 60.0$ s, t^*=0.13) in the case of the PARI one.

Finally, Fig. 10 shows the result of the analysis carried out on a sea-level record resulting from the superposition of the Venice astronomical tide, a wave signal characterizing a rough sea state (H_{m0}=4.0 m, T_p=8.0 s), and a sine-like tsunami of always the same amplitude (a= 0.4 m) and a longer period (T=16.0 minutes). As in the preceding case, the DPC algorithm is not able to detect the sinusoidal tsunami wave, while PARI and IIR-TDF algorithm are. The PARI algorithm appears to closely characterize the sine-like tsunami in terms of both amplitude and period, while the IIR-TDF one shows the already seen and explained reduction in terms of amplitude. As far as the detection delay is concerned, the IIR-TDF algorithm is confirmed to have a better performance than the PARI one. An alarm can be spread 111 s ($\tilde{t} = 30.5$ s, t^*=0.03) and 137.5 s ($\tilde{t} = 57.0$ s, t^*=0.06) after the tsunami starting time by using the IIR-TDF or the PARI algorithm respectively.

The illustrated examples allow to catch the peculiarity of each algorithm. The higher the tsunami period, the better the performance of the PARI algorithm with respect to the IIR-TDF one in terms of characterization of the tsunami waveform. This is particularly true for tsunamis of period greater than 10 minutes. On the other hand, the IIR-TDF algorithm is shown to be capable of detecting and almost effectively characterizing tsunamis with a wider range of periods. In-fact, the IIR-TDF algorithm is capable of an effective detection of tsunamis with period greater than 1 minute. Therefore, the IIR-TDF algorithm shows a wider field of application. Furthermore, in terms of detection delay, the performance of the IIR-TDF algorithm is superior to that shown by the PARI one in all the cases. Such a performance is exceeded by that shown by the DPC algorithm only in the case of mild wave climate. It is to be stressed that the detection delay determines the time at which an alarm can be spread, and therefore the elapsed time between the tsunami warning and arrival at the coast of interest. As already stated, this parameter will be essential when tsunamis may be generated either 'at' or 'near' the coast of interest.

5. Conclusion

The present chapter has presented a review of the algorithms to be implemented in the software of either bottom pressure recorders (BPRs), or tidal (TGs) or wind wave gauges (WWGs) in order to automatically perform the automatic, real-time detection of a possible tsunami within recorded signals. In-fact, although a great effort has been recently undertaken by the scientific and engineering community in developing new technologies capable of increasing the awareness of potential tsunamis, at present direct detection in sea level measurements is still the main mean to confirm the actual generation and propagation of a tsunami.

When the tsunami sources are far away from the coast of interest, it will be convenient to collect sea-level measurements far out at sea, between these sources and the coast at risk. Such a location implies great water depths, and makes bottom pressure recorders

(BPRs) with autonomous power supply the more suitable devices to be used. Two already published algorithms can be used in order to perform the automatic, real-time detection in measurements collected by these devices, i.e. the one developed by Mofjeld (1997) under the NOAA's DART program, and that proposed by Beltrami (2008) and based on an ANN (artificial neural network). Comparing the two algorithms makes it possible to show that the ANN one (Beltrami, 2008) is capable of a closer prediction of both tide and other regular patterns and therefore of a slightly better filtering performance. Such an improvement may be more or less significant, depending on the range of the tide and the characteristics of the background sea noise. In particular, the higher the tidal range at the location of interest (and the lower the background sea noise), the greater the improvement in filtering performance will be. As far as the detection performance is concerned, both the DART and the ANN algorithm can detect a tsunami. Nevertheless, these algorithms can neither properly identify its waveform nor characterize it in terms of amplitude and period. In this regard, it has been shown that a possible and simple solution consists in lengthening the algorithms' prediction time. A possible alternative to the DART and ANN algorithm may be the algorithm by Pignagnoli et al. (2010). Nevertheless, although presented at the EGU General Assembly 2010, this algorithm has not been published yet.

The use of the measurements collected by tidal gauges is clearly useful both for providing a real-time confirmation that a specific coast has been actually hit by a tsunami and for a post-processing analysis of the tsunami's characteristics. Nevertheless, tsunami detection in tidal-gauge readings may suffer of the gauge location. At present - given a suitable sampling interval ($\leq$1 minute) - the best algorithm to perform the automatic, real-time detection of possible tsunamis in this kind of measurements appears to be the ANN one. A possible alternative may be the Tsunami Early Detection Algorithm - TEDA (Tinti et al., 2009; Bressan & Tinti, 2010) presented at the EGU General Assembly 2010, but not published yet.

When tsunamis may be generated either at or 'near' the coast of interest, it will be essential to collect sea-level measurements in close proximity of the sources. Such a location implies water depths in the order of tens of meters, and makes WWGs a suitable and economically convenient choice. In this case, an important issue will be whether or not the source distance from the closest inhabited location makes the elapsed time between the tsunami generation and arrival sufficient to provide a warning. As far as the algorithms designed to detect a tsunami within a WWG's signal are concerned, those already published are that developed by McGehee & McKinney (1996), that used by the Italian Department for Civil Protection (DPC) within the Stromboli's TEWS (Leva, 2004), and that developed by the Port and Airport Research Institute of Japan - PARI (Shimizu et al., 2006; Nagai & Shimizu, 2009). While the first two algorithms are based on a cascade of time domain moving average filters, the third one uses a finite impulse response (FIR) time domain filter (TDF). The algorithm by McGehee & McKinney (1996) and by PARI (Shimizu et al., 2006) are suitable for automatic, real-time detection of tsunamis with periods greater than 10 minutes. On the other hand, that by DPC is limited by its poor performance in removing wind-waves. In order to overcome the constraints of the preceding listed algorithms, Beltrami & Di Risio (2010) have developed a new algorithm based on a cascade of causal filters, the main one being an infinite impulse response (IIR) time domain digital filter (TDF). Basic tests have shown that the IIR-TDF algorithm guarantees the detection for a wider range of tsunami periods, performing effectively both in filtering out the 'disturbance' due to wind-waves from the gauge recorded signal, and in detecting the presence in this signal of a possible tsunami (although its ability to identify the tsunami waveform has been demonstrated to be slightly inferior to that shown by the PARI algorithm). Furthermore, the IIR-TDF algorithm proposed by Beltrami & Di Risio (2010) is capable of such a detection with a shorter delay with respect to almost all the recalled already existing algorithms, and therefore of spreading the most timely warning.

6. Acknowledgments

The work presented in this chapter was funded partially by the 'Ministero dell'Istruzione, dell'Università e della Ricerca' - MIUR (Italian Ministry for Formation, University and Scientific Research) under the research projects PRIN2004 (prot. 174/2004)) and PRIN2007 (prot. 2007MNBEMY), and partially by the 'Dipartimento della Protezione Civile' - DPC (Italian Civil Protection Department).

7. References

Beltrami, G. M. (2008). An ANN algorithm for automatic, real-time tsunami detection in deep-sea level measurements, *Ocean Engineering* Vol. 35(5-6): 572–587.

Beltrami, G. M. (2010). Automatic, real-time detection and characterization of tsunamis in deep-sea level measurements. Submitted for publication to *Ocean Engineering*.

Beltrami, G. M. & De Girolamo, P. (2006). Preannuncio di un evento di maremoto. Parte 1: Identificazione in tempo reale dell'evento, *XXX Convegno di Idraulica e Costruzioni Idrauliche*, Universitá La Sapienza.

Beltrami, G. M. & Di Risio, M. (2010). Algorithms for automatic, real-time tsunami detection in wind-wave measurements - Part I: Implementation strategies and basic tests. (Submitted for pubblication to Coastal Engineering).

Bishop, C. M. (1995). *Neural network for Pattern Recognition*, Oxford University Press.

Bressan, L. & Tinti, S. (2010). Test of TEDA, Tsunami Early Detection Algorithm, *Geophysical Research Abstracts, EGU General Assembly 2010*, European Geosciences Union, pp. EGU2010–6450–1.

Butterworth, S. (1930). Theory of filter amplifier, *Experimental wireless & the wireless engineer* 7: 536–541.

Corsini, A., Farina, P., Antonello, G., Barbieri, M., Casagli, N., Coren, F., Guerri, L., Ronchetti, F., Sterzai, P. & Tarchi, D. (2003). Space-borne and ground-based SAR interferometry as tools for landslide hazard management in civil protection, *International Journal of Remote Sensing* Vol. 27(12): 2351–2369.

Di Risio, M., De Girolamo, P., Bellotti, G., Panizzo, A., Aristodemo, F., Molfetta, M. G. & Petrillo, A. F. (2009). Landslide generated tsunamis runup at the coast of a conical island: New physical model experiments, *Journal of Geophysical Research* 114: C01009, doi:10.1029/2008JC004858.

Eble, M. C. & Gonzlez, F. I. (1991). Deep-ocean bottom pressure measurements in the northeast pacific, *Journal of Atmospheric and Oceanic Technology* Vol. 8(2): 221–233.

Emery, W. J. & Thomson, R. E. (2001). *Data analysis methods in physical oceanography*, Elsevier.

Gibbs, J. W. (1899). Fourier's series, *Nature* 59: 606.

Langer, H. & Falsaperla, S. (2003). Seismic monitoring at Stromboli volcano (Italy): a case study for data reduction and parameter extraction, *Journal of volcanology and geothermal research* Vol. 128(1-3): 233–245.

Leva, D. (2004). Diagramma catena di misura sensore di pressione a Stromboli - Ginostra. (Not published paper - In Italian).

Levenberg, R. (1944). A method for the solution of certain problems in least squares, *Quarterly of Applied Mathematics* 2: 164–168.

Marquardt, D. (1963). An algorithm for least-squares estimation of nonlinear parameters, *Journal of the Society of Industrial and Applied Mathematics* 11(2): 431–441.

McGehee, D. & McKinney, J. (1996). Tsunami detection and warning capability using nearshore submerged pressure transducers - Case study of the 4 October 1994 Shikotan tsunami, *Proceedings of the 4th International Tsunami Symposium, IUGG*, Kluwer Academic Publisher, Boulder, Colorado, pp. 133–144.

Mero, T.N. (1998). NOAA/National Ocean Service Application of Real-Time Water Levels, in Proceedings. *Ocean Community Conference, The Marine Technology Society Annual Conference*, Vol. 2:November 16-19, 1998, 1036-1039.

Mofjeld, H. (1997). Tsunami detection algorithm. (Not published paper (http://nctr.pmel.noaa.gov/tda_documentation.html).

Nagai, T. & Shimizu, K. (2009). Basic design of japanese nationwide GPS buoy network with multi-purpose offshore observation system, *Journal of Earthquake and Tsunami* Vol. 3(2): 113–119.

Pignagnoli, L., Chierici, F. & Embriaco, D. (2010). A new real time tsunami detection algorithm for bottom pressure measurements in open ocean: characterization and benchmarks, *Geophysical Research Abstracts, EGU General Assembly 2010*, European Geosciences Union, pp. EGU2010–10498.

Rabinovich, A. B. &F.E. Stephenson (2004). Longwave Measurements for the Coast of British Columbia and Improvements to the Tsunami Warning Capability. *Natural Hazards*, Vol. 32, 3, 313-343.

Rumelhart, D. E., Hinton, G. E. & Williams, R. J. (1986). Learning internal representations by error propagation, *in* D. J.L. McClelland & the PDP Research Group (eds), *Parallel Distributed Processing: Explorations in the Microstructure of Cognition*, Mit Press, Cambridge, Massachusstts, pp. 318–362.

Shenoi, B. (2006). *Introduction to Digital Signal Processing and Filter Design*, John Wiley and Sons.

Shimizu, K., Nagai, T., Lee, J. H., Izumi, H., Iwasaki, M. & Fujita, T. (2006). Development of real-time tsunami detection system using offshore water surface elevation data, *Proceedings of Techno-Ocean 2006 - 19th JASNAOE Ocean Engineering Symposium*, p. 24.

Smith, S. (1997). *The Scientist and Engineer's Guide to Digital Signal Processing*, California Technical Publishing.

Tadepalli, S. & Synolakis, C. E. (1994). The run-up of n-waves on sloping beach, *Proceedings of Royal Society, London* A(4): 99–112.

Tinti, S., Bressan, L., Zaniboni, F. & Armigliato, A. (2009). S3-ur9: Development of real-time algorithms for the detection of tsunami signals on sea-level records: Testing and application to PMEL/NOAA (USA) and analysis of ISPRA (Italy) data., *Progetti DPC Convenzione 2007-2009 - Progetti Sismologici* . (Not published poster presentation - In Italian).

Tinti, S., Pagnoni, F., Zaniboni, F. & E. Bortolucci, E. (2003). Tsunami generation in Stromboli Island and impact on south-east Tyrrhenian coasts, *Natuaral Hazards and Earth System Sciences* Vol. 3: 299–309.

Tuah, H. & Hudspeth, R. T. (1982). Comparisons of numerical random sea simulations, *Journal of the Waterway Port Coastal and Ocean Division* 108(4): 569–584.

28

Airborne Geophysical Investigation of Groundwater Resources in Northern Sumatra after the Tsunami of 2004

Bernhard Siemon and Annika Steuer
Federal Institute for Geosciences and Natural Resources (BGR)
Germany

1. Introduction

The earthquake and the tsunami event on December 26th, 2004, caused not only the loss of life of a huge number of people and the destruction of houses, basic infrastructure and public facilities but also large scale salt-water intrusions and destruction of thousands of shallow water wells in the coastal region of the Province of Nanggroe Aceh Darussalam in northern Sumatra, Indonesia. The supply of a sufficient amount of potable water in all populated areas was complicated as many water pipes were broken by the earthquake and a huge number of dug wells were unusable after the tsunami. Furthermore, many new drillings were not successful in finding potable water due to the lack of information about local hydrogeological conditions. Therefore, the Indonesian and German governments set up a project dedicated to re-install the public life of the people living in the coastal regions of northern Sumatra. The focal point of this project was water assessment along the shorelines of Aceh about nine months after the tsunami.

In order to get a fast overview on the remaining freshwater resources and to assist the Indonesian authorities such as the Directorate General for Geology and Mineral Resources (DGGMR), the National Development Planning Agency (BAPPENAS) and the Executive Agency for the Rehabilitation and Reconstruction in Nanggroe Aceh Darussalam Province and Nias Islands (BRR) as well as numerous aid organisations in finding suitable locations for drilling new water wells, it was decided to use airborne geophysics. The helicopter-borne surveys including electromagnetics, magnetics and gamma-ray spectrometry were conducted by the airborne group of the German Federal Institute for Geosciences and Natural Resources (BGR) from August to November 2005. As the mineralisation of water correlates with its electrical conductivity and therefore freshwater and salt water can be distinguished in general, it was hoped that electromagnetic data would reveal freshwater resources not destroyed by the tsunami, particularly close to the populated coastal areas.

The target areas (Fig. 1) of the German-Indonesian cooperation project HELP ACEH (HELicopter Project ACEH) were the city of Banda Aceh with the district of Aceh Besar on the north coast and the area on the west coast between the towns of Calang and Meulaboh in the district of Aceh Barat (Siemon et al., 2007). In addition to that, Coca-Cola Foundation Indonesia (CCFI) funded a further survey on the north-east coast around the town of Sigli in the district of Aceh Pidie (Steuer et al., 2008).

Fig. 1. BGR survey areas (red dotted lines) in the Province of Nanggroe Aceh Darussalam, northern Sumatra. The elevation map was derived from SRTM data.

2. Method

The extremely fast airborne measurements are generally carried out on parallel flight lines. BGR uses a helicopter, a Sikorsky S-76B, to carry the geophysical equipment (Fig. 2). The helicopter-borne electromagnetic (HEM) system as well as the magnetic, GPS and laser altimeter sensors are housed in a 10 m long tube, which is dragged at a mean altitude of about 40 m above ground level. The HEM system (RESOLVE, manufactured by Fugro Airborne Surveys) consists of five small transmitter and receiver coil pairs, which are separated about 8 m. The gamma-ray spectrometer as well as further GPS and altimeter sensors are installed into the helicopter. From the three geophysical methods used simultaneously electromagnetics contributes most to groundwater investigation purposes due to the dependency of a) the electrical conductivity from the salinity of the groundwater, i.e., the groundwater quality, and b) the clay content of the subsurface, i.e., the aquifer conditions.

The HEM transmitter signals, the primary magnetic fields, induce eddy currents into the subsurface which are dependent on the electrical conductivity distribution. The relative secondary magnetic fields from these induced currents are measured at the receiver coils in parts per million (ppm) as they are related to the primary fields.
The use of different frequencies ranging from 387 Hz to 133 kHz enables investigation of different depths: High frequencies resolve the shallower parts of the subsurface and lower frequencies the deeper parts. The depth of investigation also depends on the subsurface conductivity distribution: The higher the conductivity the lower the penetration of the electromagnetic fields into the subsurface. Typical maximum investigation depths of the RESOLVE system range from about 30 m (salt-water saturated sediments) to about 150 m (freshwater saturated sandy sediments or hard rocks).

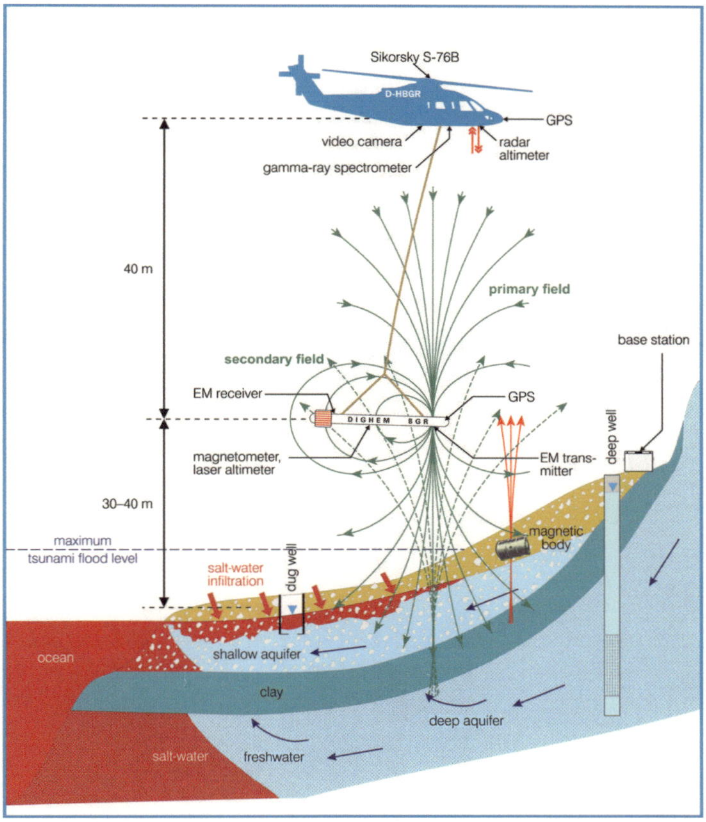

Fig. 2. Sketch of the BGR helicopter-borne geophysical system and the hydrogeological situation expected along the coasts of northern Sumatra.

At the beginning of each survey flight the HEM system was calibrated at high flight altitude (>350 m above ground level) using internal calibration coils. Phase and gain were checked several times during a survey flight at high altitude. Minor remaining calibration errors were corrected by applying correction factors obtained over the sea, where the responses of the HEM system were calculated using the conductivity of seawater and the system altitude. Siemon et al. (2006a,d,e) presented a detailed description of all survey operations.

The in-phase (I) and quadrature (Q) components of the secondary magnetic fields are generally converted to the resistivity (inverse of conductivity) of a layered half-space. The simplest one is a homogeneous half-space (Model 1 in Fig. 3). Its parameter, the apparent resistivity ρ_a [Ωm] is normally derived from the data of a single frequency (f). As two data values (I and Q) are available, a second parameter, the apparent distance D_a [m] from the sensor to the top of the conducting half-space, can be derived. The difference between apparent distance (D_a) and system height (h) measured by a laser altimeter is the apparent depth d_a (Fraser, 1978). It is a measure of how conductive the cover is with respect to the half-space: Positive or negative apparent depth values indicate resistive or conductive cover layers, respectively. In addition, the centroid depth is defined as $z^* = d_a + p_a/2$ (Siemon, 2001), where $p_a = 503.3\ (\rho_a/f)^{1/2}$ is the apparent skin depth. The centroid depth can be regarded as a centre depth of the homogeneous half-space. The resulting sounding curves, $\rho_a(z^*)$, provide an initial approximation of the vertical resistivity distribution. They are used to derive appropriate starting models for the one-dimensional (1D) inversion. A Marquardt-Levenberg inversion procedure iteratively calculates the model parameters, resistivity ρ and thickness t of the layers (Model 2 in Fig. 3), from the data of all frequencies available (Sengpiel & Siemon, 2000). The inversion procedure stops when a given threshold (e.g. 10%) is reached, which is defined as the differential fit of modelled and measured HEM data.

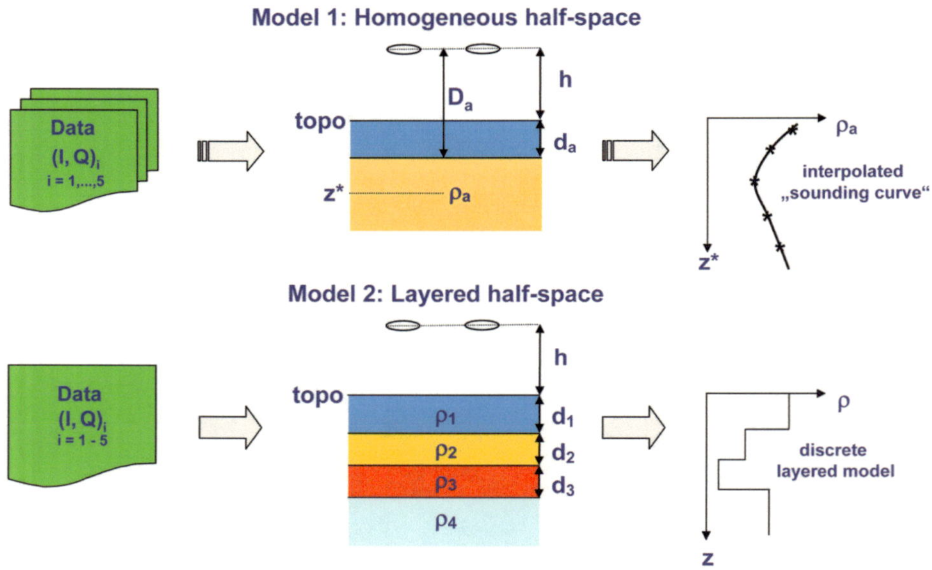

Fig. 3. Sketch of HEM inversion based on a homogeneous half-space or a layered half-space.

The results of the 1D inversion are presented as vertical resistivity sections (VRS) and as resistivity maps. The VRS are constructed by placing the resistivity models for each sounding point along a survey profile next to each other using the topographic relief in metres above sea level (m asl) as base line. The resistivity maps are derived from the apparent resistivities of single frequencies or from the 1D inversion models at selected depths below ground level (bgl). VRS in combination with resistivity maps provide a three-dimensional image of the resistivity distribution in the subsurface. All sections and maps are plotted with respect to UTM coordinates (WGS 84, zone 46N).

3. Airborne geophysical surveys

The survey areas were jointly selected by the German and Indonesian project partners in order to assist Indonesian authorities and international aid organisations in finding freshwater resources not affected by the tsunami that are located close to areas where the people suffering the catastrophe were displaced. The focus was set on populated coastal areas which were heavily destructed by the earthquake and tsunami and where water supply was a major concern.

The airborne surveys funded by BGR covered a) the city of Banda Aceh on the north coast including the valley of the Krueng Aceh as well as two coastal areas on the north-west and north-east coast (survey area Banda Aceh) and b) a broad coastal strip comprising the towns of Calang and Meulaboh as well as a number of villages on the west coast (survey area Calang-Meulaboh). Directly after completing the HELP ACEH surveys another airborne survey funded by CCFI was conducted covering the Sigli Coastal Plain on the north-east coast (survey area Sigli).

All airborne surveys were accompanied by hydrogeological reconnaissance surveys conducted by BGR and funded by the German Federal Ministry for Economic Cooperation and Development (BMZ).

3.1 Banda Aceh

Regional geology and hydrogeology

The Krueng Aceh valley is bounded to the south-west by the Aceh Fault and to the north-east by the Seulimeum Fault (Fig. 4). The West Coastal Range to the south of the Aceh Fault is built up by the pre-Tertiary limestones. The volcanics of Pleistocene to Holocene age cover most of the North Coast Foothills on either side of the Seulimeum Fault. Within the upper Krueng Aceh valley, the Tertiary rocks are covered by up to 500 m thick Plio-Pleistocene semi-consolidated calcareous and tuffaceous sandstones. From upstream of the town of Jantho to downstream of the town of Indrapuri, the Pleistocene coarse-grained partly volcanic sands and gravels form a prominent terrace surface on either side of the Krueng Aceh. These older terrace deposits may attain a thickness of up to 75 m. The alluvium near the coast of the city of Banda Aceh extends to a depth of more than 200 m below ground level becoming thinner upstream (Bennett et al., 1981a; IWACO, 1993).

The Hydrogeological Map Banda Aceh (Soetrisno, 1993a) indicates the alluvium as the main productive aquifer system within the Krueng Aceh valley. Downstream of Indrapuri, the alluvial deposits can be subdivided into a shallow aquifer system and a deep aquifer system. The top 20 metres, ranging from sandy clays to sands and gravels, are considered to be the shallow aquifer system that is directly recharged from rainfall and is still the major traditional resource for domestic water supply. The deep aquifer system comprises a few thin sand-gravel horizons below thick clay layers within the coastal belt. The thickness of the freshwater-bearing aquifer ranges from 3 to 15 m. At a depth ranging from 75 to 140 m bgl, these sandy to gravelly layers bear highly confined fresh groundwater. Upstream of Indrapuri, the alluvial sandy-gravelly deposits in the vicinity of the river courses, the older terrace sand-gravel deposits and the semi-consolidated sandstones are assumed to constitute the main aquifers of the upper part of the Krueng Aceh valley (Ploethner & Siemon, 2006a).

Helicopter-borne survey

The northernmost survey area, Banda Aceh, is bounded towards the north and west by the Andaman Sea and the Indian Ocean, respectively. It comprises the city of Banda Aceh and the valley of the Krueng Aceh (red dots in Fig. 4). The 20 km by 50 km wide area was surveyed with 24 flights within three weeks from August 23rd to September 12th, 2005. The nominal flight-line spacing of the 156 SW–NE lines was 300 m, increased to 900 m in the southern quarter of the survey area, and the 24 NW–SE tie lines were 1000 m apart, resulting in a total flight-line length of about 4000 km. The small area of 2 km by 14 km along the north-east coast was covered with one flight on September 13th, 2005, with a line spacing of 200 m for the 11 NW–SE lines. The survey flights commenced from Blang Bintang airport south-east of Banda Aceh (Fig. 1).

HEM survey results

The most important task of the project was to map the tsunami-affected salinization of shallow groundwater resources and to outline remaining shallow occurrences of potable water. Due to the dependence of the electrical conductivity on the pore-water salinity, this

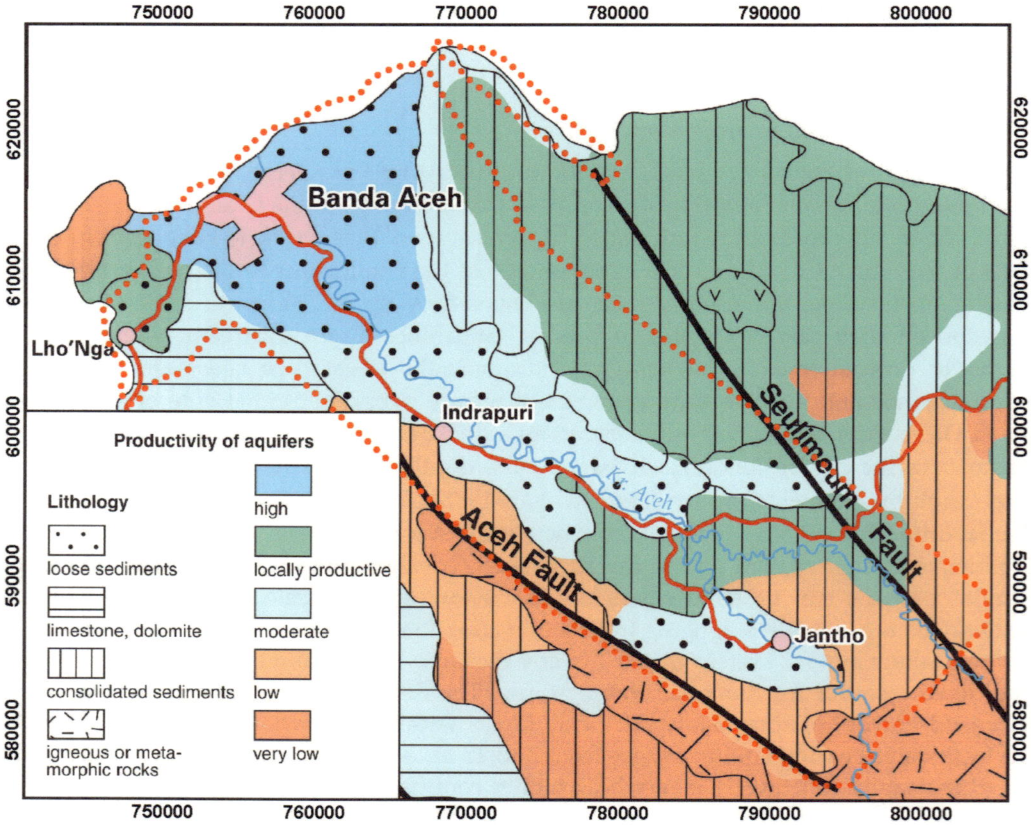

Fig. 4. Hydrogeological situation in the northernmost part of Sumatra (after Soetrisno, 1993a and Setiadi, 2004); main roads, rivers and faults are marked by red, blue and black lines, respectively, and the boundary of the airborne survey area is indicated by red dots.

task could successfully be realized by calculating and displaying the resistivity at several depth levels. About four million 1D resistivity models were calculated from five-frequency HEM data. Maps displaying the resistivity distribution at selected depths below ground level were derived from these models. Fig. 5 shows exemplarily the resistivity distribution at 10 and 30 m bgl for the entire survey area. These maps reveal very conductive areas indicated by red colours near the coasts which are obviously due to seawater saturated sediments. Freshwater resources can be expected where green and blue colours occur which represent fairly resistive areas. Low to medium resistivities (orange to yellow colours) may be caused by brackish groundwater or clayey sediments. Some of the small blue spots on the resistivity map at 10 m bgl, however, are due to laser altimeter readings affected by palm trees. The automatic procedure used for detection and correction of such misleading laser altimeter readings failed in case of widespread dense palm forests resulting in reduced laser altitude values and, thus, in elevated (virtual) ground-level values.

In order to outline shallow freshwater resources an apparent resistivity map being not affected by misleading laser altitude measurements was produced based on the data of the highest frequency (133 kHz). The very low apparent resistivities (less than 3 Ωm) along the coast close to the city of Banda Aceh clearly demonstrate that the salinization of the shallow groundwater was apparent up to several kilometres inland still nine months after the tsunami. This area, however, is definitely smaller than the extent of the tsunami flooding denoted by the dashed line in Fig. 6 (DLR, 2005a,b). It is also evident from Fig. 6 that the shallow salt-water intrusion on the north-west coast as well as that on the north-east coast is restricted to small areas close to the shoreline.

About 5 km inland the apparent resistivity map outlines an about 3 km wide strip of elevated apparent resistivities indicating freshwater resources within the Krueng Aceh valley. One is situated about 10 km to the east of the city of Banda Aceh. The lateral and vertical extents of this freshwater lens on top of saline water are clearly delineated by resistivity maps and vertical sections, e.g. along profile 29.1 in Fig. 7. The simplified lithology of borehole (B) shows exemplarily that shallow and deep aquifers are separated by clayey layers at a depth where saline water obviously occurs.

Electrical conductivity (EC) values (coloured dots and squares in Fig. 6) derived from water samples of mostly shallow boreholes, dug wells and surface waters sampled by Planéte Urgence (2005) in February 2005 and by BGR from August to October 2005 (Ploethner & Siemon, 2006a) are in good agreement with the airborne data. Discrepancies occur where water samples were taken from deep (sometimes artesian) wells. Some BGR EC samples are located within the salt-water area. They indicate deep freshwater occurrences below the salt water, not revealed by the airborne data due to the limited penetration depth of the HEM fields in highly conducting salt-water areas.

Resistivity maps, particularly at 30 m bgl (cf. Fig. 5), also outline areas of deep salt-water occurrences not directly affected by the tsunami. Saline water was found in at least two boreholes (located less than 2 km to the south of borehole (B) in Fig. 6) at 30–50 m depth before the tsunami, confirming that deep salt water occurs several kilometres inland. Therefore, all red to pink coloured areas can be regarded as affected by salt water. Freshwater resources are restricted to the hard-rock area and along the bed of the Krueng Aceh palaeoriver.

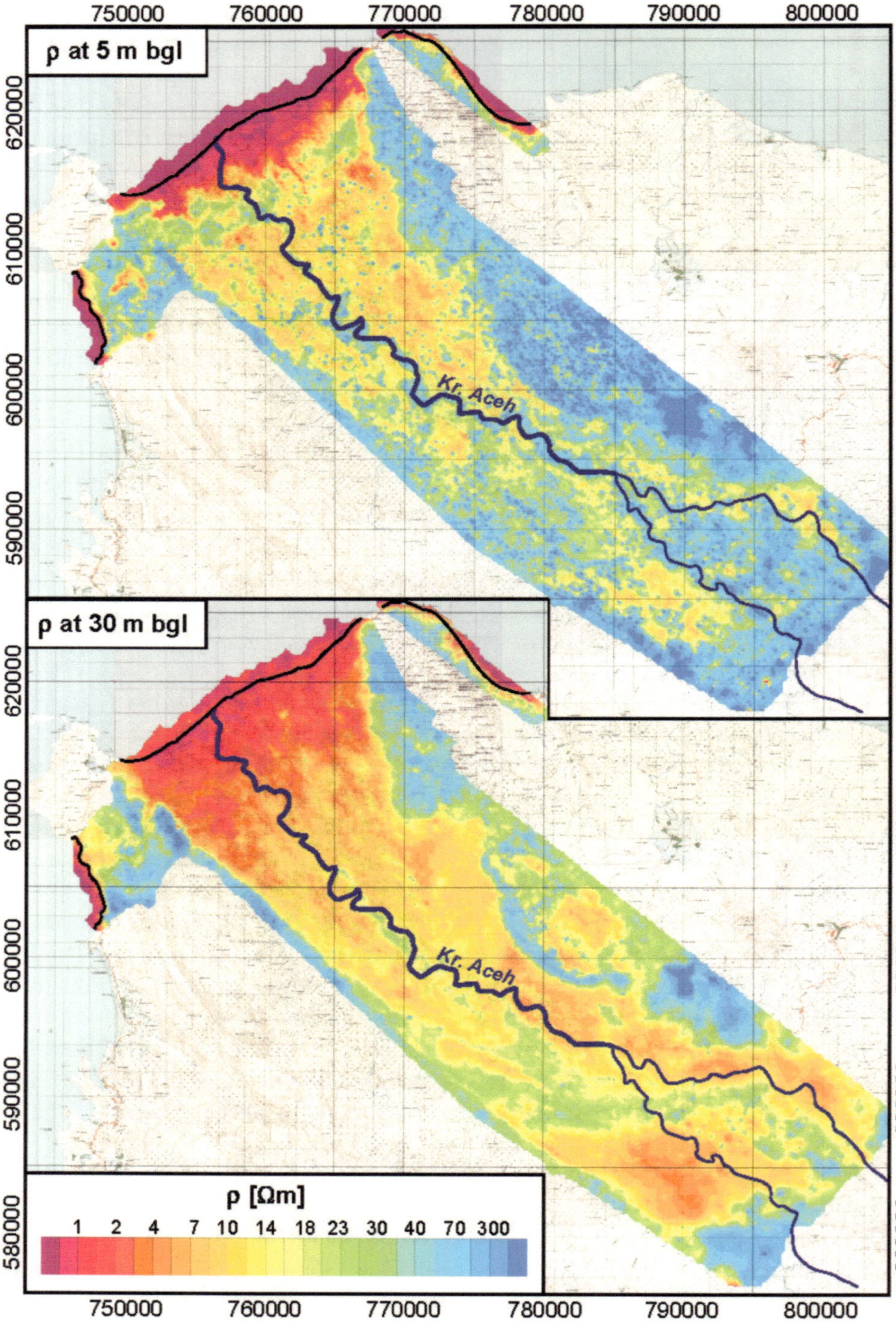

Fig. 5. Resistivity distribution at 5 and 30 m bgl derived from 1D inversion models of the Banda Aceh survey area with coast line (black) and rivers (blue). Background: TMI (1978).

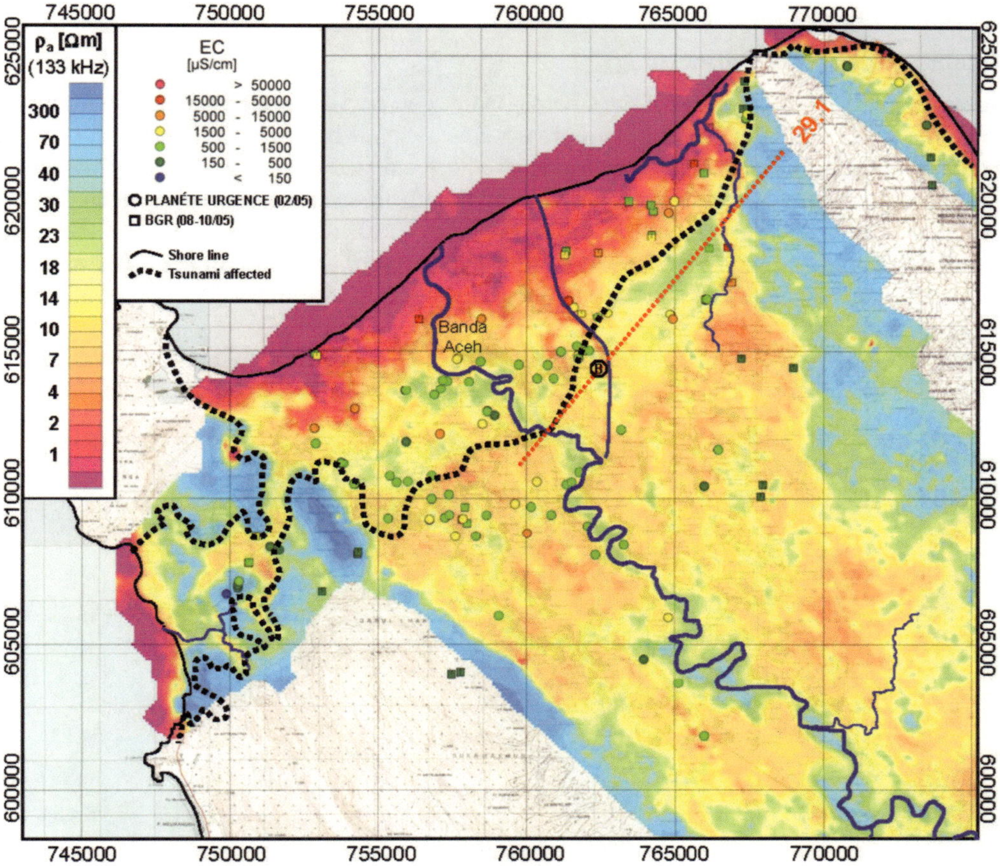

Fig. 6. Apparent resistivity at a frequency of 133 kHz revealing the lithology and the salinization of the shallow groundwater in the northern part of the Banda Aceh survey area. Water conductivity samples (coloured dots and squares), maximum extent of the tsunami flooding (dashed black line) and main rivers (blue lines) are plotted on top.

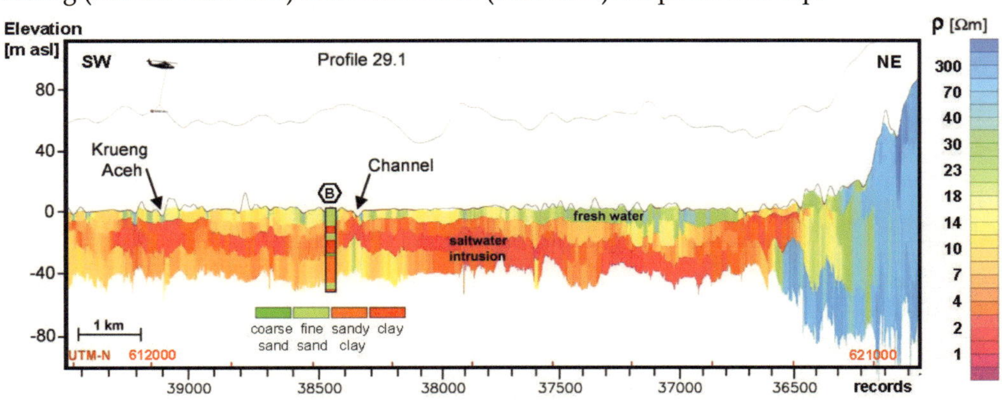

Fig. 7. Resistivity section along a profile 29.1 (cf. Fig. 6) showing a freshwater lens on top of salt-water saturated sediments; simplified lithology of a borehole (B) is plotted on top.

3.2 Calang - Meulaboh

Regional geology and hydrogeology

The central and southern parts of the Calang-Meulaboh survey area belong to the Meulaboh Embayment, an extensive coastal plain of very low relief. It has a maximum width of 50 km at Meulaboh and rarely exceeds elevations of 100 m asl (see Fig. 1). To the north-west, the coastal belt gradually becomes narrower and, to the north-east, the Meulaboh Embayment is terminated by the major westwards throwing Batee Fault whose scarp defines the north-eastern edge of the embayment. The Meulaboh Embayment is underlain to a great extent by sediments such as conglomerates, sandstones, claystones and lignites (Bennett et al., 1981b; Cameron et al., 1983).

Several rivers continuously transporting a substantial volume of suspension load cut the coastal belt (Fig. 8). By the process of sedimentation of the suspension load in the lower

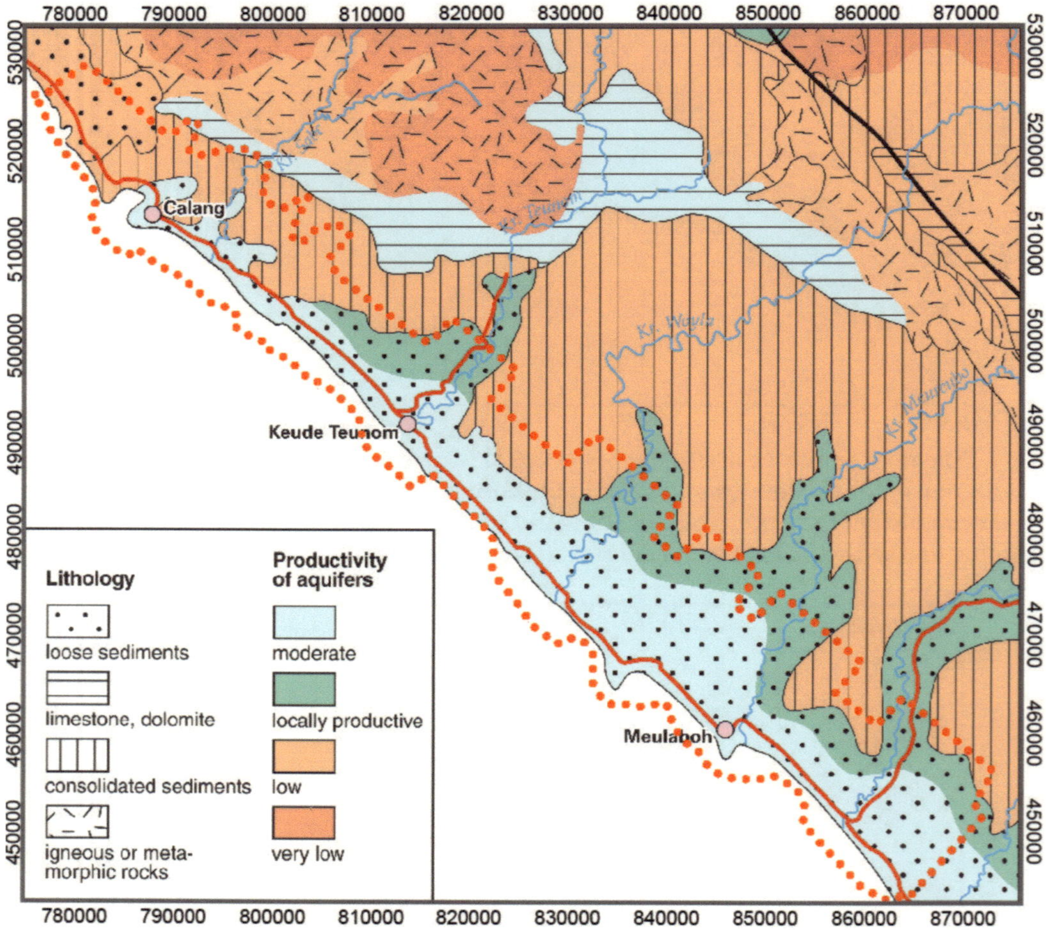

Fig. 8. Hydrogeological situation along the west coast of northern Sumatra (after Setiadi, 2004); main roads, rivers and faults are marked by red, blue and black lines, respectively, and the boundary of the airborne survey area is indicated by red dots.

reaches of the river valleys, these valleys were elevated compared with the deep-lying interfluvial swamps. In the upper 10 to 20 metres of the fine to coarse sands, a shallow unconfined aquifer is traditionally tapped by dug wells. A deep aquifer exists about 50 m below the clay-rich sediments. In the surroundings of Calang, Tertiary volcanic rocks and sedimentary rocks, like sandstones to siltstones, constitute poorly productive fissured aquifers (Ploethner & Siemon, 2006b).

Helicopter-borne survey

The survey area along the west coast (red dots in Fig. 8) in the district of Aceh Barat covers a 10–20 km by 120 km wide area. Thirty-two survey flights including 151 SW–NE profile lines and 25 NW–SE tie lines were flown, totalling about 4715 line kilometres. The nominal flight-line spacing was 500 m for both the profile lines and the tie lines. The tie lines were flown first in a 10 km wide strip parallel to the coast, and then the lines were flown perpendicular to the shoreline in the vicinity of the rivers. The helicopter-borne survey covered the area of 1700 km^2 within five weeks, from September 14th to October 21st, 2005. The survey flights commenced from a small local airport to the south-east of Meulaboh.

HEM survey results

It is obvious from the resistivity distribution at 10 m bgl (Fig. 9) that shallow freshwater occurrences can be expected in the entire survey area due to resistivity values above 40 Ωm, which represent sand, gravel and hard rock. Only close to the coast and along river-beds, where low resistivity values (below 5 Ωm) occur, contamination of the groundwater with salt water is apparent.

At greater depths (e.g. at 30 m bgl), lower values caused by clayey material dominate the resistivity maps. Areas where the top of the clay is lowered due to erosion by former rivers appear more resistive (green colours) as they were refilled with more sandy material. The blue coloured zones occurring in the central and southern parts of the survey area are correlated with swampy areas, where the top of the clay is generally lowered.

Along the coast, the conductive zone is broader than at shallow depths, correlating with a normal coastal salt-water intrusion. Unlike the situation along the coastal area close to the city of Banda Aceh (Fig. 6), shallow salt-water occurrences are mostly restricted to a small strip close to the shoreline. The two detail maps of coastal areas close to the towns of Calang and Meulaboh illustrate that, about ten months after the tsunami event, the salt-water area was smaller than it had been at the end of December 2004 (DLR, 2005c,d).

Most of the EC values derived from water samples collected in boreholes, dug wells and rivers (Ploethner & Siemon, 2006b) represent freshwater and confirm the airborne results. Only along the river valleys of the Krueng Rigaih and Krueng Meureubo (Fig. 9), resistivity and EC values indicate the presence of salt water.

3.3 Sigli

Regional geology and hydrogeology

The near-surface sediments of the Sigli Coastal Plain (Fig. 10) consist of Holocene alluvial sediments which are predominantly silty clays with intercalations of sand layers. The surrounding foot hills are built up of mainly clastic sediments of the Tertiary succession. In the north-western part of the Sigli Coastal Plain semi-consolidated tuffaceous and calcareous sandstones and coralline limestones are exposed in a horst structure (Bennet et al., 1977).

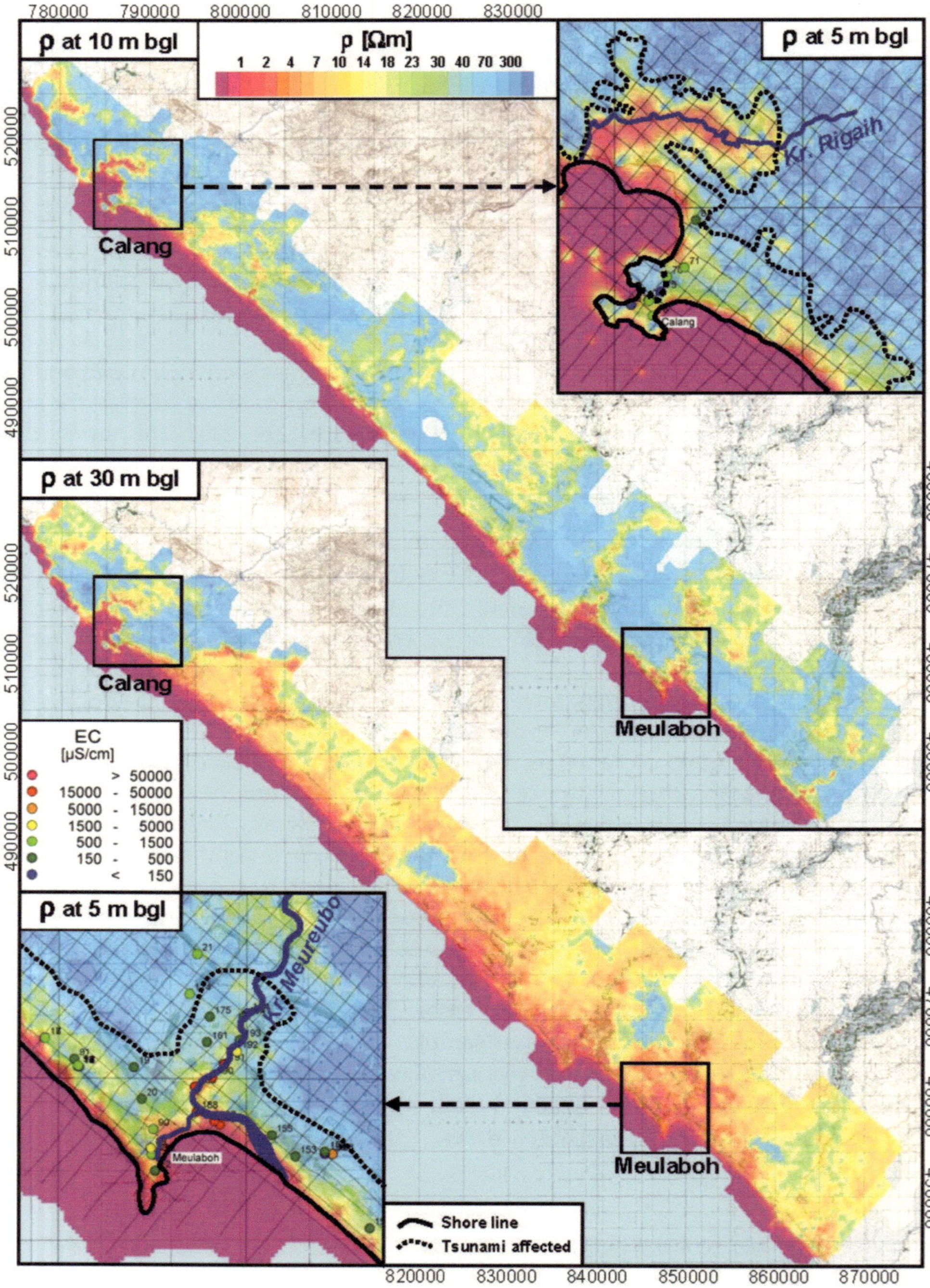

Fig. 9. Resistivity distribution along the west coast at 5, 10 and 30 m bgl derived from 1D inversion models of the Calang-Meulaboh survey area. Water conductivity samples (coloured dots) and maximum extent of the tsunami flooding (dashed line) are plotted on top of the Calang and Meulaboh detail maps. Background: TMI (1978).

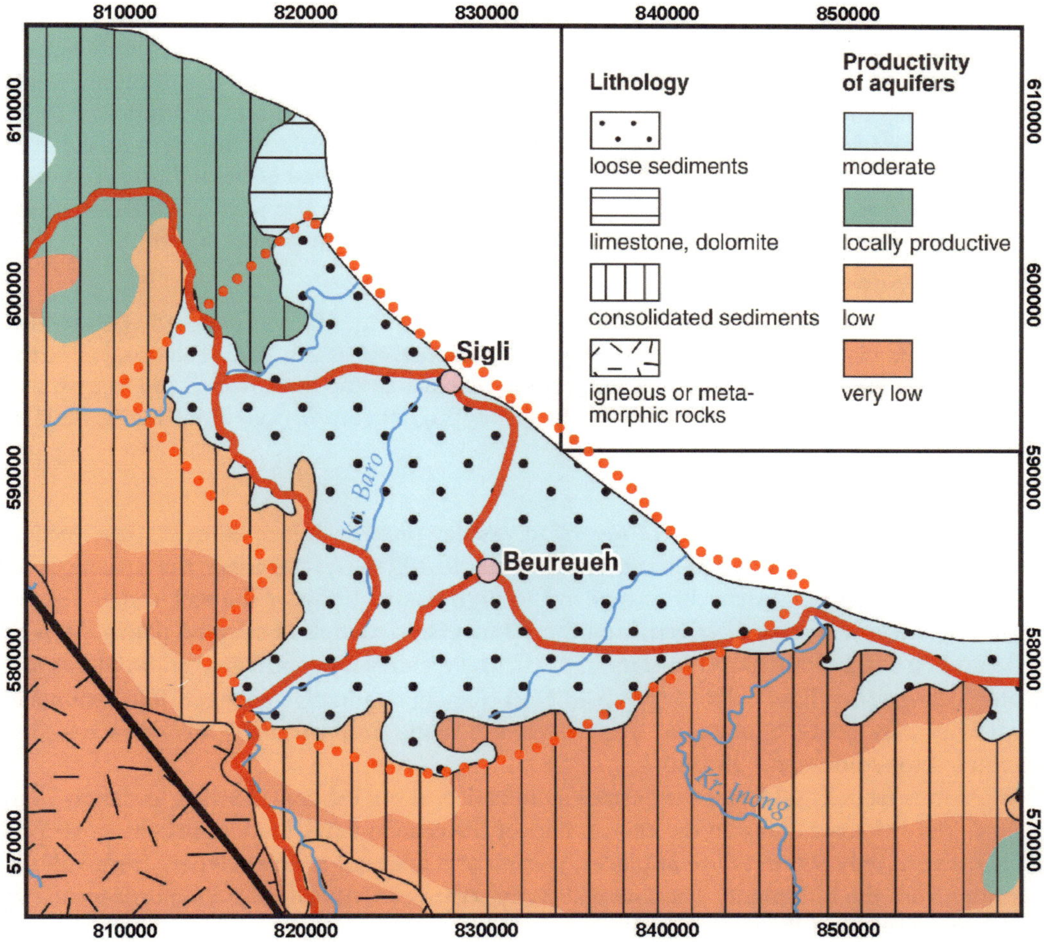

Fig. 10. Hydrogeological situation in the Sigli Coastal Plain (after Soetrisno, 1993a,b and Setiadi, 2004); main roads, rivers and faults are marked by red, blue and black lines, respectively, and the boundary of the airborne survey area is indicated by red dots.

According to IWACO (1993) and Soetrisno (1993a,b), the Tertiary sediments of the foothills at the southern and south-western margin of the plain are generally low productive. Limited shallow groundwater resources can be obtained in weathered zones. The carbonate rocks and the tuffaceous and calcareous sandstones at the north-western margin of the plain are moderately or locally productive aquifers. The alluvium of the coastal plain forms a multilayer aquifer of low to moderate transmissivity and piezometric heads of groundwater near or above land surface. It can be separated into a shallow and a deep aquifer system.

The top 10 to 20 metres of the alluvium are considered to be the shallow aquifer system within the coastal plain. The aquifer is sandy near the coast and primarily loamy further inland. It is directly recharged by rainfall. The shallow aquifer is still the major traditional resource for domestic water supply from numerous shallow dug wells. Before the tsunami, villages lying directly along the coast were also supplied with drinking water from shallow wells. In February 2005, two months after the tsunami, Planète Urgence (2005) revealed high water conductivities in most of the affected wells due to the infiltration of salt water into the groundwater. Control measurements in November 2005 in dug wells in the south-eastern corner of the plain indicated that the shallow groundwater in this area was still highly saline (Eberle et al., 2006). The deep groundwater of the Sigli Coastal Plain is supposed to be recharged in the surrounding foothills where the coastal aquifers are hydraulically in contact with older Tertiary sediments. Here, the piezometric head is being built up, which sustains the flow in the deep lying aquifers towards the coast. The groundwater along the coastal strip is largely artesian because of the apparently continuous clay layers.

Helicopter-borne survey

The survey area of approximately 700 km^2 (red dots in Fig. 10) was covered with twelve survey flights carried out between October 25th and November 7th, 2005. The spacing of the SW-NE lines was 500 m; tie lines running perpendicular to the lines were 3,000 m apart. In total 1,300 line kilometres were flown. The survey flights commenced from Blang Bintang airport, Banda Aceh.

HEM survey results

In the Sigli Coastal Plain, the 1D inversion results provide information on the vertical resistivity distribution to a maximum depth of about 100 m. At the coast, the investigation depth is limited to less than 30 m due to the high conductivity of the salt water. Fig. 11 shows the resistivity maps derived from the 1D inversion models at selected depth levels of 5, 15 and 60 m bgl. Pink to red colours ($\rho < 3$ Ωm) indicate very conductive material like salt water; orange to yellow colours ($\rho = 3\text{-}18$ Ωm) display conductive material like sediments with high clay content; potential freshwater, limestone and volcanics are more resistive (green to blue colours, $\rho > 18$ Ωm).

It is obvious from Fig. 11 that the surface salinization along the coast extents less than 3 km inland (red coloured area). In the central part of the coastal zone, the low resistivity strip is smallest with only 500 m width at places, particularly adjacent to rivers where freshwater is running from the hinterland. The low resistivity strip is widest towards the north-west and south-east corners of the survey area.

An extended system of ponds used for fish breeding lying immediately behind the shore line is thought to enable the permanent influx of seawater through tides. The discovered surface salinization covers for the most parts the area of the pond system. Therefore it is not obvious that the observed shallow salinization was produced by the tsunami waves - in opposite to the situation in the survey areas Banda Aceh and Calang-Meulaboh (Siemon et al., 2007).

In the hinterland, resistivity is generally decreasing with increasing depth. It is strongly supposed that this phenomenon can be explained by the occurrence of thicker clayey Quaternary/Tertiary sediments (mudstones) with a decreasing number of calcareous and sand intercalations. Focusing on depth levels greater than 15 m bgl (Fig. 11), it becomes evident that a resistivity high extending in the central plain over a distance of more than 20 km from the south-west down to the coast is quite consistent with increasing depth. It is considered the most favourable in terms of its potential for freshwater resources. The results of the resistivity-depth mapping infer that quite a strong groundwater flow to the coast must be assumed as the coastal low resistivity strip is smallest in the central part. The boundary between seawater and freshwater even appears to be pushed back off-shore at depths of about 60 m bgl (Fig. 11).

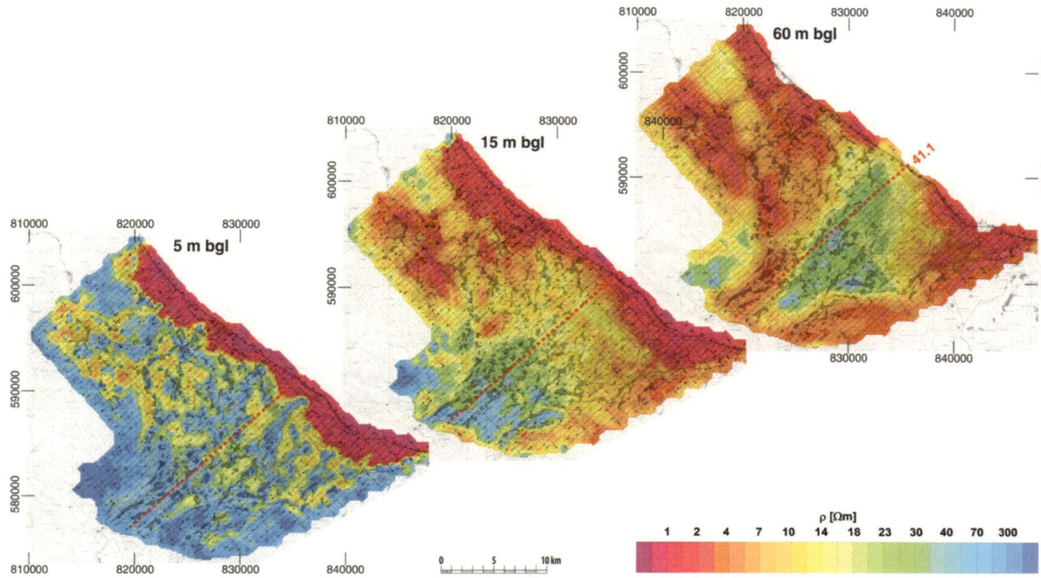

Fig. 11. Resistivity maps at selected depth levels of 5, 15 and 60 m bgl derived from HEM 1D inversion models. Additionally flight line 41.1 (cf. Fig. 12) is indicated by red dots. Background: TMI (1978).

The VRS of flight line 41.1 (Fig. 12) clearly supports this assumption of a strong groundwater flow to the coast. In the south-west resistive sediments, presumably sandstone and conglomerates, are underlain by conductive siltstone and mudstone. In the central part of the flight line, the conductive substrate dives down and the thickness of the sandstone and conglomerates grows. Towards the coast, this potential freshwater aquifer is covered by clayey material that obviously shields this aquifer against seawater intrusion.

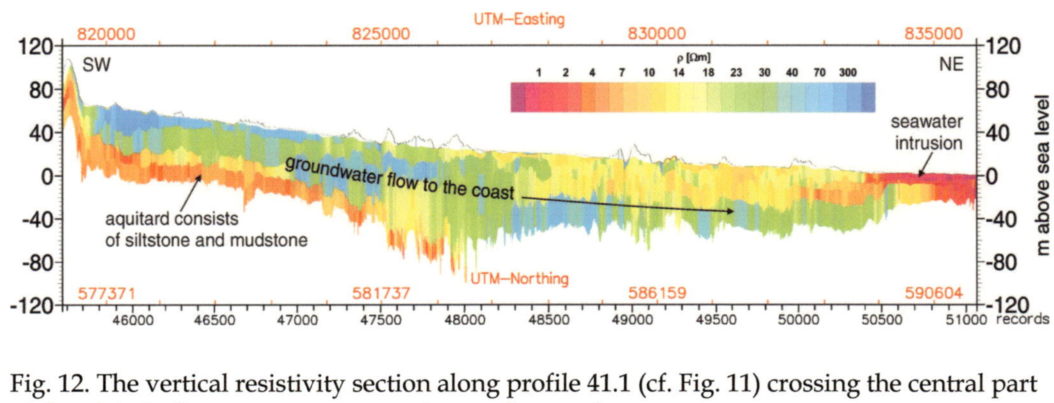

Fig. 12. The vertical resistivity section along profile 41.1 (cf. Fig. 11) crossing the central part of the plain indicates a strong groundwater flow to the coast.

4. Benefit of the airborne survey results

As the aim of the project was to assist Indonesian authorities, which were responsible for planning and realizing the sustainable reconstruction of community infrastructure, the focus was set on groundwater surveys in order to map the remaining salt-water contamination and to outline potential freshwater resources. Particularly the airborne electromagnetic results helped to get a better understanding of the hydrogeological situation as the derived resistivity maps and sections clearly showed where salt water was present and where potential freshwater resources could be expected.

On the north coast (survey area Banda Aceh) shallow salt water was found in a broad strip (about 3 km) whereas deep salt water occurred at around 30 m depth up to more than 10 km inland. On top of these deep salt-water occurrences potential freshwater lenses were mapped in an area 5-10 km from the coast (see Fig. 6) and in the palaeoriver bed of the Krueng Aceh.

The salt-water occurrences along the west coast (survey area Calang-Meulaboh) were mostly restricted to a small strip close to the shore line or some river valleys flooded by the tsunami, and thus, shallow freshwater occurrences could be expected in the entire survey area.

On the north-east coast (survey area Sigli) seawater intrusion appeared to be not strengthened by the tsunami event. The reason for that may be that in the central part of the shoreline groundwater flow from the hinterland is strongly supposed to push back seawater even to off-shore areas. A major potential freshwater occurrence was located beneath the central Sigli plain from the hilly hinterland down to the coast.

Several aid organizations drilled water-wells in order to provide the population of the tsunami affected areas with potable water. Often these drillings were not successful in finding potable water due to the lack of knowledge on the local hydrogeological situation. Many drillings ended up in salt-water aquifers or were stopped before a freshwater aquifer was reached. During and after the airborne survey BGR received many requests for appropriate drilling sites or, if a site had already been chosen, for information about the lithology to be expected. As the airborne survey was accompanied by a hydrolgeological survey those requests could be evaluated based on both geophysical and hydrogeological data. From the numerous requests evaluated for aid organisation (Siemon et al., 2006a,b), one example is shown in Fig. 13. It is evident from the vertical resistivity sections close to the proposed groundwater well (4) located about 35 km to the south-east of the town of Calang on the west coast that salt-water saturated sediments have to be expected below 15 m depth. Therefore it was recommended to drill a shallow well to tape the fresh groundwater of that sandy aquifer.

Besides requests for suitable locations for groundwater wells, a number of potential sites for a new sanitary landfill were evaluated in the vicinity of the city of Banda Aceh with respect to the existence of appropriate geological barriers. Clay as such a geological barrier is detectable, particularly in the vertical resistivity sections, due to the corresponding low resistivity of 3–30 Ωm. As a result of this evaluation, the pre-selected sites (Krause, 2005) had to be re-ranked.

Most of the requests could be answered successfully but not all – particularly if the sites for planned drillings were close to the coast where the high conductivity of salt water reduced the maximum penetration depth of the helicopter-borne electromagnetic fields. In those cases the use of deep-penetrating time-domain electromagnetics is necessary. This method was successfully applied in a ground-based follow-up survey on the north-east coast and it revealed more than 70 m deep potential freshwater resources along the shoreline close to town of Sigli (Steuer et al., 2008).

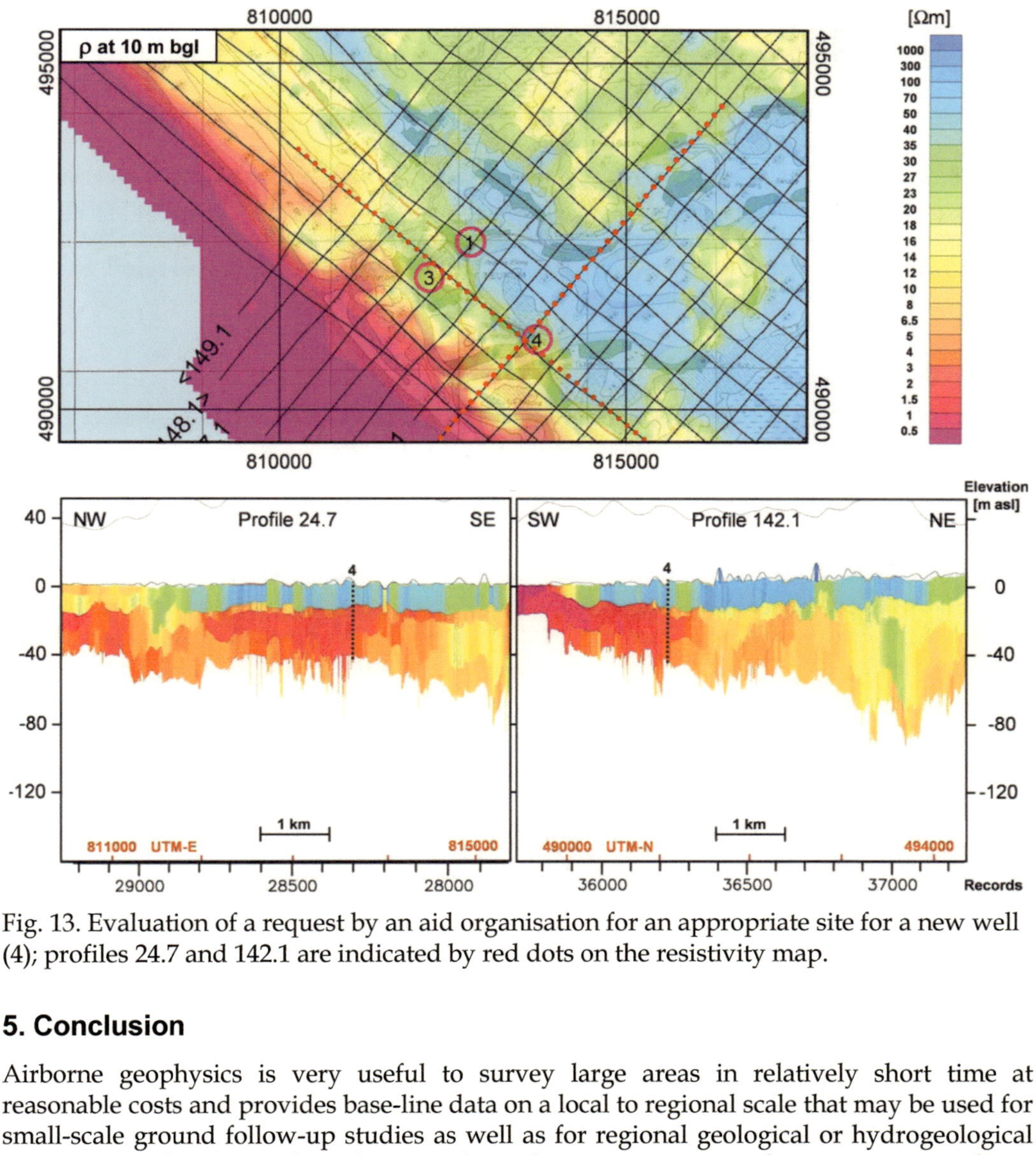

Fig. 13. Evaluation of a request by an aid organisation for an appropriate site for a new well (4); profiles 24.7 and 142.1 are indicated by red dots on the resistivity map.

5. Conclusion

Airborne geophysics is very useful to survey large areas in relatively short time at reasonable costs and provides base-line data on a local to regional scale that may be used for small-scale ground follow-up studies as well as for regional geological or hydrogeological mapping. As the electrical conductivity depends on parameters such as pore-water salinity and clay content, emphasis in the groundwater surveys along coasts of northern Sumatra was placed on the evaluation of the airborne electromagnetic data. Magnetic and radiometric data, which may also contain hydrogeological information, e.g. on fault systems and the lithology on the surface of the earth, respectively, were also evaluated but these results have not been considered in this chapter.

The electromagnetic surveys revealed several potential freshwater resources and areas of salt-water occurrences close to the coasts were mapped in detail. Many requests by aid organisations for information on the local geological and hydrogeological conditions for planned water wells were evaluated and could be successfully answered in most cases.

Close to the coasts, however, the investigation depth of the HEM system was constrained due to highly conductive near-surface salt water and, thus, ground-based time-domain electromagnetic measurements were necessary to reveal deeper coastal freshwater resources. The combination of airborne and ground-based electromagnetic techniques increased the efficiency to estimate the freshwater potential.
The airborne geophysical surveys helped to close the gaps between task-force measures and long-term planning as well as between spatial surface mapping and local borehole data. Particularly airborne electromagnetics proved to be a very efficient tool to assess the newly established post-tsunami groundwater situation and to supply hydrogeological baseline data for rehabilitation programs.

6. References

Bennet, J.D., Bridge, D.McC., Cameron, N.R., Djunuddin, A., Ghazali, S.A., Jeffery, D.H., Keats, W., Rock, N.M.S., Thompson, S.J. & Whandoyo, R. (1977). *Explanatory notes to the geologic map of the Banda Aceh quadrangle, North Sumatra.* Jakarta.

Bennett, J.D., Bridge, D.McC., Cameron, N.R., Djunuddin, A., Ghazali, S.A., Jeffery, D.H., Kartawa, W., Keats, W., Rock, N.M.S., Thomson, S. J. & Whandoyo, R. (1981a). *Geologic Map 1:250,000 of Banda Aceh Quadrangle, Sumatra.* Geological Research and Development Centre, Bandung.

Bennett, J.D., Bridge, D.McC., Cameron, N.R., Djunuddin, A., Ghazali, S.A., Jeffery, D.H., Kartawa, W., Keats, W., Rock, N.M.S. & Thompson, S.J. (1981b). *Geologic Map of the Calang Quadrangle, Sumatra.* Geological Research and Development Centre, Bandung.

Cameron, N.R., Bennett, J.D., Bridge, D.McC., Clarke, M.C.G., Djunuddin, A., Ghazali, S.A., Harahap, H., Jeffery, D.H., Kartawa, W., Keats, W., Ngabito, H., Rock, N.M.S. & Thompson, S.J. (1983). *Geologic Map of the Takengon Quadrangle, Sumatra.* Geological Research and Development Centre, Bandung.

DLR (2005a). *Indonesia/Sumatra – Banda Aceh – Map Sheet 042152, 1:40000.* Center for Satellite based Crisis Information - German Remote Sensing Data Center / German Aerospace Center (DLR).

DLR (2005b). *Indonesia/Sumatra – Lho'Nga – Map Sheet 042123 (041124), 1:40000.* Center for Satellite based Crisis Information - German Remote Sensing Data Center / German Aerospace Center (DLR).

DLR (2005c). *Indonesia/Sumatra – Calang – Map Sheet 042061, 1:40000.* Center for Satellite based Crisis Information - German Remote Sensing Data Center / German Aerospace Center (DLR).

DLR (2005d). *Indonesia/Sumatra – Meulaboh – Map Sheet 052011, 1:40000.* Center for Satellite based Crisis Information - German Remote Sensing Data Center / German Aerospace Center (DLR).

Eberle, D., Siemon, B., Klinge, H., Meyer, U., Pielawa, J., Rehli, H.-J., Schmidt, H., Steuer, A. & Voß, W. (2006). Interpretation of HEM and TEM data – Helicopter-borne and ground geophysical investigation in the province Nanggroe Aceh Darussalam, northern Sumatra, Indonesia, survey area Sigli (Pidie District) 2005. *BGR Report Vol. B-3,* Archives No. 0126196 B-3, Hanover.

Fraser, D.C. (1978). Resistivity mapping with an airborne multicoil electromagnetic system. *Geophysics*, Vol. 43, No. 1, pp. 144–172.

IWACO (1993). Kabupaten Aceh Barat: Planning for water supply development and raw water sources allocation, In: *Study of water sources allocation for water supply for D.I. Aceh Province*, prepared by Consultants for Water and Environment IWACO Ltd in association with COWI Consulting Engineers and Planners AS, PT YODYA KARYA Architects & Consulting Engineers & PT GAMMA EPSILON Consulting Engineers for Government of Indonesia, Ministry of Public Works, Directorate General of Human Settlements, Directorate of Water Supply, Banda Aceh.

Krause, P. (2005). Pre-selection of a sanitary landfill site for the city of Banda Aceh, Volume I. *GTZ-ProGReS*, Banda Aceh.

Planète Urgence (2005). Identification inventory and analysis of drinking water points in the region of Aceh (province of Sumatra). *Development mission following the tsunami of December 2004*, Banda Aceh.

Ploethner D. & Siemon B. (2006a). Hydrogeological reconnaissance survey in the province Nanggroe Aceh Darussalam, Northern Sumatra, Indonesia – Survey area Banda Aceh / Aceh Besar 2005, Helicopter Project Aceh (HELP ACEH). *BGR Report Vol. C-1*, Archives No. 0126196 C-1, Hanover.

Ploethner D. & Siemon B. (2006b). Hydrogeological reconnaissance survey in the province Nanggroe Aceh Darussalam, Northern Sumatra, Indonesia - Survey area Calang-Meulaboh 2005, Helicopter Project Aceh (HELP ACEH). *BGR Report Vol. C-2*, Archives No. 0126196 C-2, Hanover.

Sengpiel, K.-P. & Siemon, B. (2000). Advanced inversion methods for airborne electromagnetic exploration. *Geophysics*, Vol. 65, No. 6, pp. 1983-1992.

Siemon B. (2001). Improved and new resistivity-depth profiles for helicopter electromagnetic data. *Journal of Applied Geophysics*, Vol. 46, pp. 65–76.

Siemon, B., Eberle, D., Klinge, H., Rehli, H.-J., Schmidt, H., Steuer, A. & Voß, W. (2006). Technical report on the helicopter-borne and ground geophysical investigation in the province Nanggroe Aceh Darussalam, northern Sumatra, Indonesia, Survey area: Sigli (Pidie District) 2005, *BGR Report VOL. A-3*, Archives No. 0126196 A-3, Hanover.

Siemon B., Ploethner D. & Pielawa J. (2006b). Interpretation of electromagnetic data – Helicopter-borne geophysical investigation in the province Nanggroe Aceh Darussalam, Northern Sumatra, Indonesia, survey area Banda Aceh / Aceh Besar 2005, Helicopter Project Aceh (HELP ACEH). *BGR Report Vol. B-1*, Archives No. 0126196 B-1, Hanover.

Siemon B., Ploethner D. & Pielawa J. (2006c). Interpretation of electromagnetic data – Helicopter-borne geophysical investigation in the province Nanggroe Aceh Darussalam, Northern Sumatra, Indonesia, survey area Calang – Meulaboh 2005, Helicopter Project Aceh (HELP ACEH). *BGR Report Vol. B-2*, Archives No. 0126196 B-2, Hanover.

Siemon B., Röttger B. & Rehli, H.-J. (2006d). Helicopter-borne geophysical investigation in the province Nanggroe Aceh Darussalam, Northern Sumatra, Indonesia – Survey area Banda Aceh / Aceh Besar 2005, Helicopter Project Aceh (HELP ACEH),

Technical Report. *BGR Report Vol. A-1*, Archives No. 0126196 A-1, Banda Aceh, Hanover.

Siemon B., Steuer A., Voß W. & Rehli H.-J. (2006e). Helicopter-borne geophysical investigation in the province Nanggroe Aceh Darussalam, Northern Sumatra, Indonesia - Survey area Calang - Meulaboh 2005, Helicopter Project Aceh (HELP ACEH), Technical Report. *BGR Report Vol. A-2*, Archives No. 0126196 A-2, Banda Aceh, Hanover.

Siemon, B., Steuer, A., Meyer, U. & Rehli, H.-J. (2007). HELP ACEH - A post-tsunami helicopter-borne groundwater project along the coasts of Aceh, northern Sumatra. *Near Surface Geophysics*, Vol. 5, No. 4, pp. 231-240.

Setiadi, Hendri (2004). *Hydrogeological Map 1:1,000,000 of Indonesia*. Directorate of Geological and Mining Area Environment, Bandung.

Soetrisno, S. (1993a). *Hydrogeological Map 1:250,000 of* Indonesia, Sheet 0421 Banda Aceh, Directorate of Environmental Geology, Bandung.

Soetrisno, S. (1993b). *Hydrogeological Map 1:250,000 of Indonesia, Sheet 0521 Leikseumave/ Simpangilum*, Directorate of Environmental Geology, Bandung.

Steuer, A., Siemon, B. & Eberle, D. (2008). Airborne and ground-based electromagnetic investigations of the fresh-water potential in the tsunami-hit area Sigli, northern Sumatra. *Journal of Environmental & Engineering Geophysics*, Vol. 13, No. 1, pp. 39-48.

TMI (1978). *Topographic Map of Indonesia 1:50,000*. JANTOP TNI AD, Jakarta.

29

Application of Hydro-Geochemical and Geo-Electrical Techniques to Identify the Impact of Tsunami in the Coastal Groundwater

Chidambaram S[1], Ramanathan AL[2], Prasanna MV[3] and Manivannan R.[1]
[1]*Department of Earth Sciences, Annamalai University*
[2]*School of Environmental Sciences, JNU, New Delhi*
[3]*Department of Applied Geology, School of Engineering and Science, Curtin University of Technology, Sarawak*
[1,2]*India*
[3]*Malaysia*

1. Introduction

In general, the Coastal zones are dynamically active because of multivariate tectonic, fluvial marine and Aeolian geomorphic process which may vary in their dynamism with space of time. The tectonically active coasts show a greater influence on the land ocean interactive process. Tsunamis are ocean waves produced by earthquakes or underwater landslides. The word is from Japanese and means "harbor wave," because of the devastating effects these waves have had on low-lying Japanese coastal communities. On 26th December 2004, Indian subcontinent experienced the most devastating tsunami in its recorded history. The phenomenon was triggered by a submarine earthquake located at 3.4° N, 95.7° E off the coast of Sumatra (Indonesia) with an intensity of 9RSU. Tsunamis are most often generated by earthquake-induced movement of the ocean floor. Such massive earthquakes only occur in subduction zones where two of the rigid tectonic plates that comprise the earth's surface are converging, and one plate, usually composed of heavier oceanic material, dives beneath another, usually composed of lighter continental material. Aftershocks appear to be active over a 1300 Km section of the zone stretching from the Andaman Islands in the north to the earthquake epicentre below the northern tip of Sumatra, the seismic waveform data appears to indicate that only a 450 Km length of the subduction zone off northern Sumatra ruptured (Cumins & Leonard 2005). September 2004 AusGeo News article surmised, the greatest tsunami threat in the Indian Ocean appears to be posed by great subduction zone earthquakes of Sumatra.

Studies of the Indian ocean tsunami has been carried out by several workers (Table 1), it reveals that the signature of tsunami has been clearly demarcated in the sediments and only a few studies were carried out in the groundwater. Still the impact of tsunami in the groundwater by both geochemical and geophysical methods has been carried out by very few researchers. This chapter concentrates mainly on the application of the geophysical (Chidambaram et al 2008) and geochemical nature (Chidambaram et al 2010) of the ground water to characterize the tsunami signatures and to bring out the process/mechanism after the event.

2. Study area

The study area extents from Parangipettai to Pumpuhar and lies in between 79° 46'E to 79° 51' E longitudes and 11°07' N to 11° 30' N latitudes (Fig 1). It occurs within the survey of India toposheets of 58 M/15 & 16. The Vellar and Coleroon are the major rivers flowing in the study area they form a an Estuary with marshy mangrove environment at Pichavaram.

3. Geological succession

The maximum temperature ranges between 27.9°C and 36.9°C, with mean ranging from 20.8°C to 27.1°C. The long-term analysis of the rainfall data from January to May indicates that average rainfall is in the order of 1,162.35 mm/year with a maximum contribution from the NE monsoon (53.01% of the total rainfall).

Era	Age	Formations	Lithology
Quaternary	Recent to Sub-Recent	Alluvium& Laterite	Soils, alluvium and coastal sands, Clays, kankar and Laterite
------------------------ UNCONFORMITY ------------------------------			
TERTIARY	Mio-Pliocene	Cuddalore Sandstones	Argillaceous and pebble bearing grits, Clays (variegated) with lignite seams and pebble beds.
------------------------ UNCONFORMITY ------------------------------			

4. Hydrogeology

Aquifers in the study area are confined to unconfined. Water table in majority of the study area ranges from 5 – 20 mbgl (meter below ground level) during premonsoon. Whereas, depth to water table during post monsoon period it ranged between 2 and 10 mbgl. In Coleroon alluvial belt the shallow water table conditions exist less than 2 mbgl. The areal extent of this region has been further extended occupying major part of Vellar and coastal alluvial formations in east indicating extent of water logging conditions during post monsoon period. Water level observations for certain locations in the study area (Fig 2) during the period of tsunami indicate that there is considerable increase of water level in certain location accompanied by fall in few regions. Yield of bore wells tapping permeable granular zones, older sediments of study area ranges from 1176 and 29481 for drawdown ranging from 2.15 to 9.6m.Transmissivity is also higher and it ranges from 295 - 838m^2/day. Movement of water from one region to other is also governed by hydraulic conductivity (k) ranges from 13.6 to 23.6m/d.

5. Methodology

Around 40 samples collected during November 2004 from Pondicherry to Velankanni it showed that nearly eleven locations in shallow groundwater were uncontaminated by seawater intrusion along the coast and then eleven samples were identified/selected for post tsunami impact studies (In total eleven sampling locations were selected for targeting the changes after tsunami). Eleven ground water samples were collected (Fig 3a,b), during each period of sampling in the month of November 2004, January 2005, March, 2005 and August, 2005. The samples were collected from shallow aquifers through tube wells depths of 20mbgl . Samples collected in month of November 2004 (Prefix 'pr'), Jan 2005 (Prefix 'J'), March, 2005 (Prefix 'M') and August, 2005 (Prefix 'A').The samples were collected mainly from shallow aquifers (40-100 feet depth). The inundated region ranges from 0.5 to 1 Km from the coast (Table 1). These wells were selected in such a way that they are in the tsunami inundated region ranging at least about 0.2 to a maximum of 3Km from the coast and also were not earlier affected by salt water intrusion. The chemical parameters like EC (Electrical conductivity), pH, TDS, Na, K, Ca, Mg, HCO_3, Cl and SO_4. These parameters were used for the hydro geochemical interpretations. The collected samples were analyzed for major cations like, Ca and Mg by Titrimetry, Na and K by Flame photometer (CL 378); anions, Cl and HCO_3 by Tirimetry, SO_4 by Spectrophotometer (SL 171 minispec). EC and pH were measured in the field using electrode. All the analyses were carried out by adopting standard procedures (APHA, 1998). The results of the analysis were checked by the cation anion balance which was within ±8%. Heavy metal analysis were carried out by using ICPMS at NGRI, Hyderabad.

The soundings were performed by the DDR III Resistivity meter by adopting Schlumberger electrode array with maximum spacing (AB/2) in the range of 2 - 90m. Pre-tsunami resistivity was done up to 50m depth during August, 2004 and post-tsunami survey was done in 2005 at different locations (Fig 4). Interpretation of resistivity curves was done by curve matching technique using standard graphs (Rijkswaterstaat, 1969) and master curves (Ornella & Mooney, 1966) for determining thickness and resistivity of corresponding layers. Impact of saline water due to Tsunami is expected to be in shallow waters and hence the geoelectrical variations of shallow ground waters were only considered.

6. Geochemistry

Dominance of anions in the study area are in the following order; $Cl>HCO_3>SO_4$ in both pre-tsunami and post-tsunami. Though the order of preference is maintained the value of Cl has increased in Post tsunami. The order of dominance of cations in the study area is Ca>Na>Mg>K in pre-tsunami and Na>K>Ca>Mg in post-tsunami (Table 2). The figure 4 shows that there is a considerable increase in the EC, Ca, Mg, Na along with Cl and HCO_3. Increase of EC, Na and Cl are more prominent during the period of the study than other ions.

7. Heavy metals

Heavy metals like Fe, Al and Mn were analysed for these samples during post tsunami . Iron ranges from 0.12 mgl^{-1} to 0.72 mgl^{-1} with an average of 0.37 mgl^{-1} in the month of March and from 0.03 mgl^{-1} to 0.58 mgl^{-1} with an average of 0.39 mgl^{-1} in the month of August. Manganese ranges from 0.05 to 1.52 with an average of 0.52 in the month of March and from 0.009 mgl^{-1} to 1.86 mgl^{-1} with an average of 0.42 mgl^{-1} in the month of August. Aluminium ranges from 0.03 to 0.08 mgl^{-1} with an average of 0.055 mgl^{-1} in the month of March and from 0.04 mgl^{-1} to 0.08 mgl^{-1} with an average of 0.056 mgl^{-1} in the month of August. It is evident that there is no significant variation in the heavy metals subsequent to tsunami (Table 3).

8. pCO_2 and ionic strength

Partial pressure of CO_2 (pCO2) in rivers are commonly out of equilibrium with atmosphere. Two possible explanations for this apparent paradox are: (1) River waters (particularly perennial rivers) contain a significant fraction of high CO_2 groundwater and (2) the rate of re-equilibrium with the atmosphere (by releasing the excess CO_2) is relatively slow (Stumm & Morgan, 1996; Holland,1978). The Log pCO_2 (Raymahashay, 1986) for each samples are determined to study its relation to recharge (Prasanna et al,2009, Chidambaram et al,2008). Log PCO2 value, during ranges from -0.83 to -3.62 (Fig 5). The ionic strength does not show any linearity with log pCO_2 values during January and Pre-tsunami. This may be due to higher pH or lesser neutralizing ions like Ca^{2+} (Raymahashay, 1986). The samples collected during the January and pre-tsunami show lesser values of pCO2, the higher values are noted in the month of August and March. The average values of pCO2 shows the order of dominance March> August> January> Pre-tsunami. This reveals that recharge is predominant during the pre-tsunami and the January months. The Atmospheric interaction becomes lesser during March and August with more matrix interaction. The South west monsoon showers during the month of August has decreased the pCO_2values than that of March.

Ionic strength (IS) is a measure of total concentration of ions which emphasizes increased contribution of species with charges greater than one to solution non-ideality (Domenico & Schwartz, 1990).

$$I= 0.5 \sum mz^2(1) \qquad (1)$$

Where m is the concentration of a given ion in moles per liter and z is the charge on that ion. The terms in the summation include one for each ionic species present. A monograph, which simplifies calculation of ionic strength from analytical data in milligrams per liter, has been published by Hem (1961).

The approximate value of Ionic Strength can also be computed from the specific conductance of the solution if this has been measured (Lind, 1970). However, calculation should not be made unless one should have some knowledge of what the principal dissolved species present in the solution. If the composition is unknown for water with a specific conductance of 1,000 μ mhos, calculated value of IS ranges from 0.0032 to 0.134. The distribution of the ionic strength in the samples varies from lower to higher ranges. There is no significant demarcation of the IS with sampling interval except during August and March . Higher ionic strength is noted during the month of August, March, followed by

January and then Pre-tsunami. The increase of ionic strength shows linearity during the month of March and August

9. Intensity of Tsunami

There are two different tsunami intensity scales proposed by Soloviev (1970) and Ambraseys (1962). Soloviev pointed out that Imamura-Iida's scale is more like an earthquake intensity scale rather than a magnitude. He also distinguished the maximum tsunami height **h** and the mean tsunami height **h**. He then defined tsunami intensity **I** as **I = $\log_2(h)$**. Sieberg tsunami intensity scale - a descriptive tsunami intensity scale which was later modified into Sieberg-Ambraseys tsunami intensity scale (Ambraseys,1962); which describes tsunami from light tsunamis (Level 1) to disastrous tsunami (level 6) based on physical distraction caused by tsunami. Based on the calculation made with the average wave height (Table 4) in different states along the east coast of India, viz., Andhra Pradesh, Kerala, Puducherry and Tamilnadu, it is evident that the Tamilnadu and the Puducherry are the regions with higher Tsunami intensity.

10. Spatial distribution

Cl/HCO_3 ratio is generally used to identify the extent of salt water intrusion in the aquifers. An attempt has been made (Fig 6.1) to find out the relation of EC with the ratio, this indicates that EC increases in the month of March and August, it is also observed that certain amount of linearity is noted between EC and the Cl/HCO_3 ratio during all the periods of study. But higher ratios were noted only during the month of March and August 2005. The Spatial distribution of this ratio (Fig 6.2 a, b and c) during different periods of study indicates that area covered by higher ratios of these ions was noted along the river mouth and in few pockets where the water got stagnated during the event. There is a considerable increase noted in the distribution of this ratio in the month of August 2005. The average composition of the groundwaters of this region is shown by the figure 4. It indicates that the Na concentration is higher in cation and Cl in anion. Hence, the variation of these ions determines the overall water chemistry of the region.

11. Factor analysis

The statistical analysis of the data helps us to decipher the major chemical process responsible for the water chemistry of the region. The factor analysis was carried out by using Varimax rotation. Factor analysis carried out by Chidambaram et al (2010) for the study area indicate that three prominent factors were extracted for pre-tsunami and January and two factor for the month of August and March. The first component of factor loading (Figure 7a) in Pre-tsunami was represented by HCO_3, Mg, and Na indicating the process of weathering. The first component of January was represented by Cl, HCO_3, K, Mg, and Na indicating the mixing effect of seawater and fresh water. But during the month of March the first factor was represented by Cl, Ca, K, Mg, and Na indicating the evaporation mechanism, the second factor was represented by Cl and Na with negative representation of Ca,HCO_3,Mg, and SO_4 which indicates that there is a removal of these ions from the system by the precipitation of salts in pore spaces and on the surface due to precipitation. Hence there is an enrichment of Cl and Na. In August its represented by Cl, Na, and SO_4 where the

impact of saline water is clearly noted as the first factor and there is an association of Ca, HCO_3, Mg and K as second factor (Fig 7b). This is due to the leaching and dissolution of salt precipitation during the month of March. The rain during June and July has considerably leached and added these ions into the system. The representation of HCO_3 is noted during August and not in March indicate that dissolution of the precipitated salts were prominent only during the month of August after the onset of monsoon. The rains during June and July, has increased the migration of ions either by weathering or leaching of salts precipitated in summer.
The chemical compositions of different samples of the study area for different seasons are compared with the chemical composition of the sea water in the study area and the leached waters from the salt precipitate (Xue et al. 2000). The samples of August identified to have good correlation with seawater composition and that of salt precipitate. Their corresponding pre-tsunami compositions were used to obtain the mixing proportion (Chidambaram et al 2010). So the pre-tsunami water chemistry of the region is taken for the representative locations and the amount of mixing has been calculated.

Pre tsunami sample + X % Seawater → Post tsunami

The X % is determined by the mixing calculation

The percentage of sea water mixed with the groundwater after the tsunami event was obtained by the above said equation in the previous study. The results of the mixing percentage of the sea water with groundwater (Fig 8) shows the following order, A4 > A5 > A10 > A6 > A8 > A1 = A2. The higher rate of mixing is noted in the regions with more distribution channels of rivers where the water has entered and stagnated, precipitated and leached into the system. It is also noted that the bathymetry is gentle in the southern part of the area, where the tsunami water might have entered a long distance inland, where the inundation was favoured by the distributary's channels and river mouth. Chidambaram et al (2010) inferred that these processes (Fig. 9) might have been controlling the hydro-geochemical changes of ground water after the tsunami. After the tsunami, seawater entered the water table through the open wells or tube wells (M1), and the entrapped water (M2) got infiltered into the water table of the coastal alluvium during January 2005. Later high temperature in the summer months might have resulted in the formation of salt precipitates (M3) (due to evaporation of infiltered/stagnated water by impermeable clay layers) near the surface or in pore spaces and subsequent dilution in the end of March 2005 by sparse rain. After precipitation and dissolution, salt leached out from surface to the shallow groundwater zones (M4), enhances the EC and TDS of the groundwater of August 2005. Thus, the impact of direct infiltration or direct mixing of tsunami waters (M1) has a relatively lower effect than the subsequent precipitation and dissolution of salts formed by the entrapped sea water, which happened after the tsunami event.

12. Geophysical study

The geophysical survey conducted in 20 specific location in the study area (Table 5.1 and 5.2) which indicates that three to four layers were identified with resistivity for 1st layer raging from 0.012Ωm to 42Ωm, 2nd layer 0.05 Ωm to 120 Ωm and 3rd layer 0.02 Ωm to 360 Ωm. Thickness of the formation of 1st layer rays from 1 to 17, 2nd layer 2.5 to 85 and 3rd layer 7 to α. VES all four types of curves were found in the study area (Chidambaram et al 2008).

13. Curve types

There are different curve types inferred during the geophysical interpretations. The general trend of the saltwater intruded coastal aquifer is expected to be Q or K where in the resistivity values keep on decreasing with depth or decreases in the shallow and deeper aquifer respectively indicating higher conductivity. Hence, when the deeper aquifers are contaminated with saltwater they show lesser resistivity values. When the shallow aquifer are affected by the presence of saltwater they tend to give an A or K type curve, Where the first layer has low resistivity values than the under laying layers. In the study area nearly 40% of the locations have A type 15% have K type curve indicating the possibility of the recharge of tsunami waters in the shallow aquifers. All these locations with A & K type curve (Chidambaram et al 2008) are reported to have more causalities in the Tsunami event and sea water has also invaded inland through several inlets backwater, river mouths, distributaries channel etc., the entered water has left the signature into these aquifer by infiltrating into the system.

14. Formation resistivity & factor

It was found that the aquifers of shallow depths were more affected during the event of tsunami and hence the values of the 3m, 5m, 10m and 15m are, taken for discussion. Study on resistivity of the formation for the shallow depth of 3m indicate pre-tsunami samples range from moderate to saline, with almost 50% of locations showing salinity (Table 6). After the event almost all locations show salinity reflexes from resistivity values. Formation factor is obtained by ratio between formation resistivity and water specific conductance (Van Overmeeren, 1987). It is suggested that a value of 4 is for fresh water aquifer, almost all samples in post event falls in saline nature and 6 samples of pre-tsunami falls in this category. EC for rest of the samples in the Pre tsunami was not available.

15. Spatial distribution of apparent resistivity

The apparent resistivity values for specific depth have been distributed spatially trended to develop an isoresistivity map and to get and overall perspective view of the study area, three specific depth were selected for 3m,5m and 10m isoresistivity map were prepared using GIS. The 3m (Fig. 10a) isoresistivity map indicates that more region in the central part of the study area are affected with high conductivity which is interpreted with low resistivity values. The 5m (Fig.10b) isoresistivity map indicates that certain regions in the central part of the study area are affected. The 10m (Fig. 10c) isoresisitivty contour also helps us to identify only certain location along the coast show low resistivity values. The study of the resistivity values of different depths establishes that the shallow aquifers are more saline than the deeper ones.

In general due to the presence of higher intensity of tributary channels, and allows the rivers mouths (Vellar, Patinatharu etc,.) reveals the association of highly affected region, but the region with mangroves were not much affected. More over the bathymetry study in this region shows that the shelf is gentle in the southern region than in the north, which helps the giant wave to easily progress through inland.

16. Geoelectrical section

Location and thickness of fresh water lens obtained by GECS was used to delineate fresh water zones of island and thematic contour map was prepared, depicting the variation of thickness of fresh water A similar study for delineating the fresh water lenses in an Atoll (Ajay kumar, 1966) has also been carried out by researchers. The geoelectical cross section (GECS) for pre and the post Tsunami by Chidambaram et al (2008) (Fig 11.A and 11B) indicates that the lower resistivity values are noted in the north of Perunthottam. It is also suspected with a fresh water lens at a depth of 5-8m , with resistivity ranging from 100-60 ohm meter, below Kulaiyar and it extends to south of Palayar. During the Post tsunami it is noted that the southern part of the study area has been more contaminated by saline water, from Pumpuhar to Kulaiyar. It is also interesting to note that the lens has moved further north due to the compression by dense saline water from Thirumullaivasal towards north. The lesser value of this lens may also indicate that the saline water has also contaminated this perched aquifer.

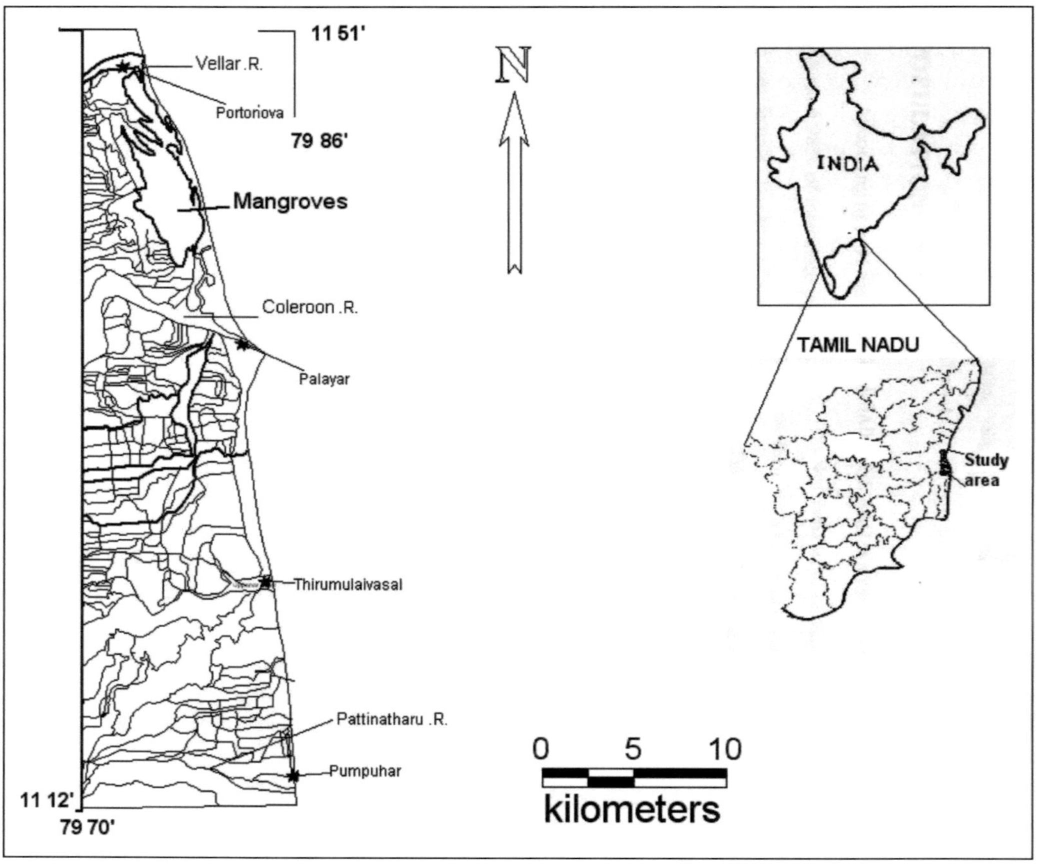

Fig. 1. Location of the study area

17. Conclusion

As the study area falls in the coastal region the water chemistry in generally expected to be saline. Hence, eleven sampling location not contaminated by Sea water intrusion were selected and the impact was noted in the shallow unconfined aquifers that too in the region where the water got stagnated and near the mouth of the river. The impact of beginning of South west Monsoon is also noted in the month of June, July and August. Few region with dilution of water chemistry is also noted after the monsoon. The gentle nature of the bathymetry has helped the tsunami wave to propagate to a greater distance inland through the river mouth and the distributary channels. The study area basically has regions with backwater recharge, impact of aquaculture, salt pans, sea water intrusion and paleo connate waters. A detailed study has to be done by using the isotope data and by biomarker approach to identify the aquifer vulnerability regions. This will help us to take necessary preventive measures in case of future calamities.

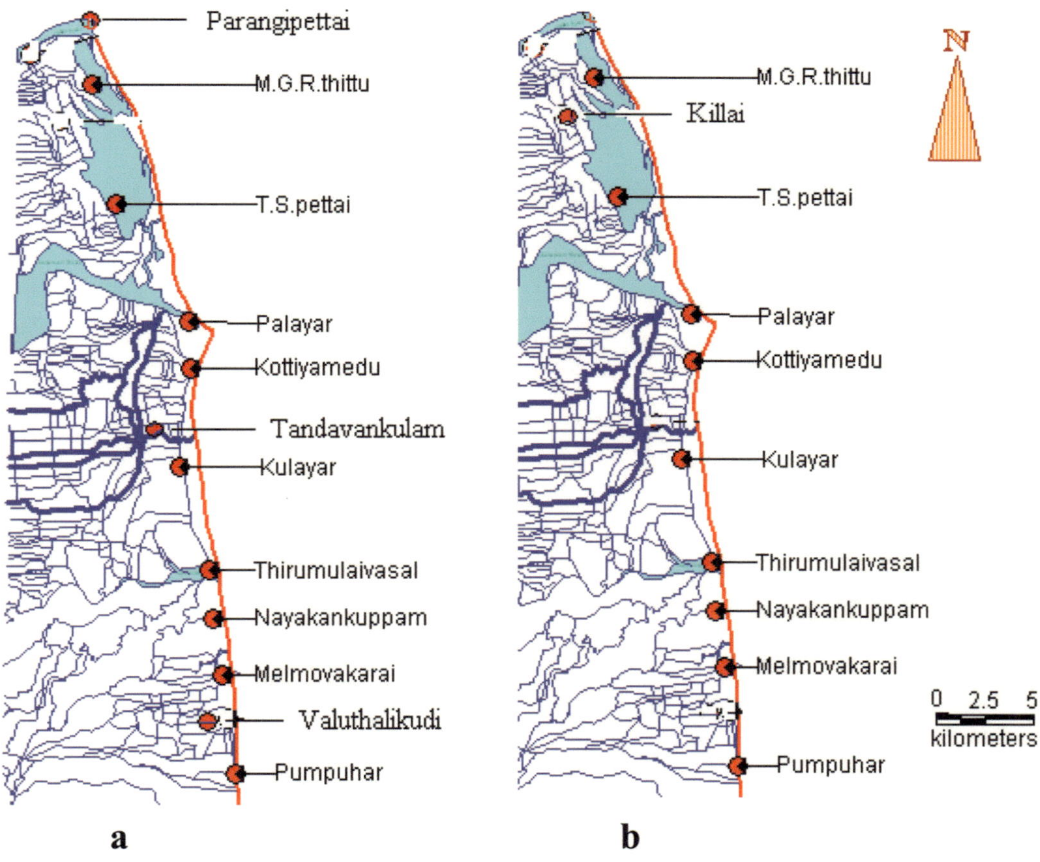

Fig. 2. Sampling location of the study area a) water sample location b) resistivity sample location

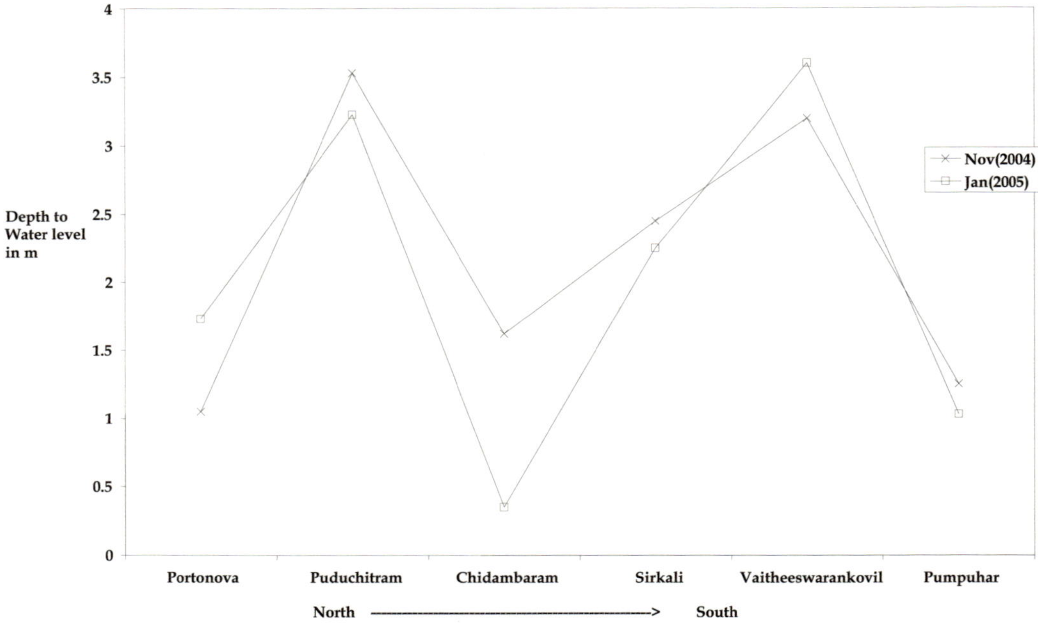

Fig. 3. Waterlevels of the study area for Pre and Post Tsunami

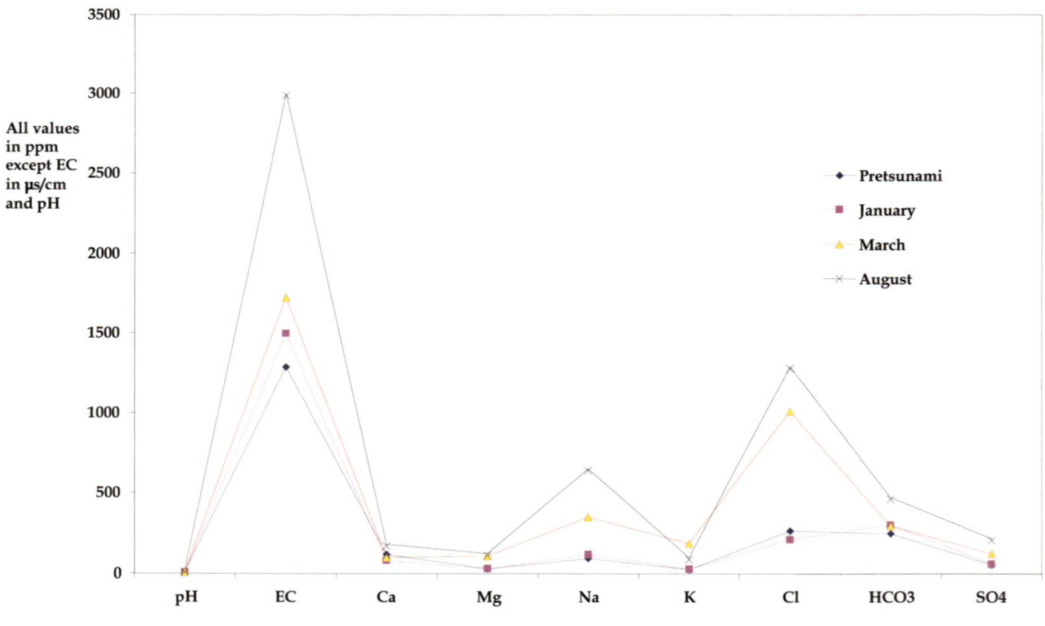

Fig. 4. Short term variation of chemical parameters after tsunami

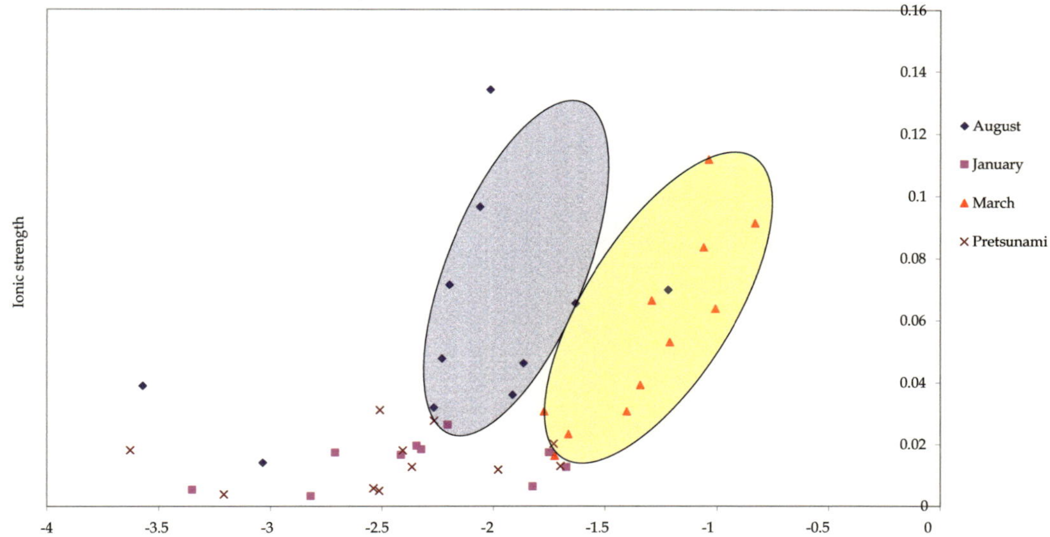

Fig. 5. Ionic strength Vs Log pCO

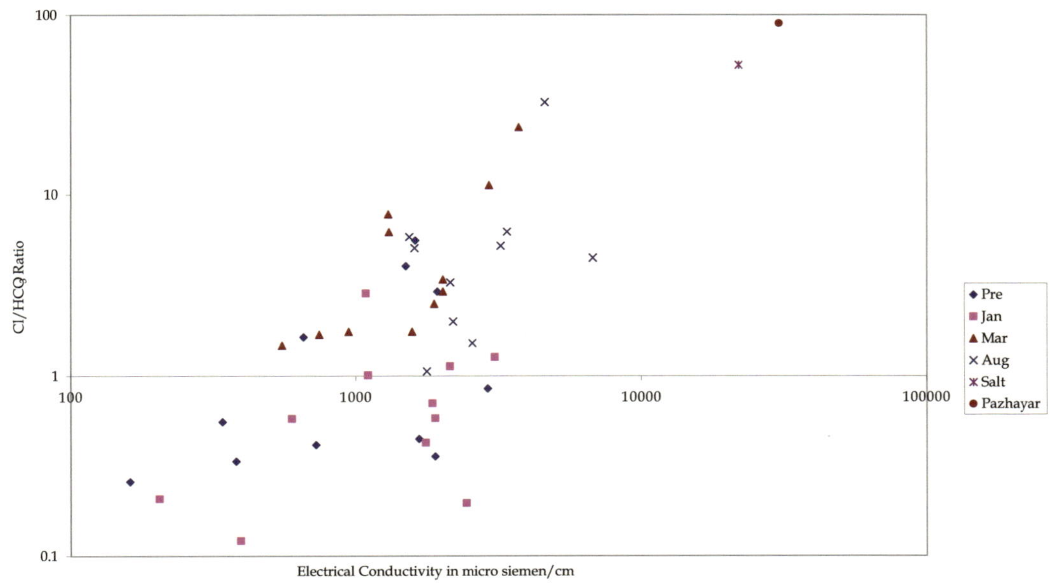

Fig. 6.1. Realtionship of the conductivity to Chloride bicarbonate ratio

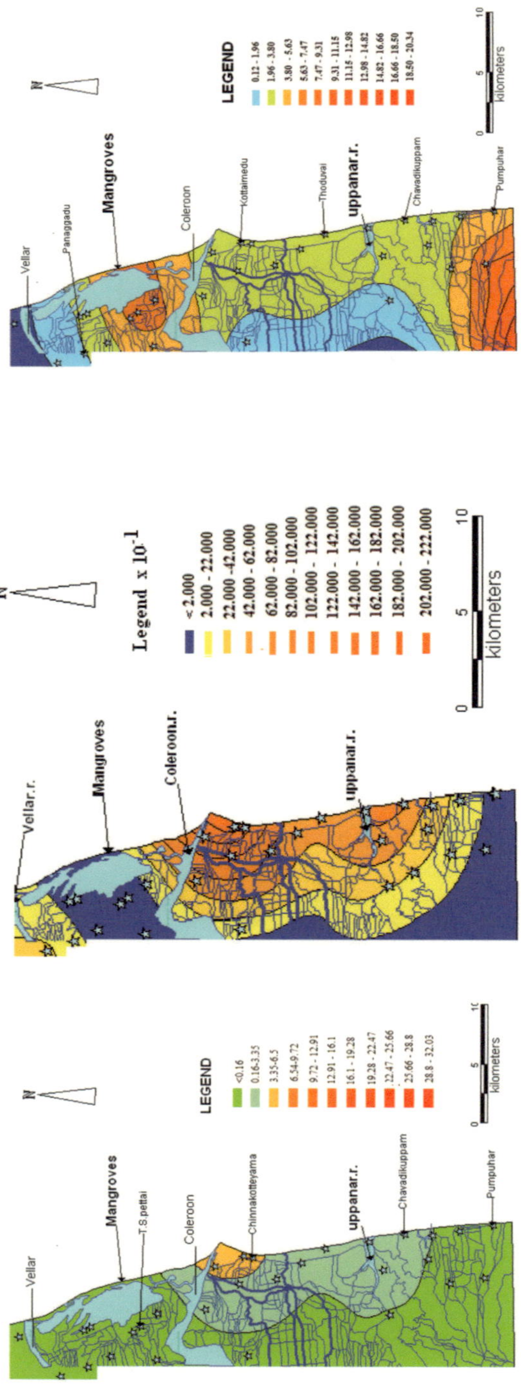

Fig. 6.2. Spatial distribution of Cl/HCO_3 ratio during a) January b) March and c)August, 2005

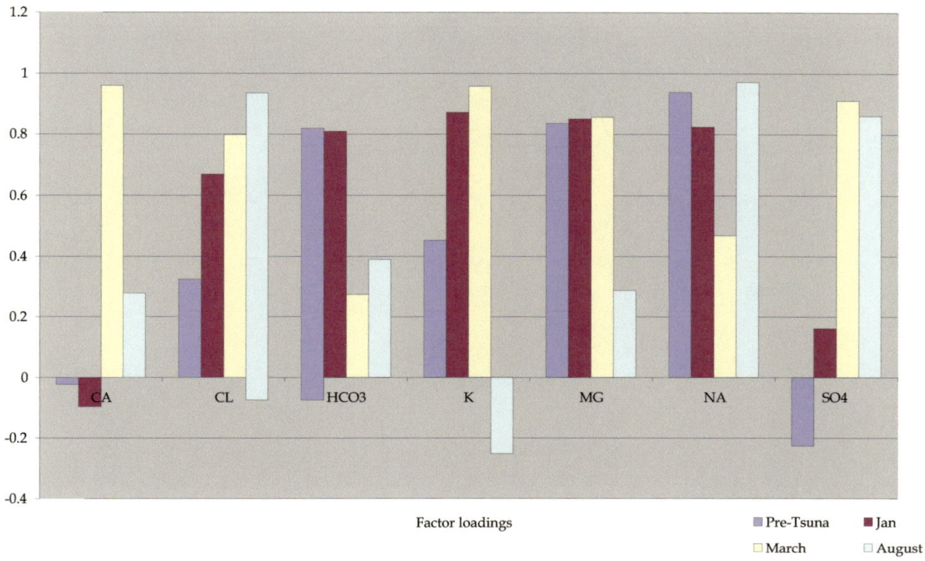

Fig. 7a. Represetation of First factor

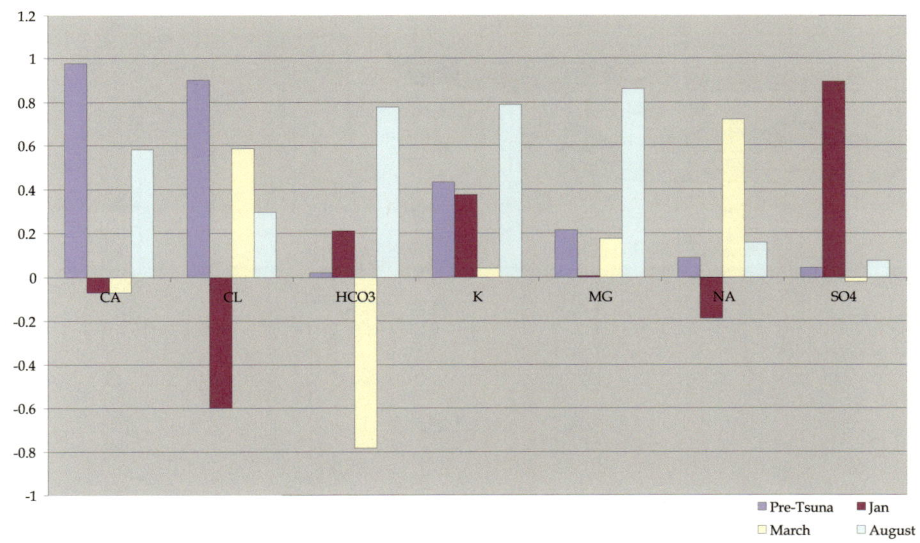

Fig. 7b. Representation of Second factor

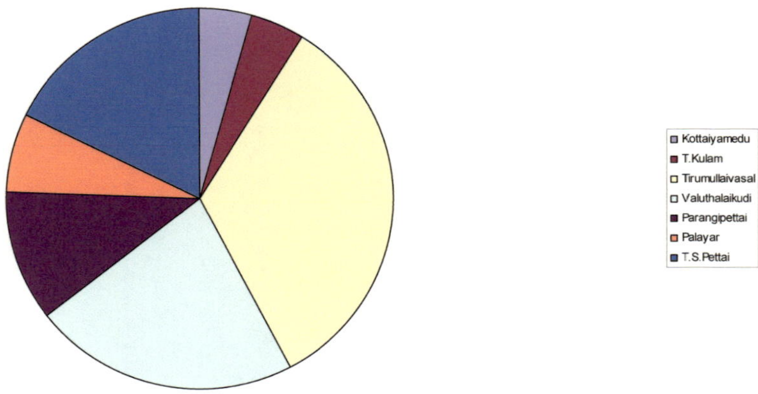

Fig. 8. Mixing proportion of sea

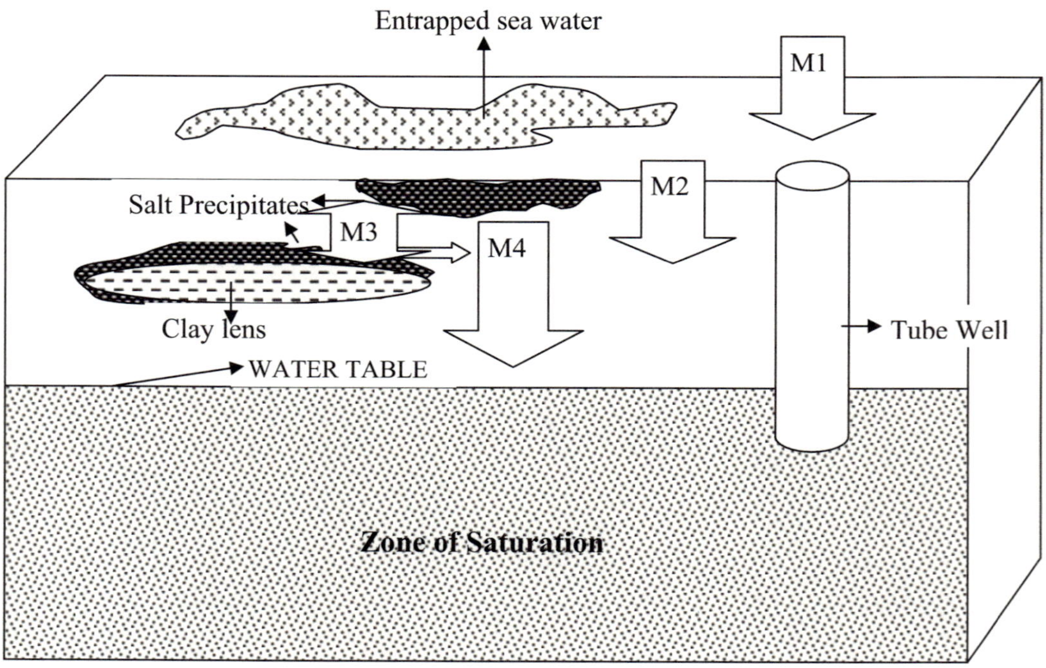

Fig 9. A Schematic representation of the possible factors for the hydrogeochemical change, after the Tsunami (Chidambararam et al 2010).

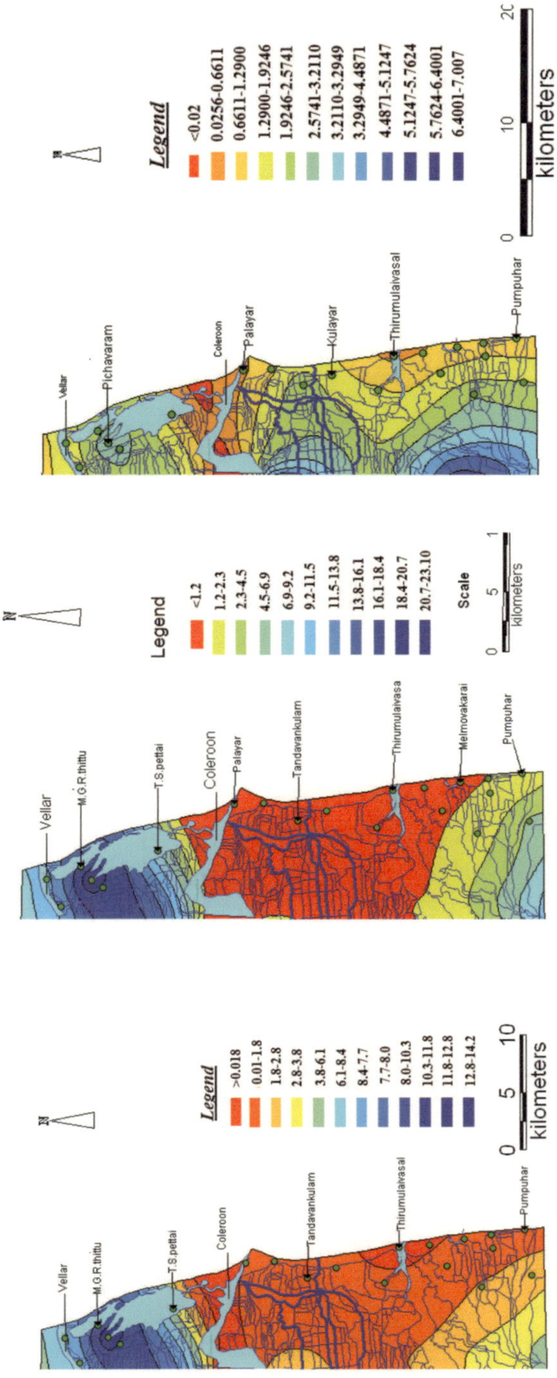

Fig. 10. Spatial distribution of apparent resistivity during a) 3m b) 5m and c)10m after Tsunami

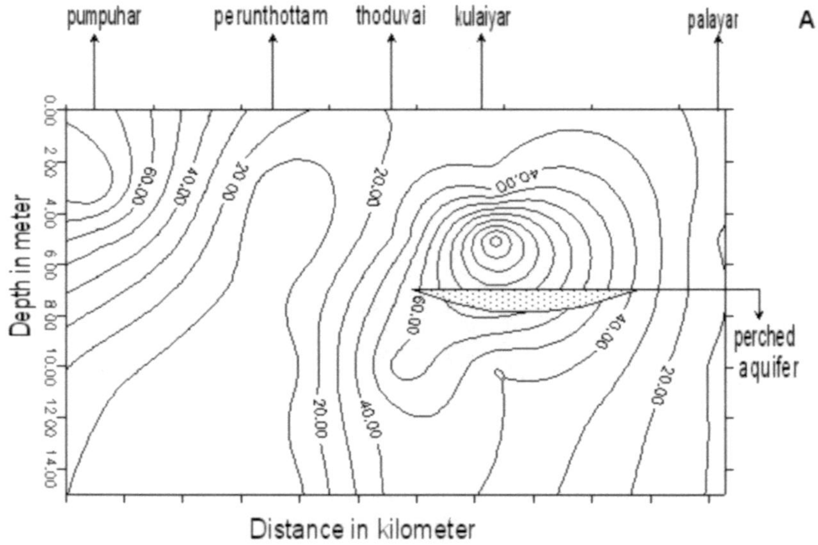

Fig. 11a. Nature of Perched aquifer during Pre-Tsunami (Chidambaram et al,2008)

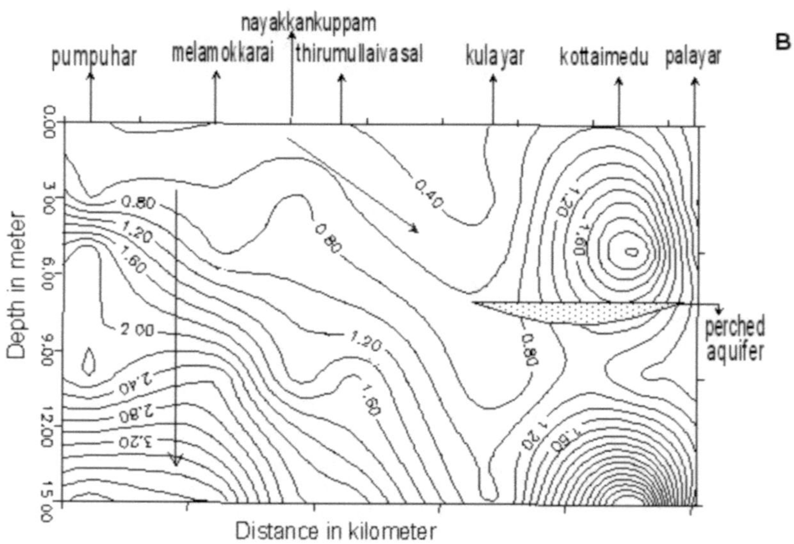

Fig. 11b. Nature of Perched aquifer during Post-Tsunami (Chidambaram et al,2008)

S. No	Authors Year	Area	Remark
1	Rai et al (2002)	Kanpur	Earthquake future
2	Scheffers and Kelletat (2003)	World wide	Tsunamigenic Sediments
3	Jain et al (2004)	General note on civil Engineering structures	Effects of M 9 Sumatra earthquake and tsunami
4	Cisternas et al (2005)	Chile	Predecessors of the giant 1960 Chile earthquake
5	Villholth et al. (2005)	Sri lanka	EC level variation
6	Kawata (2005)	Srilanka	FieldSurveyaroundGalle,SriLanka
7	Pal (2005)	India	Life loss due to tsunami
8	World Health Organization (2005)	India	Piped in water supplies and groundwater pump
9	Muraluideran et al. (2005)	Hyderabad	Earthquake had induced hydrological changes
10	Kanakasabai and Rajendran (2005)	Portonova India	Micrometeorological parameters
11	Sato(2005)	Srilanka	field investigations carried out along the west and south coasts ofSriLanka
12	Szczucinski et al. (2005)	Thailand	Tsunamigenic Sediments
13	Narasimha et al.(2005)	Nagapattinam,Tamilnadu	Six months after the tsunami.
14	Narayana et al. (2005)	Kerala coast, India	Tsunamigenic Sediments
15	Cummins and Leonard (2005)	General note	Boxing Day 2004 Tsunami
16	Altaff (2005)	Chennai, India	Impact of tsunami
17	Thangadurai (2005)	Ennore,Chennai,India	Tsunamigenic Sediments
18	Chadha et al.(2005)	East coast of India	Study on the Impact on the east coast of India
19	Nagendra et al(2005)	Nagapattinam coast	Tsunamigenic Sediments
20	Liu et al.(2005)	Srilanka	Observations by the international tsunami survey team in Sri Lanka.
21	Willams e al(2005)	Washington	Multiple sources for late-Holocene tsunamis at Discovery Bay, Washington State
22	Arya et al. (2006)	India	Some aspects of tsunami impact and recovery in India
23	Hawkes et al (2006)	Malaysia-Thailand	Sediments deposited by the 2004 Indian Ocean tsunami along the Malaysia–Thailand Peninsula
24	Nanayama and shigeno (2006)	Southwest Hokkaido.	Inflow and outflow facies from the 1993 tsunami in southwest Hokkaido
25	Goff (2006)	Srilanka	field survey after the 2004 Indian Ocean tsunami

S. No	Authors Year	Area	Remark
26	Kench et al. (2006)	Maldives	Geological effects of tsunami on mid-ocean atoll islands: the Maldives before and after the Sumatran tsunami.
27	Palanivelu et al. (2006)	Chennai India	TDS variation
28	Illangasekare et al.(2006)	Srilanka	Impacts of the 2004 tsunami on groundwater resources in SriLanka
29	Alpa Sheth et al (2006)	Indian Ocean	Tsunami on the Indian Mainland Earthquake Spectra
30	Richmond, et al (2006)	Indonesia, Sri Lanka, and The Maldives	Geologic Impacts Of The 2004 Indian Ocean Tsunami
31	Rasheed et al (2006)	South Kerala Coast and West Coast India	Tsunami Impacts On Morphology Of Beaches
32	Pilapitiya et al(2006)	Srilanka	Effect of tsunami on waste management in Sri Lanka.
33	Singarasubramanian et al (2006)	Tamilnadu coast, India	Tsunamigenic Sediments
34	Mastronuzzi et al (2006)	impacts on coastal environment	Risk assessment of catastrophic waves
35	Weiss (2006)	Indonesia	Numerical modelling of generation, propagation and run-up of tsunamis caused by oceanic impacts: model strategy and technical solutions
36	Chidambaram et al (2006)	Portnovo to Poompuhar,Tamilnadu, India	Impact of Post Tsunami in groundwater (water chemistry)
37	Rengalakshimi (2007)	Nagappattinam District	Reclamation and status of tsunami damaged soil in Nagappattinam District
38	Jaffe and Gelfembuner (2007)	Phuket, Thailand	A simple model for calculating tsunami flow speed from tsunami deposits.
39	Stein et al (2007)	Indian Ocean	Implication for regional tectonic and subduction zones
40	Chandrasekharan et al (2008)	Nagapattinam district, Tamilnadu	Variability of soil-water quality due to Tsunami-2004 in the coastal belt
41	Ranjan et al. (2008)	Thailand	Tsunamigenic Sediments
42	Martin et al (2008)	Poompuhar,Tamilnadu	Impact of Tsunami
43	Chidambaram et al. (2008)	Portnovo to Poompuhar,Tamilnadu, India	Impact of Pre and Post Tsunami in groundwater (Geoelectrical techniques)
44	Pierre leclerc et al. (2008)	Ampara district in Eastern Sri Lanka:	Tsunami impact on shallow groundwater, Conductivity measurements and qualitative interpretation
45	Sangeeta sonatz et al (2008)	India	Green reconstruction of the tsunami-affected areas in India using the integrated coastal zone management concept
46	Ravisanker and Pongathai(2008)	Sirkazhi taluk, Nagapattinam district, Tamilnadu	A study on groundwater quality

S. No	Authors Year	Area	Remark
47	Srinivasalu(2008)	South east coast of India	Tsunamigenic Sediments
48	Singh (2008)	Neill Island (SouthAndaman)	Impact of the earthquake and tsunami on the groundwater regime
49	Takashi et al.(2009)	Nagapattinam district	Impact of the tsunami on soil, groundwater and vegetation
50	Vithanage et al.(2009)	Eastern sri lanka	Effect of the Indian ocean tsunami on groundwater quality in coastal aquifers
51	Singarasubramanian et al (2009)	Tamilnadu coast, India	Geomorphology and Sedimentological
52	Rajesh et al (2009)	Pichavaram, Tamilnadu coast, India	Study on the sediments Mangrove environment after Tsunami
53	Pignatelli(2009)	Southern Apulia coastline,Italy	Evaluation of tsunami flooding using geomorphologic evidence
54	Wijetunge (2009)	East-Coast of Srilanka	Field measurements and numerical simulations of the 2004 tsunami impact on the east coast of Sri Lanka
55	Chidambaram et al (2010)	Portnovo to Poompuhar,Tamilnadu, India	Impact of Pre and Post Tsunami in groundwater (water chemistry)

Table 1. List of references for the related to Tsunami

	Pre Tsunami			January		
	Maximum	Minimum	Average	Maximum	Minimum	Average
pH	8.8	7	7.630769	8.06	6.95	7.522727
EC	2897	161	1290.923	3060	204	1497.948
Ca	340.68	28.99	117.1231	227.66	2.72	80.65909
Mg	72.16	7.98	31.04692	93	1.7	30.92818
Na	280.19	11.34	93.31923	377.4	10.2	118.8418
K	48.56	1.61	25.42462	54.4	2.04	29.22182
Cl	744.42	26.7	264.8938	538.9	30.13	210.7218
HCO_3	549.12	44.18	247.0108	693.11	51.85	302.5645
SO_4	110.24	16.9	53.49308	125.8	7.82	59.83818
	March			August		
	Maximum	Minimum	Average	Maximum	Minimum	Average
pH	7.62	7.11	7.324545	8.1	7.1	7.563636
EC	3718.63	548.87	1721.518	6743.839	1532.123	2988.282
Ca	184.34	40.75	98.42818	311.99	60	170.3609
Mg	189.85	34.69	107.92	304.8	16.8	112.7455
Na	574.71	206.9	348.9636	1837	69.7	641.8909
K	426.23	56.54	186.2827	230.7	4	87.35455
Cl	1772.42	431.31	1011.051	3217.09	177.25	1281.854
HCO_3	610.13	75.05	293.9991	1043.1	24.4	463.4
SO_4	296.18	38.13	123.6609	507.5	100.5	206.1818

Table 2. Maximum, Minimum and Average of the ion concentration all values in mg/l except EC in ms/cm and pH

S.No	Location	Fe	Mn	Al
M1	Thirumullaivasal	0.7231	1.2036	0.0623
M2	Melmokkarai	0.2931	0.0532	0.0625
M3	Pumbuhar	0.3214	0.6231	0.0321
M4	Vazhuthalaikudi	0.3897	0.7963	0.0432
M5	Palayar	0.456	1.521	0.0487
M6	Tandavankulam	0.6543	0.2356	0.0895
M7	Kulayar	0.3	0.0658	0.0823
M8	T.S.Pettai	0.3012	0.2589	0.0412
M9	Portonovo	0.4	0.1258	0.0564
M10	Kottaiyamedu	0.2136	0.4836	0.0512
M11	Nayakarpalayam	0.1256	0.456	0.0426
A1	Thirumullaivasal	0.5283	0.1065	0.0721
A2	Melmokkarai	0.3698	0.0658	0.0521
A3	Pumbuhar	0.036	0.0092	0.042
A4	Vazhuthalaikudi	0.3789	0.8357	0.0456
A5	Palayar	0.523	1.867	0.059
A6	Tandavankulam	0.5894	0.1236	0.0658
A7	Kulayar	0.302	0.0595	0.0835
A8	T.S.Pettai	0.2896	0.2136	0.0587
A9	Portonovo	0.4557	0.186	0.0428
A10	Kottaiyamedu	0.5012	0.2365	0.0563
A11	Nayakarpalayam	0.527	0.1319	0.05
R1	Palayar Sea water	0.034	0.06	0.003

Table 3. Heavy metals for post-tsunami

State	Incursion of water into the land	Average height of waves	Average intensity of Tsunami
Andhra Pradesh	500m to 2Km	2m to 5m	1 to 2.32
Kerala	1Km to 2Km	3m to 5m	1.64 to 2.32
Pondicherry	300m to 3Km	8m to 10m	3 to 3.32
Tamil Nadu	1Km to 1.5Km	7m to 10m	2.80 to 3.32

Table 4. Intensity of Tsunami along the eastern coastal states of India

S. No	Location	I Layer		II Layer		III Layer		Curve type
		Resistivity (Ωm)	Thickness (m)	Resistivity (Ωm)	Thickness (m)	Resistivity (Ωm)	Thickness (m)	
1	Kuliyar	1.24	2	1.25	5	1.28		A
2	M.G.R Thittu	16	2	240	30	96		K
3	Kavarapattu	8	2	10.5	1	2.23	∞	K
4	Perunthotum	2.2	1	1.8	5	1.44	∞	K
5	Palayar	1.6	1	0.5	7	2.25		H
6	Killar	2.52	2	1.43	5	23		H
7	Kulaiyar	28	2	1.33	4	27		H
8	T.S pettai	5.82	1	2.84	2	3.56		H
9	Thoduvai	180	3	36	5	35	9	Q
10	Pumpuhar	75	2	37.5	8	2.625	3	Q

Table 5.1. Resistivity survey values - Pre-tsunami (Chidambaram et al 2008)

S. No	Location	1st Layer		2nd Layer		3rd Layer		Curve type
		Thickness (m)	Resitivity (Ωm)	Thickness (m)	Resitivity (Ωm)	Thickness (m)	Resitivity (Ωm)	
1	Kulaiyar	5	0.8	25	1.8	∞	50	A
2	Thirumullaivasal	2	1.34	4	2.1	∞	23	A
3	Killai	5	3	35	120	∞	360	A
4	T.S.Pettai	10	1.15	18	5.2	∞	12.8	A
5	Melmuvakkarai	1	2.4	2.5	9.6	∞	48	A
6	Puthukkuppam	2	1.2	26	2.4	∞	48	A
7	Portonova	1	4.6	2.5	12.3	9.8	25.3	A
8	Thandavankulam	1	2.2	8	6.6	11.2	14.2	A
9	M.G.R Thittu	5	1.9	9.5	5	∞	2.3	K
10	Poombugar	1	2.78	6	4.89	∞	2.86	K
11	Kottayamedu	5	2.6	55	6.5	∞	3.75	K
12	Nayakkan Kuppam	1	3.35	8	1.7	22	6.1	H
13	Palaiyar	5	7.5	35	5	∞	15	H
14	Manampadi	5	1.4	5.5	1.2	∞	0.8	Q
15	Kilperumpalam	1	20.4	3	10.2	8	1.53	Q
16	Perunthotam	1	1.8	16	1.4	∞	0.244	Q
17	Thiruvenkadu	1	3.2	3	2.25	33	1.1	Q
18	Neithalvasal	1	42	16	2.52	∞	0.23	Q
19	Vazhuthalaikudi	1	10.1	3	6.2	7	2.57	Q
20	Pitchavaram	1	20.4	4	10.4	7	5.1	Q

Table 5.2. Resistivity survey values - Post-tsunami (Chidambaram et al 2008)

Category	Formation Resistivity			Formation Factor		
		Pre(10)	Post(20)		Pre(10)	Post(20)
Fresh	>300	0	0	>0.50	0	0
Slightly Fresh	150 - 300	0	0	0.25 – 0.50	0	0
Moderate	50 - 150	3	2	0.10 – 0.25	0	0
Slightly Saline	25 - 50	2	0	0.05 – 0.10	0	0
Saline	< 25	5	18	<0.05	6	20

Table 6. Variation in Formation Resistivity and formation factor, before and after tsunami

18. References

ABDUL RASHEED , A., KESAVA DAS, V., REVICHANDRAN, C., VIJAYAN P. R. and TONY THOTTAM, J.(2006) Tsunami impacts on morphology of beaches along south Kerala coast, west coast of India. *science of tsunami hazards*, vol 24, no. 1, page 24-34.

AJAYKUMAR VARMA ,R. and RAMACHANDRA K,K. (1996) Resstvity survey for describing the fresh ater lenses of Agatti Atoll Lakshadweep, *India.Jour;Assoc. Expl. Geophys*,28 .pp.11-16

ALPA SHETH, SNIGDHA SANYAL, ARVIND JAISWAL, and PRATHIBHA GANDHI (2006) Effects of the December 2004 Indian Ocean Tsunami on the Indian Mainland, *Earthquake Spectra*, Volume 22, No. S3, pages S435–S473.

ALTAFF, K. (2005) Impact of tsunami on meio fauna of Marina Beach, Chennai, India *CURRENT SCIENCE*, VOL. 89, NO. 10,25.

AMBRASEYS, N.N (1962) Data for the investigation of seismic sea waves in the eastern Mediterranean. *Bulletin Seismological Society Am*. Volume 52. Pages 895-913

APHA (1998). Standard methods for the examination of water and waste water (19th ed.). *APHA, Washington*, DC: USASS

ARYA, A.S., MANDAL, G.S., MULEY, E.V. (2006) Some aspects of tsunami impact and recovery in India. *Disaster Prevention and Management* 15(1), pp.51–66.

BRUCE M. RICHMOND, BRUCE E. JAFFE, GUY GELFENBAUM AND ROBERT, A., MORTON, Z. and GEOMORPH, N.F. (2006) Geologic Impacts of the 2004 Indian Ocean Tsunami on Indonesia, Sri Lanka, and the Maldives. *Suppl.*-Vol. 146.pp. 235 - 251

CHADHA, R.K., LATHA, G., YEH, H., PETERSON, C., KATADA, T.(2005) The tsunami of the great Sumatra earthquake of M 9.0 on 26 December 2004—Impact on the east coast of India. *Curr. Sci.* 88 (8), pp.1297–1301.

CHANDRASEKHARANA,H., SARANGIA, A., NAGARAJANB, M., SINGHA, V.P., RAOA,D.U.M ., STALINC,P., NATARAJANC, K., CHANDRASEKARANC,B., ANBAZHAGAND, S. (2008) Variability of soil-water quality due to Tsunami-2004 in the coastal belt of Nagapattinam district, Tamilnadu .*Journal of Environmental Management* 89.pp. 63–72.

CHIDAMBARAM,S. PRASANNA,MV., RAMANATHAN, AL., ANANDHAN,P., SRINIVASAMOORTHY, K., LOGANATHAN. AND SENTHILKUMAR,G (2006) Impact of Tsunami on coastal grndater-A case study from Portnova to Pumpuhar,Tamilnadu. *Ecochronicle* Vol.1,N0.2,pp.73-76

CHIDAMBARAM,S., RAMANATHAN,A.L., PRASANNA,M.V., LOGNATHAN,D., BADRINARAYANAN,T.S.,SRINIVASAMOORTHY,K. and ANANDHAN,P. (2008) Study on the impact of tsunami on shallow groundwater from Portnova to Pumpuhar, using geoelectrical technique-south east coast of India. *Indian journal of marine sciences*, Vol 37(2),pp.121-131.

CHIDAMBARAM,S.,RAMANATHAN,AL. and PRASANNA,MV. (2010) Study on the hydrogeochemical characterstics in groundwater post and pre-tsunami scenario,from port novo to poomphuhar,South east coast of India,Environmental monitoring assessment,169:553-568

CISTERNAS, M., ATWATER, B.F., TORREJON, F., SAWAI, Y., MACHUCA, G., LAGOS, M., EIPERT, A., YOULTON, C., SALGADO, I., KAMATAKI, T., SHISHIKURA, M., RAJENDRAN, C.P., MALIK, J.K., RIZAL, Y. and HUSNI, M. (2005) Predecessors of the giant 1960 Chile earthquake. *Nature* 437 (7057), pp.404–407.

COASTA PJM LEROY SAG, KERSHAW, S. and DINIS, J. Tsunamis causes behavior and sedimentary signature. Studies on Ad 1755 (Portugal) Intsities and magnitude scale for tsunami http:/tierra.rediris:es/public papers/tsunami pdf

CUMMINS AND LEONARD. (2005) The boxing day 2004 tsunami —a repeat of 1833? issue 77

DOMENICO, P.A and SCHWARTZ,W.(1990) Physical and Chemical hydrogeology. John Wiley, New York,824p

GOFF, J. LIU, PL., HIGMAN, B., MORTON, R., JAFFE, B.E., FERNANDO, H., LYNETT, P., FRIETZ, H., SYNOLAKES,C.,FERNANDO,S. (2006) Sri Lanka field survey after the 2004 Indian Ocean tsunami. *Earthquake Spectra* 22 (S3), S155–S172.

HAWKES,A., BIRD,M., COWIE, S.,GRUNDY-WARR, C., HORTON, B., TAN SHAU HWAI, A., LAW, L., MACGREGOR, C., NOTT, J., EONG ONG, J., RIGG, J., ROBINSON, R., TAN-MULLINS,M., TIONG SA, T., YASIN, Z. and WAN AIK, L. (2007) Sediments deposited by the 2004 Indian Ocean tsunami along the Malaysia–Thailand Peninsula. Mar. Geol. 242, 169–190.

HEM (1961). Calculaton and Use of Ion activity *U.S Geological survey Water Supply* pp 1532-Cp17

HOLLAND HD (1978) The chemistry of the atmosphere and Ocean .*Wiley Inter Sciences* 351p,

HORI, K., KUZUMOTO, R., HIROUCHI, D., UMITSU, M., JANJIRAWUTTIKUL, N. and PATANAKANOG, B.(2007) Horizontal and vertical variation of 2004 Indian tsunami deposits: an example of two transects along the western coast of Thailand. *Mar. Geol.* 239, pp. 163–172.

ILLANGASEKARE, T., TYLER, S.W., CLEMENT, T.P., VILLHOLTH, K.G., PERERA, A.P.G.R.L., OBEYSEKERA, J., GUNATILAKA, A., PANABOKKE, C.R., HYNDMAN, D.W., CUNNINGHAM, K.J., KALUARACHCHI, J.J., YEH,W.W.-G., VAN GENUCHTEN, M.T. and JENSEN, K. (2006) Impacts of the 2004 tsunami on groundwater resources in Sri Lanka. art. no. W05201.*WaterResour*. Res. 42 (5).

JAFFE, E.B. and GELFENBAUM, G. (2007) A simple model for calculating tsunami flow speed from tsunami deposits. *Sediment. Geol.* 200, pp.347-361.

JEAN-PIERRE LECLERCA, CHRISTOPHE BERGERA, ANTHONY FOULONB, REMI SARRAUTEB and LAURIANE GABETB (2008) Tsunami impact on shallow groundwater in the Ampara district in Eastern Sri Lanka: Conductivity measurements and qualitative interpretations J.-P. LeClerc et al. / Desalination 219 pp. 126-136

KAWATA,Y.,IMAMURA,F.,TOMITA,T.,ARIKAWA,T.,YASUDA,T (2005) TheDecember26,2004SumatraEarthquakeTsunami:FieldSurveyaroundGalle,SriLanka./http://www.drs.dpri.kyotou.ac.jp/sumatra/srilanka/galle_survey_e.htmlS.

KENCH, P.S., MCLEAN, R.F., BRANDER, R.W., NICHOL, S.L., SMITHERS, S.G.,FORD, M.R., PARNELL, K.E. and ASLAM, M.(2006) Geological effects of tsunami on mid-ocean atoll islands: the Maldives before and after the Sumatran tsunami. *Geology* 34 (3), pp 177-180.

LIND(1970) Specific Conductance as a means of estimating ionic strength,U.S.Geological Society,Professional paper,700-D,P.D272-280

LIU, P.L.-F., LYNETT,P., FERNAND,H., JAFFE,B.E., FRITZ,H., HIGMAN,B., MORTON,R., GOFF,J., SYNOLAKES,C.. (2005) Observations by the international tsunami survey team in Sri Lanka. *Science* 308, 1595

MARTIN DEVA PRASATH,P. and HIDAYATHULLA KHAN,T. (2008) Impact of Tsunami on the Heavy Metal Accumulation in Water, Sediments and Fish at Poompuhar Coast, Southeast Coast of India. ISSN: 0973-4945; CODEN ECJHAO,E-*Journal of Chemistry* Vol. 5, No. 1, pp. 16-22.

MASTRONUZZI, G. and SANSÒ, P.(2006). Risk assessment of catastrophic waves impacts on coastal environment. *Geogr. Fis. Din. Quat.* 29 (1), 83-91.

METHTHIKA VITHANAGE, KAREN G. VILLHOLTH3,KUSHANI MAHATANTIL, PETER ENGESGAARD1 AND KARSTEN H. JENSEN(2009) Effect of the indian ocean tsunami on groundwater quality in coastal aquifers in eastern sri lanka. *Science of Tsunami Hazards*, Vol. 28, No. 3, page 219 (2009)

MONTRI CHOOWONG, NAOMI MURAKSHI.,KEN-ICHIRO HISADA.,PUNYA CHARUSIRI., THASINEE CHAROENTIRITRIAT., VICHI CHUTAKOSIKANON., KRUAWAN JANKAEW., PITSANPONG KANJANAPAYANT AND SUMET PHANTUWONGRAJ (2004) Indian Ocean tsunami inflow and outflow at Phuket, Thailand. *Marine Geology* 248 (2008) 179-192

NAGENDRA, R., KAMALAK KAMMAN, B.V., SAJITH, C., SEN, G., REDDY, A.N., SRINIVASALU, S. (2005) A record of foraminiferal assemblage in tsunami sediments along Nagappattinam coast, Tamil Nadu. *Curr.Sci.* 89 (11).

NANAYAMA, F. and SHIGENO, K.(2006) Inflow and outflow facies from the 1993 tsunami in southwest Hokkaido. *Sediment. Geol.* 187, pp.139-158.

NARASIMHA, R., MENON, S., BHATTACHARYA, P., RAJANI, M.B.(2005). Six months after the tsunami. *Curr. Sci.* 89 (9), pp.1459-1461

ORELLANA, E. and MOONEY, H.M.(1996) Master table and curves for vertical electrical sounding over layered structures, Interception. Madrid, Spain,

PIGNATELLI ,C., SANSÒ,P. and MASTRONUZZI,G. (2009) Evaluation of tsunami flooding using geomorphologic evidence. *Marine Geology* 260. pp.6-18

PILAPITIYA S., VIDANAARACHCHI C. and YUEN S. (2006). Effect of tsunami on waste management in Sri Lanka. *Waste Management*. 26, pp.107-109

PRASANNA, M.V., CHIDAMBARAM ,S., SHAHUL HAMEED, A. and SRINIVASMOORTHY, K. (2009) Study of evaltion of groundwater in Gadilam Basin usng hydrogeochemical and stope data. *Environmental monitoring and assessments*, DOI 10.1007/s 10661-009-1092-5

RAI, D. C. AND MURTY, C. V. R., Reconnaissance Report on North Andaman (Diglipur) Earthquake of 14 September 2002, Department of Civil Engineering, Indian Institute of Technology Kanpur, Kanpur, 2003, p. 49. Available a www.nicee.org/EQ_ reports/Andaman/Andaman_report.htm). Inadequate seating of bridge deck over piers and abutments is a serious concern for its safety during a stronger earthquake in future.

RAJESH KUMAR RANJAN., AL RAMANATHAN., GURMEET SINGH and CHIDAMBARAM,S. (2008), Environment monitoring and assessment, Vol.147, Issue:1-3, Pp.389-411. Assessment of metal enrichments in tsunamienic sediments of Pichavaram mangroves, southeast coast of India

RAVISANKAR, N., POONGOTHAI, S., (2008) A study on groundwater quality in Tsunami affected areas of Sirkazhi taluk, Nagapattinam district, Tamilnadu, India. *Ecol.Env. & Cons.* 14 (2-3), pp.403–408.

RAYMAHASHAY,B.C (1986) Geochemistry of bicarbonate in river water,rnalf. *Geologcal Society of India*,V.27,pp.114-118.

RENGALAKSHMI, R., SENTHILKUMAR, R., SELVARASU, T., THAMIZOLI, P. (2007) Reclamation and status of tsunami damaged soil in Nagappattinam District, TamilNadu. *Curr. Sci.* 92 (9), pp.1221–1223

RIJKSWATERSTAAT. (1975) Standard graphs for Resistivity prospecting, published by *European Association of Exploration Geophysicists*, the Netherlands,

SANGEETA SONAK., PRAJWALA PANGAM., ASHA GIRIYAN (2008) Green reconstruction of the tsunami-affected areas in India using the integrated coastal zone management concept. *Journal of* Environmental Management 89. pp.14–23

SATO,S.,KOIBUCHI,Y.,HONDA,T.,WELHENA,T.,RANASINGHE,S.(2005) Tsunamion26 December 2004:fieldinvestigationscarriedoutalongthewestandsouth coasts ofSriLanka. /http://www.drs.dpri.kyoto-u.ac.jp/sumatra/srilanka-ut/ SriLanka_UTeng.htmlS.

SHIBAYAMA, T., AKIO OKOYASU., JUN SASAKI,NIMAL WIJAYARATNA.,TAKAJUKI SUZUKI., RAVINDRA JAYARATNE, MASIMIM, ZOUHRAWATY ARIFF,RYO MATSUMANU. (2006) Disaster survey of Indian Ocean tsunami in south coast of Sri Lanka and Ache, Indonesia. In: Proceedings of the 30th International Conference Coastal Engineering, San Diego, *ASCE*, 1469–1476.

SINGARASUBRAMANIAN, S.R (2006). Sediments characteristics of the M9 tsunami event between Rameswaram and Thoothukudi, Gulf of Manner, Southeast coast of India. International. *Jour. Of Sciences of Tsunami hazards*.V.25No.3. PP.160-172

SINGARASUBRAMANIAN, S.R (2009) Geomorphology and Sedimentological changes during and after the December-2004 Indian Ocean Tsunami near the Vellar river and the M.G.R island area of the central Tamilnadu coast. International. *Jour. Of science of Tsunami hazards* V.28.No.1.PP.67-74

SINGH, V.S.(2008) Impact of the earthquake and tsunami of December 26, 2004, on the groundwater regime atNeill Island (SouthAndaman). *J. Environ.Manag.* 89 (1), pp.58–62.

SOLOVIEV,S.L (1970) Recurrence of ''Tsunamis in pacific Ocean ''W.M Adams, Edition, Honolulu. pages .pp.149-164

SRINIVASALU, S., THANGADURAI, N., JONATHAN, M.P., ARMSTRONG-ALTRIN, J.S., AYYAMPERUMAL, T., RAM-MOHAN, V.(2008) Evaluation of trace-metal enrichments from the 26 December 2004 tsunami sediments along the Southeast coast of India. *Environ. Geol.* 53, pp.1711–1721.

STEIN, S., OKAL, E.A. (2007). Ultra-long period seismic study of the December 2004 Indian Ocean earthquake and implications for regional tectonics and the subduction process. *Bull. Seism. Soc. Am.* 97, S279–S295.

STUMM WERNER and MORGAN JJ. (1996) Aquatic chemistry. *John Wiley and sons Inc*, New York,1022p.

SUDHIR K. JAIN, C. V. R. MURTY, DURGESH C. RAI, JAVED N. MALIK, ALPA SHETH and ARVIND JAISWAL. (2004) Effects of M 9 Sumatra earthquake and tsunami of 26 December *CURRENT SCIENCE*, VOL. 88, NO. 3, pp.357-359

TAKASHI KUMEA, CHIEKO UMETSU A, K. PALANISAMI. (2009) Impact of the December 2004 tsunami on soil, groundwater and vegetation in the Nagapattinam district, *India Journal of Environmental Management* 90 3147–3154

THANGADURAI N, SRINIVASALU, S., JONATHAN, MP., RAJESHWARA RAO, N. and SANTOSH KUMAR, R. (2005) Pre tsunami chemistry of sediments along the inner continental shelf off Ennore, Chennai, south east coast of india. *Indian journal of marine sciences,* Vol. 34(3) pp 274-278.

UMITSU, M., TANAVUD, C. and PATANAKANOG, B. (2007) Effects of landforms on tsunami flow in the plains of Banda Aceh, Indonesia, and Nam Khem, Thailand. *Mar. Geol.* 242 (1–3), pp.141–153.

VAN OVERMEEREN R.A. (1987) Freshwater bearing sandy creek beds explored by electromagnetic measurements in a mainly saline coastal areas of Netherlands In; Garland GD (ed) proc 3rd Dec-ennial Int Conf Geophysical and Geochemical Exploration for Minerals and Grundwater,3pp.716-728

WEISS, R., WÜNNEMANN, K., BAHLBURG, H. (2006) Numerical modelling of generation, propagation and run-up of tsunamis caused by oceanic impacts: model strategy and technical solutions. *Geophys. J. Int.* 167 (1), pp.77–88.

WIJETUNGE, J.J.(2009) Field measurements and numerical simulations of the 2004 tsunami impact on the east coast of Sri Lanka. *Pure Appl. Geophys.* 166 (4), 593–622.

WILLIAMS, H.F., HUTCHINSON, I. and NELSON, A.R. (2005) Multiple sources for late-Holocene tsunamis at Discovery Bay, Washington State, USA. *Holocene 15* (1), pp.60–73.

XUE, Y., WU, J., YE, S., and ZHANG, Y. (2000). Hydrogeological and Hydrogeochemical studies for saltwater intrusion on the south coast of Laizhou Bay, China. *Groundwater, 38*(1), 38–45.

30

Generation and Propagation of Frequency-Dispersive Tsunami

Claudia Cecioni and Giorgio Bellotti
DSIC, University of Roma TRE
Italy

1. Introduction

Tsunami are long water waves generated by sudden disturbances of the sea floor or sea water surface, which is usually caused by earthquakes, landslides or volcanic eruptions. Once triggered the tsunami can propagate over long distances, carrying destruction even on far coasts, hours after the impulsive generating event. The tragic consequences of the tsunami occurred the 26th December 2004 in the Indian Ocean, involved the scientific community to develop models able to reproduce the tsunami generation and their evolution, with the aim of building Tsunami Early Warning Systems.
A single model is not able to treat adequately the generation, propagation and inundation phase of tsunami scenarios, because it can not be at the same time accurate and computational efficient. The tsunami generation most of the times requires the solution of the full three dimensional equations of the hydrodynamics (Grilli et al., 2002; Liu et al., 2005), in order to accurately reproduce the complex sea floor motion and therefore the consequent wave field.
A mathematical problem which solves the three dimensional equations is especially needed when tsunami are generated by landslides or small submarine earthquakes. An other important feature which has to be taken into account when modelling the tsunami generation, are the nonlinear terms, which allow the reproduction of waves with a wave height of the same order of the water depth. The nonlinear equations have to be solved when the generating seismic event occurs close to the coast in shallow water, which represents the most dangerous threat for people and structures.
When the tsunami propagates far from the generating source it can reach high celerity, of the order of thousands of m/s. Tsunamis in deep water are long waves (wave length of the order of hundreds kilometres) with wave height of the order of one meter; therefore, if compared to the water depth (thousands of metres), can be considered small amplitude water waves. The tsunami propagation phase can therefore be accurately modelled using linear equations. When the wave propagation is to be studied over large geographical areas it appears natural to apply simplified equations in order to reduce the computational costs. These are the depth-integrated equations which reduce the full three-dimensional problem to a two-dimensional one, making the resulting model applicable over oceanic length scales. The most widely equations used are the long wave equations, named also Nonlinear Shallow Water Equations (NSWE). One weak point of the NSWE is that they are not able to reproduce properly the celerity at which each component of the wave field propagates.

According to those models one single celerity ($c = \sqrt{gh}$) governs all the components. However it has been demonstrated (Kulikov et al., 2005) that even tsunamis generated by large earthquakes are wave packets whose propagation can be relevantly influenced by the frequency dispersion. In the last years it has therefore become well accepted that the Boussinesq Type Equations (Peregrine, 1967; Lynett & Liu, 2005) are the most suitable tool for the computation of tsunamis. As far as the tsunami approaches the coastal areas the wave height increases and the nonlinear terms become again not negligible. In this phase tsunami can properly be reproduced using the shallow water approximation.

In this work we present a possible approach for the simulation of tsunami over large geographical areas. The model makes use of linear equations, because it considers the wave propagation in deep water, where the wave amplitude is much smaller than the water depth. The full frequency dispersion of the waves is reproduced, which in our opinion plays an important rule in the tsunami propagation. A further point of interest is that, although developed to be used in the far field, the model seems to provide reasonable results also in the near field, and is able to reproduce the earthquake/landslide generation of the tsunamis. We believe the model to be useful for practical purposes because the computational procedure makes it suitable to support in real time a tsunami early warning system.

As it will be explained the model can be used to run a database of different tsunami scenario and then to elaborate in real time the tsunami waveform at some target points, by means of water level and seismic measurements while tsunami is occurring.

The model is based on the solution of the mild slope equation (MSE for short). The MSE has been originally developed by Berkhoff (1972) and it solves the wave field propagation at the undisturbed water free surface, by assuming a vertical structure of the wave motion as similar as that of the linear first order Stokes waves. The MSE in its original form is an elliptic equation which reproduces steady state situations for purely harmonic waves. Mathematically an elliptic equation defines a problem which is in general properly posed only if boundary conditions are specified, and it needs to be simultaneously solved over the whole computational domain. Copeland (1985) and Madsen & Larsen (1987) developed a pair of first-order equations, which constitutes a time dependent hyperbolic system, which is used to achieve the steady-state wave field induced by periodic waves. Kubo et al. (1992) and Kirby et al. (1992) studied the applicability of the time-dependent equations to random waves. However they concluded that only very narrow frequency wave spectra are properly reproduced by the hyperbolic versions of the MSE, since the equations coefficients are calculated using the frequency of the carrier wave. None of the above mentioned models seems able of dealing with broad spectrum wave field, such as that resulting for tsunamis. The novel approach followed here is inspired to the method used by Kirby et al. (1992), who separated the whole wave spectrum into several narrow bands and for each of these solved the time-dependent MSE. The proposed model solves the MSE in the frequency domain for all the wave frequencies, then it superimposes all the monochromatic solutions to obtain the broad banded spectrum wave field.

Although the model is originally derived for the wave propagation in the far field we also introduce a source term in the equations, which takes into account of the seafloor variation in time. We obtain the Mild Slope Equation with a generation term that describes the effect of the bottom movements on the surface waves. In order to incorporate the effects of the sea-floor movements into depth integrated equation, it is often assumed that the bottom movements instantaneously transfer to the free water surface, which is treated as identical to the vertical component of the sea-floor deformation (Kajiura, 1970). This is especially valid when the extension of the sea-floor rupture is very large in comparison to the water depth. However when the tsunami source is smaller, or of the same order of magnitude of the water depth (this is mostly the case of submerged landslide), the problem becomes more complicated. Tinti et al. (2006) proposed a filter function which attenuates the effects of the sea bottom movement in the water surface as the ratio between the water depth and the bottom movement extension becomes large. Starting from their work a slightly different filter function, applied to the MSE has been obtained (equation's derivation in the following section).

The chapter is structured as follows: section 2 describes the model equations; section 3 presents two set of experiments (physical and numerical) used to validate the model and it shows the application results on simulating landslide generated tsunami; section 4 describes the model abilities to work in a Tsunami Early Warning System and it shows an example of application to the South Tyrrhenian Sea (Italy, Mediterranean Sea). Finally in section 5 the conclusions are given. Part of the research presented in this chapter has been described in Cecioni & Bellotti (2010a, 2010b) and in Bellotti et al. (2008)

2. The proposed model

2.1 Model equation

The proposed model solves one elliptic MSE for each frequency of the wave spectrum, and then it applies the Inverse Fourier Transform to all the solutions in the frequency domain in order to recover the wave field time series. This procedure guarantees accurate reproduction of the frequency dispersion, since potentially all the components of the wave field are modelled using an appropriate elliptic equation.

In the following the MSE is derived, following the procedure of Svendsen (2005), which achieves the formulation of the time-depended hyperbolic MSE. In the present equation's derivation, it is included a generation term by considering a time variation of the water depth function, in order to reproduce a bottom movement due to submarine earthquake or landslide. We start from the linear (small amplitude) water wave equations for an incompressible irrotational fluid on an uneven bottom

$$\nabla_h^2\phi + \phi_{zz} = 0\,; \quad -h(x,y,t) < z < 0 \qquad (1)$$

$$\phi_z + \frac{1}{g}\phi_{tt} = 0\,; \qquad z = 0 \qquad (2)$$

$$\phi_z + h_t + \nabla_h\phi \cdot \nabla_h h = 0\,; \qquad z = -h(x,y,t) \qquad (3)$$

where $\phi(x,y,z,t)$ is the velocity potential in the fluid, $h(x,y,t)$ is the water depth and g is the gravity acceleration, while ∇_h is the differential operator which means the divergence in the horizontal coordinates (x,y). All these variables are real and scalar. Eq. (1) is the Laplace equation, Eq. (2) includes the linearized dynamic and kinematic boundary conditions at the free water surface, while Eq. (3) is the bottom boundary condition which reproduces the seafloor movements allowing h to vary in time. The solution of the given problem is assumed to be of the form

$$\phi(x,y,z,t) = \varphi(x,y,t)f(z) \qquad (4)$$

where $\varphi(x,y,t)$ is the velocity potential at the undisturbed free water surface $z = 0$, which can be complex and it includes the effects of reflected waves; $f(z)$ is a function that describes how the kinematic field varies along the water depth and can be chosen as that resulting from the linear wave theory valid for harmonic waves propagating in constant depth, which however still holds locally for uneven bottom, i.e.

$$f(z) = \frac{\cosh[k(h+z)]}{\cosh(kh)} \tag{5}$$

where k is the wave number, defined as $2\pi/L$ with L the wave length. In the cases of not constant depth, h and k vary with the horizontal coordinates, therefore $f = f(x,y,z)$. However if the mild slope assumption, $\frac{\nabla_h h}{kh} << 1$, is here introduced the variation of the function f with the horizontal coordinates can be neglected if compared with the vertical ones.
From the assumption (4) it comes that

$$\phi_{zz} = k^2 \varphi \frac{\cosh[k(h+z)]}{\cosh(kh)} = k^2 \phi \tag{6}$$

therefore the Laplace equation (1) can be written as

$$\nabla_h^2 \phi + k^2 \phi = 0 \tag{7}$$

The following considerations are made:

$$f(z) = 1 ; \qquad z = 0 \tag{8}$$

$$f_z = k \tanh(kh) = \frac{\omega^2}{g} ; \qquad z = 0 \tag{9}$$

$$f_z = 0 ; \qquad z = -h \tag{10}$$

In order to depth integrate the field equation (1), here it is made use of the Gauss's Theorem, which states for one dimensional domain

$$\int_a^b \left(\Phi_1 \nabla^2 \Phi_2 - \Phi_2 \nabla^2 \Phi_1\right) dx = \left[\Phi_1 \nabla \Phi_2 - \Phi_2 \nabla \Phi_1\right]_a^b \tag{11}$$

where Φ_1 and Φ_2 are arbitrary differentiable scalar functions. For the present purpose Eq. (11) is used with $x = z$, $\Phi_1 = f(z)$ and $\Phi_2 = \phi(x,y,z,t)$, therefore

$$\int_{-h}^{0} \left(f \frac{\partial^2 \phi}{\partial z^2} - \phi \frac{\partial^2 f}{\partial z^2} \right) dz = \left[f \frac{\partial \phi}{\partial z} - \phi \frac{\partial f}{\partial z} \right]_0 - \left[f \frac{\partial \phi}{\partial z} - \phi \frac{\partial f}{\partial z} \right]_{-h} \tag{12}$$

Substituting the Laplace Eq. (1) in the first term at the left hand side (LHS), and the boundary conditions at $z = 0$ and $z = -h$ (Eqs. 2 and 3) and using Eqs. (9) and (10) in the right hand side (RHS) terms, then, after changing the sign, Eq. (12) becomes

$$\int_{-h}^{0}\left(f\nabla_h^2\phi+k^2 f\phi\right)dz=\frac{1}{g}\varphi_{tt}+\varphi\frac{\omega^2}{g}-\left[fh_t\right]_{-h}-\left[f\nabla_h h\cdot\nabla_h\phi\right]_{-h} \quad (13)$$

the LHS terms can be seen as the integration over the depth of the field equation. Considering that

$$\nabla_h\phi=\nabla_h\left(\varphi f\right)=f\nabla_h\varphi+\varphi\nabla_h f \quad (14)$$

and

$$\nabla_h^2\phi=f\nabla_h^2\varphi+2\nabla_h\varphi\cdot\nabla_h f+\varphi\nabla_h^2 f \quad (15)$$

using the expression (14) for the last term of the RHS and expression (15) for the first term of the LHS, Eq. (13) becomes

$$\begin{aligned}&\int_{-h}^{0}\left(f^2\nabla_h^2\varphi+2f\nabla_h f\cdot\nabla_h\varphi+f\varphi\nabla_h^2 f+k^2f^2\varphi\right)dz=\\&=\frac{1}{g}\left(\varphi_{tt}+\omega^2\varphi\right)-\frac{1}{\cosh(kh)}h_t-\left[f\nabla_h h\cdot\left(f\nabla_h\varphi+\varphi\nabla_h f\right)\right]_{-h}\end{aligned} \quad (16)$$

Now incorporating the first two LHS terms of Eq. (16) it follows

$$\begin{aligned}&\int_{-h}^{0}\nabla_h\left(f^2\nabla_h\varphi\right)dz+\left[\nabla_h hf^2\cdot\nabla_h\varphi\right]_{-h}+\varphi k^2\int_{-h}^{0}f^2dz=\\&=\int_{-h}^{0}\varphi f\nabla_h^2 fdz-\varphi\nabla_h h\cdot\left[f\nabla_h f\right]_{-h}+\frac{1}{g}\left(\varphi_{tt}+\omega^2\varphi\right)-\frac{1}{\cosh(kh)}h_t\end{aligned} \quad (17)$$

Applying the Leibniz's rule for the first two LHS terms and knowing that

$$\int_{-h}^{0}f^2dz=\frac{cc_g}{g} \quad (18)$$

where c and c_g are respectively the phase and the group velocities, by multiplying Eq. (17) for g it results

$$\begin{aligned}&\nabla_h\cdot\left(cc_g\nabla_h\varphi\right)+\varphi k^2cc_g-\varphi_{tt}-\omega^2\varphi+h_t\frac{g}{\cosh(kh)}=\\&-g\varphi\left\{\int_{-h}^{0}f\nabla_h^2 fdz+\nabla_h h\cdot\left[f\nabla_h f\right]_{-h}\right\}\end{aligned} \quad (19)$$

Rigorously the identity of Eq. (18) is obtained for a single frequency of the wave spectrum, consequently Eq. (19) is valid for monochromatic waves, or can be seen as representative of narrow banded spectra sea state around a carrying frequency.

The RHS terms of Eq. (19) can be shown to be $O((\nabla_h h)^2, \nabla_h^2 h)$. Therefore, given the mild-slope assumption, $\nabla_h h \ll kh$, it can be argued that the RHS terms << LHS terms. Similarly, $\nabla_h^2 h \ll \nabla_h h$, which is a natural additional assumption because $\nabla_h^2 h = O(\nabla_h h)$ can only occur over short distances without changing $O(\nabla_h h)$. This means that the RHS terms are << of all the others terms, we therefore get

$$\varphi_{tt} - \nabla_h \cdot \left(cc_g \nabla_h \kappa\right) + \left(\omega^2 - k^2 cc_g\right)\varphi = -h_t \frac{g}{\cosh(kh)} \tag{20}$$

which is the **hyperbolic version of the MSE** in terms of fluid velocity potential and is usually referred to as the 'time-dependent mild-slope equation', allowing the simulation in the time-domain of the wave propagation. The so formulated MSE (20) solves the fluid velocity potential time series for a specified single wave frequency, which unambiguously determines the equation parameters, c, c_g, ω and k. The RHS term represents a source term, which takes into accounts the effects on the wave field of the bottom movement.

In terms of the free surface elevation η the MSE becomes (using the dynamic boundary condition at the free surface)

$$-\eta_{tt} + \nabla_h \cdot \left(cc_g \nabla_h \eta\right) - \left(\omega^2 - k^2 cc_g\right)\eta = -h_{tt} \frac{1}{\cosh(kh)} \tag{21}$$

Note that if the phase and group velocities are evaluated in the shallow water limit as $c = c_g = \sqrt{gh}$, then Eq. (21) reduces to the governing equation for forced long waves, with a filter function applied to the source term in order to attenuate its effect as bigger becomes kh. (Cecioni & Bellotti 2010b)

$$-\eta_{tt} + \nabla_h \cdot \left(gh \nabla_h \eta\right) = -h_{tt} \frac{1}{\cosh(kh)} \tag{22}$$

The **elliptic version of the MSE** can be obtained by taking the Fourier Transform of Eq. (21), then it comes

$$\nabla_h \cdot \left(cc_g \nabla_h N\right) + k^2 cc_g N = -fft(h_{tt}) \frac{1}{\cosh(kh)} \tag{23}$$

Where $N(x,y,\omega)$ is the Fourier Transform of $\eta(x,y,t)$, as $fft(h_{tt})$ is the Fourier Transform of h_{tt}.
It can be noted that without allowing the bottom movement, i.e.: bottom boundary condition (3) without the term h_t, the resulting equation would have been

$$\nabla_h \cdot \left(cc_g \nabla_h N\right) + k^2 cc_g N = 0 \tag{24}$$

which is properly the MSE as derived by Berkhoff (1972).
The RHS of Eq. (23) represents the source term of generating waves due to the bottom movement. It can be noted that this source term is formed by the water depth time variation and by a function which depends on the wave frequency, through the wave number. This function, $1/\cosh(kh)$, works as a low-pass frequency filter function between the movements of the bottom and those of the water free surface. In the proposed model, where the MSE is solved in the frequency domain, the inclusion of this frequency-depended source term appears natural. Once Eq. (23) is solved, one for each frequency ω, with the appropriate boundary conditions, the result in the time domain can be achieved by taking the Inverse Fourier Transform of $N(x,y,\omega)$ to obtain $\eta(x,y,t)$. For details see Bellotti et al. (2008).

2.2 Boundary conditions

The solution of a elliptic equation requires the setting of all the boundaries of the numerical domain. The boundary conditions can allow the waves to be reflected on the boundary (reflection condition) or to freely exit the domain (radiation condition). The fully-reflection boundary condition can be expressed by imposing the fluid velocity in the direction orthogonal to the boundary to be zero. In the frequency domain the reflection condition at a boundary is expressed as follows

$$N_n = 0 \tag{25}$$

where n is the vector normal to the boundary.

The radiation boundary condition can be obtained by using a mathematical formulation that allows the waves propagating toward the open boundaries to freely exit the computational domain. This condition can be easily formulated for progressive outgoing waves

$$\eta_t + \frac{c}{\cos(\theta_n)}\eta_n = 0 \tag{26}$$

where θ_n is the angle the wave direction forms with the outgoing normal to the considered boundary. The Fourier Transform of equation (26) provides the radiation condition in the frequency domain

$$N_n + ik\cos(\theta_n)N = 0 \tag{27}$$

Please note that the equation (27) is nonlinear in the sense that θ_n is not known a priori and depends on the solution itself. Iterative techniques can therefore be applied or a reasonable estimate of this parameter can be used to solve the indeterminacy.

The wave generation in the numerical domain can be modeled with two different approaches: by means of a source term included in the field equation, as derived previously (Eq. 23), or by means of a wave-maker boundary condition. A wave-maker boundary condition can be seen as a wave paddle used in physical models (although here evanescent modes are not reproduced) and can be applied when the wave generation mechanism occurs close to a boundary of the numerical domain. The wave-maker boundary condition is conveniently formulated in terms of the velocity potential at $z = 0$ as follows

$$\phi_n = u^I, \qquad z = 0 \tag{28}$$

where u^I is the velocity at $z = 0$ of the desired wave field orthogonal to the wave-maker boundary. In order to obtain a mathematical expression involving η and consequently N we make use of the dynamic boundary condition at the free surface which, if transformed in the frequency domain, results as follows

$$i\omega\Phi = -gN, \tag{29}$$

providing the usual relationship between Φ and N:

$$\Phi = -\frac{g}{i\omega}N \tag{30}$$

which allows rewriting of the wave-maker condition as

$$N_n = -\frac{i\omega}{g} U^I, \tag{31}$$

being U^I the Fourier Transform of the desired time series of velocity at $z = 0$. It is worth to remind that the proposed wave-maker condition is used to specify the fluid velocity at $z = 0$, and that the velocity field for $-h < z < 0$ is assumed to vary as the function f.

3. Validation against laboratory and numerical data

Here it is described the model validation, carried out by comparing the model results with those obtained from numerical and physical experiments, chosen as reference solution. The presented tests aim at reproducing tsunami generated by landslides. When a landslide occurs under the water in general it produces a sea floor perturbation which generates a complex three dimensional water flow in the proximity of the landslide itself. These perturbations are transferred to the water free surface depending on the water depth: if the landslide occurs in shallow water conditions, i.e. the landslide extension is larger compared to the water depth, then it can be assumed that at the water free surface occurs the same displacement of the sea bottom; this is not true when the landslide occurs in deep water conditions. In this case the column of water above the acts as a filter attenuating the bottom movement effects at the free surface.
The tests here presented will focus only on the reproduction of waves generated by landslides which propagate in deep water conditions.
In the following we present two sets of validation experiments: first the inclusion of the source term into the MSE (23) is analyzed trough numerical experiments (see also Cecioni & Bellotti 2010a), which compare the proposed model solutions with the results of a three dimensional numerical solver, taken as reference. Then the results of a physical experiment are used to further validate the proposed model (see Cecioni & Bellotti 2010b). The physical model reproduces the landslide generated waves using a realistic model of the landslide, and a wave tank large enough to study the wave propagation in the far field. These experiments reproduce the particular case of a landslide falling at the flank of a conical island and they allow the analysis of the generated wave field close to the island and in the far propagation field.

3.1 Tsunami waves generation - numerical experiments (Cecioni & Bellotti 2010a)

Here a numerical experiment which reproduces tsunami generated by submerged landslide is described. The experiment has been carried out in order to validate the correct inclusion of the source term into the MSE, and to investigate on the ability of the proposed model, based on a depth integrated equation, to simulate the wave generation due to a sea bottom movement. The results of the proposed model are compared with those of a three dimensional one, which, as long as the linearized boundary conditions can be considered to be an acceptable approximation, can be regarded as the reference (i.e. the true) solutions. The three dimensional model solves the Laplace equation within the boundary conditions. It uses the same numerical scheme of the depth integrated model, thus the equations are solved in the frequency domain and are formulated as follows

$$\nabla_h^2 \Phi + \Phi_{zz} = 0 \tag{32}$$

$$\Phi_z - \frac{\omega^2}{g}\Phi = 0 \tag{33}$$

$$\Phi_n = fft(h_t) \tag{34}$$

$$\Phi_n + ik\cos(\theta_n)\Phi = 0 \tag{35}$$

where $\Phi(x,y,\omega)$ is the Fourier transform of $\phi(x,y,t)$. Eq. (32) is the Laplace equation, Eq. (33) is the free surface dynamic and kinematic boundary condition, Eq. (34) represents the moving bottom boundary condition and Eq. (35) is the radiation condition.
We have performed many computations, varying the water depth, the landslide length and kinematic. Here however, for the sake of brevity we limit to show results of one experiment, which can be considered representative of most of our simulations. In the sketch of Fig. 1 the numerical domain is represented, which is *10 m* long and with a *1:3* slope; the landslide is a semi-ellipse, *4.21 m* long and *0.1 m* thick. The landslide suddenly starts to move after *10 s* with a velocity of *1 m/s* for *2 s* and a total displacement of *2 m*.

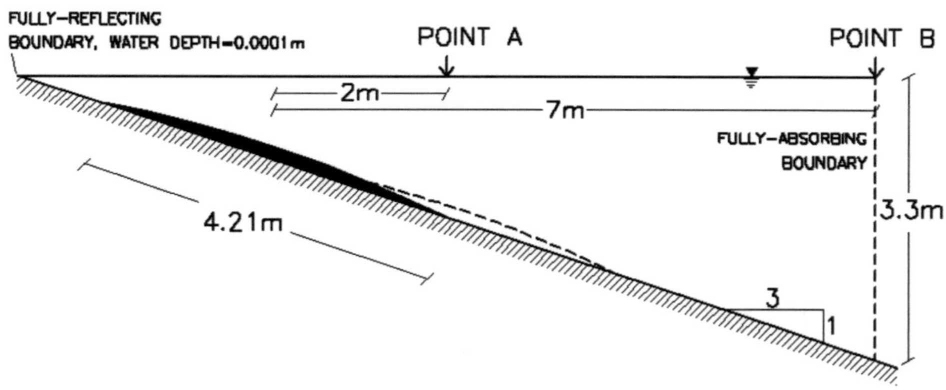

Fig. 1. Sketch of the computational domain

Numerical simulations were run with both the depth integrated model and the three dimensional one. The depth integrated model is applied in a one-dimensional domain, representative of the free water surface. At the left boundary ($x=0$ *m*) a reflecting boundary condition (Eq. 25) is imposed, by using a very small water depth (*0.0001 m*). At the right boundary, the water depth is equal to *3.3001 m* and the waves are allowed to exit the domain, by applying the radiation condition (Eq. 27). The numerical simulations reproduce a time series of *100 s*, using a Δt of *0.1 s*. This results in a total of *1000* time steps, that in the frequency domain corresponds to the same number of angular frequencies, in the range between $2\pi \cdot 10^{-2}$ and $2\pi \cdot 10$. In the practice the MSE (23) is solved just for the frequencies $2\pi \cdot 10^{-2} \le \omega \le 2\pi \cdot 2$, where there is a significant content of wave energy. The higher frequency component reproduced has, for the considered water depth, a wave length of *4 m*, thus the maximum distance between the computational nodes is *0.04 m*, since a resolution of *10* nodes per wave length is required.
The three dimensional model is solved in a two dimensional domain which covers the fluid field represented in the sketch of Fig. 1. The imposed boundary conditions are: a fully

reflecting condition at the left boundary (*x=0*), where a small water depth is imposed as for the depth integrated model; a radiation condition along the right boundary (*x=10 m*); then at the free surface the Eq. (33) is applied, which combines the dynamic and kinematic conditions, while at the bottom boundary the wave generation condition (Eq. 34) is imposed. The same component of the wave spectrum of the depth integrated model are solved. The finite element grid is composed of triangular elements, with the maximum distances between the nodes of *0.04 m*, as for the depth integrated model.

Fig. 2 shows the results of the numerical simulations in terms of surface elevation obtained with the depth integrated model (continuous black line) and the three dimensional one (dashed red line). The left and the right panels refer to the results at point A and B, where the water depth is about *1.6 m* and *3.3 m* respectively. The results of the two models appear to be in very good agreement both in the generation (left panel) and in the propagation areas (right panel).

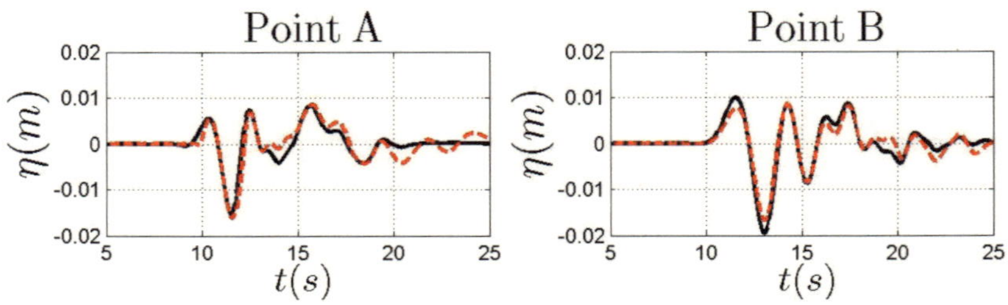

Fig. 2. Comparison of the free surface elevation obtained from the depth integrated model (continuous black line) and that obtained from the three dimensional model (dashed red line). The presented results are relative to the points A (left panel) and B (right panel) of Fig. 1.

These considerations can be supported by the analysis in the frequency domain. In Fig. 3 the left and right panels refer again to results picked up at the point A and B respectively. The thick black line represents the reference amplitude spectrum achieved with the three dimensional model, and the thin black line that obtained without using any filter in the depth integrated model, i.e.: the source term of the MSE is equal to the Fourier transform of the second derivative in time of the water depth function. The red line shows the amplitude spectrum obtained applying the MSE with the source term filtered by the wave frequency filter *1/cosh(kh)*, Eq. (23).

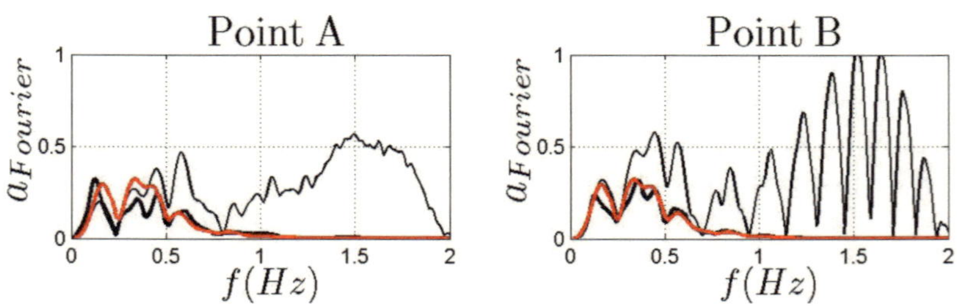

Fig. 3. Absolute values of the Fourier transform coefficients of the water surface elevations, computed with the three dimensional model (thick black line) and with the depth integrated model, without any filter function (thin black line), and computed with the proposed depth integrated model (red line).

It can be noted that from a physical point of view, the filter function has the effect of a low pass filter between the movements of the bottom and those of the free water surface. Translating the bottom deformation directly to the free surface introduces a large amount of spurious energy in the high frequency range, leading to unrealistic/wrong results.

3.2 Conical island experiments (Cecioni & Bellotti 2010b)

The results of physical experiments have been used in order to further validate the tsunami generation in the numerical model. The physical model was built in a large wave tank at the Research and Experimentation Laboratory for Coastal Defence (LIC) of the Technical University of Bari, Italy, in cooperation with the Environmental and Maritime Hydraulics Laboratory Umberto Messina (LIAM) of the University of L'Aquila, Italy. The laboratory experiments (Di Risio et al., 2009) simulate a landslide body falling on the flank of a conical island, built in order to approximately reproduce in scale *1:1000*, the Stromboli island, south Tyrrhenian Sea, Italy (Tinti et al., 2006). The physical model consists in a wave tank, *30.00 m* wide, *50.00 m* long and *3.00 m* deep; at the centre of the tank is placed a conical island, built using PVC sheets (thickness *0.01 m*) and sustained by a steel frame, with a radius of *4.47 m* at the tank bottom level. The slope of the flanks of the island is *1:3* (I vertical, 3 horizontal, see the picture on the right of Fig. 4). Experiments have been carried out varying the water depth, and consequently the shoreline curvature radius, and by varying the initial distance of the landslide from the undisturbed shoreline. The landslide model is a rigid body, with the shape of an half of the ellipsoid described by the equation $x^2/a^2 + y^2/b^2 + z^2/c^2 = 1$, where $a = 0.2\ m$, $b = 0.4\ m$ and $c = 0.05\ m$, for a total volume $V = 0.0084\ m^3$. The landslide is constrained to slide down the inclined surface by means of rails. Traditional resistive gauges were employed to register the instantaneous vertical displacement of the free surface. All the signals have been acquired simultaneously at a frequency of *1000 Hz*. The relative positions of all the gauges can be found in Fig. 4.

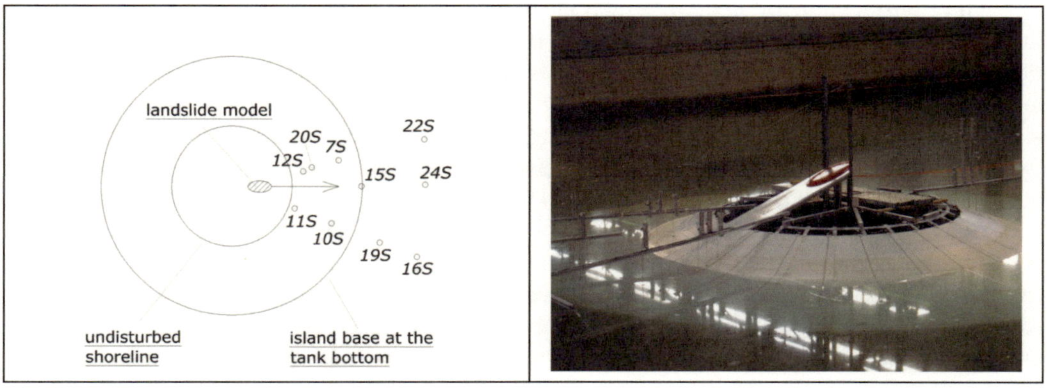

Fig. 4. Left: Layout of the sea-level gauges positions in the 3D tests at LIC. Right: Picture of the physical model.

The numerical computations have been carried out on a two-dimensional domain, sketched in Fig. 5. The numerical simulation here presented reproduces just one experimental case, defined by the off-shore constant water depth of *0.80 m*, and the shoreline radius of *2.07 m*, and characterized by an aerial landslide which falls from a distance of $\zeta = 0.30\ m$ from the undisturbed shoreline. In order to save computational costs, not all the wave tank was numerically reproduced: the domain is circular around the island up to *8.00 m* from the island centre; only half of the circular domain is reproduced given the symmetry of the problem (Fig. 5). At the internal circular border (the undisturbed shoreline) the reflection condition (25) is imposed. Along the external circular boundary the waves are allowed to freely exit the computational domain (Eq. 27). The landslide falls down in the direction of the right-bottom border, in Fig. 5. At that boundary the fully reflection condition is imposed in order to take into account the symmetric half of the domain which is not simulated.

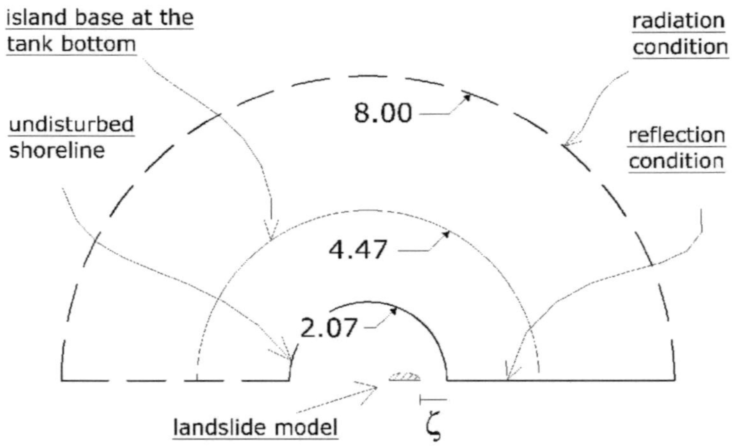

Fig. 5. Numerical domain of the depth integrated model. The numbers *2.07*, *4.47* and *8.00* express the radii in meters of respectively the undisturbed shoreline, the island base at the tank bottom, and the external circular boundary.

The water depth function $h(x,y,t)$, which takes into account the sea floor motion, due to the landslide, is calculated by knowing the landslide shape and movement. The second derivative in time of the function, $h_{tt}(x,y,t)$, is carried out using a numerical approximation and its Fourier transform is applied, in order to insert it into the field Eq. (23). The numerical simulation has been carried out using triangular linear elements, whose maximum size is *0.05 m*. Fig. 6 and 7 show the comparison between the laboratory measurement of the water surface elevation (red dashed lines) and the numerical simulations (solid black lines) at the gauges located as shown in Fig. 4.

As it can be seen from the figures, the model gives reliable results; this is evident especially for the gauges located in the off-shore area (refer to Fig. 4 for gauges position). The comparison shows that the model is not able to exactly reproduce the water level oscillations close to the shoreline, see the numerical results at gauges *S12* and *S11*. This behaviour can be explained because the linearity of the equation allows the reproduction of small amplitude landslide (or bottom movement), while in the entering phase the landslide is thick as the water depth. Moreover the numerical model equations are valid for submerged landslides, and no reproduction of the piston-like effect is made, which, which as seen in the laboratory experiment induces a deformation of the shoreline and it is responsible of the generation of waves propagating along the coast. Another point to be considered is that the numerical model does not reproduce any dissipation at the interface between water waves and the island.

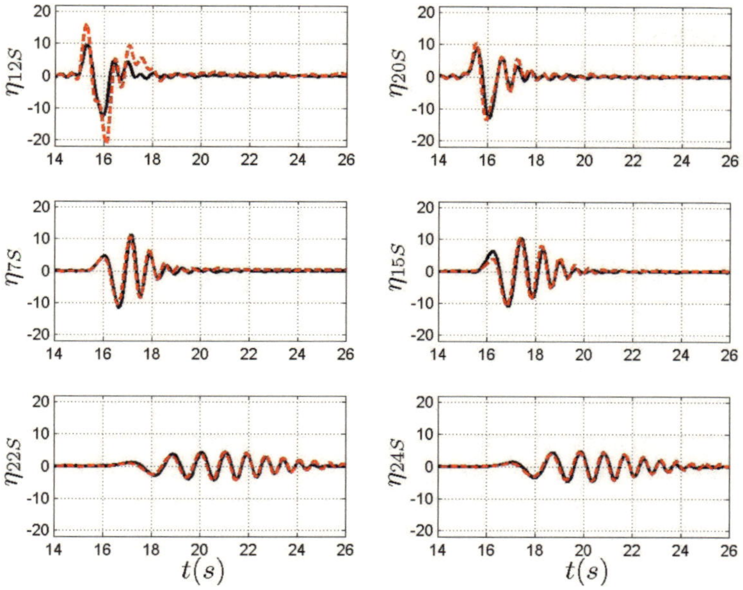

Fig. 6. Comparison of the free surface elevations (measures are in *mm*) at six gauges as measured in the physical model (red dashed line) and as obtained from the numerical model (black solid line).

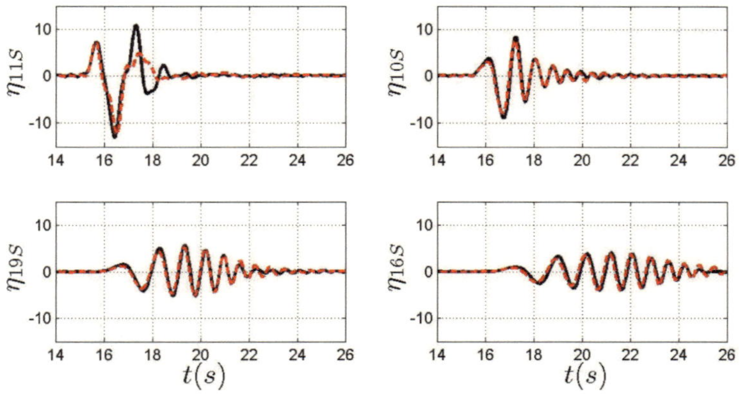

Fig. 7. Comparison of the free surface elevations (measures are in *mm*) at the other four gauges as measured in the physical model (red dashed line) and as obtained from the numerical model (black solid line).

4. Use of the model in a Tsunami early warning system

After the tragic event of December 2004, when an earthquake generated a tsunami that devastated the region of the Indian Ocean and killed about 250,000 people, the scientific community has been involved in the development of Tsunami Early Warning Systems

(TEWS for short). These systems should be triggered by seismic events, which are constantly monitored by worldwide efficient seismic networks. Once the occurrence of a submarine earthquake (or other tsunami sources, as landslides or volcanic eruptions) has been detected, rough estimates of the location of the epicentre and of the magnitude of the event are available after a short time. At that point a TEWS should decide if, given the properties of the earthquake, a tsunami alarm is to be spread. If positive, it should also be able to limit the alert to those people living at the coasts that are likely to be attacked by the waves.

It is also important that TEWS should be based on measurements of tsunami itself, in order to confirm or not that the earthquake has generated dangerous waves. It is therefore clear that, besides the earthquake and the water level measurements, a preliminary analysis is to be carried out in order to calculate the inundation level that any possible tsunami source is expected to induce along the surrounding coasts.

Many numerical models and different approaches can be used for performing the above mentioned analysis in order to build the database. Wei et al. (2003) proposed a model which solves the linear long wave equations. They have proposed an inverse method, which, thanks to the linearity of the equations, is able to infer seismic source parameters and tsunami wave form given the model simulations and water level data.

The presented model can be applied with the same scope to prepare a tsunami generation-propagation database, which results suitable in real time to forecast tsunami waveforms. The idea behind TEWS application, is to run preliminary computations without knowing the exact tsunami source, but assuming its location and here imposing a unitary wave generation term. The results of such preliminary computation are referred to as $N'(x,y,\omega)$. A second phase is applied in real time and it consists on estimating, by means of an inversion technique, the true solution. In view of the linearity of the problem the true solution in the frequency domain $N(x,y,\omega)$ can be obtained by multiplying N' for the Fourier Transform of the unknown source term, indicated as $S(\omega)$

$$N\,(x,y,\omega) = S(\omega)\,N'\,(x,y,\omega). \tag{36}$$

The model makes use of the sea level data (η_p) at one observation point P, in order to estimate the source term $S(\omega)$, by inverting Eq. (36) as follows

$$S(\omega) = N_P(\omega)/N'_P(\omega) \tag{37}$$

where N_P is the Fourier Transform of η_P and N'_P is the result of the unit source term computations at the same point P. If the records at more than one point are available, two alternative uses can be made of the data. On one hand it can be assumed that the source term is identical for all the generation areas/boundaries, and an optimization procedure can be used to find the value that best fits the data. On the other hand it can be assumed that each generation area/boundary has its own value of the source term and it is possible to write a linear system to be solved for these unknown source terms. Alternatively an over-determined system (the number of records available is greater than the number of source terms to be found) can be solved by means of an optimization procedure.

Here it is described a model application to Stromboli and the South Tyrrhenian sea. Stromboli is a volcano island in the deepwater Aeolian archipelago, Italy (Fig. 8). The NW flank of the island (named 'Sciara del fuoco') is unstable and frequently landslides occur at the coast, often related with volcanic activity. Some of these are able to generate tsunami waves which represent a real hazard for the surrounding islands and continental coast, as it

occurred on 30th of December 2002 (Tinti et al., 2005). Here we do not aim at exactly reproduce one specific tsunami event, but we build a numerical scenario just to show how the model can be applied in TEWS. Therefore, without having a tsunami water level registration at hand, we have assumed to measure a tsunami event characterized by a N-wave with period of *1 min* and height of *30 m* in deep water in front of Stromboli. This wave form is chosen after careful evaluation of the conclusions of Tinti et al. (2006), who analyzed the December 2002 tsunami event .
Fig. 8 shows the computational domain used, which encloses the islands of Stromboli, Panarea and Basiluzzo. A wave-maker condition, as that expressed by equation (31), is imposed at the boundary of the Stromboli island along the coast of the 'Sciara del fuoco'. For the preliminary computation at this boundary a unit value is assigned at the free surface fluid velocity U^I. The boundaries of the islands are modeled as impermeable and fully-reflecting using Eq. (25), while in the circular outer boundary a fully-absorbing condition is imposed to allow the outgoing waves radiation (27). The results of the preliminary computation are saved at four points of interest: point 1 is just at the wave- maker boundary; point 2 is South-West of the generation area, in proximity of the Ginostra village, along the coast; point 3 is about at the South corner of the island, in proximity of Punta Lena; finally the results in proximity of Basiluzzo, *14 km* away from the island are presented at point 4.

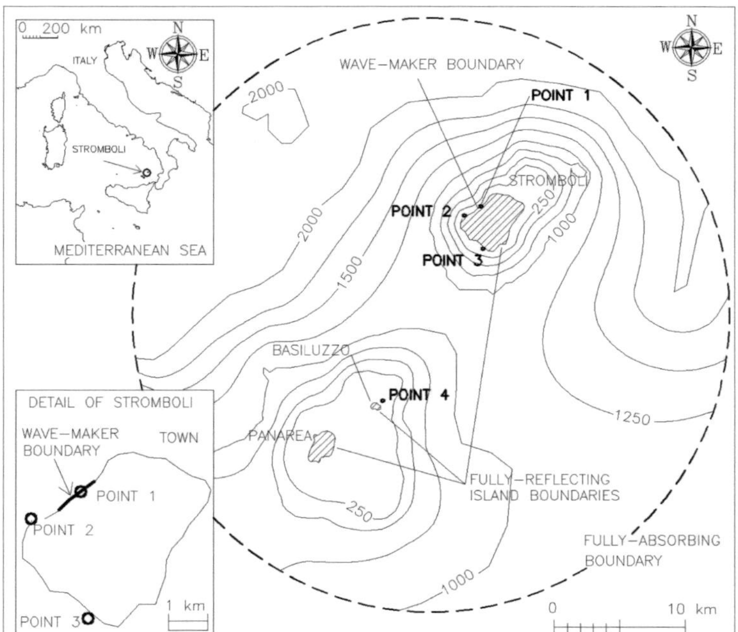

Fig. 8. Computational domain of the Stromboli-Panarea tsunami simulations.

Once the preliminary computation is carried out, we suppose that a landslide falls at 'Sciara del Fuoco' and it produces a tsunami which induces at point 1 a sea level elevation exactly as the N-wave. Thanks to the linearity of the model equations it is possible to apply Eq. (37) at the point 1 and therefore estimate the source term which has produced the N-wave at point 1. In Fig. 9 we can see the results in term of sea level elevation at the four points. It can be noted that as the distance from the 'Sciara del Fuoco' source grows, the height of the first positive wave reduces, and it appears other waves following. It can be noted at the other points 3 and 4, that the first incoming wave is not the highest. In particular at the point 4, in front of Basiluzzo Island (the most distant point from the generation area considered here) the first wave arrives for both models after about *240 s*, which is consistent with the shallow water wave celerity in a constant water depth equal to the mean water depth between Stromboli and Basiluzzo (about *600 m*).

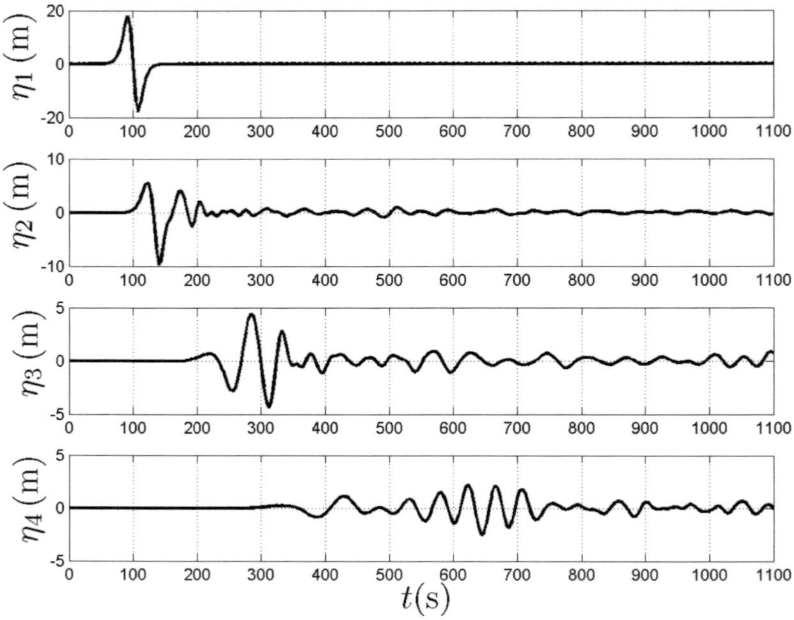

Fig. 9. Numerical results of the Stromboli-Panarea tsunami computation. Time series of the water free surface elevation as measured at point 1 and as forecasted at the other points by the model.

In Fig. 10 it is shown how the model works even with truncated input time series, in order to show which is the tsunami prediction at the target point (point 4) as the tsunami is being measured at the observation point (point 1). Several computations have been performed, assuming that the input time series is available up to a given time (t_{known}):

$\eta_{input}(t) = \eta_1(t)\,,\ t \leq t_{known}$

$\eta_{input}(t) = 0,\ t > t_{known}$

The results depend therefore on the value of t_{known}, as longer is the duration of the tsunami registration available at point 1, as better is the tsunami prediction at point 4. However it is clear that is unrealistic, given the limited time available for spreading the warning, to wait until the tsunamis has been completely measured. Few sample results are reported in Fig. 10, where the time series at the point 4 (named Panarea) are presented for t_{known} = *484, 496, 512* and *556 s*.

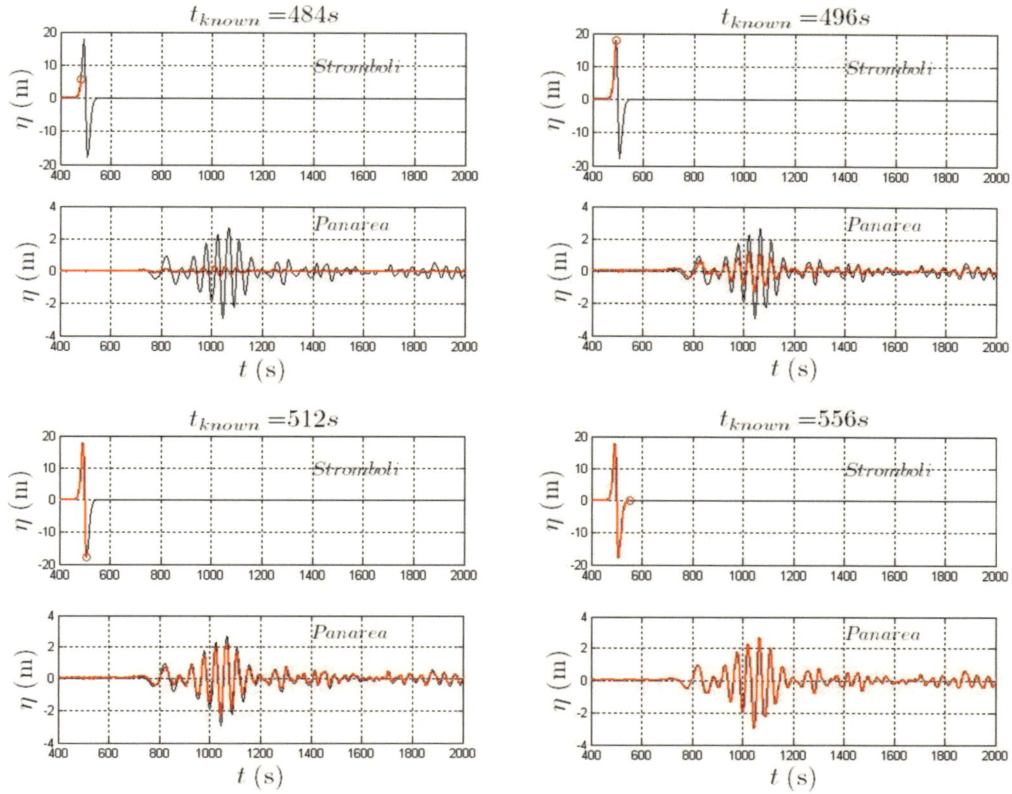

Fig. 10. Example of application of the model using truncated input time series. In each subplot the upper panel refers to the mareogram at point 1 of Fig. 8 (Stromboli) and the lower panel refers to point 4 of Fig. 8 (Panarea). The red line represents the truncated input data registered at Stromboli and predicted at Panarea, while the black lines refer to the whole mareogram registration.

On the upper panel of each subplot the input time series (η at point 1, in front of Stromboli) is represented using a thick red line, while on the lower panels it is shown the predicted η at point 4, in front of Panarea island. It is interesting to note that the use of a truncated input time series introduces very high frequency components, that are revealed by spurious short waves in the results. For t_{known} = *496 s* the crest of the first (highest) wave is used in the input time series and the prediction at Panarea appears to be already comparable to the final one. For t_{known} = *512 s*, 3/4 of the first wave in front of Stromboli is used with very good results at Panarea. Then the results tend to those that would be obtained using the whole time series. The most important parameter predicted by the model is certainly the maximum elevation at the point 4 (Panarea), referred to as ηmax. As the t_{known} increases the prediction tends to the final one, as shown in the Fig. 11. The vertical line in the figure represents the time at which the crest of the first wave is measured at point 1 (around t = *500 s*). The horizontal line indicates the final value of η_{max}. It is interesting to note that due to spurious high frequency components, induced by the truncation of the input time series, for some values of the t_{known}, η_{max} may be larger than the final one. However it can be concluded that the model is able of dynamically provide estimates of the desired parameter without suffering of instabilities. It is certainly desirable to carry out further research in the future in order to better understand if this good behavior of the model can be considered to apply under more general conditions.

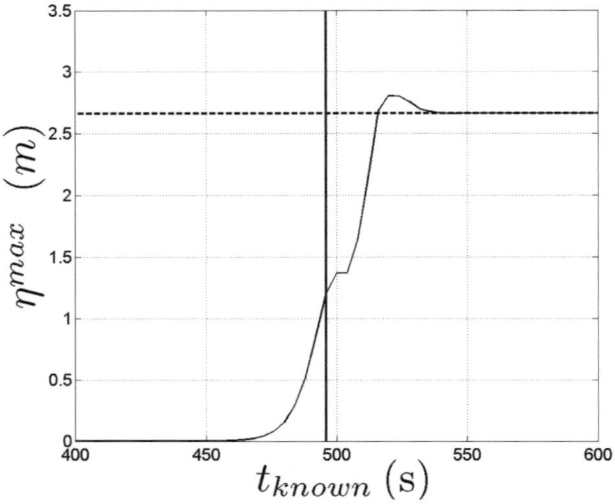

Fig. 11. Example of application of the model using truncated input time series; maximum surface elevation in front of Panarea island (point 4 of Fig. 8) as the time series at Stromboli (point 1) becomes available.

5. Conclusion

In this chapter we have presented a numerical model able to reproduce small amplitude transient waves. The model is based on linear equations, therefore it can properly reproduce waves generated and propagating in deep water. The wave generation is reproduced by forcing the mild slope equation, in order to include the effect at the free water surface due to sea floor movements. The comparison with the results obtained from a completely three dimensional laboratory experiments have proven the ability of the model to simulate wave generation and propagation.

An important feature of the presented model is that it considers the full frequency dispersion of waves, and in this sense in similar to the Boussinesq equations, when applied to small-amplitude waves. It has been demonstrated that tsunamis, although considered long wave, propagate with a celerity which is strongly influenced by the frequency. The ability of predicting with good accuracy the celerity is of the utmost importance when estimating the arrival time of the tsunamis. However the frequency dispersion has also a further very important effect on the transient waves: it induces a decrease of the wave height, especially for the first waves of a wave packet. This behaviour appears very clear in the TEWS application at Stromboli presented in this chapter, where the comparison of the results of the present model with those obtained using the linear shallow water equations has allowed to highlight this effect. It has resulted that only due to this, the wave height predicted by the LSWE can overestimate the actual height, for the experiments presented here also of a factor of 2. On the one hand it could be stated that the wave conditions predicted by the SWE are more severe than the real ones, so that safety is guaranteed. However, on the other hand, predicting tsunamis much more severe than the true ones in the context of a tsunami early warning system may lead to false alarms, which induce people not to trust the system, with possible catastrophic consequences.

6. Acknowledgements

This work was carried out under the research projects PRIN 2007 ("Development and validation of hydraulic and geologic tools for supporting a Tsunami Early Warning System. Implementation to the Stromboli Eolie-landslide case.") and FIRB 2008 ("Design, construction and operation of the SMO - Submarine Multidisciplinary Observatory-experiment"), both funded by the Italian Ministry for University and Scientific Research (MIUR).

7. References

Bellotti, G., Cecioni, C. & De Girolamo, P. (2008). Simulation of small amplitude frequency-dispersive transient waves by means of the mild slope equation. *Coastal Engineering,* 55 (6), 447-458.

Berkhoff, J.C.W. (1972). Computation of combined refraction-diffraction. *Proceedings of the 13th International Conference on Coastal Engineering ASCE,* Vancouver, Canada.

Cecioni, C. & Bellotti, G. (2010a). Modelling tsunami generated by submerged landslides using depth-integrated equations. *Applied Ocean Research,* 32(3), 343-350.

Cecioni, C. & Bellotti, G. (2010b). Inclusion of tsunamis generation into a depth integrated wave propagation model. *Natural Hazard and Earth System Sciences,* 10, 2259-2268

Copeland, G.J.M. (1985). A practical alternative to the mild-slope wave equation. *Coastal Engineering,* 9, 125-149.

Di Risio, M., De Girolamo, P., Bellottti, G., Panizzo, A., Aristodemo, F., Molfetta, M.G. & Petrillo, A.F. (2009). Landslide generated tsunamis runup at the coast of a conical island: New physical model experiments. *Journal of Geophysical Research,* 114.

Grilli, S. T.; Volgemann, S. & Watts, P. (2002). Development of a 3D numerical wave tank for modelling tsunami generation by underwater landslides. *Engineering Analysis with Boundary Elements,* 26, 301-313.

Kajiura, K. (1970). Tsunami source, energy and the directivity of wave radiation. *Bulletin Earthquake Res. Inst., Univ. Tokyo,* 48, 835-869.

Kirby, J.T., Lee, C. & Rasmussen, C. (1992). Time-Dependent Solutions of the Mild-Slope Wave Equation. *Proceedings of the 23th International Conference on Coastal Engineering -ASCE,* Venice, Italy, 391-404. By Billy L. Edge, (editor), New York, 0-87262-933-3.

Kubo, Y., Kotake, Y., Isobe, M. & Watanabe, A. (1992). Time-Dependent Mild Slope Equation for Random Waves. *Proceedings of the twenty-third International Conference on Coastal Engineering - ASCE,* Venice, Italy, 419-431. By Billy L. Edge, (editor), New York, 0-87262-933-3.

Kulikov, E.A.; Medvedev, P.P. & Lappo, S.S. (2005). Satellite recording of the Indian ocean tsunami on December 26, 2004. *Doklady Earth Sciences,* 401A, 3, (2005), 444-448.

Liu, P. L.-F.; Wu, T.-R.; Raichlen, F.; Synolakis, C. E. & Borrero, J. C. (2005). Runup and rundown generated by three-dimensional sliding masses. *Journal of fluid Mechanics,* 536, 107-144.

Lynett, P. & Liu, P.L.-F. (2002). A numerical study of submarine-landslide-generated waves and run-up. *Proc. Royal Society London,* 458, 2885-2910.

Madsen, P.A., Larsen, J., 1987. An efficient finite-difference approach to the mild-slope equation. *Coastal Engineering,* 11(4), 329-351.

Peregrine, D.H. (1967). Long waves on a beach. *Journal of Fluid Mechanics,* 27, 815-827.

Svendsen, Ib.A. (2005). Introduction to nearshore hydrodynamics. *Advanced Series on ocean Engineering,* 24, World Scientific, ISBN 981-256-204-4.

Tinti, S., Pagnoni, G. & Zaniboni, F. (2006). The landslide and tsunamis of the 30th of December 2002 in Stromboli analyzed through numerical simulations. *Bulletin of Volcanology,* 68 (5), 462-479.

Tinti, S., Manucci, G., Pagnoni, G., Armigliato, A. & Zaniboni, F. (2005). The 30 December 2002 landslide-induced tsunamis in Stromboli: sequenze of the events reconstructured from the eyewitness accounts. *Natural Hazard and Earth System Sciences,* 5, 763-775.

Wei, Y., Cheung, K.F., Curtis, G.D. & McCreery, C.S. (2003). Inverse algorithm for tsunami forecasts. *Journal of Waterway, port, Coastal and Ocean Engineering, ASCE,* 129(2), 60-69.

31

HF Skywave Radar Performance in the Tsunami Detection and Measurement Role

Stuart Anderson
Defence Science and Technology Organisation
Australia

1. Introduction

The severity of the threat posed to coastal communities by tsunamis has come into prominence following the event of Boxing Day 2004 when some 230,000 people lost their lives. Many countries have set up tsunami warning systems, or augmented existing systems, but overwhelmingly these derive their information from seismic monitoring stations which detect and locate earthquakes, not measurements of any resulting tsunamis. As a consequence, the false alarm rate is high, which leads to a loss of credibility of issued warnings and an inclination on the part of the public to ignore them.

Given our limited inability to predict whether a given seismic event will generate a tsunami, let alone provide quantitative information concerning its amplitude, researchers have sought to develop sensors which can provide reliable and accurate answers to these questions. These sensors fall into two categories. First, there are in situ devices, such as tsunameter stations and tide gauges, which monitor the ocean surface elevation at a single location. The Indian Ocean presently has only a handful of tsunameter stations, the Pacific Ocean more than thirty, as shown in Figure 1. These instruments use bottom-mounted pressure recorders to obtain time series data which can be subjected to spectrum analysis to separate low frequency, long wavelength signals of tsunamis from the much stronger responses due to wind waves and swell. Claimed sensitivity for advanced systems approaches ± 1 cm in 6000 m water depth, however given that the amplitude of a potentially dangerous tsunami may be only a few centimetres in the deep ocean, while waves and swell often exceed several metres in height, the potential for error is high. Moreover, the cost of a single deep sea tsunameter station, together with its deployment may easily exceed $250,000. Accordingly, the overall cost of a reasonably dense network would be enormous.

The second category of sensors embraces those which employ remote sensing techniques, either from space, in the form of satellite-borne altimeters, or from land, in the form of HF radars. Each of these remote sensing technologies has demonstrated that it is capable of measuring one or more of the various signatures which characterise a tsunami, with a precision which offers meaningful quantitative estimates of the amplitude of the disturbance. For example, the JASON-1 and Topex/Poseidon satellites which passed over the Boxing Day tsunami while it was propagating across the north-east Indian Ocean were able to detect the associated change in surface elevation and, in the case of Jason-1, a correlated variation in surface roughness inferred from the microwave radar scattering coefficient (Godin et al, 2009).

Allowing that there are several types of sensor capable of detecting a tsunami and measuring some parameter indicative of its magnitude, there remains the issue of spatial and temporal sampling. Ideally real-time detection, monitoring and characterisation of tsunamis requires ocean basin scale spatial coverage at a resolution significantly finer than the spatial scale of the disturbance itself (which has a wavelength typically of the order of 100 km), together with persistent observation with a temporal sampling interval measured in minutes rather than hours. This does not mean that the instantaneous field-of-view of the sensor system need span the entire ocean basin; rather, it should have the ability to survey the entire region of interest, preferably with a 'random-access' capability so that it can respond immediately to cueing and enable it to follow the propagating disturbance. Apart from an impractically large constellation of space-borne sensors, or an even denser network of tsunameters, only HF skywave radar appears to possess the desired combination of spatial and temporal sampling (Anderson, 2008). Overlaid on Figure 1 are the nominal coverage sectors of the three major Australian HF skywave radars.

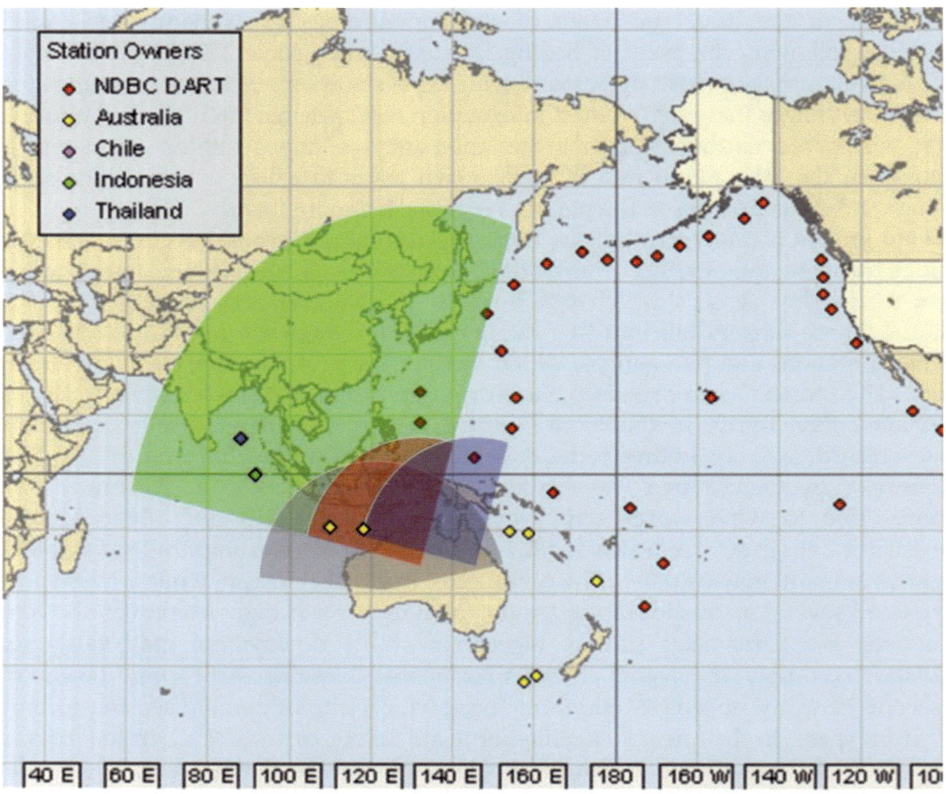

Fig. 1. The geographical distribution of deep-ocean tsunameters (NOAA, 2010) compared with the nominal coverage of the major Australian OTH radars (small sectors) and the low resolution frequency management system coverage for the central radar (large sector)

The possibility that tsunamis might be detected via quasi-vertical HF sounding of the disturbances they cause in the ionosphere was first raised in the 1970's (Hines, 1972; Najita et al, 1974), though in neither case were any of the practical difficulties of implementation addressed. Some of these issues were subsequently examined (Peltier & Hines, 1976), again in the context of a network of vertical incidence sounders, concluding that a tsunami warning capability was feasible, subject of course to suitable geographical disposition of the sounders. The idea of employing HF skywave radar in a tsunami warning role was proposed much later (Anderson, 1994), when it was noted that the oblique propagation geometry afforded a very wide coverage from a single site, eliminating the need for a network of sounders and islands on which to situate them, and introducing the prospect of mapping the tsunami-induced ionospheric disturbances continuously in space and time.

Following the Boxing Day Sumatran tsunami of 2004, interest in developing new techniques for tsunami detection increased enormously. One line of enquiry focussed on the application of tomographic inversion of total electron content (TEC) data from dense networks of Global Positioning System (GPS) receivers to image ionospheric disturbances arising from seismic phenomena via the mechanism of upwardly-propagating atmospheric gravity waves (Arteru et al, 2005; Lognonne et al, 2006). In conjunction with this TEC research, improved techniques were developed to model the temporal development of the electron density distribution in the ionosphere (Occhipinti et al, 2006). While TEC inversion techniques yield very informative 3-D images of the perturbations, these are restricted to the regions possessing such networks and in general extend no more than 200 – 300 kilometres offshore.

The possible use of HF skywave radar as a seismometer emerged again in 2004 when the possible use of the French Nostradamus radar to monitor the ionospheric signatures of earthquake-generated Rayleigh waves was proposed (Occhipinti et al, 2004; Occhipinti et al, 2010). Although the spatial resolution of Nostradamus is not high, it shares with other HF skywave radars very high Doppler resolution which is the key to detecting small perturbations in the ionosphere. It was a natural step from this to consider the application of this radar to tsunami detection (Coisson et al, 2008), based on models of the generation and propagation of the AGW produced by the expanding tsunami waveform. Around the same time, a study by a German group (Marquardt, 2007) identified HF skywave radar as a candidate technology for the tsunami detection mission, while a comprehensive assessment of HF radar signature mechanisms was reported (Anderson, 2008).

This chapter presents an assessment of the performance of HF skywave radar in the role of a tsunami warning system, with reference to all the signature mechanisms identified in the cited literature. That is not to say that other signatures are absent – the scale and energy of tsunamis are such that it would be surprising if other possibilities did not exist. First, though, we describe the unusual properties of HF skywave radar.

2. General characteristics of HF skywave radar systems

Radars operating in the HF band (3 – 30 MHz) are distinctive in that they can exploit modes of electromagnetic wave propagation other than line-of-sight. Two modes in particular are commonly exploited – skywave propagation, which involves reflection from the ionosphere, and surface wave propagation, which refers to electromagnetic waves localised at the air-sea boundary and diffracted by the earth's curvature to illuminate the sea surface beyond the horizon. Radars exploiting the former mechanism are often referred to as over-the-horizon radars (OTHR) but as this description also fits radars employing the surface wave mode, it is preferable to use the terms 'skywave radar' and 'surface wave radar'. Both types are being considered for tsunami warning (Anderson, 2008; Coisson et al, 2008; Barrick, 1979; Lipa et al, 2006; Heron et al, 2007; Dzvonkovskaya et al, 2009). Here we focus mainly on skywave radar, which can exploit tsunami signatures not accessible to surface wave radar; in addition it provides far greater spatial coverage.

The main physical differences between HF radars (of either variety mentioned) and the familiar microwave band radars result from the relative wavelengths. Whereas microwave frequencies correspond to wavelengths of 1 - 30 cm, HF frequencies equate to wavelengths of 10 - 100 m. As a consequence, antennas tend to be 10^2 — 10^3 times larger. Figure 2 shows the 2.8 kilometre receiving antenna array of the Jindalee skywave radar in Central Australia.

Fig. 2. Aerial view of the 2.8 km receiving array of the Australian Jindalee skywave radar, operational since 1982.

Representative values for some of the key radar parameters relevant to the tsunami warning mission are listed in Table 1. Although HF skywave radar is the subject of this chapter, HF surface wave radar parameter values have been included because such systems are much more common and have been widely promoted as tsunami sensing systems. A fairly comprehensive description of HF radars of both classes can be found in a recent survey (Headrick & Anderson, 2008b). Note that each of the Australian radars is essentially two radars, one a high power, high resolution system designed to detect aircraft and ships, the other a low power, low resolution system which provides frequency management advice (FMS). The latter has a significantly greater geographical coverage, as shown in Figure 1. For clarity, only one radar's FMS coverage is drawn.

Parameter	Typical values for HF skywave radar	Typical values for HF surface wave radar
total area within coverage (km^2) (subject to propagation availability)	~10^7	~10^5
range coverage (min→max) (km)	800 → 3500	5 → 180
angular coverage arc (deg)	90 - 360	60 - 180
angular beamwidth (deg)	0.3 - 3	3 - 12
resolution cell dimension (km)		
range	3 - 30	0.3 - 5
cross-range (at 2000 km)	10 - 100	n/a
number of spatial resolution cells monitored simultaneously		
range	20 - 250	30 - 250
beams	10 - 30	1 - 32
area observed simultaneously (km^2)	2 ×10^5	1 ×10^5
time taken for one observation (s)	16 - 64	50 - 300
unambiguous velocity band (ms^{-1})	± 20 - 50	± 10 - 30
effective velocity resolution (ms^{-1})	0.3 - 1	0.05 - 0.5

Table 1. Representative coverage and sampling properties of HF skywave radar when configured for tsunami detection. Optimum parameter choice varies with type of signature.

3. Tsunami signatures observable with HF skywave radar

3.1 Signature domains and mechanisms

The internal gravity wave-mediated ionospheric response is not the only tsunami signature which can be observed, at least in principle, by an HF skywave radar. In fact, it has been pointed out (Anderson, 2008) there are no fewer than six distinct mechanisms whereby tsunamis produce geophysical effects which could manifest themselves in HF skywave radar returns. It is relevant to point out that, of the identified mechanisms, HF surface wave radars observe only one - the advection, by the tsunami wave', of the shorter gravity waves which contribute most strongly to HF sea echo. Straightforward calculations reveal that a scheme for tsunami detection via the resulting Doppler shift is eminently feasible where the bathymetry is suitable (Barrick, 1979) and, after the Boxing Day tsunami, quite a number of HF surface wave radar groups conducted modelling studies of this approach (Lipa et al, 2006; Heron et al, 2007; Dzvonkovskaya et al, 2009). The other candidate tsunami signature domains identified as potentially amenable to detection by HF skywave radar are sea roughness modulation (Godin, 2004; Godin, 2008, Troitskaya & Ermakov, 2004), geomagnetohydrodynamic effects (Iyemori et al, 2005; Anderson, 2008) and infrasonic wave modulation of the lower ionosphere (Le Pichon et al, 2005; Koshevaya et al, 2004).

While it would seem that a considerable palette of potential tsunami signatures may be available, this diversity of opportunities is confronted by the realities of HF skywave radar operational performance. First, although the *potential* spatial coverage of HF skywave radar is commensurate with the perceived requirements of a tsunami monitoring system, not all of this vast region can be interrogated at any given time. The ionosphere, on which skywave propagation depends, experiences large variations over an extremely wide range of spatial and temporal scales, so the surveillance volume which can be illuminated with adequate signal strength at any given time is only some fraction of the nominal coverage. Second, even if signal strength is adequate, most prospective tsunami signatures are degraded by fluctuations in the signal phase due to dynamical processes in the ionosphere. Third, HF skywave radar is vulnerable to interference and unwanted echoes which can obscure some of the desired signals. Fourth, some natural geophysical phenomena produce Doppler signatures which are locally indistinguishable from those expected from a tsunami, so the probability of false alarm could be high. It is for these reasons that we need to assess the prospective role of HF radar as a component of a tsunami warning system in terms of the statistics of availability of propagation meeting the specific threshold requirements of the tsunami signatures of interest, taking into account the likelihood of coincident events which could result in tsunami-like signatures.

The geophysical domains in which tsunamis 'write' a signature which is plausibly observable with HF radars of one form or another fall into three categories :

- changes to ocean surface geometry and dynamics
- changes to ionosphere geometry and dynamics
- phenomena associated with tsunami run-up and landfall

In order to predict and interpret HF radar signatures of tsunamis, we need a physical model for each mechanism.

3.2 Signatures involving scattering from the ocean surface

Scattering of HF radio waves from the time-varying ocean surface results in a complex phase modulation which manifests itself in the signals received by an HF radar as an imposed Doppler spectrum. The structure of this spectrum is determined by the spatio-temporal covariance function of the surface or, equivalently, by the distribution of surface gravity waves and currents in each resolution cell. The former are conventionally represented by the directional wave spectrum, spanning wavelengths in the range $10^0 — 10^2$ m, the latter by a three-dimensional mean field associated with quasi-stationary large-scale flows. Both waves and currents are influenced by the passage of a tsunami, leading to distinct signatures as discussed below.

Signature 1 : The advection signature

Tsunamis are global scale phenomena, and hence a rigorous treatment requires the full Navier-Stokes equation on a rotating earth,

$$\rho\left[\frac{\partial v}{\partial t}+(\vec{v}\cdot\nabla)\vec{v}\right]=-\nabla p+\rho\vec{g}-\Omega\times(\Omega\times\vec{r})-2\Omega\times\vec{v}+\frac{1}{3}\mu\big(\nabla(\nabla.\vec{v})\big)+\mu\nabla^2\vec{v} \tag{1}$$

together with the general equation of continuity,

$$\frac{\partial\rho}{\partial t}+\big(\nabla\cdot(\rho\vec{v})\big)=0 \tag{2}$$

and the associated boundary conditions. Nevertheless, many essential characteristics can be derived from a simplified model obtained by assuming incompressible, irrotational, inviscid flow, ignoring Coriolis and centrifugal effects, setting aside consideration of acoustic waves in the fluid, side-stepping the possibility of significant ambient currents and noting that the wave amplitude will be small except in the immediate coastal zone. In this case, setting $\vec{v} = \nabla\varphi$, the Navier-Stokes equation reduces to the equation for potential flow:

$$\nabla_{xy}^2\varphi + \frac{\partial^2}{\partial z^2}\varphi = 0, \quad -h \le z \le \eta \tag{3}$$

which we can solve after imposing kinematic boundary conditions at the free surface and on the bottom, and the dynamic boundary condition at the free surface. As is well known, to first order, (1) reduces to

$$\frac{\partial^2}{\partial t^2}\varphi_1 + g\frac{\partial \varphi_1}{\partial z} = 0 \tag{4}$$

with solution

$$\begin{aligned} \eta_1 &= a\exp\left\{i\left(\omega t + \vec{K}\cdot\vec{r}\right)\right\} \\ \varphi_1 &= \frac{iag}{\omega\cosh(Kh)}\exp\left\{i\left(\omega t + \vec{K}\cdot\vec{r}\right)\right\}\cosh\left[\left|\vec{K}\right|(z+h)\right] \end{aligned} \tag{5}$$

Substituting in (4) yields the dispersion relation

$$\omega_1^2 = gK\tanh Kh \tag{6}$$

For tsunamis, where the wavelength is much greater than the depth, (6) reduces to

$$\omega_1 = K\sqrt{gh} \tag{7}$$

so the wave celerity is independent of wavelength and a function only of depth,

$$v = \frac{\omega}{K} = \sqrt{gh} \tag{8}$$

A feature of the solution (5) is that as the depth parameter Kh decreases, the flow becomes increasingly horizontal, and essentially in phase throughout the water column. In the case of tsunamis, the fluid motions have horizontal length scales of the order of 10^5 m, the depth parameter is small (~ 0.25 for 4000 m depth) and vertical shear is negligible except near the sea floor. Consider now the effect on the short gravity waves which are responsible for HF sea clutter. The hydrodynamic modulation amplitude is of the same order as the ratio of current velocity to the wave group velocity in the reference frame of the current. In the deep ocean, this ratio is ~ 5×10^{-5} for tsunamis so the short surface gravity waves are advected uniformly by the bulk fluid motion as the tsunami progresses. The observed monostatic radar Doppler spectrum will have a frequency offset commensurate with the orbital velocity $\vec{u}$ of the tsunami wave. In the case of the two first-order Bragg lines which are dominant at HF, the large-scale spatial modulation of the Doppler shift is given by

$$\omega_d = 2\frac{F}{c}\left[\pm\sqrt{\frac{g}{K}} - \vec{k}\cdot\vec{u}\right] \tag{9}$$

Differentiating (5) and evaluating at the surface, the advection velocity field structure is given by

$$u(\vec{r}) = \frac{g\eta(\vec{r})}{\sqrt{gh}} \tag{10}$$

with maximum value

$$u_{\max} = \frac{ag}{v} \tag{11}$$

The Boxing Day tsunami achieved an amplitude of ~ 0.7 m over the deep ocean, while the celerity was ~ 200 m.s^{-1}. Substituting,

$$u_{\max} \approx 0.035\, m.s^{-1} \tag{12}$$

This is below the minimum current velocity that can be measured with HF radar under favorable conditions so, prima facie, skywave radar cannot detect the deep ocean Bragg scatter advection signature.

For a tsunami entering water of variable depth, in the absence of dissipation, conservation of energy requires that the wave amplitude increases as the wave slows,

$$energy\ flux = \frac{1}{2}\rho g v a^2 \propto h^{\frac{1}{2}} a^2 \tag{13}$$

implying that $a \propto \sqrt[4]{h}$. According to this relation, a tsunami of amplitude 0.7 m in 4000 m depth will arrive at the 100 m isobath with an amplitude of 1.75 m. Therefore, as the wave crosses the continental slope, the modulation will increase as the tsunami slows, first becoming evident as enhancement of the bulk transport of the surface gravity waves described above, then via progressively more nonlinear effects such as wave bunching and induced breaking.

Figure 3 shows the synthetic HF radar Doppler advection signature for a tsunami approaching a coastline with a bathymetric gradient of 10 metres per kilometre. The radar is set back 1200 km from the coast and a sporadic-E layer at 100 km altitude used to set the scattering geometry. The HF radar frequency is 15 MHz, with a 50 kHz bandwidth waveform corresponding to a range resolution of 3 km. The onshore wind speed used for this example is 5 ms^{-1}. It is clear that the advection is concentrated in the last 10 - 15 km from the shoreline, so warning time would be very limited, even allowing for the reduced tsunami speed. Moreover, ocean currents and mesoscale eddies have Doppler signatures with comparable shifts, so discrimination is nontrivial.

Signature 2. The surface roughness signature

The most immediate form of interaction is aerodynamic coupling with the adjacent neutral atmosphere. Conceptually, the progressive tsunami wave modifies the wind field above the surface, the modified wind field acts on the prevailing wave distribution, and this may cause a measurable change to its radar signature via the wind stress, given by

$$\vec{\tau} = \rho_{air} C_D \left| \vec{u}_{air} - \vec{u} \right| \cdot \left(\vec{u}_{air} - \vec{u} \right) \tag{14}$$

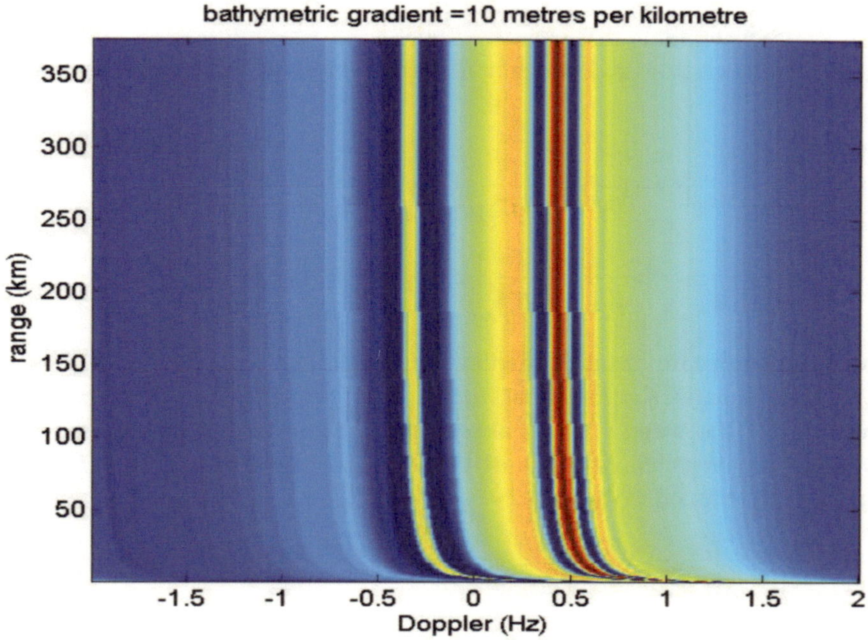

Fig. 3. A synthetic range-Doppler map for the case of a tsunami approaching a coastline with a bathymetric gradient of 10 metres per kilometre. The HF radar frequency is 15 MHz, with range resolution 3 km, and the onshore wind speed 5 ms^{-1}.

where ρ_{air} and $\vec{u}_{air}$ are the density and velocity of the air just above the surface, C_D is the drag coefficient and $\vec{u}$ is the water surface velocity given by (11). It was shown in the preceding section that the orbital velocity of the water particles at the surface is very small in deep water, of the order of a few centimetres per second. On ocean basin scales, $|\vec{u}_{air}| \sim 10$ ms^{-1} so the tsunami-induced change in wind stress is ~ 0.7%, which is insignificant. Even in the equatorial region, where $|\vec{u}_{air}| \sim 1$ ms^{-1}, the corresponding modification to the wind stress may reach ~ 7%, which is starting to appear promising. Unfortunately, in such calm conditions the horizontal scale length of surface conditions is generally low, wave development is equally variable, equilibrium seldom prevails and, as a consequence, the sea clutter Doppler spectrum is less informative.

Despite these arguments, so-called "tsunami shadows" have been reported, both visually and in microwave radar returns. This motivated a study to investigate the physics of atmosphere-ocean coupling using a more sophisticated argument (Godin, 2004; Godin, 2008). This analysis showed that, by parametrising the near-surface Reynolds stresses in terms of a turbulent viscosity which varies with height across the logarithmic boundary layer, the Navier-Stokes equation (1) can be reduced to an ODE taking the form

$$\begin{aligned}\frac{d}{dy}\left(\gamma^2\frac{dh}{dy}\right)-\kappa^4\alpha^2\gamma^2 h &= -4i\kappa^4\alpha^2\frac{d}{dy}\left[y\frac{d}{dy}(\gamma h)\right]\\ &+i\left(\frac{d^2}{dy^2}+\kappa^4\alpha^2\right)\left[\frac{y}{\gamma}\frac{d}{dy}\left(\gamma^2\frac{dh}{dy}\right)+\kappa^4\alpha^2\gamma y h\right]\end{aligned} \tag{15}$$

Under reasonable assumptions, this equation possesses wave solutions h_k, one class viscous, the other inviscid. Combining these to satisfy the boundary conditions on the surface yields expressions for the vertical and horizontal perturbations to the mean wind within the boundary layer:

$$v_H = u_0 \frac{\gamma(y)}{B_1\left(\beta^{-1},\beta^{-1}\right)}\left[B_1\left(\beta^{-1},y\right)+\frac{kD}{\kappa^2\alpha}B_2(y)\right]$$
$$v_V = iu\frac{\gamma(y)}{B_1\left(\beta^{-1},\beta^{-1}\right)}\left[\kappa^2\alpha B_3(y)-kDB_1\left(y,\beta^{-1}\right)\right] \tag{16}$$

where the B_k are known functions defined in terms of the h_k.
Analysis of these equations reveals that v_v is of the order of kDu_0, comparable with the vertical velocity of the water surface, a very small quantity. In contrast, the horizontal surface wind is driven much more strongly: v_H can be comparable with the unperturbed wind velocity, as illustrated in Figure 4.

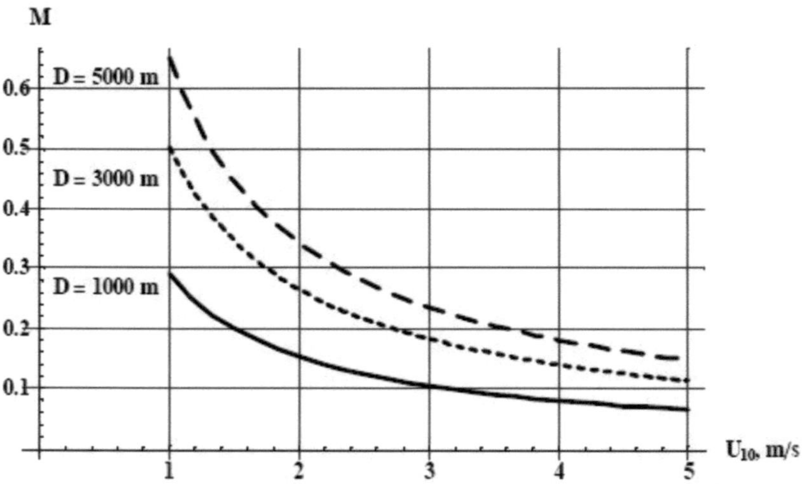

Fig. 4. Fractional change ξ in surface wind speed as a function of the unperturbed wind speed (Godin, 2008)

Under this model, the criterion for HF radar detectability should be gauged by comparing the spectrum computed for wind speed u with that computed for wind speed u′ given by

$$u' = \left(1 \pm \xi(u)\right)u \tag{17}$$

Figures 5 shows the predicted HF Doppler spectra for wind speeds u and u′ as defined in the text, for u = 4 ms^{-1}, for the case of an opposing orientation; the difference between the two is plotted in Figure 6. The corresponding plots for the aligned orientation are included as Figures 7 and 8. It can be seen that the difference is no longer negligible and may well be observable. This is not the case for higher prevailing wind speeds where the Doppler spectrum tends to saturate.

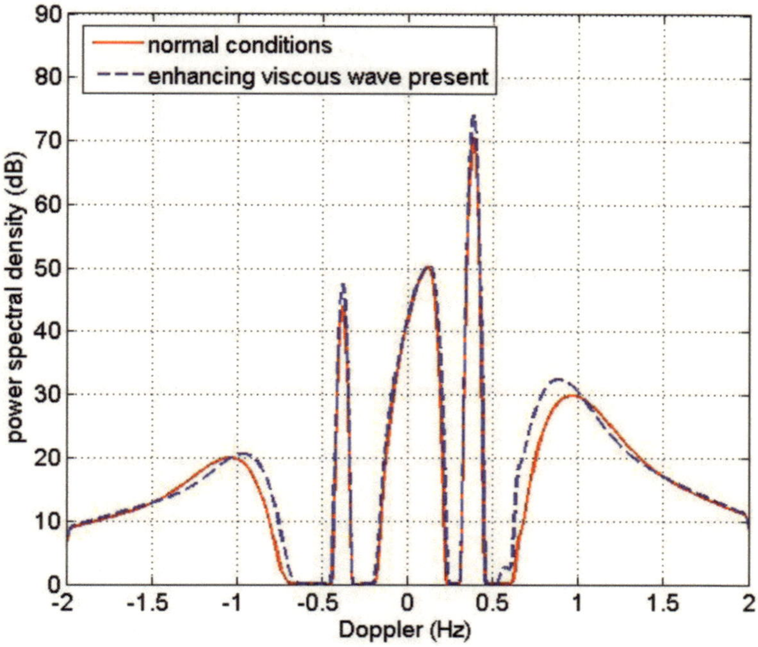

Fig. 5. Predicted HF sea clutter spectra for an ambient wind speed of 4 ms^{-1} with the viscous wave aligned with the ambient wind. Radar frequency 15 MHz

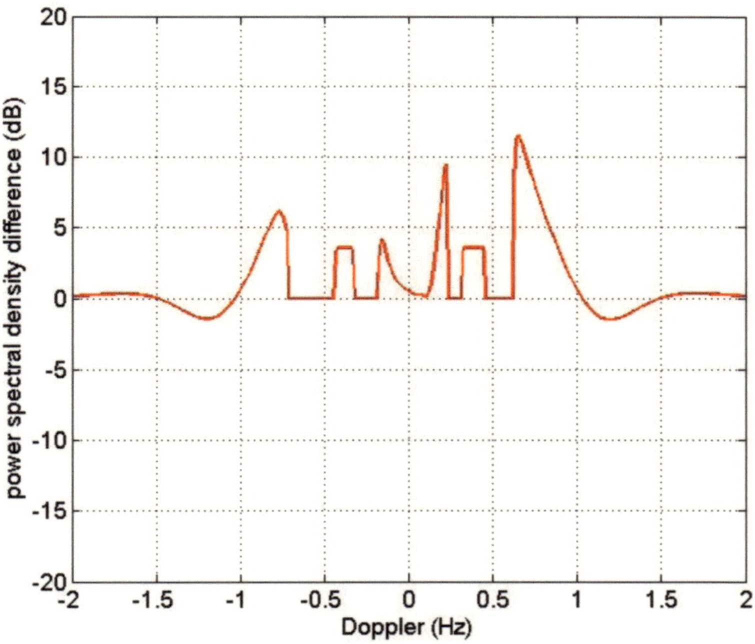

Fig. 6. The incremental change to the Doppler spectrum arising from the viscous wave for the aligned case.

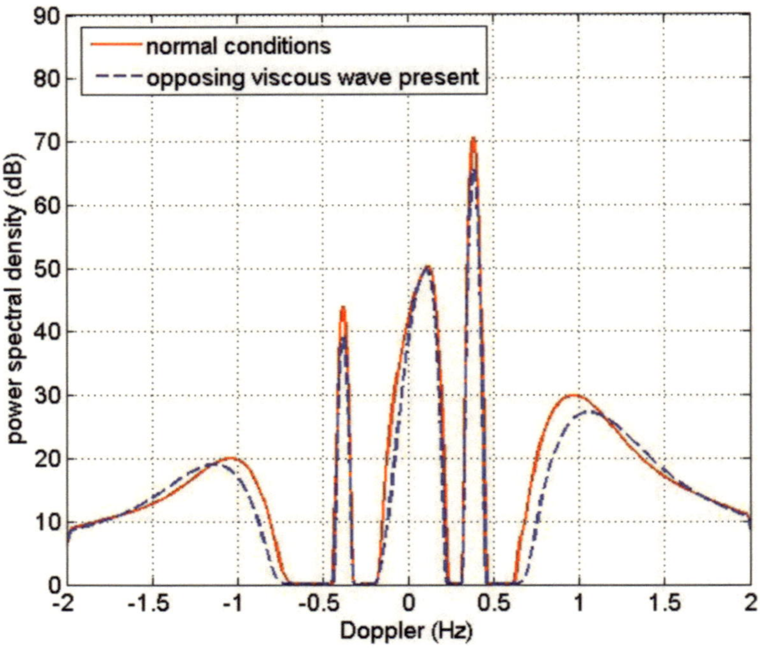

Fig. 7. Predicted HF sea clutter spectra for an ambient wind speed of 4 ms^{-1} with the viscous wave opposing the ambient wind. Radar frequency 15 MHz

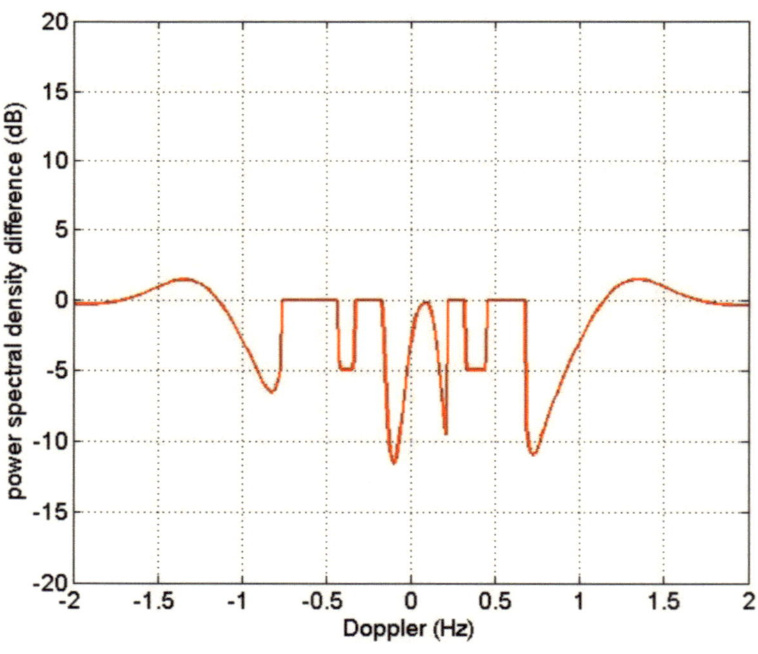

Fig. 8. The incremental change to the Doppler spectrum arising from the viscous wave for the opposing case

Experimental observations of changes to the sea roughness across the Boxing Day tsunami were made by the Jason-1 satellite at C- and Ku- bands (Troitskaya & Ermakov, 2006; Godin et al, 2009). Their swathe widths are well matched to measurement with HF radar, that is, a few tens of kilometres. Of course HF radar scattering mechanisms involve longer waves than those preferentially sampled by Jason-1, but the fact that changes were observed supports the viscous wave hypothesis.

3.2 Signatures involving the ionosphere

The ocean is coupled to the ionosphere by several physical and chemical processes of which two offer the prospect of generating a tsunami signature - mechanical coupling via the atmosphere and magnetic coupling via the geomagnetic field. The more obvious of these is the mechanical forcing due to the change in elevation of the water surface associated with the tsunami profile $\eta(\vec{x},t)$. As noted earlier, tsunami wavelengths are much greater than the water depth so the normal wave modes are essentially non-dispersive and the profile is, to first order, unchanging as the tsunami progresses, so it can be written as $\eta(\vec{x}-\vec{v}t)$ where $\vec{v}$ is the tsunami velocity. The shape of the profile is governed by the nature of the seismic displacement. It follows that the lower boundary of the atmosphere is subjected to a spatially and temporally varying forcing so the Navier-Stokes equation takes the form

$$\rho\left[\frac{\partial v}{\partial t}+(\vec{v}\cdot\nabla)\vec{v}\right]=-\nabla p+\rho\vec{g}+F \tag{18}$$

where the forcing terms correspond to compressibility, buoyancy and tsunami forcing respectively. An expression for F follows from considering the moving tsunami profile as a generalised piston, which performs work on the atmosphere according to the relation

$$\frac{\delta W}{\delta t}=-P_0\nabla\eta(\vec{x}-\vec{v}t)\cdot(\vec{v}-\vec{u}) \tag{19}$$

where P_0 is the atmospheric pressure at sea level and $\vec{u}$ is the wind velocity.

In the absence of external (tsunami) forcing, the Navier-Stokes equation, together with the equation of continuity,

$$\frac{\partial\rho}{\partial t}+\left(\nabla\cdot(\rho\vec{v})\right)=0 \tag{20}$$

and the equation of state for adiabatic behaviour,

$$\frac{\partial p}{\partial t}+\vec{v}\cdot\nabla\rho-C_0^2\left[\frac{\partial\rho}{\partial t}+\vec{v}\cdot\nabla\rho\right]=0 \tag{21}$$

describes the characteristic modes of the neutral stratified atmosphere (Hines, 1960). Here C_0 is the speed of sound,

$$C_0^2=\frac{\gamma p}{\rho} \tag{22}$$

Letting Ψ denote the vector of normalised perturbations of density and pressure from their mean values, together with the horizontal and vertical components of the fluid velocity vector,

$$\vec{\Psi} = \begin{bmatrix} \rho'/\rho_0 \\ p'/p_0 \\ v_z \\ v_x \end{bmatrix} \tag{23}$$

and setting $H = -\rho_0 \cdot \left(\frac{\partial \rho_0}{\partial z} \right)^{-1}$, the scale height,

we have

$$\begin{vmatrix} i\omega & 0 & -ik_x & -1/H - ik_z \\ 0 & -ik_x C_0^2/\gamma & i\omega & 0 \\ g & -C_0^2(1/H + ik_z)/\gamma & 0 & i\omega \\ -i\omega C_0^2 & i\omega C_0^2/\gamma & 0 & (\gamma - 1)g \end{vmatrix} \cdot \vec{\Psi} = 0 \tag{24}$$

whence we obtain

$$\omega^4 - \omega^2 C_0^2 \left(k_x^2 + k_z^2 \right) + (\gamma - 1) g^2 k_x^2 + i\gamma g \omega^2 k_z = 0 \tag{25}$$

that is, the dispersion relation for atmospheric waves. For an upward propagating wave we require that

$$k_z = k_z^{'} + i\frac{\gamma g}{2C_0^2} = k_z^{'} + i\frac{1}{2H} \tag{26}$$

We can then rewrite the dispersion relation as

$$\omega^2 C_0^2 k_z^{'} = \omega^4 - \omega^2 \left(C_0^2 k_x^2 + \frac{\gamma^2 g^2}{4C_0^2} \right) + (\gamma - 1) g^2 k_x^2 \tag{27}$$

For a given horizontal wavenumber k_x, a real solution for $k_z^{'}$ exists only when the right hand side of ( ) is non-negative. The roots of the quadratic obtained by setting it to zero, ω_1 and ω_2, divide the frequency domain into three bands of possible wave solutions:

i. $\omega < \omega_1 < \sqrt{(\gamma - 1)} \frac{g}{C_0} \equiv \omega_b$

This band corresponds to internal gravity waves, driven by buoyancy forces; ω_b is the Brunt-Vaisala frequency.

ii. $\omega > \omega_2 > \frac{\gamma g}{2C_0} \equiv \omega_a$

This band corresponds to acoustic waves, driven by compressibility forces; ω_a is the acoustic cut-off frequency.

iii. $\omega_1 < \omega < \omega_2$

This band corresponds to Lamb waves, which propagate only in the horizontal direction with $k_z^{'}$ purely imaginary.

In the lower atmosphere, $\frac{\omega_b}{2\pi} \sim 2.9 \times 10^{-3}$ Hz while $\frac{\omega_a}{2\pi} \sim 3.3 \times 10^{-3}$ Hz. At 400 km altitude these become 9.5×10^{-4} Hz and 1.1×10^{-3} Hz respectively.

From the viewpoint of tsunami detection, the most important property of these atmospheric waves is the amplification of the disturbance produced as they rise and the air density decreases. In the absence of attenuation, conservation of energy dictates that the amplitude of the vertical displacement is proportional to $\rho_0^{-1/2}(z)$. Thus, between sea level and an altitude of 125 km, the density decreases by a factor 10^{-8} so the wave amplitude grows by 10^4. A tsunami-induced surface displacement of amplitude of 10 cm may generate a displacement of ~1 km in the lower E-region. For F-region heights, the idealised amplification factor becomes 10^5 but by this height, kinematic viscosity exerts considerable damping, with peak displacement occurring at ~ 300 km. The collisional interaction of the neutral gas species with the ionised components leads to the perturbations in electron density which manifest themselves as travelling ionospheric disturbances, and it is these which have been proposed by various researchers as a possible tsunami signature. Particular interest has been shown in the three-dimensional imaging of the electron density distribution via tomographic processing of total electron content data. The time development of ionospheric perturbations over Japan, extracted from the GEONET satellite network, was found to correlate extremely well with a tsunami originating near Peru (Artru et al, 2005; Lognonne et al, 2006), and further support for the concept of TEC imaging of tsunami-generated ionospheric disturbances has come from modelling of those induced by the 2004 Boxing Day tsunami (Occhipinti et al, 2006).

The strength of the coupling between the time-varying surface displacement and any particular atmospheric wave component depends on a number of factors including not only the amplitude of the displacement but, just as importantly, its waveform. In principle this consists of both the forcing seafloor Rayleigh wave as mapped to the water surface and the freely-propagating hydrodynamic response, that is, the tsunami (Yamashita & Sato, 1976). The actual response of the neutral atmosphere at ionospheric heights to a particular tsunami event may be estimated by integration over the tsunami profile using an appropriate Green's function for the propagator in the stratified atmosphere. Various approximations to the Greens' function have been reported (Francis, 1974). Equivalently, one can employ a modal expansion in terms of the characteristic inertial-gravity and acoustic wave modes, matching the boundary conditions at the surface. The associated response of the electron density distribution in the ionospheric plasma involves an additional term in the equation of motion to account for the electromagnetic effects,

$$\rho\left[\frac{\partial v}{\partial t} + (\vec{v}\cdot\nabla)\vec{v}\right] = -\nabla p + \rho\vec{g} + \vec{j}\times\vec{B}_0 \tag{28}$$

where $\vec{j}$ depends on the Pedersen, Hall and parallel conductivities in the geomagnetic field, so the response sensed by HF radio waves is both location and direction dependent. The inclusion of Coriolis effects adds yet more complexity to the formulation. While these phenomena demand attention for a full parametric description, here we are concerned only with the broad issues of measurement by HF skywave radar.

An important consideration with the internal gravity wave disturbance is that it propagates obliquely upwards not at the acoustic velocity but at a much lower speed. As a consequence, the energy injected by the tsunami into the atmosphere will take 1 - 2 hours to reach the ionosphere where it can produce ionospheric effects detectable by an HF radar. The time taken is influenced by the ratio of the vertical and horizontal wavelengths, which in turn are governed by the tsunami parameters

Signature 3. The ground clutter Doppler spectrum modulation signature

When the internal gravity wave reaches the ionosphere, it perturbs the electron density distribution, predominantly via ion-neutral collisions and the Lorentz force. HF radar signals reflected from the perturbed ionosphere are modulated by the disturbance and this is manifested in the Doppler spectrum of ground clutter observed via the skywave path. Figure 9 presents a 'snapshot' of the Doppler shift associated with an unusual monocycle disturbance which was tracked by the Jindalee radar for over two hours, during which it changed form only slightly. Propagating towards the equator, the front half-cycle showed a

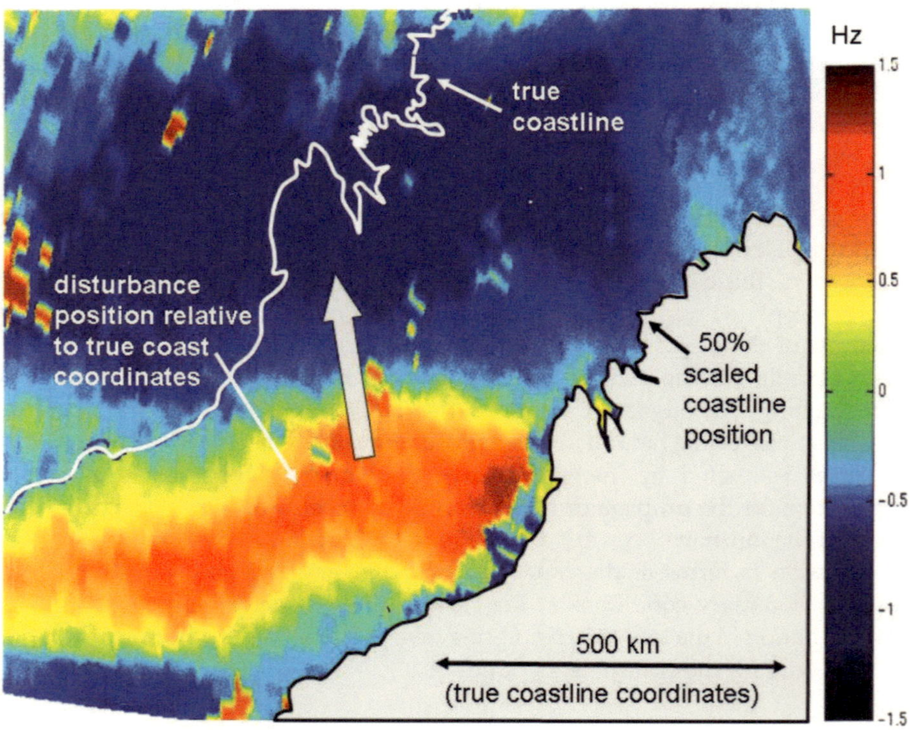

Fig. 9. A large-scale disturbance mapped with the Jindalee radar in 1986. The Doppler shift was measured for skywave paths to the ocean surface off NW Australia, then the resulting radar data geographically re-positioned so that the path Doppler shifts lie not at the corresponding ground reflection points but at the mid-points (control points) of the ray trajectories, where the phase modulation due to the disturbance is concentrated

significant negative Doppler shift, consistent with the ionosphere being displaced upwards by a gravity wave, followed by a positive Doppler restoring phase. The origin of this disturbance is not known, but it is included here because it resembles what we might expect from a tsunami.

Signature 4. The ground clutter intensity modulation signature

In addition to modulating the signal phase and hence imposing a Doppler shift on the radar echoes, internal gravity waves distort the virtual height distribution over the affected area, effectively producing a curved reflecting layer in the ionosphere. This results in focusing and defocusing of the radar signals, manifesting itself on the radar displays as a strong intensity modulation of the ground clutter. A good example of this is shown in Figure 10, a backscatter ionogram recorded with the Jindalee radar. Several cycles of a quasi-periodic disturbance are clearly visible.

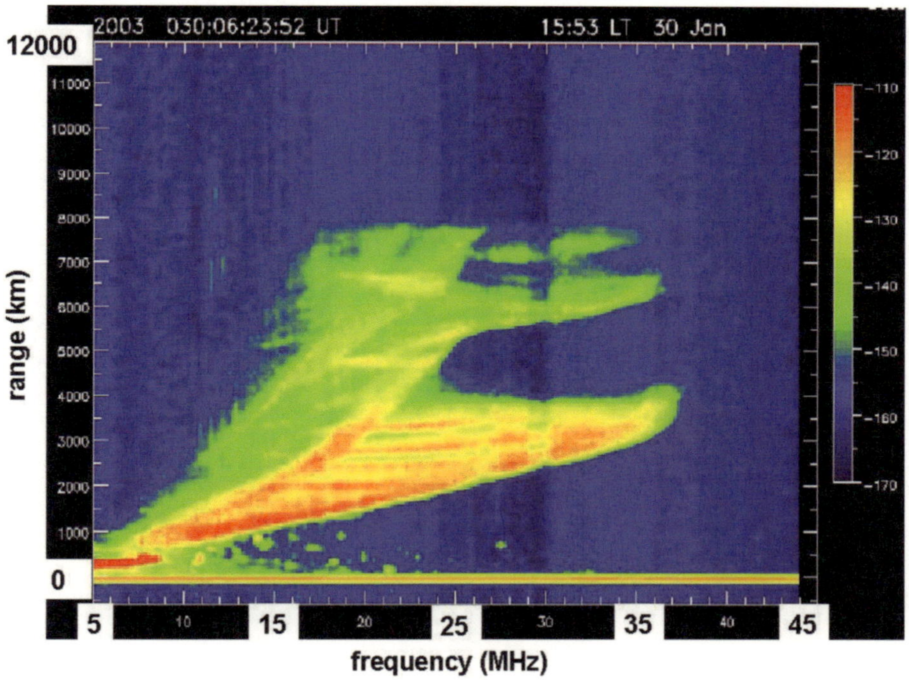

Fig. 10. A backscatter ionogram recorded during the passage of an internal gravity wave packet with 4 or 5 cycles. The bands of focusing and defocusing are almost independent of sounding frequency, supporting the interpretation. The fringe spacing is 470 km

Signature 5. The infrasonic ionosphere modulation signature

As noted above, the dispersion relation admits compression-dominated acoustic modes as well as buoyancy-dominated internal gravity waves, so there is a priori some prospect that these too might modulate the ionosphere in a way which impacts on HF radar signals. On closer examination, though, the outlook is not hopeful. At ionospheric heights, kinematic viscosity increases rapidly and this effectively imposes a low-pass filter on the wave (Spitsyn & Taraschuk, 1994). Figure 11 (adapted from Najita et al, 1974), shows the combined effect of amplitude growth and viscosity versus acoustic frequency, as a function of altitude. Clearly the spectrum of the tsunami-generated infrasonic emissions will govern its influence on a given HF skywave signal path because the fan of rays representing skywave radar illumination at a given frequency leaves substantial regions of the ionosphere unsampled. A second consideration is the likely radiation pattern of the emissions : over land, Rayleigh earthquake waves propagate at a speed much higher than the acoustic velocity C, so the source is horizontally in phase and effective upward radiation is achieved; moreover, this will be augmented by refractive focussing in stable atmospheric conditions. Over the ocean, the tsunami speed is typically ~ $0.7C_0$ so one might expect infrasonic waves excited by a tsunami to be radiated preferentially at a depth-related angle,

$$\theta_{eff} = \frac{\pi}{2} - ar\cos\left(\frac{C}{\sqrt{gD}}\right) \tag{29}$$

which corresponds to ~ 53° in water of depth 4000 m. Apart from the obvious signal loss due to increased atmospheric absorption along oblique paths, refractive signal trapping in the presence of atmospheric temperature inversions will be exacerbated.

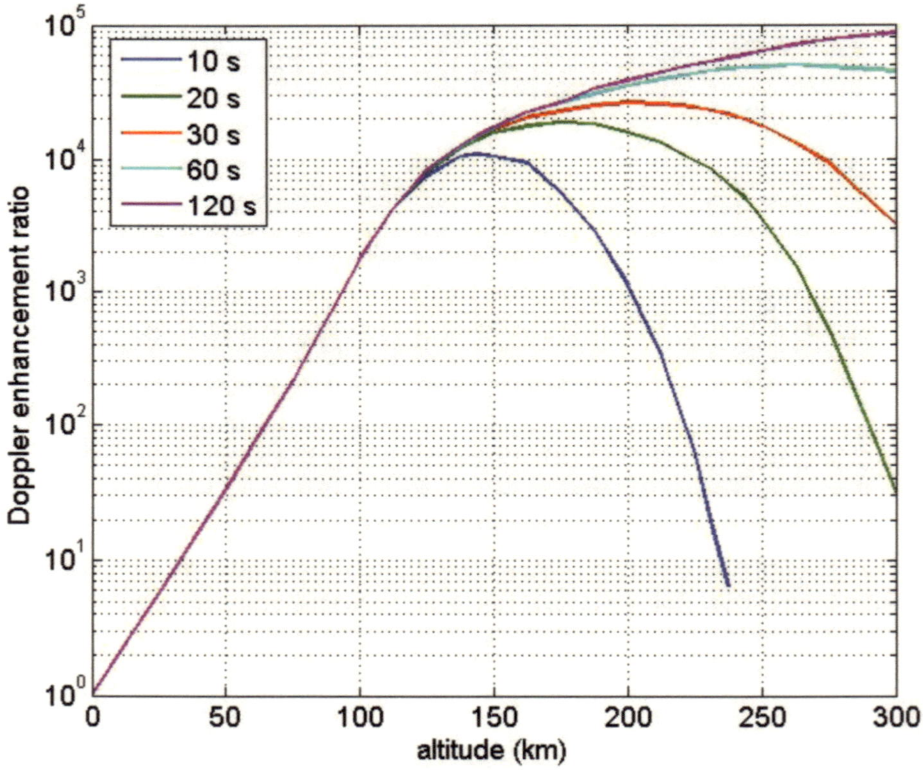

Fig. 11. The ratio of effective particle displacement at altitude to the displacement at sea level taking into account kinematic viscosity and density profiles.

In addition to these considerations, there is the issue of the coupling mechanism. Under 'frozen-in' field conditions, as apply in the ionosphere, infrasonic frequencies excite ULF magnetohydrodynamic (MHD) disturbances, that is, Alfven and fast mode oscillations at low latitudes and field line resonances at higher latitudes. These phenomena modulate the phase of HF radio waves via three mechanisms (Sutcliffe & Poole, 1976), namely, magnetic field changes, advection of the bulk plasma and compression of the plasma. There is also the issue of microbaroms, infrasonic radiation from standing waves in the ocean at long swell frequencies, though potentially these may be separable in the frequency domain.

On the other side of the ledger, infrasonic waves propagate to the lower ionosphere in about 10 minutes, in contrast to internal gravity waves which may take up to 150 minutes to reach the same height. Further, HF skywave radar has shown a remarkable ability to extract ULF phase modulations from ground and sea backscatter (Anderson & Abramovich, 1998). It should be helpful for discrimination that the azimuthal wavenumbers of ULF waves of magnetospheric origin are observed to be low, whereas tsunami-generated ULF disturbances are likely to be much more localised. Once the phase modulation sequence has been estimated, information about the cause may be inferred from the geographical distribution of the sequence structure and its temporal evolution. That this discrimination is possible can be seen from Figure 12 which shows the demodulation sequences of two specific ionospheric perturbations. Figure 12a shows the sequences for 4 range cells separated by 50 km, that is, every tenth range cell in the particular radar dwell.

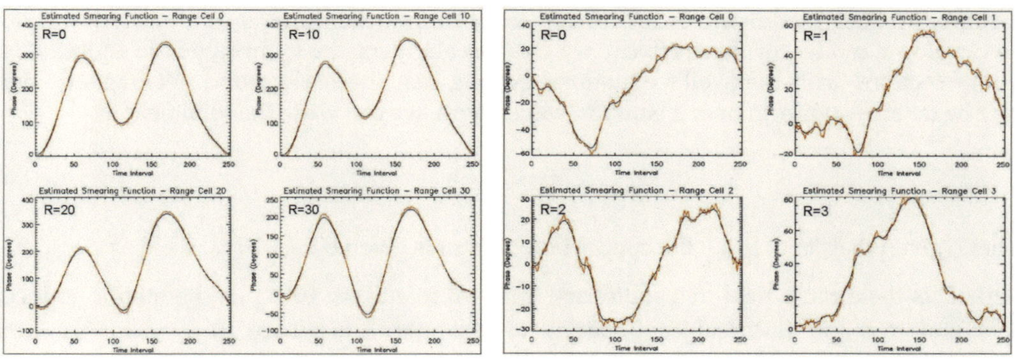

(a) (b)

Fig. 12. (a) Demodulation sequences for every 10th range cell for the case of an internal gravity wave; (b) demodulation sequences for adjacent range cells during a transient event of unknown origin. Data obtained with the Jindalee radar

The demodulation sequences are almost identical, indicating the high spatial coherence of the ionospheric perturbation which was identified by other means as a travelling ionospheric disturbance, that is, a manifestation of an internal gravity wave. In contrast, Figure 12b shows the behaviour observed during a different event; in this case the 4 range cells chosen were consecutive in range, a mere 5 km spacing, yet the sequences are far less correlated. This event was not formally identified with any particular cause.

Signature 6. The geo-magnetohydrodynamic tsunami-ionosphere coupling signature

The magnetic field within the terrestrial environment is the sum of contributions from a number of sources : (i) dynamo currents within the earth's core, (ii) magnetised rocks and other material in the earth's crust, (iii) electric currents in the ionosphere, and (iv) electric currents produced by the sea water motions. The last of these has recently been proposed as a candidate mechanism for tsunami detection and measurement (Anderson, 2008). Typically (ii) and (iii) contribute an amount less than 1% of the primary geomagnetic field due to (i), so they can often be neglected. The term of interest, (iv), is much smaller again in the case of normal ocean waves and swell but for tsunamis, with representative length scales some 10^3 times greater, a commensurate gain in the generated field perturbation might be expected. An attractive feature of this mechanism, if it is observable, is that such a disturbance propagates at the speed of light below the ionosphere.

Magnetic disturbances in the ionosphere are routinely observed with HF skywave radar when those disturbances come from above, that is, from the magnetosphere. The question addressed here is whether or not tsunamis might produce a detectable ionospheric signature from below. The mechanism involves the propagation to ionospheric heights of the magnetic field disturbances generated within the ocean body; there they can manifest themselves as Alfven waves which, under the prevailing 'frozen-in' field conditions, modulate the electron density distribution and hence impose a phase modulation on transiting radiowaves, very much as the infrasonic waves discussed earlier. Unlike the latter, though, there is little impediment to the disturbance propagating to F-region heights.

Evidence for this kind of geo-magnetohydrodynamic tsunami-ionosphere coupling appears to be available from some magnetometer records (Iyemori et al, 2005). Of course, the key limitation via-a vis tsunami warning would be discrimination of such effects from the background of similar disturbances arising from other mechanisms.

To develop a model for these effects, we need to augment the hydrodynamic equations of earlier sections with Maxwell's equations. Noting that the background geomagnetic field may be taken as uniform over a tsunami wavelength, we can write the total field as

$$\vec{B} = \vec{B}_{geo} + \vec{b} \tag{30}$$

where, given that $|\vec{b}| << |\vec{B}_{geo}|$ the conduction current is given by $\vec{J} = \sigma(\vec{E} + \vec{v} \times \vec{B}_{geo})$

Here $\vec{E}$ is the electric field in a stationary frame of reference, $\vec{v} \times \vec{B}_{geo}$ is the motion-induced field and σ is the electrical conductivity of seawater, taken here as 4 mhos.m^{-1}. From Maxwell's equations,

$$\nabla \times \vec{b} = \mu_0 \vec{J} + \varepsilon \mu_0 \frac{\partial E}{\partial t} \tag{31}$$

but at tsunami wave frequencies, the displacement current in seawater is negligible, so the magnetic field perturbation $\vec{b}$ caused by the current is given simply by Ampere's law,

$$\nabla \times \vec{b} = \mu_0 \vec{J} \tag{32}$$

Further, the electric and magnetic fields are related by Faraday's law,

$$\nabla \times \vec{E} = -\frac{\partial \vec{b}}{\partial t} \tag{33}$$

The modest conductivity of sea water is essentially ionic, and hence the Lorentz body force on the fluid parcels is negligible compared with the pressure and buoyancy forces, so there are no hydromagnetic forces to be taken into account.

Taking the curl of (31), substituting for E from (33) and using the identity $\nabla\times\left(\nabla\times\vec{b}\right)=\nabla\left(\nabla\cdot\vec{b}\right)-\nabla^2\vec{b}$, we have

$$-\nabla^2 b=\sigma_0\left(-\frac{\partial\vec{b}}{\partial t}+\nabla\times\left(\vec{v}\times\vec{B}_{geo}\right)\right)-\varepsilon\frac{\partial^2\vec{b}}{\partial t^2} \quad (34)$$

where, from Section II, the fluid velocity associated with the tsunami, as modelled by a harmonic shallow water wave, is given by

$$\vec{v}\equiv\begin{bmatrix}v_x\\ v_y\\ v_z\end{bmatrix}=\begin{bmatrix}\dfrac{-ag\cosh\left(K\left(z+h\right)\right)}{\cosh\left(Kh\right)}\cos\left(K\left(x-ct\right)\right)\\ 0\\ \dfrac{ag\sinh\left(K\left(z+h\right)\right)}{\cosh\left(Kh\right)}\sin\left(K\left(x-ct\right)\right)\end{bmatrix} \quad (35)$$

Inserting (35) into (34) yields a linear PDE which can be solved directly by Fourier transform. Thus we have a numerical model for the magnetic field perturbation caused by the tsunami. For normal sea waves, the field decays fairly rapidly above the surface with a decay factor related to the water wavelength. Given the relative length scales of tsunamis and wind waves, we may expect a tsunami to generate in-phase contributions up to the height of the ionosphere.

3.3 Signatures involving coastal processes

The arrival of a tsunami at a coastline will generate transient responses in some, though not all, of the signatures mentioned so far. These could be difficult to interpret but may not be hard to detect. Depending on the bathymetry, the advection signature will shift in Doppler before broadening as spatial refraction inhomogeneity and nonlinear wave development distort the near-unidirectional, non-dispersive deep water waveform prior to it breaking close to the shore. Prevailing current patterns will be disturbed on large and small scales (Reddy et al, 2009). Strong fixed scatterers on the shore may have their radar cross section changed drastically. Infrasound generation may be intensified, first because as the water depth decreases and the wave grows, large quantities of surface air are being compressed and rarefied, producing infrasonic waves of the observed frequency, around 0.03 Hz. A second mechanism is the interaction of the tsunami wave with the shoreline and regions of steep bathymetry(Yeh et al, 1994). Standing waves set up by reflection constitute an acoustic ‘oscillator’, similar to that responsible for microbaroms.

The relevance of all this to tsunami warning arises when a landmass engendering these transient responses lies closer to the source than the region to be alerted. In effect the landmass serves as a ‘canary’ inserted into the tsunami’s near field to produce an amplified reaction. Thus it was reported that atmospherically ducted infrasonic radiation measured at monitoring stations in the Indian Ocean region two or three hours after the Boxing Day tsunami was highly correlated with the tsunami wave landfall on the coasts of Burma and Bangladesh (Le Pichon et al, 2005). Contingent on the presence of suitable islands or other landmasses in the source region, this kind of sampling should not be overlooked. At the very least, opportunities to observe magnified advection signatures, as illustrated in Figure 3, should be exploited.

4. Limits to the performance of HF skywave radars in the tsunami warning role

The ability of HF skywave radar to detect and quantify any of the signatures discussed above is heavily constrained by factors such as (i) radar design, (ii) radar siting relative to the geographical area of concern, (iii) the instantaneous, diurnal, seasonal and inter-annual properties of the ionosphere insofar as it influences the skywave propagation channel, and (iv) the existence and prevalence of other geophysical phenomena whose signatures resemble those associated with tsunamis. Some of these issues have already been mentioned, many others are obvious. The severity of these limiting factors has often been under-estimated by those without direct experience with HF skywave radar systems, so it is apposite to provide some quantitative guide to what is achievable in practice, though leaving aside radar system issues.

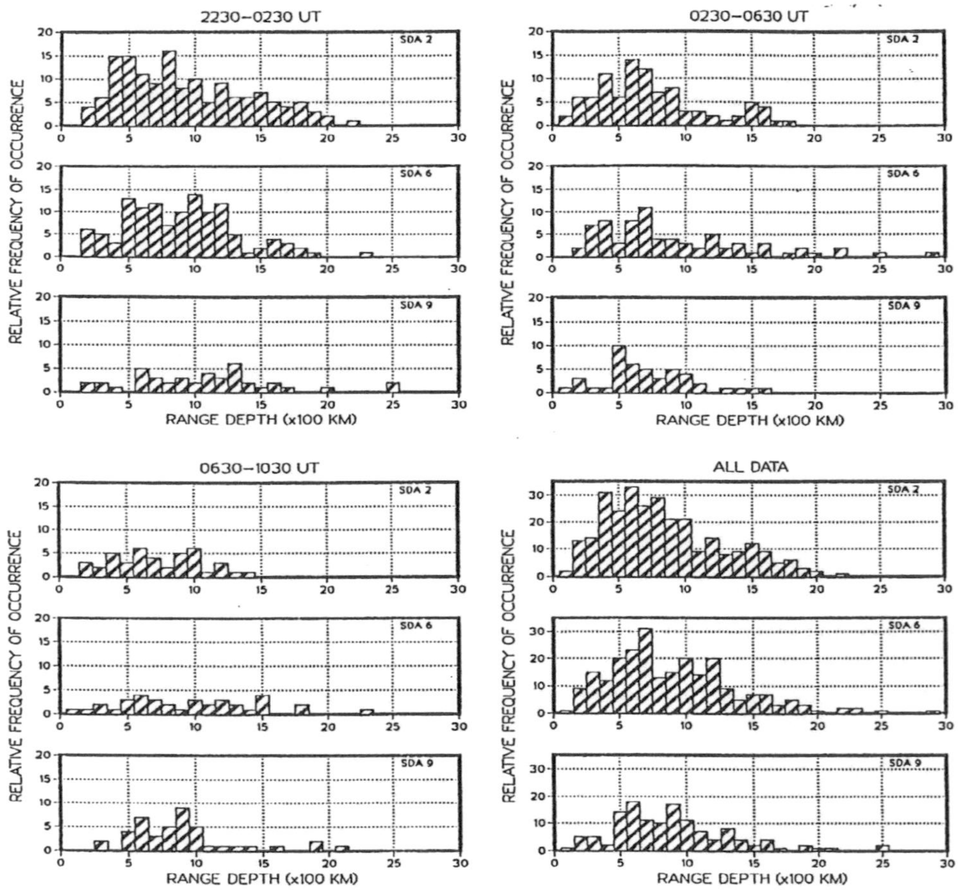

Fig. 13. Statistics of data quality relevant to observability of different tsunami signatures

First, with regard to skywave propagation, Figure 13 shows some statistics of data quality as measured on the SDA scale (Anderson, 1992). Here a rating of 9 represents near-perfect conditions, with no discernible Doppler shift, spectral broadening or multimode, whereas a rating of 0 applies to data with very low sub-clutter visibility and highly unstable phase paths. The data is binned according to instantaneous range coverage at the optimum frequency. The data used for this analysis is particularly realistic in the sense that it is not extracted from historical synoptic databases derived from frequency management systems but, instead, from records of specific missions undertaken by the Jindalee radar when the radar operators were doing their best to optimise the radar in real-time. Three levels of SDA index have been chosen for presentation here: 9, 6 and 2. At SDA=9, even the subtle roughness signature (Signature 2) should be readily observable and measurable, while at SDA=6, that would not be the case, though most ionospheric gravity wave signatures should be reasonably accurately measured. At SDA=2, large-scale ionospheric perturbations would be detectable on most occasions but would not be measurable with the accuracy needed to quantify tsunami parameters. Three time intervals have been selected : 0800 - 1200 local time, 1200 - 1600 local time and 1600 - 2000 local time. By and large, conditions at night are less favourable for the tsunami warning mission.

Figure 14, showing typical ionospheric activity as manifested in the Doppler domain, illustrates the challenge of discrimination between natural fluctuations and tsunami-generated disturbances. It seems likely that reliable discrimination will demand the use of physical models of the latter.

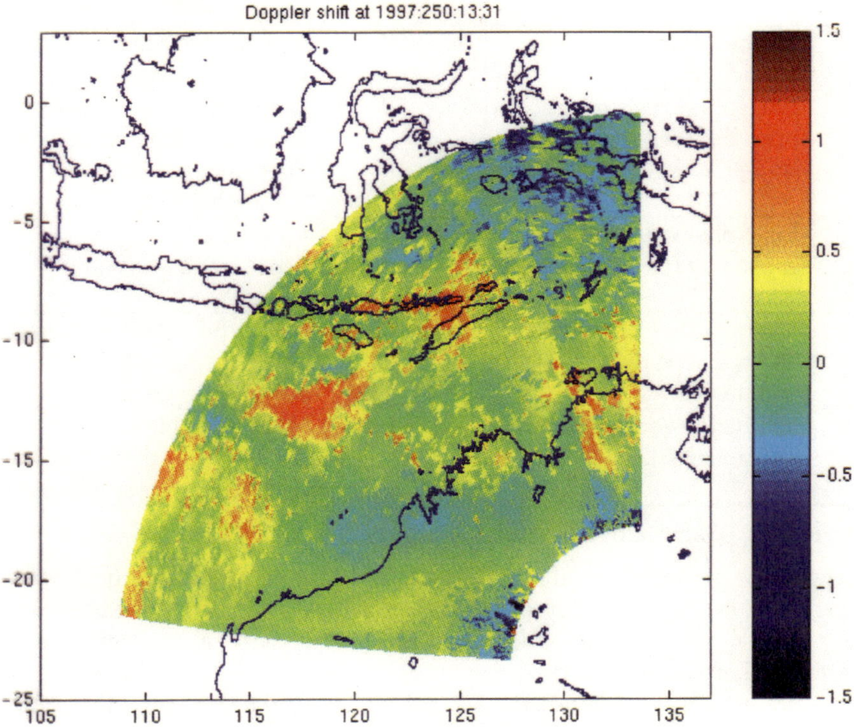

Fig. 14. A snapshot of the Doppler shift imposed on HF skywave backscatter signals under typical propagation conditions.

While detailed modelling of the physics of the signatures postulated above has yet to reach maturity, it is evident that the spatial and temporal scales of most of the identified phenomena are within the corresponding sampling resolution of HF skywave radar, while the radar Doppler resolution is exceptionally fine and likely to be the key to quantitative characterisation of tsunami amplitude. The utility of the various signatures in terms of warning and characterisation is far from uniform, as seen from the latencies listed in Table 2.

5. Conclusion

The survey reported here confirms that tsunamis may register signatures on HF radar systems via a number of different geophysical mechanisms. The mechanism cited frequently in the literature for the case of HF surface wave radar – that of advection of Bragg-resonant waves by tsunami-generated flows on the coastal shelf – is perhaps the most direct and the most accessible, but provides a useful warning only where the coastal shelf is sufficiently broad. Some other mechanisms may yield measurable signatures over much wider expanses of the ocean, though experience has shown that radar performance for this kind of diagnosis drops sharply beyond the one-hop zone.

CLASS	**PHYSICAL PHENOMENON**	**LATENCY (min)**	**VIABILITY FOR SKYWAVE RADAR**
1	Bragg-resonant gravity wave advection	1 - 2	Difficult via skywave Only possible on continental shelf
2	Viscous wave effects on surface roughness	30 – 100	Not yet demonstrated Latency an issue
3	Ionospheric disturbance caused by internal gravity waves	75 – 150	Detection well established Inversion yet to be demonstrated Poor latency Somewhat sensitive to geometry
4	Ionospheric disturbance caused by infrasonic waves	7 – 15	Limited effect Sensitive to geometry
5	Modulation of geomagnetic field by tsunami flows	1 – 2	Magnitude of effect not established Other causes exist Sensitive to geometry
6	Anomalous echoes from distant coastlines	150 – 200	Many variables Excessive latency Alternative explanations

Table 2. Summary of alternative signature latency and feasibility

While there can be no doubt that the reaction of the geophysical environment to the passage of a tsunami produces a number of effects which are potentially measurable with HF skywave radar, practical issues such as latency and false alarm rate are important considerations. There is also the matter of time line access – most if not all operational HF skywave radars have heavy operational tasking which cannot co-exist with a synoptic scale tsunami watch mission. The suitability of HF skywave radar as a tsunami sensor has been assessed here by examining the alternative mechanisms in enough detail to give some understanding of the physics involved, and interpreting modelled signatures in the context of genuine operational radar performance statistics. The evidence seems unequivocal – more often than not, the signal quality of HF skywave radar echoes will not support the identification and interpretation of the more subtle tsunami signatures, even when parametric signature models achieve higher fidelity than prevails today. Given the nature and magnitude of the candidate signatures, and the variety of alternative causative mechanisms, the outlook can scarcely be regarded as promising, but neither does it rule out a possible contribution of HF skywave radar to a tsunami warning system.

6. References

Anderson, S.J. (1992). Adaptive Remote Sensing with HF Skywave Radar, Proceedings of the IEE, Part F, Vol.139, No.2, 182-192

Anderson, S.J. (1994). Tsunami detection with HF radar. *Proceedings of National Tsunami Workshop* 1994, Brisbane, Bureau of Meteorology

Anderson, S.J. & Abramovich, Y.I. (1998). A unified approach to detection, classification and correction of ionospheric distortion in HF skywave radar systems. *Radio Science*, Vol.33, No.4, 1055–1067

Anderson, S.J. (2008). Prospects for tsunami detection and characterisation with HF skywave radar. *Proceedings of the IEEE International Radar Conference*, Adelaide, September 2004

Arteru, J., Ducic, V., Kanamori, H. & Lognonne, P. (2005). Ionospheric detection of gravity waves induced by tsunamis. *Geophysics Journal International*, Vol. 160, 840-848

Barrick, D.E. (1979). A coastal radar system for tsunami warning, *Remote Sensing of Environment*, Vol. 8, 353-358

Coisson, P., Occipinti, G., Rolland, L., Lognonne, P. and Harmel, T. (2008). Can OTH radar help tsunami warning systems ? *Proceedings of the AGU 2008 Fall meeting*, San Francisco, December 2008

Dzvonkovskaya, A. & Gurgel, K.-W. (2009). Future Contribution of HF Radar WERA to Tsunami Early Warning Systems. *European Journal of Navigation*, Vol. 7, No. 2, 17-23

Francis, S.H. (1975). Global propagation of atmospheric gravity waves : a review. Journal of Atmospheric and terrestrial Physics, Vol.37, 1011-1054

Godin, O.A. (2004). Air-sea interaction and feasibility of tsunami detection in the open ocean. *Journal of Geophysical Research*, Vol. 109, C05002

Godin, O.A. (2008). Wind over fast waves and feasibility of early tsunami detection from space. *private communication*

Godin, O.A., Irisov, V.G., Leben, R.R., Hamlington, B.D. and Wick, G.A. (2009). Variations in sea roughness induced by the 2004 Sumatra-Andaman earthquake. *Natural Hazards Earth Systems Science*, Vol.9, 1135-1147

Headrick, J.M. & Anderson, S.J. (2008). HF Over-the-Horizon Radar. *Chapter 20, Radar Handbook*, Merrill Skolnik, (Ed.), 3rd edition, McGraw Hill

Heron, M.L., Prytz, A., Heron, S.F., Helzel, T. & Schlick,T.(2007). *Application of HF coastal ocean radar to tsunami observations.*

http://www.helzel.com/helzelmed/literature/Tsunami-Australia-6912.pdf

Hines, C.O. (1960). Internal atmospheric gravity waves at ionospheric heights. *Canadian Journal of Physics*, Vol.38, 1441-1481

Hines, C. O. (1972). Gravity waves in the atmosphere. *Nature*, Vol. 239, 73– 78

Iyemori, T., Nose, M., Han, D., Gao, Y., Hashizume, M., Choosakul, N., Shinagawa, H., Tanaka, Y., Utsugi, M., Saito, A., McCreadie, H., Odagi, Y. & Yang, F. (2005).

Geomagnetic pulsations caused by the Sumatra earthquake on December 26 2004. *Geophysical Research Letters*, Vol. 32, L20807,

Kinsman, B. (1967). *Wind Waves: Their Generation and Propagation on the Ocean Surface*, Prentice-Hall, Englewood Cliffs, NJ

Koshevaya, S.V., Grimalsky, V.V., Siqueiros-Alatorre, J., Perez-Enriquez, R. and Kotsarenko, A.N. (2004). Acoustic and acousto-gravity wave pulses caused by sources of seismic origin. *Physica Scripta*, Vol.70, 72-78

Le Pichon, A., Herry, P., Mialle, P. & Verguoz, J., Brachet, N., Garces, M., Drob, D. & Ceranna, L. (2005). Infrasound associated with 2004-2005 large Sumatra earthquakes and tsunami. *Geophysical Research Letters*, Vol. 32, L19802

Lipa, B.J., Barrick, D.E., Bourg, J. & Nyden, B.B. (2006). HF radar detection of tsunamis, *J. Oceanography*, Vol. 62, 705-716

Lognonne, P., Garcia, R., Crespon, F., Occipinti, G., Kherani, A. and Artru-Lambin, J. (2006). Seismic waves in the ionosphere. *Europhysics*, Vol.37, No.4, 11-14

Marquart, N. (2007). *Concept study about new ground-based and space-borne HF and microwave systems for tsunami detection.* DLR Microwaves and Radar Institute Project GITEWS Concept Study Report, GITEWS-HF-DLR-001

Najita, K., Weaver, P. & Yuen, P.C. (1974). A tsunami warning system using and ionospheric technique. *Proceedings of the IEEE*, Vol.62, No.5, 563-567

NOAA (2010). http://www.ndbc.noaa.gov/dart.shtml

Occipinti, G., Dorey, P., Bazin, V. & Lognonne, P. (2004). Nostradamus : a new ionospheric seismometer, *Proceedings of the IEEE International Radar Conference*, Toulouse, 2004

Occipinti, G., Lognonne, P., Kherani, E.A. & Hebert, H. (2006). Three-dimensional waveform modelling of ionospheric signature induced by the 2004 Sumatra tsunami. *Geophysical Research Letters*, Vol. 33, L20104

Occhipinti, G., Dorey, P., Farges, T. & Lognonne, P. (2010). Nostradamus: The radar that wanted to be a seismometer. *Geophysical Research Letters*, doi:10.1029/2010GL044009, in press.

Peltier, W.R. & Hines, C.O. (1976). On the possible detection of tsunamis by a monitoring of the ionosphere. *Journal of Geophysical. Research*, Vol. 81, No. 12, 1995-2000

Reddy, N., Aung, T. & Singh, A. (2009). Effect of 2004 'Boxing Day' tsunami on water properties and currents in the Bay of Bengal. *American Journal of Environmental Sciences*, Vol.5, No.3, 247-255

Spitsyn, V.G. & Taraschuk, Y.E. (1994). Filtering in the upper atmosphere of acoustic waves generated by earthquakes. *Geomagnetism and Aeronomy*, Vol.34, No.1, 125-126

Sutcliffe, P. R., & Poole, A.W.V. (1984). Low latitude Pc3 pulsations and associated ionospheric oscillations measured by a digital chirp ionosonde. *Geophysical Research Letterrs*, Vol.11(12), 1172–1175

Troitskaya, Y.I. & Ermakov, S.A. (2006). Manifestations of the Indian Ocean tsunami of 2004 in satellite nadir-viewing radar backscatter variations. *Geophysical Research Letters*, Vol. 33, 2006

Yamashita, T. & Sato, R. (1976). Correlation of tsunami and sub-oceanic Rayleigh wave amplitudes: possibility of the use of the Rayleigh wave in tsunami warming system. Journal of the Physics of the Earth, Vol.24, 397-416

Yeh, H., Liu, P., Briggs, M. and Synolakis, C. (1994). Propagation and amplification of tsunamis at coastal boundaries. *Nature*, Vol.372, 353-355

Atmospheric Gravity Waves and Effects in the Upper Atmosphere Associated with Tsunamis

Michael P. Hickey
Embry-Riddle Aeronautical University
USA

1. Introduction

Tsunamis propagate at the surface of the deep ocean horizontal phase speeds of approximately 200 m/s, which is about two-thirds of the lower atmospheric sound speed. They have large horizontal wavelengths that are typically of a few hundred kilometers, and they remain coherent over large propagation distances. They also have large horizontal extents (sometimes a few thousand kilometers) parallel to their wave fronts. They can traverse great distances over a span of several hours, so that large areas of the ocean-atmosphere interface are impacted. Typical dominant wave periods associated with tsunamis are a few tens of minutes. In the deep ocean their amplitudes are usually quite small with surface displacements being only a few centimeters, but occasional large events can have amplitudes of a few tens of cm.

The speeds, wavelengths and periods of tsunamis lie within the range of those of atmospheric gravity waves. These are vertically transverse waves with motions of air parcels mainly influenced by gravity and buoyancy. The vertical displacement of the water acts like a moving corrugation at the base of the atmosphere and so very effectively generates atmospheric gravity waves. In general a spectrum of waves will be produced by a tsunami. Most of the power in the spectrum resides in internal gravity waves, with acoustic waves and evanescent waves being less efficiently generated. Internal waves can transport energy and momentum vertically through the atmosphere. Due to the decrease of mean atmospheric density with increasing altitude, the amplitude of these waves increases as they propagate upward in order to conserve wave energy. At sufficiently high altitudes molecular viscosity and thermal conductivity damp the waves, and their amplitudes then decrease with increasing altitude. Because the waves have high phase speeds (commensurate with the tsunami speed), they are deep waves with vertical wavelengths of ~ 100 km. This allows them to reach the middle thermosphere (~ 250 km altitude) before the molecular dissipation becomes severe.

Atmospheric winds also influence the upward propagation of atmospheric gravity waves. Because the winds vary with height the waves may be propagating with the wind at some heights and against the wind at other heights. In the former case the vertical wavelengths are shortened, which increases the velocity shears and thereby increases the viscous damping rate. In the latter case the vertical wavelengths are increased, which decreases the velocity shears and decreases the viscous damping rate.

At these heights the tsunami-driven atmospheric gravity waves have large amplitudes so that their interaction with the ionosphere is likely to produce detectable traveling ionospheric disturbances (TIDs). These TIDs have been observed in GPS measurements of total electron content and have exhibited a strong correlation with individual tsunami events. Modeling studies have helped the interpretation of such observations. The interaction between the neutral particles and the ions also leads to a further loss of wave energy through ion drag, but viscous effects dominate over ion drag for the typical gravity wave periods associated with tsunamis.

In addition to perturbations of the ionization, the chemistry of the thermosphere is also affected by atmospheric gravity waves. Chemiluminescent airglow emissions associated with neutral and ion chemistry are then affected. These emissions that emanate from a deep region centered at ~ 250 km altitude can be observed from the ground and/or from space. Oblique viewing allows the tsunami-driven disturbances in the thermosphere to be seen from considerable horizontal distances.

Internal atmospheric gravity waves can also transport energy and momentum over considerable vertical distances. The importance of this is due to the considerable decreases of atmospheric density with increasing altitude. In the tenuous upper atmosphere the deposition of small quantities of wave momentum can have a profound effect. Simulations suggest that for large tsunami events the dissipating gravity waves can accelerate the middle thermosphere by ~ 100 m/s within a few hours. These modeling predictions are yet to be confirmed by observations.

In this article observations and modeling of tsunami-driven atmospheric gravity waves are reviewed. Particular emphasis is given to the theory of atmospheric gravity waves, their vertical propagation, and their interaction with the ionosphere and thermospheric chemistry. The ability to measure a disturbance at a large distance from the tsunami offers a means of early detection to augment other early warning measurements. The utility of modeling and simulation in the interpretation of such tsunami-driven disturbances in the thermosphere is demonstrated.

2. Atmospheric acoustic-gravity waves

Around the dawn of the twentieth century with the advent of radio wave communications came the subsequent discovery of the ionosphere. New research was initiated related to ionospheric physics and chemistry and many new discoveries followed. One of these was the existence of traveling ionospheric disturbances (TIDs), which were ionospheric irregularities propagating rapidly in the horizontal direction with speeds of many tens to hundreds of m/s, and often with large horizontal wavelengths of 100s of km (Munro, 1948). Martyn (1950) was the first to explain the TIDs in terms of internal atmospheric gravity waves, and Hines (1960) finally produced a correct and unambiguous description of the phenomenon in terms of these waves.

Atmospheric gravity waves are fluctuations of the neutral atmosphere, usually triggered by events that cause a lifting of localized regions of the atmosphere. In the lower atmosphere sources of gravity waves include severe weather systems (Georges, 1968), and winds blowing over mountains (Scorer, 1949). A more complete discussion of these and other sources is given by Fritts and Alexander (2003). Gravity waves are the low frequency component of the more general class of acoustic-gravity waves. The acoustic waves, of higher frequencies than gravity waves, are primarily driven by events that produce a rapid compression of a region of the atmosphere. We remark that compressional effects are also important for high frequency gravity waves. At very high altitudes the oscillatory motion of air particles associated with gravity waves causes the ionospheric ions (and electrons) to oscillate due to neutral-ion collisions. The ion-electron motions are constrained by the geomagnetic field, chemistry, and electric fields.

The initial theory of acoustic-gravity waves of Hines (1960) was based on an idealized atmosphere (windless, isothermal, adiabatic wave motions). This important study showed that internal atmospheric gravity wave amplitudes will grow exponentially with increasing height as the waves propagate upward in order to conserve wave energy. If mean pressure and density decrease exponentially as $e^{-z/H}$ (where H is the atmospheric scale height), then wave amplitudes grow as $e^{z/2H}$. Due to the fact that the mean atmospheric density decreases by a factor of ~ 10^6 between sea level and the lower thermosphere (near 100 km altitude), wave amplitudes will increase by a factor of ~ 10^3 during the upward propagation of the waves over this region. A dispersion relation was derived relating the vertical wavelength ($\lambda_z = 2\pi/m$) to the horizontal wavelength ($\lambda_x = 2\pi/k$) and wave period ($T = 2\pi/\omega$)

$$m^2 = k^2\left(\frac{N^2}{\omega^2} - 1\right) - \frac{\left(\omega_a{}^2 - \omega^2\right)}{C^2} \tag{1}$$

Here, ω is the wave frequency, m is the vertical wavenumber, and k is the horizontal wavenumber. Also, C ($\sqrt{\gamma g H}$) is the sound speed, N is the Brunt-Väisälä frequency (for an isothermal atmosphere $N^2 = (\gamma - 1)g^2/C^2$), and ω_a $(= \gamma g/2C)$ is the acoustic cut-off frequency. Also, γ is the ratio of specific heats and g is the gravitational acceleration. Inspection of Equation (1) reveals that internal waves (those with real m) exist for either $\omega < N$ (gravity waves) or for $\omega > \omega_a$ (acoustic waves). For intermediate wave frequencies $N < \omega < \omega_a$ only evanescent waves exist. Internal waves ($m^2 > 0$) can transport energy and momentum vertically through the fluid in which they propagate, but evanescent waves ($m^2 < 0$), of which surface waves are a subset, cannot. Typically, gravity waves have periods ranging from ~ 10 min to a few hours, and horizontal wavelengths ranging from a few tens of km to several hundreds of km (occasionally even larger). They propagate with phase speeds of tens to hundreds of m/s, and are subsonic. Because of this, the slower gravity waves tend to have lower atmospheric sources (where the sound speed is lower), while fast gravity waves have high altitude sources where the sound speed is high. Gravity waves are vertically-transverse waves primarily influenced by the competing effects of gravity and buoyancy.

To help better understand the propagation of gravity waves we next examine the atmospheric mean state. Figure 1a shows altitude profiles of the mean temperature (T) obtained from the MSIS-90 model (Hedin, 1991), and the derived Brunt-Väisälä period (τ_B) and the sound speed C. The conditions are appropriate for the equator at times of low solar and geomagnetic activity (see Hickey et al., 2009). The temperature increases rapidly in the lower thermosphere, and asymptotes to a value of ~ 753 K in the upper thermosphere. In the thermosphere τ_B increases rapidly with increasing height and asymptotes to a value of ~ 12 min at high altitudes. Hence gravity waves having periods exceeding ~ 12 min will be internal at all heights ($\omega < N$ everywhere). The altitude variation of τ_B suggests that waves with periods satisfying 4 min ≤ T ≤ 12 min will be evanescent over some altitude range and

so would have diminished amplitudes at high altitudes. Hence, gravity waves with periods exceeding ~ 12 min should dominate the F-region (near 300 km altitude) response to an upward propagating gravity wave disturbance. The sound speed profile suggests that gravity waves of lower atmosphere origin will have phase speeds of less than about 300 m/s.
The corresponding mean winds used in our model, obtained with the HWM93 model (Hedin et al., 1996), are shown in Figure 1b. The meridional winds remain relatively small at most altitudes, with a maximum value of ~ 30 m/s (northward) near 250 km altitude. The zonal winds are much larger in the thermosphere, and approach 70 m/s (eastward) at high altitudes. Because the tsunami-generated gravity waves are fast (~ 200 m/s), we expect them to be relatively unaffected by the meridional winds. On the other hand, the zonal winds are expected to influence the gravity waves.

3. Tsunami-related upper atmosphere observations

Hines (1972) first postulated that tsunamis may be capable of producing atmospheric gravity waves that could subsequently propagate to high altitudes. Later, Peltier and Hines (1976) demonstrated the plausibility of this idea using a simple model of a gravity wave disturbance. The subject of tsunami-generated gravity waves remained relatively unexplored until the observations of Artru et al. (2005). They used Global Positioning System (GPS) measurements of total electron content (TEC) following the earthquake of Peru on June 23, 2001, and showed that the observed ionospheric disturbance had similar characteristics to the expected tsunami-generated gravity wave. In particular, they derived a horizontal propagation velocity of ~ 150 m/s in a direction consistent with the source location.
Ionospheric observations showing evidence of gravity waves following the December 26 2004 Sumatra tsunami were reported by Liu et al. (2006a, b) and by DasGupta et al. (2006). Lee et al. (2008) observed large ionospheric variations with accompanying spread F over Arecibo on the evening of December 26 2004. They postulated that tsunamigenic gravity waves associated with the Sumatra event were responsible for the observations.

4. Modeling gravity wave propagation

4.1 Previous modeling

Artru et al. (2005) used the basic theory developed by Hines (1960) to model gravity wave propagation and to provide a match to their observations. Ionospheric perturbations were not specifically modeled. However, by deriving the group velocities for the disturbances they were able to obtain good agreement between modeled and observed propagation times for the disturbances.
Occhipinti et al. (2006) performed time-dependent modeling of the December 26th 2004 Sumatra tsunamigenic gravity waves assuming that the atmospheric wave motions were inviscid and adiabatic. The ocean sea surface displacement was defined using a model of Hébert et al. (2007). A nonlinear ionospheric perturbation model was also included in order to be able to simulate the observed TEC observations. They found that the ionospheric response depended sensitively on the magnetic latitude (as described in the earlier work of Hooke, 1968). The relative TEC perturbation amplitudes were about 10%, which was noted to be in agreement with the observations of Liu et al. (2006b). Their results were also found to be largely insensitive to the effects of ion chemistry and diffusion. Occhipinti et al. (2008) continued from their previous work with a strong focus on the latitudinal variation of the

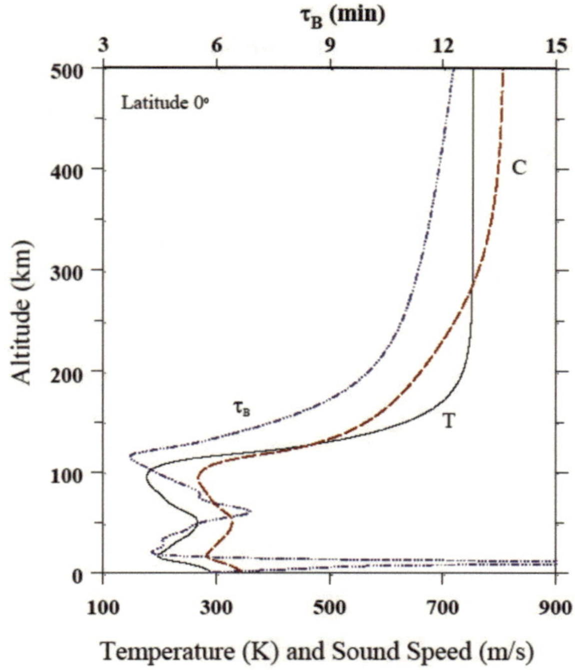

Fig. 1a. Mean temperature (T, in K), sound speed (C, in m/s) and Brunt-Väisälä period (τ_B , in min). After Hickey et al. (2009)

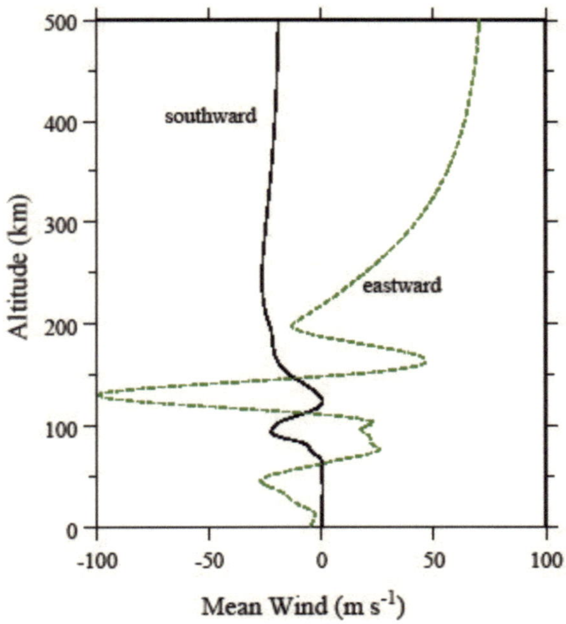

Fig. 1b. Mean southward and eastward winds (m/s). After Hickey et al. (2009)

ionospheric response to a tsunamigenic gravity wave. A clear anisotropic response was noted in their derived TEC perturbations. Note that their model used the Boussinesq approximation which applies to low frequency gravity waves for which the perturbations can be assumed to be incompressible; hence the model is essentially the same as that of Occhipinti et al. (2006).

4.2 Full-wave model description

The full-wave model, described below, has been previously used to model acoustic wave (Hickey et al., 2001; Schubert et al., 2005; Walterscheid and Hickey, 2005) and gravity wave (Hickey et al., 1997; Walterscheid and Hickey, 2001) propagation in Earth's atmosphere. A Cartesian coordinate system is used with x, y and z directed southward, eastward, and upward, respectively. The unperturbed mean state atmosphere is assumed to vary only in the vertical direction and is specified using the MSIS model (Hedin, 1991). For the simulations discussed later the lower boundary of the model is at the ground (z=0) and the upper boundary is at z=600 km.

The model solves the Navier-Stokes equations for steady, linear waves propagating in an atmosphere that is horizontally homogeneous and stratified vertically. The mean, undisturbed state is non-isothermal, and mean winds are a function of height. The model includes the eddy and molecular diffusion of heat and momentum, ion drag, Rayleigh friction and Newtonian cooling, and the Coriolis force. The momentum equation is

$$\rho\frac{D\vec{v}}{Dt}+\underline{\nabla}p-\rho\vec{g}+2\rho\vec{\Omega}_E \times\vec{v}+\underline{\nabla}\cdot\underline{\underline{\sigma}}-\underline{\nabla}\cdot\left(\rho\eta_e\underline{\nabla}\vec{v}\right)-\rho\nu_{ni}\left(\vec{v}_i-\vec{v}\right)+\rho K_R\vec{v}=0 \tag{2}$$

In this equation ρ is mass density, $\vec{v}$ is velocity, p is pressure, $\vec{g}$ is the gravitational acceleration, $\vec{\Omega}_E$ is the Earth's angular frequency, $\underline{\underline{\sigma}}$ is the viscous stress tensor, η_e is the eddy diffusivity, ν_{ni} is the neutral-ion collision frequency, $\vec{v}_i$ is the ion velocity, and K_R is the Rayleigh friction coefficient. Also, $D/Dt=\partial/\partial t+\vec{v}\cdot\underline{\nabla}$ is the substantial derivative, where t is time. The viscous stress tensor is defined as

$$\sigma_{ij}=-\mu_m\left(\frac{\partial v_i}{\partial x_j}+\frac{\partial v_j}{\partial x_i}-\frac{2}{3}\delta_{ij}\underline{\nabla}\cdot\vec{v}\right) \tag{3}$$

where μ_m is the dynamic viscosity and δ_{ij} is the Kroneka delta.

The energy equation is

$$\rho c_v\frac{DT}{Dt}+p\underline{\nabla}\cdot\vec{v}+\underline{\underline{\sigma}}:\underline{\nabla}\vec{v}-\underline{\nabla}\cdot\left(\lambda_m\underline{\nabla}T\right)-\frac{c_pT}{\theta}\underline{\nabla}\cdot\left(\rho\kappa_e\underline{\nabla}\theta\right)-\rho\nu_{ni}\left(\vec{v}_i-\vec{v}\right)^2+\rho c_v K_N T=0 \tag{4}$$

Here T is temperature, λ_m is the coefficient of thermal conductivity, c_p and c_v are the specific heats at constant pressure and constant volume, respectively, κ_e is the eddy diffusivity of heat, and K_N is the Newtonian cooling coefficient. In these studies we set K_N equal to K_R . At the upper boundary they serve as a sponge layer and are set equal to the wave frequency; at lower altitudes they decrease exponentially away from the upper boundary with a characteristic scale height of 50 km ($K_N=\omega\exp\left[(z-600km)/50km\right]$. The

potential temperature is defined as $\theta = T\left(p_{00}/p\right)^{R/c_p}$, where R is the gas constant and p_{00} is a reference pressure (taken to be the sea-level value).
The continuity equation is

$$\frac{D\rho}{Dt} + \rho\underline{\nabla}\cdot\bar{v} = 0 \tag{5}$$

The linearized equation of state for an ideal gas ($p'/\bar{p} = \rho'/\bar{\rho} + T'/\bar{T}$) completes the set. Here primes denote perturbation values and the over-bar signifies the mean state.
The solution procedure involves first linearizing Eqs. (2) through (5), and then eliminating density perturbations (ρ') using the linearized ideal gas equation of state. Next, plane wave solutions of the form $\exp i(\omega t - kx - ly)$ are assumed, where ω is the wave frequency, t is time, and k and l are the wavenumbers in the x and y directions, respectively. This allows the terms involving time and horizontal derivatives to be replaced by algebraic terms. The remaining system of equations is essentially a set of coupled second order differential equations in the vertical coordinate, z. A complete listing of these equations is provided in the appendix of Schubert et al. (2003) and by Hickey et al. (2004). These equations are then written in their finite-difference form and solved using a method described by Bruce et al. (1953), as modified by Lindzen and Kuo (1969). At the upper boundary (z=600 km) a radiation condition is assumed with solutions provided using a dispersion equation described by Hickey and Cole [1987]. The lower boundary (ground) condition allows the vertical velocity to be specified. The equations are solved on a high resolution grid with the vertical spacing chosen to be typically several meters. The input required for the model is the altitude profile of the mean state (temperature, density, molecular weight, and horizontal winds), the wave amplitude, period, horizontal wavelength and direction of propagation. The model produces altitude profiles of amplitude and phase for the perturbations (three velocity components, temperature and pressure). From these other quantities of interest can be calculated, such as momentum, sensible heat and energy fluxes and their divergence.

4.3 Spectral full-wave model

The approach of Peltier and Hines (1976) is used to model the propagation of a tsunami-generated gravity wave disturbance. The following description is also provided in Hickey et al. (2009). The initial (t=0) displacement Z at the sea surface (z=0) is prescribed by

$$Z(x,z=0,t=0) = A\left\{Ai(-x+1)\frac{x}{2}\exp[-(x-2)/2]\right\} \tag{6}$$

where Ai is the Airy function, and x is the horizontal position in units of 100 km. The amplitude of the forcing is A (in meters). The Fourier transform of (6) provides the wavenumber (k) spectrum of the forcing

$$\hat{Z}(k,0,0) = \frac{1}{2\pi}\int_{-\infty}^{\infty} Z(x,0,0)e^{ikx}dx \tag{7}$$

Our simulations are based on long wavelength water waves propagating on the ocean surface. They are nondispersive and propagate with the shallow water phase speed

$c=\sqrt{gh}$, where g is the gravitational acceleration and h is the ocean depth (Lighthill, 1978). The tsunami is taken to propagate in the negative x-direction (but this is later altered to accommodate propagation in any arbitrary direction) in an ocean of depth 4 km. Hence $c=-200$ m/s, whence $\omega=-200k$. The vertical velocity spectrum is then given by

$$\hat{W}(k,0,0)=i\omega\hat{Z}(k,0,0)=-i200k\hat{Z}(k,0,0) \tag{8}$$

We use a discrete Fourier transform to evaluate the spectrum of the surface displacement $\hat{Z}$ and the vertical velocity spectrum $\hat{W}$. Figures 1a and 1b show the surface displacement (Z) and the vertical velocity spectrum ($\hat{W}$).

Occhipinti et al. (2006) reported sea surface displacements measured by Jason-1 and Topex-Poseidon that were characterized by a dominant horizontal scale size of 400 km and an amplitude of ~ 0.5 m. Wilson (2005) reported similar values (~ 500 km and 0.3 to 0.6 m, respectively). In Hickey et al. (2009) and here we adopt the former set of values. The amplitude of the surface displacement shown in Figure (2a) initially increases rapidly and diminishes after several cycles. The vertical velocity spectrum shown in Figure 2(b) is characterized by maxima at ± 400 km, corresponding to a dominant period of 33 min. This is in general agreement with the analysis of tide gauge measurements in and near the Indian Ocean that revealed periods ranging from 10 to 60 min and a dominant period of 40 min for the Sumatra tsunami event (Abe, 2006).

The full-wave model output is obtained for each discrete wave (ω -k) in the spectrum, with the vertical velocity at the lower boundary (z=0) specified to be unity. If we denote this output by $\psi'_j(\omega,k,z)$ (here the subscript j denotes each of velocity, temperature or pressure), then the response to the forcing is given by

$$\Psi'_j(x+vt,z)=\frac{1}{2\pi}\int_{-\infty}^{\infty}\hat{W}(k,0,0)\psi'_j(\omega,k,z)e^{-ik(x+vt)}\,dk \tag{9}$$

Note that the product $\hat{W}(k,0,0)\psi'_j(\omega,k,z)\Delta k$ represents the response of a single wavenumber-dependent perturbation to the spectral forcing, where Δk is the bandwidth. Because the Brunt-Väisälä period (τ_B) exceeds ~4 min everywhere (see Figure 1), waves of period ~3 to ~4 min will experience evanescence over much of the lower and middle atmosphere and so will be unable to propagate to the thermosphere. Waves of acoustic periods will not be efficiently generated by the tsunami, and moreover they will be evanescent because their phase speeds are less than the sound speed throughout the atmosphere. Hence we truncate the spectrum and include waves having periods of 3 min (wavelength 36 km) and longer. Equation (9) was solved using 800 waves (400 positive k and 400 negative k) for values of x ranging from -1000 km to +14400 km (and Δk = 4.36 x 10^{-7} m^{-1}).

4.4 Ion-electron perturbation model

The modeled electron-ion response to a linear gravity wave includes the effects of dynamics and chemistry. We use a solution procedure similar to that previously used to model the response of minor species and related airglow emissions in the mesospause region to gravity wave forcing (Walterscheid et al., 1987; Hickey, 1988).

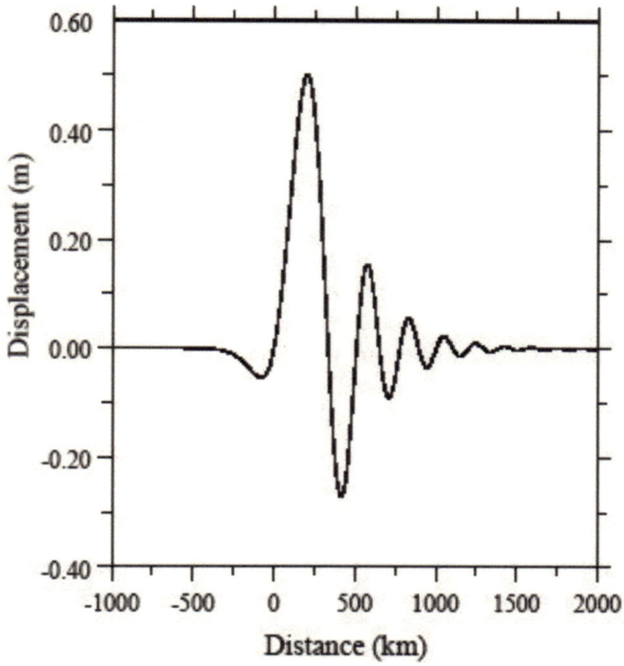

Fig. 2a. Surface displacement (after Hickey et al., 2009)

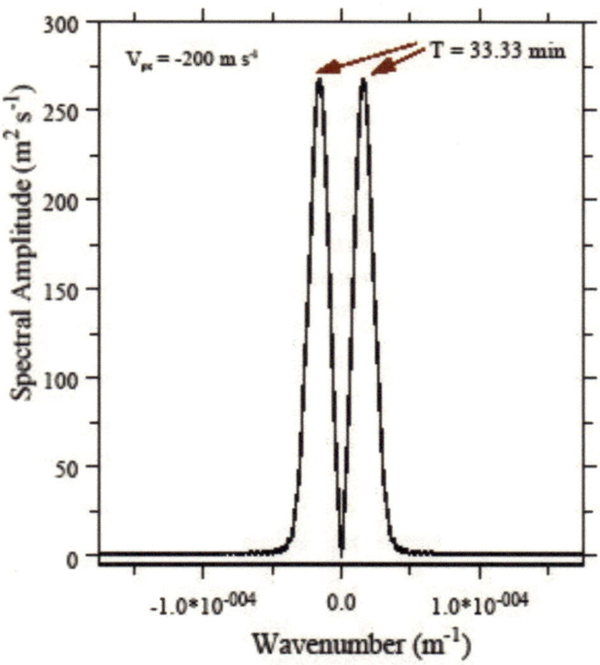

Fig. 2b. Vertical velocity spectrum at model lower boundary (after Hickey et al., 2009)

The linearized perturbation ion continuity equations for each of the ion species O^+, O_2^+, N_2^+ and NO^+ (n_j) are

$$i\omega n'_j + w'_j \frac{d\bar{n}_j}{dz} + \bar{n}_j \nabla \cdot \underline{v}'_j = P'_j - \bar{n}_j L'_j - n'_j \bar{L}_j \tag{10}$$

Here, P and L are the chemical production and loss rates, respectively. They are calculated based on the set of chemical equations of Schunk and Sojka (1996), given in Appendix 1 (as in Hickey et al., 2010a). The electron density profile is based on a Chapman layer, with a maximum value of 10^{12} m^{-3} at 300 km altitude.

For the ion momentum equation we follow the approach of McLeod (1965) and neglect polarization electric fields, gravity and diffusion:

$$\bar{v}_i = \left(\omega_i^2 + \nu_{in}^2\right)^{-1} \left\{\nu_{in}\,\omega_i\,\bar{v}x\hat{B} + \omega_i^2(\bar{v}\cdot\hat{B})\hat{B} + \nu_{in}^2\bar{v}\right\} \tag{11}$$

Here $\omega_i = q_i B/M_i$ is the ion gyrofrequency and $\nu_{in} = 2.6x10^{-15}(n_n + n_i)A^{-1/2}$ (*Kelley*, 1989) is the ion-neutral collision frequency, and where q_i and M_i are the charge and atomic or molecular weight, respectively of the ion, B is the magnetic field strength in Gauss (Wb/m^2), and n_n and n_i are the neutral and ion number density, respectively. Given the gravity wave perturbation velocity $\underline{v}'$, Eq. (11) is used to determine $\underline{v}'_j$ for each of the four ions. These ion velocities are then substituted in Eq. (10), and the system of four equations is inverted to yield the perturbation ion number densities. Other details of the calculations are provided in Hickey et al. (2009), and are not repeated here.

5. Results

5.1 Electron fluctuations

We begin by showing the results for the dominant wave in the spectrum. This wave has a period of ~ 33 min and a horizontal wavelength of 400 km. It is forced at the lower boundary of the full-wave model with a vertical velocity amplitude of 1.17×10^{-4} m s^{-1}. Mean winds, ion drag and Coriolis force are not included in this simulation, and so the propagation occurs isotropically in the horizontal direction.

The vertical velocity and temperature perturbation amplitudes are shown as a function of height in Figure 3a for two cases. In the first case the nominal viscosity and thermal conduction parameters are used, while in the second case these parameters are reduced by several orders of magnitude in order to mimic negligible dissipation (the adiabatic case). In the adiabatic case wave amplitudes increase by a factor of 10^6 over the altitude range shown in order to conserve wave energy. In the case of nominal dissipation the waves eventually dissipate and achieve a maximum amplitude near 200 km altitude. The phase of the vertical velocity fluctuations is shown as a function of height in Figure 3b. This figure demonstrates that at high altitudes the dissipation causes an increase in the vertical wavelength of the waves. Up to about 250 km altitude the vertical wavelength is ~ 100 km in both cases, but at high altitudes the vertical wavelength becomes very large in the non-adiabatic case.

The resulting O^+ velocity components for this wave are shown in Figure 4a. In this case we have included ion drag and the Coriolis force, and simulations are performed for the wave propagating either northward or eastward at the equator. The largest ion response occurs

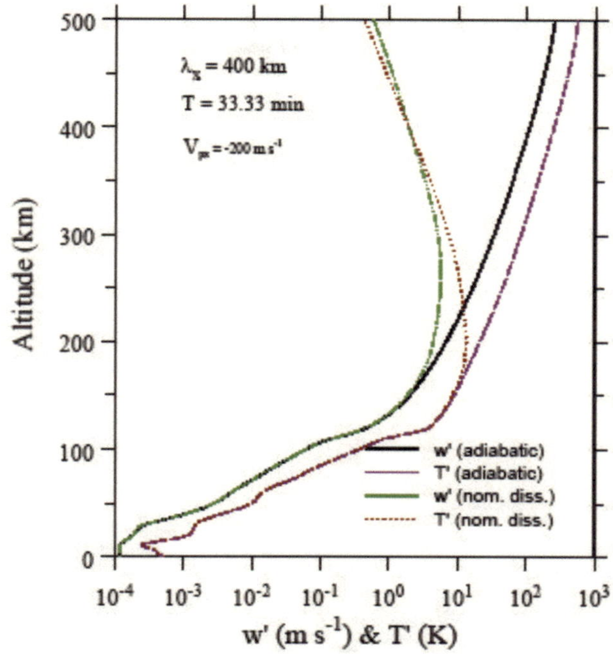

Fig. 3a. Vertical velocity and temperature perturbation amplitudes for adiabatic and non-adiabatic waves of 400 km horizontal wavelength and 33 min period (after Hickey et al., 2009)

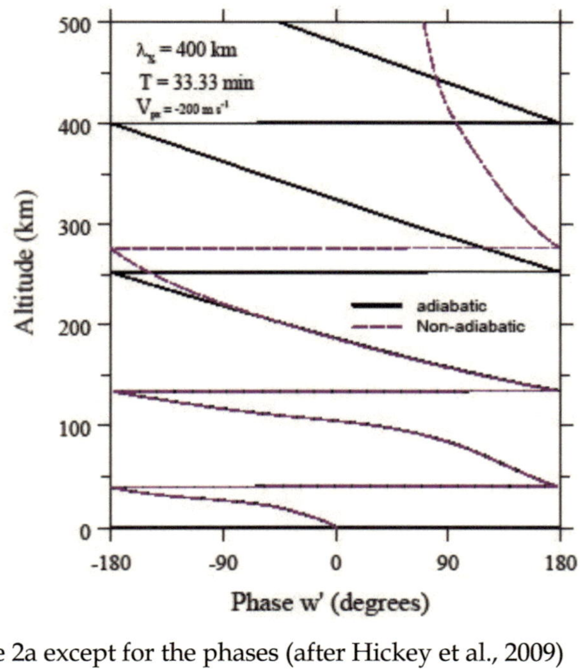

Fig. 3b. As in Figure 2a except for the phases (after Hickey et al., 2009)

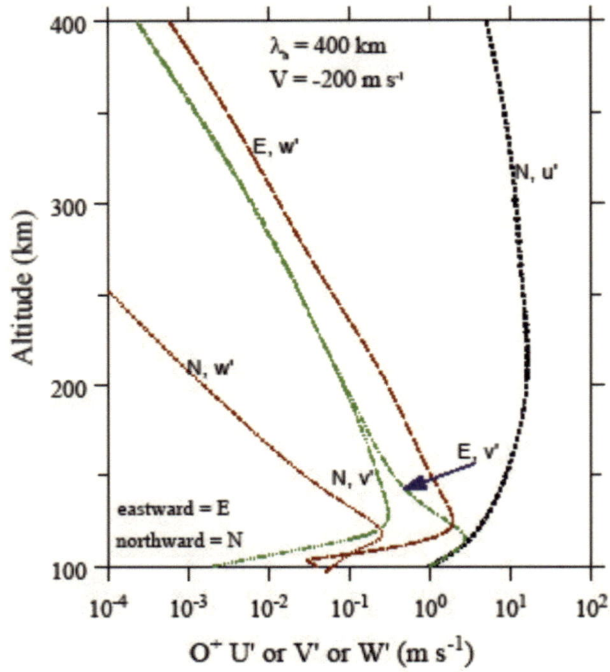

Fig. 4a. Ion velocity components for O^+ for northward and eastward propagating waves at the equator (after Hickey et al., 2010a)

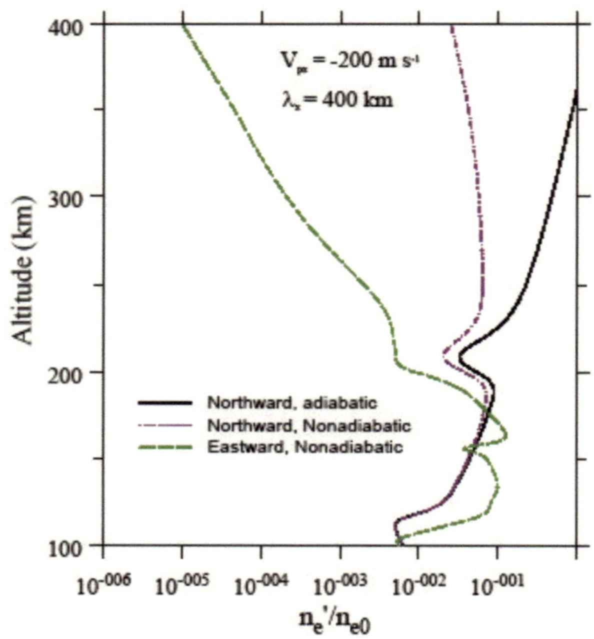

Fig. 4b. (after Hickey et al., 2009)

for northward wave propagation and for the meridional perturbation velocity (of ~ 15 m/s at 200 km altitude). By comparison, the ion velocity response is small for an eastward propagating wave, and is largest for the vertical ion velocity (~ 0.2 m/s at 200 km altitude). The associated electron density response is shown in Figure 4b. For realistic dissipation (nonadiabatic case) the largest response (of ~ 7% of the mean at 300 km altitude) occurs for the northward propagating wave. With small dissipation and for the northward propagating wave the electron density response is about an order of magnitude larger than this.

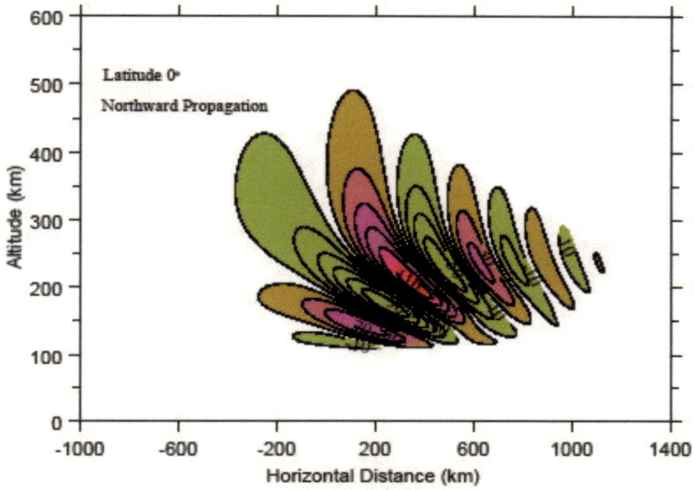

Fig. 5a. Temperature fluctuation (in K) for a northward propagating gravity wave disturbance (after Hickey et al., 2010a)

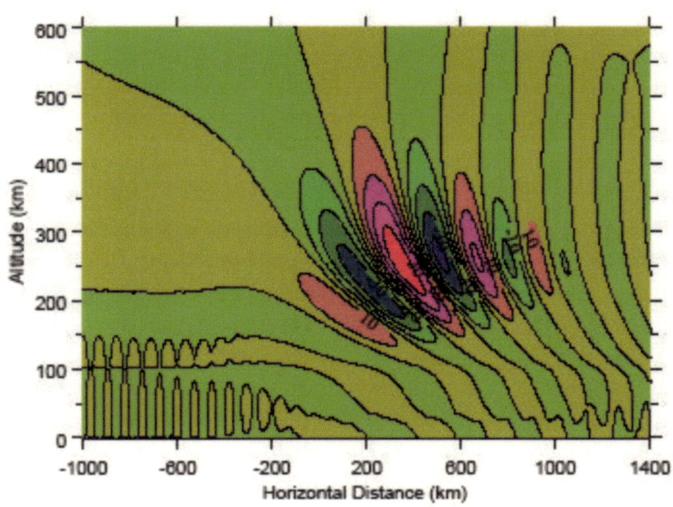

Fig. 5b. Vertical velocity fluctuation (in m/s) for a northward propagating gravity wave disturbance

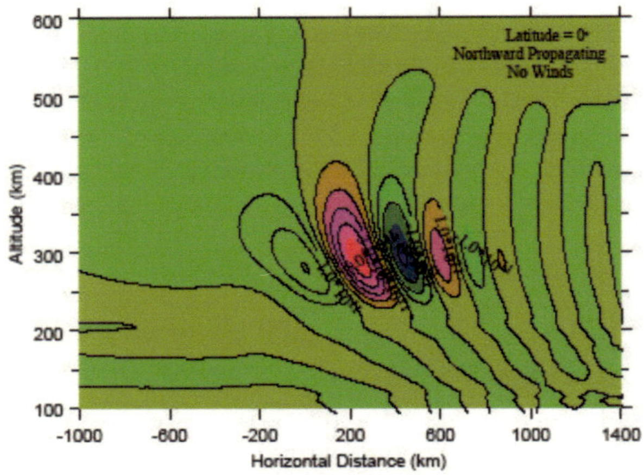

Fig. 6a. Electron density perturbations for northward propagation including dissipation and without mean winds (after Hickey et al., 2009)

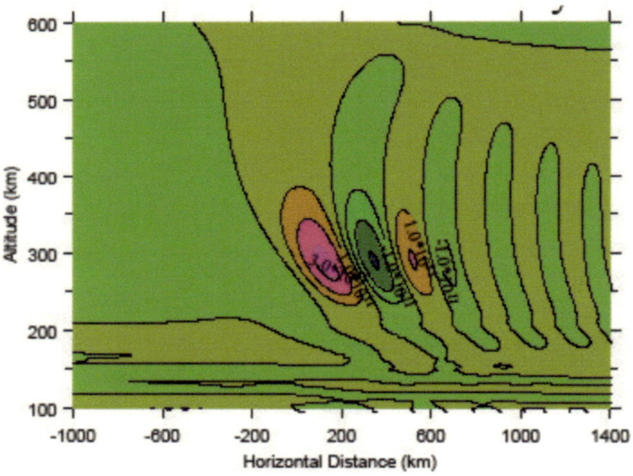

Fig. 6b. As in Figure 6a, except with the inclusion of mean winds (after Hickey et al., 2009)

The results of the spectral full-wave model are now discussed. Figure 5a shows the temperature fluctuation as a function of height and horizontal position for a northward propagating gravity wave disturbance with realistic dissipation included. The maximum perturbation of ~ 130 K (which is about 20% of the mean, undisturbed temperature) occurs over the 200 to 250 km altitude region. At greater heights the phase fronts become more vertical as a consequence of the increasing dissipation due to viscosity and thermal conduction (Hines, 1968). Figure 5b shows the corresponding vertical velocity fluctuation. The maximum of ~ 30 m/s occurs near 300 km altitude. About two to three dominant cycles are evident in these results.

Next we examine the effects of mean winds on the electron density response for a northward propagating gravity wave disturbance with realistic dissipation. Figure 6a shows the electron density response as a function of height and horizontal position without mean winds included. The maximum perturbations occur at the F2 peak near 300 km altitude and are localized to a horizontal region of ~ 800 km extent (which is twice the dominant horizontal wavelength). The maximum perturbation is ~ 60% of the mean at x ~ 200 km. With the further inclusion of mean winds (Figure 6b) the maximum perturbation is ~ 30% of the mean, while the horizontal extent of the disturbance has been reduced.

In order to simulate GPS observations, we have performed vertical integrations of the electron density perturbations to obtain the total electron content (TEC) perturbations. The integration of the mean, undisturbed electron density profile gives a TEC value of 17 TECU (where 1 TECU = 10^{16} electrons/m^2). For northward propagation (Figure 7a), TEC fluctuations are largest for the quasi-adiabatic, windless case. The inclusion of dissipation reduces the TEC fluctuation amplitude and also alters the phase of the disturbance. The further inclusion of mean winds alters the phase again, and causes a further reduction in the amplitude of the perturbations. In this latter (and more realistic) case the maximum TEC perturbations are about 3 TECU, which corresponds to an approximate 20% perturbation about the mean.

The results for an eastward propagating wave disturbance are shown in Figure 7b. Immediately evident is the significantly smaller response than for the northward propagating disturbance. As before, the smallest response occurs when dissipation and mean winds are included, and the maximum amplitude is ~ 0.01 TECU. It should be noted that these results apply to the equator. At mid-latitudes the ionospheric response to a zonally propagating gravity wave disturbance will be much larger, and can be comparable to that obtained for a meridionally propagating gravity wave disturbance (Occhipinti et al., 2008). This is because the vertical velocity fluctuation has a component parallel to the geomagnetic field at mid-latitudes, which can thus impart ion motion in that direction.

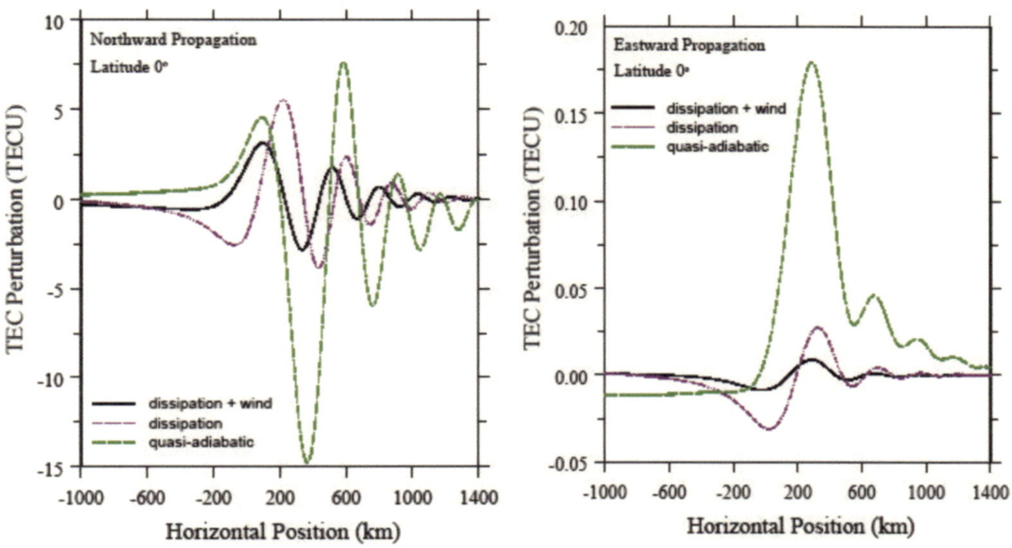

(a) (b)

Fig. 7. (a) TEC fluctuations for northward propagation (after Hickey et al., 2009). (b) TEC fluctuations for eastward propagation (after Hickey et al., 2009).

5.2 Airglow fluctuations

Atmospheric gravity waves propagating in the thermosphere are known to produce airglow variations associated with the wave-perturbed chemistry. Gravity wave motions have been detected in the 6300 A (red-line) nightglow (e.g., Sobral et al., 1978; Mendillo et al., 1997; Kubota et al., 2001), and also in the far-ultraviolet (FUV) O 1356 A emission (Paxton et al., 2003; DeMajistre et al., 2007). The chemistry associated with these emissions is included in Table 1 in Appendix 1. The mean undisturbed electron density profile is shown in Figure 8. Also shown are the altitude profiles of the mean undisturbed volume emission rates (VER) for these two emissions. The O 1356 Å VER peaks near 300 km altitude with a value of ~ 7.2 x 10^5 photons/s/m^3, and the OI 6300 Å VER peaks near 254 km altitude with a value of ~ 8.6 x 10^7 photons/s/m^3. The former profile is broader in its altitude extent than the latter.

The resulting OI 6300 Å VER perturbations for northward propagation at the equator, with nominal dissipation and in the absence of mean winds are shown as a function of height and horizontal position in Figure 9. For this particular gravity wave disturbance the perturbations maximize in the vicinity of the altitude where the mean undisturbed OI 6300 Å VER (shown in Figure 8) maximizes. The maximum value of the perturbation VER is ~6 x 10^7 photons/s/m^3, which is about 70% of the mean undisturbed value. Two complete cycles of the disturbance are evident in the VER fluctuations. Note that although we do not show the O 1356 Å VER perturbations, they resemble those for the OI 6300 Å except the maxima occur closer to 300 km altitude, in the vicinity of the peak undisturbed O 1356 Å profile (see Hickey et al., 2010a).

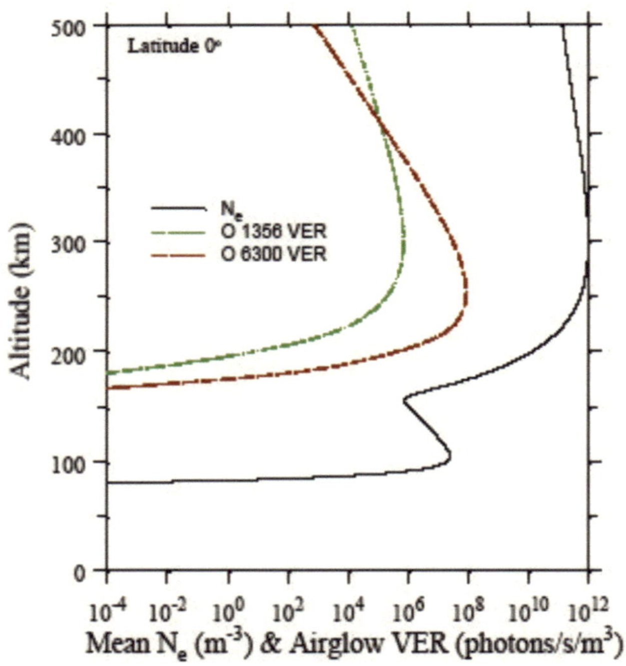

Fig. 8. Mean electron density (m^{-3}), mean OI 6300 Å VER (photons/s/m^3), and mean O 1356 Å VER (photons/s/m^3) (after Hickey et al., 2010a)

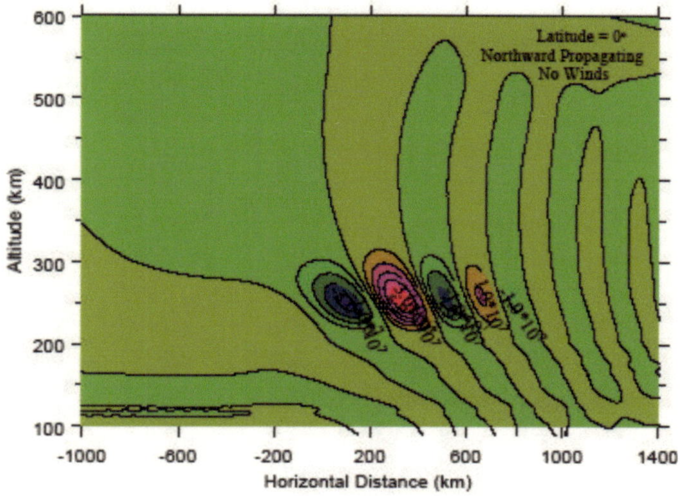

Fig. 9. OI 6300 Å VER fluctuations as a function of altitude and horizontal position (after Hickey et al., 2010a)

The brightness fluctuations are now obtained by integration over altitude of the VER fluctuations. We have done so for both the OI 6300 Å and the O 1356 Å emissions. We have also performed two integrations, one vertically, and the other along an oblique slant path inclined at an angle of ~29° from the horizontal. In each case we have then normalized the brightness fluctuations to the mean undisturbed brightness to obtain the relative brightness fluctuations (see Figure 10). The maximum relative brightness fluctuations for the OI 6300 Å emission (Figure 10a) are ~50% for overhead viewing and 25% for oblique viewing. The maximum occurs almost 400 km earlier for the oblique viewing, which equates to a time difference of ~ 30 minutes. Hence, for oblique viewing the disturbance could be observed approximately 30 minutes earlier compared to overhead viewing.

Similar results are seen for the O 1356 Å emission (Figure 10b). The relative brightness fluctuation is smaller for oblique viewing (~18% versus 43%), but appears ~ 400 km (corresponding to ~30 minutes) earlier than for overhead viewing. Also shown in this figure are the relative TEC fluctuations for overhead and oblique viewing. They strongly resemble the O 1356 Å fluctuations in terms of amplitude and phase. For overhead viewing both waveforms look strikingly similar to the original lower boundary/tsunami disturbance shown in Figure 2a. The maximum relative TEC fluctuations are ~ 33% and 8% for overhead and oblique viewing, respectively.

5.3 Momentum transport

Atmospheric gravity waves transport momentum and sensible heat. The respective fluxes per unit volume of these quantities are given by $F_m = \overline{\rho}\langle u'w' \rangle$ and $F_{sh} = \overline{\rho}c_p\langle w'T' \rangle$, where all symbols are as previously defined, and where the brackets denote an average over a complete horizontal wavelength of the wave disturbance. Here we will focus primarily on the momentum flux and its possible impact on the thermosphere. The impact of the sensible heat flux will be briefly summarized towards the end of this section. A more complete description of these wave fluxes associated with tsunamis is given by Hickey et al. (2010b).

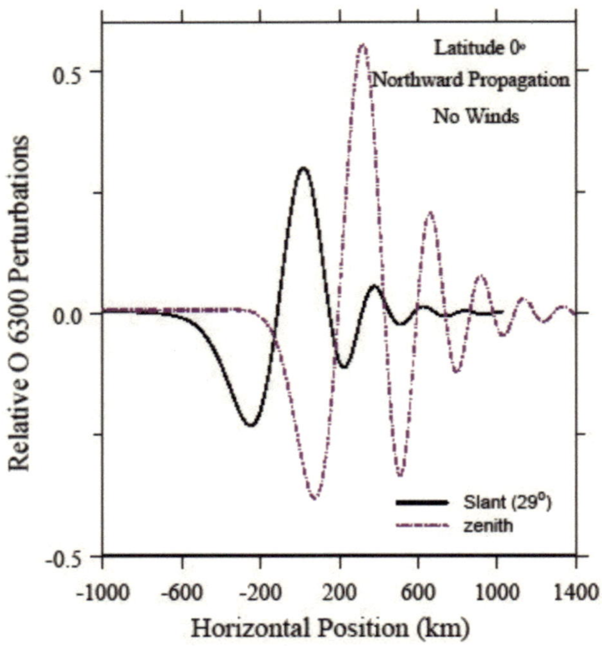

Fig. 10a. OI 6300 Å relative brightness fluctuations for overhead and oblique viewing (after Hickey et al., 2010a)

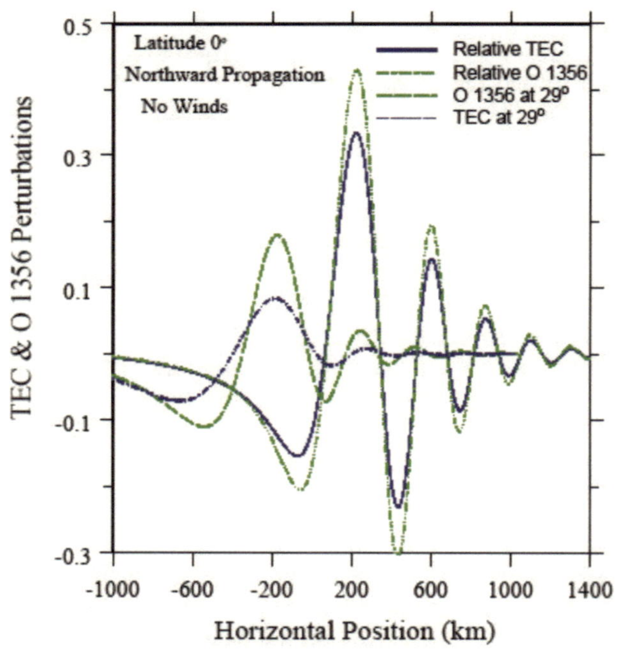

Fig. 10b. Relative total electron content and O 1356 Å fluctuations for overhead and oblique viewing (after Hickey et al., 2010a)

The non-acceleration conditions are that the waves be steady, linear, non-dissipating, and do not encounter critical levels (Walterscheid, 1995). For waves that satisfy the non-acceleration conditions the wave fluxes remain constant as the waves propagate upward. If these conditions are violated, the wave fluxes will change with height and the atmosphere can be irreversibly altered (eventually relaxing to a new steady-state). In particular, waves propagating into the thermosphere will eventually be viscously damped and a convergence of the momentum flux will occur, with a concomitant acceleration of the mean state given by

$$\frac{\partial \overline{u}_2}{\partial t} = -\frac{1}{\overline{\rho}}\frac{d}{dz}\left(\overline{\rho}\langle u'w'\rangle\right) \tag{12}$$

Here $\overline{u}_2$ is the second order, horizontally averaged change in mean state velocity associated with gravity wave forcing. This forcing is important in the terrestrial mesosphere, but in this case it is the wave amplitudes becoming nonlinear with subsequent wave breaking that leads to a momentum flux convergence. Fast gravity waves, such as those associated with tsunamis, should experience a large amplitude growth as they propagate upward through the thermosphere because they have large vertical wavelengths and so are far less dissipated than slower waves. Because they can attain large amplitudes in the middle thermosphere, their eventual dissipation might lead to large values of $\overline{u}_2$.

The major part of the gravity wave disturbance shown previously in Figure (5a) lies in a horizontal region of approximately 800 km extent. We calculate horizontal averages of wave fluxes over this horizontal extent in order to determine the second-order forcing of the atmosphere. Specifically, for a physical quantity ψ', we calculate the horizontally averaged vertical flux by evaluating

$$\langle w'\psi'\rangle = \frac{1}{N}\sum_{j=1}^{N}\mathrm{Re}(w'_j)\mathrm{Re}(\psi'_j) \tag{13}$$

In Equation (13), j is an index denoting the horizontal position of the physical variables w'_j and ψ'_j calculated from an inverse discrete Fourier transform over all waves in the spectrum. Also, N is the number of points at which the variables are evaluated over the 800 km horizontal extent of the disturbance.

The disturbance momentum flux per unit mass ($<u'w'>$) is shown in Figure 11a. At all altitudes shown it is negative and in the same direction as the phase propagation. Its magnitude increases with increasing height in the thermosphere, achieving a maximum value of ~ 1800 m^2 s^{-2} near 230 km altitude, and decreasing at greater heights due to viscous dissipation. The associated mean state acceleration is shown in Figure 11b. It maximizes in the middle thermosphere (near 250 km altitude) with a value of ~ 230 m s^{-1} h^{-1} (in the same direction as the phase velocity) and with a full-width at half maximum of ~ 120 km. Hence the dissipating disturbance drives a deep region of the thermosphere. Because the dominant part of the disturbance is of ~ 800 km horizontal extent (which is twice the dominant horizontal wavelength in the spectrum), the forcing would occur for ~ 1 h (two wave periods). Hence the acceleration shown in Figure 11b can be interpreted as the estimated change in velocity of this region of the thermosphere. This forcing would be applied over a large area (8000 km along-track for 10 h of propagation, and over a broad lateral extent). The viscous relaxation time was estimated by Hickey et al. (2010b) to be ~ 10 h near 200 km altitude and a few tens of minutes near 300 km altitude. Hence the high altitude (> 250 km) acceleration shown in Figure 11b is probably an overestimate. Note that because the wave forcing is second order, it is proportional to the square of the wave amplitude. Hence, only the largest amplitude tsunamis may be able to produce such large mean state forcing of the thermosphere.

The sensible heat flux has been previously discussed in the literature (Walterscheid, 1981, 1995) and was calculated by Hickey et al. (2010b) for the tsunamigenic disturbance related to the Sumatra 2004 event. In this case the atmospheric heating and cooling due to the dissipating disturbance (not shown) was found to be ~ +25 K h^{-1} near 180 km altitude, and -25 K h^{-1} near 290 km altitude. The expected temperature changes are relatively small compared to the undisturbed mean state temperature (~ 753 K) and may be too small to be unambiguously identified in observations.

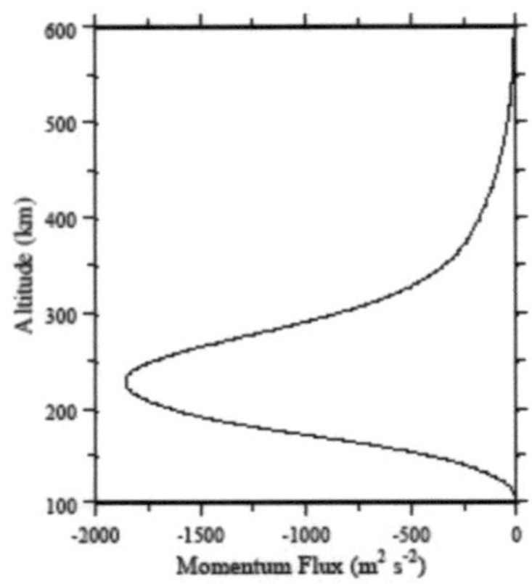

Fig. 11a. Disturbance momentum flux ($m^2 s^{-2}$) (after Hickey et al., 2010b)

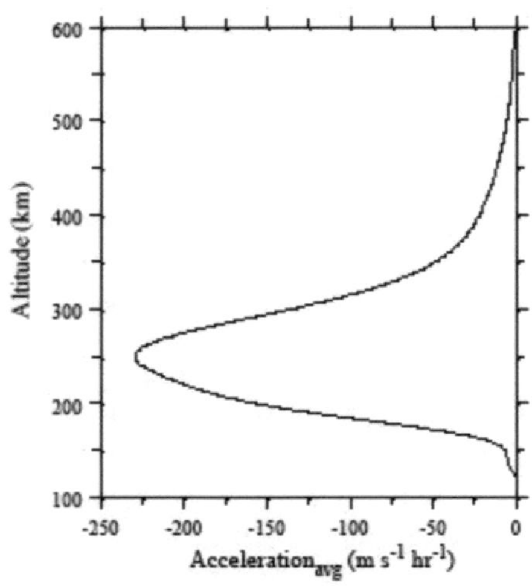

Fig. 11b. Horizontally averaged mean state acceleration due to the divergence of the disturbance momentum flux (after Hickey et al., 2010b)

6. Discussion

We have used a linear model to describe the gravity wave disturbance. For the motions of the neutral atmosphere considered here this is a reasonable assumption. Orlanski and Bryan (1969) showed that a criterion for the onset of convective instability is $u' > c$, where u' is the horizontal velocity perturbation and c is the horizontal phase speed of the wave. We find that the maximum value of u' is ~ 100 m/s, while c is 200 m/s. However, the ionospheric perturbations we have obtained are quite large, up to ~60% of the mean for realistic dissipation included, for which the assumption of linearity is violated. Nonetheless, we expect the results obtained to provide an overall good description of the expected ionospheric response. Future work should include a nonlinear treatment of this aspect of the model.

The work described herein was based on an assumed shallow water phase speed of 200 m/s and an ocean depth of ~ 4 km. Further numerical experimentation (not shown) has investigated the effect of changing this phase speed. Decreasing the phase speed results in a decrease in the vertical wavelength of the gravity waves in the spectrum. This leads to an increase in the viscous dissipation rate of the waves in the thermosphere. Because of this, these waves do not penetrate as far into the thermosphere as the faster waves do, and so achieve smaller amplitudes. Increasing the phase speed brings the waves closer to evanescence over much of the mesosphere (see discussion in Hickey et al., 2009), and so somewhat impedes their propagation to greater altitudes. Hence, we find that 200 m/s is close to an optimum phase speed for efficient gravity wave propagation from the lower atmosphere to the thermosphere. This in turn implies that the ionospheric response to tsunamis should maximize in regions where the ocean depth is ~ 4 km.

7. Conclusions

Tsunamis are a relatively rare event, but nonetheless in some instances they may have large, dramatic effects on the upper atmosphere. In addition to causing large fluctuations in electron density and in the total electron content, they will also cause comparable fluctuations in natural airglow emissions, such as the OI 6300 A and O 1356 A emissions. These airglow emissions could be observed from the ground and from space, and could help provide additional information leading to the early detection of tsunamis. The disturbances may, at times, carry significant wave momentum to F-region altitudes leading to strong accelerations of the mean flow. The observations of such events may offer an excellent opportunity of testing gravity wave theories and models, and we suggest that dedicated observations should be considered for future programs to study lower-upper atmosphere coupling and to better understand the role of tsunamis in this process.

8. Acknowledgements

This work was supported by NSF grant ATM-0639293 to Embry-Riddle Aeronautical University. The author appreciates conversations with colleagues Drs. Richard Walterscheid, Gerald Schubert, Attila Komjathy, David Galvin, and Tony Mannucchi over the course of this research.

9. References

Abe, K. (2006). Dominant periods of the 2004 Sumatra tsunami and the estimated source size, *Earth Planets Space*, *58*, 217-221.

Artru, J., V. Ducic, H. Kanamori, P. Lognonne, and M. Murakami (2005a). Ionospheric detection of gravity waves induced by tsunamis, *Geophys. J. Int., 160*, 840-848.

Artru, J., P. Lognonne, G. Occhipinti, F. Crespon, R. Garcia, E. Jeansou, and M. Murakami (2005b). Tsunami detection in the ionosphere, *Space Res. Today, 163*, 23-27.

Bruce, G. H., D. W. Peaceman, H. H. Rachford, Jr., and J. D. Rice (1953). Calculations of unsteady-state gas flow through porous media, *Petrol. Trans. AIME, 198*, 79-92.

DasGupta, A., Das., A., Hui, D., Bandyopadhyay, K. K., and Sivaraman, M. R. (2006). Ionospheric perturbation observed by the GPS following the December 26th, 2004 Sumatra-Andaman earthquake, *Earth Planets Space, 35*, 929-959.

DeMajistre, R., Paxton, L. J., and Bilitza, D. (2007). Comparison of ionospheric measurements made by digisondes with those inferred from ultraviolet airglow, *Adv. Space Res., 39*, 918-925.

Fritts, D. C., and Alexander, M. J. (2003). Gravity wave dynamics and effects in the middle atmosphere, *Rev. Geophys.*, 41(1), 1003, doi:10.1029/2001RG000106.

Georges, T. M. (1968). HF Doppler studies of traveling ionospheric disturbances, *J. Atmos. Terr. Phys., 30*, 735-746.

Hébert, H., Sladen, A., and Schindelé, F. (2007). Numerical modeling of the great 2004 Indian Ocean tsunami: Focus on the Mascarene Islands. *Bull. Seismol. Soc. Am., 97*, 1A, S208-S222.

Hedin, A. E. (1991). Extension of the MSIS thermosphere model into the middle and lower atmosphere, *J. Geophys. Res., 96*, 1159.

Hedin, A. E., et al. (1996). Empirical wind model for the upper, middle and lower atmosphere, *J. Atmos. Terr. Phys., 58*, 1421.

Hickey, M. P. and Cole, K. D. (1987). A Quartic Dispersion Equation for Internal Gravity waves in the Thermosphere. *J. Atmos. Terr. Phys., 49*, 889.

Hickey, M. P., Walterscheid, R. L. and Schubert, G. (2004). Erratum to "Gravity Wave Heating and Cooling in Jupiter's Thermosphere", *Icarus, 170*, 524-525.

Hickey, M. P., Walterscheid, R. L., Taylor, M. J., Ward, W., Schubert, G., Zhou, Q., Garcia, F., Kelley, M. C., and Shepherd, G. G. (1997). Numerical Simulations of Gravity Waves Imaged over Arecibo during the 10-day January 1993 Campaign, *J. Geophys. Res., 102*, 11,475.

Hickey, M. P., Schubert, G., and Walterscheid, R. L. (2009). The Propagation of Tsunami-Driven Gravity Waves into the Thermosphere and Ionosphere, *J. Geophys. Res., 114*, A08304, doi:10.1029/2009JA014105.

Hickey, M. P., Schubert, G., and Walterscheid, R. L. (2010a). Atmospheric airglow fluctuations due to a tsunami-driven gravity wave disturbance, *J. Geophys. Res., 115*, A06308, doi:10.1029/2009JA014977.

Hickey, M. P., Walterscheid, R. L. and Schubert, G. (2010b). Wave mean flow interactions in the thermosphere induced by a major tsunami, *J. Geophys. Res., 115*, A09309, doi:10.1029/2009JA014927.

Hines, C. O. (1960). Internal atmospheric gravity waves at ionospheric heights, *Can. J. Phys.*, 38, 1441.

Hines, C. O. (1972). Gravity waves in the atmosphere, *Nature, 239*, 73-78.

Hines, C. O. (1968). An effect of molecular dissipation in upper atmospheric gravity waves, *J. Atmos. Terr. Phys., 30*, 845-849.

Hooke, W. H. (1968). Ionospheric irregularities produced by internal atmospheric gravity waves. *J. Atmos. Sol. Terr. Phys., 30*, 795-823.

Kelley, M. C. (1989). *The Earth's Ionosphere*, Academic Press Inc.

Kubota, M., Kukunishi, H., and Okano, S. (2001). Characteristics of medium- and large-scale TIDs over Japan derived from OI 630-nm nightglow observations, *Earth Planets Space, 53*, 741-751.

Lee, M. C., Pradipta, R., Burke, W. J., Labno, A., Burton, L. M., Cohen, J. A., Dorfman, S. E., Coster, A. J., Sulzer, M. P., and Kuo, S. P. (2008). Did tsunami-launched gravity waves trigger ionospheric turbulence over Arecibo? *J. Geophys. Res., 113*, A01302, doi:10.1029/2007JA012615.

Lighthill, J. (1978). *Waves in Fluids*, Cambridge University Press, Cambridge.

Lindzen, R. S., and H. L. Kuo (1969). A reliable method for the numerical integration of a large class of ordinary and partial differential equations. *Mon. Wea. Rev.*, 97, 732-734.

Liu, J., Tsai, Y., Chen, S., Lee, C., Chen, Y., Yen, H., Chang, W., and Liu, C. (2006a). Giant ionospheric disturbances excited by the M9.3 Sumatra earthquake of 26 December 2004. *Geophys. Res. Lett., 33*, L021103, doi:10.1029/2005GL023963.

Liu, J., Tsai, Y., Ma, K., Chen, Y., Tsai, H., Lin, C., Kamogawa, M., and Lee, C. (2006b). Ionospheric GPS total electron content (TEC) disturbances triggered by the 26 December 2004 Indian Ocean tsunami, *J. Geophys. Res., 111*, A05303, doi:10.1029/2005JA011200.

Martyn, D. F. (1950). Cellular atmospheric waves in the ionosphere and troposphere, *Proc. Roy. Soc.*, A201, 216.

McLeod, M. A. (1965). Sporadic E Theory. I. Collision-Geomagnetic Equilibrium, *J. Atmos. Sci., 23*, 96-109.

Mendillo, M., Baumgardner, J., Nottingham, D., Aarons, J., Reinisch, B., Scali, J., and Kelley, M. C. (1997). Investigations of thermospheric-ionospheric dynamics with 6300-Å images from the Arecibo Observatory, *J. Geophys. Res., 102*, 7331-7343.

Munro, G. H. (1948). Short period changes in the F-region of the ionosphere, *Nature*, 162, 886.

Occhipinti, G., P. Lognonne, E. A. Kherani, and H. Hebert (2006). Three-dimensional waveform modeling of ionospheric signature induced by the 2004 Sumatra tsunami, *Geophys. Res. Lett., 33*, L20104, doi::10.1029/2006GL026865.

Occhipinti, G., E., A. Kherani , and P. Lognonne (2008). Geomagnetic dependence of ionospheric disturbances induced by tsunamigenic internal gravity waves, *Geophys. J. Int.*, 173, 753-765.

Orlanski, I., and Bryan, K. (1969). Formation of the thermocline step structure by large-amplitude internal gravity waves, *J. Geophys. Res., 74*, 6975-6983.

Parish, H. F., Schubert, G., Hickey, M. P., and Walterscheid, R. L. (2009). Propagation of tropospheric gravity waves into the upper atmosphere of Mars, *Icarus*, doi:10.1016/j.icarus.2009.04.031.

Paxton, L. J., Morrison, D., Strickland, D. J., McHarg, M. J., Zhang, Y., Wolven, B., Kil, H., Crowley, G., Christensen, A. B., and Meng, C.-I. (2003). The use of far ultraviolet remote sensing to monitor space weather, *Adv. Space Res., 31*(4), 813-818.

Peltier, W. R., and Hines, C. O. (1976). On the possible detection of tsunamis by a monitoring of the ionosphere, *J. Geophys. Res., 81*, 1995-2000.

Schubert, G., Hickey, M. P., and Walterscheid, R. L. (2003). Heating of Jupiter's thermosphere by the dissipation of upward propagating acoustic waves, *Icarus., 163*, 398-413.

Schubert, G., M. P. Hickey, and R. L. Walterscheid (2005). Physical processes in acoustic wave heating of the thermosphere, *J. Geophys. Res., 110*, D07106, doi:10.1029/2004JD005488.

Schunk, R. and J. Sojka (1996). USU model of the global ionosphere, in *STEP: Handb. of Ionospheric Models*, Ed. R.W. Schunk, SCOSTEP.

Scorer, R. S. (1949). Theory of waves in the lee of mountains, *Q. J. Roy. Met. Soc., 75*, 323, 41-56.

Walterscheid, R. L., G. Schubert, and J. Straus (1987). A dynamical-chemical model of wave-driven fluctuations in the OH nightglow, *J. Geophys. Res., 92*, 1241-1254.

Walterscheid, R. L., and M. P. Hickey (2001). One-gas models with height-dependent mean molecular weight: Effects on gravity wave propagation, *J. Geophys. Res., 106*, 28,831-28,839.

Wilson, M. (2005). Modeling the Sumatra-Andaman earthquake reveals a complex, nonuniform rupture, *Phys. Today, 58*, 19.

10. Appendix 1.

	Ion Reaction	Rate ($cm^3\ s^{-1}$)
1	$O^+ + N_2 \rightarrow NO^+ + N$	$k_1 = 5 \times 10^{-13}$
2	$O^+ + O_2 \rightarrow O_2^+ + O$	$k_2 = 2 \times 10^{-11}\ (T/300)^{-0.4}$
3	$O^+ + NO \rightarrow NO^+ + O$	$k_3 = 8 \times 10^{-13}$
4	$O^+ + e \rightarrow O(^5P) + \hbar\nu_{1356}$	$k_4 = 7.3 \times 10^{-13}$
5	$O_2^+ + N_2 \rightarrow NO^+ + NO$	$k_5 = 5 \times 10^{-16}$
6	$O_2^+ + NO \rightarrow NO^+ + O_2$	$k_6 = 4.4 \times 10^{-10}$
7a	$O_2^+ + e \rightarrow O + O(^3P)$	$k_{7a} = 0.62 \times 1.6 \times 10^{-7}\ (300/T)^{0.55}$
7b	$O_2^+ + e \rightarrow O + O(^1S)$	$k_{7b} = 0.08 \times 1.6 \times 10^{-7}\ (300/T)^{0.55}$
7c	$O_2^+ + e \rightarrow O + O(^1D)$	$k_{7c} = 1.30 \times 1.6 \times 10^{-7}\ (300/T)^{0.55}$
8	$N_2^+ + O \rightarrow NO^+ + N$	$k_8 = 1.4 \times 10^{-10}\ (300/T)^{0.44}$
9	$N_2^+ + O \rightarrow O^+ + N_2$	$k_9 = 1 \times 10^{-11}\ (300/T)^{0.23}$
10	$N_2^+ + O_2 \rightarrow O_2^+ + N_2$	$k_{10} = 5 \times 10^{-11}\ (300/T)^{-1}$
11	$N_2^+ + NO \rightarrow NO^+ + N_2$	$k_{11} = 3.3 \times 10^{-10}$
12	$N_2^+ + e \rightarrow N + N$	$k_{12} = 1.8 \times 10^{-7}\ (300/T)^{0.39}$
13	$NO^+ + e \rightarrow N + O$	$k_{13} = 4.2 \times 10^{-7}\ (300/T)^{0.85}$
14	$O(^1S) \rightarrow O(^1D) + \hbar\nu_{5577}$	$A_{5577} = 1.06\ s^{-1}$
15a	$O(^1D) \rightarrow O(^3P) + \hbar\nu$	$A_{6300+6364} = 9.34 \times 10^{-3}\ s^{-1}$
15b	$O(^1D) \rightarrow O(^3P) + \hbar\nu_{6300}$	$A_{6300} = 7.1 \times 10^{-3}\ s^{-1}$
16	$O(^1D) + N_2 \rightarrow O + N_2$	$k_{16} = 2.0 \times 10^{-11}\ \exp[107.8/T]$
17	$O(^1D) + O_2 \rightarrow O + O_2$	$k_{17} = 2.0 \times 10^{-11}\ \exp[67.5/T]$
18	$O(^1D) + O \rightarrow O + O$	$k_{18} = 8.0 \times 10^{-12}$

Table 1. Chemical reactions for ions (after Hickey et al., 2010a)

33

Autopsy and Identification Techniques

Jean-Pol Beauthier MD PhD[1], Philippe Lefèvre MSC PhD[1]
and Eddy De Valck[2]
[1]*Forensic Unit – Laboratory of Anatomy, Biomechanics and Organogenesis (LABO), Université Libre de Bruxelles, Campus Erasme CP 629 – 1070 Brussels*
[2]*Odontology – Beigem*
Belgium

"The whole of science is nothing more than a refinement of everyday thinking"
Albert Einstein (1879 – 1955)

1. Introduction

Autopsy – different steps

1.1 Introduction – the Belgian DVI team

To illustrate the different activities of the reconstructive identification process of the Tsunami victims, we refer to the activities of the Belgian Disaster Victim Identification (DVI) Team, working at "Wat Yan Yao" (Buddhist temple in Takuapa village, near Khao Lak beach, Phang-Nga province) as part of the world wide DVI community which brought assistance to the Thai government (see Fig. 1, 2 and 3).

Fig. 1. General view of the disaster (Phang-Nga province).

Fig. 2. Wat Yan Yao Site 1a – Khao Lak: the entry to buddish temple.

Fig. 3. The pictures: exposition of the decedents.

For a proper post mortem (PM) examination and from a logistic point of view, it is recommended that a DVI team be multidisciplinary so that, immediately after deployment and regardless of local conditions and facilities, it can operate autonomously. This requires not only the presence of specialized forensic experts on the team but also a strong added value of auxiliary personnel. The Belgian DVI team is composed of (i) policemen specialized in recovery/exhumation, logistics and/or administration, scientific laboratory and photography, (ii) forensic medical examiners (ideally with forensic anthropology specialization), (iii) forensic odontologists, (iv) civil protection members for body handling, transport, and assisting with the use of heavy equipment or machinery, (v) a psychologist trained in Post Traumatic Stress and available for the team all times, (vi) and a medical doctor and registered nurse attending to all team members' medical needs. We recently published this organization in Table 1.

	1 Team leader	
	1 assistant team leader	
	1 logistics member	
	1 AM and Identification coordinator	
	1 Officer for international Management Center	
	1 Medical doctor	
	1 paramedical member (nurse)	
	1 psychologist	
PM line 1		**PM line 2**
1 chief line officer		1 chief line officer
1 pathologist + 1 DVI policeman (to help write documents and reports)		1 pathologist + 1 DVI policeman (to help write documents and reports)
1 odontologist + 1 DVI policeman (to help write documents and reports)		1 odontologist + 1 DVI policeman (help in writing documents and reports)
1 science laboratory policeman (for fingerprinting)		1 science laboratory policeman (for fingerprinting)
1 DVI policeman (for washing and identifying clothing, documents, jewelry…)		1 DVI policeman (for washing and identifying clothing, documents, jewelry…)
1 DVI policeman for completing PM Interpol documents		1 DVI policeman for completing PM Interpol documents
1 Belgian Civil protection member		1 Belgian Civil protection member
	1 DVI photographer	
	1 DVI member for collecting all information/documents/samples on final desk	

Table 1. Belgian DVI Team deployment (Beauthier et al., 2009).

Ideally, such kind of PM DVI team will be self-supporting and can be operational within a few hours after arrival at a mass disaster scene.
Note that - in theory - the collection of ante mortem (AM) material is classically handled by police officers, medical examiners and forensic odontologists who try to obtain informations from the missing person's families and practitioners. The AM DVI team is usually composed of 5 policemen, but also includes Red Cross members, psychologists and social assistants.
The first Belgian DVI Team arrived in Phuket on January 2, 2005 and a second one followed two weeks later after the first team returned home. A few team members worked during several months in data base organization at the Thai Tsunami Victim Identification Information Management Centre (TTVI-IMC) in Phuket.

1.2 External examination

1.2.1 Recovery

The recovery of the bodies was carried out by the local Thai population prior to the time of the international teams' arrival in Thailand (Fig. 4).

Fig. 4. Wat Yan Yao Site 1a – Khao Lak: innumerable bodies protected by dry ice (before autopsy).

The bodies were stored at Wat Yan Yao (Site 1A) for the identification process to start. After the first logistic problems were solved (Fig. 5) – electricity and water supply, airconditioning, protected space for the autopsy room – the PM identification line was divided into four work sections: fingerprinting, medical examination, odontology examination and DNA sampling (Fig. 6).

Fig. 5. Wat Yan Yao Site 1a – Khao Lak: first examinations' activities.

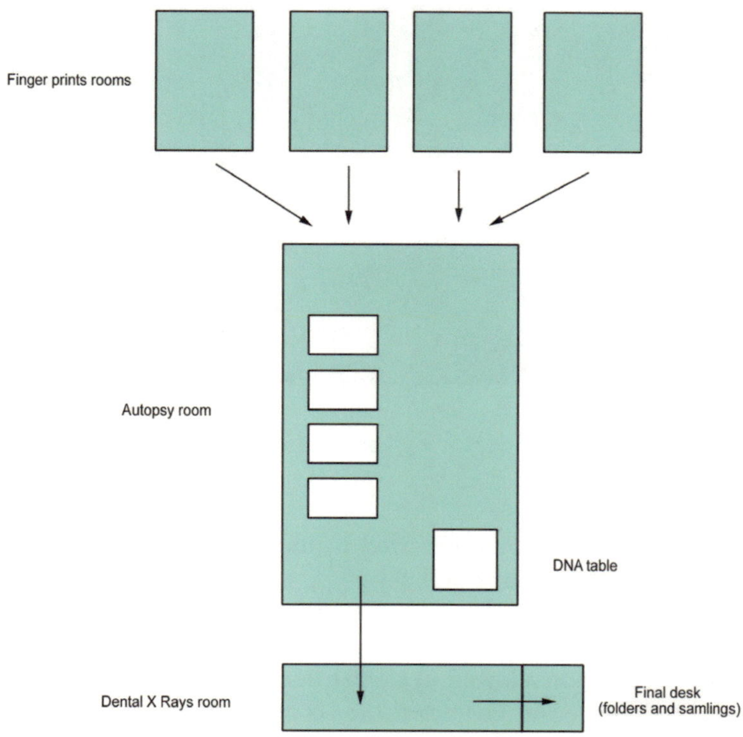

Fig. 6. Organization of the activities (first autopsy room).

1.2.2 Fingerprinting

Fingerprinting was very difficult due to the condition of the bodies but was successfully realized by specialists of the technical and scientific laboratories of the different national law enforcement agencies. Finger skin was removed by "degloving" and placed on the fingertip of one of the two operators. After powdering, the fingerprints were then transposed onto paper (Fig. 7). The use of separated fingerprinting rooms proved to be very practical and useful.

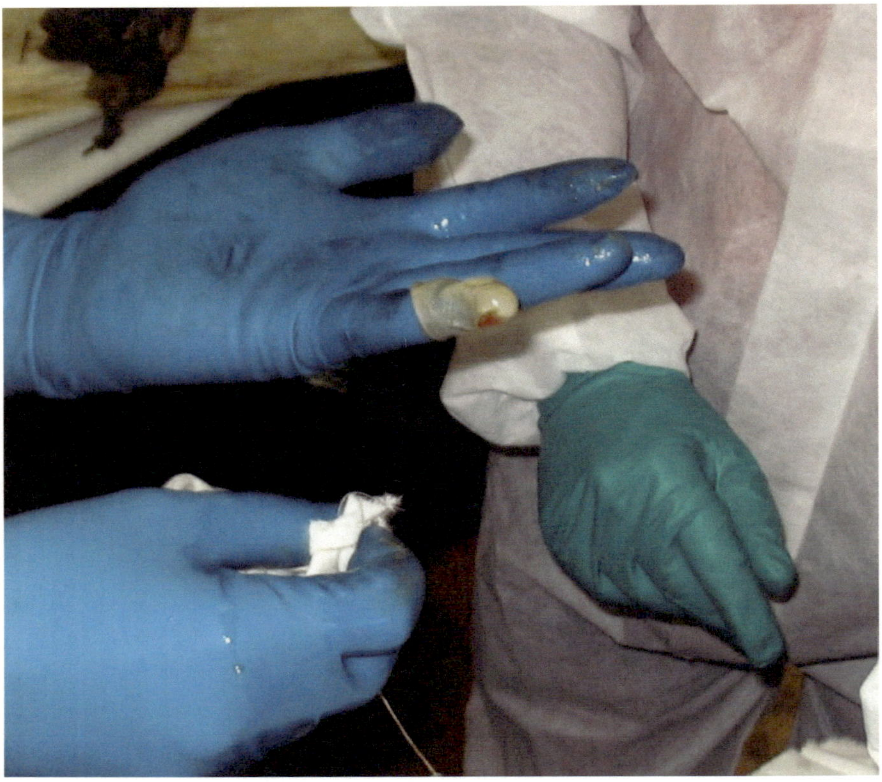

Fig. 7. Fingerprinting technique: "degloving" and "glove-on" method.

1.2.3 Autopsy room

After fingerprinting the body was transferred to the autopsy room, where external and internal descriptions took place (Fig. 8 and 9).

Clothing, jewelry and other personal belongings were systematically cleaned, described and photographed (with a metric scale and the corresponding body number in place) (Fig. 10).

The external body examination allowed (i) stature determination; (ii) description of tattoos, scars, anatomical and/or sequel characteristics, congenital abnormalities, recent or old traumatic elements; (iii) phenotypic characteristics (e.g. cranial hair color and thoracic-, axillar-, arm- and pubic hair description…).

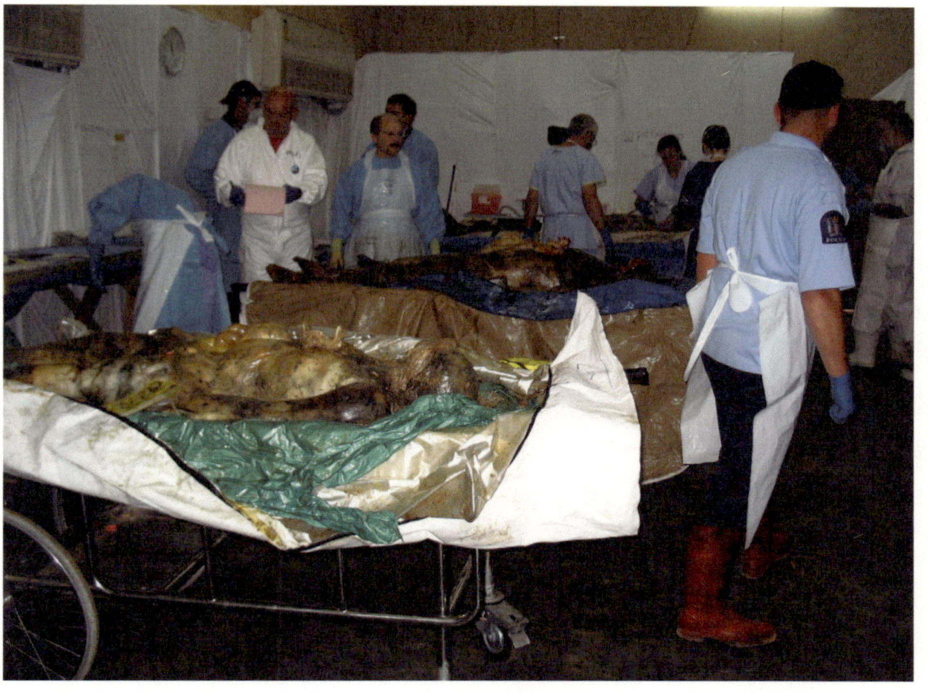

Fig. 8. First autopsy room.

Fig. 9. Second autopsy room.

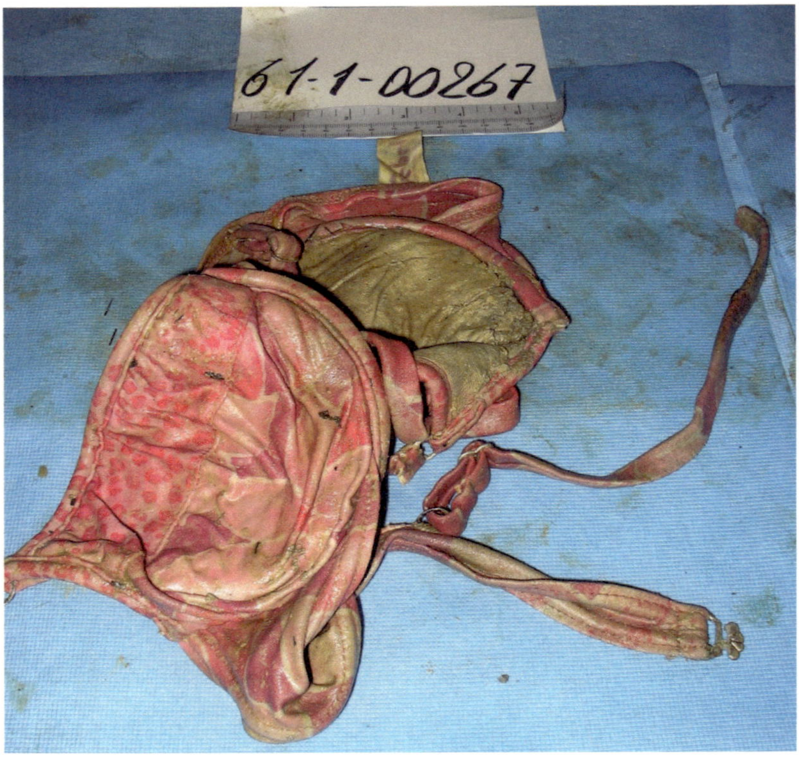

Fig. 10. Clothes.

1.3 Internal examination

According to international agreements and recommendations decided at the staff meetings, the internal autopsy in Thailand was limited to the abdominal cavity, checking for the presence/absence of the gall bladder, appendix, genitalia, etc.

Noted abnormal anatomic findings were subjected to a more extensive examination (e.g. dissection of a uterine tumor, description of tubar clips, intra-uterine devices, etc.) and description.

We estimate the necessity of an initial X-ray study, which was not realized during our Thai activities.

1.4 Facial dissection and interest of mandibular approach

To facilitate the work of the odontologists during the further part of the autopsy, it was essential to create a mandibular access.

We used a dissecting technique by making a semicircular incision, from ear to ear, behind each mandibular angle and passing through the anterior cervical region, under the submental area [modification of the Keiser-Nielsen method (Keiser-Nielsen, 1980)] (Fig. 11).

The cutaneous flap was inclined upwards to the face by using a Senn-Miller retractor allowing the exposure of the temporomandibular joints.

A second incision was made along the internal border of the mandibular body, cutting the muscles of the buccal floor, making it possible to release the lingual block (Fig. 12).

The muscles were sectioned on the mandible (*mandibula*). The *m. masseter* was cut at the level of the goniac angle and the *corpus mandibulae* (Fig. 13).
The *m. temporalis* was sectioned at the *processus coronoideus* of the mandible (Fig. 14), while the *mm. pterygoidei* were progressively dissected with the blade of the scalpel following the articular structures, when the mandible was separated at the level of the temporomandibular joint (Fig. 15).
The mandible was thus completely released (Fig. 16), without any destruction, allowing an easy repositioning in order to obtain a complete correct facial reconstruction (Beauthier et al., 2009).

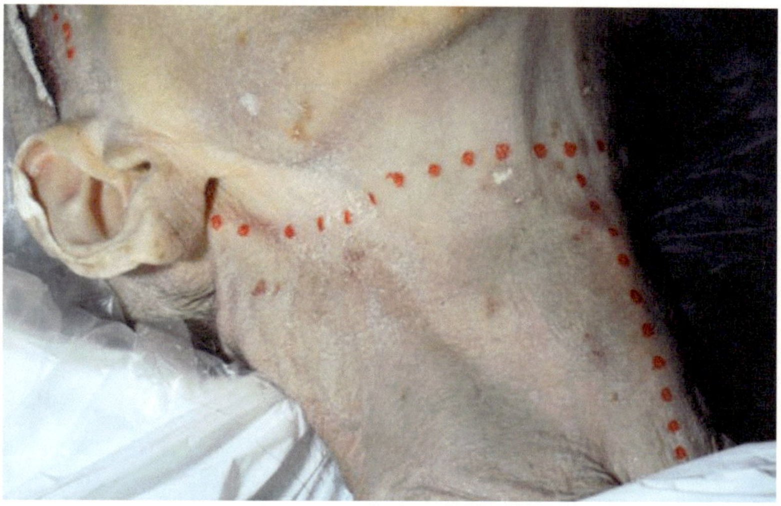

Fig. 11. Mandibular dissection, first step: cutaneous incisions.

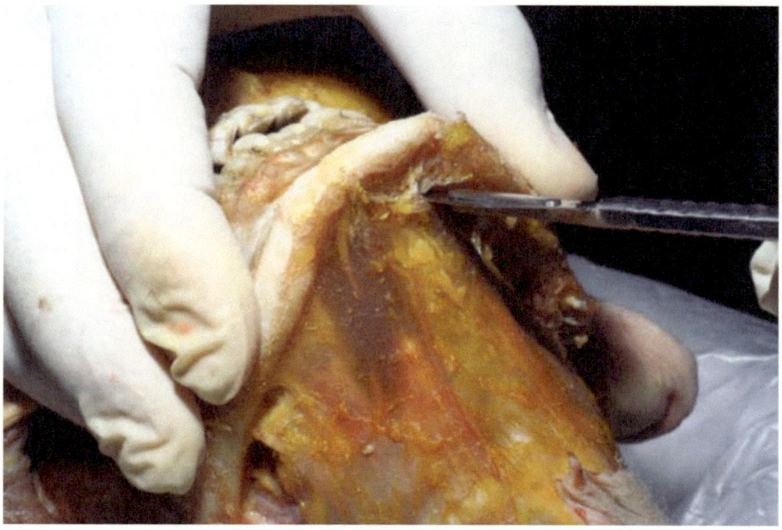

Fig. 12. Mandibular dissection, second step: buccal floor section along the inner border of *mandibula*.

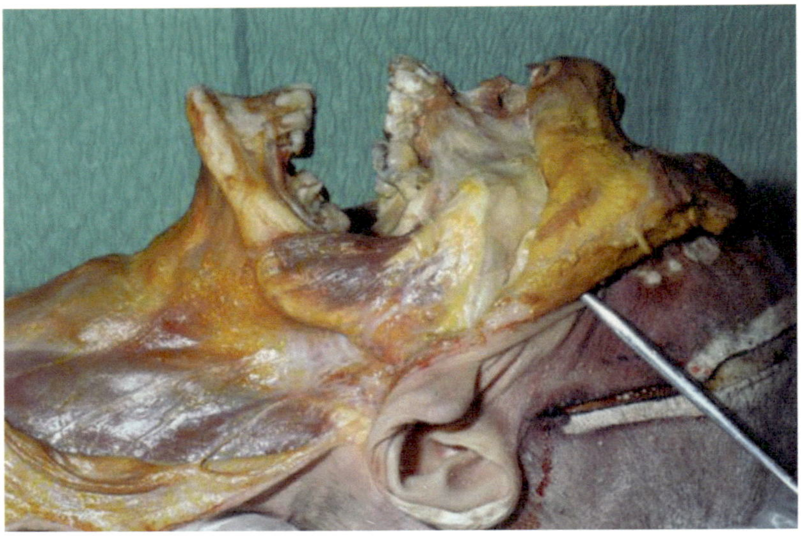

Fig. 13. Mandibular dissection, third step: facial cutaneous flap and section of masticator muscles (*m. masseter*).

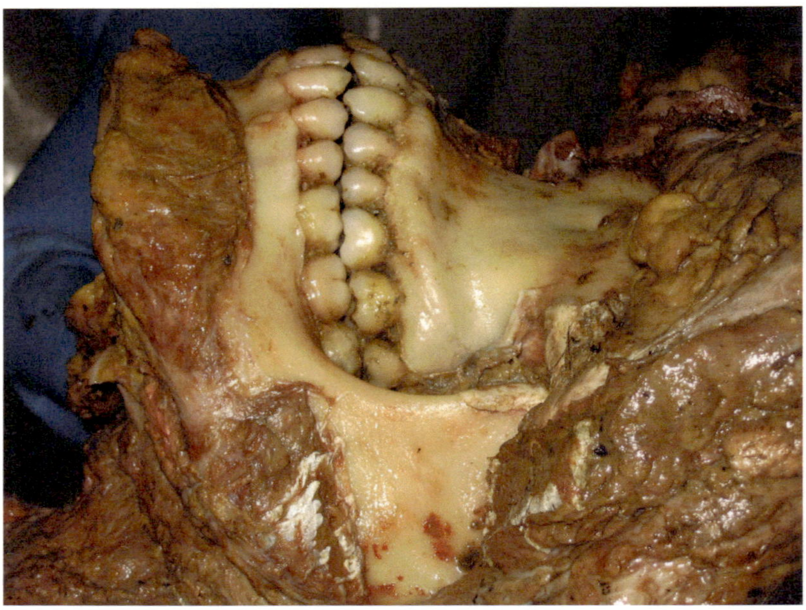

Fig. 14. Mandibular dissection, fourth step: complete cranial evidence of the *mandibula*.

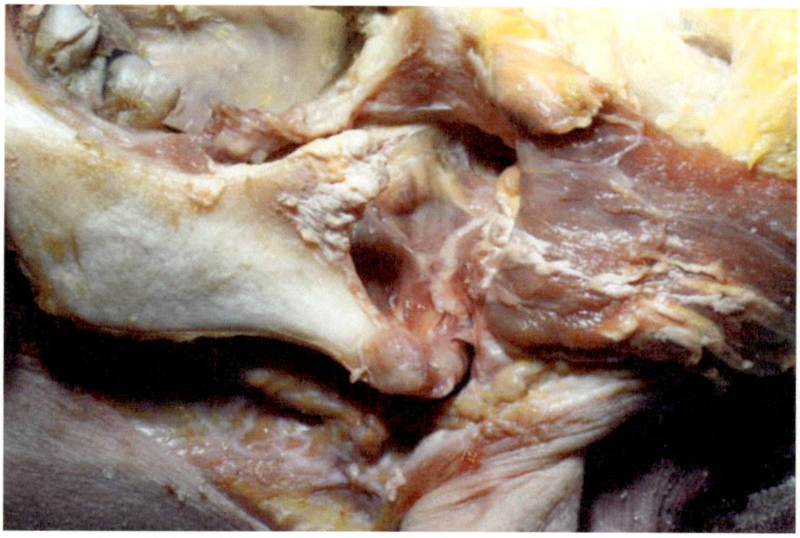

Fig. 15. Mandibular dissection, fifth step: lateral aspect of the mandibular condyle and coronoid section of the *m. temporalis.*

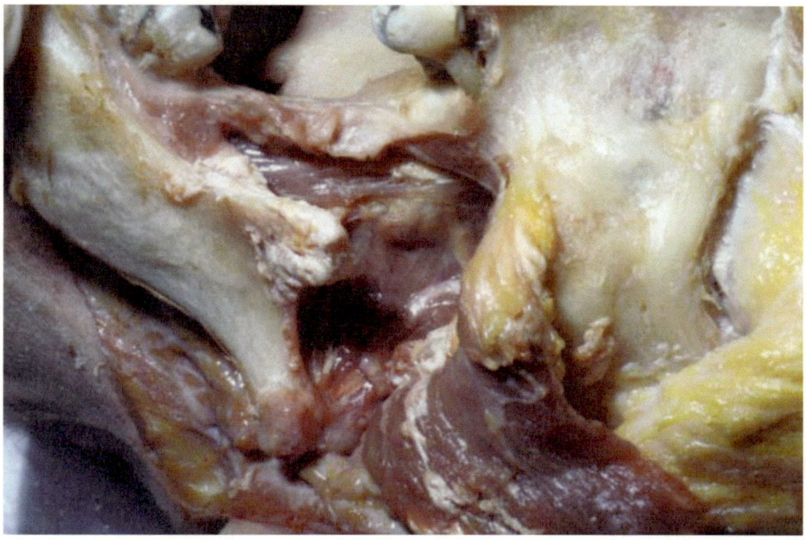

Fig. 16. Mandibular dissection, sixth step: complete disarticulation of the temporomandibular joint and muscles sections along the medial border of *mandibula* (left and right dissections) which can be deposited.

1.5 Odontological examination

This anatomical technique facilitated the forensic odontologist's PM work, with an easy access to both the maxillar and mandibular dentition (Fig. 17). Separate pictures were taken of the upper and lower jaws and teeth in occlusal position, as well as with articulating jaws (Fig. 18). The complete dentition was described using Interpol protocols, nomenclature and forms (Bajaj, 2005; Sweet, 2006) (sound teeth, pathological teeth, restored teeth, absent AM or PM teeth, attrition and oral anatomical abnormalities) (Fig. 19).

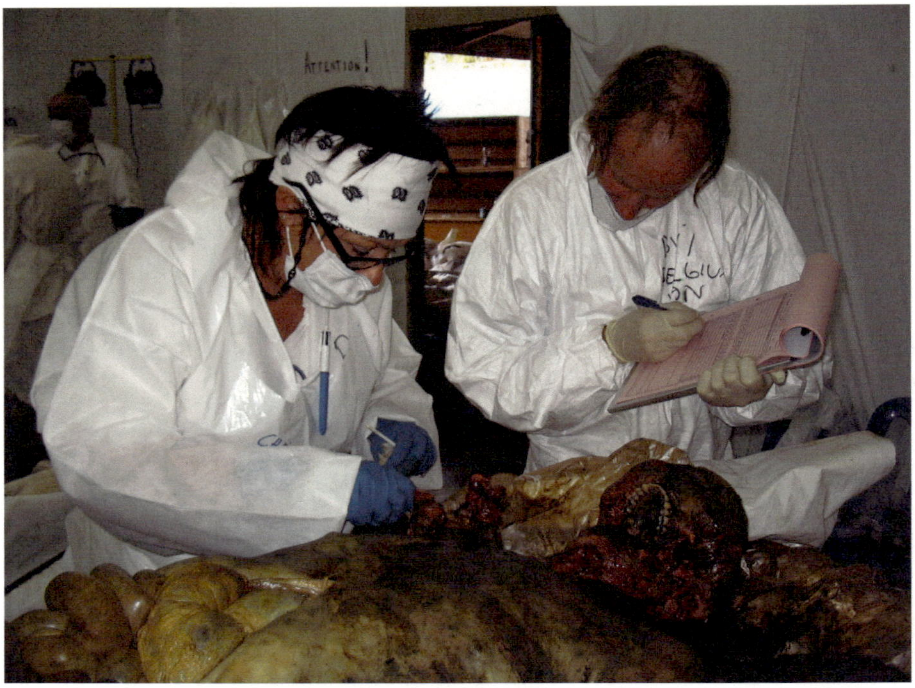

Fig. 17. Dental first examination.

All the observations were carefully registered on pages F1 & F2 and marked on the odontogram of the PM pink form[1].

The body was then transferred to the next room, where dental radiographies were carried out (bite wing radiographies, specific X-rays related to dental particularities) (Fig. 20). During the initial weeks only conventional dental X-ray equipment was present, but in a later stage a Nomad™ handheld digital dental X-ray machine was made available, facilitating the odontologist's task for X-raying considerably.

The PM work was very intense, both the number of victims and the time restrictions for PM examination, allowing only 20 min for each work section: fingerprinting, medical examination, odontological examination.

---

1 Accessible on the Interpol Internet website.

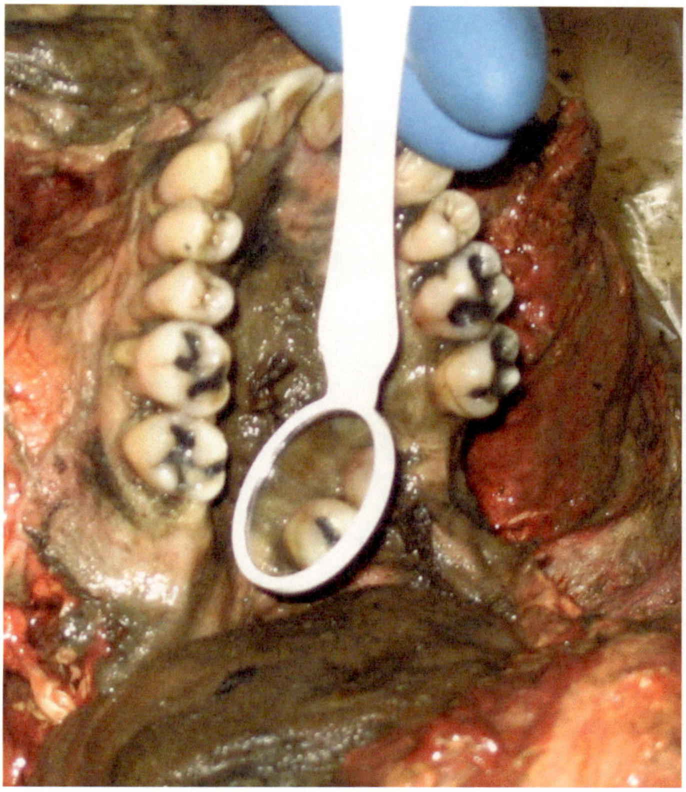

Fig. 18. Dental details.

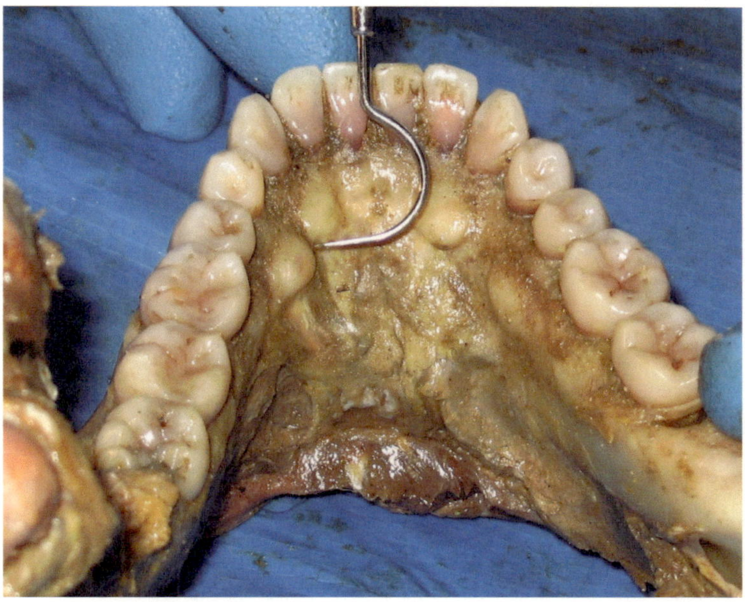

Fig. 19. Particularities - *torus mandibularis*.

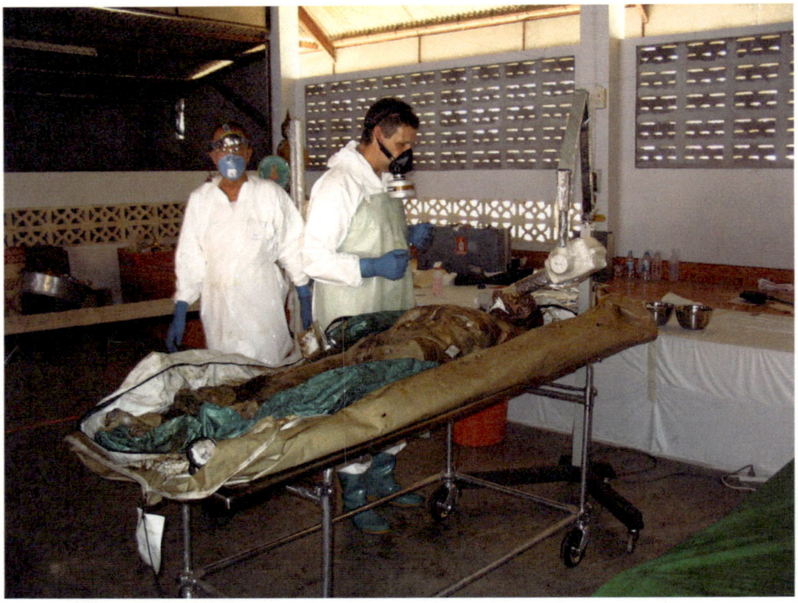

Fig. 20. Dental X-ray examination.

1.6 Reconstruction post autopsy

The particular dissection technique we explained enables also an easy reconstitution of the person's face by simply folding back the cutaneous facial tissues. This allows an eventual recognition of the victim by family members in a post identification stage if requested.

1.7 Particular considerations

Depending on the kind of PM examinations to be performed on the bodies, it might be very useful to conduct the medico-legal and anthropological examinations as near to the disaster site as possible. This helps to avoid destruction or loss of evidence caused by body transport (teeth loss, loss of particular bones that could be useful in criminal investigations, such as the hyoid bone...). The utility of this kind of "field mortuaries" in Kosovo was previously described by Beauthier et al. (2000).

As for the autopsy room management, all DVI teams deployed and working in that place should accept the agreed "Standard Operating Protocol" conditions (SOPs) and these conditions really are the rules of engagement.

The presence of a chief autopsy room manager, supervising the activities of the different teams and managing the arrival of new bodies, has proved to be of exceptional added value. At the same time, this person should have the power to enforce the SOPs and take any necessary actions if problems occur.

2. Identification criteria and techniques

2.1 General considerations

The proper identification of a decedent is not only important for humanitarian reasons for the next-of-kin but also for legal and administrative purposes. This identification issue represents one of the most difficult medico-legal and anthropological tasks as it must withstand legal scrutiny.

All possible means must be applied to achieve a scientific identification, which is sometimes extremely difficult, particularly in mass disaster situations (Lunetta et al., 2003; Poisson et al., 2003; Tsokos et al., 2006) or in matters of genocide crimes. For some the mourning process starts as soon as the family members are aware of the suspected loss but the positive identification of the decedent also may be needed and helpful with so-called "closure." Some families feel that the uncertainty is more difficult to cope with than the definitive identification.

From both legal and administrative points-of-view, non-identification creates additional problems.

2.2 Forensic anthropology

Regardless of the body's condition, the forensic and anthropological methods used during the different stages of the identification process must be rigorous and systematic (Djuric et al., 2007; Djuric, 2004).

Forensic anthropology is the application of the science of physical anthropology to the legal process. The identification of skeletal, badly decomposed, or otherwise unidentified human remains is important for both legal and humanitarian reasons. Forensic anthropologists apply standard scientific techniques developed in physical anthropology to identify human remains and also – if necessary – to assist in the detection of crime (Baccino, 2004; Beauthier, 2009; Cattaneo & Baccino, 2002; Cattaneo, 2007).

When discovering a skeleton, the physical anthropologist's aim is to prove its origin (by studying the rituals of burial, the ancestry, and the osseous characteristics which can provide interesting information on lifestyle, occupational activities, or hierarchical status).

The forensic anthropologist, who uses the same methods as his alter ego, works ultimately as an identifier and a medico-legal expert, looking for possible clues of criminal acts.

2.3 Identification process

Reconstructive identification is the first step needed to be able to achieve a comparative identification. During the reconstructive identification process all necessary information – physical characteristics, medical, odontological, DNA, fingerprints, etc. – is gathered from the unknown body so that an objective reconstructed profile of the unknown body can be established. This reconstructed PM profile can then be matched with missing persons' AM information in a comparative identification.

The Interpol DVI Standing Committee developed DVI Guidelines in the 1980's to promote the use of standardized AM and PM forms across personal and national preferences in all member countries.

(http://www.interpol.int/public/DisasterVictim/guide/default.asp).

These Interpol AM and PM forms are completed separately. Only in a later stage are the AM and PM information cross checked in order to lead to a final identification. These PM pink forms are completed by the multidisciplinary team during the different stages of the PM body examination as all useful elements for identification are systematically indexed on the PM Interpol form.

The AM data are collected from each presumed victim's family.

This information can be collected at their home or at a place where the family members are gathered, close to the disaster scene or at a distant location, according to the type of disaster and/or the wishes of the families.

Further and/or more specific information will then be collected from friends and acquaintances, the missing person's employer, doctor or dentist. It is also important to obtain pictures of the presumed victims and to collect their fingerprints. Hair samples, toothbrushes or other personal belongings of the missing person that can be useful to determine a DNA profile will be searched for and collected.

The family lineage will be established to determine from which relatives blood or saliva samples must be collected in case DNA analysis will be necessary for victim identification.

AM fingerprints are collected at the homes of the presumed victims from objects likely only touched by them (e.g. a perfume bottle for women or an after shave bottle for men).

The AM forms (yellow) are completed by the team contacting the alleged victim's family and collecting (i) medical and surgical antecedents, (ii) morphological characteristics, (iii) professional antecedents and characteristics that result from these, (iv) dental information (De Valck, 2006), (v) description of clothing, jewelry, and other personal objects worn by the missing person, (vi) missing person's pictures.

All collected AM data, even partial, are transmitted to the Identification Section and transcribed into the computer data file. The computer team works with the "DVI International" software program, developed by PlassData™ and used by the Interpol DVI Community worldwide.

This software program allows research and automatic comparisons between AM and PM data bases in both directions: AM to PM or vice versa. (Andersen Torpet, 2005; Brenner, 2006; Clement et al., 2006). The by-the-computer-suggested correspondences between an unidentified body and a missing person facilitate a more specific manual control and comparison of the corresponding AM and PM data by a team of specialists.

This team informs the Chief of the Identification Team about a final decision on the identity of the unidentified human remains.

The rigid application and the respect for the Interpol International Conventions (e.g. the dental classification and the symbols used for a correct registration of the various dental treatments) enabled an optimal procedure leading to the positive identification of these innumerable victims (De Valck, 2006).

The actual Interpol forms currently also contain an important DNA document, indicating the methods and the probes to be used, in order to allow a uniform comparative application at an international level (Ladika, 2005).

Note that at the end of the first week, the Thai Tsunami Victim Identification committee (TTVI) was created in order to establish Standard Operation Protocols for data collection and identification procedures (Tsokos et al., 2006).

The Thai Tsunami Victim Identification Information Management Center (TTVI - IMC) under the direction of the Royal Thai Police, coordinated the collaboration between the international Disaster Victim Identification Teams and Thai authorities.
(http://www.interpol.int/Public/asiandisaster/background/TTVI_FAQ.asp)

Once all PM and AM data were entered into PlassData™, the system conducted a daily automatic search to identify possible matches. These were then manually verified by members of the reconciliation team to confirm the eventual identity of a victim. All information used in the comparison process was then compiled and presented to the

Reconciliation Board, composed of specialists from all areas of forensic and investigative expertise, for final confirmation.
Once identification was confirmed, a death certificate was issued and the body released (http://www.inet.co.th/tsunami/procedures.php).
The authorities of the victim's country were then contacted so that the victim's family could be informed about the positive identification and that repatriation of the body could take place.
In concordance with the literature, we resume in Table 2, the different identification's criteria (Beauthier, 2008; Beauthier et al., 2008; Beauthier & Lefèvre, 2008; Beauthier et al., 2009; Beauthier & Lefevre, 2007; Quatrehomme, 2003).

Qualities of Identification	Examples
Certitude	X-ray comparisons
	Dental comparisons
	Fingerprint comparisons
	DNA
Probable Identification	Identity documents
	Tattoos[2]
	Compatible dental formula
Possible Identification	Scars[3]
	Pathological antecedents
Identification excluded	Incompatible dental formula
	Pathological antecedents (e.g. bone fracture) incompatible

Table 2. Medicolegal Identification possibilities (Quatrehomme et al., 1998).

2.4 DNA techniques

Because of the rapid decomposition and subsequently muscular alterations, a sampling of the femoral shaft (recommended for DNA analysis) was carried out by the anthropologist. This happened in a separated area, using clean tools different from those used for autopsies. Another possibility for DNA sampling was the extraction of two sound teeth, preferably two premolars or canines, because of the presence of a lot of pulp material inside these teeth.

3. Discussion – guidelines

From an administrative point of view we believe though that it is imperative to adopt, from the beginning of an operation, a single and final numbering system for each unidentified body. This should be determined by the management in a Standard Operating Protocol (SOP). For the same reason, the teams must be deployed to the disaster site as soon as possible. Early body examinations, a fast and unique numbering system, and the possibility of in situ pictures will considerably improve the chances of a positive identification.
A chief DNA manager can operate away from the main autopsy room. This DNA manager would be responsible for the sampling of the DNA femoral shafts and making sure there is no permutation of the Interpol forms corresponding to the bodies.

---

[2] In certain cases, the identification can be certain.

[3] With the same considerations than tattoos.

On a DNA level, it is useful to systematize the sampling of the dental and femoral diaphysis parts. In Thailand costal samples were taken from numerous bodies in the early days, but at a later stage instructions were given to start all over and instead collect the SOP approved femoral samples and/or teeth.
Only performing an external body examination has limitations and we all were well aware of that. In this particular case and because of the body conditions the search for scars for example proved to be inadequate, but tattoos however were more easily detected and of great value.
The assistance of the local Thai auxiliary personnel in interpreting or translating some of the tattoo material was of great added value.
The benefits of a separate X-ray room (Ludes et al., 1994) where radiological examination of the body by X-ray amplification[4] would allow to locate the presence of foreign bodies (e.g. osteosynthesis, prosthesis, tubar clips, vascular stents or pacemakers) are without any doubts and this idea should be taken into consideration at all times.
Radiology could be complementary - or even replace - the limited internal autopsy which provided very little helpful information as to validly evaluate anatomical characteristics under these difficult conditions (Hirsch & Shaler, 2002).
An important place in the PM information gathering was located at the final desk (Fig. 21) by (i) checking if all examinations were carried out; (ii) checking concordance of all

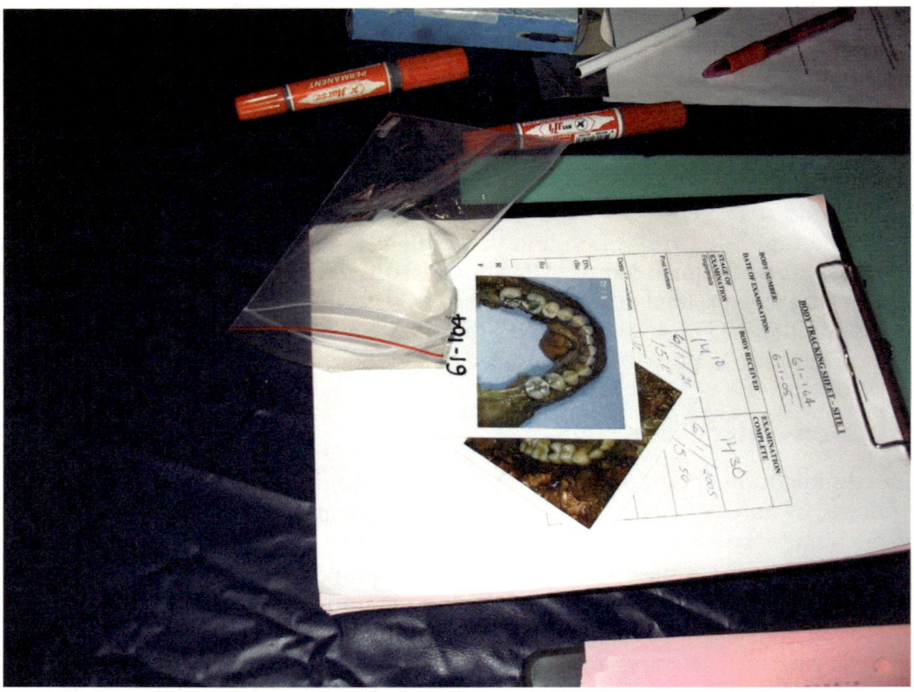

Fig. 21. Final desk.

[4] Or CT-Scan.

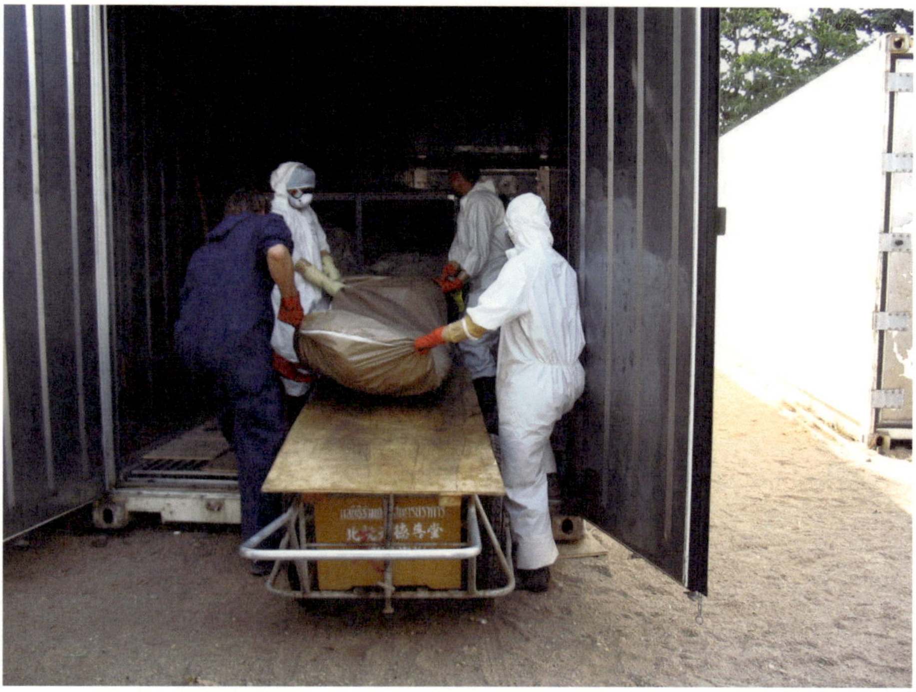

Fig. 22. Bodies' conservation (with repertory) in containers - after all PM techniques.

numbered samples with the given body number ; (iii) assuring quality control of the dental X-rays (Goodman & Edelson, 2002; Kieser et al., 2006) and (iv) quality control of all documents.

Only after meeting all these quality criteria can the body be carried out of the technical autopsy rooms and stored in an appropriate refrigerated container, where its exact location is registered by the body handler (Fig. 22).

The efficiency of the adopted methodology is strictly depending on the kind and nature of the mass disaster. Each disaster is different and the approach adapted accordingly.

The final identification results depend mainly on the presence and quality of the AM information (physical evidence such as X-rays elements (Goodman & Edelson, 2002; Messmer, 1982; Rutty et al., 2007), fingerprints if possible[5], dental records (De Valck, 2006; Kieser et al., 2006; Poisson et al., 2003) and of course DNA (Gill, 2006).

In 2006 Perrier mentioned - when referring to the Phuket identifications - the effectiveness of the following methods: 73% of identification by odontology, 24% by dactyloscopy and 3% by DNA analysis (Perrier et al., 2006). This shows the need to first focus on simple, inexpensive and quickly efficient methods, without diminishing of course the value of fingerprints and the excellence of the positive DNA identifications (Andersen Torpet, 2005; Biesecker et al., 2005; Brenner, 2006; Deng et al., 2005; Hinchliffe, 2007; Leclair et al., 2004; Leclair et al., 2007; Salo et al., 2007; Schuller-Gotzburg & Suchanek, 2007; Tack et al., 2007; Tsokos et al., 2006).

---

5 See: http://www.inet.co.th/tsunami/procedures.php.

Sribanditmongkol et al. reported similar conclusions[6]. The predominant role of forensic odontology was also underlined by Kirsch et al (2007). This highlights the need to first focus on simple, inexpensive and time-efficient methods, of course without minimizing the value of fingerprints and the excellence of DNA which are decisive in the identification process (Andersen Torpet, 2005; Biesecker et al., 2005; Brenner, 2006; Deng et al., 2005; Hinchliffe, 2007; Leclair et al., 2004; Leclair et al., 2007; Salo et al., 2007; Schuller-Gotzburg & Suchanek, 2007; Tack et al., 2007; Tsokos et al., 2006).
In December 2008, The Thai Tsunami Victim Identification and Repatriation Center mentioned that 388 bodies remain unidentified from a total of 3,696 examined bodies on the Wat Yan Yao Site 1a (Beauthier et al., 2009)[7].

4. Conclusion

Adequate victim identification management goes through strict methodology and protocols. This requires adopting and applying standard operating protocols, in agreement with the Interpol DVI Standing Committee guidelines. Such requirements are even more amplified when faced to a mass disaster, such as the Tsunami of December 26, 2004, with numerous victims of different nationalities and ethnic origin.
Such a huge operation requires a multidisciplinary team like the Belgian DVI Team, which in addition to scientific experts, includes a significant number of police officers able to multitask and capable of efficiently assisting the team's scientific personnel.
The members of the civil protection, bringing in their technical knowledge and skills to improve the overall working conditions also proved to be of great value.
In addition, the Belgian DVI team is probably one of the only teams in the world to have a physician, a nurse, a psychologist and a stress team as regular team members.
The expertise in relation to disaster victim identification is enhanced when team members - working together in an interdisciplinary way, share mutual respect - are professionally trained and skilled, and have multi-functional capacities.

5. Acknowledgments

Many thanks to Laurence Genevrois and Denise De Valck - Crenwelge (USA, BS Journalism, 1980, Texas A&M University) for assisting as translator and editor respectively.
Please note that all the photographs are the property of Prof. J.-P. Beauthier. Also note that the figures 11-13 & 15-16 illustrate the steps of the technique of mandibular dissection performed - with the helpful and friendly collaboration of Emile Godefroid - on the Laboratory of Anatomy, Biomechanics and Organogenesis (LABO) of the Université Libre de Bruxelles (ULB - Belgium) (Dir. Prof. M. Rooze)[8].

---

[6] See : http://www.who.int/hac/events/tsunamiconf/presentations/2_16_forensic_pongruk_doc.pdf., pp., www.paho.org/English/DD/PED/PungrukWHO_Lima.ppt.,
[7] See also : http://www.upi.com/Top_News/2008/12/24/Unidentified-dead-in-Thai-tsunami-at-388/UPI-85381230142146/.
[8] http://homepages.ulb.ac.be/~labo/

6. References

Andersen Torpet, L. (2005). DVI System International: software assisting in the Thai tsunami victim identification process. *J Forensic Odontostomatol* 23, pp. 19-25, 0258-414X.

Baccino, E. (2004). Forensic Anthropology Society of Europe - A section of the International Academy of Legal Medicine [newsletter]. *Int J Legal Med* 118, pp. N1, 0937-9827.

Bajaj, A. (2005). Disaster victim identification: Tsunami. *Br Dent J* 198, pp. 504-505, 0007-0610.

Beauthier, J.-P.; Boxho, P.; Crèvecoeur, J. M.; Leclercq, M.; Lefèvre, P. & Vogels, L. (2000). Mission du team belge au Kosovo, science et justice à la rencontre du drame humain - premiers résultats. *Biom Hum Anthropol* 18, pp. 43-48, 1279-7863.

Beauthier, J.-P. (2008) *Traité de médecine légale*, De Boeck Université, 9782804147983, Bruxelles.

Beauthier, J.-P.; De Valck, E.; De Winne, J. & Lefèvre, P. (2008) Catastrophes de masse, In: *Traité de médecine légale*, J-P Beauthier, (Ed.), pp. 517-531, De Boeck Université, 9782804147983, Bruxelles.

Beauthier, J.-P. & Lefèvre, P. (2008). Rôle des médecins légistes, anthropologues et odontologues lors d'activités au sein du team belge d'identification des victimes (Belgian DVI team). Revue de l'organisation et de la gestion de ces situations dans le but d'une identification optimale des victimes. *Biom Hum Anthropol* 26, pp. 45-56, 1279-7863.

Beauthier, J.-P. (2009) Contribution à l'approche anthropologique et médico-légale des sutures viscérocrâniennes utiles dans l'estimation de l'âge au décès (sutures palatines, fronto-naso-maxillaires et zygomatiques). Thèse de Doctorat en Sciences médicales (MS Thesis), Université Libre de Bruxelles, Bruxelles.

Beauthier, J.-P.; Lefèvre, P.; De Valck, E. & De Winne, J. (2009). Mass Disaster Victim Identification: the Tsunami Experience (December 26, 2004). *The Open Forensic Science Journal* 2, pp. 54-62, 1874-4028.

Beauthier, J. P. & Lefevre, P. (2007). [Guidelines in mass disaster victims identification through the Tsunami experience (December 26, 2004)]. *Rev Med Brux* 28, pp. 512-522, 0035-3639.

Biesecker, L. G.; Bailey-Wilson, J. E.; Ballantyne, J.; Baum, H.; Bieber, F. R.; Brenner, C.; Budowle, B.; Butler, J. M.; Carmody, G.; Conneally, P. M.; Duceman, B.; Eisenberg, A.; Forman, L.; Kidd, K. K.; Leclair, B.; Niezgoda, S.; Parsons, T. J.; Pugh, E.; Shaler, R.; Sherry, S. T.; Sozer, A. & Walsh, A. (2005). Epidemiology. DNA identifications after the 9/11 World Trade Center attack. *Science* 310, pp. 1122-1123, 0036-8075.

Brenner, C. H. (2006). Some mathematical problems in the DNA identification of victims in the 2004 tsunami and similar mass fatalities. *Forensic Sci Int* 157, pp. 172-180, 0379-0738.

Cattaneo, C. & Baccino, E. (2002). A call for forensic anthropology in Europe [newsletter]. *Int J Legal Med* 116, pp. N1-N2, 0937- 9827.

Cattaneo, C. (2007). Forensic anthropology: developments of a classical discipline in the new millennium. *Forensic Sci Int* 165, pp. 185-193, 0379-0738.

Clement, J. G.; Winship, V.; Ceddia, J.; Al-Amad, S.; Morales, A. & Hill, A. J. (2006). New software for computer-assisted dental-data matching in Disaster Victim Identification and long-term missing persons investigations: "DAVID Web". *Forensic Sci Int* 159, pp. Suppl 1: S24-29, 0379-0738.

De Valck, E. (2006). Major incident response: collecting ante-mortem data. *Forensic Sci Int* 159, pp. Suppl 1: S15-19, 0379-0738.

Deng, Y. J.; Li, Y. Z.; Yu, X. G.; Li, L.; Wu, D. Y.; Zhou, J.; Man, T. Y.; Yang, G.; Yan, J. W.; Cai da, Q.; Wang, J.; Yang, H. M.; Li, S. B. & Yu, J. (2005). Preliminary DNA identification for the Tsunami victims in Thailand. *Genomics Proteomics Bioinformatics* 3, pp. 143-157, 1672-0229.

Djuric, M.; Dunjic, D.; Djonic, D. & Skinner, M. (2007). Identification of victims from two mass-graves in Serbia: a critical evaluation of classical markers of identity. *Forensic Sci Int* 172, pp. 125-129, 0379-0738.

Djuric, M. P. (2004). Anthropological data in individualization of skeletal remains from a forensic context in Kosovo--a case history. *J Forensic Sci* 49, pp. 464-468, 0022-1198.

Gill, J. R. (2006). 9/11 and the New York City Office of Chief Medical Examiner. *For Sci Med Pathol* 2, pp. 29-32, 1547-769X.

Goodman, N. R. & Edelson, L. B. (2002). The efficiency of an X-ray screening system at a mass disaster. *J Forensic Sci* 47, pp. 127- 130, 0022-1198.

Hinchliffe, J. A. (2007). Disaster dentistry. *Br Dent J* 202, pp. 493-494, 0007-0610.

Hirsch, C. S. & Shaler, R. (2002). 9/11 through the eyes of a medical examiner. *J Investig Med* 50, pp. 1-3, 1081-5589.

http://www.inet.co.th/tsunami/procedures.php., pp.

http://www.interpol.int/Public/asiandisaster/background/TTVI_FAQ.asp.

http://www.interpol.int/public/DisasterVictim/guide/default.asp.

http://www.upi.com/Top_News/2008/12/24/Unidentified-dead-in-Thai-tsunami-at-388/UPI-85381230142146/.

http://www.paho.org/English/DD/PED/PungrukWHO_Lima.ppt.,

http://www.who.int/hac/events/tsunamiconf/presentations/2_16_forensic_pongruk_doc .pdf.

Keiser-Nielsen, S. (1980) *Person identification by means of teeth: a practical guide*, John Wright and Sons Ltd, 9780723605577, Bristol (UK).

Kieser, J. A.; Laing, W. & Herbison, P. (2006). Lessons learned from large-scale comparative dental analysis following the South Asian tsunami of 2004. *J Forensic Sci* 51, pp. 109-112, 0022-1198.

Kirsch, H.-P.; Rötzscher, K.; Grundmann, C. & Lessig, R. (2007). The Tsunami Disaster in the Kingdom of Thailand 2004. *Int Poster J Dent Oral Med* 9, pp. 370, 1612-7749.

Ladika, S. (2005). South Asia tsunami. DNA helps identify missing in the tsunami zone. *Science* 307, pp. 504, 0036-8075.

Leclair, B.; Fregeau, C. J.; Bowen, K. L. & Fourney, R. M. (2004). Enhanced kinship analysis and STR-based DNA typing for human identification in mass fatality incidents: the Swissair flight 111 disaster. *J Forensic Sci* 49, pp. 939-953, 0022-1198.

Leclair, B.; Shaler, R.; Carmody, G. R.; Eliason, K.; Hendrickson, B. C.; Judkins, T.; Norton, M. J.; Sears, C. & Scholl, T. (2007). Bioinformatics and human identification in mass fatality incidents: the world trade center disaster. *J Forensic Sci* 52, pp. 806-819, 0022-1198.

Ludes, B.; Tracqui, A.; Pfitzinger, H.; Kintz, P.; Levy, F.; Disteldorf, M.; Hutt, J. M.; Kaess, B.; Haag, R.; Memheld, B.; Kaempf, C.; Friederich, F.; Evenot, E. & Mangin, P. (1994). Medico-legal investigations of the Airbus, A320 crash upon Mount Ste-Odile, France. *J Forensic Sci* 39, pp. 1147-1152, 0022-1198.

Lunetta, P.; Ranta, H.; Cattaneo, C.; Piccinini, A.; Niskanen, R.; Sajantila, A. & Penttila, A. (2003). International collaboration in mass disasters involving foreign nationals within the EU: medico-legal investigation of Finnish victims of the Milan Linate airport SAS SK 686 aircraft accident on 8 October 2001. *Int J Legal Med* 117, pp. 204-210, 0937-9827.

Messmer, J. M. (1982). The use of mammographic equipment in mass disaster identification. *J Forensic Sci* 27, pp. 723-725, 0022- 1198.

Perrier, M.; Bollmann, M.; Girod, A. & Mangin, P. (2006). Swiss DVI at the tsunami disaster: Expect the unexpected. *Forensic Sci Int* 159, pp. Suppl: S30-32, 0379-0738.

Poisson, P.; Chapenoire, S.; Schuliar, Y.; Lamant, M. & Corvisier, J. M. (2003). Four major disasters in Aquitaine, France: use of odontologic techniques for identification. *Am J Forensic Med Pathol* 24, pp. 160-163, 0195-7910.

Quatrehomme, G.; Cotin, S.; Alunni-Perret, V.; Garidel, Y.; Grévin, G.; Bailet, P.; Ollier, A. & N., A. (1998). La superposition, la restauration et la reconstruction faciales: une aide à l'identification médico-légale. *J Méd Lég Droit Méd* 42, pp. 11-22, 0999-9809.

Quatrehomme, G. (2003) Anthropologie médico-légale, In: *L'identification en médecine légale. Aspects actuels,* D Malicier, (Ed.), pp. 81-128, ESKA Editions, 2-7472-0409-X, Paris.

Rutty, G. N.; Robinson, C. E.; Bouhaidar, R.; Jeffery, A. J. & Morgan, B. (2007). The role of mobile computed tomography in mass fatality incidents. *J Forensic Sci* 52, pp. 1343-1349, 0022-1198.

Salo, S.; Salo, H.; Liisanantti, A. & Reponen, J. (2007). Data transmission in dental identification of mass disaster victims. *J Forensic Odontostomatol* 25, pp. 17-22, 0258-414X.

Schuller-Gotzburg, P. & Suchanek, J. (2007). Forensic odontologists successfully identify tsunami victims in Phuket, Thailand. *Forensic Sci Int* 171, pp. 204-207, 0379-0738.

Sweet, D. (2006). Solving certain dental records problems with technology-The Canadian solution in the Thailand tsunami response. *Forensic Sci Int* 159, pp. Suppl: S20-23, 0379-0738.

Tack, L. C.; Thomas, M. & Reich, K. (2007). Automated forensic DNA purification optimized for FTA card punches and identifiler STR-based PCR analysis. *Clin Lab Med* 27, pp. 183-191, 0272-2712.

Tsokos, M.; Lessig, R.; Grundmann, C.; Benthaus, S. & Peschel, O. (2006). Experiences in tsunami victim identification. *Int J Legal Med* 120, pp. 185-187, 0937-9827.